To Kathy and Angela
For Partridge, Puffa, and Charlie Tuna

INTRODUCTION
TO
PHYSICAL SCIENCE

234567890 DODO 898765432

This book was set in Helvetica Light by Progressive Typographers.
The editors were John J. Corrigan and James W. Bradley; the designer was Nicholas Krenitsky; the production supervisor was Phil Galea.
The drawings were done by J & R Services, Inc. The cover photograph was taken by Lawrence Tamaccio.
R. R. Donnelley & Sons Company was printer and binder.

Library of Congress Cataloging in Publication Data

Riban, David M
Introduction to physical science.

Includes index.
1. Science. I. Title.
Q158.5.R52 500.2 80-23947
ISBN 0-07-052140-9

FREQUENTLY USED CONSTANTS

SPEED OF LIGHT IN VACUUM	c	3.00×10^{8} m/s
CHARGE ON ELECTRON	e	1.60×10^{-19} coulomb (C)
GRAVITATIONAL CONSTANT	G	6.67×10^{-11} N·m²/kg²
ACCELERATION DUE TO GRAVITY	g	9.81 m/s²
PLANCK'S CONSTANT	h	6.63×10^{-34} J·s
BOLTZMANN'S CONSTANT	k	1.38×10^{-23} J/K
COULOMB'S LAW CONSTANT	Q	8.99×10^{9} N·m²/C²
MASS OF THE ELECTRON	M_e	9.11×10^{-31} kg
MASS OF THE PROTON	M_p	1.673×10^{-27} kg
AVOGADRO'S NUMBER	N_0	6.02×10^{23} molecules/mole

INTRODUCTI
TO
PHYSICA
SCIENC

David M. Riban

Professor of Physics
Indiana University of Pennsylvania

McGRAW-HILL BOOK COMPANY

New York | St. Louis | San Francisco | Auckl
Bogotá | Hamburg | Johannesburg | Londo
Madrid | Mexico | Montreal | New Delhi
Panama | Paris | São Paulo | Singapore
Sydney | Tokyo | Toronto

Contents

Preface xi

CHAPTER 1
THE BEGINNINGS OF SCIENCE
1

WHAT IS SCIENCE? 1
THE EMERGENCE OF SCIENCE 2
THE RISE OF NATURAL PHILOSOPHY IN GREECE 6
SOME ACCOMPLISHMENTS OF GREEK SCIENCE 8
THE ARISTOTELIAN UNIVERSE—A THEORETICAL SYSTEM 13
THE PHYSICAL WORLD OF ARISTOTLE 17
Natural Motion 19
Forced Motion 19
Heavenly Motion 22
WHAT WENT WRONG? 23

CHAPTER 2
THE EMPIRICAL PROCESS—
Obtaining Order from Numbers
25

INTRODUCTION 25
THE ART OF MEASUREMENT 26
Measuring Systems and Units 27
Significant Figures 29
OBTAINING EQUATIONS FROM GRAPHS 31
A REVIEW OF GRAPHING 31
The Slope-Intercept Form of the Equation of a Straight Line 33
Change of Variables 37
SUMMARY 46

CHAPTER 3
THE THEORETICAL PROCESS—
Exploring the Implications of Ideas
51

INTRODUCTION 51
GALILEO'S OBSERVATIONS ON MOTION 53
THE DESCRIPTION OF MOTION 56
Defining Speed and Acceleration 56
Speed vs. Velocity—A Note on Vectors 60
The Addition of Vectors 61
Graphs of Motion—Instantaneous Speed 64
Distance from a *v*-vs.-*t* Curve 68
Speed and Distance with a Uniform Acceleration 70
SUMMARY 73

CHAPTER 4
AN EXAMPLE OF MOTIONS DEFINED—
Astronomical Motions as Viewed from the Earth
77

INTRODUCTION 77
APPARENT MOTION OF THE SUN 78
MEASURING MOTIONS 81
RATE OF SOLAR MOTION 87
* VISUALIZING EARTH-SKY RELATIONSHIPS 87
MOTION OF THE STARS 91
* SOLAR MOTIONS AND SEASONS 94
APPARENT MOTIONS OF THE MOON 98
APPARENT MOTION OF THE PLANETS 102

Venus and Mercury 103
Mars, Jupiter, and Saturn 104
SUMMARY 106

CHAPTER 5
THE CONFLICT OF TWO SYSTEMS TO EXPLAIN ASTRONOMICAL MOTIONS
109

INTRODUCTION 109
THE ANCIENT SOLUTION—ORIGINS OF THE PTOLEMAIC SYSTEM 110
THE PTOLEMAIC SYSTEM 111
INTRODUCTION TO THE COPERNICAN SYSTEM 117
THE COPERNICAN SYSTEM 118
HOW TO DECIDE?—TYCHO BRAHE AND PRECISION 122
PIECING IT TOGETHER—KEPLER AND A NEW ASTRONOMY 128
GALILEO AND THE STRUGGLE FOR THE POPULAR MIND 133
AND YET IT MOVES—THE PHYSICAL EVIDENCE FOR MOTION OF THE EARTH 135
SUMMARY 139

CHAPTER 6
NEWTON AND A SCIENCE OF MOTION
143

INTRODUCTION 143
THE FIRST LAW OF MOTION 145
THE SECOND LAW OF MOTION 147
WEIGHT VS. MASS 152
NEWTON'S THIRD LAW OF MOTION AND MOMENTUM 153
CIRCULAR MOTION 159
GRAVITATION 160
TRIUMPH AND EROSION—NEWTON'S LAWS AND ASTRONOMY 163
SUMMARY 166

CHAPTER 7
WORK AND ENERGY—
What Does Nature Preserve?
169

INTRODUCTION 169
WORK 170
A Unit for Work 172

The Work to Lift Objects—Potential Energy 174
The Work Required to Speed Up a Mass—Kinetic Energy 178
CONSERVATION OF MECHANICAL ENERGY 180
SUMMARY 184

CHAPTER 8
THERMAL ENERGY
187

INTRODUCTION 187
PROPERTIES OF HEAT 190
TEMPERATURE VS. HEAT 194
A UNIT FOR HEAT 196
* A "GOOD" BAD THEORY FOR HEAT 199
* A BORING EXPERIMENT—RUMFORD'S CANNON WORKS 202
GAS LAWS AND A LOWEST TEMPERATURE 203
HEAT AND MOTION 205
HEAT ENGINES 211
* HEAT ENGINES AND SOCIETY 212
THE LAWS OF THERMODYNAMICS 216
SUMMARY 219

CHAPTER 9
WAVES AND THE TRANSFER OF ENERGY
225

INTRODUCTION 225
WHAT IS A WAVE? 226
VARIETIES OF WAVES 227
WAVE BEHAVIOR 232
CHANGE OF MEDIUM AND SPEED 234
BOUNDARIES AND SPEED—REFRACTION 236
WHEN WAVES MEET—INTERFERENCE 239
REPEATING OR PERIODIC WAVES 241
PRODUCING SOUND 245
WAVES ON STRINGS 247
WAVES IN PIPES 249
WAVES FROM MOVING SOURCES 251
SUMMARY 253

CHAPTER 10
OTHER SOURCES OF FORCES—
Electricity and Magnetism
257

INTRODUCTION 257
NATURAL MAGNETISM 258
MAGNETIC FIELDS 260

ELECTRICITY 264
THE ELECTROSCOPE 266
FORCES BETWEEN CHARGES 269
THE WORK REQUIRED TO TRANSFER CHARGES 272
ELECTRICITY IN MOTION 273
THE FLOW OF CHARGES 275
THE INTERACTION BETWEEN ELECTRICITY AND MAGNETISM 281
PRODUCING A CURRENT WITH A MAGNETIC FIELD 285
SUMMARY 288

CHAPTER 11
LIGHT AND ELECTROMAGNETIC WAVES
293

INTRODUCTION 293
LIGHT AND MATERIALS 295
REFLECTIONS IN CURVED MIRRORS 298
REFRACTION 303
OPTICAL USES OF REFRACTION 308
SPECTRA 310
THE NATURE OF LIGHT 313
ELECTROMAGNETIC WAVES 320
SUMMARY 324

CHAPTER 12
THE BEGINNINGS OF CHEMISTRY
327

INTRODUCTION 327
ELEMENTS 332
MULTIPLE PROPORTIONS AND THE ATOMIC THEORY 334
THE SEARCH FOR PATTERN 339
RELATIVE CHEMICAL ACTIVITIES 341
A PATTERN EMERGES—THE ORDER OF THE ELEMENTS 346
SUMMARY 350

CHAPTER 13
THE BONDING OF ATOMS
355

INTRODUCTION 355
A PIECE PROVIDES A PUZZLE 357
A MODEL FOR THE ATOM 359
TOWARD A MODEL OF CHEMICAL ACTIVITY 363

CHEMICAL REACTIONS AND ENERGY 364
COVALENT BONDING 369
THE PROCESS OF DISSOLVING 372
ACIDS AND BASES 374
REACTION RATES AND CATALYSTS 377
SUMMARY 378

CHAPTER 14
THE CHEMISTRY OF CARBON AND OF LIFE
381

INTRODUCTION 381
HYDROCARBONS 383
ORGANIC GROUPS 388
THE SUBSTANCES OF LIFE 392
REQUIREMENTS OF THE BODY 393
LIPIDS OR FATS 399
PROTEINS 400
HEMOGLOBIN, A PROTEIN MOLECULE 404
THE ATTACK ON LIFE'S PLAN 408
SUMMARY 412

CHAPTER 15
THE MATERIALS OF NONLIFE—
The Matter of the Earth
415

INTRODUCTION 415
COMPOSITION OF THE EARTH 416
RESTRUCTURING THE SURFACE—THE ROCK CYCLE 420
AN EXAMPLE—THE FUTURE HISTORY OF A GRANITE TOMBSTONE 421
EROSION, HERO OR VILLAIN? 424
SEDIMENTARY ROCKS AND THE INTERPRETATION OF EARTH HISTORY 429
EARTH HISTORY 434
SUMMARY 446

CHAPTER 16
SHAPING THE CRUST
449

INTRODUCTION 449
THE EARTH'S DEEP INTERIOR 450
UPS AND DOWNS—ISOSTASY AND GEOSYNCLINES 453
AN EVEN LARGER PATTERN EMERGES 458

INDEPENDENT EVIDENCE FOR CONTINENTAL DRIFT 463
SMALLER STRUCTURAL FEATURES OF THE EARTH 466
IGNEOUS INTRUSION 472
FOLDING AND FAULTING 473
SUMMARY 478

CHAPTER 17
THE ATOMIC NUCLEUS
481

INTRODUCTION 481
A NEW FORM OF RADIATION 483
RAYS FROM ROCKS 484
THE NATURE OF RADIOACTIVITY 489
RADIOACTIVE DECAY AND TRANSMUTATION 493
NUCLEAR STABILITY 495
MASS AND ENERGY—THE THEORY OF RELATIVITY 497
RELATIVE MOTIONS 499
NUCLEAR REACTIONS AND MASS 505
* THE DISCOVERY OF NUCLEAR FISSION 508
* BUILDING A BOMB 510
SUMMARY 514

CHAPTER 18
THE ATOMIC REALM
517

INTRODUCTION 517
RADIANT ENERGY 518
THE PHOTOELECTRIC EFFECT 521
ORDER IN THE SPECTRUM 523
THE ORBITS OF ELECTRONS 524
THE DE BROGLIE HYPOTHESIS AND MATTER WAVES 530
THE QUANTUM AND CHEMISTRY 533
QUANTUM LEVELS IN SOLIDS 540
THE NATURE OF PARTICLES 544
SUMMARY 545

CHAPTER 19
THE MODERN VIEW OF THE UNIVERSE
547

INTRODUCTION 547
THE MOON 547
A VIEW OF THE PLANETS 549
THE DISTANCES OF THE STARS 558

MOTIONS OF THE STARS 560
SPECTRAL TYPE AND THE CLASSIFICATION OF STARS 562
THE EXPANSION OF TECHNIQUES—RADIO ASTRONOMY 568
THE ORGANIZATION OF STARS—THE GALAXY 571
THE EXPANDING UNIVERSE 575
SUMMARY 577

CHAPTER 20
SCIENCE AND MODERN LIFE
581

INTRODUCTION 581
SCIENCE AND POPULATION 582
ENVIRONMENTAL QUALITY 584
THE NEED FOR ENERGY 588
PATTERNS OF ENERGY USE 591
SOLAR ENERGY 597
FACT VS. FICTION—A PERSONAL STATEMENT ABOUT NUCLEAR POWER 601
RADIATION, CANCER AND MUTATION 605
NUCLEAR SAFETY 605
A NOTE TO END ON 611

APPENDIX ONE
MATHEMATICAL REVIEW
615

APPENDIX TWO
ANSWERS AND SOLUTIONS
TO ODD-NUMBERED PROBLEMS
625

APPENDIX THREE
GLOSSARY
631

Index 645

Preface

The physical sciences include such a vast array of knowledge that any attempt at covering them in a brief summary is immodest. All texts are compromises, with large numbers of "goodies" (interesting applications or fascinating pieces of information) lurking behind each page, but without enough space to mention them. Most texts respond to the task by compacting information as tightly as possible in trying to include as much as can be managed. This usually results in a text which is to be studied page-by-page rather than read. In the opinion of the author this neglects the nature of the audience for a course in physical science. Most students are having their last formal contact with the sciences in this course. They represent a wide variety of majors outside of the area of the sciences. Such students need a continuity of approach throughout a textbook which will take their eyes off the trees long enough to see the forest.

Throughout this text great emphasis has been placed on making it readable. Readability is partly measured by a formula based on such items as sentence length and word size. Probably as important, however, are the psychological aspects of readability. Wherever possible, science is not made to look like some revealed truth. The pains and missteps in its evolution are considered. The intuitive feeling for the need for a term is usually developed before the term is used and defined. Whenever possible, knowledge is applied to problems that can be perceived as real by the student. Frequently, these applications are selected to be deliberately far afield from the area being considered, to illustrate the range of a scientific principle. Science is portrayed as an endeavor of real human beings rather than as the pronouncements of an establishment of oracles. Finally, the text does not avoid some degree of editorialization and humor. Some texts are written with all the sobriety of an article for the scientific journals, intended to be read and evaluated by the dedicated experts in some subspecialty of science. In the opinion of the author, there is no area of human discourse so serious that its approach would not be helped by some good-natured humor along the way. Humor is an aid to help the human spirit along in what must sometimes be a wearying task. To individuals dedicating their professional lives to their study, the sciences can be fascinating, inspiring, and thoroughly enjoyable even though they also involve a great deal of work. A text concentrating on the niceties of scientific work without catering to one's senses lacks appeal for the nonscientist. Conversely, a text ignoring technical detail fails to provide a background for appreciating science. It is hoped that a decent balance has been achieved in this text which will speak well to students.

The text is organized into groups of chapters which are as distinct as possible. This allows a wide variety of teaching situations to be accommodated. Optional sections in various chapters are preceded by an asterisk. Chapters 4 and 5 on the development of a new astronomy may be pursued in several ways. Chapter 4 offers a description of motion in the heavens as viewed from the earth. It does not interpret these data by any scheme. Chapter 5 presents the Ptolemaic and Copernican interpretations and goes on to develop Kepler's production of a correct picture of the solar system. As a group, these chapters include much more detail for a careful consideration of this problem than most texts. The optional problem set in Chapter 5 allows students to simulate the method of Kepler in obtaining orbits. If this area is not to be emphasized, Chapter 4 may be ignored as data, and Chapter 5 simply read as an introduction to the scientific revolution. Similarly, Chapters 12, 13, and 14 develop chemistry and Chapters 15 and 16 geology in such a manner that if these areas are to be deemphasized they may be excluded. No topic essential to other portions of the text will be omitted in this process.

Chapters 2 and 3 develop the mathematical skills for the course, and the emphasis placed on these chapters will vary with the goals of the particular course. Thereafter, as much mathematics as possible is kept out of the text proper. Worked problems are included in the text when an example is necessary for continuity, but other worked examples are included with the answers and solutions to odd-numbered problems in Appendix Two. Generally, one of the text problems per chapter is fully worked out in Appendix Two. This procedure provides necessary examples for students working the numerical problems while avoiding the visual threat of a great deal of numbers in the text for less mathematically oriented courses. The questions and problems for the course generally avoid the mere repetition of information from the text and present a wide range of difficulty for the student.

In general, footnotes are used in the text to provide a space for personal observations, editorial comment, and humorous asides rather than to provide scholarly references for documentation. Most chapters include one or more inserts which are kept visually distinct from the text. These serve to present pertinent biographies or to develop interesting applications of the content of the chapter.

The author would like to acknowledge the wide range of support and encouragement he has received in preparing this book. His 11 colleagues at the Physics Department of Indiana University of Pennsylvania have been consistently supportive. In particular, Dr. Dennis Whitson and Mr. Patrick McNamara have been of specific help in areas of their expertise. Dr. Gary Buckwalter, chairman, has been unusually understanding and supportive of the problems caused by writing. Thanks also are due to several individuals who offered constructive suggestions and criticism on all or part of the manuscript. These include: Prof. Gordon O. C. Besch, Wisconsin State University, Superior; Prof. Paul R. Wignall, Brevard Community College; Prof. William Hunziker, California State University, Chico; Prof. Fred Searcy, Jr., Itawamba Junior College; Prof. Margaret B. Feero, El Camino College; Prof. R. H. Grossman, Hunter College; Prof. Frederick P. Cranston, Humboldt State University; Dr. Jessie O. Betterton, Jr., University of New Orleans; Prof. Allen P. Bonamy, Daytona Beach Junior College; Prof. Richard S. Masada, Santa Monica College; Dr. Richard E. Berg, University of Maryland; and Prof. Keith R. Honey, West Virginia Institute of Technology. While responsibility for problems or errors within the text reside solely with the author, hopefully, much has been avoided with the considered opinions and suggestions of this group.

Finally, I would like to thank my family for the patience and consideration they have shown for their part-time member. My children have had the adult decency not to run amok just to gain more time and attention during this period, which is unusual. My wife, Kathleen, provided invaluable service in typing my scrawled manuscript and in not allowing me to drop out from the human race during the writing. I sincerely hope that the student will find the text as interpretable and even enjoyable as it was intended to be.

David M. Riban

CHAPTER ONE

The Beginnings of Science

What is Science? Defining science is a difficult task. As an example for comparison, try to define music. If you have heard enough of it, you clearly have a feeling for what it is. If you have not heard music, someone's definition, while possibly correct, would not mean as much as hearing some would. Once you recognize intuitively what it is you are trying to define, there is still room for argument. For example, in a definition of music, it would be difficult to include Beethoven and acid rock in the same set of words. It also would be difficult to cover the range from a little boy proudly playing his first piece on a harmonica to a group of 100 adult professionals who have trained all their lives to act as a perfect team in generating music. Trying to define science presents a similar problem. If you think this example through, it will not surprise you that even scientists have trouble agreeing on a definition.

Science is, by one definition, "the human attempt to understand the workings of the natural world." This definition, however, does not exclude enough—activities of witch doctors, philosophers, and even adolescents can be included. Before adding to the definition, however, let us examine what the previous words do and do not mean. Science is, first of all, a *human* endeavor. It is a human creation as much as art, music, or literature. Just as form, light, and sound exist in ways we would perceive as beautiful prior to an artistic interpretation of them, the natural world similarly precedes science. Only when the human mind tries to order or communicate our experiences do we produce the arts and the sciences. Second, science attempts understanding, but success is not guaranteed. Understanding may be partial, or it may elude us entirely. No amount of endeavor by the ancient Greeks, for example, would have allowed them to understand the science underlying a transistor radio if one were dropped into their midst. One might state on faith that enough study would lead to eventual understanding; however, the statement would be exactly that—a statement of faith. Furthermore, we are never certain when our understanding of a part of nature is extremely close to being correct or when it provides only a partial explanation for the types of situations we are examining. A single exception is enough to show us that our understanding only approximates the working of nature.

The process of *understanding* implies at least three separate abilities: the abilities to explain, to predict, and to suggest new ideas. In terms of a transistor radio, for example, complete understanding would include an explanation of how it works. From this understanding, we could also predict the result of replacing its 9-volt (-V) battery with a 90-V battery, for example. Likewise, this understanding might lead us to suggest the modifications necessary to produce stereophonic sound, or a picture as well as sound.

Adding to our definition, we might say that science consists of an agreement or consensus of the informed on the way things are or how they operate. To obtain this consensus of the informed, scientists begin by making as few assumptions as possible. From there the process proceeds by reasoning and methods that can be agreed upon until its points are demonstrated. Thus, science requires a free interaction among people, careful and exact definitions of its terms, and the use of scientific methods or procedures. The latter is defined as those procedures which can be agreed upon as being correct by the informed individuals who must reach a consensus. From this process we obtain the best description of nature available at a given moment.

The Emergence of Science

The first sustained development of science occurs in ancient Greece. The earliest work was stimulated by discovery of the scholarship of older civilizations, particularly of Egypt and the area around Babylon. In both of these societies, after a promising start, learning had slowed greatly and was making little progress as the Greek era in science began.

The ancient Egyptians are known to most of us today mainly through their obsessive preparations for death. Only temples and tombs were made to endure in Egypt (Fig. 1-1), and even the most splendid of the palaces were made primarily of mud-brick. Egypt was also the most militarily secure nation of antiquity. With deserts on three sides and the sea on the fourth, it took a determined opponent to invade Egypt. Behind these natural barriers developed one of the least oppressive of the ancient civilizations. The land was highly productive and became a major supplier of food for the Mediterranean lands throughout Greek and

FIGURE 1-1
An ancient temple at Thebes in southern Egypt, the capital of the Egyptian empire (near modern Luxor). The mud left by the Nile has been cleared away down to the ancient pavement. Note that this has caused the doorway of the modern structure on the left to be isolated high in the air. (*Photograph by the author.*)

Roman times. Consequently, the learning of Egypt was devoted to very practical concerns with nature.

The Nile flows through Egypt on the top of a sunken block of the earth. Geologically, this formation is called a *graben* (Fig. 1-2). This natural trench confines the river to a narrow, flat plain bordered by low cliffs on either side. Thus, the line of fertility which is the Nile Valley stabs through the deserts of northern Africa. Rain was (and is) an unusual event in Egypt. All life depended (and depends) upon the water in the river. Once a year the Nile would become swollen with the extra runoff of the wet season at its source in central Africa, and expand to a gigantic flood. The waters would spread across the surface of the Nile graben and flood the land (Fig. 1-3). In a flat land the flood would have spread over the plain and been absorbed. The narrow valley and cliffs prevented this in Egypt. The entire country flooded until the waters spread over the delta where the river broke out of the graben to join the sea. When the waters receded, a fresh layer of fertile silt covered the land to renew the minerals in the soil. Thus, both the security and fertility of Egypt were accidents of its geology.

Much of Egyptian learning was also determined by the flood. Maintaining the social order depended upon the development of accurate surveying methods. After the flood, the country had changed. Land marks were frequently washed away or buried. Sometimes the river

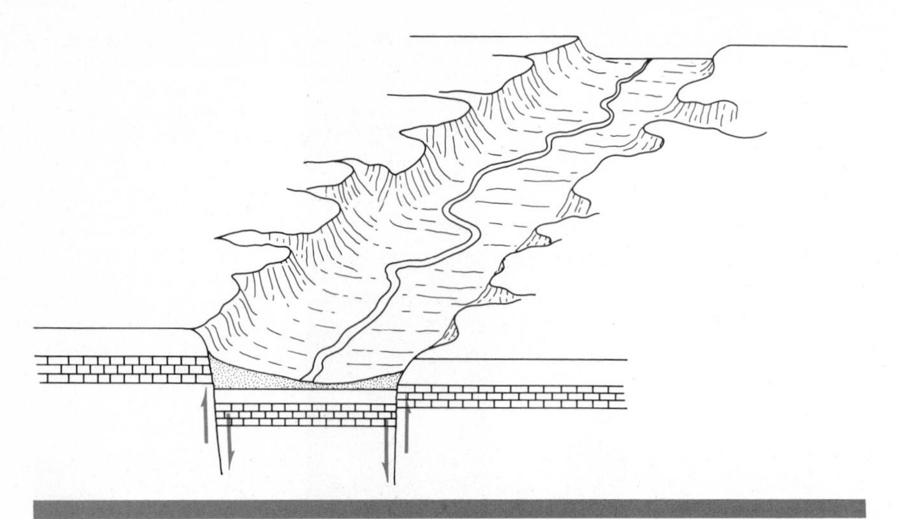

FIGURE 1-2
A graben is a block of the earth's crust between two faults or cracks which moves downward relative to the blocks on either side. For much of its course the Nile flows confined to the top of such a graben.

FIGURE 1-3
Looking across the Nile Valley in central Egypt, one can see the fertile band flooded by the Nile. In the distance is the row of cliffs marking the end of the graben. (*Photograph by the author.*)

would change its channel. Each year the Egyptian surveyors had to reestablish the limits of plots of land so that the planting could begin. Thus, the Egyptians produced a highly developed system of practical *geometry.* (The word "geometry" means "earth measurement" in Greek.) Egyptian geometry was not the formal, logical system developed later by the Greeks. It was instead a sophisticated system of methods and approximations that allowed the actual measurement of the land. Although the engineering accomplishment of the Egyptians in building the pyramids is often noted, perhaps more impressive as an engineering feat is a canal from the Nile to the Red Sea. The canal, its distance measured by an 8-day march, was cut through bedrock with copper tools and begun in about 2000 B.C. The Greek Herodotus, called the "Father of History," writes that the canal was reopened by the pharaoh Necho (about 600 B.C.). He goes on to describe a Phoenician expedition financed by the pharaoh which sailed completely around Africa, demonstrating that it was not connected to any landmass except to Asia through the Sinai Peninsula. It may be coincidence that this period of exploration and expansion comes immediately before the first great steps made by Greek science.

The Babylonians also were a very practical people. They were interested in commerce and developed an extensive system of weights and measures before 3000 B.C. Their number system was a double system. They used both a system based on 10s and one based on 12s. The system based on 12s made computations with fractions so much easier that it eventually dominated. The number 60 was magic since it was common to both systems. We preserve this by having 60 minutes (min) in 1 hour (h) and 60 seconds (s) in 1 min. We also use the Babylonian division of 360° in a circle. We preserve some of their other "number magic" in the idea of an "unlucky 13" or "lucky 7s." (As an example of how this persists, Admiral "Bull" Halsey flew into a rage when ordered to conduct the second minor American attack in the Pacific following Pearl Harbor. He had been given the code name "task Force 13" and ordered to sail on Friday the thirteenth. The number and the orders were changed!) The Babylonians were great believers in magic and mysticism. Not only did numbers have mystical properties, but so did the movements of the heavenly bodies. The Babylonians developed the science of astronomy, but only as a servant to their real interest—astrology. They believed that the positions of the planets influenced events on earth. This led them to establish very careful nightly observations of the planets and the recording of their positions. By the time of the Greek expansion of knowledge, almost 3000 years of Babylonian records were available.

To keep records through time the Babylonians used a 60-day calendar. For shorter intervals they used the moon's cycle—the "moonth," which became our month. Since there were seven special objects in the sky, they used a 7-day week—calling the first day after the sun, the second after the moon, and the remaining days after the planets. (We still use Sun-day, Moon-day, and Saturn's-day. The other days are named for the Norse gods Tieu, Woden, Thor, and Fria.) The Babylonians were compulsive timekeepers. They invented and refined the sundial and water clock. But this work also acted to serve the purposes of their fortune-telling. When the Persian Empire fell to Alexander the Great, the Babylonian contribution most widely received by the world was the system of astrology. There are more people making a living from astrology today than in ancient Babylonia. The system flourishes whenever people want soft and simple answers to hard and complex questions.

Whatever its shortcomings, astrology as a commercial enterprise was (and is) a success. In the United States today there are five professional astrologers for every astronomer. Historically, astrology provided the money for the development of astronomy.

The Rise of Natural Philosophy in Greece

Three major elements seem to come together to stimulate the development of science in Greece. The first is a basic change in the Greek life-style. The early Greeks were farmers, but (around 700 B.C.) population pressures in many Greek city-states forced an expansion to new lands (see Fig. 1-4). Some cities, such as Miletus in Asia Minor (modern Turkey), expanded to form colonies to the north along the Black Sea. Others, such as Corinth, expanded into the western Mediterranean. With the expansion came wealth. However, there also came a need to rethink the nature of the world. New experiences compelled some to move beyond the simple views of a farming society. From the east and south came the second element, the learning of more mature civilizations. However, the really significant change occurring at this time in Greece constitutes the third element—attitude.

Earlier attempts at understanding nature always had ended in a supernatural interpretation. The concept of a world running by mechanisms that a human being could understand—a world governed by physical laws—was an incredible idea that simply *did not occur* in most early civilizations. This is one of the reasons why ancient Greece is so special. By 600 B.C. we already find the philosophers at Miletus assuming that events in nature occur because of natural causes. The seeds of a mode of thought are planted in ancient Greece that eventually—after several false starts—will lead to modern science. To us it seems "quaint" to see an Indian rainmaker hired to "dance up a storm" for a group of drought-stricken farmers. This would not seem strange at all to almost any person in almost any civilization but our own. This type of behavior is the norm of human experience throughout history. Magic, mysticism, and the belief in the direct intervention of the supernatural into even the simplest of life's events governed human thought prior to the Greeks (and most of it afterward). Why is a pack of mosquitoes "bugging you" at night? Because someone wishes you evil, or in retribution for some nasty thing you did? The thought rarely occurs that they are driven by hunger and you happen to be the local supermarket.

In Greece a new way of thinking emerged. By interpreting the natural world as a series of events based on cause and effect—governed by physical laws—the Greeks initiated the movement to modern thinking. Ideas could accumulate. Wrong ideas could be discarded when they would not work; ideas that seemed to work could be preserved for the next generation. This new way of thinking was not necessarily popular in Greece, however. Anaxagoras was eventually driven into exile for his teachings. Socrates was executed. The Milesian philosophers were regarded as suspicious atheists. All was not quiet on the Greek front—it

FIGURE 1-4
Much of Greece is quite different from the setting of other early civilizations, as may be seen in this view of the southern peninsula of Greece. Since Greece lacked the large, stable rivers of Egypt and Babylon, agriculture was much more difficult. (*Photograph by the author.*)

never is during periods of conspicuous advances in human thinking. Quiet times seem to produce dull people. Change and intellectual achievement seem greatest when the majority opinion of the citizenry is that the world (and all that is good, useful, or right) is falling apart around them. And, if there is one thing the Greeks produced in abundance, it was intellectual achievement! The question of why this intellectual expansion occurs in Greece, or why it occurs at all, is interesting but not central to this book. The belief that the world operates in a manner that can be uncovered by humans engaged in rational thought is central. This is the starting point of science.

One clarification is essential. Not all Greeks were wise and intelligent nor were they all original thinkers. Most were too busy scratching out a living to be involved in an expansion of knowledge. (Indeed, when asked what it took to be a philosopher, Aristotle gave as his first criterion: "Be independently wealthy.") Most Greeks were very conservative in their views. However, Greece was such a diverse place that new ideas were usually tolerated. Open argument over ideas was widely practiced as an art in itself.

Individuals concerned with ideas were called *philosophers,* or "lovers of wisdom." Those philosophers attempting to describe what nature was like were called *natural philosophers.* With natural philosophy, the foundations of science are begun.

Some Accomplishments of Greek Science

Greek science is usually taken to begin around 600 B.C. with Thales of Miletus. Thales had traveled widely throughout Egypt and to Babylon. According to tradition, Thales began the study of geometry in Greece, based on knowledge he brought back from Egypt. The respect he won from other citizens was probably based less on this than on his shrewd business deals, which made him a fortune. Thales is best known for his successful prediction of an eclipse (possibly in 610 B.C. or 585 B.C.), and yet 2000 years later an eclipse was still an event that caused panic among the general population. Thales reportedly had studied astronomy in Babylon. He understood the cycle of eclipses worked out by Babylonians from their records. Perhaps it was the application of this knowledge once back in Miletus that promoted the idea among the Greeks that nature could be understood.

Miletus was located in a 100-kilometer (-km) (70-mile) stretch of coastline the Greeks called Ionia (Fig. 1-5). The Ionian school produced a number of distinguished natural philosophers. The last one of great fame was the physicist Anaxagoras who lived during the time of Soc-

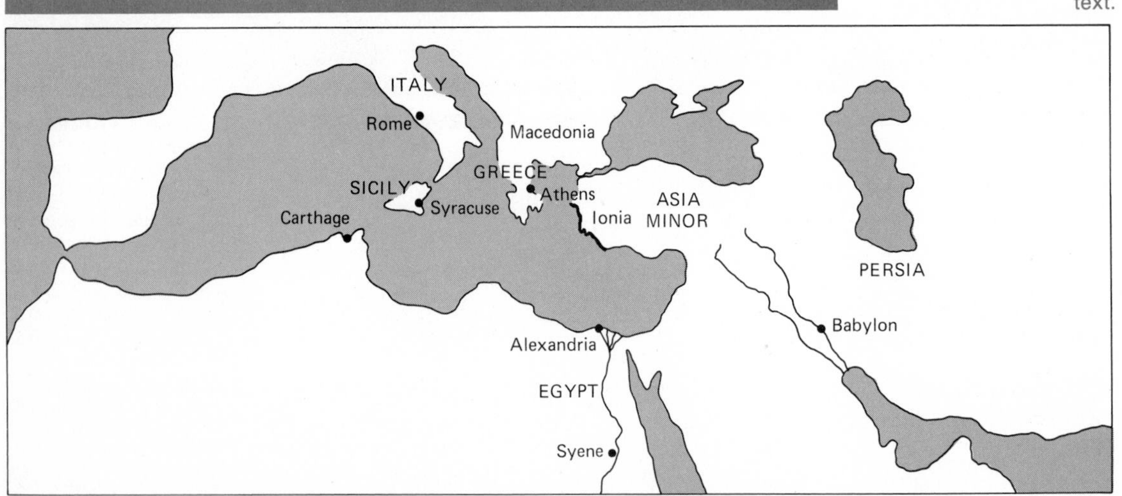

FIGURE 1-5 The eastern Mediterranean Sea indicating locations mentioned in the text.

rates. The line of philosophers from Thales to Anaxagoras was long, and much changed during the interval. Democritus, for example, maintained on philosophical grounds that all matter was composed of very small, indivisible pieces which he called *atoms.* Although rejected by other Greeks (notably by Aristotle), this idea was the basis for modern chemistry when reintroduced on experimental grounds in 1808, some 22 centuries later.

Only a generation after Thales a second school of Greek science began with Pythagoras. He, too, was born in Ionia, but he spent most of his life in southern Italy. Pythagoras is best known for the theorem in geometry named for him: For any right triangle, $a^2 + b^2 = c^2$, where a and b are the lengths of the sides making up the right angle, and c is the length of the hypotenuse, or the side opposite the right angle. He also discovered what might be called the first mathematical principle in the harmonies produced by stringed musical instruments. These two relationships convinced him of a mystical, mathematical order to nature. Although doing much important work, the followers of Pythagoras formed a closed and secret society in time. Pythagoras was the first important scientist to maintain that the earth was a sphere. In less than 200 years this became the accepted view in Greek science. The reasons usually given for accepting the roundness of the earth were:

1 The shadow of the earth falling on the moon during a lunar eclipse is always circular. However, the only solid that can always cast a circular shadow is a sphere. Thus, the earth is a sphere.

2 When ships sail away over a smooth sea, the hull disappears from view well before the sail. This could be accounted for if the earth were a sphere, and the ship was disappearing from view around the curve.* (See Fig. 1-6.)

During the 200 years between 400 and 200 B.C. scientific accomplishments multiplied. A method was devised by Hipparchus to measure the distance to the moon. This indicated that the moon was somewhat over 30 earth diameters away, or about 400,000 km in modern measurement. His value was within 5 percent of the accepted value today (382,260 km). Aristarchus earlier had devised a method to determine the distance to the sun in terms of the distance to the moon. He concluded from his measurements that the sun was at least 17 times as far away as the moon. The sun is actually nearly 400 times as far away as the moon, but the method used by Aristarchus was correct. The lack of instruments to measure extremely fine angles kept him from an accurate determination. However, even this number (over 6.5 million kilometers) was so vast that the Greeks were startled. Aristarchus also maintained that the sun

* This was the main reason for posting lookouts in a crow's nest in sailing ships. If the sea were perfectly flat, a person could see just as far from the deck surface as atop the mast. This is a different case than on land, where a high viewpoint allows you to see over nearby obstacles. There are no obstacles on a smooth sea.

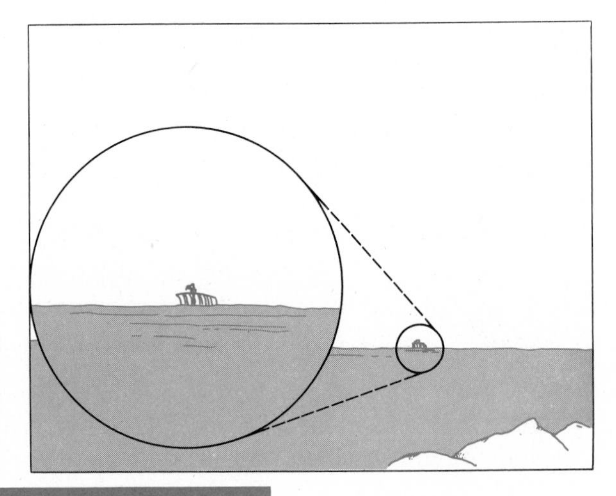

FIGURE 1-6
Sailing around the curvature of the earth a ship would disappear from view from the bottom up. This effect was well known in the ancient world.

was the center of the universe, and that the earth circled the sun as well as rotating on its own axis. His calculations had indicated that while the moon was somewhat smaller than the earth, the sun was much, much larger. The Greeks reasoned that if the earth was moving around the sun, in 6 months we should be halfway around in our orbit, or twice the distance to the sun from where we are now. The Greeks reasoned that being at a vastly different spot in space, we should obtain a greatly different view of space. In particular, to an observer on earth, nearby stars should shift in the sky against the background of more distant stars as the earth moves through space (Fig. 1-7). This shift is called *parallax.*

You can see parallax for yourself quite easily. Hold up a finger at arm's length. With one eye closed, line the finger up with same object across the room. Now, holding your finger steady, close the eye you sighted with and open the other eye. The finger will no longer line up with the object. It will appear to have jumped to a new position. Blinking rapidly from a left-eye view to a right-eye view, back and forth several times, will cause the finger to appear to jump between two different positions. This is parallax: the apparent movement of a nearer object against a

FIGURE 1-7
Parallax. As seen from point *A* the stump lines up with the distant oak tree. When viewed from point *B* the stump appears to shift its position against the distant forest and line up with the pine tree.

more distant background object due to a change in the position of the viewer.

While most Greeks immediately rejected Aristarchus' idea of a moving earth, several observed the stars carefully for any sign of parallax. In particular, Hipparchus, the greatest observational astronomer of ancient Greece, who lived just after the time of Aristarchus, searched for this effect. Hipparchus developed spherical trigonometry to allow him to determine and describe stellar position. He painstakingly measured the positions of almost 1000 stars to produce the best star map known to the world before the telescope. Still, Hipparchus could not detect the parallax demanded by the theory of his contemporary. He concluded that the earth was motionless, and went on to develop most of what became the Ptolemaic System of the universe (which is described in Chap. 5).

The last great accomplishment of the main period of Greek astronomy was the determination of the size of the earth by Eratosthenes shortly before 200 B.C. Before you read the separate description of his work at the end of this chapter, ask yourself how you would go about determining the earth's circumference using a road map, a calendar, and a protractor.

The most modern of the Greek scientists, and certainly the best physicist of the ancient world, was Archimedes. He worked in relative isolation from other Greek scientists at Syracuse, on the island of Sicily, where he developed most of his ideas. He considered himself a mathematician, and in the application of mathematics to physical problems he had no equal. His most famous discovery was the concept of density, buoyancy, and displacement in fluids. Hiero, the tyrant (king) of Syracuse, put a problem before Archimedes. He had just had a new crown made shaped like a wreath of leaves. However, he suspected that the goldsmith had cheated him and had added some silver to the gold.

Hiero wanted Archimedes to tell him whether the crown was pure gold or not without harming the crown in any manner. Tradition has it that the solution came to Archimedes as he lowered himself into his bath. As he entered the water the level of the water rose and overflowed the tub. He reasoned that for every cubic foot of his body that slipped under the water, a cubic foot of water was squeezed out of position and flowed over the edge of the tub. Seeing the application at once, he jumped from the tub

Ancient detective work

Was Hiero's crown pure gold or had the artisan cheated? Archimedes realized that gold packs into space more densely than other materials. Thus, if 200 g of gold is dropped into water, it will push out of the way, or displace, some 10.4 cubic centimeters (cm^3) of the water. The same mass of silver takes up some 19.0 cm^3. If a block of each of these materials is slowly lowered into a full container of water, the overflow could be measured to determine the volume of the block.

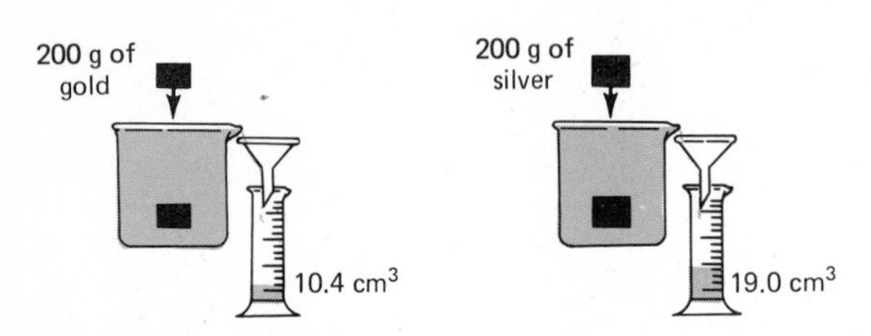

If Hiero's crown was pure gold and had the same mass of 200 g, it too should displace only 10.4 cm^3 of water. *Density* is defined as the mass of an object divided by its volume ($D = M/V$), and gold is the densest material known on earth (then or now). Thus, any material added to it would make the crown take up more space for its mass.

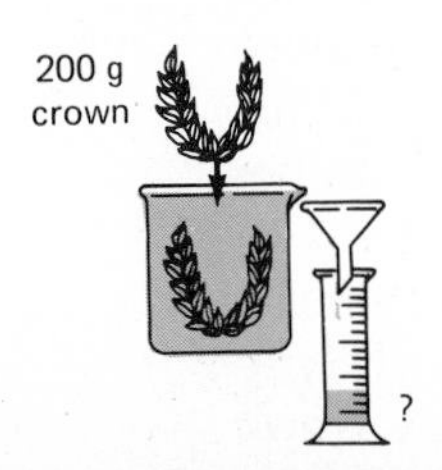

Actually, the precise measurement of the volume of a liquid is rather difficult. Archimedes pondered for a way to measure the crown with precision as he lowered himself into his bath. The fundamental idea that occurred to him was that objects lose weight when placed in water. For example, ships and people lose all of their weight and actually float in water, but even heavy objects seem lighter underwater. This can be summarized by *Archimedes' principle: An object immersed in a fluid loses an amount of weight equal to the weight of the fluid it displaces.* For example, the 200-g block of gold in the first illustration could be balanced by 200 g in the air. If we rebalance it with the

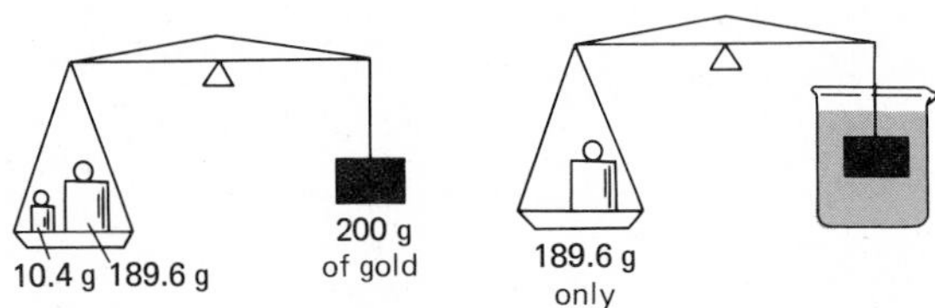

gold under the surface of the water, it will appear lighter. The block will have displaced 10.4 cm^3 of water with a mass of 1.0 g for each cubic centimeter, or 10.4 g total. Thus, the gold should now be balanced by only 189.6 g (200 g − 10.4 g). Balancing was a well-developed skill of measurement in the ancient world and allowed Archimedes to determine the authenticity of the crown by balancing it in air while suspended in water.

and ran home shouting "Eureka!" ("I have found it."), supposedly forgetting to put his clothes on in the process and thereby becoming history's first official streaker.*

What Archimedes realized was that for equal weights, silver has a greater volume than gold. Thus, if the crown weighed 200 grams (g) and if it displaced the same amount of water as 200 g of pure gold, it was made of pure gold. If it displaced more water, the gold was mixed with silver, which takes up more space for the same weight, or is less *dense*.

Archimedes' greatest claim to fame was in the early application of proper methods. Given the problem, he reasoned out the principle which was operating. He then expressed the principle mathematically, deduced the consequences, and tested them experimentally. Throughout, he refused to get involved in *why* the principle should be working or what part it played in some grand scheme of nature. These he left to lesser intellects more skilled at debate than in producing verifiable principles. In this sense, Archimedes is the first true, modern scientist. He was less concerned about second-guessing the motives of creation of science. During a trip to Egypt he designed a screw inside a tube which raised water when turned. The device is still used today. In addition, in a classic study, Archimedes developed the mathematical principles of the law of levers.

The Aristotelian Universe— A Theoretical System

In the last section we reviewed some specific Greek accomplishments in science, but the Greeks had problems as well. (See Fig. 1-8.) Most frequently, having discovered some particularly pleasing relationship, the Greeks were inclined to try to extend this idea into a grand scheme for all of nature. Having found and examined a single tree, the Greeks immediately wanted to infer the size and shape of the forest. These overextensions from the available information produced great discussions but limited knowledge. Roughly four periods can be identified in the development of Greek science.

1 (600 to 400 B.C.) *The Early Schools of Science.* Best illustrated by the Ionian tradition started by Thales or by the Pythagoreans. The early schools were mixtures of brilliant insights and wild speculation.

2 (400 to 300 B.C.) *The Emergence of the Great Philosophic Systems.* These systems were based on the accumulation of much previous good work and tried to weave knowledge and observation together into a sweeping interpretation of the grand design of the universe. The best example is the Aristotelian System, which we are about to consider.

* In this period, bathing was often in public baths in cities, not in private bathtubs. The public baths were a social as well as hygenic institution. Even though this story is ancient, a lack of authenticity is possibly indicated by its attitude. The ancients were not particularly unused to nudity, since most heavy labor was done unclothed in hot climates, and bathing was communal although loosely segregated by sex.

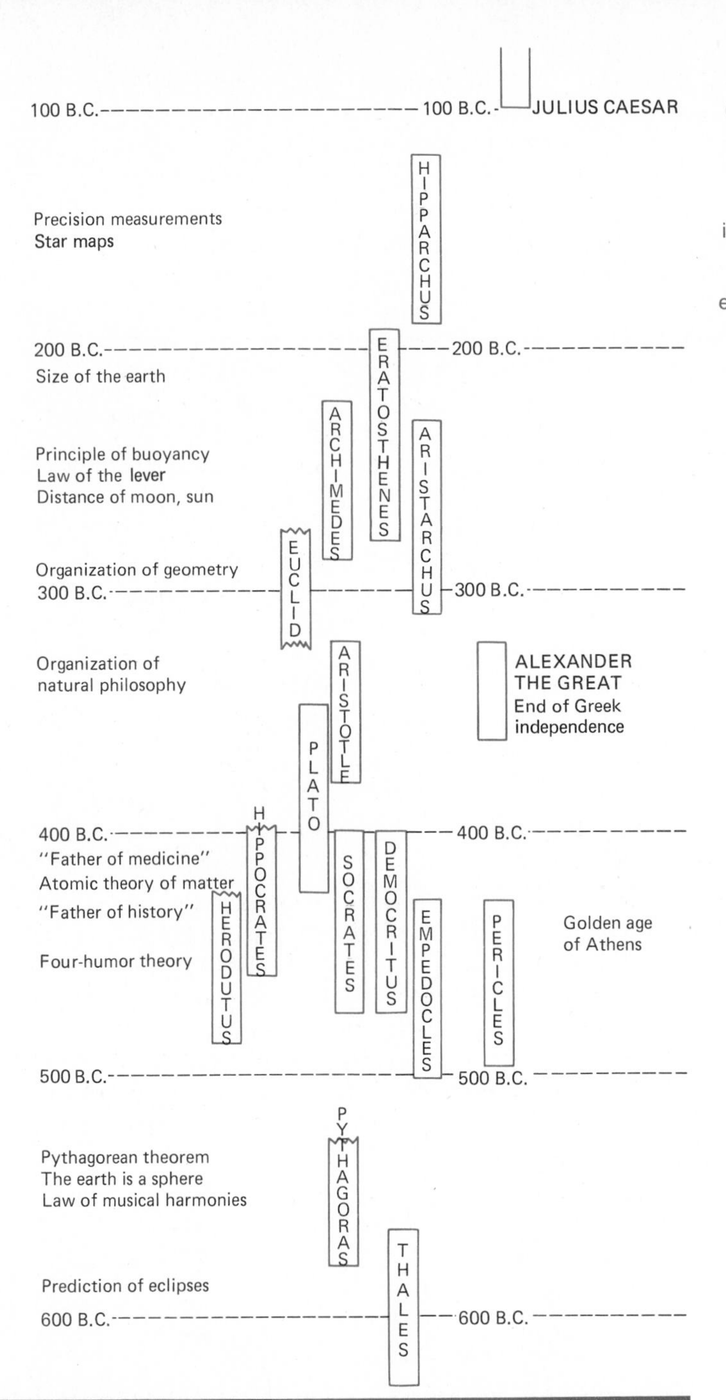

FIGURE 1-8 Greek science extended over a time 5 times as long as the existence of the United States. The life spans of prominent individuals mentioned in the text for the first 5 centuries of Greek science are indicated in the figure.

3 (300 to 100 B.C.) *The Age of the Specialist.* This period is characterized by individuals working within one branch of the sciences and no longer trying to summarize all of nature. Much of the most enduring work was done in this period by people such as Euclid, Aristarchus, Hipparchus, Eratosthenes, and Archimedes.

4 (Beyond 100 B.C. to the end of classical civilization with the fall of Rome, 440 A.D.) *The Summarizers.* This period overlaps with that of the specialists. Increasingly, however, individual scholars within a field studied what had been done before them and pieced together the best of it into a polished summary of the field. The advantage of this group was that they had the last word. The best example is Claudius Ptolemy in astronomy, to be discussed in Chap. 5.

Geographically, the first period centered in the outlying areas of the Greek world. The second period was focused in Athens, while the last two were dominated by Alexandria in Egypt.

Having considered several individual discoveries, you should be willing to admit that many Greek achievements have withstood the test of time and can correctly be called works of science. The methods used in many Greek measurements in astronomy, for example, have been held correct from the time they were originally performed to the present. However, these bright spots were not typical of the majority of Greek work in natural philosophy.

Much of this work resembles the well-known ancient Indian proverb of the blind men and the elephant. Paraphrased, this is as follows:

A number of blind men lived in the kingdom who had no concept of what an elephant was like. In his kindness, the rajah had them led to an unusually docile beast so that they could experience the elephant for themselves. The first blind man, in reaching out, grabbed the elephant's tail. After exploring its general size and texture, this man concluded: "The elephant is very much like a rope." The second man contacted the elephant's leg. After sampling its surface and contour, he concluded: "The elephant is very much like a tree." The next man touched the elephant's side. After groping with his hands across the side, he concluded: "The elephant is very much like a wall." (And so on, through several more men, each concluding something quite different.)

So it was with most of the Greek natural philosophers. From a limited amount of information they tried to conclude the nature of the entire physical universe. If you view Greek science as a series of unconnected, brilliant accomplishments along with many equally unconnected "bad guesses," you can never appreciate the hold which it had on the minds of human beings. The Greeks were very prone to conclude the entire "size and shape" of the natural world from an examination of its "tail."

To give you some idea of the range of this type of conclusion, we will examine a single system describing the nature of the physical world. Even though it will appear quite long, our description is an extremely abbreviated form of this system. The example presented is the most enduring view of the world yet produced in Western history. Although proposed in the fourth century B.C., these views were not seriously challenged on all fronts until almost 2000 years later. This system is the view

The size of the earth

How large is the earth? This question was first answered for the ancient world by Eratosthenes (276–194 B.C.). His accomplishment was spectacular enough and so simple in its conception that it is worthwhile to discuss it even today. By the time of Eratosthenes, the center of scholarship, and of science in particular, had moved to Alexandria in Egypt. Supposedly, Alexander himself had walked out the outlines of the city when he ordered it built. The actual expansion of Alexandria came under Alexander's general Ptolemy and his son Ptolemy II. The city was the capital of their third of Alexander's empire. In the royal compound, two institutions were built which became the focal point of ancient learning. The first was the museum. Technically, a museum was a temple dedicated to the muses—the gods of history, art, dance, astronomy, etc. There had been many museums, but this became *the museum*, the one which gave new meaning to this word in dozens of later languages. Today we would probably call it a scientific research institute. The second institution of learning at Alexandria was the great library. This became the largest and most famous library of the ancient world. The director of the library was always one of the foremost men of letters of the day. Eratosthenes was the only scientist ever to head the great library. His interests included so many fields that he was nicknamed "pentathlos" (after the pentathlon event in the Olympics, where athletes compete in five separate events.)*

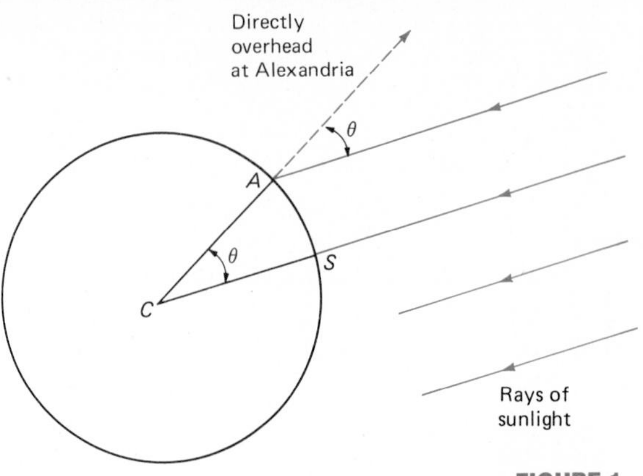

FIGURE 1
The angle θ between the directly overhead direction at Alexandria (A) and the sun's rays was the same as the angle between the direction to Alexandria and that to Syene (S) as seen at the center of the earth (C). (Why?) Measuring θ gave Eratosthenes the size of the earth.

Eratosthenes heard that in Syene, near modern Aswan, in southern Egypt, on the longest day of the year, the sunlight at noon fell straight down a deep well shaft, illuminating the surface of the water. This clearly implied that on that date (June 21) the sun was directly overhead at noon at Syene. But, Eratosthenes knew that on that date in Alexandria, in northern Egypt, the sun was not directly overhead. A stick placed in the ground left a shadow indicating that the sun was 7.2° from being straight up at noon on that date. But there was only one sun, and Aristarchus' measurements had indicated that the sun was incredibly far away. This difference in the height of the sun had to be due to the differing positions on the curved surface of the earth. The sun was in the same direction in space for both Syene and Alexandria. Alexandria was not directly beneath the sun, but some 7.2° around the curve of the earth from that point. Thus, Eratosthenes knew that Syene and Alexandria were separated by what we would call 7.2° of latitude. But how far was this in kilometers, or miles (he actually used the Greek measure of *stadia*)? Here, Eratosthenes was fortunate. Syene was almost due south of Alexandria, and between them was the most intensively surveyed strip of land on earth—the Nile Valley. Eratosthenes obtained 5000 stadia as the distance. Thus, 7.2°, which is exactly one-fiftieth of a circle, was 5000 stadia in length.

$$\frac{\text{Measured part of the circle}}{\text{Total circle}} = \frac{\text{measured distance}}{\text{total distance around the circle}}$$

$$\frac{7.2°}{360°} = \frac{\text{5000 stadia}}{\text{circumference of the earth}}$$

Eratosthenes eventually settled on 252,000 stadia after repeated measurements. This would be 39,700 km in our measuring system. The currently accepted value for the circumference of the earth measured across the poles is 39,940 km. This startling agreement is mostly luck. The measurements were not that well known. But, this *was* a measurement. Earlier numbers were guesses.

* Actually, this was ridicule, not good-natured kidding, for not sticking to one thing he was best in. Eratosthenes was good in everything, from literature to filing systems, but was the world's first *great* physical geographer.

of the physical world proposed by the philosopher Aristotle. Note that this single system contains the basis for explanations in geology, chemistry, physics, astronomy, and much else. Aristotle undertook to synthesize all of human knowledge. Behind him were centuries of Greek achievement, while in his own day Greece was losing its political identity and would never completely recover the independent spirit that produced its greatest achievements. Thus Aristotle, in addition to his own contributions, became the great summarizer of Greek knowledge. Later Greek work was done largely by specialists within particular fields. Some of this work was so successful that Aristotle did not have much impact in those areas—for example, astronomy or geometry. But in most areas of human knowledge his impact became so great that the late Middle Ages referred to him simply as "the philosopher."

The Physical World of Aristotle

In examining matter, the Greeks recognized two sets of properties which they believed were most important: hot vs. cold and wet vs. dry. The combinations of these properties led the Greeks to assert that there were four basic substances, called *elements* or *essences* (Fig. 1-9). A typical rock was cold and dry and an example of the element earth. Of course, not all rocks were alike, so few, if any, were pure earth. Rocks were mostly composed of the element earth but also could have some of the other three elements. These four building blocks of *earth, air, fire,* and *water* made up, in varying combinations, all of known materials.

Furthermore, each of these elements had a "natural place" in the design of the universe. Earth, being the heaviest of these elements, migrated to the center of the universe. A rock released anywhere would immediately proceed toward the center and pack down as tightly as possible on the other rocks in its way. In this way, the ball of rock upon which the Greeks found themselves had accumulated. In the same manner, water migrated toward the center. Being less dense, the water had to pack down over the top of the rock. Also, being a fluid, water could

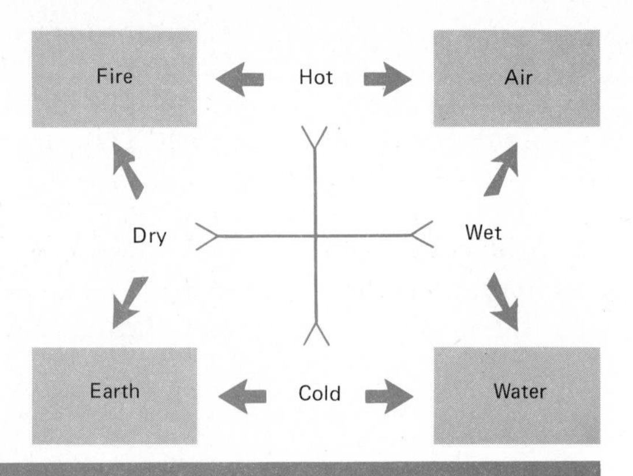

FIGURE 1-9
Two sets of opposite characteristics combined to produce the four basic elements which the Greeks believed made up all matter on earth.

change its shape in response to its downhill mission and flow to the lowest possible level. Solid rocks could not do this and thus formed bulges and humps which stuck up through the water into the air. Above the water was the air. Air, too, attempted to reach the center. It filled cavities within the earth when water was excluded. Above all of this was fire, the least capable of displacing the others in the struggle to achieve the center of the universe. Thus, the volume of the known universe—the sublunary sphere or roughly everything below the orbit of the moon—ideally would be made of four spheres (Fig. 1-10). Clustered around the center of the universe would be the sphere of earth, then a spherical shell of water, then air, and lastly fire. All of these would be separate and unmixing in an ideal case. But the sublunary sphere was not ideal. It was the sphere of corruption or change, as opposed to the sphere of perfection located from the moon outward. Mixing did occur on the earth. The sun sent in a stream of fire to the surface. Some of this was stored by plants for later release in cooking and heating over the earth. One can picture the Greeks convincing themselves of the correctness of their ideas around an evening campfire. In the upward rush of the flames, the fire, suddenly released from the wood, races to its natural place in the universe. In exactly the same manner, a bubble of air released underwater is displaced upward to its appropriate place in the universe.

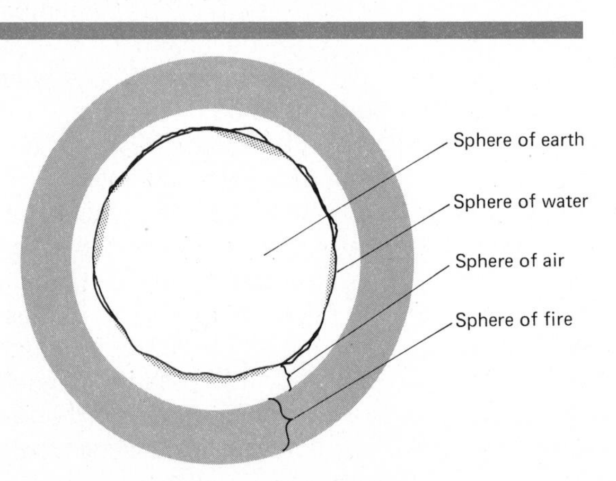

FIGURE 1-10
According to the Greeks, the differing densities of the materials crowding toward the center of the universe led to a separation of matter into four distinct spheres.

Perhaps most awe-inspiring of all is the action of the fire of the sun at the surface of the water and air. There, a mixing occurs and the fire bound to the water forms a vapor. This vapor, balanced between the tendency of the fire to rise in air and that of the water to fall, ascends to form the clouds. When finally the separation of the elements does occur, the fire retires to its loft, while the released water cascades upon the surface in a fall of droplets, renewing the vegetation. These droplets, in their eternal quest for the center of the universe move ever downward, first forming small trickles, then streams, and finally raging rivers on their way to the sea—the sphere of water. We may summarize this more formally as follows.

NATURAL MOTION

Natural motion is a spontaneous motion at least some part of which is upward or downward and occurs in response to matter seeking its natural place in the universe. (For example, water moves sideways yet always downward.)

In an era when measuring instruments were primitive and inaccurate, the Greeks relied mostly on reasoning to uncover the nature of the universe. A large rock weighed more than a small rock. The natural force pulling it downward was greater. Therefore, the Greeks reasoned that it should fall at a greater velocity than the smaller rock. Philosophical considerations led the Greeks to conclude that objects fell at speeds proportional to their weight and that the speed during the fall was constant. Observing the rise of an air bubble in water or the dropping of metal spheres in water would seem to confirm this. Thus, a 5-kilogram (-kg) bowling ball would be expected to fall through the air 10 times as fast as a $\frac{1}{2}$-kg baseball. As the weight W increased, the speed of falling v would increase. If you double one, the other will also be doubled. Today we would indicate this as:

$$v \propto W$$

where the symbol $\propto$ means "is proportional to." This was *not* the way Aristotle indicated it, however. First of all, algebra had not been invented, and second, Aristotle did not believe in mathematical laws as descriptions of nature (as did his great teacher Plato). To Aristotle, relationships in nature were approximate, not exact, and mathematics could be misleading.

FORCED MOTION

Of course, not all motion is in response to seeking out a natural place in the universe. Motion also may be in the horizontal plane, or a rock may be lifted. These types of motion differ from natural motion in that they require an agent to produce the motion. Someone or something must exert a *force* on the object to be moved. To lift a rock requires that an upward force be applied. If the force is large enough to overcome the natural force of the rock's weight, the rock can be raised. However, once this force is removed, the rock falls instantly in response to its natural tendency.

An extension of Greek thinking—The art of speculation

The Four-Humor Theory of Disease

What does the four-element theory of matter imply about the nature of the human body? In spite of the reference to some of us as windbags, the fact is that if you drop us, we fall. Thus, there seems to be more earth and water in the human body pulling us down than fire pulling us up. On the other hand, there is clearly some air—we don't do well if we stop breathing. While humans do float, they don't do it nearly so well as some animals. May we conclude that there is a great deal of water in the human composition? We "know" there must be some fire in humans since they tend to be warmer than their environment. (Although it isn't necessary that fire display its presence this way—consider a lump of coal, which must have considerable fire in it but can be quite cold.) Furthermore, can we conclude that the upward pull of the air and fire in a human body just about balances the downward pull of the earth in us, accounting for our floating behavior in water?

If we accept this line of argument it becomes clear that humans are a fairly delicate balance of the elements (consider that we don't *always* float). From this it is a small step to consider what the consequences might be of changing the balance of the four elements in the human body. This suggests where the problem may be when the human body starts to work badly.

Think about this for a minute!

Consider that humans give off solid (earthy) and liquid (watery) wastes. We are constantly exchanging air in breathing, and we radiate heat (fire) continuously. Could these be attempts to regulate the balance of the four elements? If so, you might expect four separate regulatory systems within the body. Could this be the job of all of those strange-looking blobs of flesh that humans are stuffed with? But organs are stationary; how, for example, would the water-level-processing organ collect excess water from all over the body? Ancient physicians believed they had located four fluids, or *humors*, in the

On a horizontal surface, a heavy cart will move forward if a sufficient pulling force is exerted by a horse. But, in the absence of such a force, the cart will just sit there at rest. If the horse stops pulling the cart, the motion of the cart stops. Thus, Aristotle believed that the normal situation is for objects to be at rest unless acted on by some outside force. If the force is large enough, motion will result. However, this *forced motion* (or *violent motion*) differs from natural motion in that when the force stops, the motion stops. Natural motion will persist until the object has found its natural place in the universe. Furthermore, in the example of the cart and horse, if the horse pulls with a constant force, the cart will move at a constant speed. All of this sounds quite reasonable until it is very closely examined. For centuries this was the opinion of the majority of thinking human beings trying their best to find order in the world around them.

Before we leave forced motion, we should note that there is one obvious problem, which concerns what is called *projectile motion.* Consider the motion of a bullet fired from a gun. The bullet sits quietly inside the gun until the gunpowder behind it is ignited. Then, the expanding body—one associated with each of the elements. They believed that an imbalance of one or more of these could result in anything from serious disease to minor ailments—such as an "evil humor" (no pun, this is the origin of the phrase).

To their credit, physicians in Greece seem to have been reluctant to tinker with the human body if they weren't very sure of what was wrong. But, they did love to speculate, as did their philosopher friends. The most influential medical writer of the ancient world was the physician Galen. He left studies in Alexandria in the second century A.D. to specialize in the diseases of the very rich in Rome. The influence of his writings through history has been enormous. Renaissance artists, such as Michelangelo, over 1000 years later found themselves studying the anatomy books of Galen. (By that time dissection of humans had long been banned, and you had to rely on secondhand sources to understand what was inside the human body causing all of the bumps, bulges, and lines we see on the surface.)

Physicians based treatments on the four-humor theory, since the theory did seem reasonable and few alternatives were available. For example, if a patient has a clear excess of fire, what could be more logical than to seek to remove some of it? Thus, you could drain some of the fluid associated with fire, the blood, from the body. A more reasonable treatment for high fever would be hard to imagine for someone who believed in the theory. Fortunately, many patients were even strong enough to survive the treatment. The common nature of this practice still can be seen all over the world. The red-striped barber pole marks the one person in any town who will definitely have the sharp instruments needed to open a vein. The practice of bleeding is where the pole got its red stripe. The practice isn't entirely dead. In 1953, when Joseph Stalin, the dictator of the Soviet Union, was dying, he was treated with blood-sucking leeches. This was interpreted in the West as indicating that his death was inevitable, and the bleeding was a concession to unite the traditional elements of his country. Wrong ideas die very hard!

gases generated by the chemical reaction push outward against all of the walls confining them. The only movable wall is the bullet, which is violently forced down the barrel of the gun. The forces acting against the bullet are extremely great, and it emerges from the gun at high speed. But now what happens? The bullet emerges into the air and no longer has the force of the expanding gases pushing on it. In the absence of a force, violent motion should cease according to the Aristotelian view of forced motion. The bullet should stop in midair and quietly fall to the ground somewhere near the end of the barrel. While this might be a great relief to whatever was the intended target of the gun, it obviously does not happen.

While the ancients did not have guns, there is no evidence that Aristotle advised Alexander to leave behind his 1000 Cretan archers for his attack on Persia because they might all become casualties with foot wounds. The ancients had enough experience with projectiles to realize that the violent motion of a spear or arrow persists long after the force is removed. They attributed this continuity of motion to the action of the air itself. As the arrow moves rapidly forward, it leaves vacant a volume of space behind it. The air rapidly rushes in to fill this void, and, in so doing, the air exerts a forward push on the arrow. According to the ancients, this forward force accounted for the persistance of motion of a projectile after release.* A modern observer might object that normal projectile motion could not exist in a vacuum. The ancients, however, denied that a vacuum could exist at all. The reason that the air would rush in behind the arrow in the first place was to prevent a vacuum from forming. The existence of space without matter filling it presented serious philosophical problems to Greek thought, and they rejected the possibility of a vacuum. This was summarized in the popular slogan, "Nature abhors a vacuum."

HEAVENLY MOTION

Natural motion and forced motion summarized the everyday occurrences of motion which the Greeks observed around them. The objects in the heavens presented a different case altogether. These objects did not fall nor did they recede from the earth. Day after day and year after year, the planets (which included the sun and the moon for the Greeks) and the stars moved continually about the earth. Unlike the earth, which represented the *sphere of corruption* or change, the heavens were the *sphere of perfection.* Beyond the sphere of fire the heavens began, and here new rules were called for. The heavens were the purest handiwork of creation, and out there everything was perfect, ceaseless, and unchanging.

This view has several implications. First of all, the heavens are not made of the four elements found on the earth. They must be composed of some "fifth element," which Aristotle called the *ether.* Since no "chunk" of ether could leave a place that was perfect and unchangeable, none was available for analysis. However, the ether filled all of the heavens, since nature abhorred a vacuum out there, too. This wasn't too large a task, since the heavens consisted of the separate shells containing the planets and then the large sphere of the fixed stars enclosing all of the universe. Beyond this there was nothing—not the nothing of empty space or a vacuum, merely nothing. The universe was of a fixed size and bounded by the sphere of the fixed stars. All of this structure was centered, of course, on the earth.

The second implication of the nature of the heavens was that heavenly motions must be perfect. To the Greek mind, the motion of an automobile in traffic would not be perfect—speed up, slow down, brake hard as you get cut off, give it gas again, more gas when the horns behind you start blowing. Perfect would be much more like the flight of a Concorde jet through the stratosphere—smooth, vibrationless, at high constant speed. The fastest motions experienced by the Greeks were in galloping on horseback or in sailing with a stiff breeze on their sail. The passenger in both cases would hardly describe these motions as "perfect." The

* There were several competing explanations of projectile motion in ancient science. This is the easiest one to visualize, and all are wrong.

Greek concept of perfection demanded an unchanging, eternal motion. Such a motion must be at constant speed and on a path of constant curvature. The only path that returns on itself that has constant curvature is a circle. Therefore, *all heavenly motion must be at constant speed in a circle.* There were very few points on which the independently minded Greeks were more in agreement than this. Any other motion would offend their sense of beauty, symmetry, and perfection.

In this section you have had a brief exposure to the structure and motions of the universe as perceived by Aristotle. His synthesis attempted to pull together all physical phenomena and relate them in a single coherent structure. This was a monumental work of the human intellect. Of course, it is incorrect on each and every point listed, but this does not detract from it as a great theoretical system. A flawed idea which is only partly workable at least provides a starting point for discussion. The intuitive power of this system can be seen from the fact that it was held as more or less correct for 2000 years.

What Went Wrong?

An overview of Greek science is incomplete unless we ask, "Where did they go wrong?" They did not produce the fundamental changes in thinking which we call the scientific revolution. That came almost 2000 years after Aristotle, and mainly as a reaction against Greek views.

Several items seriously hampered Greek work. One point which is often suggested is a Greek reluctance to do serious experimentation, which was equated with manual labor. Even in democratic Athens the large majority of the male population was composed of slaves and free, but nonvoting, men. (The female half of the population had even fewer rights.) High-class Greeks did have a bias toward thinking and discussing rather than measuring and tinkering. Aristotle is often criticized for asserting that human males and females had different numbers of teeth. With a wife, a daughter, and numerous slaves available to him he really could have counted. Still, Aristotle had better attitudes toward direct experimentation ("getting his hands dirty") than did most Greeks.

A second problem was that the role of mathematics to the physical sciences was still not clear. Aristotle, fundamentally a biologist, regarded relationships in science as approximate rather than mathematically precise. Had he lived a century or two later he might have changed his mind when reviewing the greatest scientific accomplishments of the Greek mind. However, it was Aristotle's broad description of the world rather than the sharp penetrations of an Archimedes, for example, that held the human mind. Also hampering direct mathematical experimentation was the number system itself. If you try long division of Roman numerals (where do you think the Romans got them?), you will appreciate why the Greeks avoided calculation whenever it was possible.

Probably the most critical problem with Greek science was the same throughout its development: The Greeks asked the wrong questions about nature—questions which they could scarcely be expected to answer correctly. The Egyptians and Babylonians had questioned the natural world by asking "Who?" The Greeks moved one step closer by asking "Why?" The beginning questions to pose to the world around us are much more modest. The appropriate questions are "What?" and "How?" If you cannot define "love," for example, you will have precious little likelihood of being able to explain it. The Greeks were more interested in explaining nature than in describing it. It required more experience with the depths of humanity's ignorance to force us back to this more modest beginning.

Problems

1 What could the ancients conclude about the relative proportions of the elements fire, earth, and water in the Goodyear blimp?

2 You have a friend at a university which the map indicates to be exactly 500 km north of your campus. If you are experimentally to determine the size of the earth as a class project, what measurement must you ask your friend to make?

3 A careful study indicates that less money is won at a roulette table on the number 13 than on the number 7. Does this show that the former is an unlucky number and the latter a lucky one?

4 Why do so many Greek ideas sound so reasonable when they are completely incorrect?

5 Which of the following statements would Aristotle object to?

(*a*) A rock falls at a constant speed determined by its weight.
(*b*) Without forces, a moving object would remain moving in the same manner.
(*c*) The earth exerts a downward pull or force on the moon.
(*d*) Given a large enough sideways push in space, a rock could remain orbiting the earth forever.
(*e*) A bubble of air released underwater should rise.

6 Why was the idea of an earth moving around the sun rejected in the ancient world?

7 Density is defined as mass divided by volume. If 200 g of gold occupies a volume of 10.4 cm^3, what is the density of gold? If 200 g of silver occupies 19.0 cm^3 of volume, what is the density of silver?

8 Algol, "the Demon Star," in the constellation of Perseus in the sky, changes its brightness notably over time. Why would the Greeks maintain that Algol was not a true star but some object high above the earth and nearer to us than the moon?

9 What could account for the power that the Aristotelian view of the physical world had over serious thinking for such a long period of time?

10 The word "quintessence" literally means "fifth element." Look up its meaning and see if you can figure out how it acquired this meaning. What was the quintessence to the ancients?

11 Can you reread the section on the Aristotelian universe and convince yourself that each part of it is incorrect?

CHAPTER TWO

The Empirical Process—Obtaining Order from Numbers

Introduction Eighteen years before the birth of Galileo, Tartaglia demonstrated that a cannon achieved its maximum range when fired at an angle of 45°. Tilt the cannon more, or tilt the cannon less, and the cannonball strikes closer to the cannon. Tartaglia did not know why this was true, but it was. He had experimented by firing cannons at various angles of elevation, and this was the result.

Later, Galileo experimented with rolling balls down ramps. From a lengthy study he concluded two things. First of all, a horizontal motion persists unless something stops it. Second, all objects when released fall at the same constantly increasing speed (at the same acceleration). From these two ideas he could conclude, even without firing a cannon, that a cannonball must attain its greatest range when fired at an angle of 45°.

Twenty-some years after Galileo's death, Newton wrote down three general laws describing the nature and causes of all motion and the form of the law of gravitation by which objects with mass are attracted to other objects with mass. From these laws an individual who had never set foot on any planet, nor seen a cannon, could have predicted the angle of firing to obtain the greatest range. All of Galileo's results, as well as additional results, also could have been predicted.

The art of producing understanding where it was previously absent is quite complex. At one time it was fashionable to talk about "the scientific method." In fact, there are really many approaches which have led to great understandings. Similarly, traditional approaches to a given problem have frequently produced only repeated failure. High school texts used to outline the steps of scientific method as follows: Define the problem; gather pertinent information; formulate hypotheses (suggested explanations); design experiments to confirm or reject each hypothesis. This is an excellent blueprint for conducting an investigation. Unfortunately, there is no master plan for the inevitable discovery of new ideas in science. As a simple example, it is difficult for an astronomer who is trying to understand how stars are formed to set up an experiment to check hypotheses. In reality, there are multiple methods by which science proceeds. There are, however, two basic processes or modes of operation going on in science continually. These are the *empirical process* and the *theoretical process.* Science requires both of them.

The word "empirical" comes from the Greek word meaning "experience." Thus the empirical process is the process of deriving understanding from observation or experiment. Empirical knowledge comes from what we can directly measure about the world. It leads to the ability to describe what will happen in a given situation. The theoretical process seeks to find broad principles that can be applied to explain a wide variety of relationships. Theory usually proceeds out of a base of known information gathered by empiricism. While empiricism produces a continuous growth of knowledge, the great leaps of science come from theory. Theoretical breakthroughs occur which produce a new picture of the ways by which nature is working. The application of these new ideas to numerous problems in different areas triggers a flood of new understanding.

The majority of specific Greek achievements in science were related to that branch of mathematics which the Greeks knew best, geometry. Eratosthenes' measurement of the size of the earth and Aristarchus' determination of the distance of the moon and sun all depended upon the Greek knowledge of geometry. That this early work was best in astronomy was no accident. The mathematics of space is geometry. Greek work was less enduring where the relationships involved number but not necessarily spatial relationships. If any moral has emerged from the 5 centuries of modern sciences, it is that *the natural world is very highly ordered in both structures and in actions. Furthermore, the best and only practical way to describe this order is with mathematics.*

In this book we will not emphasize the role of mathematics in the physical sciences. To ignore that role, however, would be dishonest. Science relies on measurement and on the ability to make exact, mathematical statements which can be either verified or rejected.

The Art of Measurement

One of the major conclusions to be made from the study of science is that the behavior of nature is essentially quantitative. That is, the workings of nature are best described by numbers. As with many ideas, this

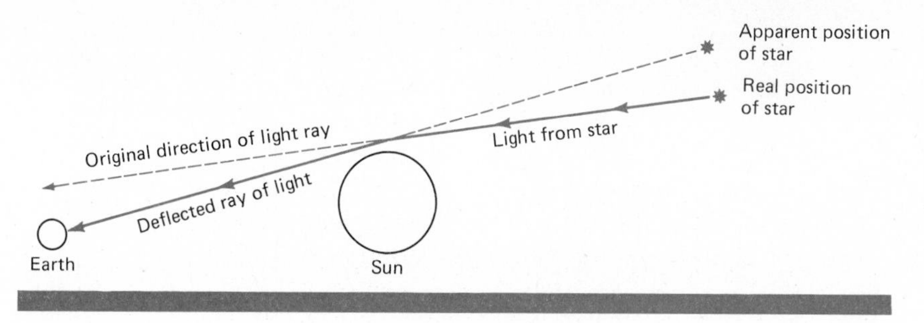

FIGURE 2-1
Einstein predicted the deflection of starlight passing near the sun. The scales are exaggerated. If the sun were 1 inch (in) across in the diagram, the earth would be a period placed 3 yards (yd) to the left and the nearest star would be almost 400 mi away.

was expressed best by Galileo, who said that science "is written in that great book which ever lies before our eyes—the universe—but we cannot understand it if we do not learn the language and grasp the symbols in which it is written. This book is written in the language of mathematics."

One of the main advantages of the mathematical description of nature is that ideas may be demonstrated easily to be wrong. For example, in 1915 Einstein asserted that as a consequence of his general theory of relativity the light from distant stars would be deflected or bent by as much as 1.75 seconds of arc (1.75″) when it skimmed near the surface of the sun on its way to the earth (Fig. 2-1). This type of assertion definitely allows a direct measurement to show that an idea is incorrect. The measurement in this example is quite delicate. A second of arc is roughly the width of a dime as seen by the naked eye at a distance of 4 miles (mi)! No one had previously suggested that the path of light should be bent at all by passing near to a large mass, much less indicated by how much it should be bent for a given mass. If starlight was not deflected by passing close to the sun, the prediction would be wrong, and so would Einstein's theory of relativity. When astronomers established that starlight really was deflected, by observing stars during a total eclipse, Einstein's theory had correctly predicted a totally unsuspected phenomenon. While this, by itself, did not "prove" that the general theory of relativity was correct, the theory had passed a very important test and had to be considered very seriously. Thus, if we phrase ideas mathematically, not only do we tend to reflect an exactness required by the way nature behaves, we also provide an easy opportunity for proving ourselves wrong.

Mathematical description obliges us to measure quantities and assign numbers to them. For some variables, such as length or weight, this is easy. (For others, such as love or honesty, it is difficult or impossible.) One of the successes of the physical sciences is that we can usually agree on what must be measured to describe a particular variable. Physical scientists rely very much on measurement and mathematical descriptions because nature is essentially mathematical in the way it behaves.

MEASURING SYSTEMS AND UNITS

The essential requirement of any measuring system is the use of units which are the same for all investigators and are of a known, agreed-upon

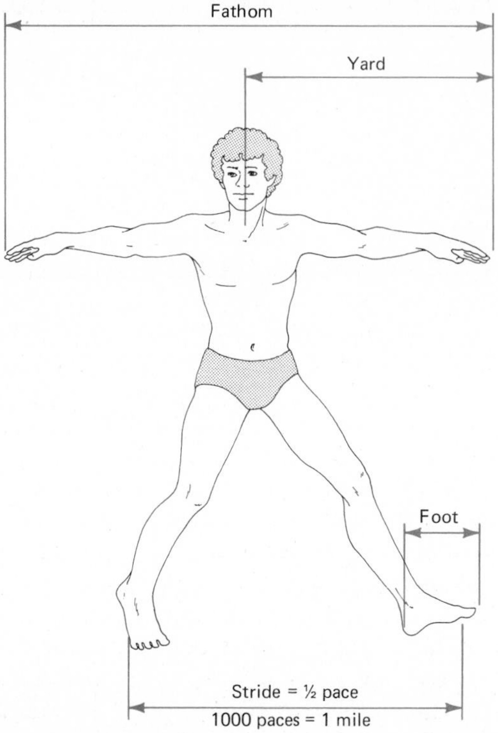

FIGURE 2-2
Before the standardization of units, most measures were defined as part of the human body so that people could carry their system around with them.

size. Beyond this, the choice of units is largely one of convenience. Some systems are unnecessarily hard to work with. The clearest example of an inconvenient system is probably the measurement of length in the English system of units (which is why the English had the good sense to abandon it). Inches are customarily divided into eighths, sixteenths, thirty-seconds, and sixty-fourths. Twelve inches make up a foot, three feet a yard, two yards a fathom, two and two-thirds fathoms a rod, one hundred and ten yards a furlong, eight furlongs a mile, three miles a league, and so on. Conversion between these units is tedious, and the likelihood of making an error during the calculations is quite high. (For example, how many sixteenths of an inch are there in 7 leagues?) Nature provides no obvious standard for most units, including length. Thus, early societies based their measuring systems on parts of the body because they could be remembered easily (Fig. 2-2). (The inch was the length between the first and second joints of the index finger. The origin of the foot was obvious. The other measurements are also "built in," and the origin of each makes interesting reading.) Each of these units originally violated the first requirement of a measuring system in that they were not the same for all individuals. This problem eventually was solved by passing laws standardizing each unit and fixing their relative sizes. However, when this was done, the greatest advantage of a system based on the human body was eliminated—most people no longer had the legal measuring system as part of their body.

TABLE 2-1
Commonly Used Prefixes

SYMBOL	SI PREFIX	MULTIPLE
G	giga	1 billion, or 10^9
M	mega	1 million, or 10^6
k	kilo	1000, or 10^3
c	centi	one-hundredth, or 10^{-2}
m	milli	one-thousandth, or 10^{-3}
μ	micro	one-millionth, or 10^{-6}

The current system standard throughout the world is the metric system, originally developed in 1795. The only major nation on earth *not* using the metric system for everyday measurements is the United States. The pace for changeover to the metric system has been increasing in the United States over the past decade, and conversion seems inevitable. All of the units used throughout this book will be part of the international standard system of metric units (SI units). If you are not already familiar with some of these units, it is about time you learned something about the world standards. A few SI units, such as the volt or the watt, are already known to you since there are no English-system equivalent units in common use. One considerable advantage to the metric SI units is that only one unit is used to measure one property. Thus, the SI unit of length is the *meter* (which is slightly larger than a yard at 39.37 in). When a larger or a smaller unit is more convenient for measuring in a particular case, these are always derived from the basic units in a standard way which eliminates awkwardly sized conversion factors (Table 2-1). Such a larger or smaller unit is indicated by placing a single-letter prefix in front of the abbreviation of the basic unit. Thus, one millimeter (mm) is one-thousandth of a meter (m) or 10^{-3} m. Whenever the prefix m is used with a unit, the prefix m may be replaced by the multiplier 10^{-3}. Thus, 217 mm = 217×10^{-3} m, or 0.217 m. Likewise, a megawatt (MW) is one million watts, and a kilovolt (kV) is one thousand volts. In this manner, only one standard unit need be defined for a single physical variable, while differently sized units may be derived from it with ease.

SIGNIFICANT FIGURES

In trying to draw conclusions from numbers scientists have learned to be very careful with their handling of numbers. The handling of a number obtained as the result of a measurement is quite different from the standard treatment of numbers in a mathematics class. As an example, consider a square sheet of paper. We wish to measure the area of the paper. For this task we have been provided with a ruler marked in centimeters (cm). Since our ruler has only centimeter marks, we can determine the length of a side of the paper only to the nearest centimeter without estimating. We do this and determine that the length of each side of the square, to the nearest centimeter, is 7 cm. Now, how are we to report the area of the sheet of paper? Clearly the area is the length times the width, or 7 cm × 7 cm. What is this product? Normally, 7 × 7 is 49, but if you report this as the answer, you have made an important error. If we had 7 boys and each had 7 apples, clearly we would have 49 apples. However, in this case, the possibility of having 7.1 boys doesn't exist, and an apple with a bite out of it is still likely to be called "an apple." The point is that in the boy-apple example the numbers are arrived at by the process of counting. Numbers arrived at from counting whole units are handled differently from the numbers that result from a measurement.

In the case of our square of paper, the measurement indicated a length of 7 cm. However, the length could have been somewhat less or somewhat more and we would still have had to call it 7 cm since we can determine the length only to the nearest centimeter with our ruler. Reporting the length at 7 cm means that we determined that the length was closer to 7 cm than to 6 cm, and that it was closer to 7 cm than to 8 cm. Thus, reporting a measurement as 7 cm means that we determined the length to be between 6.5 and 7.5 cm. But this range introduces considerable uncertainty in our determination of area. Consider:

$$\begin{array}{r} 6.5 \\ \times 6.5 \\ \hline 42.\not{4}\not{2} \end{array} \qquad \begin{array}{r} 7.5 \\ \times 7.5 \\ \hline 56.\not{4}\not{5} \end{array}$$

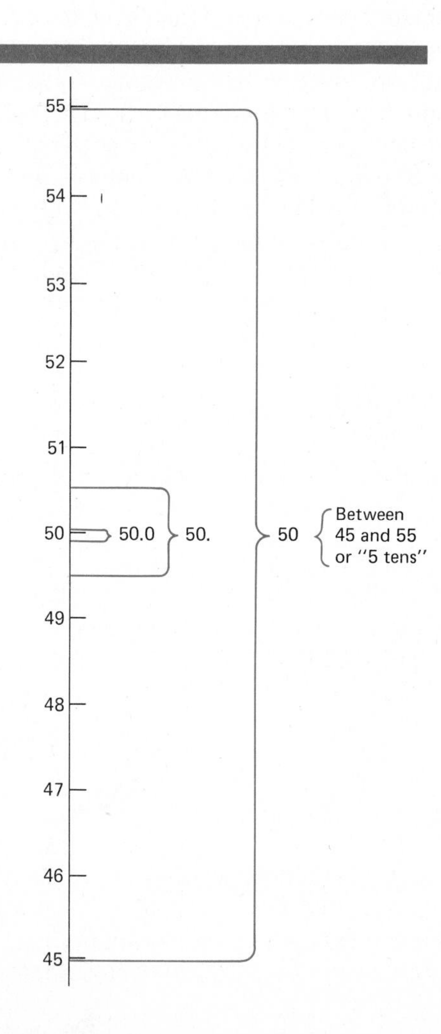

FIGURE 2-3
The numbers 50, 50., and 50.0 imply greatly different possible ranges for the result of a measurement.

Thus, the correct area for our square of paper could be as low as 42 cm^2 or as high as 56 cm^2. In either case we would have had no choice but to call the side length 7, to the nearest centimeter. Thus, in our example 7 cm × 7 cm $\neq$ 49 cm^2. To report the result as 49 would be misleading. We cannot know that this area is closer to 49 than to 48, or closer to 49 than to 50. Yet, *that is what reporting a number as 49 asserts.* Instead, we report our final product here as 50 cm^2. When a number is written in this manner in science, it means "five 10s to the nearest 10." If we meant 50—not 49, not 51, but 50—we would write 50. cm^2. The placement of the decimal point after the zero indicates that it represents a measured zero, not a place-marker (Fig. 2-3).

Of course, it is difficult to write 7 cm × 7 cm = 50 cm^2. Our mathematical training generally objects to this result because of our experience with counting numbers. Yet, there is a great danger in handling numbers otherwise. Ignorance is our normal state, and from experience we can cope well with our not knowing something. (We have no choice but to cope with ignorance, since most of us know only the slightest fragments of what there is to be known about almost any subject.) It is when we *think* we know something that, in fact, is not correct that we get in trouble. Action based on a "fact" that is really not a fact is the source of much trouble in science, as elsewhere. Built into the reporting of every measured number is an assurance of how well that number is known. The correct reporting of numbers is covered under the title "significant figures." Violations of the rules for handling significant figures can easily lead you to assert more than you really know. Sooner or later this can lead to conclusions unsupported by the measurements. See the discussion of rules for significant figures appearing in Appendix One at the end of the book.

Obtaining Equations from Graphs

Graphs allow the relationship between variables to be pictured more easily than an algebraic expression allows. We will make extensive use of graphs in our work. You will need not only the ability to plot data to obtain graphs, but also the ability to obtain the equation of a straight line from its graph. This skill isn't as difficult as adjusting a recipe for an 8-in pie to a square 10-in pan. If you are not sure of your ability to handle graphing, pay particular attention throughout this section and refer to it if you need help later.

A Review of Graphing

In graphing we deal with numerical quantities that can take on different values, or vary in size. These quantities—such as length or time—are called *variables.* A graph is a method of representing the corresponding values of two variables simultaneously. One variable is called the *x variable* and is represented along the bottom of a sheet of paper having both

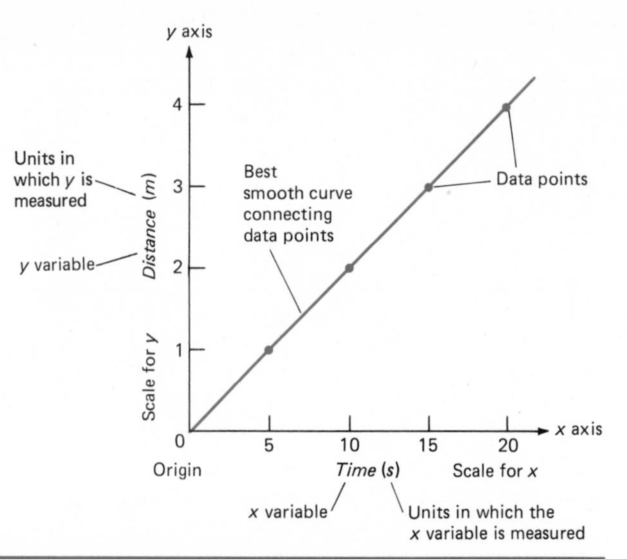

FIGURE 2-4
Illustration showing the parts of a graph. Note that the x and y variables are not usually called x and y in the sciences unless we would be graphing the number of xylophones vs. the number of yo-yos, or referring to a purely general case.

horizontal and vertical lines (Fig. 2-4). The second or *y variable* is represented along the left side of the graph paper. Some point, usually in the lower left-hand corner of the graph, is chosen as the *origin.* The origin is the one, and only, point where both the value of x and the value of y are zero. A data point at the origin would indicate that both x and y are zero simultaneously. A horizontal line running through the origin is called the x axis. Along this line the x variable may take on different values, but the value of y on the x axis is zero everywhere. Along the x axis we indicate a *scale* for x. That is, we choose a fixed amount of distance along the x axis to represent a fixed change in the size of x. Thus, if we move right 1 cm from the origin, we might choose to have x increase from zero at the origin to, for example, 5 at this new point. Once we have chosen this scale we must stick with it, in simple graphing, so that if 1 cm indicates that x increases by 5, 2 cm from the origin, x must be 10, and 3 cm to the right x must be 15, and so on. Such a scale is said to be *linear*—that is, equally sized distances along the scale mean equally sized increases in the value of x.

We now may draw a vertical line through the origin for the *y axis,* and determine a scale for y. This is also to be a linear scale for simple graphing. Everywhere on the y axis, the value of x is taken to be zero. Similarly, if x is 5 at a distance 1 cm to the right of the origin, x is also 5 for any point directly above or below this point on the graph. Thus, a certain distance to the right or left of the y axis indicates a certain change in size for the x variable from zero. Likewise, a change in distance above or below the x axis indicates a definite size change in the value of y. Each point on the graph paper now has been assigned a particular value for x and a particular value for y. Data plotted to a particular point indicate a value for x as well as a value which the y variable had *at the same time.* Thus, a series of plotted points indicates how changes in one variable are related to changes in the other variable. Consider the graph in Fig.

2-4. Starting at the origin, as we move to each successive point plotted, we move upward and we move rightward. Thus, each successive point is getting larger in its value of y (more upward) and is getting larger in its value for x (more rightward). A relationship between x and y where the values of the two variables always get larger together, such as that shown, produces a straight line on the graph. The straight line is the simplest form of order to interpret on a graph, and we will consider it first.

THE SLOPE-INTERCEPT FORM OF THE EQUATION OF A STRAIGHT LINE

To draw a line on a sheet of graph paper, we need two pieces of information. First of all, we need to know where to put the pencil down on the paper to begin. Second, we need to know in which direction to head from our starting point. A particularly useful choice for a starting point is that place where the graph will cross the y axis. This is called the *y intercept,* and its value is usually indicated by the letter b. Thus, the y intercept is the value of y when x is zero (as it is anywhere on the y axis). The second item of information needed to draw the graph is the heading we are to take from our starting point, called the *slope* of the line, which is usually indicated by the letter m. As can be imagined from its name, slope is a measure of the tilt or inclination of a line. We can give both a rough (but accurate) definition of slope and a more formal definition.

If you start on the line and move to the right so that x has increased by 1 unit, the slope of the line is the number of units of y you must go up (positive) or down (negative) in order to get back on the line. For example, in the graph shown in Fig. 2-5, if we begin at the point labeled A, moving right 1 unit in x brings us to the point labeled B. To get back to

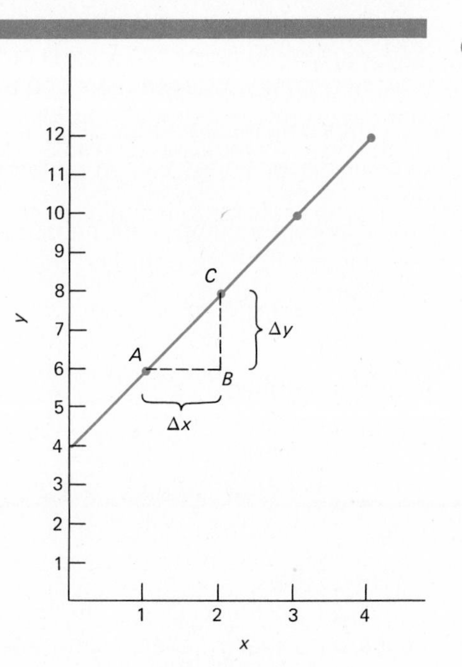

FIGURE 2-5
Graph for the determination of the equation of a line.

the line we must move upward to point C. The value of y at B is 6, while the y value at point C is 8. Thus, we will have to move upward 2 units in y to once again be on the line. The slope of this line is $+2$.

To formalize the definition of slope it is convenient to introduce the symbol Δ (Greek letter capital *delta*). The term Δx is read as "the change in x." To find the value for Δx we would take the final value of x in the interval we are considering and subtract from it the original or initial value of x in that interval. Thus,

$$\Delta x = x \text{ final} - x \text{ initial}$$

(or $x_f - x_0$, for short.) Consider Fig. 2-5 again. The interval we are interested in is that between point A on the line and point C on the line. The variable x began this interval (at point A) with a value of 1. It ended the interval (at point C) with a value of 2. Thus,

$$\begin{aligned}\Delta x &= x_f - x_0\\ &= 2 - 1\\ &= 1\end{aligned}$$

After you work through that example, you are ready for the formal definition of slope. The slope of a straight line is a number obtained by dividing the change in y between any two points by the change in x between those same two points.

$$\text{Slope} = m = \frac{\Delta y}{\Delta x}$$

In the example of Fig. 2-5, $\Delta y = 8 - 6 = 2$, while $\Delta x = 2 - 1 = 1$. Thus,

$$m = \frac{2}{1} = 2 \qquad \text{our slope}$$

Some caution is necessary since the variables used in physical sciences normally have units. Likewise, slopes will, in general, have units as well as a numerical size.

Both the slope and the intercept are important because, if we know these, we can write the equation of the straight line immediately. *The slope-intercept form for the equation of a line is*

$$y = mx + b$$

where: y is whatever variable is graphed along the y axis; x is whatever variable is graphed along the x axis; m is the slope of the line; and b is the y intercept. If we note that the y intercept of the graph in Fig. 2-5 is 4, we can write the equation of this line immediately. Substituting into the form $y = mx + b$, we get $y = 2x + 4$ as the equation of this line.

Example: Temperature Conversions

Let us apply this to a physical problem. Currently, the Celsius (°C) scale of temperature measure (formerly called centigrade) is coming into more widespread common use. During the transition to the use of the Celsius scale, it might be necessary to convert from the traditional Fahrenheit (°F) system. What is the relationship between the two scales? You may recall that water freezes at 0°C, which is 32°F, and that it boils at 100°C or 212°F. If you are average, this is about all you remember about the relationship between the two temperature scales (Fig. 2-6). Still, we have enough information to obtain the relationship by graphing. Figure 2-7 graphs the information we have. We are justified in connecting our two data points by a straight line since both scales increase uniformly from freezing to boiling. (After all, the same glass tube can be given 100 evenly spaced marks for Celsius or 180 marks for Fahrenheit.) Looking at our graph, we can see that the change in the x variable between our two points is

$$\begin{aligned}\Delta x &= \Delta \mathrm{C} \\ &= 100 - 0 \\ &= 100\end{aligned}$$

Similarly,

$$\begin{aligned}\Delta y &= \Delta \mathrm{F} \\ &= 212 - 32 \\ &= 180\end{aligned}$$

Thus the slope is

$$\begin{aligned}m &= \frac{\Delta \mathrm{F}}{\Delta \mathrm{C}} \\ &= \frac{180}{100} \\ &= \frac{9}{5} = 1.8\end{aligned}$$

We can now read off the equation for this graph from the form $y = mx + b$. Here, we have graphed F on the y axis, so F is our y variable. Likewise, C is the x variable. Inspection of the graph shows that the y intercept is 32. Thus, the equation of this graph is

$$y = mx + b$$

or $\mathrm{F} = \frac{9}{5}\mathrm{C} + 32$

This is a form which you may recognize as being the conversion formula from degrees Celsius to degrees Fahrenheit. As you can see, it is quite easily derived from a knowledge of graphing and a few simple facts about the two temperature scales.

You will need to know how to obtain the equation of a straight line using the slope-intercept form we have just discussed.

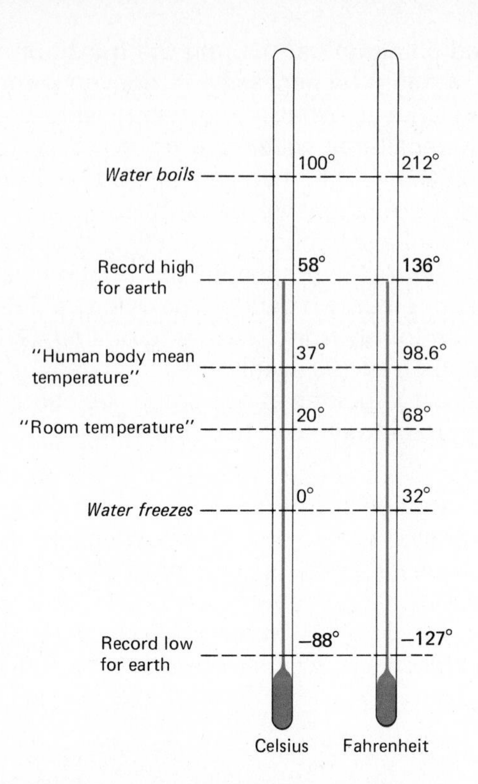

FIGURE 2-6
A comparison of some values on the Celsius and the Fahrenheit temperature scales.

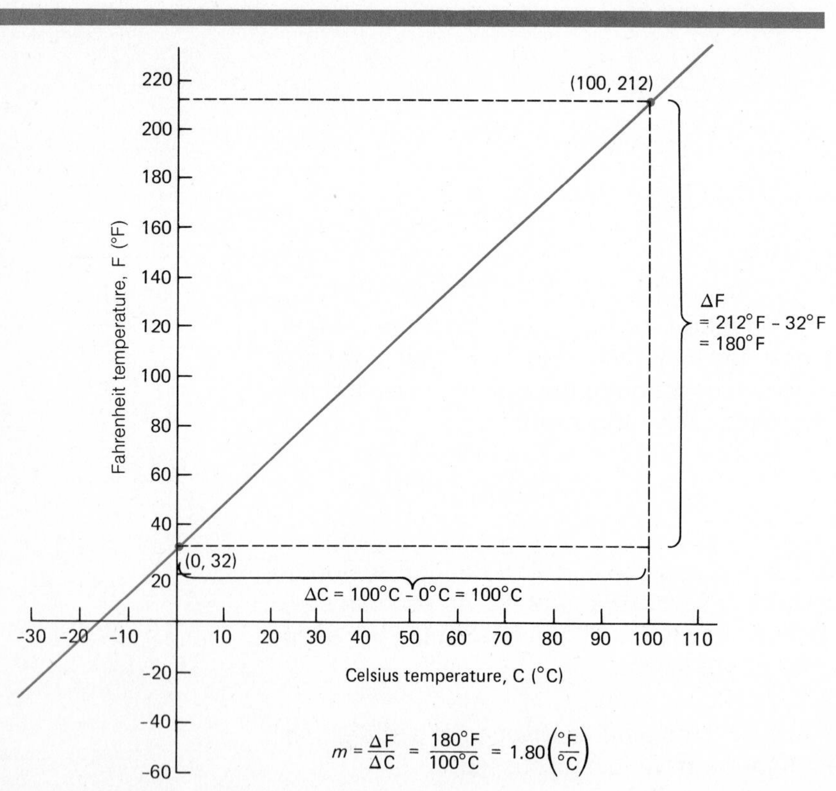

FIGURE 2-7
A graph of Fahrenheit temperature vs. Celsius temperature to determine the relationship between the temperature scales.

CHANGE OF VARIABLES

We have now learned to obtain the equation of a straight line from its graph. While the goal of this chapter is to allow us to obtain relationships from data, we still have a problem. Straight-line relationships generally are not the most useful or interesting ones in science. Nature tends to favor simple relationships, but not necessarily this simple. How are we to find a relationship between variables which presents something other than a straight line when graphed? The answer we present here, and use frequently throughout the book, is one that depends upon the bias of natural relationships toward simple mathematical forms. *If a graph isn't a straight line, make it into one!* By saying this, we certainly are not suggesting that we hide from reality. This history of science already has too many unhappy examples of neglecting facts. The strategy being suggested is both a positive one and is quite "legal" mathematically. *Here is the strategy:* If a graph between x and y is not a straight line, examine the line to determine why it is not. Try to determine what mathematical change would have to be made to x, for example, to make it grow in step with y. Apply the change to the values of x. If a straight line results, determine its equation from the form $y = mx + b$. That series of steps might appear to be confusing, so let us work through an example.

FIGURE 2-8
In the J-tube method, the difference in height h of the mercury in the two columns is a measure of the pressure added. In the hypodermic-syringe method, the pressure added is directly proportional to the number of standard weights supported by the piston.

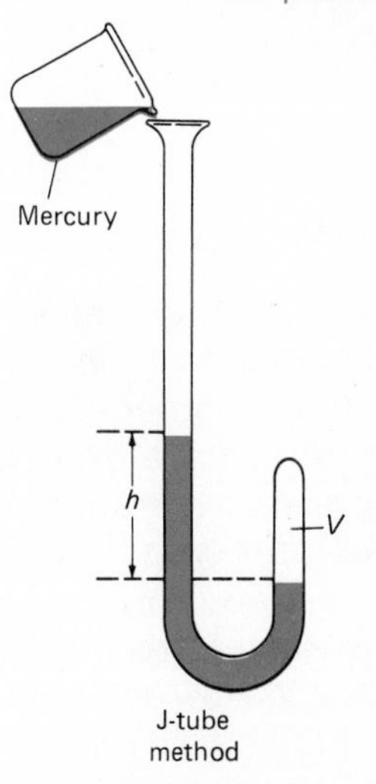

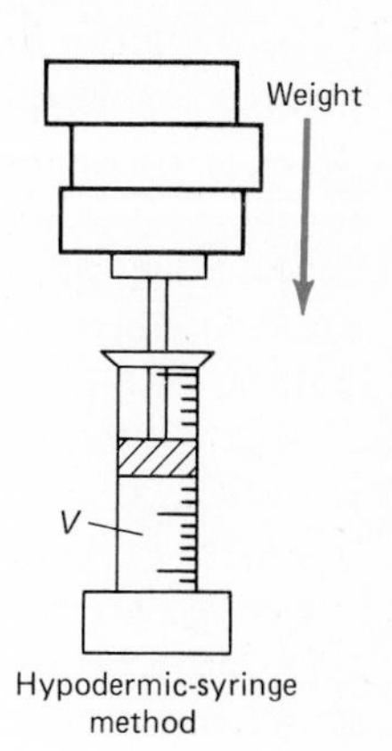

Example: Pressure and Volume

Consider the case of an enclosed gas, such as we might find in a balloon. As we push harder on the walls of the balloon, its volume shrinks. As we reduce our pressure on the balloon, the volume increases to resume its original value. The two variables involved are the volume V which the gas occupies and the pressure P, or force per unit area, with which we push on the gas. To investigate the relationship between these two variables, it is more convenient to use apparatus which will allow the easy measurement of volumes and pressures. Such apparatus usually consists of some cylindrical enclosure for the gas with one end sealed and the other a movable wall, or piston. Two varieties of such devices are shown in Fig. 2-8. Data obtained in measuring pressures applied and the corresponding volume of gas in each case is shown in Table 2-2. It can easily be seen from the numbers that as the pressure increases, the volume of the gas decreases, as we expected. Graphing this data does not produce a straight line. Instead, the curve illustrated in Fig. 2-9 results. The implication of the curve seems to be that as P becomes very large, V becomes quite small, but it does not shrink to zero. Examination of

TABLE 2-2
Volume of an Enclosed Gas at Various Pressures Applied to the Gas, in Arbitrary Units

PRESSURE	VOLUME
1.0	2.0
1.5	1.3
2.0	1.0
3.0	0.67
4.0	0.50

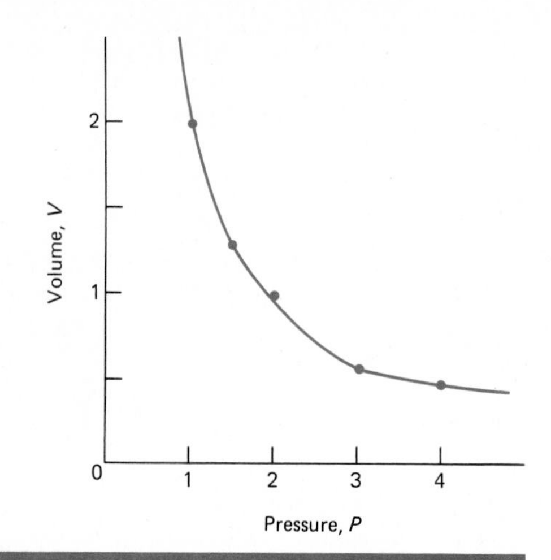

FIGURE 2-9
Graph of pressure applied vs. volume for an enclosed gas.

the graph indicates that each time we double the pressure, the volume is cut in half. Conversely, as the pressure drops to a small value, the volume increases. The curve seems to imply that if we could make the pressure drop to a very low value (by reducing the natural air pressure surrounding the apparatus by placing it in a vacuum chamber, for example), the volume would grow to be very large.

Now, how are we to obtain the equation of such a relationship? Clearly, a great deal of order is present in the relationship between P and V. How are we to summarize it? If we refer back to the strategy for obtaining an equation presented just before this example, we can try to apply the steps given.

(a) *Determine why the graph is not a straight line.* In this case, the reason is quite clear. One variable is growing in size while the other is shrinking. However, this shrinkage and growth is not proportional. As P goes from 1 unit to 2 units, V decreases to half its original value. However, as P again increases by 1 unit from 2 to 3 units, V decreases only by a half of its previous drop, from 1.0 to 0.67 on the graph.

(b) *Determine the mathematical change necessary to make the variables grow together in step.* In our example, as the value of V increases, P becomes increasingly small. To allow both variables to grow and shrink together we need a mathematical operation that will take large values of P and produce small numbers, and that will take small values of P and produce large numbers. We need to *invert* the size of the values of P. Such an operation is to take the inverse of the values of P. The *inverse* of a value is defined as 1 divided by that number. Thus, the inverse of 3 is $\frac{1}{3}$, or 0.33, and the inverse of 10 is 0.1. Note that the larger a number is, the smaller its inverse. Thus, $\frac{1}{2}$ is larger than $\frac{1}{50}$, although 2 is smaller than 50. Let us proceed with the steps in the strategy.

(c) *Apply the mathematical change necessary to the values of the x variable.* This is reasonably easy to accomplish. Taking previous

INVERSE PRESSURE, $1/P$	VOLUME, V
1.0	2.0
0.67	1.3
0.50	1.0
0.33	0.67
0.25	0.50

TABLE 2-3
Volume of an Enclosed Gas vs. the Inverse of the Pressure Applied, in Arbitrary Units

values for P from Table 2-2, we can calculate the values for $1/P$, as shown in Table 2-3. Thus, if P is 1.5, $1/P$ becomes 1/1.5 or 0.67, and so on. The values of V remain unchanged during this operation.

(*d*) *Regraph with the new variable to determine if a straight line results.* In Fig. 2-10 we have graphed the values of $1/P$ as the new x variable, while keeping the original data for V as the y variable. As can be seen, a straight line does result from this procedure. We have found the basic order within our data and may now proceed to the final step.

(*e*) *Determine the equation of the line if a straight line results.* We may use the form $y = mx + b$ to determine the equation of the line that has been produced and, therefore, the equation of the relationship between P and V. Let us proceed carefully. First of all, we may observe that the line passes through the origin. Thus, the value of the y intercept b is zero. Measuring from the graph, we may verify that the line increases by 2 units vertically on the y axis for each unit we proceed to the right along the x axis. From this we may conclude that the slope is 2.0. (In this example, we have eliminated units for clarity, but the slope normally would have units indicating what V and P must be measured in to have the equation valid.) We note that our y variable is V. The variable we have graphed on the x axis is $1/P$. Thus, $1/P$ is our x variable. Substituting these values into the slope-intercept form for the equation of a line we obtain:

$$y = mx + b$$

$$V = 2.0(1/P) + 0$$

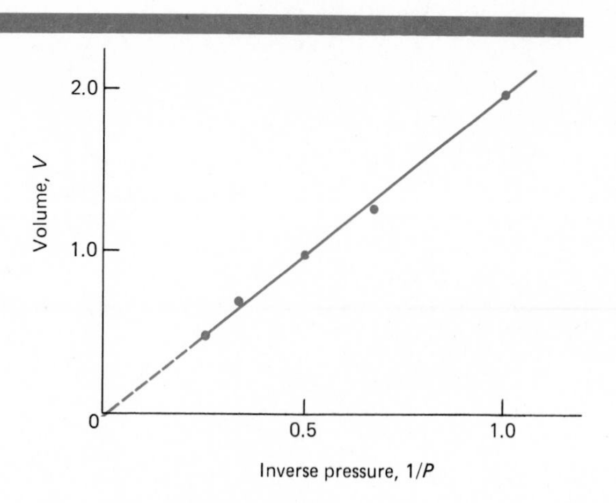

FIGURE 2-10
Graph of inverse pressure, or $1/P$, vs. the volume of an enclosed gas from the same data as in Fig. 2-9.

This may be simplified as $V = 2.0/P$ or as $PV = 2.0$. Thus, the product of pressure times volume is a constant, and we have determined the relationship.

If this process strikes you as being a form of trial and error, relax; it is! If you are worried that there are just too many different relationships to try, you can relax again. Aside from straight-line relationships, we will concern ourselves with only four other frequently occurring forms. If a relationship is not covered by one of this small group, we will either give up and simply tell its form or derive it separately.

The first form we will consider is graphed in Fig. 2-11. We easily can see that x and y are not growing at the same rate. Consider the dashed line drawn in Fig. 2-11. For small values of x and y, near the origin, the graph appears to be quite close to the trend of the line. As the variables get larger, however, the graph pulls more and more away from the dashed line. As x gets larger, the y variable gets larger, but at a much faster rate. The value for y is growing faster than x, and the graph is being pulled off in the positive y direction. We need a mathematical strategy to make the variables grow together. This strategy must not change by much the smaller values of the variable, since the early part of the curve is nearly a line. However, large values of the variable will need to be changed a great deal. What mathematical process changes small values relatively little, while altering large values a great deal? One such process is raising to a power—for example, squaring a variable. If we square 1, the result is 1, for no numerical change. Squaring 2 produces a 100 percent increase to 4. But, for a large number, the change is quite large. The number 10 increases 10 times, or 1000 percent, on squaring. Raising x to a power will have the effect of "stretching" the graph in the

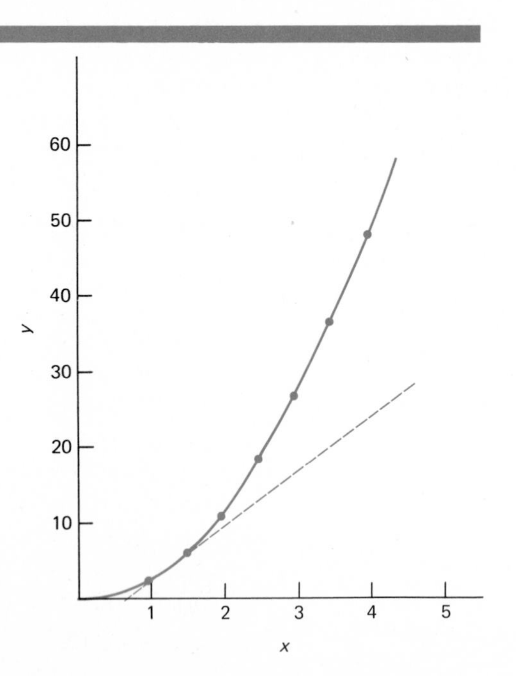

FIGURE 2-11
Graph of a direct-square relationship between x and y.

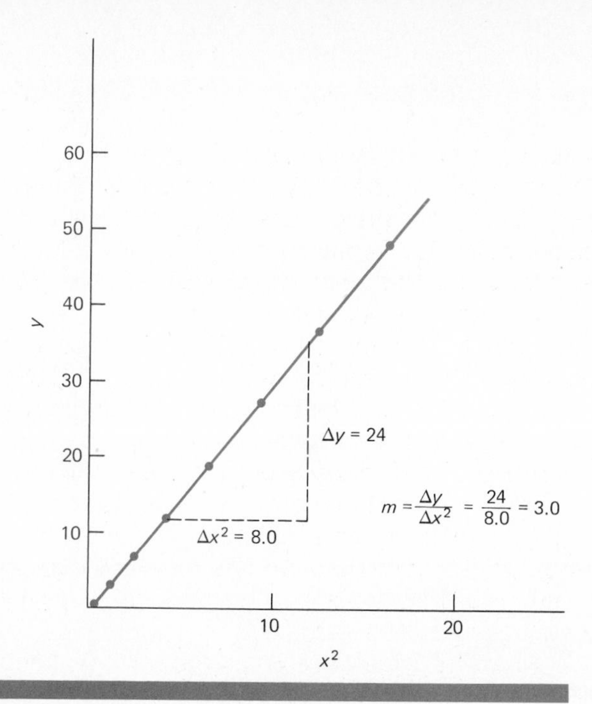

FIGURE 2-12
Graph of the square of x, or x^2, vs. y for the same data used to generate Fig. 2-11.

rightward direction. The "stretch" is not even. Regions closer to the origin are less affected. Squaring all values for x accompanying Fig. 2-11, we may now regraph y vs. x^2 as shown in Fig. 2-12. This is indeed a straight line with an intercept of zero and a slope of 3 as measured on the graph. Thus, the equation of this graph may now be written since it is a straight line. In the form $y = mx + b$, $y = 3x^2$ is the final equation.

Had the squaring of x corrected some, but not all, of the deviation from a straight line, we could have had the case where y is proportional to the cube of x. A graph of y vs. x^3 would yield a straight line from which we could obtain the equation of the relationship. Actually, cubic relationships are not particularly common, and most curves that we encounter which look like Fig. 2-11 will reveal a square relationship. In the event that neither squaring nor cubing the appropriate variable produces a straight line, *punt!* Trial and error is only a hopeful process when the number of options to be tried can be reduced to a manageable set.

The last remaining type of curve is also frequently encountered in nature. It is illustrated in Fig. 2-13. This curve is similar in appearance to our example with pressure and volume. Like that example, it represents a type of inverse variation. In this case, the y variable is inversely proportional to the square of x, or $y = k(1/x^2)$, where k is a constant determined by the choice of units for x and y. Nature abounds with situations where one variable is proportional to the square, cube, inverse, or inverse square of another variable, but this is hardly required. The fact that so very many situations will be covered by the cases outlined is part of the reason scientists are impressed with the orderliness of nature. We may now consider a real example in detail.

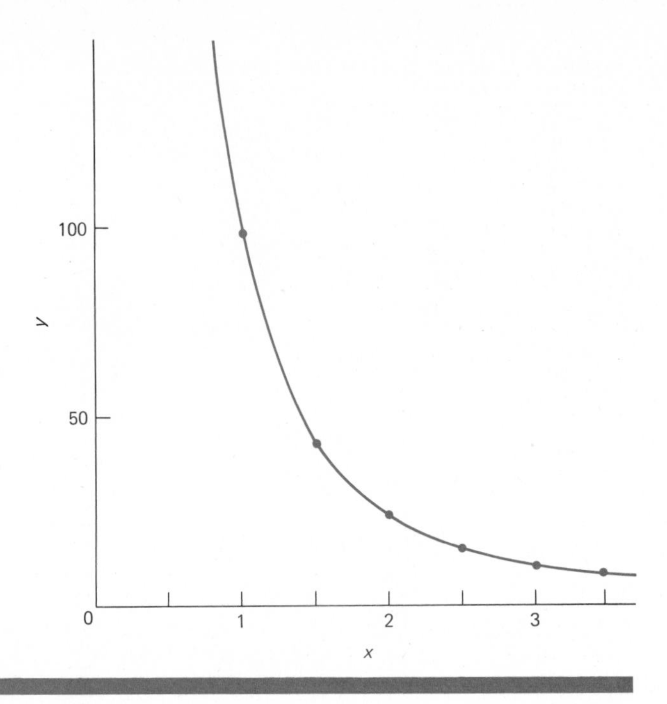

FIGURE 2-13
Graph of x vs. y for an inverse-square relationship. In this case the equation relating the variables is $y = 100/x^2$.

Example: The Distance Covered by a Falling Cannonball

An old story relates that Galileo dropped a musket ball and a cannonball off the Leaning Tower of Pisa to show that they fell together. Aristotle's physics would have required the cannonball to have startled the tourists by embedding itself in the pavement below before the musket ball had fallen one-hundredth of the distance to the ground.* Suppose, however, that Galileo not only dropped these objects, but that he could record the position of the cannonball every $\frac{1}{2}$ s. What could such data reveal? What would these times and positions tell the trained observer? We will use this as an example in obtaining relationships from numbers.

Figure 2-14 shows a sequence of seven photographs taken at intervals of 0.30 s by a repeating camera. The photographs show a striped bowling ball being dropped from a dormitory. This is a reasonable simulation of Galileo's work, except that we have the ability to measure small time intervals, which Galileo did not. The distance between the windowsill on one floor and that of the floor above is 3.15 m. This allows us to determine the distance scale of the photographs. Measurement is made easier by use of the composite photograph of these seven positions reproduced in Fig. 2-15. To deter-

* This story is one of the myths of science. It would have been a dramatic test of Galileo's ideas, but it didn't occur. The tourists were safe. The story doesn't appear until long after any paperboy on the scene would have died of old age. Besides which, cracking the plaza pavement would have been a criminal offense.

FIGURE 2-14
A sequence of seven photographs of a standard bowling ball being dropped from the seventh floor of a dormitory building. (Turn the page to properly view the sequence from left to right.) The ball has white stripes painted on it for increased visibility, and its position is indicated by a white arrow in each photograph. (*Photographs by the author.*)

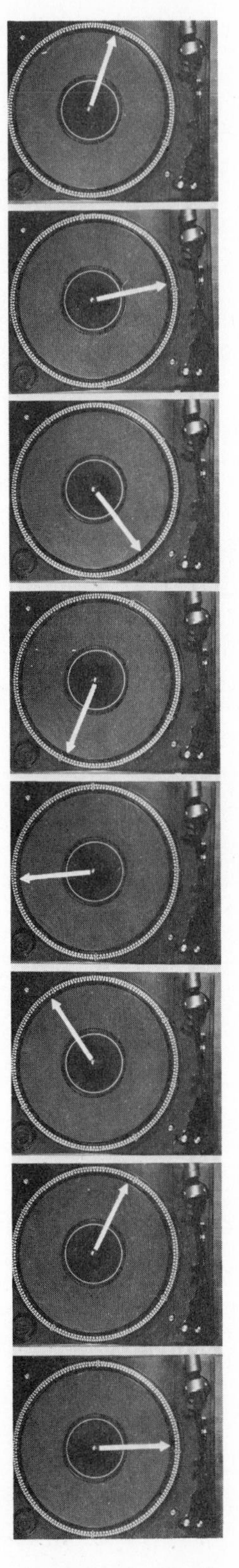

FIGURE 2-15
(*Far left*) A composite of the seven photographs shown in Fig. 2-14 showing the relative spacing of the positions of the ball during each photograph. Each photograph was taken at 1/1000 s to minimize motion of the bowling ball during the exposure. Why is the shadow of the ball on the building elliptical and not circular? (*Photograph by the author.*)

FIGURE 2-16
(*Near left*) A series of photographs of a phonograph turntable at $33\frac{1}{3}$ revolutions per minute (rpm) by the camera used to produce Fig. 2-14. Comparison of the first and seventh images shows that the turntable moves through slightly more than one revolution (366°) in six intervals between pictures. Since $33\frac{1}{3}$ turns per minute is 1.80 seconds per turn, the interval between pictures works out to 0.305 s. (*Photographs by the author.*)

mine the time between photographs, the camera was used to photograph the movement of a mark on a phonograph turntable. This sequence of photographs is shown in Fig. 2-16. You are encouraged to verify the numbers we obtained for use in the table.

Graphing d vs. t from the data in Table 2-4, we obtain the graph shown in Fig. 2-17. This graph is not a straight line, but its form is familiar as the second case of the previous section. To see whether this is really the situation, we may regraph to determine whether we obtain a straight line. Squaring the values of t above, we obtain the data in Table 2-5. Graphing these data we obtain a straight line of slope 4.9 m/s² (Fig. 2-18). Thus we may conclude that the relationship between time and the distance covered by an object falling from rest is $d = (4.9\ \text{m/s}^2)\ t^2$, and *not* $d \propto t$ as Aristotle believed.

TABLE 2-4

Distance d vs. Time t for a Falling Ball

PHOTOGRAPH	TIME SINCE RELEASE, s	DISTANCE TRAVELED, m
1	0.000	0.00
2	0.305	0.61
3	0.610	1.89
4	0.915	4.11
5	1.22	7.15
6	1.53	11.2
7	1.83	16.4

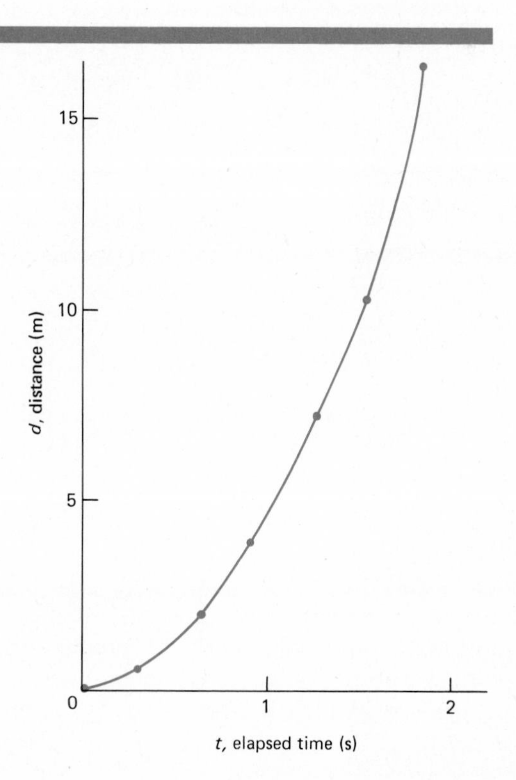

FIGURE 2-17
A graph of the distance fallen by the bowling ball in Fig. 2-14 vs. the time elapsed since the first picture was taken.

t^2, s^2	d, m
0.000	0.00
0.093	0.61
0.372	1.89
0.837	4.11
1.49	7.15
2.34	11.2
3.35	16.4

TABLE 2-5
Values of t^2 and d for a Falling Ball

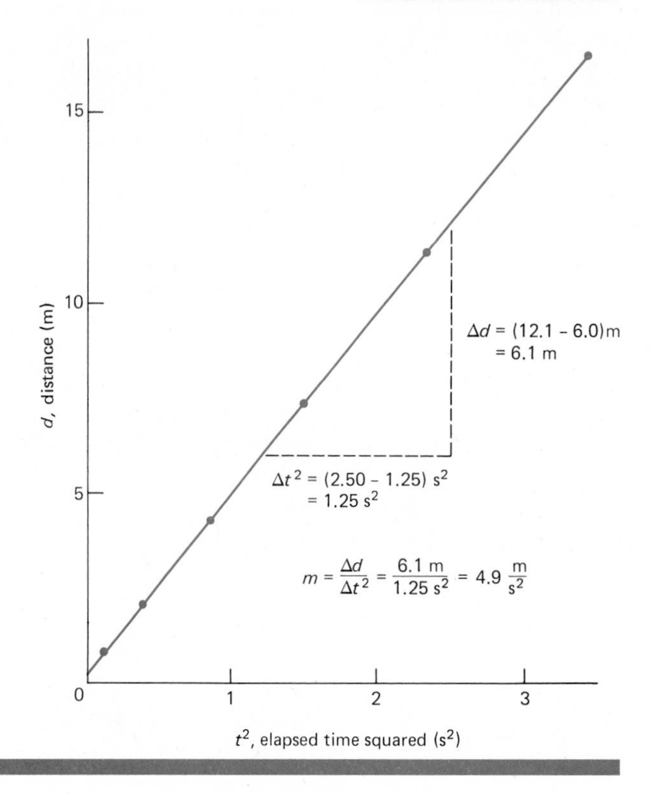

FIGURE 2-18
A graph of the distance fallen by the bowling ball in Fig. 2-14 vs. the square of the time elapsed since the first picture was taken.

Summary

Empiricism is the process of producing order from direct experience. In the sciences, this involves obtaining data by direct measurement and finding order in the numbers that result. There are several ways to summarize a relationship between variables. The table of experimental numbers itself contains the relationship, but it may be difficult to determine from the raw numbers. Furthermore, a table of data covers only the situations already measured. It is frequently not apparent how the numbers would change when measurement is extended to an untried situation. The most compact summary of a relationship is an equation relating the variables of interest. The equation allows the rapid calculation of the effect a change in one variable will have on another variable. Equations are easily written and remembered. The difficulty with equations is that they communicate little to individuals who have no mathematical experience. They are often used blindly by the mathematically naïve in some sort of number-guessing game. Also, they often are not viewed by the novice as a statement of a physical relationship between variables. Mathematical procedures remembered from an algebra class tend to dominate the picture of what is happening in the physical relationship between the variables. A midway course between dealing with only measured numbers and with the equation is provided by the graph of a relationship. Graphs allow a mental picture to be formed of the influence of one variable upon the value of another. Further, graphs can provide a useful middle stage between data and equations. Certain types of relationships between variables occur frequently enough in nature that familiarity with the graphs of these relationships provides the means for obtaining the equation between these variables. Particularly common forms of variation are the inverse, inverse square, direct square, and cubic proportionalities between variables.

Questions

1 If you observe sale prices such as $3.98, $2.49, and $1.99, what does it tell you about the average person's normal method of handling significant figures?

2 What advantages do graphs have over equations in describing physical relationships? Disadvantages?

3 In the days before the standardization of units in the English system do you suspect that most merchants selling cloth (yard goods) were large men or small men?

4 The temperature outside is 24°C. Is it more appropriate to wear shorts or an overcoat when you go outside?

5 "Clear sky at night, a sailor's delight.
Red sun in the morning; sailor take warning."
This folk verse is an empirical statement. What does that mean?

6 If you pull sideways on the tip of a car antenna it moves a distance from its normal position. What would be the shape of graph of force applied vs. distance displaced? Would it be safe to use this curve to predict what would happen for very large forces?

7 Suppose you graphed height vs. IQ for every student in a local elementary school. How would you expect the points to be distributed on the graph? If you did this for height vs. reading level, would the distribution of points be the same?

8 For graphs 1 to 3 identify the:

(a) x variable
(b) x scale
(c) y variable
(d) y scale
(e) slope of the graph
(f) Type of variation
(g) Equation of the relationship

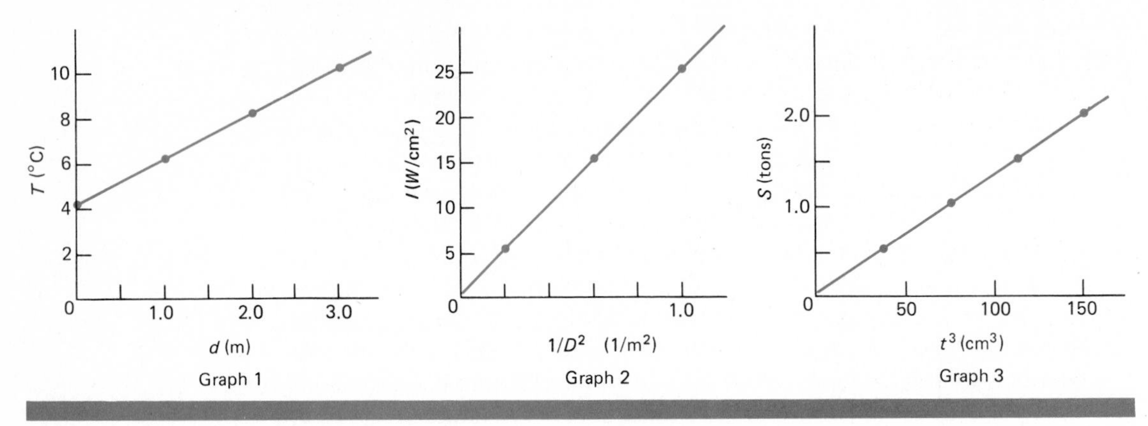

9 To manage an empire the Romans had to know just how long it would take to march an army from any one place to another. Thus, crews counted the paces as the army marched and erected a milestone every 1000 paces. Go to a long corridor with square tiles and mark a starting point. Take 20 normal walking steps forward and count the number of tiles you covered (usually 9-in square or $\frac{3}{4}$ ft each). Two steps make up one pace. How does your mile compare with that of the Roman army?

10 Aristotle asserted that falling objects fell at constant speed. Does this agree with the data in Fig. 2-15?

11 A very careful measurement and regraphing of the data of the bowling-ball drop, as in Fig. 2-18, indicate that the y intercept is not precisely zero. The intercept is instead on the order of $d = +0.04$ m. How would you interpret this nonzero intercept?

Problems

1 A mile is 1.609 km. What is the national speed limit of 55 mi/h in kilometers per hour (km/h)?

2 A speed of 80 km/h (50 mi/h) is about 22 m/s. If your reaction time is about 0.4 s, how far will you travel, in meters, before your foot can reach the brake pedal in an emergency?

3 The event in the Olympics closest to the mile run is the 1500-m run. Which is longer? If

a runner can do the mile in 3 min 54 s, how long should the runner take for the 1500-m race?

4 The distance from San Francisco to Los Angeles is 580 km. How many miles is this?

5 Nitrogen gas liquefies at −144°C. What temperature is this on the Fahrenheit scale?

6 The data in the table below give the distance from the basket vs. the percentage of successful shots during competition.

DISTANCE FROM BASKET, m	SHOOTING PERCENTAGE
1	62
2	52
3	40
4	32
5	28
6	24
7	21
8	19.5
9	18
10	17

Source: J. W. Bunn, *Scientific Principles of Coaching*, Prentice-Hall, Inc., New York, 1955, p. 225.

(*a*) What type of a relationship exists between distance D and shooting percentage p?
(*b*) What type of curve would you expect for D vs. p?
(*c*) With what do you think you would plot vs. D to obtain a straight line?
(*d*) Try the plot of D as the x variable vs. $1/p$ to see if the relationship yields a straight line. (Since this is data from measurement, a small amount of "wiggle" in the points is to be expected.)
(*e*) Write the equation relating D and p from your graph.

CHAPTER THREE

The Theoretical Process—Exploring the Implications of Ideas

Introduction The theoretical process in science has at least two meanings. The grand use of the term "theoretical process" is reserved for those spectacular flights of human mind that occur rarely but change human thinking when they do occur. This usage is associated with the term "theoretical breakthrough" and the creation of new modes of thought. You can picture the event much like Michelangelo's *Creation* on the roof of the Sistine Chapel. God stretches forth a finger and one waits, breathlessly, for the spark to be transferred. However, instead of the finger of an Adam ready to receive the spark of life, we have the mind of a Newton or an Einstein ready to advance one full rung on the ladder of the human understanding of nature. This kind of theory is denied to all but a handful of humans in the history of science.* A major new theoretical system is a scientific revolution and defines the *unusual* event in science.

Theoretical work in science is not limited to the rare cases involving scientific revolutions. The everyday use of theoretical work is to explore the implications of physical ideas. This is usually done mathematically, and the process leads to conclusions which might not have been foreseen, but may be checked experimentally. Thus, "playing with ideas" on paper leads to many new experiments suggested by the theory. A *theory*, in this usage, is a well-formed mental model of how something is working in nature

* In the entire history of physics there have been only four new theoretical systems devised: the laws of motion by Newton; electromagnetic theory by Maxwell; relativity by Einstein; and quantum mechanics by many, notably Heisenberg and Schrödinger. This averages to one theoretical revolution per century for modern physics, which has the largest number of such revolutions among the sciences.

to be confirmed by experiment. These are usually expressed mathematically since it is easier to reason out the implications of mathematical statements.

No human investigation can call itself true science unless it proceeds through mathematical demonstrations.

Leonardo da Vinci

This use of the theoretical process, then, is exploratory. It serves as a tool for guiding empirical thought, and is essential for the day-to-day progress of science. The simple fact is that the world has too much data available for empirical thought to process. Without guidance concerning what data to collect and examine, empiricism would allow our understanding of the physical world to grow only very slowly.

Consider a real example from the physical world. Naturally occurring rubber has many useful properties. One of the most useful is its water-repellent nature. Impressed by this, one young businessman in the last century developed a way to coat cloth with rubber and produced the first watertight rainwear. He sold a large order to the U.S. Postal Service, and went bankrupt when the order was returned. The problem concerned the behavior of natural rubber with temperature changes. At very low temperatures the rubber became brittle and tended to crack and flake from the cloth. This was less of a problem than the behavior at high temperature. On a hot, muggy day, the rubber would partially melt and become "sticky," and at really high temperatures it would flow. The problem was to stabilize the rubber to preserve its flexibility at low temperatures and yet retain its solidity under high temperatures. The businessman was obsessed with this problem and devoted his life to it. He lost his business, all of his personal property, and eventually his family because he devoted all of his time and resources to the empirical attempt to stabilize rubber. He knew what he wanted in a material, but he had nothing to guide his empirical thought. There was no model to suggest what internal arrangement of the rubber molecules gave them their properties or how to modify them to obtain new properties. Charles Goodyear eventually did succeed in stabilizing rubber. Supposedly by accident, he spilled rubber and sulfur together on a hot stove. The rubber that resulted had the desired properties, and Goodyear called the process *vulcanization.* This became the basis of the world rubber industry but it was too late for Goodyear. He had lost everything and was in debt to anyone who had taken pity on him. Goodyear died broke and broken. His name is remembered because his lifetime search eventually stumbled on success. However, unguided empiricism has a very small chance of success. There simply are too many possible combinations to try, or items of data to collect, to expect rapid success with empiricism. Furthermore, once you succeed, you are not sure why you succeeded, and what changes might contribute to even greater success.

FIGURE 3-1
Galileo Galilei is often called the "Father of Modern Science." He was an incredibly gifted man, stimulating the development of instrumentation in many branches of science. His sharp logic and clear writing made him both adored and hated in his time. Much of this complexity may be inferred from this portrait of about 1624. (*Courtesy AIP Niels Bohr Library.*)

Empiricism without theory produces knowledge but little understanding. Theory without empiricism (as we have seen in the case of the Greeks) can lead us so far from reality that our ideas are worthless. It is in a balance between these two modes of scientific endeavor that human understanding of the physical world can prosper.

In this chapter we will examine the extension of a few of the ideas of Galileo from his empirical studies. We will use these ideas to describe motion in general and motion which occurs under a constant acceleration in particular. The description of motion is called *kinematics*.

Galileo's Observations on Motion

Galileo (Fig. 3-1) conducted many experiments by rolling a ball down a ramp. He chose this line of experimentation because the clocks of his day were too crude to allow the timing of a falling object. The force causing the ball to roll downhill was its weight or the downward pull of gravity (Fig. 3-2A). However, Galileo reasoned that on a slightly tilted ramp, the weight acted straight downward and only a small portion of the weight acted to make the ball roll down the ramp (Fig. 3-2B). By changing the tilt of the ramp, the experimenter could experience the effects of any part of the ball's weight (Fig. 3-2C). With a flat, horizontal ramp, none of the weight contributed to the ball's motion, so it did not move. With a straight up-and-down ramp the ball was falling freely, with all of its weight contributing to motion. Between these extremes, a tilt could be selected to sample the effects of any portion of the ball's weight acting on the motion of the ball. (For example, with a ramp tilted at 30° to the horizontal, one-half of the weight is acting to roll the ball down the ramp.) By choosing ramps with a relatively small tilt, Galileo could reduce the size of the force, to the point where the motion of the ball could be measured by the simple clocks available. He had to repeat his measurements many times to be confident of his results with the crude equipment available. However, by using ramps, Galileo found it possible to "cut down" or *dilute* the influence of gravity on the ball's motion. Thus, measurement of the motion became possible.

Galileo's experiments convinced him that several of Aristotle's conclusions about motion were incorrect. The most apparent of these was Aristotle's assertion that without forces, motion would cease. Galileo noted that once started, motion persisted until something stopped it. For example, a ball rolling off the ramp onto the flat surface remained moving. Galileo determined that its motion was sufficient to climb partway a second ramp placed in its path. He experimented enough to convince himself that the ball's motion was sufficient to allow the ball to return to almost the original height of its release, regardless of the tilt of the second ramp (Fig. 3-3). The small loss in height he attributed to *friction*—small, inescapable forces retarding the motion. (For example, small ridges in the wood which the ball had either to climb over or to crush along its path, plus the air the ball had to push out of its way as the ball moved, were sources of frictional forces.) Galileo reasoned that in an *ideal case,* where the ramps were perfectly hard and smooth, and there was no air to push away, the ball would always roll up the second ramp to the same height from which it was released on the first ramp. From this Galileo drew his first major conclusion. If there was no second ramp to climb, in the absence of frictional forces, the ball would continue rolling without a stop until it encountered an obstacle (Fig. 3-4). Today we call this idea the *principle of inertia.* We define *mass* as the measure of the amount of matter in an object. All objects of the same mass have the same tendency to resist a change in their motion—they have the same *inertia.* We may state the principle as:

> ***In the absence of forces effecting motion, a mass at rest tends to remain at rest, or a mass in motion tends to remain in motion at constant speed in a straight line.****

* The author's most interesting recollection of the principle of inertia occurred one early winter day. During a lecture in a windowless auditorium a fine freezing rain had blanketed the campus. As students emerged from the building this was not apparent to them, and their first step off the dry concrete doorsill found them on a smooth ice surface with a slight downhill run for 2 m leveling off to a long, straight sidewalk. Boosted by this slight head start, whole clusters of students slid smoothly past their intersection 5 m from the door making heroic but useless efforts to turn onto the correct sidewalk. Before long a spectator's gallery inside the building was cheering each effort to turn and the equally desperate attempts to stay upright. In the absence of forces, a mass in motion tends to remain in motion at constant speed in a straight line—the principle of inertia.

FIGURE 3-2A The weight of the ball may be divided into two parts. One part pushes directly into the ramp and one part along it. Only the latter portion of the weight of the ball tends to move it down the ramp.

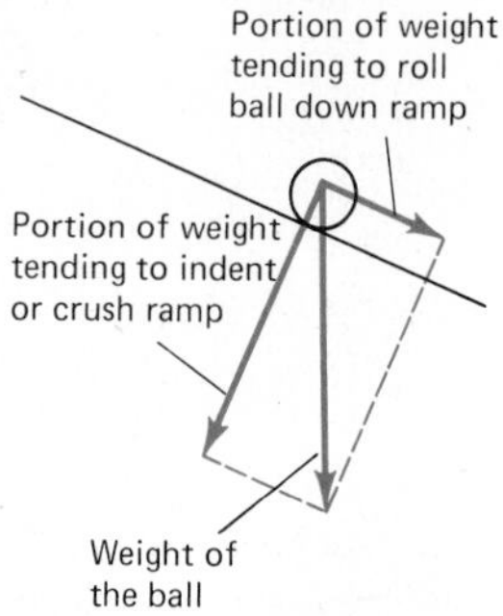

Portion of weight tending to roll ball down ramp

Portion of weight tending to indent or crush ramp

Weight of the ball

FIGURE 3-2B As the ramp becomes less steep the portion of the weight causing motion becomes smaller.

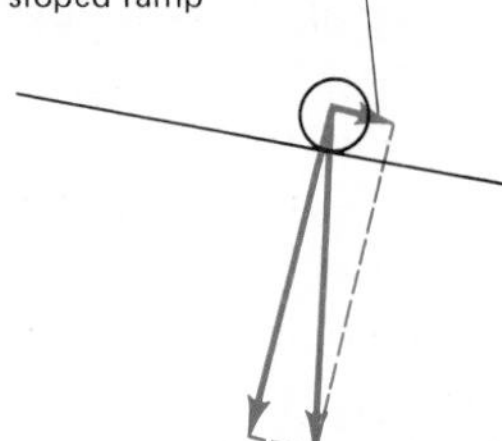
Only a small portion of the weight acts to move the ball down a gently sloped ramp

FIGURE 3-2C For very steep ramps, the force moving the ball approaches the size of the weight.

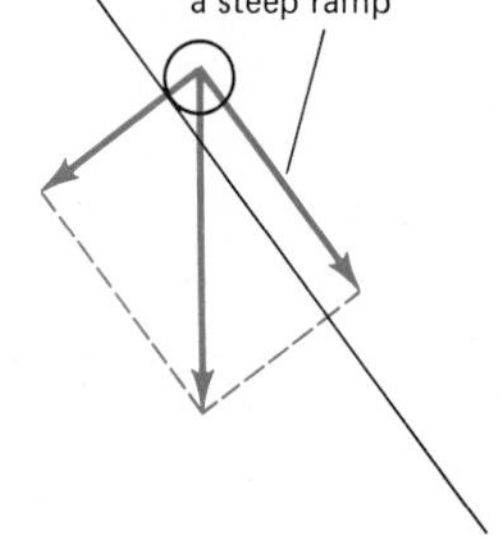
A large portion of the weight acts to move the ball down a steep ramp

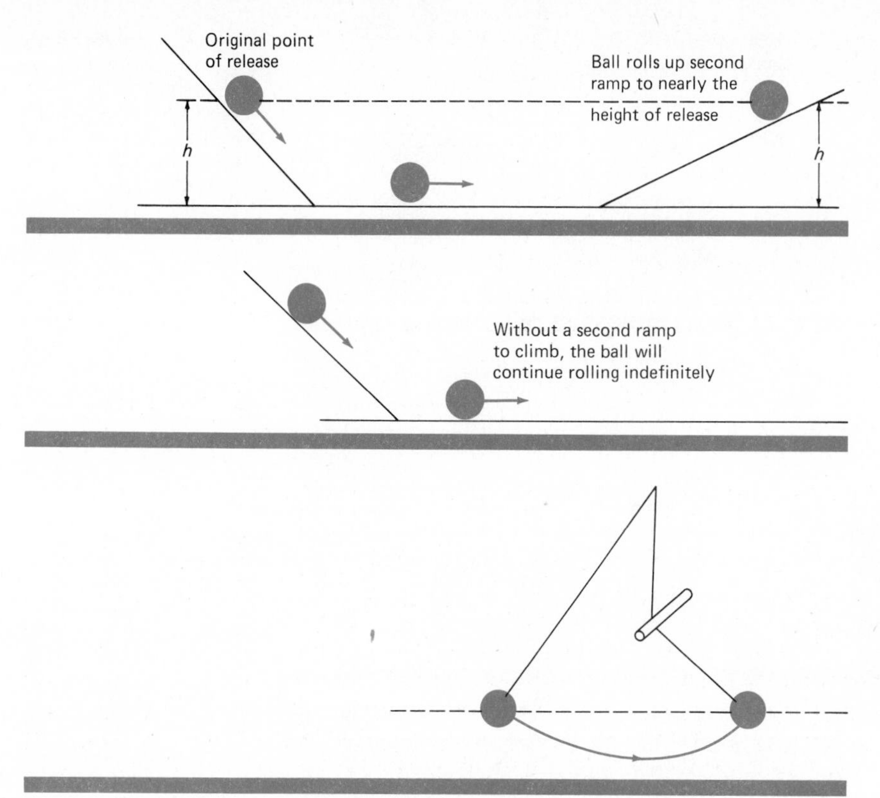
Original point
of release
Ball rolls up second
ramp to nearly the
height of release
h
h
Without a second ramp
to climb, the ball will
continue rolling indefinitely

FIGURE 3-3
A ball rolled down a ramp tends to climb a second ramp to the original height of its release.

FIGURE 3-4
The principle of inertia—in the absence of forces the ball will continue moving at constant speed.

FIGURE 3-5
The pendulum represents a nearly frictionless case of the ball and ramp since it eliminates the ramp. The swinging pendulum returns to its height of release even when a rod is used to change the arc of its swing.

Galileo checked the principle of inertia in many different experimental situations. Perhaps the best example of a nearly frictionless "ball and ramp" available to Galileo was the pendulum. By doing away with the contact with a ramp altogether, the pendulum approached the ideal case (Fig. 3-5). Galileo determined that the ball tended to return to the height of release even if the length of the pendulum was changed by contact with a fixed rod.

Thus, Galileo rejected the first part of Aristotle's ideas of motion. Aristotle asserted that the natural tendency of matter (horizontally) was to be at rest in the absence of forces. Instead, Galileo stated that the tendency of matter in the absence of forces was to continue doing exactly what it was doing, whether that was moving or being at rest. This change of view begins the modern treatment of motion.

The second major objection raised by Galileo to Aristotle's rules for motion is more well known. The simple fact was that balls that were rolled down ramps speeded up. The steeper the ramp or the longer the ramp, the greater the change in the speed of the ball (Fig. 3-6). From his studies Galileo concluded that the rate at which a falling object gains speed is constant. (Aristotle, you will remember, concluded that the speed of a falling object was constant from the instant of release and was determined by its weight—an object twice as heavy should fall at twice the speed.) Galileo stated that all objects would fall at the same

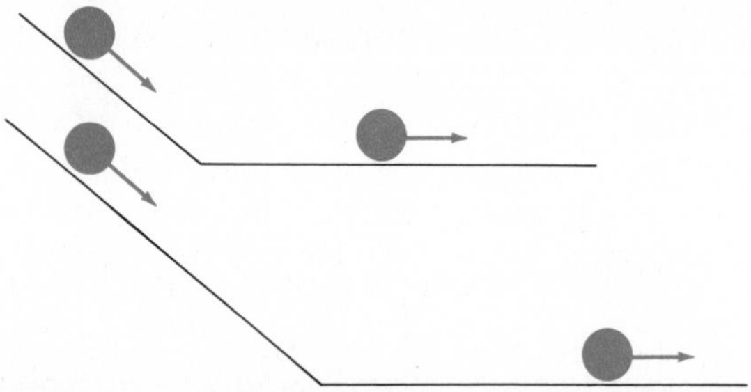

FIGURE 3-6
The ball rolling the greater distance down the ramp comes off at a greater speed. Therefore, falling objects gain speed as they fall.

rate of constantly increasing speed *in the ideal case.* In the real case, a feather cannot push the air out of the way as effectively as a coin can, so it is noticeably slower in falling. Galileo asserted that in a vacuum the fall of all objects would be similar.*

Galileo's colleagues tended to ridicule his need to talk about "ideal cases." Why not consider the real world, not some fairy tale in the mind? Furthermore, they disliked the idea of a vacuum as much as Aristotle had (and for the same reasons). To answer their objections, Galileo suggested many experiments.

You can see for yourself who is closer to being correct quite easily. Drop a marble and a golf ball from shoulder height. The golf ball is about 10 times as heavy as the marble. According to Aristotle's views, the golf ball should hit the floor when the marble has moved only one-tenth of the distance to the floor. According to Galileo both the marble and the golf ball should hit the floor at the same time. What happens? You can determine whether Galileo's concern about air resistance and his talking about an "ideal case" were justified. Drop a sheet of notebook paper and a coin from shoulder height. Is there a clear winner in several tries? Now crumple the paper into as tight a ball as you can manage. Drop the paper ball and the coin several times? Does the condition of the paper influence its action in falling? The weight of the paper was constant. What has changed?

The point of view that would eventually be held to be correct was that one which offered the best prediction of events in nature—Galileo's. Furthermore, Galileo assured that future tests would be done carefully by taking great care in defining his terms and expressing them mathematically.

The Description of Motion

DEFINING SPEED AND ACCELERATION

Up to this point in the text we have used most terms in their everyday sense and have not concerned ourselves with precise definition. In general, this is a bad way to proceed in a science. Until terms are carefully

* This was demonstrated to be correct some 330 years later when an astronaut dropped a feather and a hammer on the moon for television viewers. The two objects fell together and struck the surface at the same time.

defined, we are not sure we mean the same thing when we use the same words (Fig. 3-7). Or we may have a "feeling" for the meaning of a term but have no idea of how to go about measuring it. This is almost as bad as the use of undefined terms, since science deals in measurable quantities. This lack of precision may be excused to this point on two grounds. First of all, it is useful to become comfortable with the style and pace of a subject before encountering too much that is really new or different. Second, the physical variables used to this point have been ones you are likely to know how to measure. (Knowledge of how to measure a variable serves as one type of definition of it—a very useful type, called an *operational definition.*) Length, time—even temperature and pressure—are variables you have been using for many years. For example, it is likely that if a visiting Martian inquired, you would be able to provide enough information to allow the Martian to build a working thermometer. It is not so clear that most individuals could describe how to build a speedometer. Therefore, from this point we will try to provide more careful definitions of the terms we will use. A frequent mistake of beginners in science (or mathematics) is to pay more attention to conclusions than to definitions. The opposite approach is more appropriate, since once definitions are set the conclusions are inevitable.

Speed is the rate at which distance changes. Hiding in that simple sentence are several important mathematical notions. In a mathematical statement about the physical world, the word "is" translates to "is equal to." The word "rate" implies that we will be considering the ratio of two changes—that is, the change in one variable divided by the change in another. Usually, the variable we divide by is time unless we are told otherwise. (Thus, the rate of growth of a plant is the change in height divided by the time it took to make that change, $G = \Delta h/\Delta t$.) Rates are very important in science, so it would be useful to remember this.

The car in Fig. 3-8 is moving to the right at some speed v. To measure that speed we determine the time t_1 at which the car is at a particular position d_1. We later measure the time t_2 at which the car is at position d_2. The change in distance, Δd, divided by the time required to accomplish that change in distance, Δt, is

$$v = \frac{\Delta d}{\Delta t}$$

$$= \frac{d_2 - d_1}{t_2 - t_1}$$

where v is the *velocity,* the size of which we call *speed.*

The form of the equation also suggests that the units in which v will be measured are distance units divided by time units. Thus, speeds are measured in miles per hour, kilometers per hour, meters per second, and so on. ("Per" means "divided by.")

One concern in the measurement of speed is that the speed may change a great deal and a measurement taken across a large interval of distance and/or time does not have to represent the true speed at any

FIGURE 3-7
Motion consists of a body changing its position in space as it changes its location in time. At each instant of time the body is in a particular position in space. Thus, the description of motion involves the measurement of changes in spatial location and changes in time. (*Photograph by the author.*)

given time. For example, a car leaves Chicago for St. Louis some 450 km away. It arrives 5 h later. What was the car's speed? We may obtain an answer by using the formula

$$v = \frac{\Delta d}{\Delta t}$$

$$= \frac{450 \text{ km}}{5 \text{ h}}$$

$$= 90 \text{ km/h}$$

Of course, this number doesn't necessarily tell us how fast the car was going at any given time. The number is an *average speed* over a 5-h interval. During that time, the car speeded up and slowed down with traffic, and probably even stopped for gasoline. Thus, at any one time the car's speed could have any reasonable value, even though its average speed was 90 km/h. In scientific applications we are usually more interested in the speed at a particular instant than in the average speed. The *instantaneous speed* is more important for what is happening at that instant. After all, whether a car penetrates a brick wall depends upon its speed at the instant it hits the wall, not on its average speed for the past 2 h. To distinguish an average value over a large interval from an instantaneous value, a line or *bar* is written over the average. Thus, the average speed may be written as

$$\bar{v} = \frac{\Delta d}{\Delta t}$$

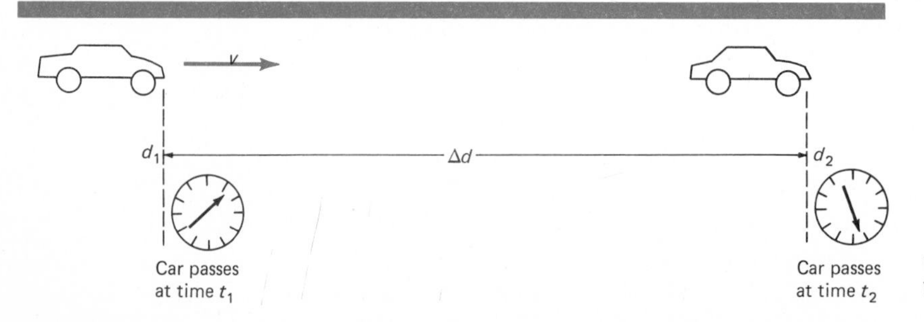

FIGURE 3-8
The measurement of speed involves measuring a change in distance and the time interval across which this change in distance took place. Here, $d_2 - d_1 = \Delta d$, the change in distance, while this change was accomplished in the time interval $t_2 - t_1$ or Δt.

which may be read as "v-bar, or v-average, equals change in d over change in t." Of course, all distance changes must be measured across some time interval. When does a number cease being an average value and become an instantaneous value? Clearly, the shorter a time interval across which the two distances d_1 and d_2 are measured, the closer the measured value is likely to be to an instantaneous value. A speed measured over a time interval of 5 s of motion is much more likely to be near the instantaneous speed of the car than one measured over 5 h of motion. Technically, a true instantaneous speed is measured only if the time interval Δt becomes vanishingly small, or

$$\overline{v}_{\text{inst}} = \frac{\Delta d}{\Delta t} \qquad \text{as } \Delta t \text{ approaches zero in size}$$

For a *constant-speed motion* we have no problem in finding the instantaneous speed since it is the same as the average speed at any point in the motion.

Galileo presented a second definition which was of great value in discussing motion. He gave us the standard definition for discussing changes of motion. The process of speeding up or slowing down is called *acceleration.* Galileo defined *linear acceleration* as the time rate of change of speed.* That is,

$$a = \frac{\Delta v}{\Delta t}$$

This is mathematically and verbally identical to the definition of speed with the change of a for v and v for d in the earlier equation. The difference between average acceleration and instantaneous acceleration is just as important as with speeds. The case of constant acceleration is called *uniformly accelerated motion.* Galileo's measurements indicated that a falling object was uniformly accelerated. That is, the object gains downward speed at a rate that is constant in time. Every second a falling object, free of any frictional restraints, will gain an additional 9.8 m/s. Thus, if the object is initially at rest and is released,

* Actually we define acceleration as the rate of change of velocity, not of speed. The difference is discussed in the next section.

1 s later it will be falling at a speed of 9.8 m/s; 2 s after its release it will have increased its speed by another 9.8 m/s and be falling at a rate of 19.6 m/s. After 3 s its speed will be 29.4 m/s, and so on. A freely falling object gains 9.8 m/s of downward speed each second of its fall. This number, 9.8 m/s each second, or 9.8 m/s^2 is called the *acceleration due to gravity* and is expressed by the letter g whenever it appears in mathematical expressions.

$$a = \frac{\Delta v}{\Delta t}$$

$$= \frac{9.8 \text{ m/s}}{1.0 \text{ s}}$$

$$= 9.8 \text{ m/s}^2 = g$$

The value of g varies only slightly for different positions on the earth's surface. In general, it is greatest at low elevations near the poles and least at high altitudes near the equator. Very small variations in the value of g have been found to be important to geologists studying the structure of the earth. Since the maximum variation at the surface of the earth is about $\frac{1}{2}$ percent, we will treat g as a constant. A more precise value for sea level at the equator is 9.7805 m/s^2. Frequently, problems may be simplified by approximating g as 10 m/s^2, which introduces only a 2 percent error in the result of calculation.

SPEED VS. VELOCITY—A NOTE ON VECTORS

It may have occurred to you that it is strange to use the letter v to represent speed. Generally, an attempt is made to use a letter that suggests the variable, such as l for length or t for time. This is not done with speed because a new consideration enters physical laws when we begin to discuss motion. This consideration is that of direction. The practical effects of walking one block south are quite different from those of walking one block north. If you are trying to get somewhere, *direction counts,* it is a major concern. Likewise, a given speed to the east is quite different in its implications from the same speed to the west. Therefore, at some point in science it is necessary for the mathematical treatment of variables to include direction as well as size. A variable that has a direction as well as a numerical size is called a *vector.* The variables you are accustomed to which indicate only size are called *scalars.* Speed is a scalar. If we wish to consider not only the speed but direction as well, we use the term *velocity.* If you say that your car is traveling 80 km/h (50 mi/h), you are reporting its speed. If you say you are moving 80 km/h to the east, you have given your velocity. In the sciences, velocity is discussed much more frequently than the speed, so the letter v is used for both variables. If speed is meant, no change is required. Should we want to discuss velocity, it is either referred to by name, or the symbol v is modified. Texts indicate vectors by having the letter printed in boldface type, **v**, rather than in normal type. Another alternative is to print an arrow over the letter representing a vector: $\vec{v}$. Vector equations are more complicated than the scalar equations you are used to. Both sides of a vector equation must agree not only in size but in direction as well. Also, the arithmetic of vectors is different from the arithmetic of numbers. (How do you multiply 2 north by 3 west and what direction will the result have?) While vectors may be new to you, the rules for handling them are not difficult to understand. We will not refer to vectors frequently in the text, but when we do, it will be important that you understand their use. The following subsection discusses the rules of operations with vectors in terms of the addition of vectors.

FIGURE 3-9A
The sum of two vectors in the same direction is just equal in size to their numerical sum.

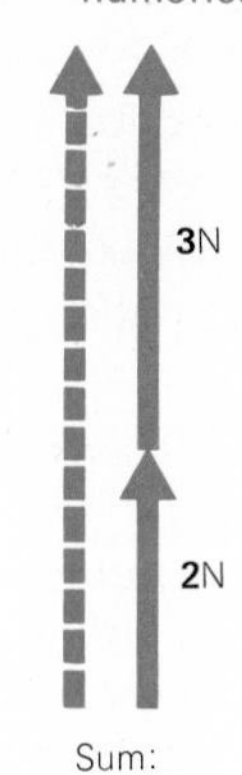

FIGURE 3-9B
The sum of vectors in opposite directions is equal in size to their numerical difference.

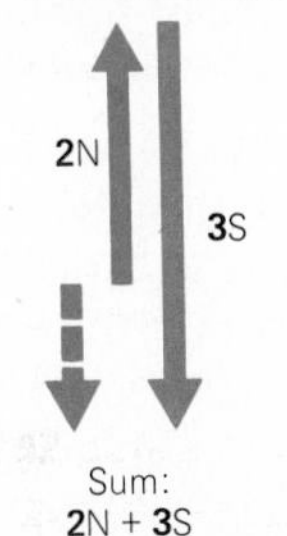

FIGURE 3-9C
In adding vectors the length and direction of each must be preserved. The tail of each vector begins where the head of the previous vector ended. The sum is the length and is in the direction from the tail of the first vector to the head of the last vector added.

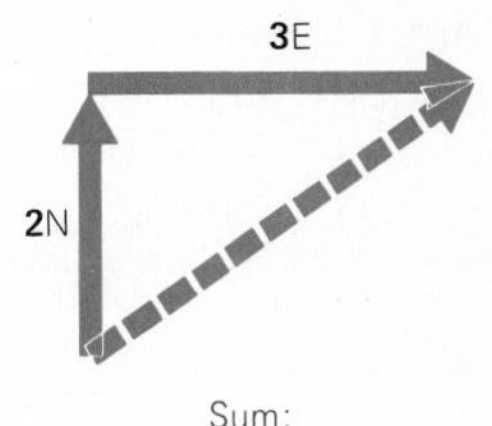

THE ADDITION OF VECTORS

With numbers, 2 + 3 = 5. It should *not* surprise you that this is not generally true with vectors. Consider this example. You walk 2 blocks north, and then you walk an additional 3 blocks north. How far are you from your starting point? (See Fig. 3-9A.) Clearly you would be 5 blocks from your starting point. Thus, in this example, **2** + **3** = **5**. However, suppose that you first walked 2 blocks north but that you then walked 3 blocks south. What would your distance now be? (See Fig. 3-9B.) Again, it would be clear that you are now 1 block south of your starting point. Thus, **2** + **3** = **1**. This result is not likely to look reasonable when written in numbers like this. We might be tempted to gloss over this difference by indicating north as positive and south as negative. We could then rewrite this second case as a scalar equation, $2 + (-3) = -1$. But, if we choose that method to reduce the problem to the familiar arithmetic of scalars, how will we handle this third case? (See Fig. 3-9C.) You walk 2 blocks north and then you walk 3 blocks *east.* How far are you now from your starting point? 2 + 3 = ? Recognizing that the two vectors in Fig. 3-9C form a right triangle with the dashed line of our distance being the hypotenuse, we could obtain the distance from the theorem of Pythagoras, $a^2 + b^2 = c^2$. Here, we obtain $2^2 + 3^2 = d^2$, or $d = 3.6$ blocks when we complete the arithmetic. Thus, for this case, **2** + **3** = **3.6**.

If you have followed this carefully, by now you are ready to admit that the arithmetic of vectors is different from the arithmetic of numbers. Indeed, the sum of two vectors (**2** + **3**) may be as large as their numerical sum (Fig. 3-9A), or as small as their numerical difference (Fig. 3-9B), or any value in between these extremes (as in the case of Fig. 3-9C). The numerical value of the distance, as well as its direction from the starting point, depends upon the direction of the two vectors as well as their size.

Now that you understand and accept that the arithmetic of vectors will be different from the arithmetic of numbers, you are entitled to ask, "How do we add vectors?" To answer this, please note that you have already witnessed the answer for three important cases in the examples just given. We can state these as formal rules:

Case 1: If two vectors *agree in direction,* their sum is the numerical sum of the vectors and the direction of the sum is the same as the direction of the two vectors (Fig. 3-9A).

Case 2: If two vectors are *in opposite directions,* their sum is the numerical difference of the two vectors and is in the direction of the larger vector (Fig. 3-9B).

Case 3: If two vectors are *at right angles* to each other, their numerical sum may be found by using the Pythagorean theorem (Fig. 3-9C).

In this third case, the direction of the sum is less easy to describe. If the direction is needed, we may proceed to the general strategy for adding vectors. Before we do this, however, you should note that the three cases given above cover the majority of situations which will arise in the simple handling of vectors.

FIGURE 3-10
So long as the direction and relative length of each vector is preserved, they may be added in any order to produce the same sum.

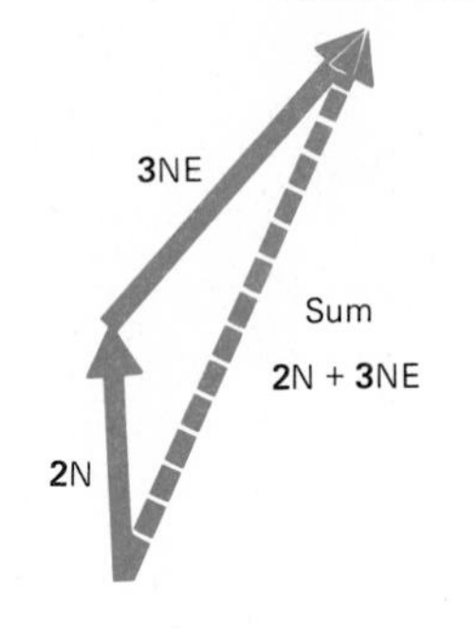

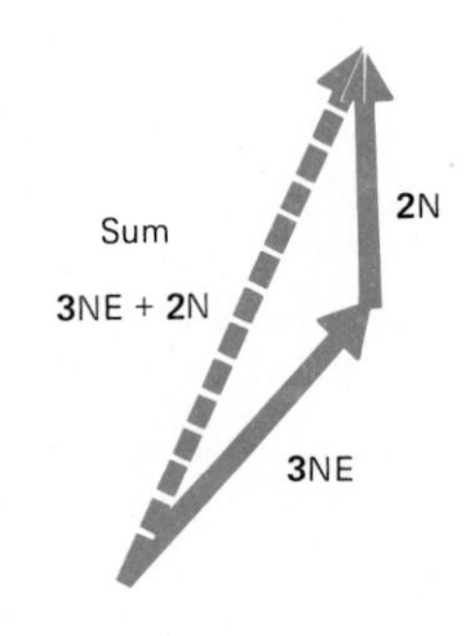

The general strategy for the simple addition of vectors has already been illustrated for you in the three parts of Fig. 3-9. Consider the third part, Fig. 3-9C. The first vector was drawn from a dot representing our starting point and points directly upward on the page. Up thus represents north, as is customary on maps. If you use a ruler you may verify that this vector is 2.0 cm long. The second vector is drawn from the ending point of the first directly to the right, or east, and is 3.0 cm long. The dashed line representing the total distance, the sum, is drawn from the tail of the first vector at our starting point to the head of the second vector at our ending point. This dashed vector represents the distance and the direction we are from our starting point when we finish the motion. Using your ruler again, you may verify that this vector is 3.6 cm long. On the scale of the other two vectors, if 1 cm represents 1 block, this means the sum is 3.6 blocks. Using a protractor on the drawing you may verify that the angle the sum vector makes with north is 56°. Thus the sum is 3.6 blocks in a direction 56° east of north. We have completely determined the sum of the two vectors from the drawing. *This procedure represents a general strategy for adding two (or more) vectors. Make a scale drawing and measure the sum from the drawing.* In Fig. 3-10 we have added by drawing two vectors representing 2 blocks north and 3 blocks northeast in two different ways. In the first drawing we began with the north vector and added the northeast vector to it. In the second drawing we reversed the order in which we drew the vectors. Measurement will verify that in both cases the sum is identical in length and in angle to the top of the page. Thus, the order in which vectors are added does not influence the sum. Or, if one vector **A** is added to another vector **B**,

$$\mathbf{A} + \mathbf{B} = \mathbf{B} + \mathbf{A}$$

This is equivalent to saying that if you walk 2 blocks north then 3 blocks east, you are at the same spot as if you walked 3 blocks east and then 2 blocks north. In addition of vectors, the only things that *must* be preserved for each vector are its size and its direction. The order of addition is unimportant.

The subtraction of vectors is quite simple once addition has been mastered. To subtract a vector we simply add its negative—that is, we add a vector equal in length but opposite in direction. Thus, to subtract 2 blocks north we would add 2 blocks *south.*

One operation with vectors remains for us to consider—the *resolution of vectors.* Just as two or more vectors can be added to produce a

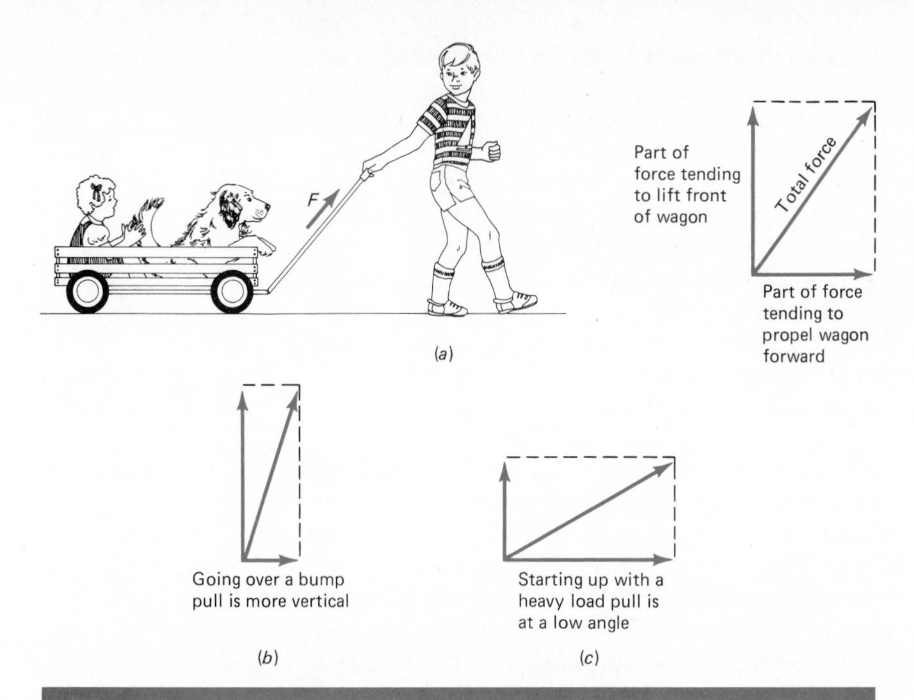

FIGURE 3-11 In pulling a wagon the angle of pull is adjusted to provide either a greater upward force or a greater forward force as the need demands.

single sum vector, a single vector may be broken down into two or more vectors that add up to it. An example of this might be the case illustrated in Fig. 3-9C. We have seen that the sum was 3.6 blocks in a direction 56° east of north. Suppose that the starting point represented your house and the destination at the head of this sum vector was the post office. If you were sitting in front of your house and a stranger asked you where the nearest post office was, you would be unlikely to reply, "It's 3.6 blocks in a direction 56° east of north." Even if you knew this to be correct, you would be more helpful (and less likely to start an argument) if you replied, "It's 2 blocks north and 3 blocks northeast." Similarly, it is often very convenient in science to break down and consider a single vector in terms of two or more vectors which add up to the original vector by the rules of vector addition.

An example of this might be to interpret pulling behavior of a wagon. (See Fig. 3-11.) You may have noted that when the pulling gets difficult, as in starting up or covering soft ground, it is easier to lower the angle of the handle to the ground. This can be understood by resolving the pull or force vector into two parts—an upward part and a forward part (Fig. 3-11*a*). The pull which is exerted on the wagon acts along the line of the handle. It has both size and a direction (it is impossible to pull in no direction); therefore this force is a vector. Clearly the vector labeled "forward pull" and the vector labeled "upward pull" add up to the total force vector. The upward part of the pull tends to raise the front part of the wagon, but it does not contribute to the forward motion. The forward part of the pull provides the force to overcome the rolling friction and to speed up the wagon. To start up a greater forward pull might be desired.

To get this, one might simply pull harder, but this would increase the upward part of the pull also. Indeed, the upward part of the force might become great enough to lift the front of the wagon off the ground and dump the occupants. Rather than increasing the pulling force, a second strategy is to lower the angle of pull (Fig. 3-11*b*). Then, even for the same total pull, a larger forward part is available to begin the motion. In this case, to discuss the problem it has been convenient to *resolve* the total pull into a *vertical component,* the upward pull (which does not contribute to the forward motion), and a *horizontal component,* the forward pull (which determines the motion). It is frequently convenient to break up or resolve a single vector into two perpendicular parts which cannot contribute to each other, much as an upward pull does not determine the forward motion.

As a numerical example of the resolution of a vector, consider the following problem. A boy throws a rock at a speed of 20 m/s and at an angle of 60° to the horizontal. At what speed is the rock gaining altitude when it is released? To solve this problem we must resolve the initial motion of the rock into two parts—an upward part and a horizontal part. We first draw the vector representing the total velocity of the rock **v** (Fig. 3-12A). We have chosen the scale of 1 cm, representing 5 m/s of motion. The vector is 4.0 cm long in the drawing. The angle of the vector is measured to be 60° from the line representing the horizontal surface of the ground. At this point the starting point and the ending point of the vector are determined. Any set of vectors added to each other, tail to head, which begin at the starting point and finish at the ending point of our vector, would therefore add up to our velocity vector. However, only one set of vectors will answer the problem. To see why, consider the set **A** and **B** drawn in Fig. 3-12B. Clearly they add up to **v**. What is the upward speed of the rock? It is not vector **A**, since **A** is mostly horizontal although it has some upward part. Vector **B** does not represent all of the upward motion. Furthermore, some of **B**'s length represents horizontal motion. The vectors of Fig. 3-12B do not solve the problem. To produce a set that does, we return to our original drawing of **v**. From the tip of **v** we drop a line perpendicular to the surface (Fig. 3-12C). The distance from the base of **v** to this line we label $\mathbf{v}_H$, the horizontal speed. The distance from the tip of $\mathbf{v}_H$ to the tip of **v** we label $\mathbf{v}_V$. This set of vectors equal **v**, or $\mathbf{v}_V + \mathbf{v}_H = \mathbf{v}$ (Fig. 3-12A). Furthermore, $\mathbf{v}_V$ is purely vertical and represents all of the upward motion of **v**. Measuring $\mathbf{v}_V$ we find it to be about 3.1 cm long. Thus it represents about 17 m/s of upward motion. This is the initial upward speed of the rock. Similarly, $\mathbf{v}_H$ is about 10 m/s of horizontal speed. In this manner we have isolated the upward and the sideways portions of the motion for consideration.

FIGURE 3-12A
The initial motion of the rock has both upward and sideways speed. It is necessary to separate these two components of the motion.

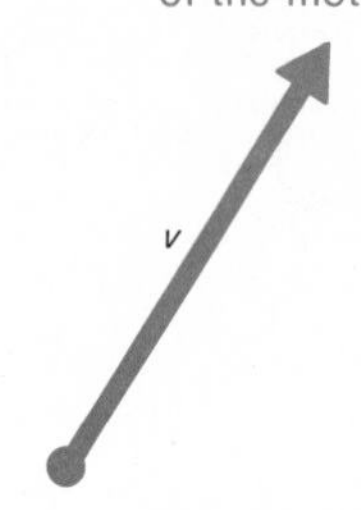

FIGURE 3-12B
While the set of vectors **A** and **B** add to the original vector **v**, they are not useful. Both **A** and **B** have upward and sideways components to them, so we cannot consider these motions separately.

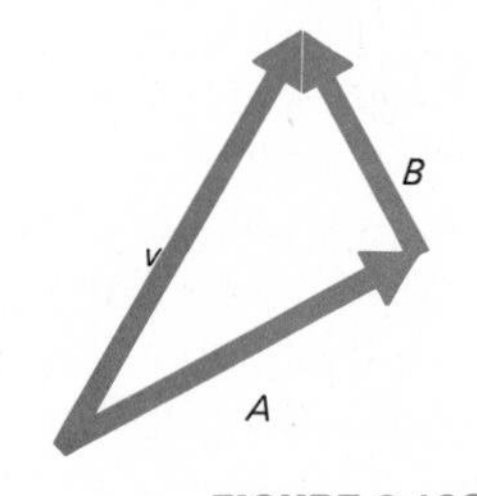

FIGURE 3-12C
The set of vectors $\mathbf{v}_H$ and $\mathbf{v}_V$ add to the original vector **v**. In addition, no part of $\mathbf{v}_H$ is vertical, so $\mathbf{v}_V$ represents the entire vertical component of the motion. Likewise, $\mathbf{v}_H$ represents the entire horizontal velocity of the rock.

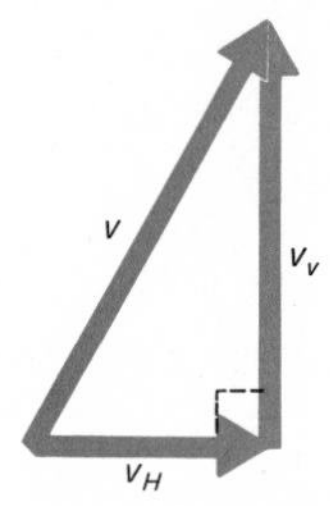

GRAPHS OF MOTION—INSTANTANEOUS SPEED

We now have two definitions:

1 Velocity is the rate at which distance changes with time.

2 Acceleration is the rate at which velocity changes with time.

$$v \equiv \frac{\Delta d}{\Delta t} \qquad a \equiv \frac{\Delta v}{\Delta t}$$

The form of each of these expressions is mathematically identical to the definition of the slope of a graph.

$$m \equiv \frac{\Delta y}{\Delta x}$$

This similarity of the forms of these definitions leads to a powerful conclusion. If we make a graph of distance along the y axis vs. time along the x axis, *the slope of the graph at any position must be the speed* v, since

$$m = \frac{\Delta y}{\Delta x} = \frac{\Delta d}{\Delta t} = v$$

Consider the simple graph of distance vs. time shown in Fig. 3-13.* What was the speed of the car at the time labeled 3.0 s? The first property of the graph that we should notice is that it is a straight line. Thus, its slope is constant. Since this is a graph of d vs. t, the value of the slope *is* the speed, which must, therefore, also be constant. Furthermore, since the speed is constant, we may measure it across any size interval in the

* In giving a graph a title, such as "distance vs. time," it is standard practice to name what we call the *dependent* variable first and graph it on the y axis. The *independent* variable is named last and graphed on the x axis. In this example, the distance covered depends upon the time that has elapsed, and is the dependent variable. Time keeps ticking away and is utterly independent of the motion indicated by the graph. Time is almost always the x variable when it is graphed.

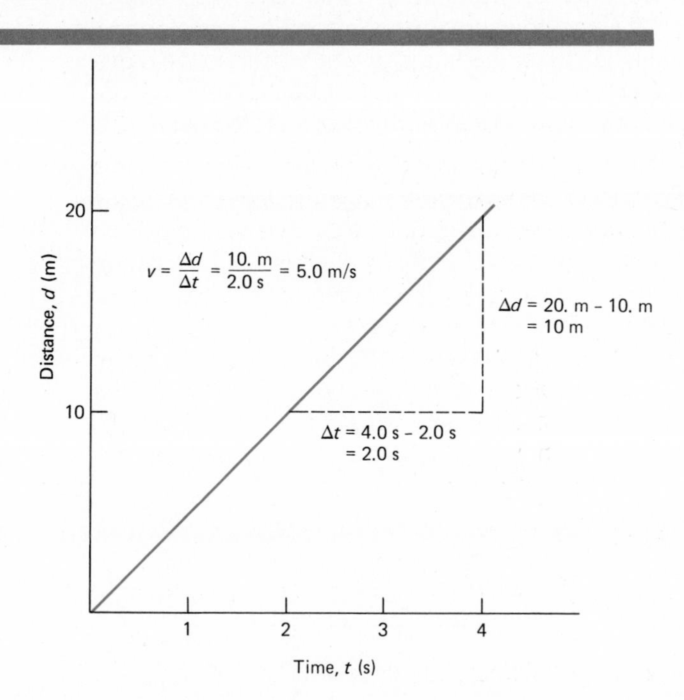

FIGURE 3-13
From the mathematical form of the definition of speed, the speed of an object at a given time *is* the same as the slope of a graph of distance vs. time for that time. Here, the slope of the curve is constant so this represents a constant-speed motion with distance increasing uniformly with time.

graph. Here, the instantaneous value of the speed at 3.0 s is the same as the average speed for a large interval. Choosing two convenient and widely separated points ($t = 2.0$ and $t = 4.0$), we can now determine the speed (Fig. 3-13):

$$v = \frac{\Delta d}{\Delta t}$$

$$= \frac{20.\ \text{m} - 10.\ \text{m}}{4.0\ \text{s} - 2.0\ \text{s}}$$

$$= \frac{10.\ \text{m}}{2.0\ \text{s}} = 5.0\ \text{m/s}$$

The motion shown in Fig. 3-14 is more difficult. Here, the speed is not constant. If the object whose motion is graphed would have collided with a wall at $t = 2.0$ s, at what instantaneous speed would it have hit the wall? To determine this is more of a problem. The slope of the line, and thus the speed, is continually changing. Study of the graph shows that the slope becomes greater as the time increases. Thus, the speed at 1.5 s, for example, is less than the speed at 2.0 s. Later, say at $t = 2.5$ s, the slope of the curve is much steeper, and we may conclude that the speed is greater. To find the speed at 2.0 s, we must determine the slope at $t = 2.0$ s, not earlier or later along the graph. Furthermore, if we chose some time interval along the graph with 2.0 s as its midpoint, in general we would not be justified in assuming that the average speed for this interval is the same as the speed at 2.0 s. Thus, obtaining the average speed for the interval $t = 1.5$ s to $t = 2.5$ s does *not* yield the value of the instantaneous speed at $t = 2.0$ s. We need to determine the slope or trend of the line at $t = 2.0$ s and nowhere else. We can find this slope by using a rule which states that the slope of a curve at a point is the same as that of a straight line tangent to the curve at that point. We may construct such a tangent line by aligning a straightedge near the curve and

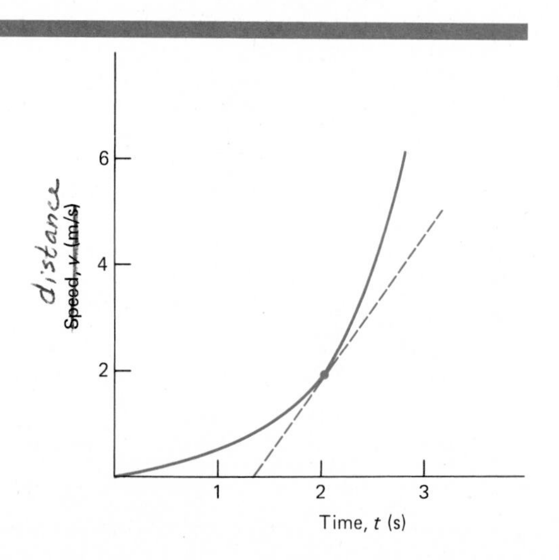

FIGURE 3-14
The speed at any time is still given by the slope of the *d*-vs.-*t* curve for the case of a more complex motion. Here, the slope of the line is continually changing. The speed at time $t = 2.0$ s, is given by the slope of the line at that time, as is indicated by the dashed line on the drawing.

then sliding it into place until it just touches the curve at the point $t = 2.0$ s (and does not touch the curve before or after that point with a curve such as this). The line formed by the straightedge is now *tangent to* the curve at the point $t = 2.0$ s. More importantly for us, the slope of the line is the slope of the curve at $t = 2.0$ s. The slope of this line measured on the scale of the graph will yield the instantaneous value of the speed at $t = 2.0$ s. Such a line tangent to the curve at $t = 2.0$ s is represented on Fig. 3-14 by the dashed line. Measurement of the slope of this line yields the value 0.5 m/s as the speed of the motion at $t = 2.0$ s.

We will not need to determine so many instantaneous speeds here as we might have to in a laboratory situation. Yet, it is important to note that we need not throw up our hands in despair if asked to obtain an instantaneous value from a graph. The value of the slope of a graph at a point is the same as the slope of a straight line constructed tangent to the point on the curve.

One final note before we leave this section. From the form of the definition of acceleration, $a = \Delta v/\Delta t$, it should be clear that *the acceleration of an object is the same as the slope of a graph of its speed vs. time at any point.* The definition of slope is identical in form to that of acceleration with the x variable replaced by time and y by v. Thus, everything we have said about speed as determined from a d-vs.-t curve applies to acceleration as determined from a v-vs.-t curve. For example, in the curve shown in Fig. 3-15, we may determine the acceleration to be

$$a = \frac{\Delta v}{\Delta t}$$

$$= \frac{(15 - 5.0)\ \text{m/s}}{(3.0 - 1.0)\ \text{s}}$$

$$= \frac{10\ \text{m/s}}{2.0\ \text{s}} = 5.0\ \text{m/s}^2$$

Verification of this figure is simple if you have followed the discussion.

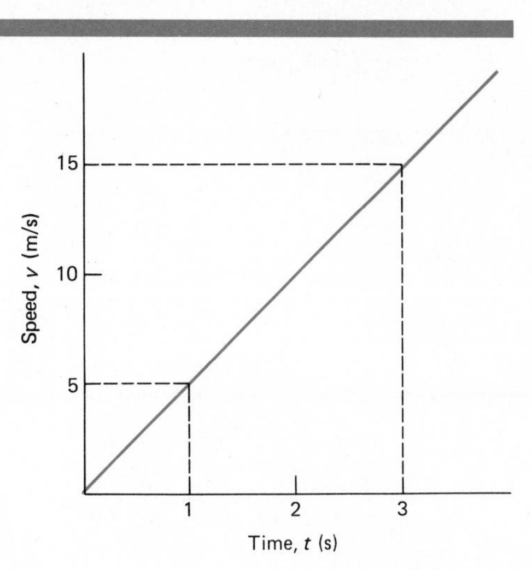

FIGURE 3-15
The acceleration may be measured from a graph of speed vs. time since it is the slope of this graph at any point in time.

DISTANCE FROM A v-VS.-t CURVE

Consider the curve of Fig. 3-16. What is the distance traveled for the 3 s graphed? In this case we can see that the speed is constant at $v = 5.0$ m/s. We can obtain the distance for the case of constant speed quite easily. Since $v = \Delta d/\Delta t$, then $\Delta d = v\ \Delta t$. Because v has a constant value, we may use this last expression with ease. The change in distance is just the value of v, 5.0 m/s, times the time interval of 3.0 s:

$$\Delta d = v\ \Delta t$$
$$= (5.0\ \text{m/s})(3.0\ \text{s}) = 15\ \text{m}$$

You will notice that this is also the area under the curve in Fig. 3-16. The curve being considered has a height of 5.0 m/s and a width of 3.0 s. Thus,

$$\text{Area} = h \times w$$
$$= (5.0\ \text{m/s})(3.0\ \text{s}) = 15\ \text{m}$$

This observation points out a perfectly general conclusion about v-vs.-t curves. *The distance traveled is always given by the area under a v-vs.-t curve, even when the curve itself is not a straight line.*

The only difficulty in using this relationship arises when the curve is not a straight line. For example, consider the curve shown in Fig. 3-17. What is the distance traveled in the time interval from $t = 1.5$ s until $t = 2.0$ s? This change in distance will be equal to the shaded area shown in the figure. Any technique that will allow us to determine this area would provide a method to obtain Δd for this time interval. Perhaps the easiest method is to approximate the shaded area as a tall, thin rectangle. Observing the graph shows that the decline in speed for the interval is fairly even, since the curve is close to a straight line in this interval. At $t = 1.5$ s, the speed indicated by the graph is 9.0 m/s. At $t = 2.0$ s, the speed is 7.0 m/s. Since the decline in speed over the interval is close to uniform, we can take the average of these two values as an approximation of the average speed for the interval:

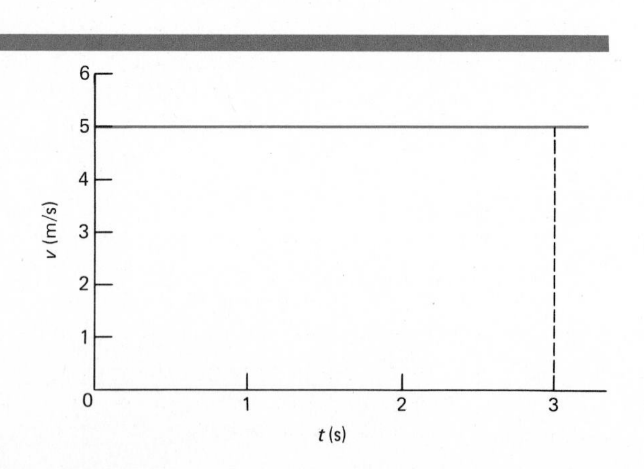

FIGURE 3-16 A graph of speed vs. time for a constant-speed motion. The distance traveled in the 3 s graphed is just the area under the curve, in this case, 15 m.

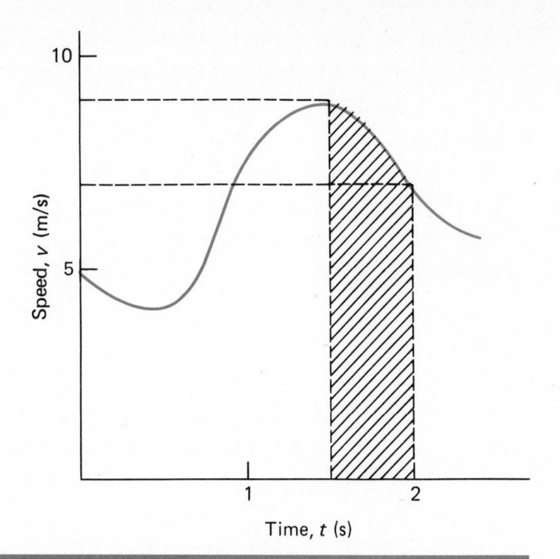

FIGURE 3-17 The distance traveled in the time from $t = 1.5$ s to $t = 2.0$ s is the area bounded by these times, the curve, and the t axis, which is indicated in the drawing.

$$\bar{v} = \frac{v_0 + v_f}{2}$$

$$= \frac{9.0 + 7.0}{2} \text{ m/s} = 8.0 \text{ m/s}$$

(Averaging by adding the initial and final values v_0 and v_f and dividing by 2 is valid *only* if the change between these values is uniform—a straight line.) With this approximate value for the average speed during the interval, we may now calculate the distance covered:

$$\text{Area} = \Delta d$$

$$= h \times w$$

$$= (8.0 \text{ m/s})(0.5 \text{ s}) = 4 \text{ m}$$

If it were necessary to obtain the total distance represented by the curve graphed, it would be possible to consider a series of small time intervals as shown in Fig. 3-18. The average speed could be estimated for each interval, and Δd for that interval could be calculated. The total distance would be simply the sum of these individual Δd's. The accuracy of this method improves as the size of the intervals used becomes smaller. This is because the average speed calculated from the endpoints of the interval approaches the true average speed for the interval.

SPEED AND DISTANCE WITH A UNIFORM ACCELERATION

We are now ready to consider a motion that is quite important in nature. This is the case of motion with a constant acceleration, or *uniformly accelerated motion.* You will recall that Galileo asserted that the motion of a falling object was uniformly accelerated. The determination of distances, speeds, and the like are important for uniformly accelerated motion because this situation is encountered frequently in nature. Thus, the relationships obtained in this section will be used often.

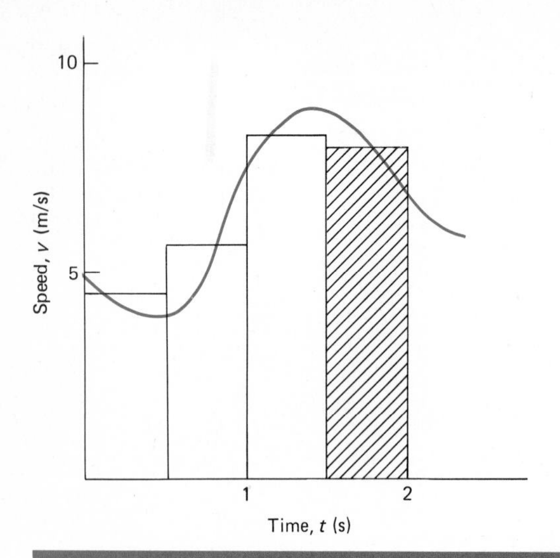

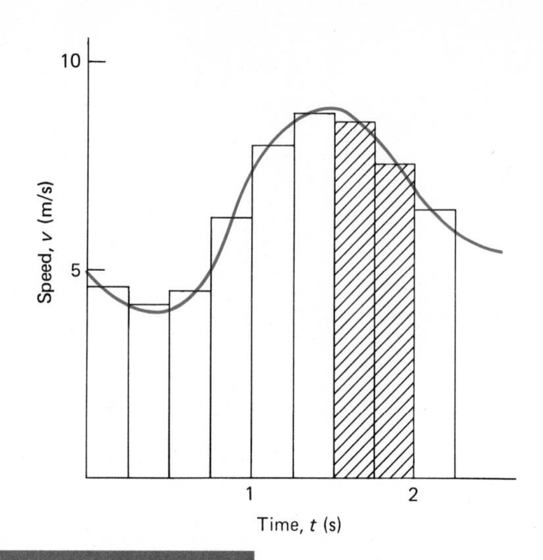

FIGURE 3-18A
The total area under the curve may be approximated by adding the areas of the bars shown. Each bar represents $\frac{1}{2}$ s of time and is taken to the height of the average speed for the $\frac{1}{2}$ s in question. While this yields the approximate area, it does not give the precise area under the graph.

FIGURE 3-18B
The precision of determining the area may be improved by using smaller time intervals, here 0.25 s each. The area of the bars is now closer to the actual area enclosed by the graph. The exact area would not be determined unless the width of the time interval was allowed to become vanishingly thin. This may be done mathematically but not graphically.

Examine the situation shown in the v-vs.-t curve of Fig. 3-19. The motion to be considered begins at $t = 0$ with an initial speed v_0. The speed increased uniformly to a final speed v_f at time t_f. Since the v-vs.-t curve is a straight line, we are assured that the acceleration is constant and this is a case of uniformly accelerated motion. Furthermore, since no numerical values are used,* it is a perfectly general case of uniformly accelerated motion, and its conclusions will apply to any case representing such motion.

We may obtain the acceleration from the slope of the curve, since $a = \Delta v/\Delta t$. Or, if we know the acceleration, the time, and the initial speed, we can use the same expression to determine the final speed:

$$\Delta v = a\ \Delta t$$

$$v_f - v_0 = a(t_f - t_0) = a(t_f - 0)$$

or $$v_f = v_0 + at \qquad (1)$$

This expression is frequently useful. Note that if we start from rest, $v_0 = 0$ and $v_f = at$.

Perhaps the easiest way to determine the distance covered by the motion is to recall that the distance is average speed times time, $d = \overline{v}t$. Under a constant acceleration, the average speed is quite easy to obtain, and we have done this before:

$$\overline{v} = \frac{v_f + v_0}{2}$$

Thus $$\Delta d = \frac{v_f + v_0}{2}t \qquad (2)$$

* The only numerical value assigned is to let the initial time for the motion, t_0, be equal to zero. This does not reduce the generality of the treatment since time is an independent variable and we can choose to begin the interval under consideration whenever we please.

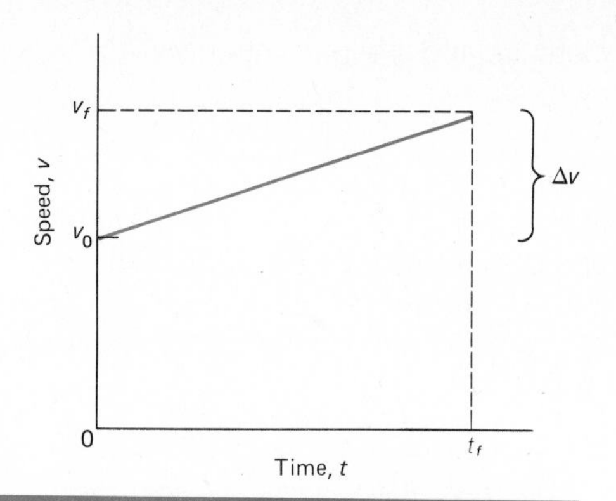

FIGURE 3-19
A general speed-vs.-time graph for a uniformly accelerated motion. Since the slope of a v-vs.-t graph is the acceleration, the straight line indicates that the acceleration of the motion graphed here is a constant.

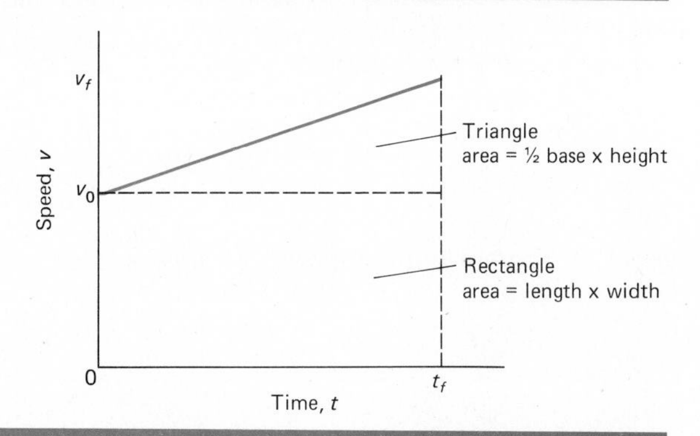

FIGURE 3-20
The area under the general speed-vs.-time graph of a uniformly accelerated motion divides into two geometrical shapes, a triangle and a rectangle, for which the area may be easily obtained. The distance traveled in this motion is just the sum of the areas of these two figures.

Frequently, v_f will not be known for the motion. In these cases it would be useful to have an expression for the distance not involving v_f. You will recall that the change in distance is just the area under the v-vs.-t curve. Examining our curve, we may see that it can be divided into two shapes, the areas of which are easy to obtain (Fig. 3-20). The lower area is a rectangle, and the upper area is a triangle. Thus

$$\Delta d = \text{area under curve}$$

$$\Delta d = \text{area}_{\text{rectangle}} + \text{area}_{\text{triangle}}$$

The rectangle has a height of v_0, and its width is t_f. Thus, its area is

$$A\square = lw = v_0 t$$

The triangle will have an area of one-half its base times its height. The base of this triangle is again t_f, or t, the total time for the motion. The height of the triangle is $v_f - v_0$ or Δv. Thus, $A\triangle = \frac{1}{2}t\ \Delta v$, and

$$\begin{aligned} \Delta d &= A\square + A\triangle \\ &= v_0 t + \tfrac{1}{2}t\ \Delta v \\ &= v_0 t + \tfrac{1}{2}t(at) \\ &= v_0 t + \tfrac{1}{2}at^2 \end{aligned}$$

or $$d_f = d_0 + v_0 t + \tfrac{1}{2}at^2 \qquad (3)$$

Note that if the initial distance and the initial speed are both zero, this reduces to the frequently used form

$$d_f = \tfrac{1}{2}at^2 \qquad (3')$$

One additional expression is useful. How can we find the final speed knowing only the acceleration, the distance, and the initial speed (if any) and not the time which the motion required? To obtain an expression, we may eliminate the time from two expressions we already know:

$$\bar{v} = \frac{\Delta d}{\Delta t} \qquad \text{and} \qquad a = \frac{\Delta v}{\Delta t}$$

(We must use the expression for average speed since we will need to use the total distance.) We obtain the following expressions on solving for Δt:

$$\Delta t = \frac{\Delta d}{\bar{v}} \qquad \text{and} \qquad \Delta t = \frac{\Delta v}{a}$$

Since both right-hand parts of these equations equal Δt, they are equal to each other. Thus,

$$\frac{\Delta v}{a} = \frac{\Delta d}{\bar{v}}$$

or

$$\bar{v}\,\Delta v = a\,\Delta d$$

$$\frac{v_f + v_0}{2}(v_f - v_0) = ad$$

$$v_f^2 - v_0^2 = 2ad$$

or

$$v_f^2 = v_0^2 + 2ad \qquad (4)$$

If the initial speed is zero, this becomes

$$v_f^2 = 2ad \qquad (4')$$

Examples

1 If a rock is dropped from a cliff, how far will it fall in 3.0 s?

In this case we know that $a = g = 9.8\ \text{m/s}^2$, and we also know that this is an example of uniformly accelerated motion. The word "dropped" indicates that $v_0 = 0$.

$$\begin{aligned} d &= \tfrac{1}{2}at^2 \\ &= \tfrac{1}{2}(-9.8\ \text{m/s}^2)(3.0\ \text{s})^2 \\ &= \tfrac{1}{2}(-9.8\ \text{m/s}^2)(9.0\ \text{s}^2) \\ &= -44.1\ \text{ms}^2/\text{s}^2 \\ &= -44\ \text{m} \end{aligned}$$

2 A cliff is 50. m high. A rock is thrown from the cliff with an initial upward speed of 10. m/s. How high is it above the ground 4.0 s later?

This is a complex problem chosen to illustrate the translation of the conditions into mathematics. Follow along, verifying each statement carefully. (1) Again, we know $a = g = -9.8\ \text{m/s}^2$ since the rock will be acting under the influence of gravity once released. (2) We may choose ground level as $d = 0$. If we do that, the rock was released at the clifftop, so $d_0 = +50$ m. The positive sign indicates the release point was above the level chosen for $d = 0$. (3) The rock was thrown upward; therefore v_0 is positive and is given as 10. m/s, $v_0 = +10.$ m/s. (4) Finally, we are to solve when $t = 4.0$ s.

$$\begin{aligned} d_f &= d_0 + v_0 t + \tfrac{1}{2}at^2 \\ &= (+50.\ \text{m}) + (+10.\ \text{m/s})(4.0\ \text{s}) + \tfrac{1}{2}(-9.8\ \text{m/s}^2)(4.0\ \text{s})^2 \\ &= 50.\ \text{m} + 40.\ \text{m} + (-78\ \text{m}) \\ &= 12\ \text{m} \end{aligned}$$

The rock is 12 m from the ground or 38 m below the clifftop at $t = 4.0$ s.

Summary

Galileo rolled balls down ramps to study the diluted action of gravity. He concluded that objects fell with constantly increasing speed. *Inertia* is defined as the tendency of matter to continue in its state of motion unless acted on by an outside force. *Mass* is the measure of the inertia of a body. Small departures from this ideal behavior could be attributable to *frictional forces* tending to oppose the motion.

Velocity v is a measure of a motion and has both a size called the *speed* and a *direction.* Variables that have both size and direction are called *vectors. Average speed* may be defined as the total distance covered divided by the time required for this motion to be completed, or $v = d$. *Instantaneous speed* is given by $v = \Delta d/\Delta t$ provided the time interval, Δt, is very small. *Acceleration* is defined as the rate at which velocity changes, or $a = \Delta v/\Delta t$. Both speed and acceleration have mathematical definitions of the same form as the definition of the slope of a line, $m = \Delta y/\Delta x$. Thus, the slope of each portion of a d-vs.-t curve is the speed for that portion of the curve. Likewise, the slope of a v-vs.-t curve gives the acceleration of the motion. Similarly, the area enclosed between the t axis of a v-vs.-t curve and the curve itself for any time interval gives numerically the distance covered by the motion in that interval. For motion starting at a speed v_0 and accelerating uniformly to speed v_f, this produces several useful formulas:

Average speed: $\bar{v} = \dfrac{v_f + v_0}{2}$

Distance traveled: $\Delta d = v_0 t + \tfrac{1}{2}at^2$

Final speed: $v_f^2 = v_0^2 - 2ad$

Problems

1 Give some examples of objects that keep moving without being pushed.

2 From 1954 to 1975 the world record for the mile run was changed 11 times. The record dropped from 3 m 59.4 s to 3 m 49.4 s in this time. At what rate was the world record dropping?

3 The world record for the 1500-m run is 3 m 32.2 s, while the record for the mile run is 3 m 49.4 s. Which of these two record-holders had a higher average speed for the run?

4 The world record for the 100-yd dash is 9.0 s, while the record for the 100-m dash is 9.95 s. Which of these runs represents a higher average speed?

5 A Chevrolet Monte Carlo goes from 0 to 50 mi/h (80 km/s) in 8.6 s. What is its average acceleration for this time in m/s? Is it likely that the instantaneous acceleration throughout this time would be the same?

6 What is the current national speed limit in kilometers per hour?

7 How fast is a rock traveling after 1 s of falling? After 5 s?

8 If a rock hits the ground 3 s after it was dropped, at what speed would it strike? What is this speed in miles per hour?

9 How far does a rock fall in the first second after its release? The first 3 s?

10 The John Hancock Tower (tallest building in Boston) is 60 stories tall, or 240 m (790 ft). How much time would it take a metal ball to fall from its top to the pavement below? How fast would the ball be traveling when it hit?

11 A fully loaded oil tanker traveling at 15 knots (7.7 m/s) requires 8 km (5 mi) to come to a stop with its engines at full reverse. What is the maximum acceleration of the tanker? How much time would it take to come to a stop? What are the implications of this for a tanker coming into dock?

12 Ralph went for a walk and the figure shows a distance-vs.-time graph for his walk. See if you can tell during which time interval Ralph did the following:

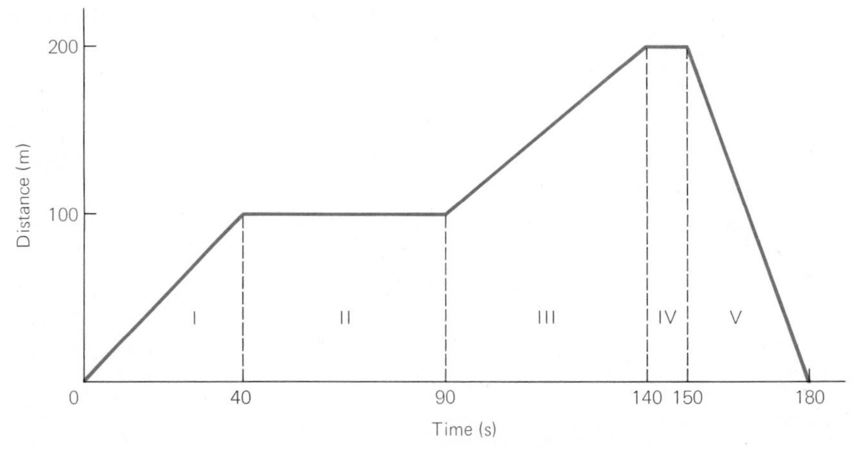

(*a*) Met a friend and talked for a while.
(*b*) Stopped to check his pockets to see if he remembered to bring his keys.
(*c*) Ran back home for his keys.
(*d*) Walked rapidly toward his office.
(*e*) Walked slowly through the park.
(*f*) Arrived back at his house.

13 In which segment of the motion I, II, or III does Ralph have the highest speed? At what speed is he walking during segment I?

14 What is the greatest distance Ralph gets from his house?

15 The world record for the 100-m dash is 9.95 s. What can you infer about Ralph's physical condition from the curve?

16 Janice drives her car through town as shown by the figure below which graphs her speed vs. time. See if you can tell during which time interval the following happens:

(*a*) Janice slowly coasts to a stop for a red light.
(*b*) She picks up speed from her stop at a light.
(*c*) She accelerates to beat the light ahead of her, but she will miss it and must stop.
(*d*) She lets up on her speed when a child looks as if he or she might run into the street, but doesn't.
(*e*) She drives steadily down the street.

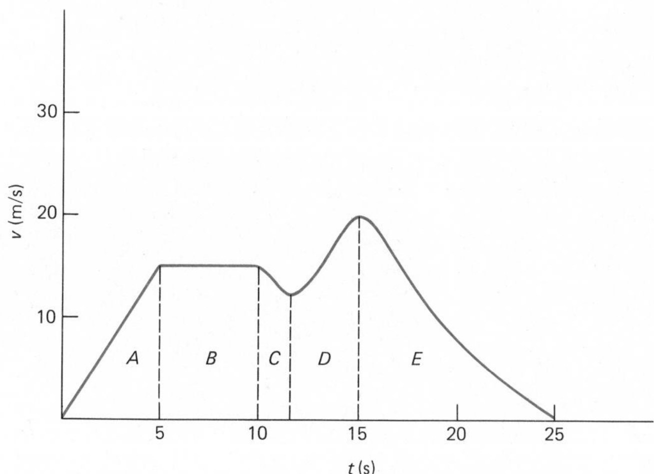

17 What is Janice's top speed in the interval?

18 About how far is it between the stoplights?

19 What is the likely speed limit for this stretch?

20 How far did Janice go in the first 15 s?

21 What is Janice's acceleration at t = 3.0 s?

22 At what time was Janice pushing hardest on the accelerator? Brake pedal?

23 The airline distance Chicago to New York is about 1100 km. If an airplane flies at 550 km/h, how long will the trip take if:

(*a*) There is no wind at all?
(*b*) There is a constant wind blowing opposite the direction of motion (a headwind) of 183 km/h?
(*c*) There is a tailwind of 100 km/h?
(*d*) Would it make sense, in terms of time, to fly 20 min out of your way to lose such a headwind and pick up the tailwind going in your direction?

CHAPTER FOUR

An Example of Motions Defined— Astronomical Motions as Viewed from the Earth

Introduction Now that we have studied motion, it is useful to examine the most interesting moving system provided by nature—the heavens. To an observer on the earth, the motions are many and complex. The sun appears to move daily as well as seasonally. The stars move in our night sky. The moon moves around the earth and among the stars. The planets "wander" through the sky, and comets occasionally move among all these objects. Yet, no matter how complex they appear, there is great order in the heavenly motions. If you could get away from the lights and distractions of our everyday life and observe these motions through time, you too would be convinced of the great order displayed. In today's world, our awareness of the sky is limited to occasional glimpses. We produce our own daylight and our own summer in our houses and increasingly neglect the natural world about us. This is particularly true of the heavens. The sky is something we tend to notice in the transition between events—between classes during the day or between a party and the dorm some weekend night. Indeed, on an urban campus, we may be denied seeing any but the most brilliant objects in the night sky. Reflections of streetlights and neon signs cause the haze of particles in the air above our cities to glow more brightly than all but the brightest stars. Children from the city are often awestruck by the brilliance and beauty of the nighttime sky on their first trip to rural areas. Reacting similarly to much of nature, adults tend to notice that the sky is there, mumble "how pretty," and then turn their attention to other matters, saving themselves the trouble of real wonderment.

There is a second reason to carefully examine the motions of the heavens as viewed from the earth. If the first is simple curiosity (or the desire to pass a course), the second is to understand our civilization. The interpretation of motion in the heavens and the process of defining modern science grew up together. The revolution in astronomy gives us an unusually clean glimpse of radical changes occurring in scientific thinking. To educated individuals in all fields, the Copernican revolution in astronomy is the prime example of science emerging from the Middle Ages. To scientists it is viewed as the starting of events in the scientific revolution that found its high point in Newton. However you want to view it, the change in human perception of the universe that occupies the sixteenth century is a critical turning point in our civilization. Modern science emerges from it and grows until its products dominate our way of life. *Science is a very human activity*—humanity's attempt to order and understand the workings of the universe. As such, it is humanmade, and subject to all of the faults of human beings. The advantage to science is that bad ideas crumble in time. Meanwhile, very good guesses can go on providing useful predictions for much longer. What is correct? How do we decide between rival ideas? By what rules do we accept or reject ideas as being truly descriptive of nature? These were the real issues being decided in the sixteenth century.

The interpretation of the motions observed in the heavens must begin with a description of these motions. In this chapter we will describe the motions of objects in the heavens as viewed from the earth. *Description must come before interpretation.* The facts given in this chapter are the observable positions of the heavenly bodies at various times, and the motions they appear to go through to get to these places at the observed times. These are the relationships and motions which any successful theory of the heavens must explain and predict. However, throughout the chapter we will avoid the issue of "causes," or what is really happening to produce the observed motions, and leave the interpretation of these motions to Chap. 5.

Apparent Motion of the Sun

At first glance, the motion of the sun seems to be very regular. The sun appears to rise in the east, set in the west, and move across the sky during the day. We may even define due east as the average rising point of the sun throughout the years. Likewise, due west becomes the average setting position. In between rising and setting the sun moves at a steady pace across the sky. In general, the sun does not rise straight up out of the horizon, but moves away from the horizon at some angle. This angle is fixed for any one place on the earth. Only at the equator does the sun rise perpendicularly out of the horizon and set perpendicularly into the horizon.

These motions lead to a number of useful ideas. First of all, it is possible to obtain your latitude on earth from the angle between the sun's path at rising or setting and the horizon. In Fig. 4-1A this angle is labeled ϕ (phi). This angle is constant for any one position on the earth, and is just the complement of the latitude—that is, the angle which, when added to your latitude, gives you 90°. Thus, at the equator, where latitude is 0°, the sun rises straight up, making an angle of 90° with the horizon. If the sun makes an angle of 55° with the horizon as it rises (or sets), you would calculate your latitude to be 35° north or south of the equator (Fig. 4-1A).

This behavior of the rising or setting sun explains the short twilight period at the equator (Fig. 4-1B). Twilight is the period of time between the first brightening of the eastern sky and sunrise. It also is the time between the setting of the sun and the last brightness of the horizon in the west. Rudyard Kipling reported the tropics as the place where "the dawn comes up like thunder/outer China 'crost the Bay!" Dawn may not be that abrupt, but it is short in the tropics.

On the other hand, when you proceed toward either pole of the earth, the sun "cuts" the horizon at a smaller and smaller angle. At the pole itself, the path of the sun for any one day just parallels the horizon (Fig. 4-1C). If the sun is up, it is up all day and at the same approximate height all day long. If the sun is below the horizon, it remains at the same distance below the horizon all day long. Since the sun is never more than $23\frac{1}{2}°$ below the horizon at the pole, only in the middle of that pole's winter—with the sun at its lowest—does it really get quite dark. During the month of October, for example, the sun is moving just a few degrees below the horizon for an observer at the North Pole, and the brightness is similar to the first half-hour after sunset in mid-latitudes. Given this motion of the sun, neither pole has daily sunrises and sunsets. The North Pole gets one sunrise, on March 21, which takes all day as the sun slowly drifts above the horizon. Once up, the sun stays above the horizon for half a year until the next sunset in late September.

Figure 4-2 also indicates why the length of night and day varies with the seasons. When the sun rises due east and sets due west, precisely one-half of its daily path around the earth is above the horizon. Thus, we will receive 12 h of daylight and 12 h of darkness on these days. In summer in the Northern Hemisphere, the sun rises to the north of east and sets to the north of west. More than one-half of its sweep of the sky is above the horizon, and the daylight period is correspondingly longer. At 45° N latitude, the daily cycle divides up to about 15 h of daylight and 9 h of darkness. However, in winter this situation is exactly reversed. Less than one-half of the sun's path around the earth is above the horizon, and we receive less than 12 h of sunlight each day.

At the equator, the horizon always cuts the sun's apparent daily path in half (Fig. 4-1B). The sun always rises 6 h before noon, sets 6 h after noon, and provides 12 h of daylight in between. Thus, over a year, every spot on earth has had the sun above the horizon one-half of the time. (This neglects minor effects, such as the bending of sunlight entering the

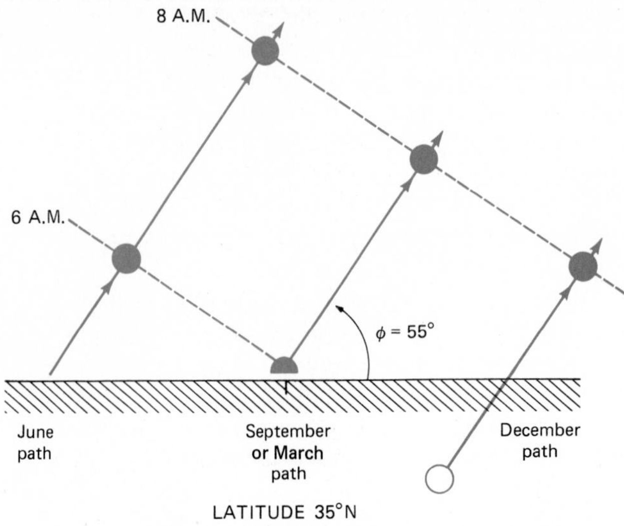

FIGURE 4-1A The track of the rising sun as seen from 35° N latitude at varying times of year. All the daily tracks of the sun are parallel, but in our summer the sun rises to the north of east. This causes part of its track to be above the horizon before the average time of dawn (6 A.M.). Thus, days are longer in the summer.

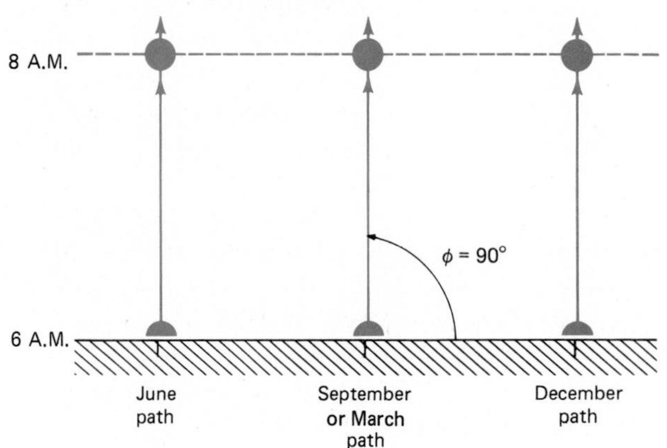

FIGURE 4-1B The track of the rising sun is shown for varying dates as seen at the equator. Since the sun rises perpendicularly from the horizon, all sunrises are at the same time each day. As with other locations on the earth, however, the sun rises straight through the due east point on the horizon only in late September or late March.

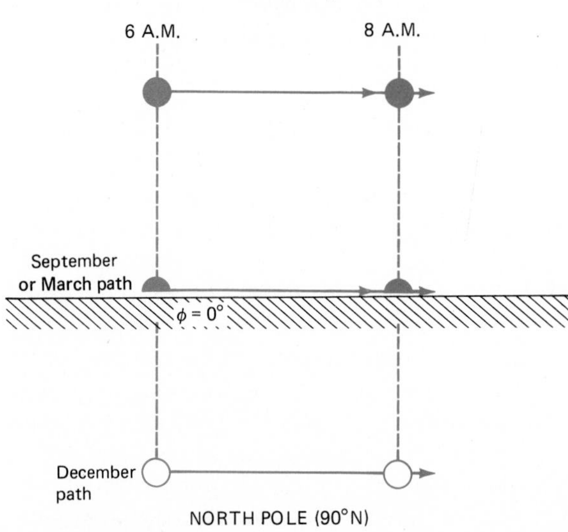

FIGURE 4-1C At the North Pole the track of the sun parallels the horizon each day. The sun rises in late March, gets as high as $23\frac{1}{2}$° in the sky by June 21, and then sets in September. Even when set, the sun is never more than $23\frac{1}{2}$° below the horizon for the pole. This leads to months of twilight conditions from scattered sunlight.

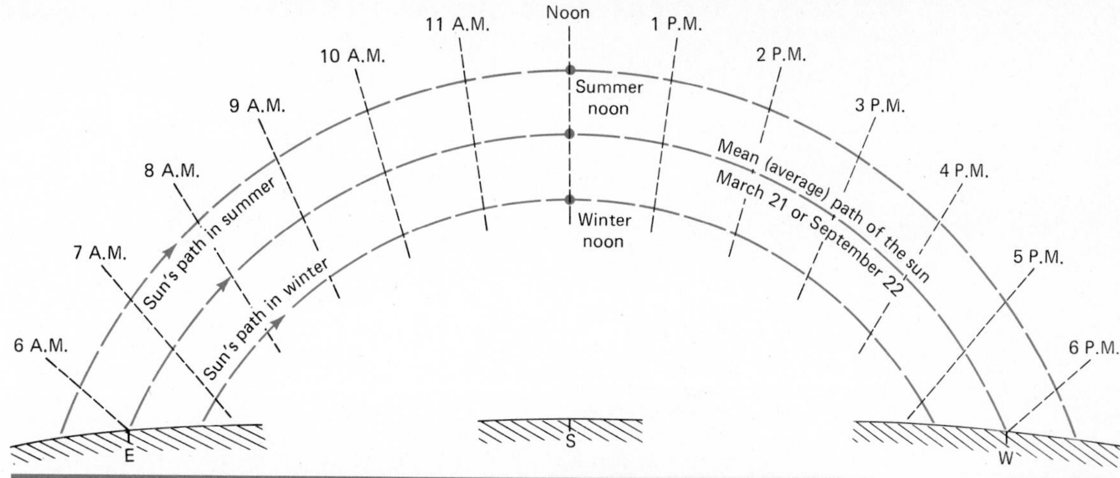

FIGURE 4-2
The visible portion of the sun's daily path through the sky is shown for various seasons for a mid-latitude location. For much of the United States the difference between summer and winter paths is even more extreme than is shown. For example, northern cities such as Chicago and New York receive about 15 h of daylight on June 21 (and only 9 h of darkness). This implies that the sun would rise about 4:30 A.M. and set at 7:30 P.M. This extreme difference is difficult to illustrate on a flat page.

earth's atmosphere, which can add 4 min of daylight per day. Obviously, local conditions make a difference—such as living in a deep canyon, or having high mountains on the eastern horizon, and so on.)

Measuring Motions

A check of the progress of the sun across the sky can be made rather easily if no great accuracy is desired. At the simplest level, we may place a stick in the ground and mark the position of its shadow every hour. For working indoors, a suitable measuring device may be made from a sheet of paper resting on cardboard with a large pin inserted in the center of the sheet. For easy measurement of angles, a sheet of polar coordinate graph paper is the best choice as a record sheet. (See Fig. 4-3.) (When you have made this device, you will have put together the modern equivalent of the ancient Babylonian gnomon.) If placed in a window facing south, the position of the pin can be marked each hour. Using this apparatus (shown in Fig. 4-3), you can determine the height of the sun in the sky at the same time that you obtain its position.

The height of the sun in the sky is measured by the length of the shadow of the pin (Fig. 4-4). If measured in degrees, this is called the *altitude* of the sun. The altitude varies from zero, if the object is on the horizon, to 90°, if the object is directly overhead at the observer's *zenith.* As an example, if you were on the equator in mid-March, the shadow of the pin would get shorter each hour until noon as the sun rose in altitude (Fig. 4-5). At noon the pin would cast no shadow since the sun would be

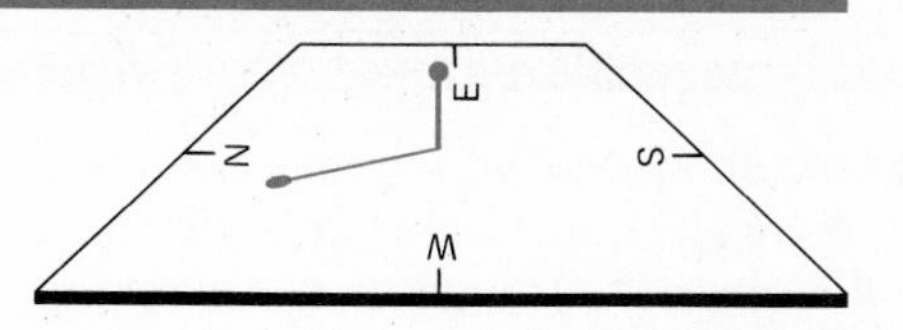

FIGURE 4-3
The simplest device to record the position of the sun in the sky is a stick in the ground, or, as here, a pin in a marked sheet of paper. The shadow points directly away from the direction to the sun, and its length indicates the height of the sun.

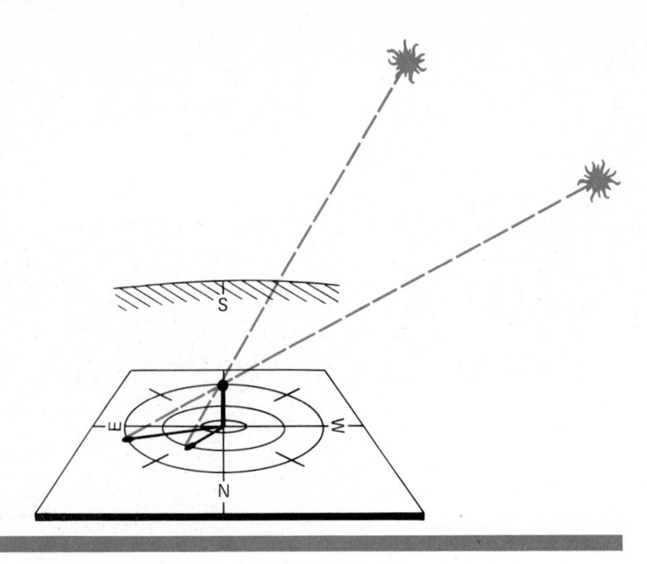

FIGURE 4-4
During the day the sun varies in both direction and height. Both changes may conveniently be recorded using a sheet of polar coordinate graph paper as a base for the pin.

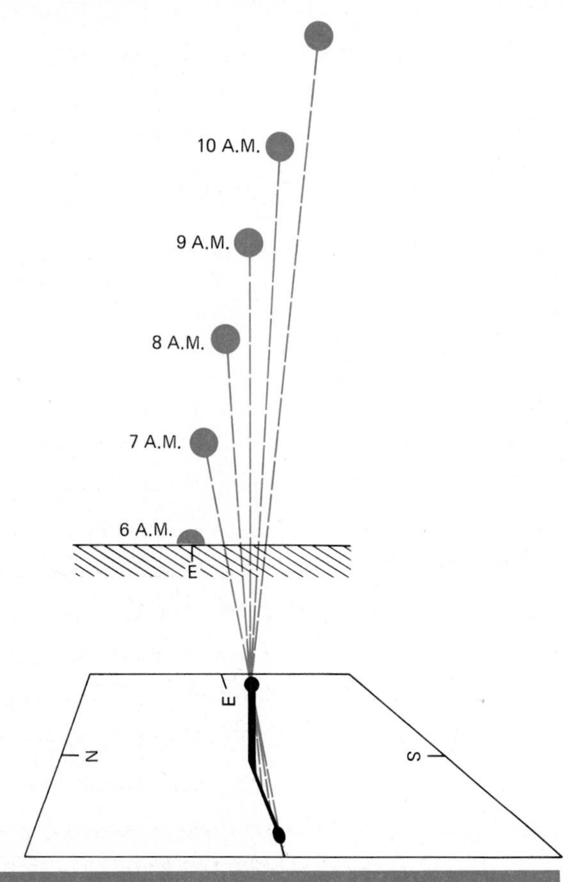

FIGURE 4-5
The daily variation in the sun's location at the equator in March would be entirely in shadow length. At noon the sun would be directly over the pin and cast no shadow.

at the zenith. After noon the shadow would lengthen in the opposite of the original direction. Figure 4-4 also illustrates that a second variable of position is measured by the gnomon. The second variable is indicated by the direction in which the shadow points along the paper. Sighting along the shadow on the paper, we find a point on the horizon which appears to be directly under the sun. This variable is called the sun's horizon angle, or *azimuth.* To record this angle, imagine the entire circle of the horizon divided into degrees, with north being 0°, east = 90°, south = 180°, west = 270°, and back around to north at 360° or 0°.* This system of recording positions in the sky by *altitude* and *azimuth* is called *the observer system.* This system is easy to learn and to use, and it clearly defines positions in the observer's sky.

The observer system has two main disadvantages in astronomy. The first objection to it is that observers at different positions on earth will obtain different measurements for the same object. Conversion of the measurements from one time and place to another is possible, but it would be more convenient to have a single system which was the same for all observers. The second objection is that the measurements of motions in the sky are needlessly complex. In Fig. 4-6 we see that the sun after sunrise is moving in both altitude (upward) and in azimuth (southward). The problem is that it does not do this all day long. If we look at Fig. 4-7, we see that around noon the altitude of the sun is approximately constant; it is moving in azimuth only. How can we compare the angular motion of the sun through the sky to determine if the rate is constant if all of the motion is in one "direction" one time, but only part of it is in that "direction" at other times? We want to determine whether the *diagonal* path of the sun's motion in Fig. 4-6 represents the same angle in an hour's time as the angle for an hour's motion along the sun's path in Fig. 4-7 near noon. To do this easily, we need a new way to measure the sun's

* Actually, when using this system, astronomers use south as zero. Navigators and surveyors use north as the zero of azimuth. Since more use is made of the observer system outside of astronomy, we have gone with the majority.

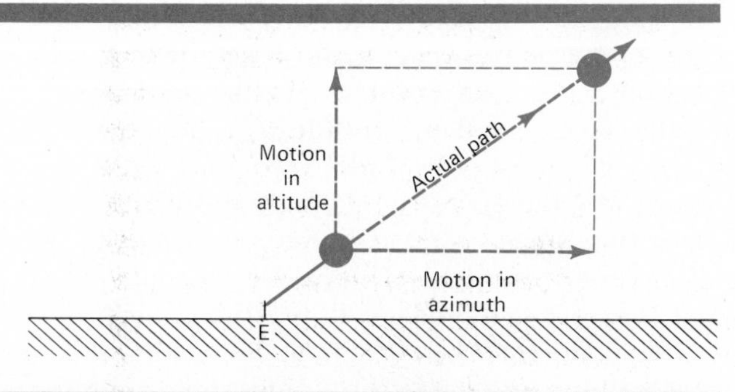

FIGURE 4-6
For most locations the sun changes both its direction along the horizon and its height above the horizon. These measures are called azimuth and altitude, respectively. This variation is most apparent near sunrise or sunset, as shown.

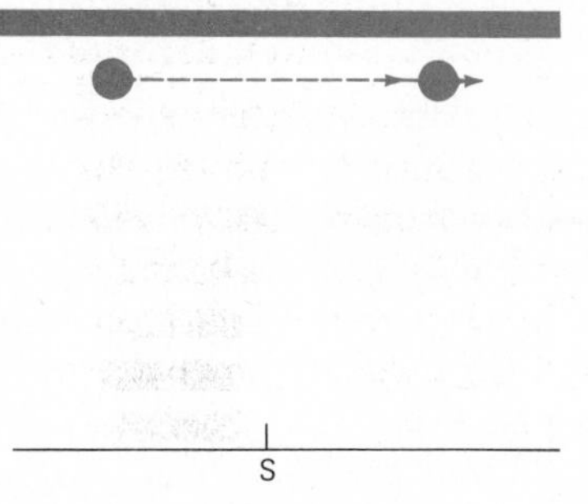

FIGURE 4-7
Around noon the sun changes its altitude very little but still has a considerable variation in its azimuth. The angle of the sun's shadow on the gnomon of Fig. 4-4 will change by 1° every 4 min around noon.

The sun vs. the clock

Are all days the same length? Clearly, the daylight period in the summer is longer than the daylight period in the winter. But, is the time between one noon and the next always the same? In the ancient world, the most dependable timepiece was the sun. In peasant societies, the people rose with the sun and quit when "the cerfew tolls the knell of parting day." The high point of the sun defined noon, time for the midday meal, and time to seek shade in the summer. Many events can be regulated quite nicely by time recorded on a sundial. More complex societies found it necessary to keep time also at night. It was convenient to be able to tell when the midnight watch was to go on duty. Since sundials have obvious defects for this type of measure, other methods were needed. Much of the populace could keep rough time from the position of the stars. For cloudy nights, a standard candle could be used, which burned down a known amount in length per hour. The Romans favored the clypsedra, or water clock, which effectively counted the drip rate of water through a standard opening. Less mechanical and more obvious in its functioning was the hourglass. However, all of these devices would allow you to slip in a few minutes now and then with no one the wiser.

The first clocks with the regularity and precision we expect from a timepiece were developed following a discovery by Galileo. He was sitting in church one day when a minor earthquake rocked the cathedral. The earthquake did no damage, but the long chandeliers suspended from the ceiling high above were set in motion. Galileo watched the long, slow swings and had an idea. It occurred to him that the time required for one swing did not depend on the size of the swing. As the swings became smaller they took just as long as the big swings had taken. He used his pulse rate to time each swing and became convinced that the time required for one swing was constant. (He does not record how he kept his heartbeat constant while discovering an important physical principle. Nor does history record the reaction of the priest to this inattention to the sermon.) Later, the constancy of the period of the pendulum became the basis for the first accurate clocks. Each swing could be used to advance the gears of a clock one notch. The backswing could be used to reset the release mechanism. These two actions represent the "tick-tock" of the pendulum clock.

The development of the first accurate pendulum clocks gave scientists the ability to perform experiments requiring the careful measurement of time. (For example, it led to the first determination of the speed of light, by Olaus Römer, within 35 years of Galileo's death.)

Given a precise clock, it is possible to show that all days are not the same in length. Clocks are adjusted to mean (or average) solar time, not to the true sun. The time between noon of one day and noon of the next is rarely an exact 24 h on the clock. The sun runs either ahead of the clock or behind the clock most of the time. (This is a "real" ahead-of or behind and has nothing to do with time zones, which is another complication.) If the navigator of a ship was using the sun to fix a position and did not realize that this variation took place, his determinations of longitude would be useless. The real sun can run as much as 16 min of time ahead of or behind the average, or clock, sun. Translated into positional error, this difference leads to being off by almost *280 mi*. The navigator might determine that he was in the Atlantic Ocean opposite Philadelphia, and really be on the outskirts of Pittsburgh. Or, in the Pacific Ocean, with the real sun running slow, he could believe himself in Los Angeles and be approaching Phoenix, Arizona. In either case, it would be rough sailing! The difference between the average sun and the true sun is called the *equation of time*. You probably have seen a correction

chart for this difference and did not realize it. Many globes print a lopsided figure eight in the Pacific Ocean, called an *analemma*. The analemma gives you two pieces of information at once. For any date, it indicates at what latitude the sun is directly overhead. Secondly, the east-west part of the figure shows how much the real sun is running ahead of or behind the clock sun (local time, not zone time) for that date. The fact that all days are not the same length is not due to an inconstant rotation. It is because the earth's orbit is an ellipse and not a circle.

motion along its path in the sky—a new coordinate system. This is not a minor step! When we adopt a new coordinate system based on motions, the motions become much easier to visualize and describe. Let us proceed slowly.

To measure angles along the path of the motion of the sun with our gnomon, all we have to do is tilt the cardboard upward toward the south. At some angle of tilt we arrive at the point where the sun appears to move along the plane of the cardboard. Of course, the sun is higher in the sky at noon in the summer, and lower in winter, so we must adjust the tilt of the cardboard to point to the average noontime altitude of the sun. Now, no matter what time of the day we sight along the cardboard, we find the sun at a constant height from the plane of the board (Fig. 4-8). One problem does arise, namely, that unless this is done in the summer, the sun will lie so close to the plane of the cardboard (or slightly below it) that it will cast no shadow of the pin on the paper—the entire paper will be shaded. We can solve this problem by using a new apparatus made from a protractor and a Tinkertoy (Fig. 4-9). This apparatus uses the same idea but allows us to record the position of the sun easily.

Once we have completed the steps needed to adequately measure the sun's motion, we will find that most astronomical motions become

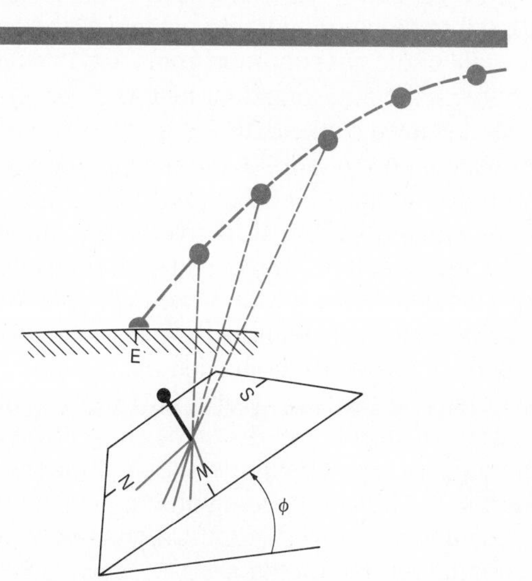

FIGURE 4-8
If we tilt our gnomon around an east-west line so that its southern edge points upward to the average height of the noon sun, measurement can be simplified. The sun will be at a fixed height from this plane on any given day, and the motion of its shadow is now a constant 1° every 4 min. The plane of our gnomon now defines a line in the sky called the celestial equator.

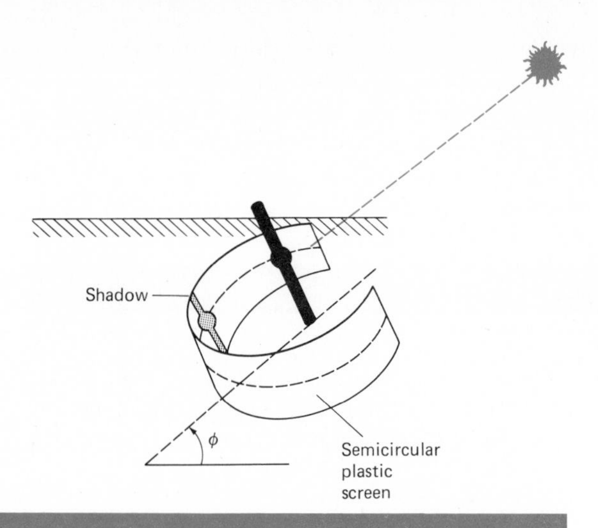

FIGURE 4-9
A slight modification of the apparatus allows us to measure the angle for the sun at any height above or below the celestial equator. If we replaced the central stick with a small ball, the height of its shadow above and below the central band on the scale would measure the coordinate called declination.

quite simple in this system. We have "fallen into" the standard measuring system of astronomers, and it is time to define some terms. The average, or mean, path of the sun across the sky is called the *celestial equator.* Another way of defining the celestial equator (CE) is as the set of all points in the sky directly overhead at the equator of the earth.

Instead of measuring altitude, we now can measure the angular distance north or south of the celestial equator in the sky. This coordinate is called *declination.* If an object has a declination of positive 10° in the sky ($\delta = +10°$), this means that it is found 10° north of the celestial equator. Negative declination indicates that the object is south of the celestial equator in the sky; declination in the sky is a coordinate to latitude on earth. In fact, any object observed directly overhead has the same declination as your latitude on earth. The main advantage in using this coordinate system can be expressed easily: *All stars have constant declination.* While the stars appear to move daily from east to west through the sky, they maintain a constant angle of separation from the celestial equator. Sirius, the brightest star in the sky, for example, has a declination of −16°38′. This means that it is always found 16°38′ south of the celestial equator in the sky. The most important star for observers in the Northern Hemisphere would have a declination of +90°, meaning that it would appear to be fixed in the sky while all of the stars rotated around its position. The North Star, or Polaris, is actually almost a full degree away from the true pole of the sky. This difference is small enough so that its daily motion in a tiny circle is not noticeable without measurement. Polaris is effectively in the same position at all times for an observer in the Northern Hemisphere. Polaris is not among the 20 brightest stars as viewed from the earth, but it has the entire area of the sky to itself without any bright stars for competition. Thus, Polaris stands out clearly and is easily located. (If you are in the Southern Hemisphere, the North Star is always below your horizon and you never see it. Unfortunately, no relatively bright star marks the South Pole of the sky.)

The measurement of position in the sky using declination as the angle above or below the CE, and another coordinate, *right ascension,* for measuring east-west separations, defines the *celestial coordinate system* for measuring astronomical positions. This system is used by all astronomers.

Rate of Solar Motion

The sun appears to circle once around us each day as it proceeds westward across the sky. The rate of this motion is a uniform 15°/h measured along the sun's path through the sky. If we define the angle along the path as θ (the Greek letter "theta"), we can say that the rate at which θ changes is constant, or,

$$\frac{\text{Change in } \theta}{\text{Change in time}} = \frac{\Delta\theta}{\Delta t}$$

$$= c$$

$$= \frac{360^\circ}{24 \text{ h}} = 15^\circ/\text{h}$$

using the method of expressing changes which we learned in the last two chapters. The expression $\Delta\theta/\Delta t$ is called the *angular speed* of the sun, or the rate at which it sweeps through an angle measured by an observer on earth. Thus, to the accuracy of naked-eye measurement and using ordinary clocks, we can say that the angular speed of the sun along its path is constant.

We have also seen that the sun moves northward and southward in the sky seasonally by as much as $23\frac{1}{2}^\circ$ from its mean position on the CE. It arrives at its highest noon position on June 21, which marks our first day of summer in the Northern Hemisphere. The lowest noontime height is achieved for us on December 21, the first day of winter.*

* Visualising Earth-Sky Relationships

Many people find it difficult to visualize the sky as seen from different places on earth. Imagine yourself standing at the North Pole of the earth. Directly overhead, at your *zenith,* would be the North Pole of the sky marked approximately by the star Polaris. Exactly on your horizon in all directions would be the CE. No stars would ever rise; none would set.

* It is no accident that civilizations in the mid-latitudes historically tend to have a very major holiday at the end of December. It was frightening to a prescientific society to watch the sun slip lower and lower in the sky each day while the weather grew increasingly dismal. In several societies, a priestly class developed just to keep an eye on the sun. When the turnaround of the sun was certain (a few days after December 21), the news went out for the great midwinter celebration. The sun was on its way back up in the sky! Our own celebration of this season, Christmas and New Year's Day, derives from this custom. Early Christians in Rome found it much healthier to celebrate the birth of Christ when the Romans were busy celebrating their feast of the Saturnalia. To have celebrated Christmas in the spring, when the birth actually occurred, would have led to a happy and expanding lion population at the expense of a diminishing supply of Christians.

They would maintain a constant height from the horizon and circle eastward forever. Of course, most of the year you wouldn't be able to see them, since the sun would rise in March. It would spiral around daily, inching its way upward in the sky. In June, the sun would reach its highest point and then spiral down 91 daily loops to its setting in September. Only in the deep winter, with the sun well below your horizon, and with the glow of its scattered rays completely gone, could you watch the silent circling of the stars—once per day on their endless journey.

Picking any direction and calling it south, you could now proceed to move away from the pole toward the equator of the earth. For every degree of latitude you went southward around the curve of the earth's surface, the CE would rise in the sky 1°. Overhead, Polaris would move away from your zenith and slide 1° northward in the sky for each degree you moved equatorward. At latitude $66\frac{1}{2}°$ N, we reach the Arctic Circle—the farthest south one can go and still have at least one full 24-h period during the year when the sun will not rise (and one 24-h period when it will never set). We have moved $23\frac{1}{2}°$ toward the equator from the pole. The celestial equator rises before us. It comes above our horizon at due east, as it does for all observers on earth. It then arcs across the sky to reach a maximum height of $23\frac{1}{2}°$ above the south point on the horizon, and finally goes below the horizon at due west, as it does for every observer on earth. A smooth sweep of our hand from east to a height of $23\frac{1}{2}°$ at south and then down to due west separates the northern half of the heavens (that portion which we could see at the North Pole) from the emerging southern portion. Meanwhile, overhead, Polaris has slipped away from our zenith and has retreated northward $23\frac{1}{2}°$. Its altitude above the north point on our horizon is now $66\frac{1}{2}°$. It will remain fixed at that position for the rest of our lives if we choose to stay at this location.

For those of us not overly fond of polar bears and nose-rubbing, we will move farther southward. The next convenient place to stop would be the Tropic of Cancer at $23\frac{1}{2}°$ N latitude. This would be easy because we have already described the sky for you. All you must do is change the numbers in the preceding paragraph and reread it. The number $66\frac{1}{2}°$ becomes $23\frac{1}{2}°$ and vice versa. In the reference to the Arctic Circle, read "Tropic of Cancer, the farthest north we can go on earth and still have the sun directly overhead at noon at least 1 day per year." Try rereading the paragraph now and then return to this point.

Actually, while the numbers $23\frac{1}{2}$ and $66\frac{1}{2}$ become special because of the seasonal migration of the noonday sun up and down in the sky, we may reread the same paragraph for any latitude on earth.* For example, Chicago is at latitude 42° N. To read the paragraph for Chicago, use the latitude of 42° and its complement of 90° − 42°, or 48°. Read "42°" wherever "$66\frac{1}{2}°$" occurs in the original statement, and "48°" wherever "$23\frac{1}{2}°$" occurs. Omit the mention of the Arctic Circle, since 42° N is a spe-

* If you want to read it for south of the equator, you must exchange "south" and "north" in the statement and remember that there is no real "South Star" comparable to Polaris.

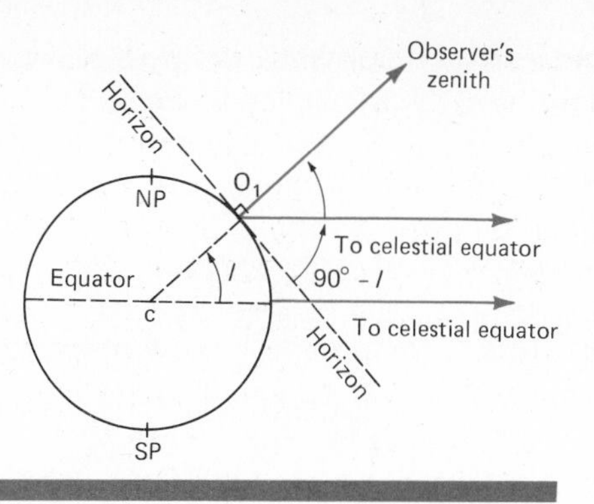

FIGURE 4-10 An observer at Chicago, O_1, would have a horizon indicated by the dashed line. Above and to the right of this line the sky could be visible. The angle between this observer's zenith, or directly overhead point in the sky, and the celestial equator is the same as the latitude angle as defined at the center of the earth.

cial latitude only for Chicagoans. The paragraph is now valid for Chicago. By similar changes we can now describe the sky for any position on earth. If you visualize relationships better by using a diagram, consider the case of Chicago shown in Fig. 4-10.

The earth is shown with the poles and equator indicated. An observer labeled O_1 is placed at the position of Chicago. The dashed line of the observer's horizon separates the half of the sky he or she can see from the part blocked by the earth itself. Directly out from the center of the earth through the observer is an arrow extending toward the observer's zenith. Since the zenith is directly overhead for any observer, it is always 90° away from the observer's horizon. The direction to the CE is marked both for the observer and for the equator of the earth. These lines are parallel since the CE can be taken as being infinitely far away. Note that the latitude l is defined at the center of the earth. The latitude is just the angle between the equator and a position on the surface as seen from the center of the earth. Because the lines to the CE are parallel, we can see that the angle between the zenith and the CE is the latitude of the observer. Another way of saying this is that the height of the CE above the south point on an observer's horizon is the complement of the latitude, $90 - l$. (Why?)

If this discussion has rekindled some of your enthusiasm for geometry, you may wish to prove our next assertion. *The height of the North Star above an observer's horizon is the same as the observer's latitude.* Figure 4-11A shows several astronomically significant directions for our observer. What tends to make the figure confusing is that two of the defined items are in the observer system—the zenith and the horizon. These change in relationship to the others for different observational positions on earth. The remaining four directions have fixed relationships to each other in celestial coordinates and are independent of the observer. Consider what happens to these directions when we redraw the situation for north of the Arctic Circle in Fig. 4-11B. The observer O_2 is now much nearer the pole, and the North Star is now high in the sky near the zenith. Meanwhile the CE is down near the horizon. Consider the same situation for observer O_3 south of the Tropic of Cancer in Fig. 4-11C. Figure 4-12 shows two directions for each of the three observers simultaneously.

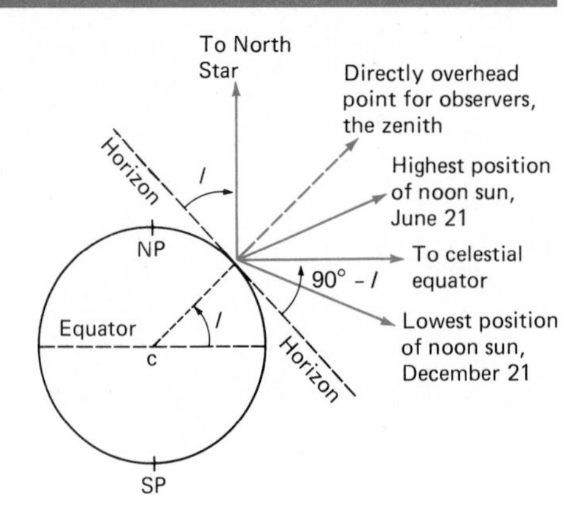

FIGURE 4-11A
The same observer at Chicago in Fig. 4-10 has several other astronomically significant directions. What is the angle between the marked highest position of the noon sun and the celestial equator? Why is the angle between the direction to the North Star and the horizon labeled l?

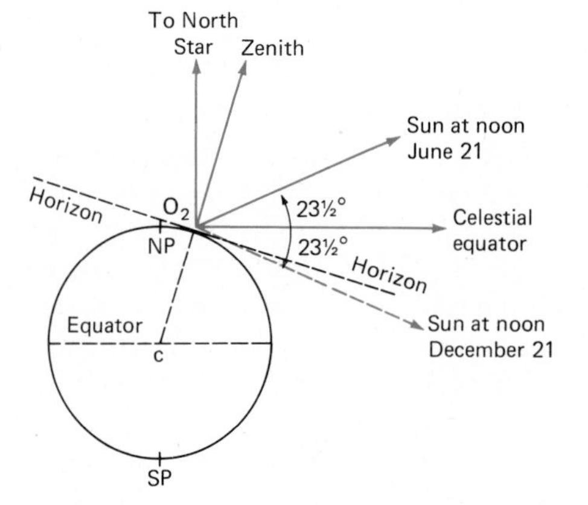

FIGURE 4-11B
The same astronomical directions as for the observer at Chicago are drawn for an observer, O_2, above the Arctic Circle. Why is the line to the sun at noon on December 21 drawn in dashed and not solid? What angles in the diagram would yield the observer's latitude?

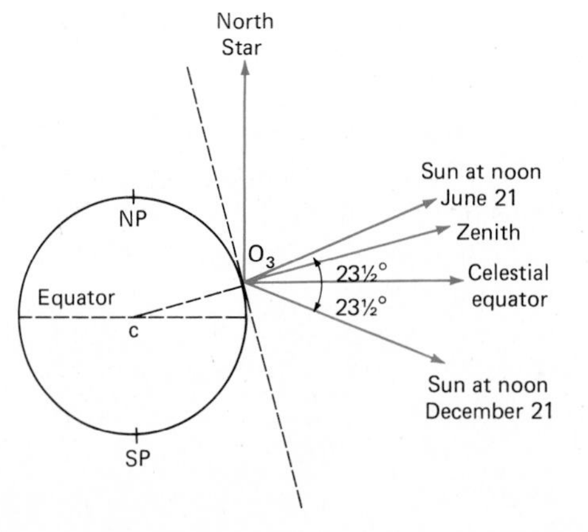

FIGURE 4-11C
Various astronomical directions for a third observer, O_3, south of the Tropic of Cancer. What position of the sun could be seen by this observer never visible to O_1 or O_2? What does the dashed but unlabeled line represent?

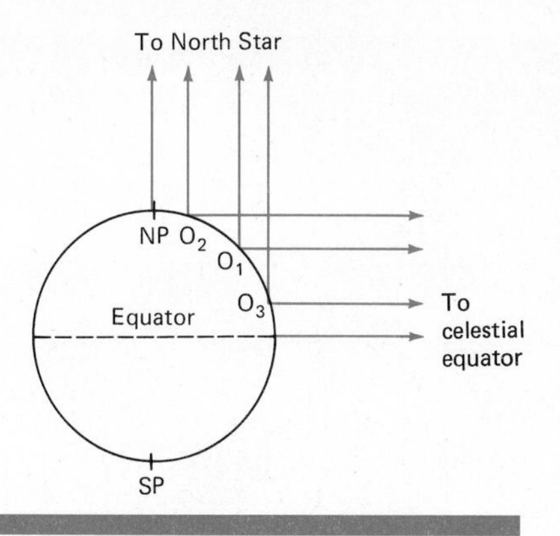

FIGURE 4-12
While each of the observers O_1, O_2, and O_3 had individual directions for horizons and zeniths, the astronomical directions remained constant in space. This illustration shows the direction to the celestial equator for each observer. Note that it is in the direction of the zenith for an observer at the equator. Likewise, the direction to the North Star for each observer is the same direction in space and parallel to the axis of the earth.

Motion of the Stars

The stars move from east to west in our night sky. If you are not aware of this, it would take only an hour or so to verify it some night. Line up some star in the southern sky with two telephone poles or the edge of a building or anything fixed to the earth. Come back and check in an hour and the star will have moved to the west.* A fast check with a protractor would disclose that the angle through which the star has moved in that hour is about 15°.

But now we must be more careful. If the stars move around the earth at the same speed as the sun, the east-west distance between the stars and the sun would be constant. If you are at all familiar with the stars, you know that there are winter groups of stars and summer groups of stars—and so on for each season. A particular group of stars is called a winter constellation, for example, if it is viewed most easily in the winter, since the sun is in the same part of the sky as that constellation during the summer. This implies that the sun has an east-west motion relative to the stars as well as its seasonal rise and fall in declination. We can easily check this. Normally, we could "run a 1-day race." Find a star in a north-south line with the sun and then send them scurrying off to the west. One day later we could observe which got back to the starting line first after going entirely around the sky. Unfortunately, it would not be quite that simple since any star near the sun is not visible to us. The sunlight scattered by the earth's atmosphere makes the sky glow much more brightly in the daytime than the stars, thus rendering them essentially invisible. (The brightest stars can be observed during the day if you know exactly where to look and can somehow cut down the glare of the rest of the sky. The brighter planets are much easier to observe during

* The author can verify from personal experience that it is possible to perceive this motion directly. Or, as we say with computers, "in real time," not by delayed action. Lying in a sleeping bag at the bottom of the Grand Canyon, he was startled at the speed with which stars appeared at one edge of the slit of sky far above and "shot the gap" over to the other wall of the canyon to disappear. The motion was clearly perceptible since the eye could judge the rate at which the separation of wall and star was growing or shrinking. Other than for this effect, however, he does not recommend the bottom of a deep canyon for stargazing.

FIGURE 4-13
The motion of stars around the north polar axis of the sky is shown in this photograph. A camera with its lens left open records the motion of the stars as blurred streaks. Near the point of rotation is the bright smear of the North Star—not quite at the polar axis. About how long did it take the camera to record this picture? (*Lick Observatory photograph.*)

the day—and even they are difficult to find!) Fortunately, we have a totally acceptable stand-in for the sun—the clock. Since the clock is adjusted to the mean length of the day, a good clock is a sun substitute.

To begin the "race" referred to above, we need to find a well-defined starting and finishing line. The best way to do this is to attach a simple pointer—a tube or pipe—to an immovable object such as a wall. Line it up with the east-to-west track of some bright star and fasten it securely. Sighting through the pipe, we note the time when the star is exactly centered. After writing the time down, we leave the tube undisturbed until the next night. Arriving a little early, we note our star coming back for its transit through the viewing area of our tube. When it is exactly centered, we again carefully note the time. If you have measured the two times carefully, you will find that they are not even close. *The star is early, al-*

most 4 min early! The star has taken only 23 hours 56 minutes 4 seconds ($23^h56^m4^s$) to go once around the sky. Its east-to-west motion around us has been completed in 3 min 56 s less than that of the sun's east-to-west motion around the sky. The star has outraced the sun in moving toward the west. Hence, the sun appears to slip slowly eastward among the stars daily. The same observation could be made for any other star in the sky. Thus, the star day or *sidereal day* is about 4 min shorter than the *solar day.*

If we multiply the difference in time between the sidereal and solar days by the number of solar days in a year,

$$3^m56^s \times 365\tfrac{1}{4}^d = 24^h00^s$$

we get 1 full day as the difference between the two motions each year. This fact can, indeed, be used to define the year. The time, for example, between a noon lineup of a certain star and the sun and the next noon lineup of that same star and the sun is 1 year. The stars appear to make about 366 "loops" around the observer each year for each 365 "loops" of the sun. All year long, the stars move westward slightly more rapidly than the sun does. Since the sun moves all the way around to line up again with a particular star in 1 year, its rate of motion through the stars must be about 1°/day.

$$\frac{\Delta\theta}{\Delta t} = \frac{360°}{365 \text{ days}} \simeq 1°/\text{day}$$

Since the sun is about $\frac{1}{2}°$ wide as seen from the earth, it moves about two apparent solar diameters per day eastward among the stars. This motion is large enough to measure easily if we can see anything near the sun to use as a reference. The fact that we cannot see any stars in the vicinity of the sun is more of a nuisance than a serious handicap. Once we have mapped the stars, we may determine the location of the sun among the stars by indirect means. For example, we could note what stars are above the due south position at midnight. The sun is exactly halfway around in the sky at midnight, and we could mark this position on our star map. By direct measurement of how high the sun was the previous noon, we could mark its position above or below the CE. Thus, we would have determined both the declination of the sun and its east-west position in the sky. Actually, the ancients had worked out the path of the sun in the sky very thoroughly. The annual path of the sun through the field of the stars is called the *ecliptic.* (It is called this because the moon must be on the ecliptic in order to have an eclipse. Usually the moon is north or south of the ecliptic in the sky as it passes the zone where an eclipse could occur, and nothing happens.) The ecliptic provides us with a convenient starting point for measuring east-west angles in the sky. As the sun rises and sinks in the sky seasonally, its path, the ecliptic, cuts the CE twice—the first intersection occurs as the sun is rising in the sky daily, headed toward our summer in the Northern Hemisphere. The second intersection of the ecliptic and the CE occurs when the sun sinks back into the Southern Hemisphere of the sky to begin our autumn. The first crossing occurs on about March 21 and is called the *vernal equinox* point in the sky. (This event takes place in the constellation of Aries, the ram, and is marked with the old symbol for this star group—♈.) This point in the sky is used as the starting point for all east-west measurement in the celestial coordinate system.

Astronomers actually measure east-west distances in a coordinate called *right ascension* (or RA for short). Unlike declination, right ascension is measured in units of time. Beginning at 0 h at the vernal equinox point and moving eastward, the sky is divided into 24 zones, each 1 h wide. Thus, a star exactly 30° to the east of the vernal equinox point would have an RA of 2^h0^m. By measuring all east-west angular distances in time units, astronomers can calculate how long it will take for objects to move to a particular position. The star mentioned above, for example, would be in the position of the vernal equinox exactly 2 h later—it would rise 2 h after the equinox point, set 2 h later, and so on. We shall not do much with the concept of right ascension, but it is part of the system used universally by astronomers. In case you develop an interest in astronomy, this brief description should allow you to interpret star charts and do further reading.

* Solar Motions and Seasons

Figure 4-14 presents a map of the stars with the celestial equator and ecliptic shown. The ecliptic is marked to allow the reader to determine where the sun would be for any date. Of course, stars in the vicinity of the sun on a given date would not be visible. Thus, you can readily determine what part of the sky would be considered the winter constellations, and so on. The band of the 12 constellations through which the ecliptic passes was named the *zodiac* by the ancients. The vernal equinox point is marked on the chart. The word "vernal" means "occurring in the spring." The word "equinox" indicates a date at which all points on the earth have equally long daylight and darkness periods of 12 h each. Since the sun is on the celestial equator at the time of the vernal equinox, it appears to rise due east for everyone, sets due west, and is above the horizon for exactly one-half of its daily loop around the sky. The other time of year at which this occurs is the *autumnal equinox,* when the sun crosses the celestial equator heading southward in the sky on September 23 or 24.

The reason for this division of night and day is easy to understand. Figure 4-15 illustrates the earth receiving light from the sun at an equinox. The sun at an equinox is on the celestial equator. It is, therefore, directly overhead at noon on the equator. The sunlight comes straight down, striking the surface at the equator perpendicularly at noon. Notice in the drawing that the terminator, the line dividing daylight from darkness, cuts all latitudes evenly on this date. All the way from the North Pole to the South Pole every latitude is exactly half in darkness and half in daylight. Hence, the term "equinox".

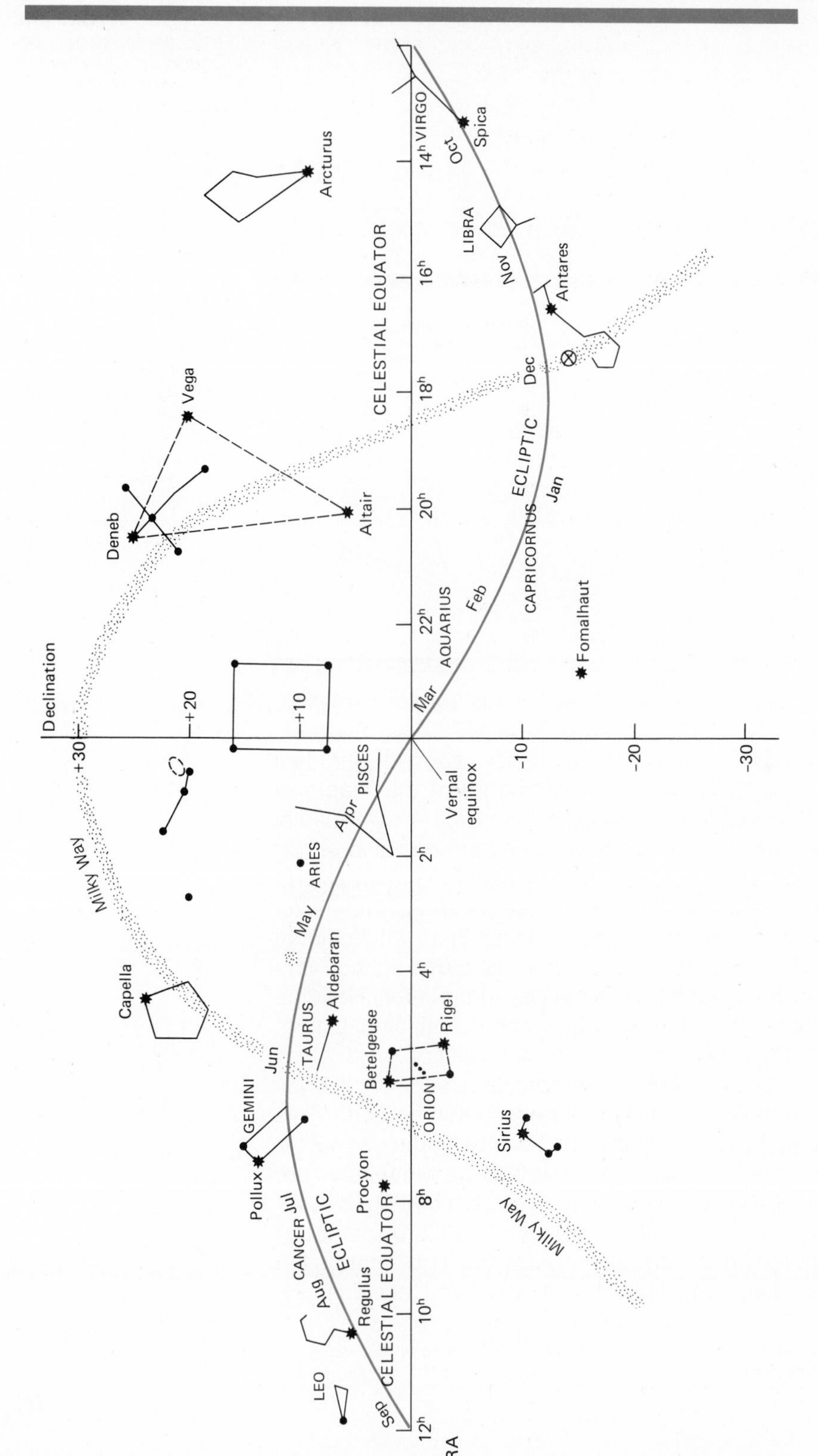

FIGURE 4-14
A map of the sky as seen from the earth. Only the brightest stars are shown. The celestial equator is used to divide the sky into northern and southern halves. The pathway of the sun through the stars, the ecliptic, is marked by the dates when the sun is in that section of the sky. Constellations along the ecliptic are named as are stars among the 20 brightest in the sky. The symbol ⊗ marks the direction to the center of our galaxy.

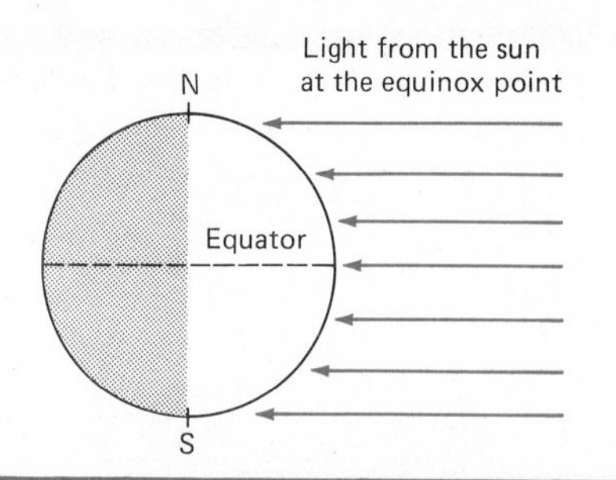

FIGURE 4-15
At the equinox every position on earth spends one-half of the day in daylight and one-half in darkness. At the poles the sun is either rising or setting on this date.

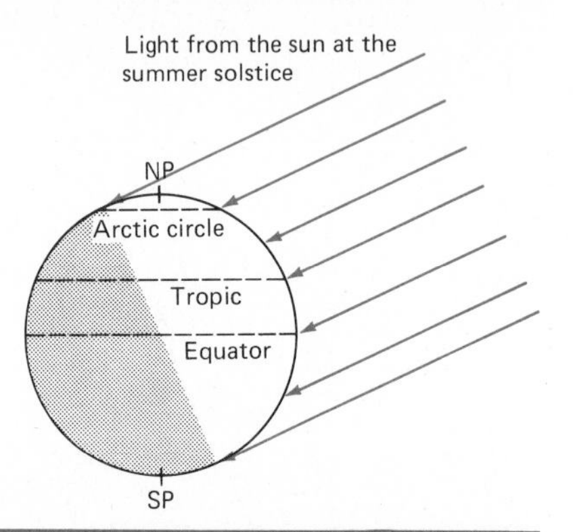

FIGURE 4-16
At the summer solstice the sun is the highest it ever gets for observers in the Northern Hemisphere. Note that sunlight "reaches over" the North Pole to illuminate the entire Arctic Circle. Similarly the zone around the South Pole is never exposed to sunlight on this date.

Compare this with the situation on June 21, the *summer solstice,* when the sun is highest in the Northern Hemisphere sky. This is shown in Fig. 4-16. Sunlight is now falling perpendicularly downward ("is incident perpendicular to the surface" in textbook language) at the Tropic of Cancer at noon.* Observers on the tropic would look directly overhead at noon to see the sun. An observer on the equator would have to look $23\frac{1}{2}°$ north of the zenith at noon to see the sun. Note that the terminator still divides the equator into equal halves of daylight and darkness. Note also, however, that this is the only latitude on earth at which there will be equal periods of night and day. South of the equator all latitudes spend more time in the night portion of the earth than in the daylight portion. Near the South Pole, there is a zone where the sun will not be visible all day long. South of $66\frac{1}{2}°$ S latitude, the Antarctic Circle, the sun will not rise at all on this day. Meanwhile, near the North Pole, note that sunlight passes above and over the pole to illuminate a zone beyond it in far north. This zone will stay in sunlight all day long. It is marked by the line at $66\frac{1}{2}°$ N latitude, the Arctic Circle. All points north of this latitude will receive uninterrupted sunlight for 24 h on this date. Before you start feeling sorry for the Southern Hemisphere, remember that the situation will be reversed in 6 months, when the sun will be as far south in the sky as it gets during the year. (For practice, see if you can draw the situation for December 21, the winter solstice.)

* Examination of the star map in Fig. 4-14 shows that the sun is located in the constellation of Cancer, the crab, on this date. The Tropic of Cancer is named from this. Similarly, the sun is in the constellation of Capricornus, the sea goat, 6 months later, giving the name Tropic of Capricorn to the southern tropic line.

While we are concluding this section with some remarks about the seasons, it might be useful to discuss the relationship of the sun's position in the sky to climate. Where is the hottest spot on earth? If the only thing that mattered was the presence or absence of sunlight, the hot spots would be the North and South Poles! After all, either pole will receive 6 months of sunlight, 24 h a day, before the sun disappears for the winter! Picture a hot summer day with the sun beating down on you. If we could "freeze" that situation and keep the sun in position, life would get unbearable quite fast. It would certainly take less than 6 months for life to become unbearable where you live. The only relief during a hot summer is that the sun sets and gives us a chance to cool down a bit before we start over at dawn. Yet, we are all aware that polar bears and eskimos do not run around scantily clad. Clearly, the length of the daylight period is only one of the factors that determines how warm it will get.

The second major factor is the intensity of the heat and light delivered by the sun to a given square meter of the earth's surface. This is usually referred to as the "angle of the sun's rays." You have probably observed that the sun feels hotter around noontime than in the morning or late afternoon. The effect of the ray angle of sunlight on the heat delivered to a given surface is easy to visualize. Consider a narrow-beam flashlight, as shown in Fig. 4-17A. The flashlight is giving off a certain amount of light. If held 1 m directly above the center of a piece of cardboard screen, the light will be spread over a certain area of the screen as shown. Now consider what happens when we tilt the flashlight angle to the screen. Figure 4-17B shows the same flashlight, still 1 m from the point where the center of the beam hits the cardboard. Note the area of the screen now in the beam. The same amount of light is spread over a much greater area. Thus, the amount of heat and light *per unit area* is much less. Thus, an area of surface at the pole receives less heat and light from a "low" sun that does a similar area near the equator from a "high" sun. The rate per unit area at which heat and light is received from the sun is called *insolation*.

A calculation of the area at the North Pole over which sunlight is spread on June 21, when the sun is highest in the sky, discloses that the sunlight is spread over 215 times as much area as it is at the Tropic of Cancer at noon on the same day. On the other hand, the pole receives

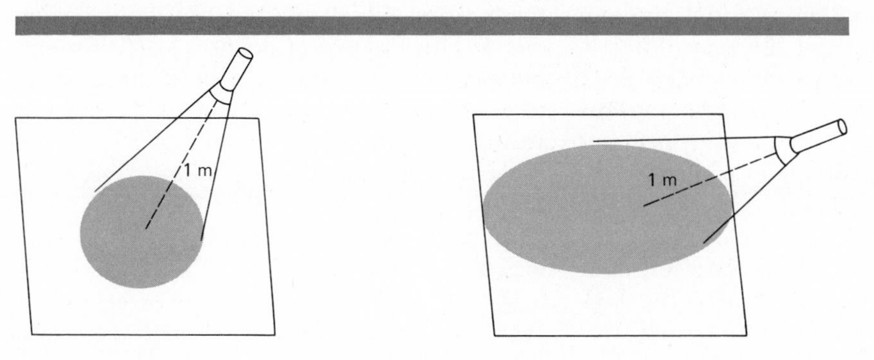

FIGURE 4-17A
(*Far left*) A flashlight beam shining directly on a screen 1 m distant leaves a small circular area of illumination on the screen. All of the heat and light from the flashlight is concentrated in this area.

FIGURE 4-17B
(*Near left*) The same flashlight, kept 1 m from the point where the center of the beam hits the cardboard, is moved to the side so the beam hits the screen at an angle. The same amount of heat and light is now spread out over a much larger area on the screen so that a square of a given size receives much less illumination than before.

24 h of sunlight on this day, and the sun remains at a constant altitude. At the tropic, the sun is up only part of the day, and its altitude varies from 0 to 90°. Calculating both the effect of the angle of the sun's rays and the length of the daylight period indicates that the tropic can accumulate more heat. While this advantage is small, it is cumulative. During the summer season tropical regions gain more heat daily than they can lose by radiation to space, and they warm up. Even though the sun is lower, it is up longer each day. Between the Tropic of Cancer and the North Pole, a maximum for heating effects is reached. Chicago or New York get hotter in the summer than Singapore, but they make up for this by being much, much colder in the winter.

Apparent Motions of the Moon

Of all the objects in the heavens moving against the background of the distant stars, the moon moves the fastest. The cycle that takes the sun a year—once around through the stars—takes the moon only one "moonth." This *is* the origin of our word "month." Timekeeping by primitive societies tended to be based on a moon or lunar cycle rather than a solar cycle. The moon was much easier to keep track of in relation to the stars. The moon appears to move eastward through the star field just as the sun does. The moon moves nightly through the stars and lines up

"Moonths" and years

The time between a lineup in the sky of the sun and the moon is $29\frac{1}{2}$ days. The real "moonth" is thus $29\frac{1}{2}$ days long. If we had half of our months with 29 days and half with 30 days, the months would keep in step with the phases of the moon. We could ensure, for example, a full moon on the fifteenth of each month. This is almost exactly the calendar in use in Moslem countries to this day. The problem is that while the month is a convenient division of time, there are not an even number of lunar cycles in the year: $29\frac{1}{2}^d \times 12 = 354$ days, or 11 days short of the year. Since it seems handy to have the months keep cycle with the year, phases lose out once a sun-centered calendar is introduced. It is handy to have January always in the winter, for example. (This appears less critical to Moslems.) Now with 11 days left over to divide between the 12 months, the task would seem easy. Slip each of the six, 29-day months an extra day and give five of the 30-day months an extra day. The remaining 30-day month of the original set could get its extra day every fourth year, when we take note of the fact that the year really takes 365.2422 days by inserting an extra day in the year. This was just the recommendation given to Julius Caesar by Egyptian astronomers from Alexandria.

The ancient Roman calendar was a lunar calendar. Each month began with the new moon on a day that Romans called the *calends* (from which we get the word "calendar"). The full moon occurred a half-month later on the *ides*. Unaware of the trouble the ides of March would get him into one day, Julius nevertheless abolished the lunar calendar and established a calendar based on the Egyptian advice. The Egyptians (who were largely Greek) determined that the year was about $365\frac{1}{4}$ days long. This called for a leap year having an extra day every fourth year. The first problem arose when the proposed calendar had a month named after Julius himself, July. Naturally, if a month is named after the emperor, it is going to be one of the big, 31-day months, not a skimpy 30-day one. This would have been fine, too, until Augustus Caesar, Julius's successor, had the next month in line named after him. Naturally, this had to be a 31-day month, too. This required "stealing" a day somewhere. January was named after Janus, a two-faced god with a special temple in Rome. Aside from inspiring the adjective "two-faced," Janus was a popular god, and held on to 31 days. The same was true of Mars, patron god of Rome, who defended all 31 days in March. To make a long story short, February had the least influential constituency, and lost a day. It shrank from 29 down to 28 days. As a consolation prize, it was given the extra day every leap year.*

Almost 16 centuries later it was becoming obvious that 0.2433 was not one-fourth. By that time the seasons were almost half a month out of agreement with the calendar. The Gregorian calendar adopted in 1585 removed 1 leap year per century and adjusted all months backward to the season they occupied in late Biblical times. We now have a leap year every year which is divisible by four, except on the first year of a century. The century year is not a leap year if it is divisible by 400. Thus, 1900 was a leap year, but 2000 will not be. Russia was the last European nation to adopt this system and did not convert until 1923. Without some well-worked-out system, calendar-keeping is a very complex business. Most primitive societies have required a priestly class just to keep track of the calendar and tell the peasants when it was time to plant the potatoes—or whatever was the major local crop. Nowadays, most of us take timekeeping for granted and rarely think much about it.

* W. M. O'Neil in his book *Time and the Calendars,* rev. ed. (Sydney University Press, 1978) concludes that Augustus' tinkering with the calendar is a fiction dating from books of the thirteenth century. Be aware of this possibility, but any story repeated for 800 years is part of the cultural background even if not authentic. Remember George and the cherry tree?

with a particular star every $27\frac{1}{3}$ days. Thus, its motion through the stars is $360°/27\frac{1}{3}$ days $= 13°/\text{day} \simeq 0.55°/\text{h}$. The moon is the same apparent size as the sun, about $\frac{1}{2}°$ in diameter. Therefore, the moon slips through the stars a distance equal to its own width each hour. This motion is large enough to be easily observed some night when you feel inclined to do so.

Consider the moon, the sun, and a star all lined up on a north-south line in the sky. Unfortunately, while the moon is moving 13°/day eastward through the stars, the sun is also moving about 1°/day eastward. This means that while the moon can line up with the same star in the sky after $27\frac{1}{3}$ days, the sun has moved ahead an additional 27° during the same time. The moon must move a little more than 2 additional days to line up again with the sun. Thus, there are two kinds of months. The *sidereal* month is the time it takes the moon to move through 360° in space and line up with a given star. The *solar* month, or *synodic* month, is the one we are more familiar with, since the moon's phases depend on its orientation with respect to the sun. The solar month is $29\frac{1}{2}$ days long. This is the time between successive lineups of the sun and the moon as seen from the earth, or the time between successive repetitions of a particular lunar phase. (The time between full moons, for example, is 1 synodic month.)

We may easily summarize the moon's separate motion in the sky.* The moon moves eastward through the field of stars about 13°/day in a path that centers on the ecliptic (the path of the sun) but deviates from it by as much as 5°, or about 10 moonwidths. Thus, the moon is always to be found within 10 of its own widths of the path of the sun through the stars.

The lunar path crosses the ecliptic in two places. These are called the *ascending* and *descending nodes* of its path. The node is ascending if the moon is moving upward from south of the ecliptic as it crosses. Similarly, it is a descending node if the moon is moving downward from north of the ecliptic as it crosses. The nodes and their positions in space are important because eclipses can occur only near the ascending and descending nodes of the moon's orbit. If the moon moved exactly on the ecliptic, we would have two eclipses every month.

There are several types of eclipses. A *solar* eclipse occurs when the shadow of the moon falls on the earth (Fig. 4-18). To an observer on the earth the events can be frightening. First, an increasingly large "bite" is observed missing from the observed disk of the sun. The size of the "missing" piece increases for about an hour as the moon slowly moves in front of the sun. Toward the end of this hour, the color of the landscape changes as sunlight from the redder outer regions of the sun dominates. If the eclipse is to be *total,* these final events occur: The landscape quickly darkens, and the edge of the moon's shadow may be seen rushing in at great speed from the west. Mountains, hills, forest, clouds are suddenly plunged into darkness as the shadow races toward the observer. Then, the night descends. It is small wonder that this set of events terrified primitive human beings. Of course, very few people see the sun totally blocked by the moon. Totality is confined to a region perhaps 100 mi wide by a few thousand miles long. Secondly, totality can last only a few minutes at any one spot on earth. Finally, the moon may be farther than average from the earth during the eclipse, and its shadow may not reach the earth. This situation is called an *annular eclipse.* On earth, the dark disk of the moon may be observed surrounded on all sides by a bright halo formed by the sun.

Whether total or annular, solar eclipses are more common than *lunar* eclipses, even though fewer people see them. The earth is a much bigger "screen" for the moon's shadow to fall on. However, when a lunar eclipse does occur, it is visible from more than half the earth. When this happens, the moon in its motion through space enters the shadow of the earth. During a lunar eclipse, the moon rarely gets completely dark. the atmosphere of the earth acts as a lens and bends or refracts some light around into the shadow. Usually, the moon dims considerably and ac-

* We use the word "separate" because everything in the heavens appears to go around or up once per day as basic motion. This motion dominates the smaller, separate motions of the sun, moon, and planets.

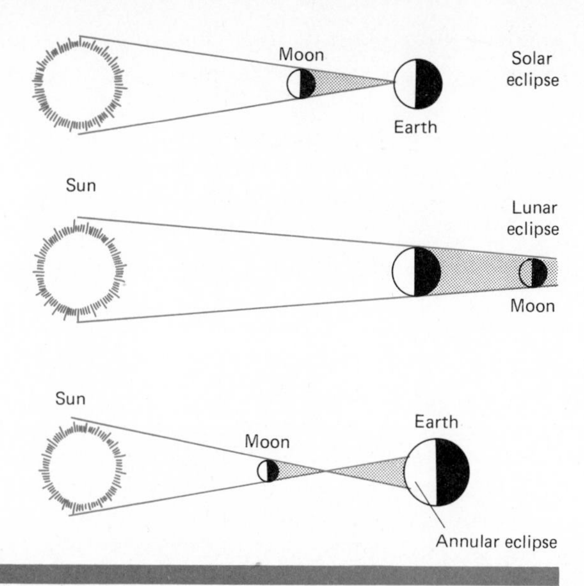

FIGURE 4-18
A solar eclipse occurs when the moon's shadow falls on the earth. In the lunar eclipse the moon moves into the earth's shadow zone in space. The annular eclipse results when the moon's shadow tapers out before reaching the earth's surface.

quires a dull, reddish cast. In any event, when the moon enters the deep shadow of the earth (the *umbra*) where all of the sun's surface is blocked, a dark edge proceeds across the moon. In a total lunar eclipse, this edge moves across the moon's surface until the moon completely enters the umbra. It may remain there for up to 2 h. On emerging from the shadow, a bright zone appears on the side of the moon which was the first to darken and proceeds across its surface. We have already mentioned the fact that the earth's shadow as it moves across the moon's face is always an arc of a circle.

Eclipses are exciting events, but they can occur only in the region of the nodes of the moon's orbit. Away from a node, the moon will pass either north or south of the place in the sky where an eclipse could take place. These nodes do not remain fixed at certain places on the ecliptic, but move westward along it. This *regression,* or "backward" motion, of the nodes is on the order of 20°/year. An interesting fact is that the forward motion of the sun and the backward motion of the nodes synchronize so that they agree after about 18 years $11\frac{1}{3}$ days. The ancient Babylonians noted from their records that cycles of eclipses repeated themselves in this time interval, which they called the *saros.* It was this cyclic behavior of eclipses and the Babylonian records which allowed Thales of Miletus to predict an eclipse in about 600 B.C.

The phases of the moon repeat themselves every month. The side of the moon facing the sun appears bright by reflected sunlight. The side away from the sun appears dark. From the earth, we are usually looking at a part of each side (Fig. 4-19). When the moon is in the direction of the sun, we are looking at the dark face and, of course, see nothing. This is called the *new-moon phase.* As the moon moves farther and farther to the east from the sun in the sky, we first see some of the illuminated side peeking around the western edge of the moon. This is the *crescent*

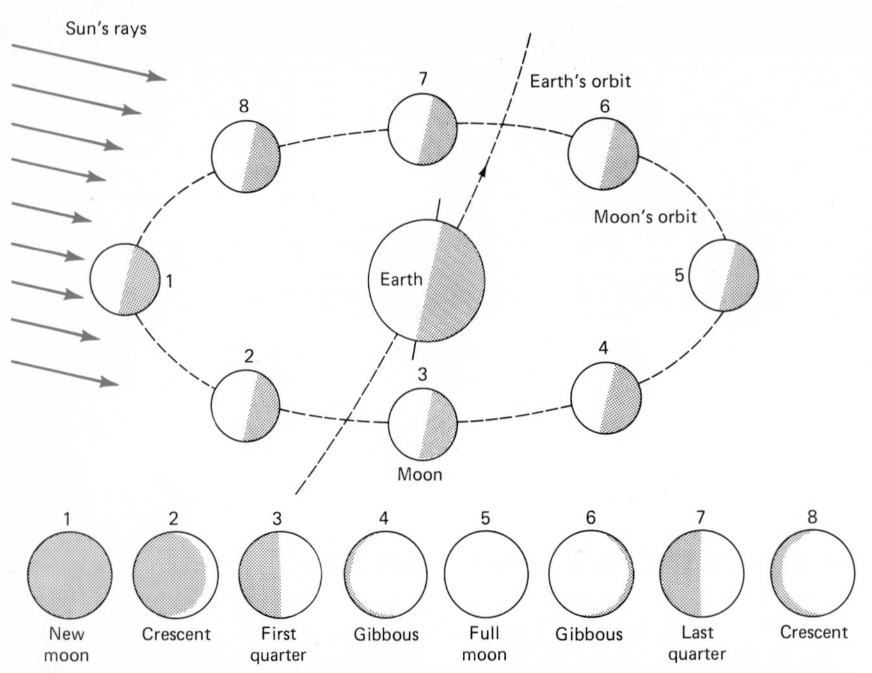

FIGURE 4-19 As the moon moves around the earth, its side toward the sun is always illuminated. When the moon is in the direction of the sun, we see none of this illuminated face, as in position 1. As the moon moves around the earth, more and more of the side facing the earth is in sunlight. Finally, at position 5, the moon is in the opposite direction from the sun for the observer, and its entire visible face is illuminated in a full moon.

phase. As the lunar phases proceed, we eventually see half (*half-moon phase*), more than half (*gibbous phase*), and finally all (*full-moon phase*) of the illuminated half of the moon which is facing the sun. As the moon continues around, the cycle of phases repeats itself in reverse order as less and less of the illuminated half is visible. This cycle has been correctly interpreted by humans for as long as there have been astronomers and written records. Its treatment in folklore is more romantic. The terms "lunacy" and "lunatic" derive from the supposed tendency of the full moon to drive humans mad.

Apparent Motion of the Planets

Up to this point, the description of the motions of the objects in the heavens has been reasonably straightforward. Both the sun and the moon proceed slowly eastward through the stars. Their rates of movement may be regarded as essentially constant. The only real complication is that their paths through the stars are tilted or inclined with respect to the daily motion of the star sphere around us. This is more a convenient nuisance (seasons are *very* convenient for life on earth) than a serious problem for description or measurement of these motions. There are, to be sure, some complications that we have not mentioned. For example, the equinox points, where the sun's apparent path crosses the celestial equator, are not fixed in space, but move or *precess.* This *precession of the equinoxes* was one of the more subtle motions discovered by the ancients. Hipparchus estimated this effect, in the second century B.C., to

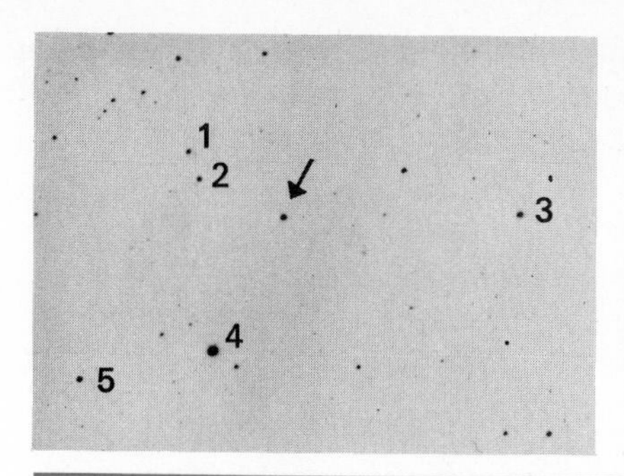

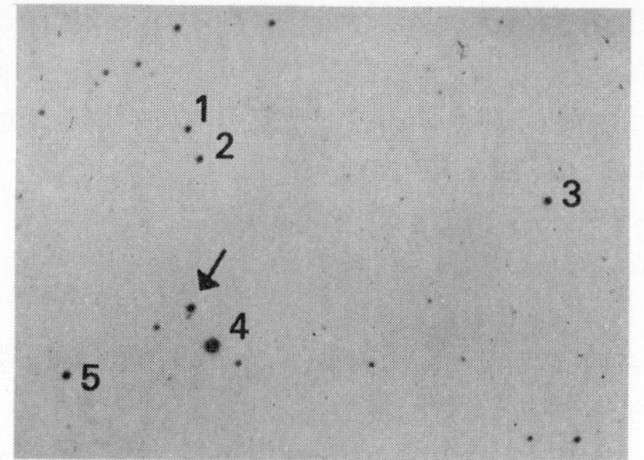

FIGURE 4-20 These photographic negatives illustrate the motions of members of the solar system relative to the stars. The arrows mark the positions of the minor planet, or *asteroid,* Ceres between the nights of (*left*) 4 April and (*right*) 8 April 1972. Several nearby stars are numbered as reference marks in the two photographs. These photographs were taken *without a telescope* by a standard 35-mm camera with a telephoto lens. Most of the objects photographed are not visible to the naked eye, yet are easily recorded by the camera. A complete system of coordinate astronomy with the 35-mm camera for colleges without access to a telescope has been developed by M. K. Gainer (St. Vincent's College, Department of Physics, Latrobe, PA). This method also allows dedicated amateurs a method of pursuing quantitative studies in astronomy without much expensive equipment. (*Courtesy of M. K. Gainer.*)

be about *36 seconds of arc per year* (36″/year)—or 1/36,000 of a circle per year. (He was not correct by the way, as the precession completes a circle in 26,000 years.) The point is that the regularity of astronomical motion is very striking, and quite small effects made themselves apparent early in history.*

We have arrived at the planets!

After the reasonable, sedate motions of the sun and moon, the planets are quite a change. (See Fig. 4-20.) First of all, the planets share the "once-around daily" motion of the star sphere as a basic movement. From this point onward, however, their motions are not that simple. The basic separate motion of each of the planets is eastward through the stars, as it is for the sun and moon. However, the planets do not move at constant rates—or even at nearly constant rates. As a matter of fact, at one time or another, each planet appears to stop and move *backward* through the stars! This backward, or westward, motion through the star field is called *retrograde motion* and is the first major complication in interpreting planetary motion. The second problem is that the planets vary in brightness as they move through the stars. There is one immediate regularity to the motions of the planets. With the entire sky to choose from, the planets nevertheless are confined to the narrow belt centered on the ecliptic called the *zodiac.* The ancients knew of five planets besides the earth: Mercury, Venus, Mars, Jupiter, and Saturn. Their motions are of two basic types, which we will describe separately.

VENUS AND MERCURY

Venus and Mercury share the fact that they are always found in the general vicinity of the sun, *never* on the other side of the sky. In fact, Mercury is never more than 28° away from the sun, while Venus may move as far as 48° from the sun. Mercury is usually so close to the sun that it is difficult to observe. However, it shares the general features of the motions of Venus, which we shall describe. As seen from the earth, Venus goes through a cycle of motions that takes 584 days. This period of time is

* Apparently quite small effects can have very large implications for humanity. The authors of an article in *Science,* 10 December 1976, concluded that small changes in the earth's orbital geometry could be the fundamental causes of ice ages. Three effects—the eccentricity of the earth's orbit, precession index, and variation in the obliquity of the earth's axis—gave scientists the ability to pinpoint past advances and retreats of the great ice sheets.

called its *synodic period* (from the word "synod," which means "meeting"). During these 584 days, Venus moves westward through the stars for 43 days. In general, this westward motion cannot be observed easily since it takes place when Venus is near the sun.

The best way to describe the motion of Venus is to picture its movement in daily jumps. Imagine that we are going to observe Venus each night just after sunset. Day by day, Venus slowly rises up out of the area where the sun just set, and each night it gets a little farther from the position of the sun just below the horizon. As it moves farther from the sun each night, Venus also becomes noticeably brighter. About a month before it nears its maximum separation from the sun (greatest *eastward elongation* is the technical term), it is at its brightest. After the sun and the moon, Venus is the brightest object visible from earth. For a considerable period of time, Venus seems to occupy almost the same position in the sky each night at sunset. It is still moving eastward through the stars since the stars all move about 1° to the west between sunsets. Venus has, however, slowed down so that its motion through the stars matches that of the sun—about 1°/day. Finally, Venus begins appearing lower in the sky each sunset. In closing the gap with the sun, Venus moves greater amounts each day until it is actually moving westward faster than the stars and is lost in the sunset. We can observe the other portion of Venus' orbit by shifting our attention to the time just before dawn (provided you have a good alarm clock and a great deal of enthusiasm to get up that early). A few weeks after Venus has disappeared into the sunset, we may observe it rising before the sun. At first, it gains in altitude rapidly in the eastern sky. Each morning it is a little higher in the sky at dawn. It gradually slows down as it approaches its maximum westward elongation of 48° from the sun. It appears to maintain this elongation for some time as the stars move a degree westward past it between dawns. Venus then picks up speed and moves with greater and greater daily eastward speed through the stars until it is lost in the brightness of the dawn. One synodic period, or 584 days, has passed. The motions described then begin over again.

It may have been difficult to follow on paper the motions of Venus and Mercury. I would suggest you read the section slowly in your dorm room, or some isolated spot, with a great deal of pointing and hand waving. Better yet, go and see a simulation of the motion in a planetarium. The motions of both Venus and Mercury are quite orderly in the sky, but are characterized by extremely variable speeds of eastward motion through the stars.

MARS, JUPITER, AND SATURN

The motion of the remaining planets as seen from the earth is entirely different from that of Mercury or Venus. The apparent motion of Mars is the largest in this group of planets so we will concentrate on Mars. Normally, Mars moves in a well-behaved fashion eastward through the stars. If we pick a condition, such as "Mars will be directly above the south point on our horizon at 9 P.M.," we can begin timing the passage of Mars around the star sphere. We would find that it requires 780 days for Mars to repeat a position such as the one stated. This is, therefore, the period of time required for Mars to complete one cycle of its motions as viewed from the earth. In other words, the synodic period of Mars is 780 days. The condition for Mars' position each night must be carefully defined in order to determine Mars' synodic period. Like all planets, Mars displays backward or retrograde motion through the stars. Thus, if the condition for one cycle is poorly defined, Mars might back up among the stars to the starting point in the sky 10 days later! The best starting point might be when Mars lines up with the sun, except, of course, that we cannot see Mars during that portion of its motion. During most of its 780-day cycle, Mars is very well behaved. To complete a cycle through the stars in 780 days, Mars must average about $\frac{1}{2}°$ of motion per night. Since this is only half the rate at which the sun moves through the sky, the sun always comes up from the west in the stars and overtakes Mars. Since our clocks are set by the sun, this means that Mars appears in the eastern sky before dawn and slowly gets higher and higher in the sky at each dawn. About a year after it is first visible before dawn, Mars will be visible all night long, centered in the sky above south at midnight. It takes the better part of a year to sink lower each night toward the west and disappear into the sunrise. Mars then cannot be seen until it emerges from the bright glow before sunrise to begin the cycle again.

AN EXAMPLE OF MOTIONS DEFINED—ASTRONOMICAL MOTIONS AS VIEWED FROM THE EARTH

In the midst of this relatively simple motion, Mars does a backward "loop-the-loop" among the stars! Each synodic period, just when Mars is up high in the nighttime sky for excellent viewing, Mars comes to a dead halt among the stars. It then begins to move backward, or to the west among the stars. It backs up for almost 2 months before halting and proceeding eastward again. This *retrograde loop* of Mars is the most well known of the special, peculiar motions of the planets. It is illustrated in Fig. 4-21. Note that Mars will actually pass some stars 3 times during the loop—once while still displaying its regular motion, once while in retro-

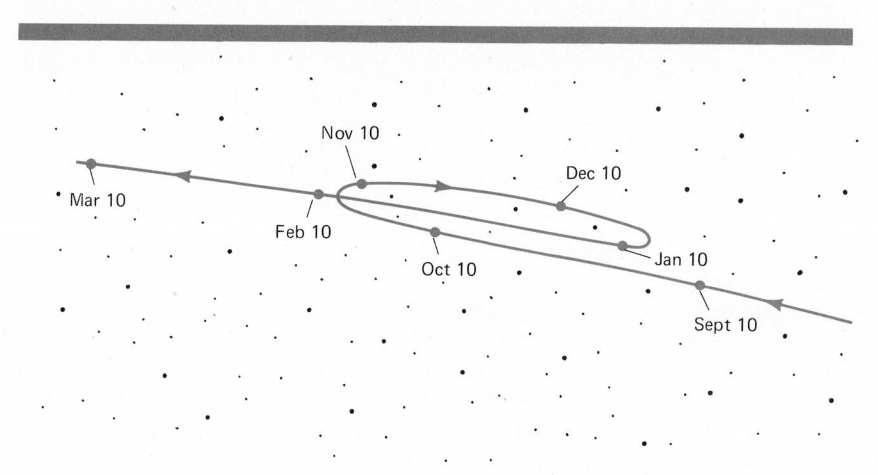

FIGURE 4-21
For almost 2 months every 2 years, Mars appears to "back up" in its motion through the stars. The exact shape of this retrograde loop varies every 2 years from an open loop to motion back along the original line of the planet's motion. During the midportion of its retrograde motion, here early December, Mars also becomes noticeably brighter than it has been or will be for the next 2 years. This is a most difficult motion for an astronomical system to predict.

grade motion, and a third time when it resumes its regular eastward motion. Instead of minor variations in speed or position, here we are confronted with a backward loop-the-loop among the stars which must be explained. If you came to this course relatively naïve about astronomy (most do), it may seem odd that the problem of explaining astronomical motion occupies such a high place in the serious work of civilization. Here you have the most important set of data demanding interpretation in the early history of science: Planets do loop-the-loops in the sky! Why? Explanations must wait for the next chapter, but perhaps astronomical motion looks a little bit more like a problem now.

Furthermore, we are not finished yet! There are two other items that must be mentioned in connection with retrograde loops. The first is that Mars noticeably *changes its brightness* during its motions. It is clearly at its brightest during the middle of its retrograde loop. A satisfactory explanation of these motions will have to account for this variation in brightness. For the second point (we have saved the strangest for the last), Mars does one retrograde loop in each passage through the stars. Jupiter does 11 backward loops per cycle, and Saturn does 28! Indeed, about a third of all Saturn's motion among the stars is much like someone scribbling flattened ovals on a sheet of paper. Perhaps it is not surprising that some of the greatest minds in human history have devoted themselves to grappling with astronomical motions.

Summary

In this chapter, we have tried to list and define various astronomical motions as seen by a naked-eye observer on earth. We have found the following motions which need interpretation:

1 A basic motion of the entire sky appears to be essentially once around the observer per day. Certain objects take slightly more or less than a day. The least time is normally taken by the stars, which complete the motion in $23^h56^m4^s$ of clock time. The stars thus make precisely one more loop around the sky per year than does the sun. Since all of the stars share this motion from east to west (around a pivot located near the North Star), we can choose this as the *basic* motion of the heavens and interpret other motions in terms of how they differ from the star motion.

2 The sun, moon, and each of the planets display motions separate from the star motion. Thus, we may say that they move or change their daily position among the stars.

3 The most fundamental fact about this separate motion is that it is essentially eastward through the stars. That is, each night the moon, for example, will be east of its position on previous nights. The planets usually move eastward through the stars, although each planet is observed to display some *retrograde,* or westward, motion among the stars at some portion of its path.

4 The sun, while moving daily westward about 1° among the stars, also moves to the north and to the south in the sky in a cycle that takes 1 year. This cycle is the basic cause of the seasons. Maximum motion of the sun from the celestial equator is $23\frac{1}{2}°$, either north or south. The extremes of this motion define four special latitudes on earth—the two tropics and the Arctic and Antarctic Circles. While

the length of time the sun is above the horizon in different seasons varies with latitude, all places on earth have the sun above the horizon for half of the time during the year, and below the horizon for one-half the time.

5 The annual motion of the sun through the stars defines a path called the *ecliptic.* The ecliptic is significant in that the moon and the planets are always found on or near the ecliptic. The intersections of the ecliptic and the celestial equator are called the *equinox points.* When the sun is in either of these positions, all places on earth have equal periods of daylight and darkness for that day. The spring (or vernal) equinox, which the sun occupies on March 21, is used as the starting point for all east-west astronomical measurement in the sky.

6 The moon moves around the earth and lines up with the sun every $29\frac{1}{2}$ days. During this time, we see varying amounts of the moon's illuminated portion. This explains the phases exhibited by the moon each month. The moon's eastward motion through the stars is the largest movement of the heavenly objects discussed at some 13° daily.

7 The path of the moon through the stars does not exactly coincide with the ecliptic, but is inclined to it at an angle of some 5°. Since the moon and the sun each appear about $\frac{1}{2}$° wide in the sky, this difference of 5° is enough that the moon does not move across the face of the sun each month. Eclipses occur only when the sun is near the places, called *nodes,* at which the moon's path crosses the ecliptic. *Solar eclipses* occur when the moon moves between the sun and the earth and its shadow falls on the earth. Solar eclipses may be either partial, if only part of the sun is blocked out; *total,* if the entire sun is covered; or *annular,* if the moon is not quite big enough to block out the entire sun. *Lunar eclipses* occur when the moon enters the earth's shadow in space, and may be either partial or total.

8 Venus and Mercury are always seen in the vicinity of the sun. They move eastward of the sun to the limit of their separation—48° in the case of Venus. They then move rapidly back toward the sun, cross its position in the sky, and are then seen on the other side of the sun. During the fastest part of this motion, both planets are moving in retrograde motion daily through the stars. The brightness of Venus varies in a complex manner depending upon where it is in its cycle.

9 Mars, Jupiter, and Saturn may be viewed at any point in their paths around the sun. They all move less rapidly through the stars than the sun does. Each of these planets normally moves eastward through the stars, but all display retrograde motion. Mars executes one backward loop in the sky for each 780-day cycle of motion. This retrograde motion lasts about 2 months and covers almost one-twelfth of Mars' total path in the sky. The other planets display more numerous retrograde loops during the cycle of their motions.

Questions

1 What is your latitude and longitude?

2 At what angle to the horizon does an observer's sun set and rise?

3 Why was astronomy so important that each ancient civilization—Greek, Egyptian, Chinese, Mayan, whatever—developed the study of the heavens?

4 Distinguish zodiac from ecliptic.

5 If the Big Dipper is positioned with the handle level in the sky, as though the imaginary "bowl" could hold water, how much later would you have to look to see it standing with the handle straight up in the sky? With the bowl upside down?

6 What is the only object in the sky that will remain in approximately the same position for the whole of your life?

7 How can you tell the difference between a planet and a bright star in the sky?

8 Do the planets appear directly overhead in the night sky for the residents of the United States?

9 What are the advantages of the celestial coordinate system used by astronomers compared to the observer system? What are its disadvantages?

10 Solar collectors for heating systems are roughly 3 times as effective in summer as in winter in western Pennsylvania. What factors determine how much heat such a collector can gather?

11 The moon is about $\frac{1}{2}°$ wide in the sky as seen from the earth. How large does the sun appear in the sky? How do you know?

12 What objects in the sky display no "backward," or retrograde, motion among the stars?

13 Where on the earth am I if the celestial equator is directly overhead? If the sun is up for 12 h on December 21?

Problems

1 What is the highest the sun ever gets in your sky? What is your lowest noontime elevation of the sun?

2 What is approximate altitude and azimuth of the sun at noon today? At sunset?

3 At about what time of day will a full moon rise and set? A new moon? A first-quarter moon?

4 In the summer a full moon is usually low in the sky at midnight (under 20° for a midlatitude). In the winter the full moon can be very high at midnight (73° altitude for an observer at 45° N). Why is the full moon's behavior opposite the sun's?

5 What is the altitude of the North Star above your northern horizon? Can you provide a rough "proof" of this from the diagram in Figure 4-11A?

6 For the latitude of Chicago (42° N) what is the highest the moon would ever get in the sky? What is the lowest the moon would ever get when it is in midsky—on the observer's zenith meridian (the line joining the zenith and the south point on the horizon)?

7 If we were on the moon viewing astronomical motions, the moon has one side permanently facing toward the earth. Imagine yourself in the middle of that side.

(*a*) How long would the day be?
(*b*) How long would the year be?
(*c*) Would the earth go through phases as the moon does for us on earth?
(*d*) How long would the stars take to go once around?
(*e*) Would the celestial equator and the vernal equinox point have any special meaning for you on the moon?

8 Take a thin cardboard tube or a pipe and fasten it down securely pointing at some very bright star in the southern part of the nighttime sky. As you watch the star through the tube, how long does it take to convince yourself that the star is moving? What mistake could you make here that could cause you to have a very long wait?

9 The ancients had measured the moon's distance and estimated it to be 30 earth diameters away. This is very close to the correct value. The moon's diameter is 3.4×10^3 km. The earth has a diameter of 12.8×10^3 km.

(*a*) How wide in degrees would the earth appear from the moon?
(*b*) Since the sun appears the same size in the sky as the moon, if we know that its distance is 1.5×10^8 km from the earth, what is the sun's diameter?

CHAPTER FIVE

The Conflict of Two Systems to Explain Astronomical Motions

Introduction The preceding chapter presented a reasonably complete picture of astronomical motions as viewed from the earth. Perhaps you noticed that some care was taken to avoid interpreting these motions. You would have to have led a very sheltered life indeed to be unaware that the earth spins on its axis and moves around the sun. But how do you know that such "common-knowledge facts" are correct? No primitive people ever assumed the earth to be an active mover as they observed the heavens. Thus, these ideas are not obvious. The conclusion that the earth moves comes from weighing evidence accumulated over a long period of time.

Imagine that this is your first day in this class. The instructor hands out an essay exam for you to complete during the hour. The problem asked of you is: "Present at least two types of physical evidence for the rotation of the earth on its axis and two for the revolution of the earth about the sun." If you are typical of students your age, you would flunk this task. Why? You have been told about rotation and revolution all of your life. But now you have an obligation as an educated individual to have reasons for accepting ideas that are not at all obvious. Stop for a minute and test yourself. Would you pass this exam? If so, pat yourself on the back; you have the first requirement of workers in science—sufficient curiosity to demand reasons for nonobvious "facts."

If you would not do so well—or if you want to succeed in your coursework—resolve to stay with us through the chapter. You will find it difficult to be neutral in the conflict of the two great world systems: an earth-centered universe vs. a sun-centered universe with a moving earth. You believe you know the results! Criticizing the design of the Wright Brothers' biplane from the cockpit of a jet leads you to certain biases. Having lived in a particular society all our lives, we tend to absorb the modes of thought dominant in that society. To think differently is certainly difficult and very nearly impossible. But let us at least give it a try.

This chapter presents two interpretations of astronomical motions. Both the ancient Ptolemaic System (completed in the second century A.D.) and the more recent Copernican System (published in 1543) are described. The highlights of the dispute concerning the acceptance or rejection of these systems are presented. The presence of two very different ways of describing astronomical motions forced scientists to clarify their ideas on the rules for accepting or rejecting theories. *Some knowledge of this struggle with ideas is a necessary part of the background of an educated individual in any field of study, since it is an important part of the intellectual development of our civilization.*

The Ancient Solution—Origins of the Ptolemaic System

Much of the knowledge of the ancient world came to a great clearinghouse in Alexandria in Egypt. This included records of the positions of the heavenly objects and the attempts to predict their future positions. How are we to interpret astronomical motions? There had been considerable discussion of this question in the ancient world, but the last of the great summarizers of astronomy would have the final word.

The culminating work of ancient astronomy was written by the Alexandrian Greek Claudius Ptolemy in the second century A.D.* Ptolemy was the last of a series of great observational astronomers, which began with Eratosthenes in the third century B.C. (You may recall from Chap. 1 that Eratosthenes was the first person to accurately measure the circumference of the earth.) A century later came Hipparchus, who was mentioned in the last chapter. Hipparchus was probably the best observational astronomer of the ancient world—an "astronomer's astronomer." However, the person who formulated the one grand, internally consistent theory of the movements of celestial bodies was Ptolemy. With Ptolemy, the "last word" in astronomy had been said by the Greeks. The Ptolemaic System influenced astronomical thought for well over 1000 years. (Harvard dropped it from the curriculum only in the nineteenth century.) When the Arabs came upon the book of Ptolemy, they recognized the treasure they had found. They gave it the name by which it has always been known since—the *Almagest.* The title in Arabic means "the greatest." This accolade exceeds anything we could say about the Ptolemaic System, so let us examine it.

* No relation to the pharaohs of the same name. Ptolemy was a very common Greek name, similar to "John Smith."

We have seen some of the ideas demanded by the Greeks in their astronomy in the first chapter. The Greek demand for order, beauty, and harmony required a system to be simple—even when it was not. In particular, majority opinion had settled on several specific requirements. First of all, the heavens were the sphere of perfection, and everything in the heavens had to conform to the Greek ideal of perfection. This meant that irregular motions were out. The only curves allowed for the paths of celestial bodies were circles, i.e., the only "perfect" geometric figures with constant curvature. Moreover, the motion must take place at constant speed, since changes are not allowed in a perfect system. Secondly, unusual ideas, such as a moving earth, had been eliminated. This was partly because such ideas violated common sense and partly because they did not seem to agree with observation. If the earth was moving around the sun, as the radical Aristarchus had suggested, the stars should appear to shift in the sky. This effect, called parallax (see Chap. 1), could not be observed by even the most careful of ancient astronomers.

Where did this leave Ptolemy? How does one account for backward loop-the-loops in the sky when it has been assumed that the planets must move at constant speeds in circles? If you think that it cannot be done, you have very little faith in the power of the human mind. (Given a powerful enough reason to want it to be done, humans could probably secure a fit to the data to describe square orbits!) Basically, human beings are clever, and if we have to fit nasty facts into the way we wish to look at something, we eventually manage to do so. The easiest thing to do is to ignore the data and assert that things are the way we wish them to be. We have always been good at this. Fortunately, this approach will not work for long in science. Mars will be somewhere in the sky a year from now, and we have either called its position correctly or we have not. Obviously, from what has been said, the Ptolemaic System was very successful. Well over 1000 years after it was put together, this theoretical system was still being used to predict planetary positions. The Ptolemaic System was indeed a very good theoretical system. Most present-day systems will do well to require so little modification in the same span of time.

The Ptolemaic System

How are we to account for the observed motions of the heavens and predict past and future positions for the various objects in the sky? Actually, one motion is explained by assuming motion at constant speed in a circle—motion of the stars. Two other motions come close, that of the sun and that of the moon. Both of these can profit from some fine adjustment, but motion at constant speed in a circle was a close approximation before precision clocks. The planets were another problem. Ptolemy "tinkered with" each of the standard Greek assumptions:

1 Motion in circles

Constant-speed motions

The immovable earth at the center of all motions

But this tinkering was done in such a way that the basic ideas were preserved and Greek sensibilities were not offended.

Let us begin with circles. Many examples of circular motion can be found in a modern amusement park if you watch the motions carefully. If you are standing in the middle of a merry-go-round and watching the motion of someone on the outside edge, you are viewing circular motion from the center of the circle. If the rate of turn is constant, the person will move through equal angles in equal periods of time. This is another way of saying that his or her speed is constant. Watching this motion for a few minutes should convince you that there is no way that this motion will by itself produce a backward looping movement. The clue here is in the motion of the horses on the merry-go-round. While they go around, they have a separate up-and-down motion. This gives them a much more complex motion as seen from the center. The ancient solution to the apparent backward motion of the planets was simple. *Have the planets move in more than one circle at once.*

Consider how this would work for Venus. In Fig. 5-1 we have an

Going around in circles

A Do-It-Yourself Guide to Epicycles

A compound circular motion may be produced using a toy called the Spirograph.* However, since the wheels in these sets are rigidly geared to the bigger rings, the wheel turns at the rate its outside edge moves along the ring. What this means is that a point on the circle cannot be rotating faster than it revolves about the disk. Astronomically, this means a planet cannot move backward in the sky. Its forward motion along the orbit is always greater than its rotation in the epicycle.

A simple way out of this for do-it-yourself epicyclists is to proceed as follows. Cut a small circle out of thin cardboard (such as a writing tablet backing). A circle 3 or 4 cm in diameter is good. With a paper punch, punch a hole in the center of the circle. For a 12-position epicycle, which is about the simplest to follow, mark it as shown in the illustration. The marking is similar to a clock face, but the numbers run counterclockwise around the edge. Note that a straight line is extended across the center of the circle from the 4 to the 10. This line allows us to keep the circle in the correct position. Call this the 4–10 line. Now, on a sheet of *lined* paper draw a circle—either with a compass or by using a peanut butter jar lid. Around the circumference of the circle, place evenly separated marks to represent successive positions of the center of the cardboard circle. The spacing of these marks will determine the path you will produce. If the marks are about two-thirds as far apart as the marks on the cardboard circle, a nice orbit results.

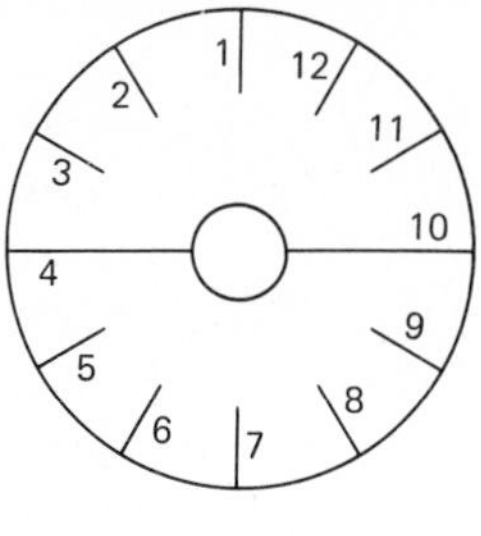

* TM, Kenner Toy Division, General Foods Corporation.

You are now ready to draw an epicyclic path. Place the cardboard disk on the paper so that a mark of the drawn circle is centered in the hole in the disc. Always line up the 4–10 line on the disc parallel to the lines on the paper. Draw a dot on the paper at the end of the 1 on the disk. Now slide the disk along the circle counterclockwise until the second mark is centered in the hole. Making sure the 4–10 line is parallel to the lines on the paper, mark a dot on the paper where the 2 on the disk meets the paper. Proceed around the orbit, advancing one mark on the epicycle for each mark on the circle. When you are finished, connect the dots in order to see what the motion of an object performing this double circular motion would be. The illustration shows parts of paths drawn in this way (to scale) with the size disk shown. If you enjoy making constructions of this sort, try to determine what size and markings on such a disk would simulate the epicyclic orbit of Jupiter, which shows 11 backward loops per circuit of the entire sky.

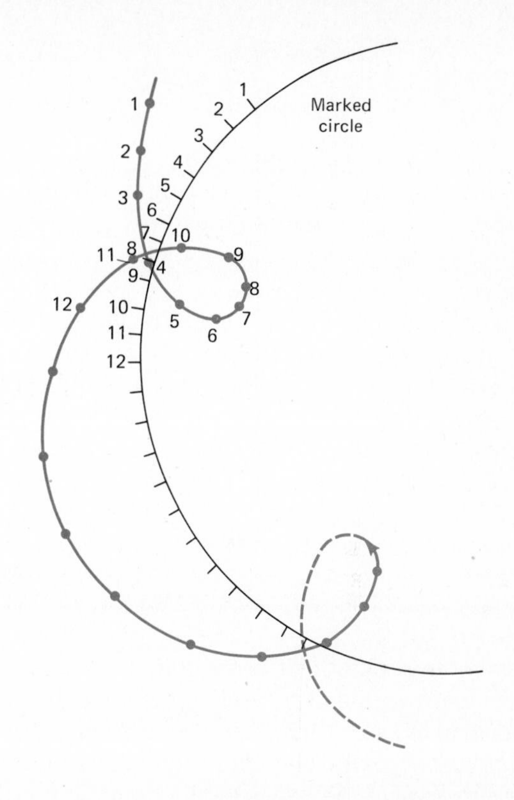

alignment of the sun, Venus, and the earth. As Venus and the sun move to the left, there are only three possible choices which maintain constant-speed motion. Choice 1 is that they stay lined up as they move. The sun would have to move faster than Venus since it is going around in a large circle. But this will not produce the observed motions at all, since if Venus stays in line with the sun, we would never be able to see it at all. Choice 2 requires that Venus move to the left faster than the sun, while choice 3 requires that Venus moves to the left more slowly than the sun. However, neither of these will work since, as times goes on, the gap between Venus and the sun will widen. Since both are required to move at constant speed, the gap would continue to grow until Venus separated from the sun by more than 48°. But Venus never gets farther than this from the sun. How did Ptolemy resolve this apparent discrepancy?*

Remember the motion of the horse on the merry-go-round? Two motions going on at once can be much more interesting than one. If circles are required, why not two circles at once? We have redrawn Fig. 5-1 with a slight change. The sun and the earth are in the same positions as before. Now, instead of Venus occupying the same position as before, that

* In this account, we are giving Ptolemy the credit for the Ptolemaic System. Actually most of the features of the system were worked out well before Ptolemy. The greatest contribution was probably that of Hipparchus.

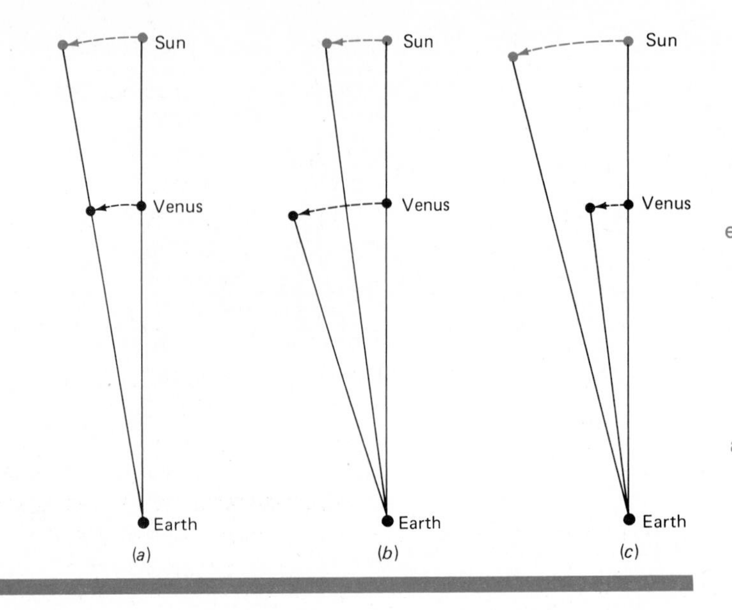

FIGURE 5-1 Three options for the relative motion of the sun and Venus as seen from a fixed earth. In all cases, the sun and Venus are originally in line as seen from the earth. (*a*) In the first case both objects move at such a rate as to stay in line as seen from the earth. (*b*) In the second case the motion of Venus is greater than that of the sun, and it would appear to pull ahead in the sky. (*c*) The third option has the sun moving through a greater angle in a given time so it would draw away from Venus. None of these options would simulate the actual motion observed in the sky where Venus moves back and forth among the stars but is never more than 48° away from the sun.

point is the center of a circle (Fig. 5-2). Venus now proceeds around in this smaller circle at constant speed. We then combine this with choice 1 shown in Fig. 5-1. The center of the small circle goes around at the same angular rate as the sun. Thus, the center of the small circle is always on the line joining the earth and the sun. So long as the speed of Venus in the small circle is greater than the speed at which the center of the circle moves to the left. Venus will appear to follow the path shown in Fig. 5-2, where several successive positions are illustrated. Note that with this system, Venus can never be seen farther from the sun than an angle determined by the radius of the small circle and the distance of its center from the earth. Thus, the most troublesome features of the apparent motion are neatly accounted for. In this scheme, the path followed by the center of the circle is called the *deferent.* The smaller circle in which Venus moves is called the *epicycle.*

It should be apparent to you that this same scheme will explain the retrograde motion of Mars. In fact, Mars' case is simpler, since it is not required that the center of the epicycle stay in line with the sun. Figure 5-3 shows the epicycle of Mars generating a retrograde loop as seen from the earth. As a bonus of this system, note that Mars is clearly much closer to the earth in the middle of its retrograde loop. This, you may recall, is exactly when Mars appears brightest in our sky.

Could Ptolemy now account for the observed motions of the planets? After all, the system must be able to predict accurately future positions, not just approximate them. Either Mars will be in its predicted position or it will not. The answer to this question is that epicycles alone will *not* do a thorough job of prediction; more is needed. To "fine-tune" this system, two additional steps are required. Both are somewhat technical, so we will describe them only briefly.

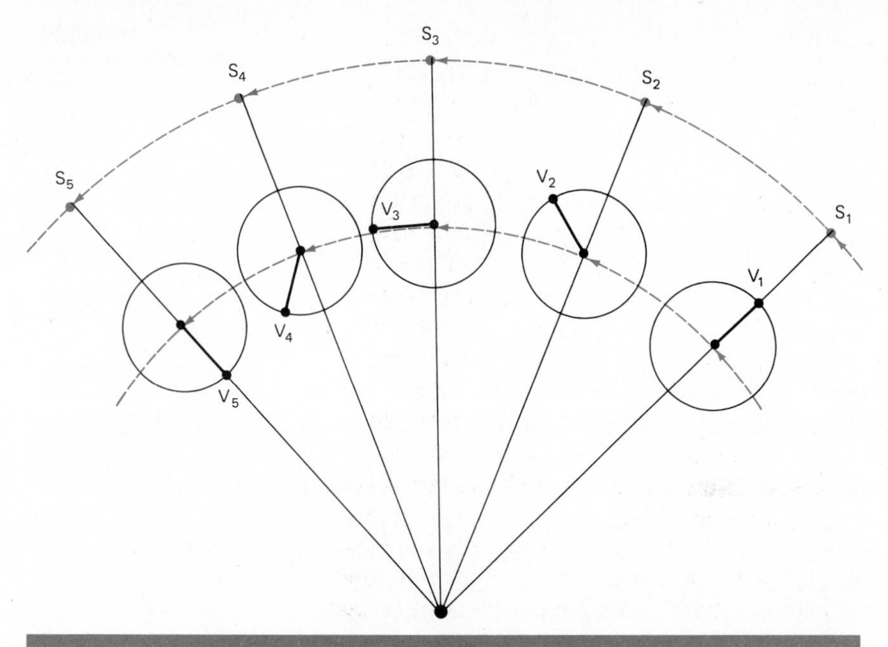

FIGURE 5-2 The principle of the epicycle—Venus moves on a small circle, called an epicycle, while this circle moves around the earth in a circular orbit. The center of the circle moves so that it remains on the line joining the earth and the sun. Thus, in the first position, the sun at S_1, Venus at V_1, and the earth are lined up. When the sun moves to position S_2, the separate motion of Venus on the epicycle causes it to move out to position V_2, and so on with the other positions shown.

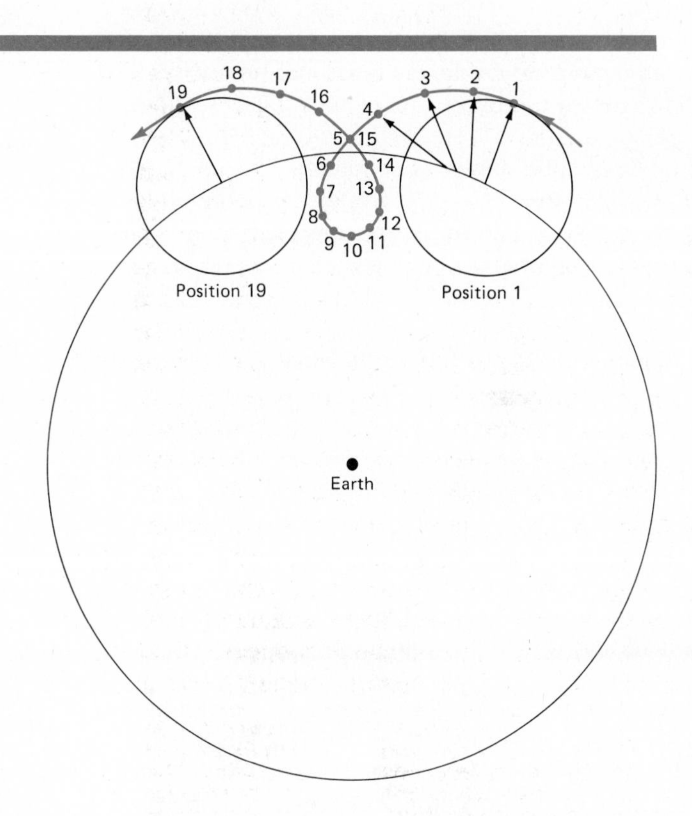

FIGURE 5-3 The generation of the retrograde loop of Mars from its motion in an epicycle can be seen. The full epicycle is shown for positions 1 and 19 only. The entire epicycle moves to the left about one-fifth of its radius between positions. In the same time, Mars moves one-sixteenth of the distance around the epicycle. The effect of this combined circular motion is that Mars makes a loop in space pointed toward the earth.

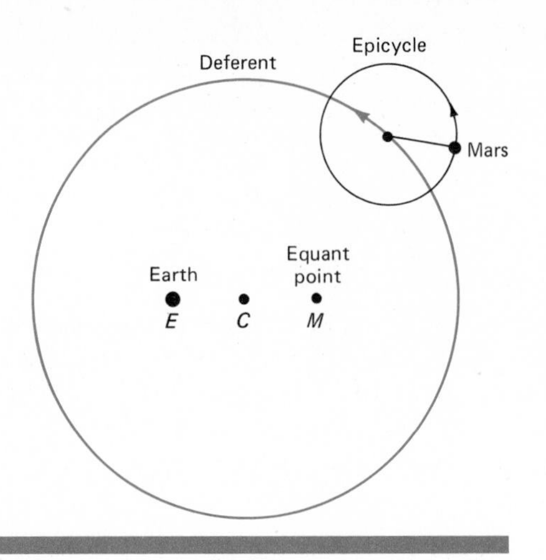

FIGURE 5-4
The Ptolemaic System not only has planets moving in epicycles, but uses two other corrections to adjust the observed speeds of motion. The earth is moved off the center (eccentric) of the orbit. A similar point M, the equant point, is the only position in space from which the motion would appear to be at constant speed. The separation of these points is greatly exaggerated in the drawing compared to the size of the orbit.

Using simple circular motions, Mars, for example, will be found to run slower in certain parts of the sky than expected and faster in others. To resolve this problem Ptolemy modified the requirement of constant speed. The motion in the circle is at a constant angular rate—so many degrees per day—but *not as seen from the center of the circle.* To observe the turning rate as constant, Ptolemy suggested that one had to be at a point M off the center of the circle. (This point was called the *equant* point.) Furthermore, Ptolemy suggested that the earth was not at the center of the orbit. It was moved off center by the same amount as the equant point, but in the opposite direction. This is illustrated in Fig. 5-4. Note that we have three centers now: the center of motions M; the geometric center of the orbit C; and the center of all interest, our earth, at E.

This essentially completes the Ptolemaic System. By carefully adjusting all distances, sizes, and speeds, a system was constructed which satisfactorily predicted planetary positions.

We may now summarize the Ptolemaic System:

1 All astronomical motions are around the earth, which is motionless.

2 All objects share a daily motion around the earth with the stars.

3 From the earth outward, the astronomical objects are the moon (closest), Mercury, Venus, the sun, Mars, Jupiter, Saturn, the stars.

4 Planetary motions are composed of a basic circular orbit plus a separate circular motion in an epicycle around a point moving on that orbit.

5 The centers of the epicycles of Venus and Mercury move around the orbit so that they always remain on the line joining the earth and the sun.

There are other parts to the complete system. For example, to answer the question, "What keeps the planets up; why don't they fall?", it was conventional to say that they were hung like lanterns on spheres of the purest transparent crystal. Christian astronomers answered the question of why the planets move continually by suggesting that angels push the crystal spheres. And so on. Some of these ideas are actually important historically. The crystal sphere idea, for example, caused problems later when *measurements* indicated that comets pass through planetary orbits.

Introduction to the Copernican System

The difficulty with any system of astronomical motion is that the objects keep moving. Any small error that accumulates will eventually become apparent. Only errors that compensate or correct themselves each cycle may remain hidden. After the Ptolemaic System had been in use for over 1000 years, it carried with it the burden of its records. If some small error required an adjustment, the change that was to be made had to take care of the problem in the night sky. But that change had to preserve the positions of objects recorded in Ptolemy's data as well. A change made to correct the sky of 1400 A.D. had to have no effect on the sky of 150 A.D. as recorded in the *Almagest*. And corrections did become necessary. Astronomers argued as to the best way to adjust the system in order to correctly predict planetary positions. Should one move the earth more off center (more eccentric), slide out the equant point, increase the speed of an epicyclic rotation, change the radius of a deferent, or what? There were a number of ways to adjust the Ptolemaic System, but most of these changes also would have changed the ancient sky.

FIGURE 5-5
Most of the nations of the Western world issued stamps to commemorate the 500th anniversary of the birth of Copernicus in 1973. The Polish set of stamps shown here reproduced most of the portraits of Copernicus. The most frequently seen portrait is at the upper right above, the stamp issued by the United States.

It is widely reported that late-Renaissance astronomers added epicycles to the original epicycles to adjust planetary positions. Recent calculations of the modern sky using the original Ptolemaic System yield about the same levels of accuracy in predicting positions as were obtained in the fifteenth century. Thus, if additional complexities were introduced, they had no practical effect. For all practical purposes, the system was adequate. The delightful fact about the human mind, however, is that it can be curious first and practical later.

Into this background came Niklas Koppernigk (1474–1543) of Poland, whom the world has always known by his Latinized name Copernicus (Fig. 5-5). He was born into exciting times. At the age of 18, he was at the Crakow Academy when Columbus discovered America. Four years later, he was a student at the University of Bologna in Italy where he first encountered discontent with the Ptolemaic System. Most of his adult life was spent as a canon to the Bishop of Warmia in Poland (Fig. 5-6). He was near the scene of action of the Florentine reformer Savanarola in Italy; the "wheeling and dealing" in Rome during the Jubilee year of 1500; the break with the church of a northern neighbor, Martin Luther; and much more.

FIGURE 5-6
Copernicus spent most of his adult working life living in the two upper floors of the left-hand tower in the wall surrounding the cathedral at Frombork. The windows led to his study, and most of his observations of the sky were taken from the top of the wall. *(Photograph courtesy of the embassy of the Polish People's Republic.)*

Copernicus was apparently working on his idea of a sun-centered universe as early as the age of 30, but it is not clear that he realized the degree of controversy it would cause. Although the world coined the word "revolution" from the title of his book *De Revolutionibus Orbium Coelestium,* there was little about Copernicus that could be considered revolutionary. He was the picture of an establishment man, and was quite successful in all that he undertook.

The Copernican System

Copernicus set out to produce a simpler system to account for the observed motions of the objects in the heavens. He was particularly disturbed by the equant used by Ptolemy. To be sure, the evidence against the Ptolemaic System was limited. Generally, it described motions well but could be quite in error when two or more motions were involved. Copernicus apparently noted the discrepancies in prediction during the planetary conjuctions of 1504. (A *conjunction* is the apparent alignment of two objects in the sky.) On this occasion, there were notable errors of prediction of as much as 2°. In order to predict a conjunction one must predict the motion of each planet correctly, or be fortunate enough for the separate errors to cancel each other.

Copernicus worked on his system during most of his life. However, the publication of his findings was delayed until the year of his death in 1543. *De Revolutionibus* is one of the most important books in the history of Western civilization. Its publication marked the start of a transition to modern science that lasted a century and a half and ended with the publication of Newton's *Principia* in 1686.

The fundamental change of the Copernican System was to place the sun as the immovable center of the solar system. The earth must then be in motion, In fact, according to Copernicus the earth must have three motions:

1. A daily rotation on its axis
2. A yearly motion in an orbit around the sun
3. A long-term "wobbling" of the axis of rotation to account for the precession of the equinoxes

Starting from these hypotheses, Copernicus had the moon moving in an epicycle around the earth, while the earth moves in a circle around the sun. The planets then moved around the sun in circular orbits. In order to account for the observed changes in speed of the planets, Copernicus was obliged to keep epicycles but he did succeed in eliminating the use of the equant. Furthermore, the epicycles were not needed to account for the retrograde motions, only to adjust speeds. Mercury and Venus would be expected to show westward motions among the stars since they were going around the sun (Fig. 5-7).

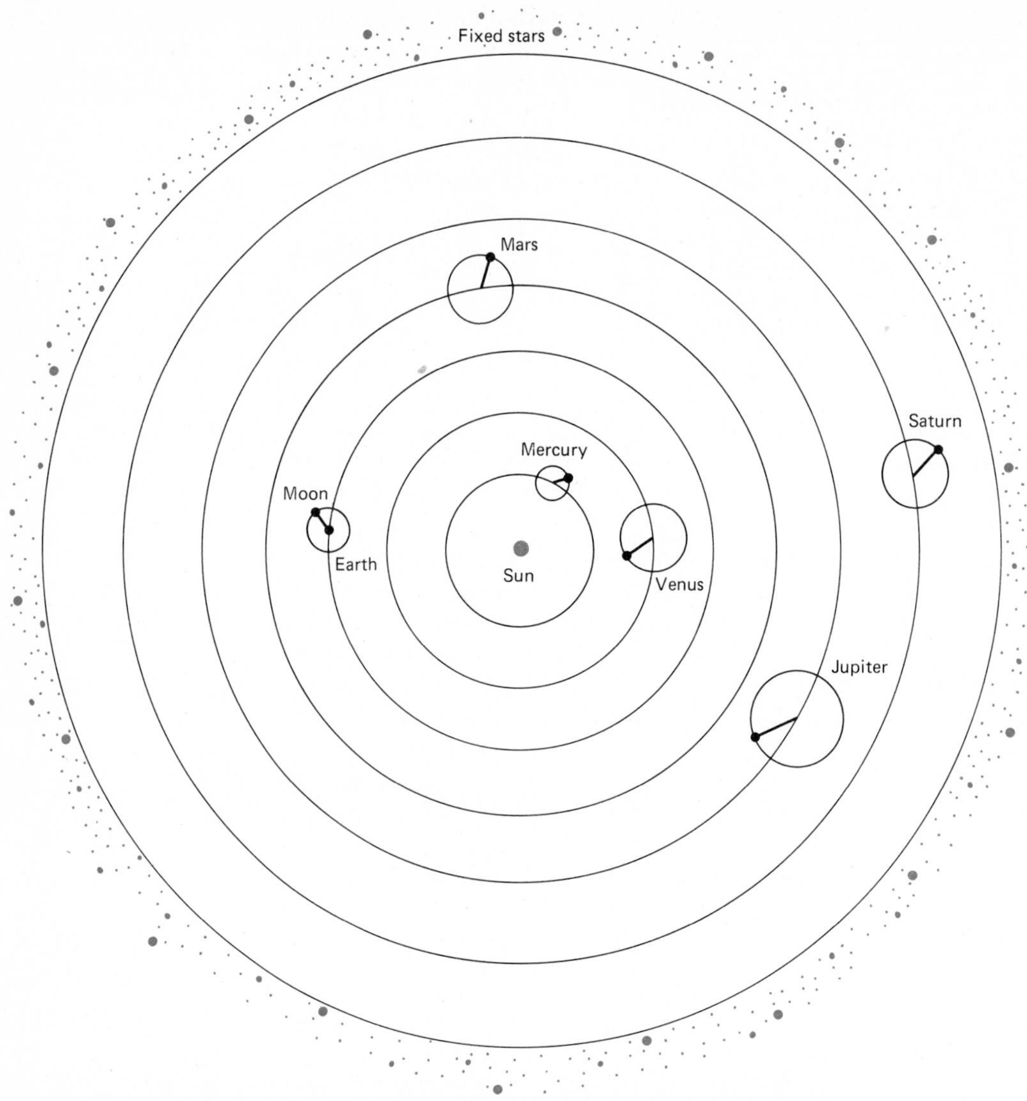

FIGURE 5-7
The original Copernican System centered planetary motion on the sun. Copernicus was obliged to keep epicycles, exaggerated in size in the illustration, to make small adjustments to fit observed motions. Likewise, all orbits were circular and all motions were at constant speed in the Copernican System.

Mars and the other planets were another matter entirely. The retrograde motion of Mars turns out to be an expected feature of this system. The earth is moving faster in its orbit than Mars is moving in the Martian orbit. As the earth "sweeps" past Mars each synodic period, Mars appears to move backward among the distant stars. This is much the same as observing a car which you pass on the highway. As you pass, the car appears to be moving backward with respect to the more distant scenery. This is an effect you should examine carefully in order to obtain an understanding of it. Figure 5-8 indicates the basic idea. To understand the idea clearly, you should carefully *study* Fig. 5-8, not merely look at it.

This in essence is the Copernican System. Note that it keeps circular orbits, constant-speed motion, and epicycles. It places the sun in the center of the system of motions with everything orbiting the sun except the moon (which orbits the earth) and the stars (which are motionless).

The Copernican System is, geometrically, less of a departure from the Ptolemaic System than is often assumed. As we have seen, Copernicus retained circular orbits and epicycles. In some ways his system is conservative since it moves back toward the Greek ideal of "perfect mo-

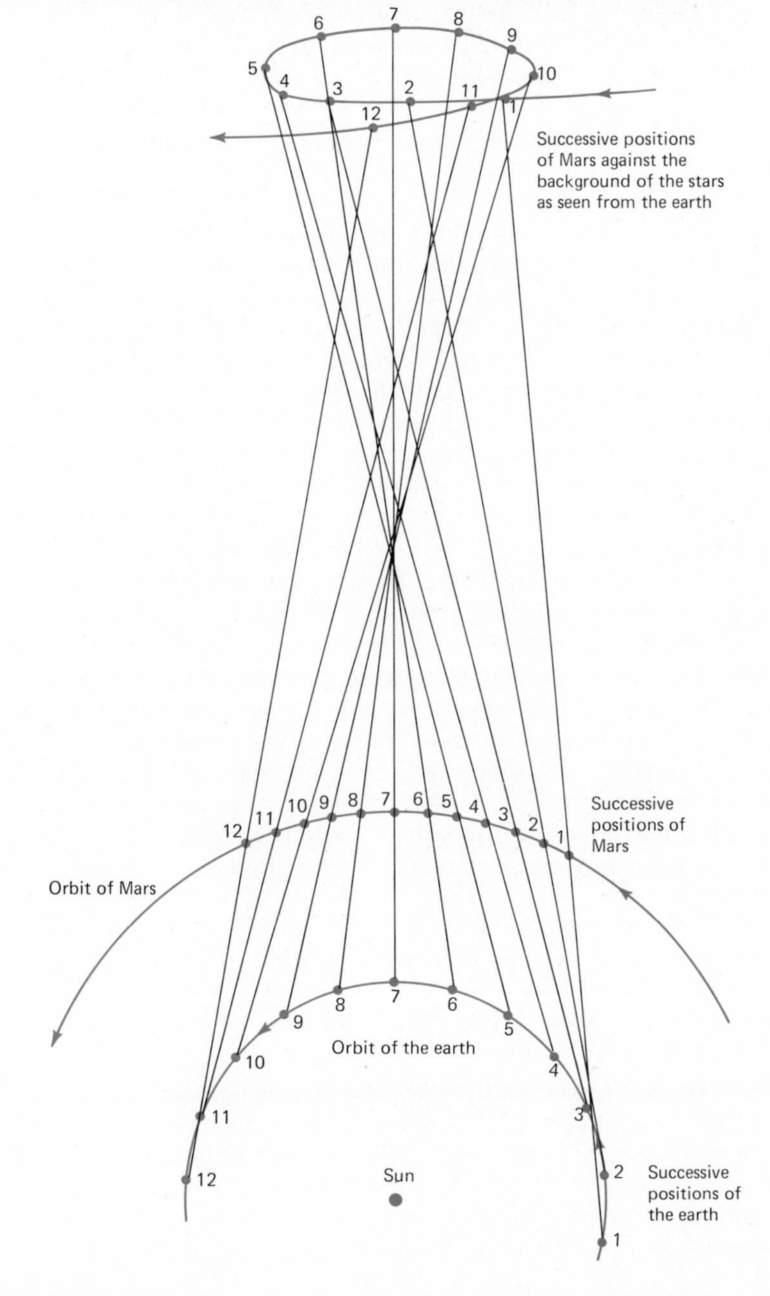

FIGURE 5-8 As the more rapidly moving earth passes Mars in its orbit, Mars appears to move backward against the stars. With a straightedge you may rapidly verify each sighting shown. For example, as we view from the earth at position 1 on its orbit, we see Mars in position 1 of its orbit. Our view extends beyond Mars to the stars, and we see the image of Mars against the star field as position 1 at the top of the drawing. While the earth moves to position 2, Mars, moving more slowly, also moves to its position 2. The first four sets of positions lead to an apparent leftward motion of the image of Mars through the star field. However, from position 5 through 10 the earth rapidly passes Mars in its orbit. This leads to an apparent rightward, or retrograde, motion of the image of Mars among the stars.

tions" in the heavens. From a practical standpoint, it is somewhat easier to calculate positions using a sun-centered system than using an earth-centered one. On the other hand, neither system predicts to the limit of observation. In general, they are about equally accurate and perform poorly only when two or more motions are involved—as in planetary conjunctions or eclipses. How then does one choose between them? A better question might be, "Why bother to choose at all?" If you want to predict positions, use the easier system, since the two are equally inaccurate. As for what to believe was really happening in the solar system, it remained to the individual's preferences. No physical proof was offered for either system. Most people rejected the concept of a moving earth as absurd and contrary to common sense. A few individuals who preferred order and balance, however, noted an intuitive appeal to the Copernican System. For example, the planets were arranged in a logical order. Mercury, with the shortest period to complete its orbit, was closest to the sun. Then the planets were arranged in the same order as the time it took them to complete one orbit—Venus, Earth, Mars, Jupiter, and Saturn.

The big problem with accepting the Copernican design of the universe as correct was that it changed humanity's entire concept of itself and its relationship to the universe. This change of view is a fundamental step in the intellectual history of the human race. No field of human thought remained untouched by this change in viewpoint. A rethinking was required not only in astronomy and science, but across the entire realm of human ideas. This change in world view is discussed in many works. Here, we will limit ourselves to its impact on scientific thought. The Copernican System required a new physics of the heavens. Under Ptolemy, the earth was fundamentally different from the substance of the heavens. Hard, dense, and massive, it occupied the middle of the universe. Everything external was totally unlike the earth, or else doomed to fall in and join the earth—like meteorites. Out there were these marvelous little torches in the sky to illuminate humanity's path, delight the senses, and stimulate the contemplation of nature—"perfect," in ceaseless motion, intellectual "hula hoops" testifying to a grand design in creation. But, the key idea—*different*—not the same stuff and substance as the earth, *not subject to the same rules.* The Copernican design changes all of this. Now the earth is just one of seven objects whirling about the heavenly torch.

How to Decide?—Tycho Brahe and Precision

The most outstanding astronomer produced by Europe in the decades after Copernicus was the Danish nobleman Tycho Brahe. With Brahe we recognize the beginning of a new kind of thinking in science. Brahe elevated observational astronomy to new heights. He constructed the most precise naked-eye observing instruments in Western history. To Brahe, prediction had to be precise. He was appalled by the inaccuracies in

FIGURE 5-9
Tycho Brahe seated in his observatory with several of his instruments. The largest instrument is the curved mural quadrant used to measure the height of objects above the southern horizon. The assistant at the extreme right sights the object through a hole in the upper left wall. The instrument itself is essentially half of a very finely divided protractor with a sighting device sliding along the scale. The illustration is from Brahe's 1598 book on astronomical instruments. Both Brahe and Ulug Bey (Fig. 5-10) produced data as good as the human eye could produce without a telescope.

prediction of both the Alfonsine tables of planetary positions (calculated using the Ptolemaic System) and the (usually lesser) inaccuracies in the Prutenic tables (calculated using the Copernican System). He realized that the basis of astronomical prediction was precise data. Neither Ptolemy nor Copernicus were especially good observational astronomers. Both tolerated inaccuracies which Brahe would not tolerate; *precision was Brahe's unceasing obsession.* For 30 years, he recorded positions in the heavens with an accuracy that previous observers had not approached. (See Fig. 5-9.) Indeed, most prior workers saw no great *need* for more accurate data. (See Fig. 5-10.) Brahe belonged to the period before the utility of a mathematical description of nature became apparent, a period when mathematical laws describing physical phenomena were not generally used. The power of mathematics in describing relationships was unrealized, but Brahe was to contribute to changing this situation. Copernicus was content with fixing the position of a planet in the sky if it were in error by no more than the diameter of a softball viewed at the length of a football field. Brahe tolerated less than half of this margin

FIGURE 5-10 The world's oldest astronomical observatory, in the modern sense of the word, was used over a century before Tycho Brahe built the first in the Western world. All that remains of the observatory of Ulug Bey at Samarkand in central Asia is this great quadrant dug into a trench two stories deep. An observing cart rolled on the two stone tracks shown with the observer moving between the tracks. At each side are staircases for assistants. The only light in this shaft comes from a single opening at the center of the curved measuring surface, as in Fig. 5-9. The observatory, along with many universities, was destroyed two generations before the discovery of America in a conservative Islamic religious revolution. (*Photograph by the author.*)

of error in routine observations. For data that were particularly critical, Brahe's error approached the width of a quarter seen at the same distance. While most astronomers made only occasional observations and spent most of their time in paperwork, Brahe observed and recorded positions for the planets whenever the weather permitted. He accumulated the most accurate data ever compiled by a naked-eye astronomer.

Analyzing this data was another problem. The carefully observed positions did not seem to fit the system of Ptolemy or that of Copernicus. Several features of the Copernican System of motions seemed to work out better than the Ptolemaic System, yet no evidence could be found in the data to support a moving earth. Brahe could not find the parallax of the stars which a moving earth seemed to demand. The inconsistencies between prediction and observation finally defeated him. Brahe never seemed to doubt that an accurate description of the motions must account *exactly* for the observed positions. This requirement of precise agreement of observation and theory was a definite contribution of Brahe. Given a disagreement between theory and careful observation, *the facts win* and the theory must be modified. Later, what Brahe only implied, Kepler was to state emphatically.

The need for data

Tycho Brahe

Tycho Brahe: Juvenile Delinquent to Wizard

What gives a life purpose? This has been a "growing-up" question for youth from the start of human leisure onward. Tycho Brahe is in many ways a prototype of the spoiled, sophisticated, suburban youth of our day. Fortunately, psychoanalysts were not available in his day or he might have become successful in the family business—thus making the world poorer.

Brahe was born into the Danish nobility but never quite fit in. His parents let a rich uncle raise him based on a promise made before his birth and after an eye-opening inspection of the uncle's bank account. As an adolescent, Tycho seemed to be without purpose. He lived the good life that most adolescents dreamed of but which only nobility could afford. Today, you might picture him as supporting his own motorcycle gang! As a student, he fought a duel—and lost! Unlike most duels (then or now), the issue did not seem to be a member of the opposite sex. The quarrel (according to one source) concerned who was the better mathematician. Tycho might have been sensitive on this point even at this early age. Anyway, during the duel part of Tycho's nose was sliced off. This is just the sort of thing a young man needs in the days before "miracle surgery" to gain self-confidence. (Tycho kept himself supplied with artificial noses for the rest of his life.) To put it mildly, Tycho was the black sheep of the Brahe family. The family, at one point, was all set to ship him off to manage their trading right in Goa, halfway around the world in India. While due south of Siberia, and having a much warmer climate, this would have served the same geographical purpose—exile.

Several things happened to completely change his life. (I hesitate to say "straighten out," since Tycho was strange all of his life.) One item was a book on mathematics; another was the appearance of a nova or "new star" in the sky. Tycho developed a passion for astronomy, wrote the definitive paper on the nova (which is still called "Tycho's star"), and established a reputation as a budding scientist. This did not all happen at once, but it was the set of events which shaped Tycho's career.

Now, back in these darker ages, there was widespread belief in a form of ancient Babylonian mysticism called astrology. Today, there would be an uproar if we had a Department of Fortune-Telling in the Government (although both economists and a weather bureau are grudgingly tolerated). In Tycho's day, it was one area where funding from the state was likely to occur for a scientist.*

Tycho, his reputation built on the patient and meticulous observation of his nova, became known as an astronomer. More importantly, he became known as a skilled observer. To make a long story shorter, he was granted funds by the king of Denmark to build the Western world's first astronomical observatory. This was before the day of the telescope, yet the building, Uraniborg on the island of Hveen, looked somewhat like a modern observatory. Better observing meant better tabulated positions for the planets.

Tycho was the world's best naked-eye

* The last Western "state astrologer" was supported by Adolf Hitler. I trust you know how well that came out.

astronomer. He had read Copernicus's book and was convinced it was wrong. But, almost alone, he realized the correct way to resolve the Copernicus-vs.-Ptolemy argument was through *better data*! This was a novel idea and is Brahe's claim to fame. Given more precise data, which system would break down? Brahe built the best instruments for measuring positions in the sky that the world had yet produced. His instruments were essentially protractors, but they were big, solid, finely divided protractors! The concept of more precision and better data in science was almost lacking before Brahe's time. If one looks at some instruments of the period, the artwork is more finely and precisely engraved than the scales from which the measurements were taken. Not Brahe's instruments! Tycho Brahe achieved an accuracy of observational measurement which is at—or beyond—the resolution of the human eye. He was an observational fanatic! Let a story suffice: When the king would come to the island for a visit, Brahe would sometimes snub him; he would come to dinner in his "painting clothes" and ignore almost everyone. Yet, when darkness would approach, he would excuse himself. He would reappear in the finest of embroidered silks ready for his night's observing. For mere mortals, Brahe made few concessions, but for "communication with the heavens" nothing but the best was sufficient.

The king was tolerant of genius but his son was not. It is not surprising, therefore, that when the king died, Brahe was encouraged to find a new home. He found a patron in Rudolph II, the Holy Roman Emperor. Rudolph wanted Brahe to produce the most precise tables of planetary positions yet made for his astrologers to work from. Brahe was more interested in the nature of the solar system than in the Rudolphine tables, but he agreed and was installed in a castle on a hill outside of Prague as the imperial astrologer. It is great fun to speculate on what part of our mental image of a wizard comes from this weird creature in his embroidered silks and silver nose roaming the battlements of the castle on the hill at night and carrying the title "imperial astrologer." Brahe would have loved it!

While in no hurry to finish the Rudolphine tables—and thus end his reason for support—Brahe was becoming disturbed by his work. It was becoming clear that although he possessed the best observational data of planetary positions ever collected in the history of the world, he was not enough of a mathematician to make sense out of them. Frustrated, he began to seek out a mathematical genius to help him in his work and, if we have any glimpse into his character, hated every minute of having to lose his nose twice!

Johannes Kepler

Johannes Kepler: Mystic to "Mathematicus"

It is hard to imagine a more different background from Tycho Brahe's than that of Johannes Kepler. While Brahe was highborn and aristocratic, Kepler was the son of an innkeeper's daughter and was concerned with his own legitimacy. While Brahe was a misfit in his family, Kepler came from a long line of misfits. The role was natural, and he gave it new stature. Brahe believed in an earth-centered universe, while Kepler became fond of the ideas of Copernicus when he first read them. Brahe was Catholic and Kepler was Protestant in a period when religious wars were politically about to tear Europe to shreds. Brahe performed feats of eyesight denied the average human being but was only a fair mathematician. Kepler was a mathematical genius, interested in astronomy, but so nearsighted that in later life he was lucky to find the moon in the night sky—much less become one of the great figures in the history of astronomy.

Both were unusual! Both were lucky to be led into their life's work. Both worked like demons at what they did. While Brahe was volatile and unpredictable except in his work, Kepler was disgustingly premediated in most of his life's activities—to the point of generating "checklists of pros and cons" in the process of selecting a suitable wife from a large field.* Like Brahe, Kepler was always in a fight in his youth. Unlike Brahe, who lost only one important fight, Kepler seems to have lost them all!

After an unusual education, Kepler became a college graduate looking for his first job. He found one in the town of Gratz. Gratz was small but prosperous and proud of its tiny university. Kepler was hired in a dual role. He was the town astrologer who cast the official horoscope for the town fathers, and he taught mathematics at the university. In the second role, Kepler was an utter failure. Students could not understand him. No student lasted through the first class, and he never taught again!

In the second role, Kepler had the habit of giving relatively specific horoscopes—sudden death to the nonlucky! History was fond of Kepler. Action on one "long-range forecast" made the town fathers rich and Kepler famous. Kepler also published several books at this stage in his life, all of which were rich in mysticism. *Kepler was a mystic.* He was always looking for a strange, significant pattern in things. And he was bright enough to find one even if it was not really there! An early obsession was the relationship of the spacing of planetary orbits to the regular solids. The order is not there, but it makes a lovely model. Later, his idea of the solar system led him to write music. What holds the planets up? They are like lanterns attached to spheres of purest crystal, rotating around the sun (he was a Copernican!). When crystal rubs against something, what does it do? It rings out in pure tones! Kepler wrote the "Music of the Spheres"—notes with which all of the spheres ring out while grinding against each other to keep the solar system functioning. Clearly, Kepler was a mystic!

Kepler stands out in human history because of several acts of great intellectual honesty which define the correct procedure in dealing with data for modern science. It is interesting that these acts are buried in the many volumes of nonsense—be it ever so much fun to read—which Kepler generated.

Following his association with Brahe, Kepler was named "Imperial Mathematician" in an effort to keep alive the Rudolphine tables. Kepler did succeed in producing the Rudolphine tables, which were based on the data of Brahe. He also used that data to his own advantage. Aside from an interlude to defend his mother in her trial for witchcraft, Kepler managed to find in Brahe's data the pattern for a new understanding of the heavens.

Sodium and Water—Brahe and Kepler

The association of Tycho Brahe and Johannes Kepler is one of the most interesting in the history of science. Working teams frequently run into problems—from Gilbert and Sullivan to Sonny and Cher! The Brahe-Kepler collaboration seemed doomed from the start.

Right down to the stump of the missing nose, the aristocractic Dane's fierce pride must have been seared at the realization that he had to seek a better mathematician to help him with his data. No matter, the mathematician came to him. Kepler was on hand, in anticipation of the kill. Kepler was all that Brahe lacked, and needed.

Obsessed with finding order in the heavens, Kepler knew he needed Brahe's data.

* It should not be implied that Kepler was a very adequate husband. He frequently neglected his family for long periods of time and saw his wife only in passing. Passing was sufficient to generate 12 children by two wives!

Nearsighted astronomers do not easily obtain their own data. Kepler combined an almost naïve belief in his own ability with ferocious persistency. (Kepler even made a bet he would crack the orbit of Mars in 8 days. It took him 8 years, but he stuck with it.) This self-confidence must have eaten away at Brahe. Each man needed the other, and resented it. They were the best of friends—they were the worst of enemies! Torn with resentment, Brahe began the systematic mental torture of Kepler. Kepler had been assigned the problem of discovering the true orbit of Mars. Brahe, however, would not let him have free access to the observational records of the planet's positions. Picture the scene in Benatek Castle. It is dinner time in the great hall. At the head table sits Brahe with visiting dignitaries. Seated farther away are the Brahe children, cousins, nephews, aunts—the whole horde of relatives which orbited Brahe. At the end of the hall are lesser mortals. Late in the meal, Brahe rises for a toast and the throng grows quiet. "Here's to Mars, noble mover of the sky, who on the night of April 14, 1584 was located at. . . ." At this point Brahe rattles off the precise position of Mars in the sky for that date. From somewhere down at the end of the hall, among the lab assistants and instrument polishers, a choking sound is heard. Then there is a stir as an awkward Kepler runs from the hall to write down the data before he forgets it. This (with only slight dramatization) was one example of how Brahe released the data to Kepler in bits and pieces. It is small wonder that the two of them argued fiercely and frequently. Particularly violent arguments usually ended with Kepler packing his bags and leaving. After a week or so, however, the horrible truth settled upon both astronomers. They needed each other! They were the perfect match of capabilities to decode the universe—and they hated it! Fortunately for us, they always did get back together, and Kepler was present at Brahe's death.

Kepler let the Brahe relatives mourn while he gathered up all the records Tycho had denied him. Soon enough the Brahes realized that the data which the emperor needed to produce his tables had been taken by Kepler. Kepler was hounded by lawsuits in an attempt to get the data back. In a politically fragmented Europe, the Protestant Kepler (who was also Imperial Mathematicus to the Holy Roman Emperor) moved from one jurisdiction to another to rid himself of Brahe's relatives. Eventually, they tired of the pursuit, and with the publication of the Rudolphine tables the usefulness of the data became scientific and not monetary.

Piecing it Together—Kepler and a New Astronomy

Although Brahe had a treasure trove of data, he jealously refused to publish it. He earnestly desired to be the one to discover the true order of the heavens. Toward the end of his life, he took on a mathematical assistant, Johannes Kepler, who had been waiting for the opportunity to work with Brahe's data. After Brahe's death, Kepler plunged into the problem of astronomical motion with access to all of Brahe's observations. Kepler spent years in calculating the best "fit" he could obtain for the orbit of Mars. Finally, he arrived at a discrepancy between Brahe's observations and the predicted position of Mars of 8 minutes of arc (8′). This was less than the inaccuracies of observation that Ptolemy or Copernicus had allowed themselves. On the other hand, it was twice the error in position to be expected from Brahe's data. Now Kepler gives us no reason to believe from his early work that he will let a few facts get in the way of an idea that he likes. To Kepler's immense credit, he immediately scrapped 5 years of calculations and started over, stating:

But for us, who by divine kindness were given an accurate observer such as Tycho Brahe, for us it is fitting that we should acknowledge this divine gift and put it to use. . . . If I had believed that we could ignore these eight, I would have patched up my hypothesis accordingly. But since it was not permissible to ignore them, those eight minutes point the road to a complete reformation of astronomy.

In one glorious moment, Kepler had thrown off his ties to the Middle Ages. In an act of incredible intellectual courage, he had required the theory to fit the facts; it could not, and he had discarded it. But what now? He had spent 5 years polishing and correcting the Copernican treatment of the orbit of Mars. Every reasonable correction had been made. The best values of circular speeds, epicycle sizes—everything—had been worked out. The Copernican System would not work! What now? Back to Ptolemy? Hardly! Kepler's improvements of the Copernican System yielded an accuracy far beyond what had been attained with an earth-centered system, yet its predictions were not quite in accord with the results of observations. Nevertheless, most scientists would have taken what he had and proudly published it. Kepler discarded the system and started over.

Something was clearly wrong, but what? Kepler decided to reexamine the shapes of the orbits. From the beginnings of astronomy as they rose out of the misty lore of astrology, planetary orbits had been assumed to be circles. Kepler was to find otherwise. Working from the data of Brahe, two relationships emerged which we now call Kepler's first and second laws.

Kepler I: *The orbits of planets are ellipses with the sun located at one focus.*

An ellipse is a "flattened" circle. Actually, the circle is a special geometrical case of the ellipse. A *circle* can be defined as the location of all points in a plane equidistant from a given point called the center. The definition of the ellipse is similar. An *ellipse* is the location of all points in a plane equidistant from two points called *foci* (singular: *focus*). An ellipse is shown in Fig. 5-11. The points f_1 and f_2 are the two foci of the ellipse. Point P_1 is a point on the ellipse. The distance d_1 from f_1 to P_1, plus the distance d_2 from f_2 to P_1, add up to some number we will call D. Thus $d_1 + d_2 = D$ (a constant). This means that if we take another point on the ellipse, P_2, the distances to each focus will be different, d_3 and d_4. But, $d_3 + d_4 = D$, the same D as before. The sum of the two distances to the foci is the same for any point on the ellipse. If the two foci are far apart for a given D, a long, flat ellipse results. As the foci move closer together, the ellipse gets "rounder" and more nearly circular. If the two foci come together, a circle results (Fig. 5-12).

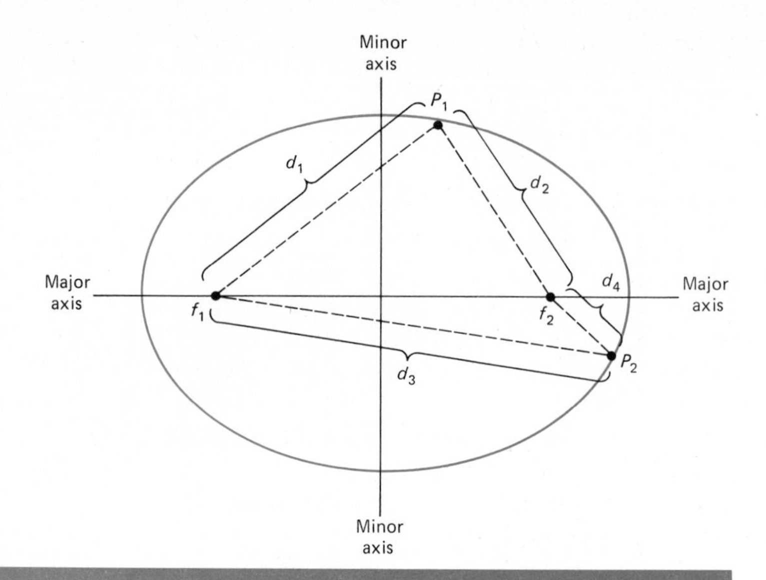

FIGURE 5-11
In an ellipse all points on the figure are a given distance from two points called foci, labeled f_1 and f_2. Point P_1 is a distance d_1 from the first focus and d_2 from the second focus. For P_2 to be a point on the ellipse, its distances d_3 and d_4 must add up to the same length as $d_1 + d_2$.

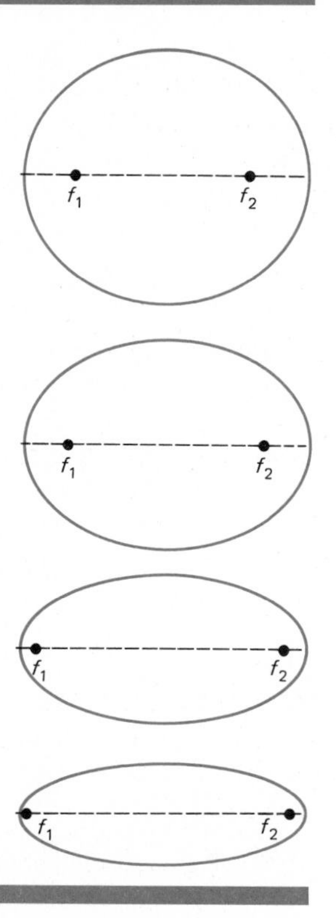

FIGURE 5-12
As the two foci of the ellipse get farther and farther apart, the shape of the ellipse becomes more flattened or eccentric. The orbits of the planets are less eccentric than even the first ellipse drawn here. In fact, a scale drawing of the orbit may appear so circular to the eye that a measurement is required to demonstrate that the orbit is not a circle.

The ellipse is usually described in terms of the greatest length across it (through the two foci), the *major axis,* and the shortest distance across it, the *minor axis.* The ratio of the distance between the foci to the length of the major axis is called the *eccentricity* of the ellipse, *e*. The circle is an ellipse with $e = 0$. For the planetary orbits, the fit of the data to ellipses was as precise as Brahe's data. However, the ellipse presented a problem because it was clear that an object such as Mars traveled on the ellipse at variable speed. Thus was eliminated the second assumption of both systems of the universe: uniform speed. Kepler found a way of describing the speed of a planet in its orbit in his second law:

Kepler II: *An imaginary line joining a planet to the sun sweeps out equal areas of space in equal periods of time (the equal-area law).*

To see what the second law implies consider Fig. 5-13. If the planet is at P_1, the line connecting this position to the sun is shown by the line SP_1. As the planet moves, this line moves with the planet. At some time later, say 1 month, the planet is at P_2. The sun-planet line has swept through the shaded area indicated. If we pick another position in the orbit, P_3, how far will the planet have moved 1 month later? Kepler's second law indicates that it will have moved from P_3 to a new point P_4 such that the shaded area SP_3P_4 just equals the previous shaded area SP_1P_2. Thus, area 1 = area 2 in the diagram, provided that the *time* between P_1 and P_2 is the same as the *time* between P_3 and P_4. Note that this implies that the farther a planet is from the sun, the more slowly it moves in its orbit.

We should point out that these results left Kepler miserable! He had devoted all of his life to finding beautiful and mysterious order in the universe. He sought to reveal the mystic order of symmetry and grand pattern in everything. Consider the titles of two of his books—*Mysterium Cosmigraphicum* (*The Cosmographic Mystery*) or *De Harmonice Munde* (*On the Harmony of the World*). The truth of the matter is that much of what Kepler published in his life was wild and speculative, based on imagined glimpses of grand design. It tends to be interesting nonsense, because Kepler is always talking to the reader and shares all of his efforts—even his failures. But now he had demolished two of the elements of symmetry agreed on since the ancient world—circular orbits and constant-speed motion. In their place, he had substituted the oblong of the ellipse and a means of determining how the speed of a planet varied with position in its orbit. The fact that he was correct was small consolation for the geometrically pleasing ideas he had dethroned; he

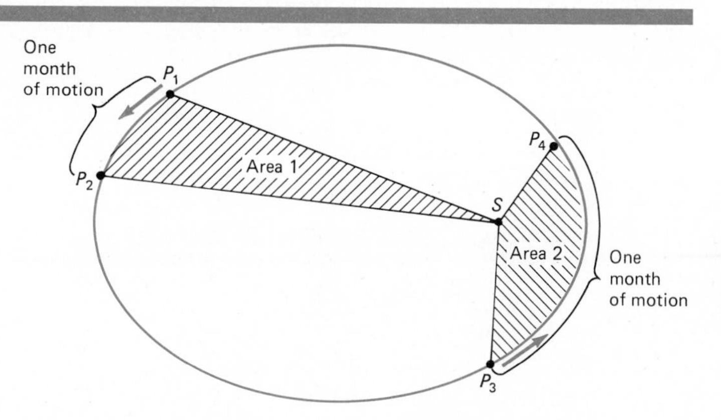

FIGURE 5-13
Since each planet continually varies in its speed along its orbit, predicting its position at some time in the future can be difficult. Kepler summarized the variation in speed in his equal-area law. If the time for the planet to go from P_1 to P_2 is the same as the time it takes to go from P_3 to P_4, then area 1 will equal area 2 in the diagram.

was miserable! He resolved to find some beautiful, new element of order in Brahe's data or die trying. He found it!

Kepler sought a relationship between a planet's distance R from the sun and the time it takes the planet to complete an orbit, the sidereal period T. Of course, he had to determine the radius of the orbits and find the sidereal period T for each planet. The only period of a planet which we may observe from the earth is its synodic period T_s. The time required for a planet to go 360° around the sun with respect to the stars cannot be measured directly by an earth-based observer. After much work and some trial and error, Kepler emerged with an empirical law. This law was important psychologically to Kepler because it revealed the order he was seeking. In addition, the law is important to science because it is a precise, mathematically expressed statement, relatively easy to confirm or reject. This law is, in many ways, the prototype of what science later became.

Kepler III: The Harmonic Law *The ratio of the cube of the radius of a planet's orbit to the square of its period is a constant, or* $R^3/T^2 = k$.

For example, the earth takes 1 year to go around the sun, and its distance from the sun is defined as 1 astronomical unit (AU). Measuring periods in years and distances in astronomical units, we obtain for the earth:

$$\frac{R^3}{T^2} = \frac{(1\ \text{AU})^3}{(1\ \text{year})^2}$$

$$= 1\,\frac{\text{AU}^3}{\text{year}^2}$$

$$= k \qquad \text{(a constant)}$$

Mars is 1.52 AU from the sun. Mars' distance from the sun cubed divided by its period (in earth years) squared must be the same constant, $1\,\frac{\text{AU}^3}{\text{year}^2}$. Thus, we have

$$\frac{R^3}{T^2} = 1\,\frac{\text{AU}^3}{\text{year}^2} = \frac{(1.52\ \text{AU})^3}{T^2}$$

Taking the square root of both sides, we get

$$T = 1.874\ \text{years} = 648\ \text{days}$$

which is the period of Mars. Thus:

$$\frac{R^3_{earth}}{T^2_{earth}} = \frac{R^3_{Mars}}{T^2_{Mars}}$$

and this equals R^3/T^2 for any other planet so long as T is in earth years and R is in astronomical units.

It should be pointed out that Kepler had no idea as to what his three laws suggested or meant. They were *empirical laws* derived from observational data. They described reality and showed great order, but they did not reveal to Kepler a grand scheme underlying nature. It is ironic that the seeker of a master design such as Kepler should become the great empiricist in our story, but not the master theorist.

The revolution in astronomy is now essentially complete. Regardless of how the facts were interpreted, the Brahe-Kepler system of elliptic orbits provided a complete description of astronomical motion. When Kepler published the Rudolphine tables and astronomers compared predictions using these tables, their resistance gradually faded away. This was not so among the general public. The Copernican revolution was a battle of new ideas against traditional modes of thought. It was bigger than astronomy, and no historical account would be complete without discussing the main line of resistance—the way nonastronomers pictured the universe.

Galileo and the Struggle for the Popular Mind

The main figure in the battle for the popular mind in the Copernican revolution was Galileo Galilei. While the struggle for the professional opinion of astronomers was nearing completion in Kepler, the same struggle for the minds of nonastronomers had barely begun. Few educated men were aware that a dry text in Latin, written in great detail and full of calculations, was about to change their world. Galileo lived at the same time as Kepler, and some of their correspondence still exists. However, they led separate arms of the movement toward acceptance of a sun-centered universe. Most Protestant groups had denounced *De Revolutionibus* by this time, as could be anticipated from their insistence on a literal interpretation of the Bible. The Roman Catholic Church had not. It had a tradition of tolerating relatively radical views among scholars. However, one point should be noted. No matter how much the Protestant clergy disagreed with the Copernican views, it lacked the political power to suppress unpopular ideas. The Catholic Church did not. With the Inquisition, it held the apparatus to secure silence and conformity.

Galileo was a Copernican, although not publicly so until 1609. While he did not invent the telescope, he was the first person to use it to observe the heavens. The excitement of what he found still comes through in the pages of his book *The Starry Messenger.* (Furthermore, this book was written in Italian for the masses to read—not in the scholarly Latin.) What he observed also dealt a severe blow to the Ptolemaic System among the educated populace but did little to directly confirm the Copernican System. Some of the observations Galileo made with his telescope were as follows:

1 The moon had mountain ranges and flat plains. Its surface was highly etched and textured like the earth's. (This was hardly a "perfect" object in a sphere of perfection as pre-Copernicans believed.)

2 The sun had spots on it. Furthermore, the motion of the spots indicated that the sun rotated on its axis in a little less than a month. (This was a motion *not* around the earth and was a blow to the conservatives.)

FIGURE 5-14 Photograph of the constellation Orion taken with a wide-angle lens with a 10-in focal length. Even though this is one of the most frequently viewed areas of the sky, not one out of a hundred of the stars seen here is visible to the naked eye. A small telescope would render thousands of additional stars visible in the same area. Sirius is just at the edge of the photograph at the lower left. The bright star at lower right-center is Rigel. Orion is labeled on the star map of Fig. 4-14. (*Photograph courtesy of the Hale Observatories.*)

3 There were thousands and thousands of stars too faint to be seen by the naked eye (Fig. 5-14). The bright band in the sky called the Milky Way was composed of innumerable faint stars so near to each other that the entire sky in their area glowed. (This was also a setback for the conservatives! If these objects were so faint they could not be seen, they could not have been made for humanity—as all the heavenly objects were supposed to have been. Giordano Bruno had been burned at the stake in 1600 for maintaining this and other unpopular ideas.)

4 Even through the telescope, the brightest of the stars were pinpoints and not little spheres. While moon, sun, and planets grew in size when observed through the telescope, stars did not (and they do not grow in size even when observed with the largest telescopes today—they are essentially geometric points). This argued that the stars could be much farther than opinion had held up to that time. The argument that there was no parallax of the stars, often used against the idea of a moving earth, was weakened.

5 Jupiter had moons! Four bright, starlike objects could be observed near the planet. Their motions could be easily explained by postulating that they orbited Jupiter. (In the Ptolemaic System, all motions are around the earth. Theologists argued that motion not around the earth was pointless and therefore could not exist.)

6 Venus showed phases like the moon as it moved in its orbit. (This constituted a serious objection to the Ptolemaic System. There is no way to observe a "full Venus" phase unless it moves around the sun.)

In *The Starry Messenger,* Galileo simply told of what he had observed in the heavens, but none of it was pro-Ptolemaic. It is also true that none of it was conclusive evidence for the Copernican System. But the wedge in the popular mind had been driven deeply. The "sphere of perfection" was dead. Some scholars offered more resistance and even refused to look through Galileo's telescope. Galileo propagandized far and wide about the wonders of the heavens (and sold quite a few telescopes on the side). What happened from this point is the subject of libraries of commentary. In case your education is incomplete in this celebrated controversy, here is a "bare-bones" account of the story.

Galileo was warned by church authorities not to be too forward in advocating a moving earth. Knowing the unhappy fate of individuals defying the Church, Galileo went on to other work for some time. However, he eventually published his *Dialogue on the Great World Systems.* In it, we find a discussion between individuals holding opposite opinions. Galileo's bias is apparent in that the Copernican is called Salviati (wise) and the Ptolemaist Simplicio (fool). Of course, Salviati wins most of the points. Galileo was brought before the Inquisition. Some scholars maintain that he was framed by false documents. Others claim that he was unwise enough to put statements and questions asked him by the Pope years earlier in the mouth of Simplicio. At any rate, he was found guilty of spreading heresy and forced to publicly confess and state that the Copernican ideas were wrong. Legend has it that on the way out he mumbled "and yet it (the earth) moves." He was sentenced to spend the rest of his life under house arrest and to refrain from publishing. He went on to continue his studies on motion, his most important scientific work. His book on motion was published outside of Roman Catholic Europe after his death. Had he been burned at the stake, like Bruno, we might have lost his most valuable contribution to the scientific revolution. In this, and in the total judgment of history, Galileo had the last laugh. He was granted "unconditional amnesty" by the Church in 1953, over 300 years after his condemnation.

And Yet it Moves— The Physical Evidence for Motion of the Earth

Initially the Copernican System was regarded as a convenient way of calculating planetary positions. Later it became regarded as truly representing the structure of the solar system. "Would not a rapidly spinning earth fly apart? Would not objects be swept off its surface? Would not the air be left behind?" Questions such as these were reexamined and found to be inconclusive as objections to a moving earth. The physics of Aristotle did not allow a decision once it was suspected of being inadequate. But how are we to know that the earth does indeed move? Most of the proof awaited the further development of the work begun by Tycho Brahe: precision measurement. It took almost 2 centuries from the death of Galileo in 1642 for astronomers to find the direct proof sought since ancient times, the parallax of stars. At the time of this discovery in 1836, the issue of a moving earth had long since been resolved by other evidence.

By 1672, it had been observed that pendulum clocks ran slower near the equator and faster toward the poles. The difference between pole and equator could be accounted for by assuming objects at the equator weighed about one-half percent less than at the poles. This agreed well with the prediction of Newton's laws (in the next chapter) for a spinning earth. Thus, the earth spins on its axis.

At about the same time (1675), the astronomer Olaus Römer obtained the first determination of the speed of light. In observing and timing the eclipses of the moons of Jupiter, he was disturbed by the fact that the predicted time of these eclipses did not agree with their observed times. It occurred to him that if the earth were moving in its orbit, it would sometimes be much farther from Jupiter than at other times (Fig. 5-15). Moreover, the delay in observing the eclipse agreed exactly with the increased distance of the earth from Jupiter for other points in the earth's orbit. Thus, the evidence confirmed that the earth was moving around the sun.

The discovery of stellar aberration by James Bradley in 1727 was a conclusive milestone in establishing the physical motion of the earth through space. Bradley was looking for stellar parallax to confirm the motion of the earth. To detect this small, seasonal shift in the positions of nearer stars due to the earth's motion was a very difficult problem. Today we can photograph a portion of the sky with a telescope and then compare the recorded positions on the photograph with a similar photograph taken 6 months earlier. A persistent back-and-forth movement in a series of such photographs is proof of parallax. Before photography, the detection of this small motion had eluded observers since ancient Greece. Bradley set about using the most precise instrument of his day—the pendulum clock. He carefully measured the time at which certain stars were directly overhead. To ensure that he was looking straight up, his telescope used the reflection of the sky in a pool of liquid mercury. Bradley reasoned that if a particular star *transited,* or crossed directly overhead, slightly earlier and then slightly later than nearby stars at 6-month intervals, it would be a confirmation of parallax. Instead, he found a totally different and completely unsuspected phenomenon. *All* stars which he observed shifted in transit time! The phenomenon was named the *aberration of starlight* (Fig. 5-16). If you are running through a rainstorm, you must tilt your umbrella forward to keep from getting wet.

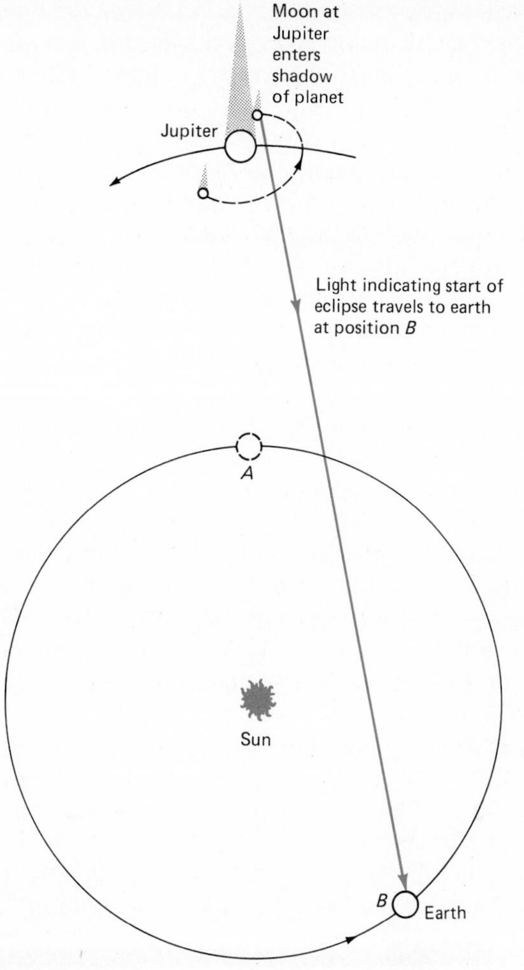

FIGURE 5-15 The first determination of the speed of light was completed in 1675 by Olaus Römer and depended on the fact that the earth moved around the sun. Jupiter has several large moons which frequently enter the planet's shadow in an eclipse. If the time of the eclipse was calculated when the earth was at position *A*, closest to the eclipse, the actual eclipse will appear to occur "late" to an observer on earth. The additional time is the time for light to travel across the earth's orbit and reach position *B*.

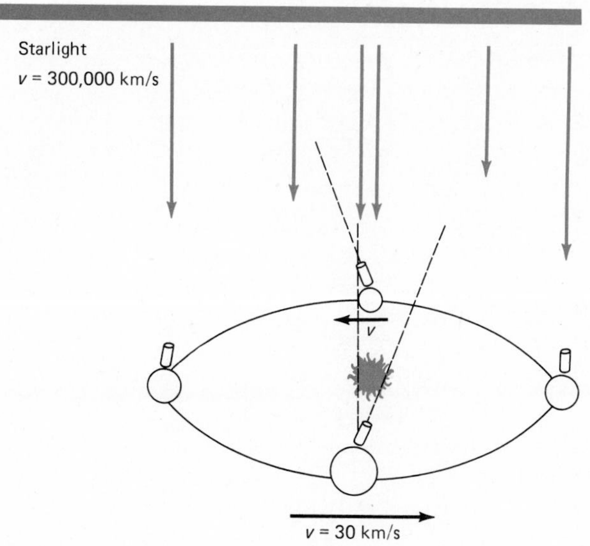

FIGURE 5-16 The aberration of starlight. Telescopes on the earth must be tilted slightly in the direction of the earth's motion to center the image of a star perpendicular to the plane of the earth's orbit. Six months later the earth is moving in the opposite direction, the telescope must be tilted that way, and the star seems to have shifted its position. This effect is much larger in size than that of parallax.

Your motion makes it appear as though the raindrops are coming at you at a slant, although they may be falling straight downward. Aberration is similar in that we must tilt our telescopes slightly in the direction of the earth's motion to have starlight come straight down the tubes. This "slanting" of starlight is a proof of our motion through space and was discovered by Bradley a half-century before the American Revolution.

Most of the strong resistance to a moving earth among academics collapsed with the announcement of the discovery of stellar parallax in 1836 by Friedrich Bessel, and later by Wilhelm Struve. This was the effect sought for over 2000 years to confirm an orbiting earth.

The conclusive experimental confirmation of the rotation of the earth was provided in 1851 by the French physicist Jean Foucault. He hung a long pendulum from the dome of the Pantheon in Paris. As the pendulum swung back and forth, Foucault observed that it slowly changed its plane of swing. Foucault suggested, however, that the pendulum was maintaining its plane of swing while the earth rotated beaneath it. This effect is easy to describe for a pendulum at the pole, but is more complex for other locations (Fig. 5-17). Picture a pendulum at the North Pole swinging back and forth so that at each swing the bob is moving toward star *A* in space. As time goes on, the earth will rotate under the pendulum, which will continue to swing toward star *A* each swing. Unless there is a twisting force in the mounting tending to pull the pendulum around with the earth, it will continue to swing in a plane containing star *A*. To an observer at the pole rotating to the east with the earth, it will appear that the plane of swing is moving to the west at the rate of 15°/h. Actually, the pendulum is swinging in a constant plane in space while the earth moves beneath it. The conclusive nature of Foucault's experiment explains why most science museums (and many college science buildings) today include a shaft for a Foucault pendulum.

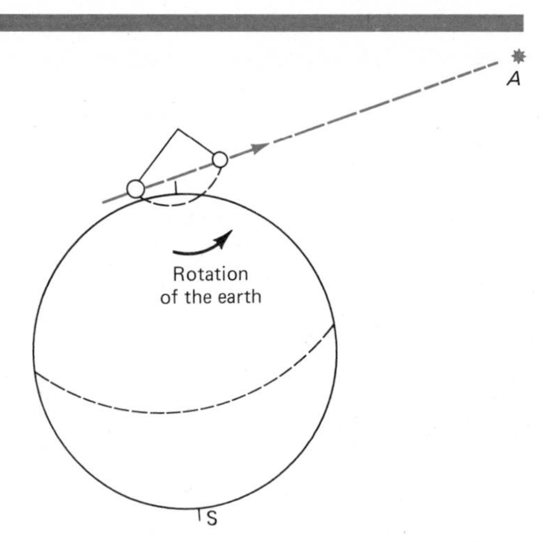

FIGURE 5-17 The pendulum at the North Pole swings so that its motion "points" at star *A*. As the earth turns toward the east under the pendulum, it will maintain its plane of swing and still point at star *A*. To an observer rotating with the earth it will appear that the pendulum swing is slowly twisting to the west.

Summary

We have now seen the main cast of characters for the Copernican revolution. Each played an important role in the development of our ideas.

Claudius Ptolemy (around 150 A.D.) Put together a mathematical system for predicting the motions of objects in the heavens, a system based on uniform circular motions around a fixed earth.

Nicolaus Copernicus (1473–1543) Published a modern sun-centered system of motions.

Tycho Brahe (1546–1601) Realized that better ability to predict in astronomy (and determination of what the solar system was really like) depended upon greater precision in the data available. He reformed naked-eye observation and collected the most accurate positional data ever taken by an astronomer prior to the invention of the telescope.

Johannes Kepler (1572–1630) Worked from the data of Brahe and eventually showed that while the system was arranged essentially as Copernicus had suggested, it differed in every detail of motion.

Galileo Galilei (1564–1642) Popularized the dispute between the rival systems and added the weight of telescopic observations to hasten the end of the Ptolemaic System.

*Problems

This set allows you to see some of the techniques of Copernicus and Kepler in working out the basics of orbits.

1 An old merry-go-round is made of two rings of horses that rotate in the same direction but at different speeds. The inner ring, which turns faster, rotates once in 60 s. If your horse on the inner ring and a particular horse on the outer ring come opposite each other every 90 s, can you figure out how long it takes the outer ring to go once around?

2 What exactly does the previous problem have to do with astronomy?

3 Copernicus calculated the sidereal period of the planets (how long they take to go 360° around the sun in space) from their synodic period (how long it is between line-ups of the sun, the earth, and the planet). Why could he not determine the sidereal period, or year, directly for the planets?

4 Copernicus used the formula $1/T_{syn} = 1/T_{inner} - 1/T_{outer}$ to calculate sidereal periods; where T_{syn} is the synodic period of the two planets, and T_{inner} and T_{outer} are the sidereal periods of the planets closer to the sun and farther from the sun, respectively. Suppose the earth ($T = 1$ year) lined up with a planet farther out from the sun every 3 years. What would the sidereal period of this outer planet be?

5 Can you use the formula of Copernicus in Prob. 4 to solve Prob. 1?

6 Observationally, how can you determine the synodic period of a planet with respect to the earth?

7 From ancient Greece onward, orbits were determined by assuming a shape and tinkering with speeds until the best fit was obtained. Kepler proceeded differently by determining the shape of orbits point by point from data. Look out of your window at night and select a distant streetlight. Can you determine its direction from your window? Can you determine its distance?

8 Suppose you had a large map of your campus and could determine the position of your window on the drawing of your dormitory.

(*a*) Measuring with a protractor, could you mark a line on the map along which the street lamp must be? Would this tell you where the lamp was?

(*b*) Suppose now that you went 10 rooms down the hall and sighted the same lamp from another dormitory window. (Most dorms are long and thin for economy of construction. If yours is not, pretend that it is.) Could you determine a line of sight from this window to the lamp and mark it on the campus map?

(*c*) Now you know the position in space where the street lamp is located? Where?

(*d*) How could this be applied to determining the orbit of a planet?

9 The earth is too small for two simultaneous lines of sight from different positions on earth to pin down the distance of a planet. Kepler had to use a slightly more complicated scheme. When will Mars be in the same position in space again as it is right at this moment? At that time, will the earth be in the same position in space as it is right now?

10 How can we tell approximately where the earth is in its orbit? Through what angle, as seen from the sun, does the earth move each day?

11 The sidereal period (year) of Mars is 687 days. How many days short of 2 earth years is this? Roughly how many degrees along the earth's orbit would separate two lines of sight from the earth to Mars the first time and 1 Martian year later?

12 (Project: You will need a compass, ruler, protractor, and a fine-point pencil.) Draw a circle of radius 5 cm to represent the earth's orbit. The earth is moving counterclockwise in this circle as we look down on its North Pole. Draw an arrow on the diagram representing this motion. The center of the orbit is, of course, the sun. Let the long axis of the paper to the right be the direction in space to the vernal equinox point (VE) in the heavens (astronomers use this direction as their zero line for angle measurement). Draw a line from the sun in the direction of VE. At what date is the earth in the position in its orbit on the line from the sun to the vernal equinox point? (Think before you answer this. The question is not, "At what date is the sun at the VE as seen from the earth?")

(*a*) On October 14 of a particular year Mars is observed in the heavens. Where is the earth in its orbit on October 14? (How many degrees past where the line from the sun to the VE crosses the earth's orbit?)

(*b*) Measure off this angle on your diagram and mark the position of the earth in its orbit carefully with a fine mark. Label its postion E_1.

(*c*) Which direction is it from position E_1 to the VE? Mark this direction with a fine dashed line from E. (Remember that the VE is infinitely far away in the heavens.)

(*d*) On this date Mars is observed to be at an angle of 27° from the VE, back in the direction from which the earth has come. Measure this angle at E_1 from the dashed VE line and draw a fine line several centimeters from E_1 in the direction Mars was sighted. Mars is somewhere along this line, but we do not know where yet.

(*e*) How long will it take Mars to go once around in its orbit and be back in the position in space we sighted it on October 14?

(*f*) Almost 2 years later, on September 1, we again take a sighting of Mars. Determine the position of the earth in its orbit for a September 1 sighting and mark this position E_2. Draw a dashed line from this position toward the VE.

(*g*) Mars is seen to be some 39° from the VE and ahead of it in space (in the direction of the earth's motion from the VE). Mark off the line of sight from E_2 to Mars.

(*h*) Since our two sightings from E_1 and

E_2 are taken 687 days, or exactly 1 Martian year, apart, what do we know about the point where these two lines of sight intersect?

(*i*) Mark this intersection point *M* on your diagram. You have determined from data one point in the orbit of Mars. By working in a similar fashion you could determine hundreds of points in the orbit to obtain its size and shape, as did Kepler. On your diagram measure the radius of the earth's orbit and the distance from the sun to Mars. How many times as far from the sun as the earth is Mars? Would this method of constructing orbits allow you to obtain the relative spacing of the planets? Would it allow you to determine the distance to Mars?

13 You now have the methods used by Copernicus and Kepler to determine the sidereal periods of the planets *T* (Prob. 4), and their distances from the sun *R* (Prob. 12). These values for the planets known to Kepler are given below. With a caculator, or by graphing, verify that Kepler's harmonic law works.

	SIDEREAL PERIOD *T*, years	**ORBITAL RADIUS *R*, AU**
Mercury	0.241	0.387
Venus	0.615	0.723
Earth	1.00	1.00
Mars	1.88	1.52
Jupiter	11.9	5.20
Saturn	29.5	9.55

14 Uranus is 19.2 AU from the sun. What should the sidereal period of Uranus be?

15 Neptune has a sidereal period of 165 years. How far should Neptune be from the sun?

16 Venus has a sidereal period of 225 days. What would the synodic period of the earth and Venus be? (Be careful of "inner" and "outer" here.)

CHAPTER SIX

Newton and a Science of Motion

Introduction Human beings have tried to determine the nature and causes of motion since ancient Greece and beyond. With the relationships found by Kepler from Brahe's data the stage was set for a resolution of this quest. The Keplerian description of astronomical motions worked. Kepler's three laws yielded predictions as good as the observations supplied for the calculations. This agreement improved with the rapid expansion of the use of the telescope in astronomy. Using the telescope as an ultrafine pointer to determine position in the sky, the predictions of positions calculated by the Ptolemaic System or the original system of Copernicus collapsed. Kepler's laws, however, continued to produce results confirmed by careful observation. For example, when an apparent error in prediction was discovered, Römer found that it was not an error at all but an independent evidence of the earth's motion. From this he obtained the first measurement of the speed of light.

It was found that the moons of Jupiter (and later those of Saturn as well) obeyed Kepler's three laws in their motions. Orbital radius cubed divided by period squared for each moon was a constant ($R^3/T^2 = k$). However, the constant was of one value for objects orbiting Jupiter and of a different value for objects orbiting the sun (and of yet another value for the moons of Saturn). The constant seemed to be determined by the object being orbited. Why? How?

Kepler's laws were *empirical laws* derived from the data of Brahe. There was no argument that these relationships correctly described what was happening. However, they did not explain anything, they merely described. A more fundamental set of relationships was hidden under these laws, but what? At least one clue came from the laws themselves. Whatever order nature was imposing on astronomical motions, this order was relentlessly and precisely mathematical. There was no "sloppiness" built into these predictions—the better the data, the better the prediction. Galileo appeared to be correct in saying that the book of nature is written in mathematics. But why should the heavens be moving like divine clockwork while motion here on earth seemed so confused? Could order be found in all motion? When, how, and why do things move? Should a different set of rules apply to the heavens than to motions on earth? The ancients had thought that the heavens were special. The precision of Keplerian prediction was convincing evidence for a moving earth. If the earth was not the center of the universe, why should it be covered by special laws of motion?

The modern age in science dates from the resolution of these questions. Galileo died in 1642 after beginning the modern treatment of motion and 100 years after the book of Copernicus appeared. In the same year of 1642 Isaac Newton was born in England. Newton was to produce the modern treatment of motion. In the work of Newton we find the completion of the scientific revolution away from the physical science of the Greeks. The development we have followed from Copernicus in 1500 ends in Newton's publication of a complete science of motion and its causes in 1687.

Sir Isaac Newton

"The Very Model of a Modern Major Scientist"

I do not know what I may appear to the world; but to myself I seem to have been only a boy playing on the seashore, and diverting myself now and then finding a smoother pebble or a prettier shell than ordinary, whilst the great ocean of truth lay all undiscovered before me.

The contribution of Sir Isaac Newton to the production of modern science was huge. Before Newton there were many correct but unconnected ideas about forces and motion along with many more incorrect ones. After Newton the world had a finished and complete science of motion, which was added to but not fundamentally altered for over 2 centuries. Little in Newton's early life indicated that he would become perhaps the greatest single figure in the emergence of the modern world. The publication of his greatest work, the *Principia*, is the most important event in the scientific revolution which led to modern civilization.

Newton's father died shortly before the birth of Isaac in 1642. When his mother remarried, she left Isaac behind to be raised by his grandparents. Some modern scholars find in this abandonment the roots of the retiring life-style Newton developed. He never married. Newton was a good but unspectacular student until his university days at Cambridge.

In 1665 the "black death" ravaged England, and Cambridge University was closed. The bubonic plague in Europe killed as much as one-third of the population. Nobody knew how to stop the spread of the plague, but everyone who could afford to do so left the cities for the isolation of the countryside. Newton spent 2 years at his home in Woolsthorpe (population: two families). These were the most scientifically productive years of Newton's life, at age 23. During these years, Newton invented and developed the field of mathematics called calculus and formulated the laws of motion. It was here that the falling apple led to a law for gravitation. Later, he presented a paper on color theory to the Royal Society (one of the earliest scientific societies, founded in 1645, incorporated in 1662, and publisher of the first scientific journal in England in 1665).

The most influential scientist in England at the time was Sir Thomas Hooke, and Newton's ideas directly contradicted Hooke's. Thus, many people (including Hooke) felt obliged to criticize Newton's ideas mercilessly. The resulting unpleasantness soured Newton on exposing his work to view, and he decided not to publish his ideas again. Fortunately he was talked out of this position by Astronomer Royal Sir Edmund Halley. Halley is best known for his study of comets. He believed that comets were repeating objects that came back every so often, not just once. He asked Newton what kind of force would be required for a repeating orbit. Without hesitation Newton indicated that an inverse-square force law was required, and he started rummaging through his miscellaneous unpublished papers for the proof he had done years before. As more and more of Newton's ideas emerged, Halley was astonished. Eventually he convinced Newton to publish a book, and Halley paid the costs himself (although he was not a rich man). Newton's *Principia* appeared in 1687, and was an instant classic in science.

Newton became famous, and, characteristically, too busy to do very much more in science. His later life was spent in reforming the English monetary system as master of the mint, serving in Parliament, being knighted, and in studies and publications in religion. (Newton's works in theology are far more lengthy and numerous than his publications in science and mathematics.) Newton himself remained modest, writing, "If I have seen further, it is by standing on the shoulders of giants." He died in 1727.

The First Law of Motion

For the first basic idea governing motion Newton used the principle of inertia. Although Galileo had used this principle several times, he had never stated it as directly as Newton did.

> ***Newton's First Law of Motion*** *A mass in motion will remain in motion at constant speed in a straight line, or a mass at rest will remain at rest, unless acted on by an external force.*

Inertia is the name given to this tendency of a mass to preserve its motion.

The commonest situation in which we become aware of the action of this law might well be when we are in a bus or car which goes around a curve rapidly. If you have been riding standing in the aisle of a bus making such a curve, you sense a tendency to move to the outside wall of the bus. In fact, one often speaks of being "flung to the outside wall" of the bus in such a situation. This sensation is so real that early treatments described a *centrifugal force*—a force away from the center of the curve—which moved you to the outside wall. Newton's first law correctly explains this situation and reveals this centrifugal force to be a fiction (Fig. 6-1). You, as a mass, tend to continue in a straight line in the direction you are moving. The bus, on the other hand, proceeds into the curve. To do this, the front tires are twisted around at an angle to the direction of motion. To keep going in the original direction the tires would have to skid sideways against the road. The friction between the front tires and the road exerts a sideways force against the tires toward the interior of the curve (a *centripetal force*). This force bends the motion of the bus around to follow the curve. Unfortunately, you, the passenger, are an independent mass and you tend to continue in constant-speed motion in a straight line unless acted on by an outside force. As you tend to move straight forward this outside force is eventually supplied by the outer wall of the bus as it is pulled around into your path and forces you into following the curve which the vehicle is making. We are so used to walls being inanimate objects that it is difficult to think of the wall coming over to pound us into following the general curve of the vehicle. Thus, the fiction of some force that "pulls" us to the outside wall is invented to explain the events. The effects in the bus are similar, but the way of thinking about the events are very different after Newton's first law. An independent observation supports Newton's view. If the vehicle is going too fast into the curve, the friction of the tires sliding across the pavement cannot supply enough sideways force to push the vehicle around into the curve. The result is that the vehicle tends to continue moving in a straight line at a constant speed, skids across the pavement, and crashes.*

The greatest idea hidden in Newton's first law is probably that *motion does not need further explanation.* Any motion, once started, will continue unchanged unless forces are acting on a mass. Indeed, a change in a motion is *proof* of the action of forces if this law is correct. Therefore, a projectile moves freely horizontally because of an absence of horizontal forces, but always is pulled downward because of a constant downward force acting on it. Immediately we may also infer that forces must be acting on the moon and the planets, for example, because their motions are not in straight lines. Indeed the first law of motion completely changes the basic question about the motions of the heavens. Since ancient times natural philosophers felt they had to explain why the moon and the planets did not fall down to the earth. After the first law the question becomes: "Why do not the planets and the moon keep moving in a straight line at a constant speed in the direction they are already going?" Why do they not just wander off into space? What holds

* Police investigators routinely estimate the speed of a car into such a crash by measuring what they call "centrifugal skid marks" left by rubber scraping off the tires as they grind against the pavement. A better term might be "inertial skid marks." You might look for such evidence at road curves as you drive. Not all such marks result in a crash. Sometimes the frictional forces just manage to overcome the car's inertial tendency in time (and frequently in the wrong lane).

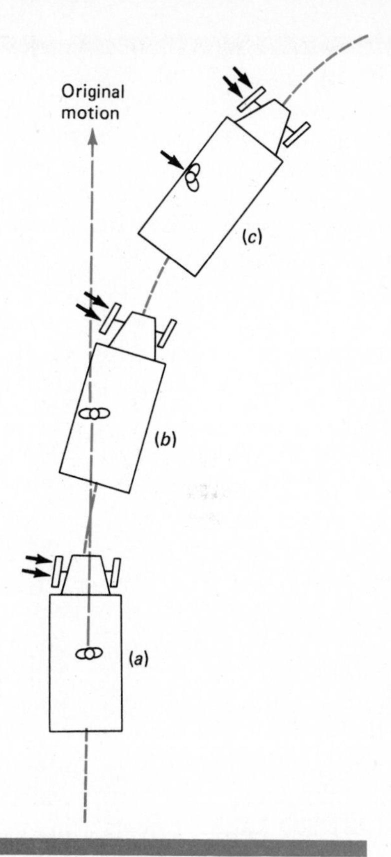

FIGURE 6-1
Bus and human being tend to move in a straight line at constant speed unless acted on by an outside force. (*a*) The bus tires are twisted at an angle to the motion, and friction with the road provides a force to move the bus into the curve. (*b*) The bus is moving into the curve, but inertia causes the person to remain in the original motion. (*c*) The wall of the bus is providing the force to accelerate the person into a new path of motion.

them in orbit? The old view looked for a *centrifugal force* to push the moon away and keep it "up." The new view required a *centripetal force* to constantly pull the moon in toward the earth and keep it from escaping due to its inertia.

The Second Law of Motion

Given this tendency for masses to maintain their motions, how do we change motion? Motion itself needs no explanation, but changes in motion do.

Obviously, the application of a force can change the state of motion of objects. For example, when a foot applies a force to a football on a kickoff, the football changes from a state of rest, or no motion, to rapid forward and upward motion. Forces are essential to change a state of motion. What is the exact relationship of forces to changes in motion? This is the question addressed by Newton's second law of motion.

One clue is found in the drop of our bowling ball. A constant unbalanced force applied to an object produces a constant acceleration. But what effect do forces of different size have on the same mass?

It is not difficult to obtain an intuitive picture of the relationship of force to motion. Imagine the following situation. We take five identical

Coriolis force—rockets and winds

One of the great strengths of America lies in consistent agricultural productivity. We enjoy much good land and favorable climates as well as farming expertise and advanced technology. In part, our climate is regulated by the principle of inertia. To understand this, consider the motion of earth on its axis each day. The earth rotates to the east, completing one turn per day. At the equator, the earth is 40,000 km in circumference. To go once around this distance in a day, an object must move at a speed of 1700 km/h (about 1000 mi/h). Thus, all objects resting at the equator have an eastward speed of 1700 km/h. The east-west distance around the earth at the equator is larger than any other location on earth, so the daily motion there is the greatest. At the latitude of Philadelphia (40° N), the distance eastward around the earth is only 31,000 km, so objects at Philadelphia are moving eastward there at the reduced rate of 1300 km/h. Human beings are not sensitive to speed, only acceleration, so we normally do not notice this motion, or the difference in motion between these two locations. Suppose, however, we started mail-rocket service between the equator and Philadelphia? If we shot a rocket from the equator and were unaware that it was moving eastward 400 km/h faster than its target, we would miss the landing site. For every hour in its flight, the rocket would appear to drift 400 km eastward of its target. Having no apparent way to explain this drift if we ignore the rotation of the earth, we invent a force which we can say "pulls" the rocket off course. This fictitious force on the earth is called the *coriolis force*. The rocket's drift is really due to its inertia, or tendency to keep its eastward motion. This drift becomes very large when shooting our rocket over the pole. Here, target and rocket have motions in different directions in space. In a 1-h flight, a rocket bound from New Orleans to New Dehli in India could wind up four countries over in Saudi Arabia owing to this failure to account for inertia. This error corresponds to aiming at Philadelphia but landing in Salt Lake City, Utah.

Even slow motions show this drift when moving north and south. Air normally flows from regions of higher pressure to regions of lower pressure. However, air moving northward to a low is deflected to the east by its own inertia. Likewise, southward-moving air appears to deflect to the west of its target. Thus, air moving into a low-pressure area does not move directly into it, but appears to swirl into the low counterclockwise in the Northern Hemisphere. This has important implications for the weather patterns of the United States. To see why, look at the map.

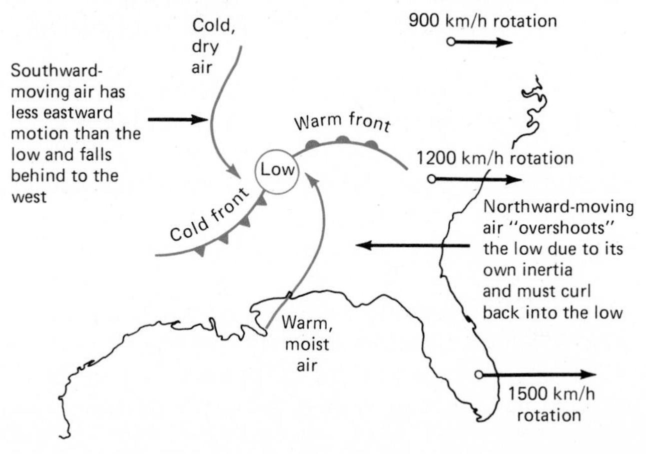

As the air swirls slowly into the low, warm, moist air from over the Gulf of Mexico curves in front of the low. Meanwhile, cold, dry air from over Canada is pulled in behind the low. Masses of air with greatly different properties do not mix well. Thus, the advancing edge of the Canadian air along the ground is called a *cold front*. This is usually marked by the sudden, heavy release of the moisture in the air which the cold air is advancing against. Cold fronts are associated with thunderstorms, or, if their advance is very vigorous, with tornadoes. The warm front associated with the advancing Gulf air is also marked by rain. In this case the rain is spread hundreds of miles north of the front and occurs as a long, steady rain. Most of the rainfall in the United States is associated with frontal activities such as this, and ultimately with the inertia of air masses.

Chevrolet Chevettes, each of which has a mass of about 1 megagram (Mg). We will use one car unchanged and call it car *C*. In the remainder of the cars we tear out the engines and replace them. To produce a larger force we can install the largest Cadillac engine in car *B*. For car *A*, we cut off the roof and drop in a Pratt and Whitney turbojet engine. This develops roughly 20 times the force and we do not have to bother connecting it to the wheels. To car *D* we add a standard Volkswagen engine, while car *E* receives the engine from a riding power mower. We now arrange the five cars so that at the push of a button all five engines will race at full power and move the cars forward. Once you have fully pictured this situation, you have little doubt about what will happen once the button is pushed. By the time it is clear that car *E* is gaining some speed, car *A* has disappeared over the horizon. We have kept the masses of these five objects constant and varied the force applied. Clearly, *the larger the force applied to a given mass, the greater the change in its motion.*

We could do a variation on this thought experiment by using the five identical Chevrolet engines. This time, however, we mount the identical engines on the power mower, the Volkswagen, the Chevy, the largest Cadillac, and in place of a jet transport, a large truck (since we do not want it to take off anyway). In this mental exercise we keep the force constant, since we are using identical engines, and vary the masses. Again, you can picture the result easily. The lawn mower will zoom off into the sunset leaving a streak of melted rubber on the pavement. It is likely that it will burst into flames from its overheated parts before we notice the truck moving. We can conclude that *the smaller the mass to which a given force is applied, the greater the change in motion which results.*

Repeating these experiments on a laboratory scale yields the basic relationships. The data for force applied vs. the acceleration of a constant mass object are graphed in Fig. 6-2. From this we may conclude that acceleration *a* is proportional to the force *F* if the mass *m* is constant. In symbols, this is

$$F \propto a]_m$$

Similarly, data on the acceleration of various masses acted upon by a

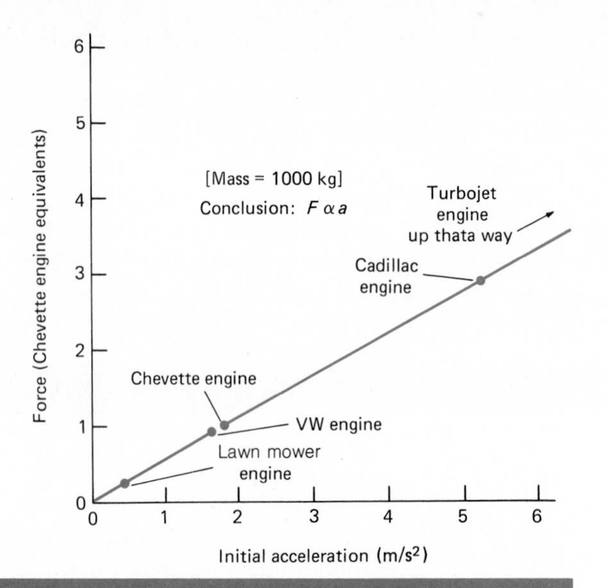

FIGURE 6-2 Graph of force vs. acceleration for five identical masses propelled by engines generating driving forces. The variables graphed are proportional, or $F \propto a$, if mass is constant.

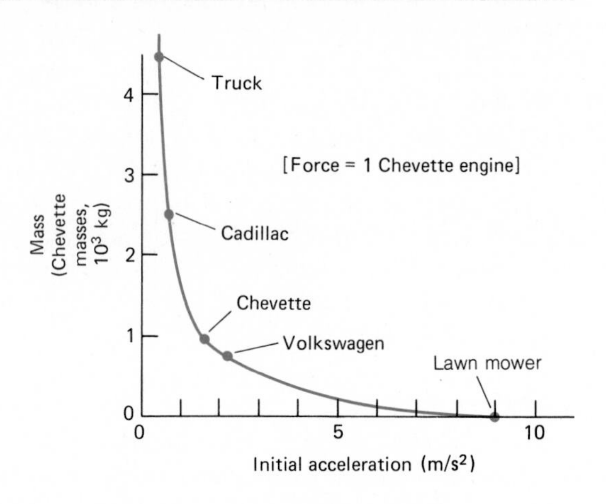

FIGURE 6-3A Graph of mass vs. acceleration for five vehicles propelled by engines generating the same force. The shape of the curve suggests an inverse relationship.

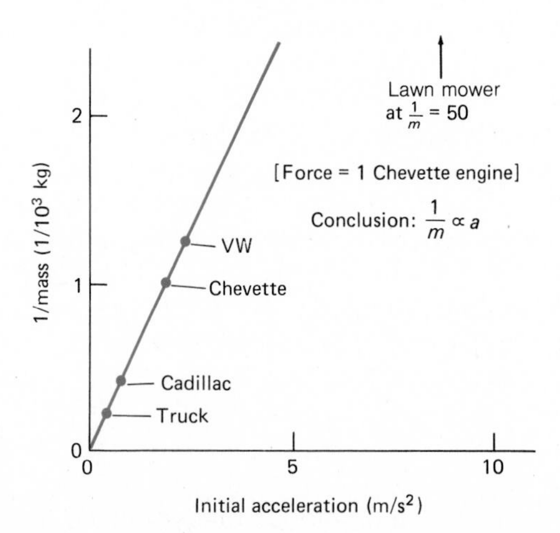

FIGURE 6-3B Graph of inverse mass ($1/m$) vs. acceleration for five vehicles propelled by engines generating the same force. The variables graphed are proportional, or $1/m \propto a$, provided the force is constant.

constant force are graphed in Fig. 6-3A. The shape of the curve is that of an inverse proportion. If we test this by regraphing a vs. $1/m$, we obtain the straight line shown in Fig. 6-3B. Thus, acceleration is proportional to the inverse of mass if the force is constant, or

$$\frac{1}{m} \propto a]_F$$

From these two proportions we may write the combined proportion

$$\frac{F}{m} \propto a \qquad \text{or} \qquad F \propto ma$$

To change this to an equation we have only to add a constant to adjust for the units in which forces, masses, and accelerations are measured. Thus,

$$F = kma$$

This is essentially the form of Newton's second law of motion. Newton defined *force* as the rate of change of *momentum*. Momentum is the product of mass times velocity, or mv. A rate, you may recall, is the ratio of two changes, with the denominator usually the change in time. Thus, Newton's second law becomes

$$F = k\,\frac{\Delta(mv)}{\Delta t}$$

There is no reason to guess that mass changes with or depends on the time, so we may treat it as a constant and rewrite the expression as

$$F = km\,\frac{\Delta v}{\Delta t} = kma$$

which is the same as our result before.

Newton's Second Law of Motion *An object experiences an acceleration which is directly proportional to the unbalanced force acting on it and inversely proportional to its mass.*

To this point we have not defined a unit in which force is to be measured. The convenient and universal choice is to select a unit which will make the constant in $F = kma$ equal to 1. We may *define* the unit of force to be that force which will produce an acceleration of 1.0 m/s² when acting on 1.0 kg of mass. We call this unit of force the *newton* (1.0 N). If force is measured in newtons, mass in kilograms, and acceleration in meters per second per second, the constant k has a numerical value of exactly 1. We may then write Newton's second law of motion as

$$F = ma$$

Example

A bowling ball has a mass of 7.3 kg (a standard 16-lb ball). It is speeded up from rest to a speed of 5.6 m/s in 0.50 s. What was the average force applied?

$$F = ma$$

$$= m \frac{\Delta v}{\Delta t}$$

$$= (7.3 \text{ kg}) \frac{5.0 \text{ m/s}}{0.50 \text{ s}}$$

$$= 73 \text{ kg} \cdot \text{m/s}^2 = 73 \text{ N}$$

Example

A bowling ball has a mass of 7.3 kg. When released it falls with an acceleration of 9.8 m/s^2. With what force is the ball being pulled downward?

$$F = ma$$

$$= (7.3 \text{ kg})(9.8 \text{ m/s}^2)$$

$$= 72 \text{ kg} \cdot \text{m/s}^2 = 72 \text{ N}$$

This downward pull of 72 N is called, of course, the *weight* of the ball. We see that 16 lb equals 72 N. Thus, 1 lb is about 4.5 N, or 1.0 N = 0.22 lb. Pounds and newtons are comparable units in that they are both measures of the same quantity—force.

Weight vs. Mass

Frequently we will see conversions from pounds to kilograms. These units measure different items, but the distinction is blurred because weight and mass are nearly proportional on the surface of the earth. Weight is the force with which an object is being pulled by gravity. *Weight is a force*—the force of gravity acting on an object. Mass, on the other hand, is a measure of the quantity of matter in an object, whether or not gravity is acting. Mass is measured by the inertia of matter, or the degree to which it resists changes in its motion. If a person weighs 200 lb (900 N) at the North Pole, he or she will weigh only 199 lb (896 N) at the equator as measured on a scale. The quantity of matter in the person has not changed, and he or she has a mass of 91 kg at both locations as measured with a balance. Scales measure a pull or force directly, while balances compare the pulls on two masses. The distinction between weight and mass becomes important in the sciences because some variables depend on one and some on the other of these two measures. An example may help. A person is lying on the pavement. A truck slowly moves over him or her, and the person is crushed. Now consider the same event deep in space. A person is lying flat on a plane of some material. A truck moves slowly over him or her. This time the situation is very different. Truck, human, and plane are weightless. There is no downward force pulling on the truck which is too great for the person's body to support. The person is not crushed—but the plane and the truck are pried apart by the person's presence. Heavy objects crush things underneath them because of the large downward pull of gravity acting on them—because of their weight.

Now consider a second set of events. A person is standing in a road when hit by a truck at the legal speed limit of 88 km/h (55 mi/h). He or she makes a rapid move toward the next county, sustaining serious, possibly fatal damage. Now, what will happen if a weightless astronaut is in the path of the now weightless truck in space as it approaches at the same 88 km/h? The answer is that the astronaut will be killed just as though on earth. In neither collision is weight a factor. In both cases the truck is a large mass in rapid motion. To stop that motion requires the application of forces larger than the human could exert. The tendency of the mass of the truck to preserve its state of motion exerts forces on the human greater than the body could withstand, and damage results. The collision situation is not a problem of weight but of the inertia of masses. On the earth's surface we exert forces to overcome both weight and the inertia of masses. It is not particularly easy to get a grand piano moving even when it is mounted on rollers, and once it is in rapid motion, don't stand in front of it!

The calculation of weights from masses and vice versa for the earth's surface is quite important and will be encountered frequently. The basic relationship between force and mass is

$$F = ma$$

In the gravitational case this becomes:

$$\text{Force due to gravity} = \text{mass} \times \text{acceleration due to gravity}$$

or $$\text{Weight} = \text{mass} \times g$$

$$W = mg$$

where g is our old acquaintance of 9.8 m/s², the acceleration due to gravity.

Newton's Third Law of Motion and Momentum

How are forces produced? What is required to generate a force? Newton's third law of motion concerns itself with the production of forces. Forces represent the interaction between two or more objects. An object must be pushed or pulled *by* something. Similarly, when you exert a push, you must exert it on some material object. When you push on a shopping cart, you and the cart are interacting. If you were in space, this might be the only interaction involved (Fig. 6-4). On earth, several other interactions are involved at the same time. For example, the floor

F

Force of shopping cart on astronaut

-F

Force of astronaut on shopping cart

FIGURE 6-4
Forces occur only in equal and opposite sets. Whenever two objects interact, both experience the same size force from the other at every instant of time, but in opposite directions.

pushes up against your feet and balances your weight, your foot pushes backward against the floor to accelerate you forward, and so on. In each case, we can define forces in terms of the interaction between two bodies. For the moment, consider only the shopping cart and yourself. You push against the shopping cart with your hands. As you do this, the flesh on your hands indents and you have a sense of pressure against the skin. The cart is resisting your force by exerting a push back against you. The same sort of situation occurs when you lean against a wall. If the wall doesn't exert a backward force on you, you will fall. If the chair or floor underneath you right now was not exerting an upward force on you this instant, you would accelerate downward in response to your weight From this kind of consideration Newton reasoned that forces always occur in pairs, and furthermore, in matched pairs. Assume body A and body B are interacting. Body A exerts a force on body B which we may indicate by F_{AB} (the force of A acting on B). Newton reasoned that this required a counterforce F_{BA}, the force of B acting on A. Furthermore, these two forces were of the same size but opposite in direction, or $F_{AB} = -F_{BA}$. Thus, if you are exerting a forward force of 25 N on the shopping cart, at the same instant the cart is exerting a backward force of 25 N on you. This may be summed up as:

> ***Newton's Third Law of Motion*** *Forces are generated in the interaction of two bodies. The bodies exert forces on each other that are equal in size but opposite in direction.*

This idea is frequently summarized by the slogan: "For every action there is an equal and opposite reaction." Careless use of this statement leads to many incorrect ideas. For example, if I put a coin in a vending machine, this is an action in the common use of the term. When (and if) the machine dispenses a can of pop, we could term this a "reaction." This is *not* the way these terms are used in science. There is a temptation to use the terms "action and reaction" like the terms "cause and effect." Among other problems this introduces the idea that one comes before the other in time. The forces referred to by Newton's third law are *at every instant in time* equal in size and opposite in direction.

The form $F_{AB} = -F_{BA}$ is perhaps the best way to think of this relationship because it provides no such mental problem of which comes first.

Closely related to the third law of motion is another principle which is very important. Picture pushing the shopping cart in space again. Suppose the push of 25 N lasts for exactly 1 s before you and the cart push away from each other. What effect will the set of forces generated have on your motions if you have a mass of 50 kg and the cart has a mass of 15 kg? This is easy to calculate.

$$F = ma \qquad \text{or} \qquad F = m\frac{\Delta v}{\Delta t}$$

Effect on you:

$$\Delta v = \frac{F}{m}\Delta t$$

$$= \frac{(25\text{ N})\ (1.0\text{ s})}{50\text{ kg}} = 0.50\text{ m/s}$$

Effect on cart:

$$\Delta v = \frac{F}{m}\Delta t$$

$$= \frac{(-25\text{ N})\ (1.0\text{ s})}{15\text{ kg}} = -1.7\text{ m/s}$$

Thus, after the interaction you are moving away from your original position at a speed of $\frac{1}{2}$ m/s. Meanwhile, the cart is moving away from the original position in the opposite direction (that's what the minus sign indicates) at 1.7 m/s. Note that your new speed times your mass is just the opposite of the cart's speed times its mass. This is a perfectly general result which always occurs in just this way for two isolated bodies. We can express it more formally. From the third law of motion,

$$F_{AB} = -F_{BA}$$

but, the force of A acting on B equals the mass of B times the acceleration of B by the second law, $F_{AB} = m_B a_B$. Similarly, $F_{BA} = m_A a_A$. Thus,

$$m_B a_B = -m_A a_A$$

Substituting from the definition of acceleration, $a = \Delta v/\Delta t$, we get

$$m_B \frac{\Delta v_B}{\Delta t_B} = -m_A \frac{\Delta v_A}{\Delta t_A}$$

The time interval Δt_B during which force F_{AB} acts is identical to the time interval during which F_{BA} acts. These forces are equal in size and opposite in direction every instant. They start together, grow or shrink together, and end together in time. Thus, $\Delta t_B = \Delta t_A$, and

$$m_B\,\Delta v_B = -m_A\,\Delta v_A$$

or, as Newton would have written it,

$$\Delta(mv)_B = -\Delta(mv)_A \quad \text{or} \quad \Delta(mv)_B + \Delta(mv)_A = 0$$

The product of mass times velocity is called *momentum.* Our last expression indicates that during this interaction, the sum of the changes in momentum is zero. If one object gains momentum in one direction, the other object gains equal momentum in the opposite direction. We have obtained this result for a system of two objects, but the idea is completely general for any number of objects which interact among themselves with no forces involving additional objects. We may state this as:

> ***The Law of Conservation of Momentum*** *The sum of the changes in momentum (mv) for a closed system is zero.*

Another way of saying this is to assert that the momentum of the universe is constant, since the universe contains all objects and is therefore a closed system. This is the first example we have encountered of a particular kind of relationship called a *conservation law.* A conservation law asserts that some variable, in this case momentum, remains constant and cannot be gained or lost—merely redistributed among the parts of a system of objects. This type of natural law is important for two reasons. Conservation laws seem to be quite basic and cut to the core of nature. Conservation laws identify items which are preserved in nature *no matter what else is happening.* Philosophically, they loom large in trying to understand nature. Practically, conservation laws are significant because we may apply them in cases too complex to treat with other principles. Even if we have no way of determining all the interactions in a system, we may assert that momentum is conserved and be right in the answers that result. For example: Two cars streak toward each other and collide, each traveling 88 km/h. One car has a mass of 2000 kg, and the other has a mass of 1200 kg. During the collision the cars crunch together and lock into a single lump of deformed metal. During the collision mul-

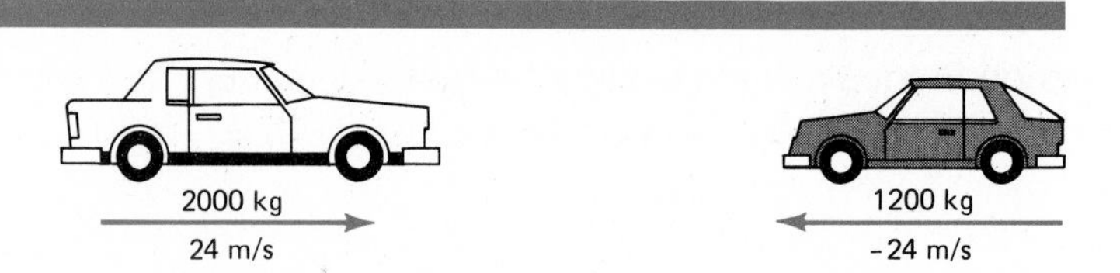

Before collision: Total momentum = + 1900 kg · m/s

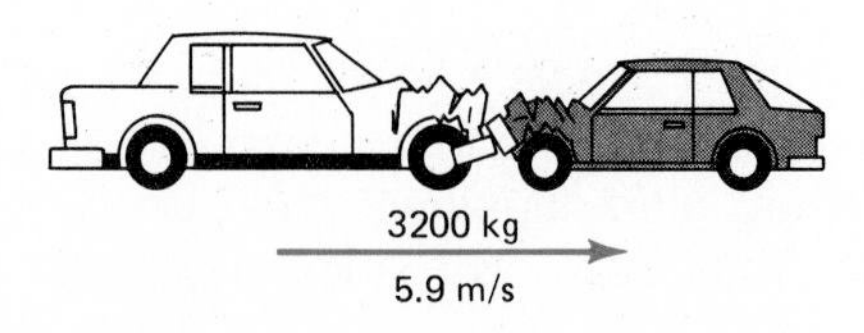

After collision: Total momentum = + 1900 kg · m/s

FIGURE 6-5 Two cars of unequal mass collide and lock together. The law of conservation of momentum indicates that the momentum after collision must equal the momentum before the collision, and thus predicts the velocity at which the cars will skid away from the impact point.

tiple forces are involved between the various parts of the cars. These forces vary in size and direction instant by instant. Even though the force applied to a particular part will equal its mass times its acceleration, the thousands of simultaneous interactions make the situation too complex to analyze. Yet, the law of conservation of momentum predicts that after the collision the cars will be skidding in the original direction of motion of the more massive car at a speed of 5.9 m/s. None of the details of the interaction is necessary to make this prediction (Fig. 6-5).

Before collision: $m_1 = 2000\ \text{kg} \qquad m_2 = 1200\ \text{kg}$

$v_1 = +24\ \text{m/s} \qquad v_2 = -24\ \text{m/s}^*$

$$\text{Total momentum} = (2000\ \text{kg})(+24\ \text{m/s}) + (1200\ \text{kg})(-24\ \text{m/s}) = 1{,}9 \times 10^3\ \text{kg} \cdot \text{m/s}$$

After collision: $m = m_1 + m_2 = 3200\ \text{kg}$

$v_f = ?$

Momentum must be conserved, and so

$$\text{Total momentum} = 1.9 \times 10^3\ \text{kg} \cdot \text{m/s} = (3200\ \text{kg})v_f$$

Therefore

$$v_f = +5.9\ \text{m/s}$$

While this may seem to be an adequate job of prediction, there is more. Momentum has a particular direction, and therefore the law of conservation of momentum (like most laws involving force) is a vector law. Not only must the size of the total momentum *mv* be preserved, its direction must also be preserved. Thus, if our cars had not hit head on but at some angle, the law of conservation of momentum would allow us to predict the one and only angle the locked-together cars would skid off at following the collision. To comprehend the power of this principle, consider the following situation. An artillery shell is approaching its target at high speed and explodes in midair. The fragments of the shell and the expanding gases from the explosion take off in all directions. If we take the mass of each fragment times the velocity of each fragment after the explosion, the sum of all these vectors—added as vectors in three-dimensional space—will add up to the original size and direction of the momentum of the shell before the explosion. What is most remarkable is that whenever we devise a clever experiment to actually measure such a complex situation, it really works—momentum is conserved in even the most complex situations measured to date.

As a final note, momentum of a system about its center of mass is also conserved. This is called *angular momentum,* given by the formula *mvr*. For a compact mass, *r* is the perpendicular distance from the center of its mass to the line of action of *v*. Although we will not use it in any cal-

* 88 km/h = 24 m/s.

FIGURE 6-6
Motion of an entire object from place to place in space is called *translational motion.* A set of forces acting at different positions on an object is called a *torque,* and tends to set up a *rotational motion* of the object about some pivot. This stroboscopic photograph of a gymnast shows both types of motion. (*Photograph by the author.*)

culations, the law of conservation of angular momentum is a powerful statement concerning nature. A classic example involving angular momentum is that of an ice skater twirling in position. The skater begins to spin with the body as spread out as possible. As the arms and legs are pulled in the skater will spin faster and faster. With the arms and one leg spread wide, the mass of the skater is relatively far from the center of mass. When the limbs are pulled in, r becomes smaller. The law of conservation of momentum states that mvr will remain a constant through all of this. Since the skater's mass is constant, as r becomes smaller, v must become larger. Thus, the rate of spin is increased and the skater enters a

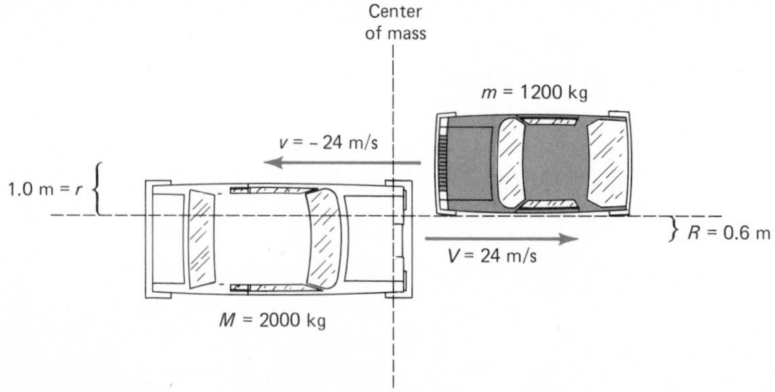

FIGURE 6-7
The center of mass of a system, here the two cars, is that point with as much mass to the left as to the right and above as below. Angular momentum about this point before the collision is $MVR + mvr$ in the diagram. Since it is conserved, in the collision it allows determination of the rate of spin of the locked-together cars after collision.

fast twirl. Aside from being a pleasant and impressive sight to watch, it is a convincing demonstration of this law. Think of this the next time you watch a skater perform (Fig. 6-6).

Angular momentum is conserved in collisions along with linear (or straight-line) momentum. In the example of the collision of cars, had the cars hit moving parallel but somewhat off center (Fig. 6-7), we could have predicted the rate at which the remains would have been spinning from the law of conservation of angular momentum, in addition to the direction and speed of their skidding away from the collision as we did above.

Circular Motion

To discuss the motion of the moon and gravitational force, we will need to know how to describe circular motion. In particular, we need an expression for the circular speed v_c and the circular acceleration a_c. *Uniform circular motion* is defined as motion in a circle at constant speed. Obtaining the speed is easy. The average speed is just the total distance traveled by the object divided by the time it takes to perform this motion. A convenient distance to consider is one full swing around the circle. During this time the object travels the length of the *circumference* of the circle, or $\Delta d = 2\pi R$. It travels this distance in the time required for one full loop, which we define as the *period* of the motion T. Thus,

$$v_c = \frac{2\pi R}{T}$$

For example, if a fly is standing in the grooves of a $33\frac{1}{3}$-rpm record some 12.0 cm from the center, we find its speed as follows: If 33.3 turns are made each minute, 1 turn is made in 1/33.3 min, or 0.0300 min. In general, the number of cycles per time period is called the *frequency* f. The time required to complete one full cycle is the *period* T. The relationship between these variables is $T = 1/f$. Thus if an object makes 5 turns per minute ($f = 5/\text{min}$), each turn takes $\frac{1}{5}$ min ($T = 0.2$ min). For our fly, $f = 33.3/\text{min}$, thus $T = 0.0300$ min. Substituting 3.14 for π, we can now calculate its speed in the circle as

$$v_c = \frac{2\pi R}{T}$$

$$= \frac{2(3.14)(12.0\ \text{cm})}{(0.0300\ \text{min})(60\ \text{s/min})} = 41.9\ \text{cm/s}$$

This speed of about 0.42 m/s is the speed that the fly would ram into the almost motionless pickup arm of the phonograph if it would be too dizzy to take off.

The acceleration of an object in uniform circular motion, a_c, may be shown to have a similar form mathematically.

$$a_c = \frac{4\pi^2 R}{T^2} \qquad \text{or} \qquad a_c = \frac{v_c^2}{R}$$

The appropriate form to use for the circular acceleration depends upon what is known about the situation.

Gravitation

Newton was sitting in an orchard pondering why the moon stays in orbit around the earth rather than move off into space when an apple fell nearby. Unlike most such famous stories, this one is true. The fall of the apple caused Newton to consider the force of gravitation. There is clearly a force on apples pulling them toward the earth. Could the same force be responsible for the motion of the moon? Since the moon goes around the earth in a nearly circular orbit, a force of nearly constant size directed always toward the center of the orbit was required—a centripetal force. Was the same force that holds us to the earth's surface providing this force? If so, what was the nature of this force? Why should masses be pulled toward the earth? That was a key idea—mass. The force of the earth must be an attraction for all mass. The force must act on mass itself. Newton reasoned that gravitational force must be an attraction of mass for mass.

Consider the moon. The earth exerts an attractive force on the moon. The more mass the moon has, the greater this force should be. Thus, the force should be proportional to the mass of the moon. Upon what else should this force depend? From the third law of motion, if the earth exerts a force on the moon, the moon must be exerting the same size force on the earth at the same time. This force exerted by the moon must be an attraction for the mass of the earth. Newton reasoned that gravitational force must be an attraction of any mass for any other mass. Masses must attract each other. If this were so, the size of the earth's attraction for the moon must depend on how much mass is possessed by the earth. The more massive the earth, the stronger its pull on the moon should be.

Upon what else should this force depend? For several reasons, the force should depend on distance. The greater the distance, the weaker the force must become. Why? First of all, we have examples of this kind of force. The attraction between two magnets gets weaker as they get farther from one another and stronger as they get closer. Thus, a large and powerful magnet far away may have much less effect on another magnet than a smaller, weaker magnet close at hand. In a similar way, the sun is a million times the volume of the earth. We could be safe in assuming that it is much more massive than the earth.* Yet, we experience an attractive force toward the earth which we call our weight, not toward the more massive sun. Some chunks of the earth are nearer to you and some are farther. But, the average distance for all the chunks of the earth's mass is your distance from the center of the earth. Thus the combined pull of all of the particles of the earth's mass is the same as if all of the earth's mass were concentrated in a single blob at the earth's center,

* The sun is actually "only" some 300,000 times as massive as the earth. The sun is a sphere of gases while the earth is a sphere or rock and metals. Rocks are generally more densely packed than gases.

some 6400 km (4000 mi) beneath your feet. Massive as it is, the center of the sun is 23,000 times as far away from you.* To have the gravitational force fall off rapidly with the distance R, the force must have some kind of inverse proportionality to distance. That is, $F_G \propto 1/R^n$, where n is some unknown power. From several lines of investigation Newton determined this power to be exactly 2.

Thus, the earth's pull on the moon should be proportional to the earth's mass and to the moon's mass, and inversely proportional to the square of the distance from the earth to the moon. Or for any two masses m_1 and m_2 where R is the distance from the center of one to the center of the other, we may write (Fig. 6-8)

$$F = G\frac{m_1 m_2}{R^2}$$

This is Newton's *law of universal gravitation*—the law of gravity. Here, G is a constant determined by the choice of units for mass, force, and distance. We can see that the force falls off as the inverse of the square of the distance. That is, if we double the distance, the force drops to one-fourth of what it was; if we triple R, the force is one-ninth, and so on—$F \propto 1/R^2$. The center of the earth is 6400 km beneath your feet and you are attracted toward it with an acceleration of 9.8 m/s²? The Greeks had determined that the moon is almost exactly 60 times as far from the center of the earth as you are (30 times the diameter of the earth). By this reasoning the force on a particular mass at the distance of the moon should drop to $1/(60)^2$ or 1/3600 of the force if it were on the earth's surface. Thus the acceleration of a mass at the distance of the moon should be 1/3600 of its acceleration on earth or $(\frac{1}{3600})(9.8\ \text{m/s}^2) = 0.0027\ \text{m/s}^2$. The distance of the moon is 3.8×10^8 m, and it completes one circuit of the earth in $27\frac{1}{2}$ days or 2.4×10^6 s. Using these numbers with the formulas for circular motion, we obtain $a_c = 0.0027\ \text{m/s}^2$ as the required acceleration to maintain an object in a circular orbit of the earth at the distance of the moon.

Working from his form for a law of gravitation, Newton was able to show that all three of Kepler's laws were the required behavior of masses

* The earth is pulling on us hard enough to give us a downward acceleration of 9.8 m/s² if we are dropped. By the formulas for the acceleration in circular motion given in the last section, we can calculate our acceleration toward the sun. The period required to complete one circuit of the sun is 1 year, or 365 days, or 3.2×10^7 s. The distance to the sun is 1.5×10^8 km (9.3×10^7 mi). Performing the calculation using $a_c = (4\pi^2 R)/T^2$ shows that our acceleration toward the sun is only about one-thousandth of our acceleration toward the earth.

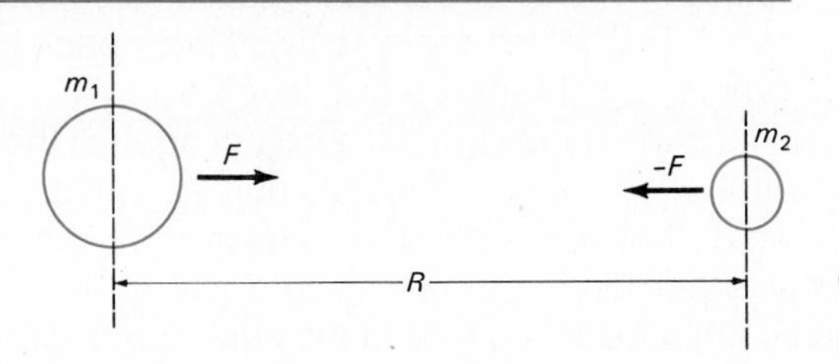

FIGURE 6-8
Any two masses, m_1 and m_2, separated by a distance R exert a gravitational attraction for each other. The size of the force of attraction is $(m_1 m_2)/R^2$ times a constant to adjust for the units of measurement.

if the force was inversely proportional to the square of the distance. Theory, based on the idea of an attraction of mass for mass, had replaced the empiricism of Kepler. One law based on a fundamental idea of how the universe was operating replaced three laws describing that operation. But gravitation did more! A useful law in science should not only explain what is known, it should suggest relationships not previously suspected. Mars, for example, is being attracted by all of the masses in the universe. This implies that the orbit of Mars is not determined only by the sun. In particular, Mars passes near enough to Jupiter, the most massive planet in the solar system, that the mass of Jupiter should have slight effect on the orbit of Mars. When instrumental astronomy had advanced enough, these small disturbances, or *perturbations,* from a pure elliptical orbit could be found in the paths of all of the objects in the solar system.

One of the earliest direct tests of Newton's laws was provided by the shape of the earth itself. If a kilogram of mass is at the North Pole, we may estimate its weight by using $W = mg$ and assuming $g = 9.800\ \text{m/s}^2$. Thus the weight of 1 kg would be 9.800 N. What would the weight of this kilogram be at the equator? A fast and simple reaction might be to guess that its weight would be the same. However, the earth is in rapid rotation, and a mass at the equator rotates through a circle every 24 h. A centripetal force is required to pull the mass into this circular motion.

$$F_c = m\frac{v^2}{R} = m\frac{4\pi^2 R}{T^2}$$

$$= (1.0\ \text{kg})\frac{4\pi^2(6370\ \text{km})}{(24\ \text{h})^2}$$

The mass moves in a circle of radius 6370 km (the radius of the earth) once each day. Thus,

$$F_c = (1.0\ \text{kg})\frac{4(3.14)^2(6.37 \times 10^6\ \text{m})}{(24 \times 3600\ \text{s})^2}$$

$$= 3.37 \times 10^{-2}\ \text{N}$$

$$= 0.034\ \text{N}$$

This force could be supplied by the gravitational attraction of the earth for our kilogram of mass. But then, the weight of the kilogram would be measured as only 9.766 N, not 9.800 N as at the pole. At the pole, all of the gravitational force acts to pull an object downward against the surface. At the equator, the attractive force of the earth must also supply the force to overcome the object's inertial tendency to fly off into space in the direction it happens to be moving. The force that is left over (by far the larger part in the case of the earth) acts to pull the object against the surface. This reduction in the effective gravitational pull as we proceed toward the equator is small, but it is enough to ensure that the earth warps out of a spherical shape and "bulges" toward the equator (Fig. 6-9). The bulge is small compared to the size of the earth but is large on

FIGURE 6-9
The equatorial bulge of the earth is too small to be seen in a photograph from space. Jupiter, however, is 11 times the diameter of the earth and rotates in only 10 h. The centripetal force required to hold objects at the equator is over 60 times as great for Jupiter, and this leads to an obvious bulge at the equator. Jupiter is thicker through the equator than at the poles by 5000 km, or almost a full earth radius. Incidentally, this is one of the finest photographs ever taken of Jupiter by an earth-based telescope. Compare it with the satellite photographs in Chap. 19. (*Courtesy of the Hale Observatories.*)

the human scale. The earth is 23 km thicker through the equator than through the poles. This difference provided a check on Newton's ideas. If the surface of the earth is a sphere, a different set of stars should be directly overhead at various points than if the earth is nonspherical. A more elegant way to state this is to say that a degree of latitude on the earth should represent a smaller north-south distance near the equator than toward the poles.

Triumph and Erosion—Newton's Laws and Astronomy

The planet Uranus is technically visible to the naked eye. If you know exactly where to look on a very good night, it will appear as one of the dimmest stars visible. Astronomers rate the brightness of objects in the sky in magnitudes. Stars of the first magnitude are very bright. The eye can see objects as dim as the seventh magnitude—the brightness of Uranus. Since Uranus takes 84 years to make one orbital cycle, its motion through the stars is too small to attract attention over a short period. When Sir William Herschel demonstrated it to be a planet in 1781, it had already been plotted on various star maps at various positions as one of the dim stars for over 100 years. Astronomers eagerly began accumulating information about the first member of the solar system not known to

FIGURE 6-10 The largest optical telescope in the world today is the 6-m (236-in) reflector at Mount Pastukhov in the Soviet Caucasus. This is 5 times the diameter, but 24 times the light-collecting area of Herschel's telescope. The 5-m (200-in) Hale telescope at Mount Palomar in California was the largest from 1949 until the completion of the Mount Pastukhov instrument in 1976. The newer instrument has yet to match the pace of discoveries maintained by the California instrument. (*Photograph courtesy of D. Kruikshank, University of Hawaii.*)

the ancients. Instruments were improving rapidly in this period. Newton had improved on Galileo's telescope by replacing the lens with a mirror 5 cm wide. Herschel used a mirror 122 cm (48 in) across in a tube 12 m long (Fig. 6-10). The precision of defining positions in the sky improved rapidly with larger, more stable instruments. Within 50 years of the discovery of Uranus it was apparent that its motion was not exactly what was expected. Even after calculating the small effects of the gravitational pulls of Jupiter, Saturn, and other bodies on Uranus, its path deviated from the prediction of Newton's laws. Were the laws wrong? After 150 years of uninterrupted triumph, it did not seem likely.

A young English astronomer, John Couch Adams, believed the small disturbance from the expected orbit was due to the presence of an additional planet. In 1846, Adams, then 26 years old, took his calculations to the royal astronomer. Adams predicted the size and location of a new planet that was perturbing Uranus' orbit using Newton's laws. He was turned away. A year later, while Adams was being ignored, a French mathematician Urbain Leverrier produced the same calculations and drew the same conclusions. Similarly, Leverrier was ignored by French astronomers with access to a telescope. Slowly both the English and the French began to take their young colleagues more seriously (Leverrier was 8 years older than Adams). It was too late. Leverrier had sent his calculations to Galle in Berlin, saying:*

* Sir Harold Spencer Jones, "John Couch Adams and the Discovery of Neptune," in James Newman (ed.), *The World of Mathematics*, vol. 2, Simon & Schuster, New York, 1959, pp. 822–839.

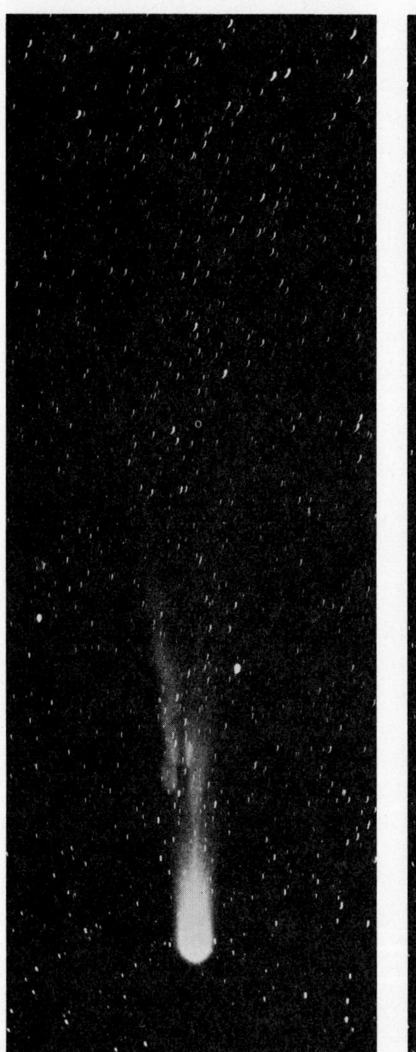

FIGURE 6-11 Newton's laws were published because Halley believed that comets could be in repeating orbits (see the section on Newton at the start of this chapter). Halley's Comet is the most spectacular of the returning comets, having been recorded for 29 returns. The photograph shows the rapid changes in the comet's tail observed during 1910. The next appearance of the comet will be in 1985 to 1986. Plans are being made to shoot a space probe to the comet for measurements. The head of the comet is many earth diameters wide, and the tails can stretch as far as the distance from the earth to the sun. (*Lick Observatory photograph.*)

Direct your telescope to a point on the ecliptic in the constellation of Aquarius, in longitude 326°, and you will find within a degree of that place a new planet, looking like a star of the ninth magnitude, and having a perceptible disc.

Galle did, and within a half-hour he had discovered the planet Neptune as predicted. This represented a triumph of prediction for Newton's laws.

Within a half-century, two more problems were encountered. Neptune's orbit was not precisely according to prediction. An attempt to find a new planet by the method of Leverrier and Adams did not work. Long after most astronomers had given up, Clyde Tombaugh in Arizona discovered Pluto in 1930. Small, out where planets were huge; dense, where its neighbors were nondense, Pluto was a misfit. Pluto's orbit is so irregular it actually crosses that of Neptune. Even if it were not what was expected, Pluto obeyed the laws of Newtonian mechanics. The problem with the little one up front was more serious. The long axis of Mercury's orbit slowly swings around in space. Newton's laws predict this effect due to the disturbance of Venus, but not its correct size. The shift or precession of the axis is 42 seconds of arc (42″) per century too large according to Newton's laws. This angle is the same as the width of a basketball in Milwaukee as seen from Chicago. Nevertheless, the discrepancy is there; it is real and must be accounted for. Astronomers suggested that yet another planet might orbit quite close to the sun's surface and be very difficult to detect. The search was fruitless. The motion of Mercury really did require a modification of Newton's laws. Thus, after $2\frac{1}{2}$ centuries of successful predictions, the total dominance of Newtonian mechanics ended. Einstein's general theory of relativity gave us a somewhat different and more general set of rules with which to analyze motions.

Summary

The first modern treatment of motion was completed by Newton with his three laws of motion. Publication of these laws was the peak event in what we call the scientific revolution and fundamentally altered the nature of the civilization. *Mass* is defined as a measure of the amount of matter in an object. The *first law of motion* asserts that forces are not necessary to produce motions, only to change motions:

> *An object will retain whatever motion it has at constant speed in a straight line unless acted on by an outside force.*

Mass may be measured from the *inertia* of an object, or its tendency to resist a change in its motion. The greater the inertia, the more massive the object is.

The second law of motion relates forces to the changes in motion they produce in the expression $F = ma$. Here, F is the unbalanced force, while m is the mass of the object acted on and a is the acceleration produced by the action of the force. In words:

> *The acceleration an object will experience is directly proportional to the unbalanced force applied and inversely proportional to the mass of the object.*

From this relationship we define the metric unit of force, the newton (N), as that force which produces an acceleration of 1.0 m/s² when applied to an object with a mass of 1.0 kg. A newton is approximately one-fifth of a pound (1.0 N = 0.22 lb). Weight is the force with which an object is being pulled downward by gravity. If released on earth, such an object would experience an acceleration of g, or 9.8 m/s², downward. Thus, the weight is given from Newton's second law of motion as $F = ma$, or $W = mg$.

The *third law of motion* asserts that forces between interacting objects are always generated in equal and opposite sets ($F_{AB} = -F_{BA}$). This in turn implies the *law of conservation of momentum.* Momentum is the product of mass times velocity. This law asserts that if we add the separate momenta of the parts of a system before an interaction and again after the interaction, the two sums must be the same. Momentum is neither gained nor lost during interactions, merely transferred from object to object.

In the *law of gravitation,* Newton asserted that all masses exert an attractive force for all other masses. This force is directly proportional to the product of the masses and inversely proportional to the square of the distance separating their centers. Mathematically,

$$F = \frac{Gm_1m_2}{R^2}$$

where m_1 and m_2 are the masses, R is the separation, and G is called the gravitational constant and was measured after Newton's death by Cavendish ($G = 6.7 \times 10^{-11}\ \mathrm{N \cdot m^2/kg^2}$). Of particular interest in the application of gravitation to astronomical systems is the situation of constant-speed motion in a circle, or *uniform* circular motion. The speed in this circular motion, v_c, is given by the formula $v_c = 2\pi R/T$, where R is the radius of the circular motion and T is the *period,* or time to complete one cycle of the motion. Circular motion is accelerated motion, even though the speed does not change, because the object must be constantly pulled out of the straight line it would move along in the absence of forces. The acceleration required in circular motion, a_c, is given by

$$a_c = \frac{v_c^2}{R} \quad \text{or} \quad a_c = \frac{4\pi^2 R}{T^2}$$

Questions

1 Children often ask what keeps the sun up or what keeps the moon up in the sky. What is wrong with such a question scientifically?

2 A car is moving 55 mi/h down a smooth, straight section of highway. What is the total force acting on the car?

3 What evidence assures you that an unbalanced force is acting on an object?

4 When you are standing still, what do you do to begin walking forward?

5 What determines the direction in which an object accelerates?

6 What happens when a person steps briskly out of a small rowboat for a pier?

7 If the engine of a car provides a constant force, why doesn't the car accelerate continuously at a constant rate?

8 Would the acceleration of a rocket be constant during the time the engine is burning?

9 If weight is the force pulling objects toward the earth as they are dropped, why doesn't a very heavy object accelerate downward at a greater rate than a lighter object?

10 Several children are playing with a playground merry-go-round. They run along the outside to get it up to speed and then hop on. What can they do at this point to increase its speed or rotation yet farther?

11 Two cars collide and skid off in a direction determined by the law of conservation of momentum. In a very short time the cars come to a stop. What happened to the momentum which was supposed to be conserved in a collision?

Problems

1 What is the weight of a 1.0-kg mass? A 2.0-kg mass? Since the second mass has twice as much weight pulling it downward, why doesn't it accelerate toward the earth at twice the rate of the first mass?

2 Calculate your weight in newtons. From this calculate your mass in kilograms. If you were on the surface of the moon where the acceleration due to gravity is 1.6 m/s², how would these numbers change?

3 A rocket has a mass of 100 kg. How much force must its engine develop before it can leave the ground? If the engine provided an upward force of 1000 N, at what rate would the rocket accelerate upward? What would the rate be for 1100 N of thrust?

4 A Chevrolet Citation can go from a standing start to 26.8 m/s (60 mi/h) in 10.6 s. What is its average acceleration during this time? If the mass of the car is 1160 kg, what is the average force that is applied?

5 A 20-kg boy is in a frictionless wagon with a mass of 5.0 kg. In the wagon he has 10 bricks with a mass of 1.0 kg each. What happens as the boy throws each brick in the backward direction at a speed of 2.0 m/s? What will be the final speed of the wagon and boy? If the boy then jumps off the wagon in the backward direction with a speed of 2.0 m/s, what will be the final speed of the wagon?

6 Carol has a mass of 50 kg. (Use $g = 10\ m/s^2$.)

(*a*) What would she weigh on a scale reading in newtons?
(*b*) What would she weigh on a scale reading in pounds?
(*c*) If she was wearing frictionless roller skates, how hard would you have to pull her down the hall so that her speed increased 2 m/s for every second you pulled?
(*d*) If you pulled on her for 4 s at this rate, what would happen when you let go?
(*e*) If you tied Carol to the end of a rope and tried to lift her to the second floor, how hard would you have to pull on the rope?
(*f*) How hard would you have to pull on the rope not just to lift her but to accelerate her upward at a rate of $1.0\ m/s^2$?

CHAPTER SEVEN

Work and Energy— What Does Nature Preserve?

Introduction Forces are important in the analysis of motion. Furthermore, we have enough experience with forces to allow us to picture the mechanism of an interaction. But, forces cannot be the entire story for interactions. The reason is simple—nature does not guard them jealously enough! Within reason, we may multiply forces by any amount that we wish. Picture a student driving a large car at highway speed. A truck pulls into the road some distance ahead. With the force developed by a single leg, our student is able to bring tons of metal to a screeching halt in a full-panic stop. This can be done even without assistance from the engine in the form of power brakes, although this would reduce the force that would have to be applied. Nonpower brakes are not like a light switch which activates an outside system. The brakes can perform perfectly well even if the engine, battery, and most of the strange looking pieces under the hood of the car were taken out and thrown away. Indeed, the brakes would perform better under these conditions since there would be less inertia to overcome. The foot on the pedal is the source of *all* the force that stops the wheels from turning. Yet, if the student is out in the highway exerting the same push against the hood of the oncoming car, the force would have a very small effect on the motion. Thus, the brake system must multiply the force of the foot on the pedal somehow. To consider a related example, our student cannot directly lift the car up off the road. Yet, with the use of a few shaped lumps of metal stored in the trunk, he or she is perfectly capable of raising the weight of the car to change a tire. Assume that the student cannot exert a force greater than about 500 N (100 lb) downward on the handle of the jack. Yet, this force is enough to raise one end of the car completely off the ground. A direct lift by a weightlifter would probably require at least 5000 N to accomplish the same thing. Here, again, we have the multiplication of force. What is constant in such situations? Expressed differently, what do we have to give up to be able to attain large forces? Any device which changes the size or direction of a force is called a *machine.* By examination of simple machines it is possible to determine what is sacrificed to obtain larger forces.

Work

Suppose a group wanted to get one large wooden keg of imported beer to the second floor of a fraternity house before a party. The second floor is 3.0 m above ground level. The keg has a mass of 150 kg and a weight

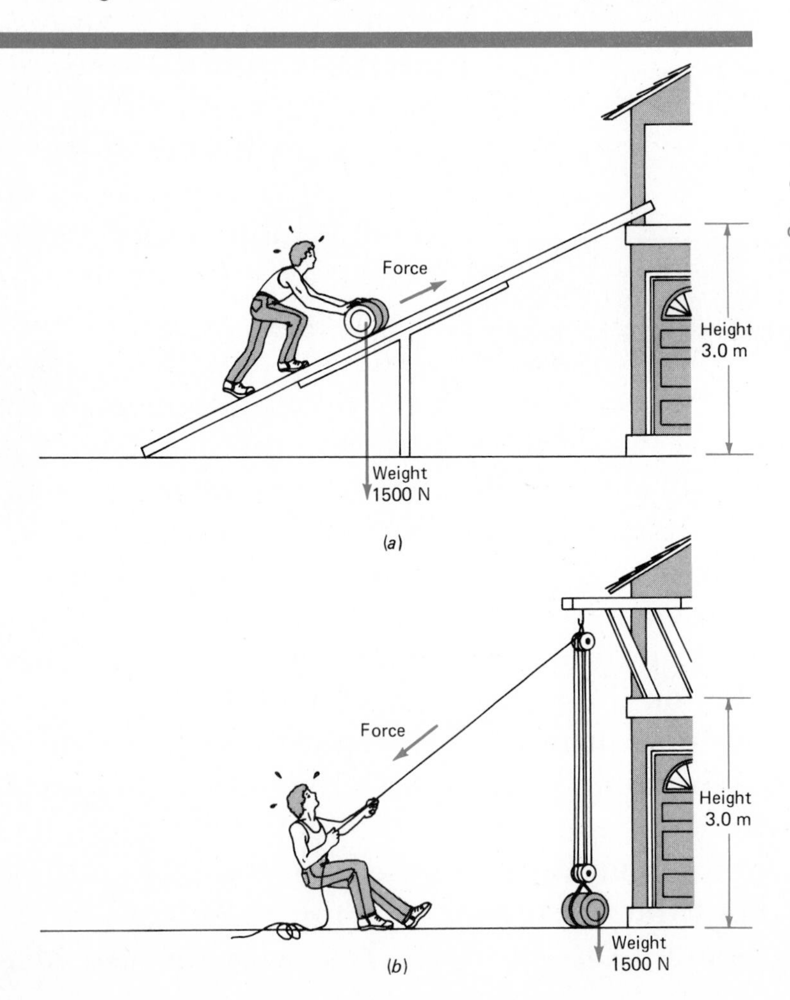

FIGURE 7-1 Machines are devices that allow us to change the size or direction of our force. To raise the 1500-N keg 3.0 m by direct lifting requires too great a force. The inclined plane in (*a*) allows a smaller force to be exerted as does the block and tackle in (*b*).

of 1500 N (330 lb). Clearly, this is too much for one member to lift directly. Consider a few of their options. Two of their largest members could lift the keg, one at each end. Together, they would have to exert a force of 1500 N. They could borrow two 4-m-long beams to make a ramp 8.0 m long to the second-floor balcony. If they roll the keg up this ramp, they would find that a force of 560 N would be required. Alternatively, they could borrow a block and tackle as shown in Fig. 7-1. Using this pulley system, they would find that a force of 380 N would be required to lift the keg.

We could continue this list of options for some time since many devices exist which could be used for lifting the keg. However, the point has been made that each of these options requires a different force to accomplish the same task. It is useful to define exactly what that task is. Here, the task is to raise an object with a weight of 1500 N through a height of 3.0 m. Clearly it is less of a task if the keg weighs less. It is also less of a task if the height through which the keg is to be raised is less. To define the size of such a task, we introduce the idea of *work*. As you have seen, the work involved in getting the keg to the second floor involves two variables, the weight of the keg and the height to which it must be raised. *Work may be defined as the product of the force exerted times the distance through which that force acts.* Or,

$$W = F \cdot d$$

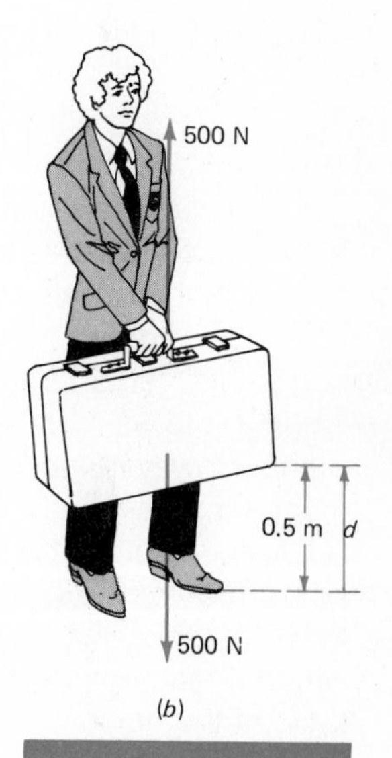

FIGURE 7-2
Work is the product of force times the distance moved in the direction of the force. In (*a*) no part of the distance is in the direction of the force exerted by the bellhop. The upward pull just balances the weight as the suitcase glides forward. As the suitcase is lifted in (*b*), work is done equaling the force (500 N) times the distance lifted (0.5 m).

Example

Note that this definition does not allow blind substitution of forces and distances to obtain meaningful values for the work done. For example, how much work is accomplished by the Statue of Liberty each day in holding her torch up? The answer is none. While large forces are involved, they do not act through any distance. The torch is not lifted; thus the work is zero. If you got a job holding up the roof somewhere you might have to expend great effort. At the end of the day you might say that you had put in a hard day's work. But, if your upward force did not lift the roof you would have done no work in the scientific sense. As used in science, work is a measure of accomplishment, not effort. Forces do no work unless there is some motion in the direction of the force. The force must be exerted through a distance to accomplish work. Another example: A bellhop in a hotel carries a heavy suitcase from the check-in desk to the elevator (Fig. 7-2). The suitcase has a weight of 500 N, and the distance involved is 5.0 m. How much work did the bellhop do? This is a more complicated problem. Clearly, some work was done to pick up the suitcase in the first place. A small amount of work had to be done to get the suitcase moving in the direction of the elevator. To produce motion we must exert a force through a distance. But, presuming the problem begins with the suitcase already in the bellhop's hand and in forward motion, no work needs to be done. The force required here just maintains the height of the suitcase. The force does not raise the suitcase. The force is purely vertical, while the motion is horizontal. No part of the distance moved is in the direction of the force. The bellhop could be replaced by a pillar on a roller skate supporting the weight while the suitcase coasted sideways to the elevator. However, if the bellhop initially lifted the suitcase, at that time he or she did some work. If the suitcase was lifted through a height of 50 cm, we may calculate the work done.

$$W = F \cdot d$$
$$= (500\ \text{N})(0.50\ \text{m}) = 250\ \text{N} \cdot \text{m}$$

A UNIT FOR WORK

Since we will use units for work quite frequently, it is convenient to define a separate unit. *If a force of one newton is exerted through a distance of one meter we say that one joule (J) of work is done.*

$$(1.0\ \text{N})(1.0\ \text{m}) = (1.0\ \text{kg} \cdot \text{m/s}^2)(1.0\ \text{m})$$
$$= 1.0\ \text{kg} \cdot \text{m}^2/\text{s}^2 = 1.0\ \text{J}$$

Thus, a joule is 1 $\text{kg} \cdot \text{m}^2/\text{s}^2$ in standard units of length, mass, and time.

We may now return to our keg of beer. One of our ways to get the keg to the second floor is to use a ramp, or *inclined plane.* The usefulness of such a plane is determined by its tilt or inclination. If the plane is purely horizontal, it takes almost no work to roll the keg along it. Of course, the keg is not being raised either. As the plane is tilted upward through a small angle we must exert an increased force on the keg to roll it up the plane. However, the force we must exert to move the keg up the plane is much less than the weight of the keg. At an angle of 10° from the horizontal, for example, we find that the force to roll the keg uphill is about 17 percent of the weight of the keg. On the other hand, when the plane is much steeper, the advantage in using it drops rapidly. The amount by which a machine multiplies a force is called the *mechanical advantage* (MA) of the machine. If a machine has an MA of 2.0, this would mean it would double your force. The inclined plane with a tilt of 10° allows us to lift (roll upward) a 100-N weight with a force of only 17 N. The plane has allowed us to increase our force almost 6 times. If the heaviest object we could lift directly was equal to our own weight, by the use of this plane we could raise an object almost 6 times as heavy. The MA of an inclined plane varies with its tilt from a very high value for a low inclination to an MA of 1 for a straight up-and-down plane where we are effectively lifting the object directly. The plane available for our keg of beer has a length of 8.0 m and rises to a height of 3.0 m. The force to roll our 1500-N keg up this plane was given as 560 N, so the plane multiplies our force 2.8 times (MA = 2.8). The work which is accomplished at the end of the problem will be to have the 1500-N keg some 3.0 m higher than the level at which it started. This represents 4500 J of work if we lifted the keg directly. As we actually use the ramp, we push on the keg in the direction parallel to the surface of the plane. We push with a force of 560 N and must exert this force through the length of the plane. The work we do on the keg is $W = F \cdot d = (560\ \text{N})(8.0\ \text{m}) = 4500\ \text{J}$. Thus, even using the inclined plane, we have done as much work at the end as if we had lifted the keg directly. The use of the plane did not save us work, it allowed the use of a smaller force to accomplish the same task. That force, however, had to be exerted across a greater distance to accomplish the same task. The work done depends upon the accomplishment—raising a 1500-N weight 3.0 m—not on how it is done. No machine requires less than 4500 J of work input to accomplish this task, although many will require more than this to make up for wasted motions. A comparison of how much work must be put into getting a particular task accomplished with the work that task represents is called the *efficiency* of a system. With the inclined plane, we accomplished a 4500-J task. To do this we did 4500 J of direct work. The efficiency would be 4500 J/4500 J = 1, or, more normally, we would multiply by 100 to express this as a percentage.

Simple machines frequently allow us to get almost as much work out as we put into them—they can have high efficiencies. More complicated devices usually have lower efficiency.

The second machine to raise the keg is a pulley system. To lift an object we must pull on the free end of the rope which winds around the pulleys. The work that we do is to exert a force F through a distance equal to the amount of rope we pull out of the system. If we pull 1 m of rope past the last pulley, there must be 1 m less rope circling the wheels in the pulley system. Since there are four segments of rope connecting the upper pulleys and the lower pulleys, each segment must become $\frac{1}{4}$ m shorter to make up for the loss of the rope from the system. Thus, when we pull through a distance of 1 m with our force, the weight is lifted by $\frac{1}{4}$ m.

Neglecting losses from work which we do but which does not contribute to lifting the keg, the work we get out should equal the work we put into the system.*

$$\text{Work in} = \text{work out}$$

$$F(1.0) = (mg)(0.25)$$

This implies that the force we must exert to lift the keg is just one-fourth the weight of the keg, $F = \frac{1}{4}mg$. Thus, this pulley would multiply our force by 4, or MA 4.

Hiding throughout this section is a very important principle which we will state here. *In all of the collected experience of the human race, no device has ever been encountered in which the work out is greater than the work put into the device.* The best we can do with a device is to get out all of the work we put in as useful work. Usually, we do less well.

* Such losses would include stretching the rope. To stretch a rope you must exert a force through a distance. Also included would be the bending of the beam supporting the pulley, overcoming the friction to allow the pulley wheels to turn, and many other types of unwanted work. Most of these are very minor for a simple device, allowing most of the work we do to be the useful, intended work we want to do. Contrast this with the situation in a complicated device. An automobile uses less than one-third of the work potential of the fuel for motion, and your muscles manage to use only one-fifth of the work potential of your fuel in doing external work.

The italicized statement conveys the correct idea about work but is far from being a completely satisfying statement scientifically. We will continue to clarify it for some time to come.

What do we lose or give up when we use a machine? Clearly we can gain force, but what must we sacrifice? To lift a keg 3.0 m we had to exert our push through the 8.0-m length of a ramp. With the pulleys we had to pull 12.0 m of rope through the pulleys to lift the weight by 3.0 m. If we gain force, we must have our force act through a greater distance. This also works in reverse in that if we will settle for a smaller force, it can act through a greater distance. In this way we can gain speed of action at the expense of force. In the act of throwing a baseball, the shoulder muscles contract only through a distance of perhaps 10 cm. This force is used to whip the arm around rapidly and accelerate the ball to a high speed. If we tried the same trick with a bowling ball, we would rip the shoulder out of its socket. Normally, we can generate forces large enough to manage a bowling ball, but in this application we have sacrificed force for speed of action. The force available is inadequate to hurl the bowling ball casually without exceeding the limits of human construction. Within reason, we can have a large force or we can have great speed of action, but not both for a single source of work. A single source of work can generally do work at a given rate. A large football player might have raised the keg to the second floor in 10 s. His rate of doing work would be 4500 J in 10 s, or 450 J/s. One of the fraternity brothers using the ramp might have taken 60 s to do the same task. He accomplished 4500 J of work in 60 s, or worked at the rate of 75 J/s. The football player is correctly described as being more powerful. *Power is the rate at which work is accomplished.*

$$\text{Power} = \frac{\text{work}}{\text{time}}$$

When discussing power the standard unit is the *joule per second,* and this is called the *watt* (W). Thus, we have a 450-W football player and a 75-W frat member. Humans are capable of bursts of work at higher power, but these numbers bracket the range of sustained adult performance fairly well.

THE WORK TO LIFT OBJECTS—POTENTIAL ENERGY

To lift a 7.0-kg bowling ball up to a closet shelf, we must do work. The work that is done is the force we exert times the distance through which that force acts. If we know that the closet shelf is 2.0 m high, we can calculate the work that must be done. To lift the ball we must exert an upward force equal to its weight, mg, or (7.0 kg)(9.8 m/s^2) = 69 N. We will have to exert that force the entire height from floor to shelf to lift the ball through this distance. Thus the work done will be $F \cdot d$ or (69 N)(2.0 m), which is 140 J.

Now what? We have done 140 J of work. Our bowling ball is resting quietly on the shelf. We could view this as the end of the problem, but it is not. A very important idea can now occur to us. What has happened to the work we have done on the bowling ball? Is it gone—lost forever? The answer has probably occurred to anyone who has stored a loose bowling ball on a closet shelf only to have it roll off and land on a foot. A bowling ball at a considerable height can become a lethal weapon. The ball is perfectly capable of exerting forces through distances on the way down. The ball possesses the capacity or ability to do work by falling from its shelf. Picture the bowling ball on the shelf for a minute. Imagine that we attached a wire to the ball and ran it over pulleys to a small weight on the floor (Fig. 7-3). As the bowling ball falls, it can raise the weight. In fact, if the pulleys were frictionless and required no work to turn, the bowling ball could lift any weight up to its own weight.* If, we attached a second, identical bowling ball at the bottom, the upper ball could balance exactly its weight. Since the forces would be balanced, we would not expect motion to occur. An unbalanced force is required to overcome inertia and start the system in motion. If, however, a poorly flying moth collided with the top of the upper bowling ball, this force could start the system moving. Once moving it would continue in motion at constant speed since all forces are balanced. Thus, in falling, our bowling ball can raise an identical weight through the same height as it falls. *The work we did to get the bowling ball up to the closet shelf was not lost, it was stored in the position of the bowling ball.* The ball does exactly as much work on the way down as we did in raising it in the first place. Namely, it lifts a bowling ball of weight 69 N through a height of 2.0 m. The work we put into a system may come back out as work. Thus, a bowling ball on a shelf possesses the ability to do an amount of work equal to its weight times its height: Potential work = mgh.

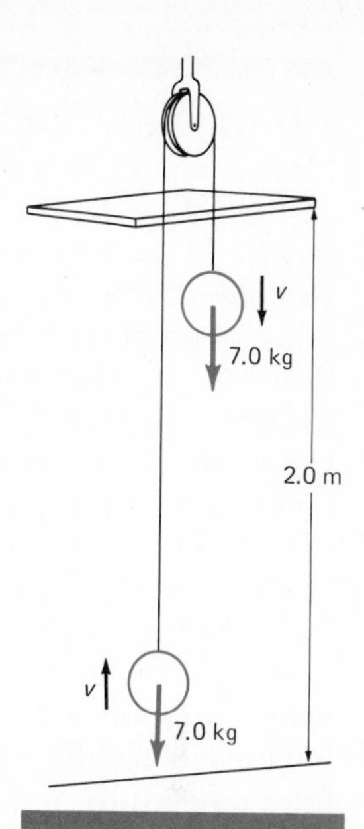

FIGURE 7-3
If we roll the bowling ball off the closet shelf and give it the slightest of starts downhill, it has the ability to raise an identical ball to the shelf as it falls. Thus, the work performed in raising the ball to the shelf is not lost, but is stored in the position of the ball, or, as potential energy.

Rather than talk about stored work we define a new term. *Energy* is the ability or capacity to do work. Anything that can do work—can exert forces through distances—is said to possess energy. Any weight capable of falling has energy. Energy which is stored in the position (or condition) of an object is called *potential energy.* Until the situation develops where the object may lose this potential energy by changing its position the energy remains stored. For example, the work done by the ancient Egyptians in hauling the topmost block to the top of the great pyramid of Cheops remains stored in the position of that block. The energy remains there until released. Once the block begins to fall it may do work in shattering stone blocks, crushing camels, and digging pits in the sand at the base of the pyramid. All of these require that forces be exerted through distances—all require that work be done. How much work is done depends upon the weight of the object times the distance through which it falls:

$$E_p = mgh$$

* In practice it is difficult to engineer a truly frictionless system on earth. But we can come close. During construction of the 5-m (200-in) Hale telescope on Mount Palomar, the work crew broke for lunch one day. By mistake they had left the telescope locking mechanism off. This was no problem in itself since the entire telescope mount was precisely balanced and stayed in place. A worker eating lunch in the dome rested a lunchbox on the huge telescope frame. When the worker looked back later, the entire telescope mount was in rotation from the added weight. Some 500 tons in motion from the force of a lunchbox. (It is difficult to resist making the conclusion that it obviously was not a light lunch.)

Life as a lever

Motion of a human being depends upon the ability of certain tissues to shorten their length by a few percent on demand. The strength of this contraction is determined by the cross-sectional area of the muscle taken through its thickest part. Thus, bulging biceps really do imply a greater force of contraction—if it's all muscle. Almost all muscles are designed to pull, not to push, a notable exception being the tongue. If your arm was one long muscle and you wanted to lift something to your mouth, you would have problems. Such an arm that began 1 m long would still be over 90 cm away from your mouth when fully contracted. Muscles contract powerfully but not through much distance. To solve the problem of obtaining considerable motion from the small distance of the contractions, all higher life forms use systems of levers. This is similar to increasing the size of your stride by running around on stilts. The levers are the *bones* of the skeletal system. Muscles attach to bones by tough, noncontracting fibers called *tendons.* During the course of his dissections, da Vinci noted that muscles start on one bone and end on the next bone in line. Actually, bones do not touch each other. Pads of smooth *cartilage* form nearly friction-free contacts between bones at the joints while the bones are held to each other by tough fibers called *ligaments.* If you picture yourself in a race with a giant amoeba, you can realize that the human system of motion on built-in stilts provides for very rapid motions compared to a less-organized blob of matter. We also pay the penalty for this levering system.

Perhaps the best known system is that of the human forearm. The large biceps muscle raises the forearm, while the smaller triceps muscle pulls to lower it. Since muscles only pull, they frequently occur in sets such as this with one undoing the action of the other. (Of course, they are not used at the same time any more than you use "drive" and "reverse" on your car at the same time.) The biceps attaches to the forearm to form what is called a *third-class lever.* In such a lever the lifting force is applied to the lever between the pivot point, or *fulcrum*, and the load which is to be lifted. Because the load is way out on the end of the lever, if the biceps contracts to raise the forearm by 1 cm, the load will be swung up by about 10 cm. If the load had a weight of 100 N, the work resulting from this contraction is (100 N)(0.1 m) = 10 J. The biceps had to do at least this much work to

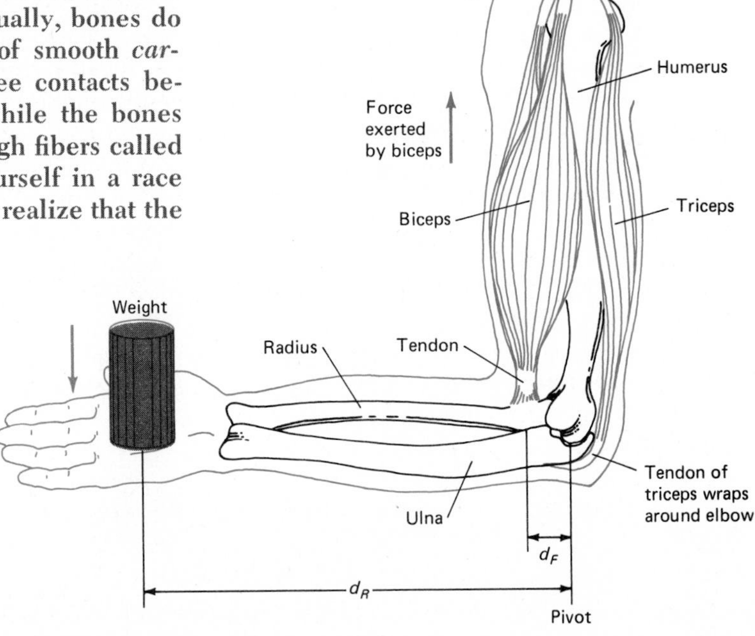

raise the load (it also had to raise the arm itself). Thus, the minimal force exerted by the biceps is $W = F \cdot d$, $F = (10\ \text{J})/0.01\ \text{m}) = 1000\ \text{N}$. The biceps had to exert a force 10 times as large as the weight being lifted! All third-class levers sacrifice force for speed, and the human body has mostly third-class levering systems. This is a price we pay for rapid, smooth motion. The biceps in your arm could lift 10 times the weight if attached directly as it can through the system of your arm. Think about that the next time you chin yourself and the force comes almost entirely from your biceps. The large forces generated by the muscles require a very strong bone structure. Bones will withstand about 10^4 N/cm^2 (6 tons/in^2) before being crushed or four-fifths of this before they are pulled apart. Bones can withstand most of the steady forces normally encountered quite well. Sudden forces, such as in a collision, may be quite large momentarily and are the cause of most bone fractures. Some muscular activities do require very large forces to be exerted when the direction in which the muscles must pull lies close to the fulcrum. For example, the simple act of holding your arm straight outward requires a large force of the muscles since they must act near the fulcrum in the shoulder. A weight placed in the outstretched hand may require as much as 100 times its weight in the force the shoulder muscles must exert to support it. Even supporting the weight of the arm is tiring. Try it by holding your arm out for a few minutes and notice the relief in the muscle when you stop. One serious problem area is in the lower back when you bend over. The forces of the bones pressing against each other easily reach 3 times your weight—provided you are not trying to lift anything but yourself!

One of the unique characteristics of humans is our erect posture. Standing places the load of our body on the bones and leaves most muscles resting. Standing also frees the arms for other things. Another unique item is our hands. The same hands that can tear a small animal apart can assemble a watch. Paleontologists argue which comes first, the human brain or the erect posture and prehensile hand. Whatever the result of such a "chicken-vs.-egg" circle, we should note that both are required to produce a civilization. The hand attains its agility with strength by getting the bulky muscles out of the way. The muscles directing your fingers are between the two bones of the lower arm and are attached to the fingers by long cordlike tendons. You can observe these tendons moving under the skin at the back of the hand. The one large muscle remaining in the hand pulls the thumb over to press against the fingers in grasping. This is the fleshy blob at the base of the thumb. Imagine what chance you would have to assemble a watch if you had 10 or more such muscles on the hand to direct the fingers. A phrase to describe the clumsy is to say they're "all thumbs." This is what we would all be without good design.

The potential energy of an object may vary depending on what level it falls to. For example, our bowling ball might have fallen from its original shelf to another shelf halfway to the floor. The potential energy released would be only one-half of its potential energy relative to the floor, or one-half of the 140 J it took to raise it from the floor to the high shelf. On the other hand, the bowling ball might just keep going all the way to the floor. Once there it might be able to crack right through the floor and fall to a lower story. If we are on the top floor of an apartment house, we may picture the ball doing work against floor after floor as it falls and winds up in the basement. If we determine the work done by the ball we will find it much greater than the 140 J we had calculated it to be originally. This is because we had calculated the work to lift the ball from the floor, our floor, to the shelf. We had not considered the work to lift the bowling ball to the top floor of the apartment house in the first place. Potential energy is always calculated from one position to another position. If a lower position is available, there is still potential energy which could be converted to work. *There is no convenient level at which we could say that a weight has no potential energy.* If we use ground level, the weight could always fall down a mine shaft. If we pick the bottom of the deepest mine, it could fall into an undiscovered cavern lower than that, and so on. Even a mass at the center of the earth possesses gravitational potential energy. If you could overcome its inertia along the earth's orbit somehow, such a mass would be attracted into the sun. The mass could exert forces through distances while falling to the sun just as our bowling ball can as it falls toward the earth. Futhermore, why should we believe the center of the sun is as low as the ball can get? It isn't! Thus, potential energy is *not absolute; it is relative.* Mass, for example, is an absolute variable. We know perfectly well what zero mass means. We do not know what the absence of any potential energy for a mass means. Therefore, whenever we calculate a potential energy it must be from some defined level to another defined level.

THE WORK REQUIRED TO SPEED UP A MASS—KINETIC ENERGY

Suppose our bowling ball rolls off the closet shelf and falls. While it is halfway down we snap a photograph of the ball as it is falling. The bowling ball is not lifting another weight in this case. The work required to move the air out of its way as it falls is negligible (at low speed). It is fine to talk about the work the ball can do against the floor when it collides. But, in our picture, it is halfway down. It has lost half its height above the floor; thus, it has lost half of its potential energy. Does the work stored as potential energy disappear during the fall only to reappear as work done against the floor? Where is the lost work at the instant of our picture? The answer is that it is hiding in the blurred image of the ball. At the time of the photograph the ball is in rapid motion. What does it take to speed up a mass? From all of your experience with shopping carts, coaster wagons, and even bowling balls (when used for their intended purpose) you know the answer to this question. To speed up a mass you must exert a force through a distance—you must do work. If we observe the middle part of a bowling alley, we can see a ball whizzing along at some constant speed on its way to its target. How much work had to be done on the ball to get it to that state of motion? The ball had to be speeded up from a state of no motion to the state in which we observe it. To do this, an unbalanced force had to applied. For simplicity, assume that this force was constant for the time during which it was applied to the bowling ball. From the second law of motion, this force was equal to the mass of the bowling ball times its acceleration, $F = ma$. If the unbalanced force was constant, so was the ball's acceleration during the application of the force. The work that was done on the ball was this force times the distance through which it acted,

$$W = F \cdot d$$

or, substituting for F,

$$W = mad$$

Here we may use an expression developed for motion under a constant acceleration in Chap. 3 (Eq. 4a), $v_f^2 = 2ad$, or $\frac{1}{2}v^2 = ad$. The work done to speed up a mass is, substituting for ad,

$$W = \tfrac{1}{2}mv^2$$

Like potential energy, this work is stored in the ball. In this case it is stored in the motion of the mass of the ball. A moving bowling ball is certainly capable of exerting forces through distances. Whenever a mass is in motion, it possesses the ability to do work; it has energy. We call the energy due to the motion of a mass *kinetic energy,* and calculate it as above,

$$E_k = \tfrac{1}{2}mv^2$$

The bowling ball in our photograph lost potential energy as it fell. This energy has gone into the increased speed of the bowling ball during falling. All the while the bowling ball has been losing height, it has been gaining speed. If all of the energy of the bowling ball is to be accounted for, the potential energy lost should equal the kinetic energy gained. Thus, an instant before the ball hits the floor it has lost almost all of its potential energy. How much motion should the ball have? What should the speed of the ball be at the collision with the floor?

$$(E_p)_{lost} = (E_k)_{gained}$$

or $$mgh = \tfrac{1}{2}mv^2$$

The variable m drops out, and

$$2gh = v^2$$

or $$v = \sqrt{2gh}$$

$$= \sqrt{2(9.8\ \text{m/s}^2)(2.0\ \text{m})}$$

$$= \sqrt{39\ \text{m}^2/\text{s}^2} = 6.3\ \text{m/s}$$

Viewing the process of falling from the standpoint of energy transfer gives us a view Galileo would have appreciated. He maintained that all objects would gain speed at the same rate in an ideal case. The small differences which were observed were attributed by Galileo to air resistance. We can see this effect clearly from the standpoint of energy. Ideally, as an object falls all of its potential energy would go into kinetic energy. In air, however, some work must be done in moving the air out of the path of the falling object. Suppose we drop two spheres, each with a diameter of 10. cm. The spheres are made of lead and wood. The lead sphere will have a weight of 58 N (13 lb), while the

wooden sphere weighs only about 4 N, one-fourteenth as much. When the lead sphere drops 1 m, it loses 58 J of potential energy, while the wooden sphere loses only 4 J for the same drop. Yet, both spheres are of identical size and shape and must do the same amount of work on the particles of air in their paths to get them out of the way. Potential energy expended in doing work against air molecules is not converted into the kinetic energy of the spheres. Consider the case in which the spheres are traveling very fast and must do 4 J of work against the air for each meter of travel. At this point, the wooden sphere expends all of the potential energy lost in falling in speeding the air out of its way. No energy is converted to kinetic energy, and the wooden sphere falls at constant speed. This is called the *terminal velocity* of the sphere and would be around 24 m/s (53 mi/h).* At this same speed, only 4 J of the 58 J of potential energy lost by the lead in each meter of dropping goes to fighting the air. The majority (93 percent) of the energy of the lead still goes into speeding up the sphere. Thus, the effects of air resistance are quite different on objects of different densities. Yet, as Galileo stated, all objects, even feathers and coins, fall at the same acceleration in a vacuum. In that case, all of the potential energy lost goes into kinetic energy, $\not{m}gh = \frac{1}{2}\not{m}v^2$ or $2gh = v^2$. The size of the mass of the object has no effect on the rate of the acceleration or on the final speed.

Conservation of Mechanical Energy

Let us examine the exchange of potential energy and kinetic energy more carefully. Consider Tarzan doing his famous vine-swinging act. As he begins, Tarzan is some height above the bottom point of the swing (Fig. 7-4). This is the greatest height he will have so we will label it with a capital *H*. Following the required bloodcurdling yell he will swing freely. As he swings his height becomes less, and he loses potential energy. By the time he has reached the lowest point in the swing, the potential energy lost has been *mgH*, his weight times the height through which he has fallen. He has also moved a considerable distance sideways, but horizontal motion does not involve doing work with or against the force of gravity. Thus, there is no potential energy change associated with the horizontal movement. Unless he lets go of the vine or it snaps, this is as low as he will get during the swing. The potential energy which has been lost has been associated with his increasing sideways speed. At the bottom of the swing he is moving as rapidly as he will at any point in the swing, *V*. After the bottom is reached the only way to go is uphill. This means that his potential energy will be increasing. The only source for this increase in potential energy is in the energy of his sideways motion. Kinetic energy is transferred back to potential energy, and Tarzan slows down as he rises on the far side of his swing. Excluding work done in de-

* The terminal velocity for a human falling through air is about 63 m/s (140 mi/h). For anyone landing on a hard surface this velocity would be truly terminal. A fall of about $\frac{1}{2}$ m onto concrete landing flat on the heels with the knees locked is adequate to snap the human bones. This is why parachutists must land with knees bent and proceed immediately into a forward roll. This procedure extends the time over which kinetic energy is delivered to the ground and keeps the forces of landing low enough to survive. Some remarkable escapes from falls have occurred in spite of all of this. In one case a pilot was ejected from his plane several miles above the surface. His parachute failed to open and he plunged to the ground, hitting at the terminal velocity. In landing he plunged into a pine forest snapping limbs as he dropped (his and the tree's). The stopping process was extended over enough distance that the pilot survived. In addition to breaking his fall, the tree managed to break some 200 bones in the pilot's body, which is essentially all of them. Even more remarkable was such a fall into a thick snowbank from which the aviator walked away without physical injury.

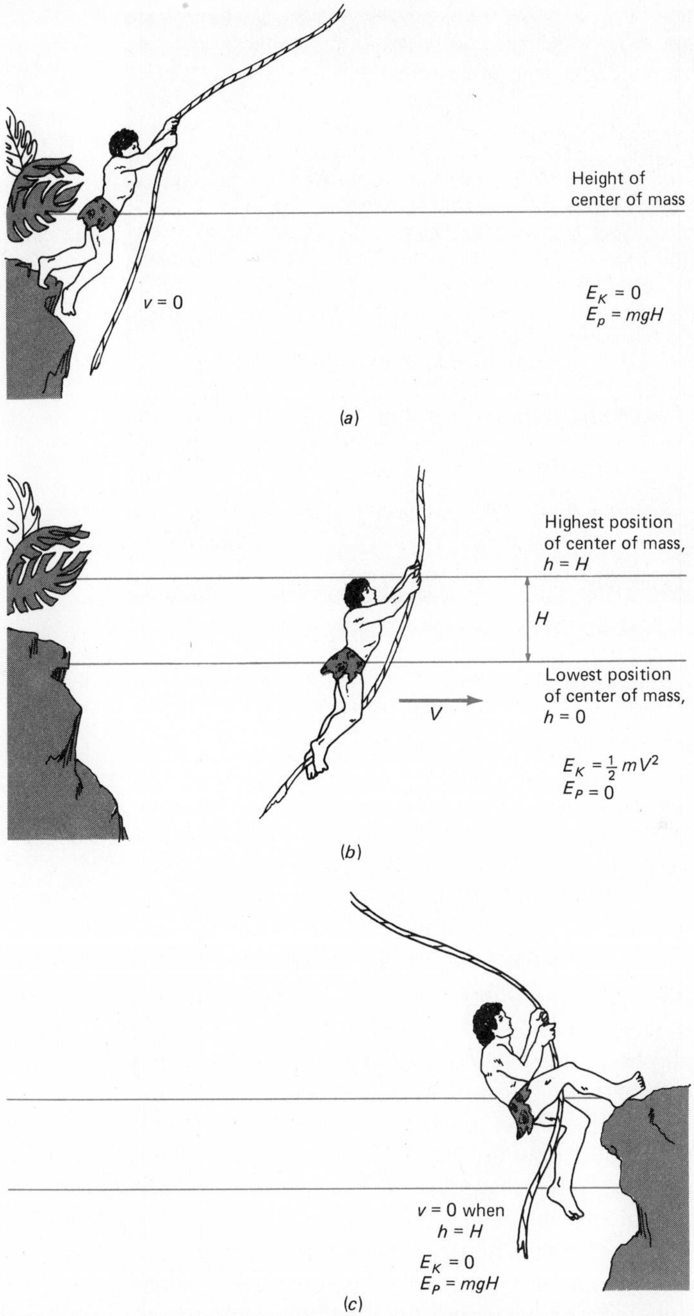

Height of center of mass

$v = 0$

$E_K = 0$
$E_p = mgH$

(a)

Highest position of center of mass, $h = H$

H

Lowest position of center of mass, $h = 0$

V

$E_K = \frac{1}{2} mV^2$
$E_P = 0$

(b)

$v = 0$ when $h = H$

$E_K = 0$
$E_P = mgH$

(c)

FIGURE 7-4
As he poises above the abyss (*a*), Tarzan's energy is entirely due to his height since he has no motion. At the low point of the swing (*b*) he loses all of his height but is traveling at his greatest speed, *V*. As he arrives at his original height on the far side (*c*) his speed drops to zero. In each case, he has the same total energy. For (*a*) and (*c*) the energy is entirely potential energy, while in (*b*) it is entirely kinetic energy. Between these positions he would have some of each form of energy, but the total energy would be the same for any of the positions shown.

forming the vine or in pushing the air out of his way, he should have enough energy to reach the same height from which he originally fell. If his target is higher than this, it would be time for a second bloodcurdling yell as he fell backward on the return swing.

The assumption which runs throughout this interpretation is that en-

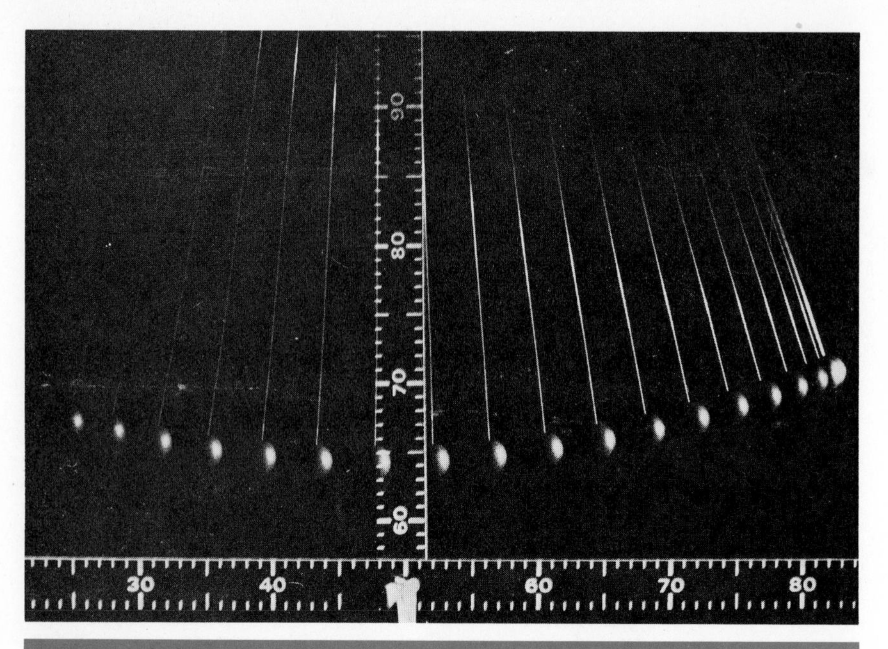

FIGURE 7-5 Stroboscopic photograph of a single pendulum swing. Pictures are taken at equal time intervals. What happens to the pendulum's speed as it nears the bottom of its swing? The top? How can you tell? At what point is the potential energy least? The kinetic energy? (*Photograph by the author.*)

ergy is a constant in this situation. The energy may be shuffled from potential energy to kinetic energy, and vice versa, but its total amount is constant. This is a rather pleasant idea, but, before making such a conclusion it is useful to examine it experimentally. We could measure the separate photographs of such a swing from a motion picture. The camera takes 24 separate still pictures per second, and when flashed on a screen at this rate the motion looks continuous.* But each separate picture records a position, and the time between pictures is known. From this we could measure Tarzan's average speed between two adjacent pictures and his average height. The problem is to judge exactly what point on Tarzan to measure as the arm and legs are extended or drawn in. It is simpler to use a system which is less complex. Figure 7-5 shows a photograph of a swinging pendulum. The photograph was made with a light source which flashed 20 times per second. Each flash lasted only 30 microseconds (μs) and has recorded the pendulum as "frozen" in a single position. We can measure how far the pendulum moved in the 1/20 s between flashes, and from the distance and the time we may obtain the average speed for that interval. Figure 7-6 shows the relationship of the kinetic energy ($\frac{1}{2}mv^2$), the potential energy (mgh), and their sum, the total energy, during the swing. The sum is constant. From careful measurements of this type of photograph we would be allowed a conclusion. *For isolated mechanical systems the sum of the potential energy and kinetic energy is constant.* This is known as the *law of conservation of mechanical energy,* and is almost correct. As you know, swinging pen-

* The motion looks continuous today. Early motion pictures were sometimes taken and shown at a slower rate, and the blank screen between pictures became perceptible as a flicker. The eye "remembers" an image for about $\frac{1}{16}$ s once it is seen. If the next picture is up on the screen before this, the eye has not recovered enough to record the blank screen and sees one image "melt" smoothly into the next in continuous motion.

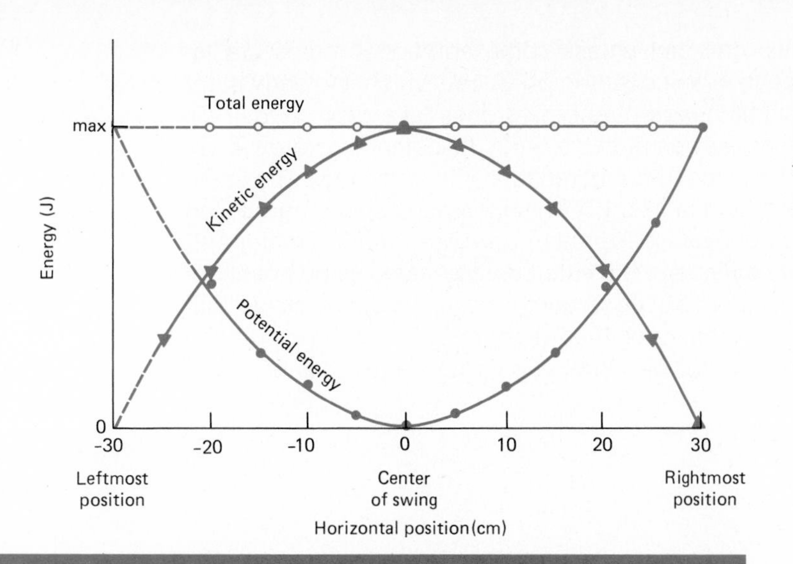

FIGURE 7-6 A graph of potential energy, kinetic energy and their sum, total energy, vs. the distance from vertical position in the photograph of Fig. 7-5. What can you conclude from the graph?

dulums run down. The swinging lasts much longer in a vacuum chamber and with some nearly frictionless attachment. Nevertheless, they do run down and stop. We can simply say that this means that the system was not "isolated," which is a condition of the law. *Isolated* means that there is no interaction with anything that is not part of the system, that all forces and exchanges of energy are between the parts of the system. The requirement that the system be "mechanical" means that no energy is stored in any form but motion or position—kinetic or potential.

One example of an isolated mechanical system is a satellite in orbit around the earth. A body in a circular orbit maintains constant "height," hence constant potential energy. From this we may conclude that its

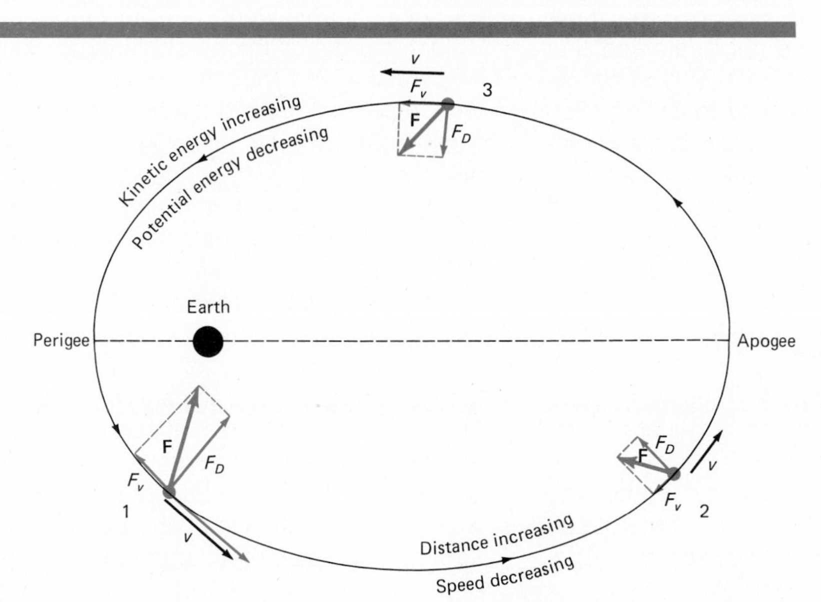

FIGURE 7-7 The gravitational force pulling a satellite for three positions in its orbit is labeled *F*. This may be divided into two parts. F_D is perpendicular to the satellite's motion *v* and can change only the direction of motion. F_v is parallel to the satellite's motion and will increase or decrease its speed. From perigee, past 1 and 2, to apogee, F_v is always in the opposite direction from *v* and thus slows the satellite down. What does this imply about energy changes?

kinetic energy is also constant unless some additional forces are involved. If our satellite is in an eccentric orbit, one which is clearly elliptical, or oblong, an interchange of energies does take place (Fig. 7-7). As the satellite approaches nearer to the earth, its potential energy is decreased. There is a corresponding increase in kinetic energy. The satellite moves at greater speed in its orbit. This increased speed carries the satellite past its point of closest approach (*perigee*) and to greater altitude. However, as the altitude of the satellite increases, its potential energy increases at the expense of its kinetic energy. The speed slows until the farthest point in the orbit is reached (*apogee*). For each point in the orbit, the sum of the potential and kinetic energies is a constant.

Summary

Nature allows the size and direction of forces to be changed or transformed quite easily. Any device which transforms a force in size, direction, or both size and direction is called a *machine.* The number of times a force is multiplied by a machine is referred to as its *mechanical advantage* (MA = force out/force in). Since force is not held constant, it is useful to inquire as to what does appear to be constant in all normal transactions. This turns out to be the variable named *work.* Work is defined, in its scientific sense, as *the product of the force times the distance through which this force is exerted* ($W = F \cdot d$). The mere exertion of a force does not guarantee that work is done; instead, this force must be exerted through some distance to accomplish work. If force is measured in newtons and distance is given in meters, work is measured in newton-meters ($N \cdot m$) or *joules* (J). For all transactions examined, *work in = work out,* although the measurement involved may be complex except for simple machines. The work out of a machine, however, may not all contribute to the task desired; thus only a fraction may be *useful work.* Thus, work done combatting frictional forces is generally not useful (or recoverable) work. The useful work out divided by the work put into a machine gives the machine's *efficiency.* The rate at which work is done is called *power* ($P = W/t$). A rate of one joule per second is called one *watt* of power.

Work done on an object can change its condition in such a way that the work may be recovered later. The work expended in raising a mass to a height may be recovered as the mass falls. The mass has stored in its position the ability or capacity to do future work. Such stored work is called *energy,* and is the capacity to exert forces through distances under certain conditions. The work stored in the increased height of mass is called gravitational *potential energy* (E_p). It is equal to the work required to raise the mass to its position. This work was the weight of the object, mg, times the height through which it was raised, h. Thus, $E_p = mgh$. Similarly, work may be stored in the motion of a mass. Such stored work is called *kinetic energy,* or energy of motion (E_k). This is equal to the work required to get the mass up to its speed and is equal to one-half its mass times its speed squared ($E_k = \frac{1}{2}mv^2$).

For a simple mechanical situation in which negligible work is done against frictional forces the sum of the potential and kinetic energies is constant ($E_p + E_k = k$). This is the law of conservation of mechanical energy. It was the forerunner of the more global law of *conservation of energy,* which was to emerge much later and become one of the key principles of all science.

Questions

1 You toil all weekend to complete three major papers and two assignments due Monday and study for a test. You are really proud of all of the work you accomplished when some science major informs you that you didn't do any work. Why?

2 A pillar is cracked in the temple and Sampson is called in to hold up the roof while it is replaced. For $\frac{1}{2}$ h he must exert a force of 2000 lb to keep the roof in place. How much work did he do?

3 If muscle fatigue is not a measure of work accomplished, what does it measure?

4 How do brakes stop a car?

5 Is a doorknob a machine? A light switch? A nail? A screwdriver? A hammer? On what basis do you judge—how can you tell?

6 What evidence can you provide that a nutcracker is a machine?

7 Assume that a 1000-kg car fell off the top of a dam and did no work until it hit the river surface. Where did its lost potential energy go? At what speed will it hit the water?

8 Explain how you would determine the kinetic energy for a position shown in the photograph of Fig. 7-5. How would you determine the potential energy?

9 Seymour Simian raises a 600-lb box of toothpicks to a loading dock in a single motion. Cuthbert Lemonfizz has his trained termites take one toothpick each and reassemble the load on the dock. Since the same work is accomplished in each case, what is the main difference between Seymour and the termites in this case?

10 At the back of the jaw near the ears you can feel the surfaces on which the jaw pivots. By lightly clenching and releasing the teeth you can feel the jaw muscles flexing forward from this position. Have a friend measure the horizontal distance from the line of the pivot to the line of the muscle. Next measure the distance from the line of the muscle to the front teeth of the lower jaw.

(*a*) From these numbers, if the front teeth exerted a biting force of 20 lb, with what force would the jaw muscles have to contract?

(*b*) In moving the jaw, which swings through a larger arc, the front teeth or the biting muscles?

(*c*) Pressure is the force exerted divided by the area across which it is applied. The front teeth (incisors) are designed to chop up food. Assume that the 20-lb biting force above is transmitted by a single set of incisors to a raw carrot. If each incisor is 0.3 in wide and 0.05 in thick at its biting surface, what is the pressure crushing through the surface of the carrot where it is contacted by the teeth?

(*d*) On either side of the row of incisors are the canines, of Count Dracula fame. Unlike the wedge-shaped incisors, these are roughly cone-shaped. Does this increase or decrease the pressure during biting?

(*e*) Farther back in the mouth are the flatter molars intended for crushing. Can a tooth farther back in the mouth exert more force or less force during biting?

(*f*) The ridge of bone under the eye (the zygomatic bone) is much heavier in proportion in a wild primate, such as the chimpanzee or gorilla, than in a human being. Why would this bone be heavier in a wild creature?

Problems

1 What force is required to lift an 8.0-kg watermelon? What work would be done if it was raised 1.0 m?

2 Sally has a mass of 60 kg and climbs up a 3.0-m rope in the gym. How much work does she do? Later, she walks up the stairs to the second floor some 3.0 m higher. How much work does this require? What is the chief difference between these two actions in terms of the work?

3 A toy car is pushed down a ramp and drops 80 cm before coming out onto a smooth horizontal surface. What is the fastest the car can leave the bottom of the ramp from a standing start at the top?

4 Little Davy has a mass of 20 kg. He is dragged 5.0 m along a floor to a bathtub at a constant slow speed. The force required is a constant 50 N, disregarding jiggles due to screaming and kicking.

(*a*) How much work was done?
(*b*) How much was Davy's potential energy changed?
(*c*) How much was Davy's kinetic energy changed?
(*d*) Work is called useful if it goes into some recoverable form. How much of this work was useful?
(*e*) Where did the work done in this case go?
(*f*) How much of a force would it take to lift Davy?
(*g*) What fraction of this force was required to drag him across the floor?
(*h*) This fraction (force to drag/weight) is called the *coefficient of friction* of the surface. Typical values are given in the table below. Was the surface leading to the bathtub quite rough or smooth?
(*i*) The harder and smoother a bathroom surface, the easier it is to clean. However, this also lowers the coefficient of friction. What problem arises here?
(*j*) What application is made of the low-value steel on ice in the table?

SURFACES	COEFFICIENT OF SLIDING FRICTION
Rubber on concrete	0.7
Leather on wood	0.5
Steel on wood	0.4
Steel on steel	0.2
Steel on ice	0.01
Bone on joint	0.003

5 A cubic meter (1 m^3) of water contains 1000 liters and has a mass of 1000 kg.

(*a*) What is the potential energy of 1.0 m^3 of water 3.0 m above the surface of a river?
(*b*) If this water is allowed to fall to the river level, how much work can it do?
(*c*) If exactly 1 m^3 of water is allowed to fall to the river each second, at what rate can work be done?

6 We may repeat the parts of the previous question for a real case. Hoover Dam is about 220 m high.

(*a*) What is the potential energy of 1 m^3 of water at the surface of Lake Mead behind the dam compared to the Colorado River at its base?
(*b*) If 2000 m^3 of water is allowed to fall from this height through the powerhouse to the river each second, how much potential energy may be converted to work each second?
(*c*) If one-third of this work may be converted to electric energy, what would be the electric output of the dam?
(*d*) The actual output of Hoover Dam is 1346 MW. Assuming the numbers given to be correct, what is the actual efficiency with which the potential energy is converted into electricity?

CHAPTER EIGHT

Thermal Energy

Introduction During the early part of the scientific era there was no clear distinction made between heat and temperature. Temperature was widely believed to be a measure of the intensity of the heat, or how "tightly packed" the heat was in an object. A similar case would be pressure, which is a measure of how concentrated a force is over an area. Pressure is the force per unit area ($P = F/A$). The pressure in a bicycle tire might be higher than that in an automobile tire, but the total outward force against the walls of the automobile tire is much greater due to the greater area. Thus, a glass of water and Lake Erie might be at the same temperature, but clearly Lake Erie would contain more heat. The heat needed to warm the glass of water by 1° would be insignificant in comparison to the heat needed to warm Lake Erie by the same amount. Thus, the heat needed to bring about a temperature change depends upon the total mass to be warmed. According to this view temperature might be a measure of the heat per unit volume, or something like that. This is *not* correct; the relationship between heat and temperature is more complex.

For the moment, we may accept your intuitive idea of what these factors are. *Heat* is whatever is delivered to a pot of water by a fire underneath it. *Temperature* is a variable measured by a device called a *thermometer*. Thermometers go back to Galileo's time, and they depend upon another property of matter. As heat is added to matter, the matter generally increases in size or *expands*.

This process is reversible so that as heat is taken away the matter *contracts*. This expansion and contraction with changes in the state of "hotness" is not the same for all kinds of matter. Galileo noted that the change in volume of air for a small change in hotness was much greater than the change in volume of glass, for example. Thus, a bulb of glass filled with air could be used to measure changes in hotness. A typical gas thermometer is shown in Fig. 8-1. As the temperature increases, the volume of the gas in the bulb increases. As the air expands the only movable wall of its enclosure is the bead of liquid in the thin stem of the thermometer, and this is pushed farther from the bulb. By making marks on the stem to record the position of the liquid bead, temperatures could be compared. Such a gas thermometer is actually more sensitive to temperature changes than most thermometers in common use today, and they still are used in some laboratories. The problem is that the gas bulb is usually the size of a baseball or larger. Most hospital patients would object to having their temperature taken with such a device. Thermometers today use the same principle but are scaled down to a convenient size. Because the stem is very thin, small changes of the volume of the fluid in the bulb are magnified and can be measured. Most such thermometers today use mercury or alcohol with a red dye as the fluid in a bulb sealed at high vacuum. Where more durability is required, one type of solid thermometer depends upon the difference in expansion and contraction of two materials. This is the *bimetallic strip* (Fig. 8-1*b*). Two strips of metal are welded together along the length. As the temperature increases, one of the metals expands faster than the other. The result is to curve the strip with the more expanded metal on the longer outside of the curve. Such a thermometer is particularly useful in *thermostats*. As the metals bend in response to temperature changes, they may make an electrical contact to turn on a furnace. Later, as the room temperature increases, the strip bends in the other direction until the electrical contact is broken and the furnace is turned off. Still more modern (but not necessarily better) types of thermometers depend upon the variation of other properties of matter with temperature. Electronic thermometers are now in widespread use in hospitals.

A scale for temperature is needed in these devices to produce numbers which can be compared. Early scales are all arbitrary; some nearly constant temperature which could be reproduced in a laboratory was assigned a number. Fahrenheit used an ice, salt, and water mixture; got it as cold as he could; and marked that as his zero. He set human body temperature at 96° and obtained the rest of the marks on the thermometer by repeatedly dividing in half the distance between marks. About the same time, Celsius originated a scale using the freezing point of water as 0°C and the boiling point as 100°C. The Celsius scale is more convenient, but both scales are arbitrary. Both get us into a problem which we try to avoid in science—*zero does not mean none.* We say that these scales are not *absolute.* Both will indicate the *relative* bigness or smallness of temperature, but 20 does not mean twice as much as 10. Eventually we will want a better scale than either of these in which to

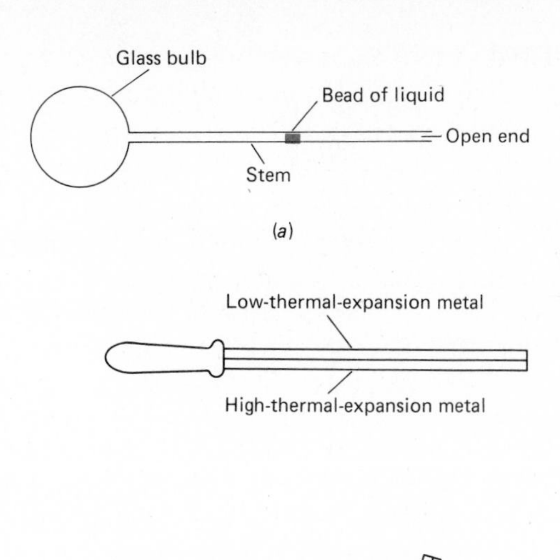

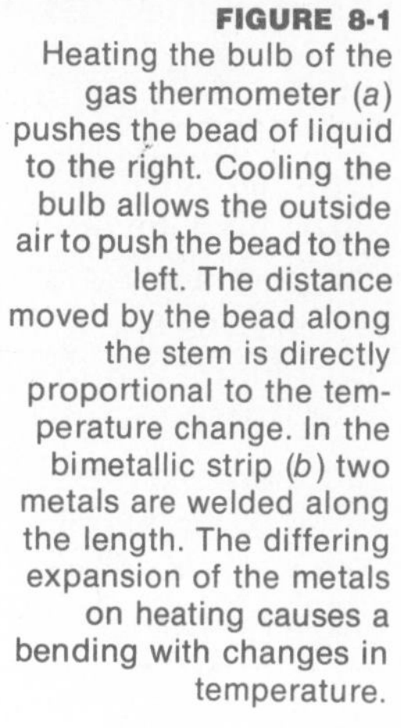

FIGURE 8-1 Heating the bulb of the gas thermometer (*a*) pushes the bead of liquid to the right. Cooling the bulb allows the outside air to push the bead to the left. The distance moved by the bead along the stem is directly proportional to the temperature change. In the bimetallic strip (*b*) two metals are welded along the length. The differing expansion of the metals on heating causes a bending with changes in temperature.

measure temperature, where zero does mean none and 20 is twice as much as 10. However, even though temperatures may not be absolute in the Celsius scale, *temperature changes* will be. To go from 10 to 30°C is to have twice the temperature change as to go from 10 to 20°C. Thus, we may explore the effects of temperature change. One that we have already mentioned is the expansion and contraction of matter with changes in temperature. For example, steel expands in length by 1 part in 10^5 parts for each degree Celsius which the temperature rises. Consider what this can mean. The trans-Siberian railroad runs some 5000 km across Asia. Winter temperatures are routinely as low as −31°C, while summer temperatures rise to 35°C. This is a temperature change of 66°C. Thus in winter the track is 3 km (≃2 mi) shorter than in the summer! This relationship is expressed by the following equation, in which E is the expansion factor (in this case, $10^{-5}/1°C$), ΔT is the temperature change, and L is the length of the railroad:

$$\Delta L = E \times \Delta T \times L$$

$$= (10^{-5}/1°\text{C})(66°\text{C})(5 \times 10^3 \text{ km}) = 3 \text{ km}$$

All structures, from the trans-Alaska pipeline to the Verrazano Narrows Bridge, pose a similar problem of change of length that must be solved during the design stage—or else! The problem is even more serious for boilers, nuclear reactors, or automobile engines, the parts of which are assembled at room temperature but will have to fit at 1000° or more.

Properties of Heat

Armed with our thermometer we may now explore the relationship between temperature and heat. We know some things about heat already from our common experiences. If two objects are placed in contact, heat flows from the object at the higher temperature to the one with the lower temperature. We say that heat is *conducted* from the warmer object to the cooler. The flow of heat from a cooler to a hotter object is never observed in nature. Even in refrigerators and freezers the flow is always from the hotter object to the cooler.* This flow occurs even when there is no contact between the objects. If a block of steel at 100°C and a second block at 20°C are placed near each other, but not touching, inside a vacuum chamber, the transfer of heat still occurs. The transfer of heat without the use of physical contact is called *radiation*. If you have ever been in a room of an older house during winter and felt very chilled even though the thermometer is registering a comfortable temperature, you may have experienced the effects of radiation. In the exchange of heat by radiation with the cold outside wall of the house you will lose heat, being the warmer object. Clothes generally do not influence this heat exchange by radiation, but shiny metallic surfaces can reflect radiated heat. Space suits for astronauts have a metallic coating to reduce heat transfer by radiation (as do their ships). Most of space is at a radiation temperature of about 260°C below zero, while the sun's surface is at 6000°C. Without protection for thermal radiation, this presents you with the options of being thoroughly frozen or very well done in space. In houses, radiation chilling is prevented by adding insulation to the outside walls. Such insulation is particularly effective if it includes a metallic foil to retard heat exchange by radiation.

Radiation is one of three mechanisms for the transfer of heat. The other two are *conduction* and *convection*. Convection is a process that depends upon a flow of matter—currents—between the hotter object and the cooler one to transfer heat. These three processes are easily pictured. For example, if you wish to pass a note to a classmate three seats ahead of you in class, you have similar options. You could pass the note from student to student down the row—conduction. As a second option, you could pass the note to the person who happened to be walking down the aisle for delivery to your friend—convection. Finally, in desperation, you could just throw the note—radiation. The only point not made by this analogy is that with convection the temperature difference of the two bodies is frequently the cause for the flow to develop. For example, a hot metal warms the air in contact with it. The air expands with its increased temperature and is now less dense than the surrounding air. Being less

* An explanation may be needed here since this is not obvious. As a gas expands it cools, and as it is compressed it heats up. The gas in the sealed system of your freezer is compressed and gets quite hot. Air is blown over the coils containing this hot, compressed gas, lowering it to room temperature. Now the gas is allowed to expand into a partial vacuum. The temperature of the gas drops to a value much lower than the freezing point of water. The cold gas is circulated near the slab of meat you wish to freeze and the warmer meat loses heat to it. The slightly warmed gas is now recompressed to attain a temperature higher than room temperature, and it delivers the heat from the meat to the outside air. Each heat transfer (meat to cold gas; hot gas to air) is from a hotter object to a colder one.

Keeping your cool

The highest forms of animal life on earth all operate their bodies at constant temperatures, about 37°C in the case of humans. The rate at which chemical reactions take place is dependent upon temperature. Reaction rates about double with a temperature increase of 10°C. The speed and force of muscular contraction, heart rate, breathing rate, even the process of thinking is linked with the rate of chemical reactions. Animals that do not regulate their temperatures become sluggish and incapable of sustained activity as the temperature falls or dangerously overactive as the temperature increases. To live at a constant rate you must regulate your internal temperature.

All bodily activities generate heat. To maintain a constant temperature you must get rid of heat as fast as it is generated—no faster or your temperature drops; no slower or your temperature rises. Additionally, when your surroundings are warmer than you are, you must manage to lose the heat you gain from this source too. The rate at which a human generates heat depends upon body structure. However, this rate divided by body area tends to be constant for humans and is defined as *metabolic rate*. This rate when lying down, or relaxing, is known as the *basal metabolic rate* (BMR) and is about 11 cal/s for each square meter of body area (or 40 kcal/m^2 · h, or about 46 W/m^2). More specifically, a student with a mass of 60 kg (weight of 132 lb) and a height of 1.7 m (5 ft 6 in) would have about 1.7 m^2 of body surface area. Resting on the sand, our student would generate energy internally at the rate of about 80 W.

Picture our student at a reservoir beach in the desert. The sun is up at an angle of 60°, and the still air has a temperature of 40°C (104°F). Since the air is hotter than the body, the air will tend to deliver heat to the body. The lowest rate for conduction with this temperature difference would be about 30 W on a naked body (the bathing suit comes close enough to this for scientific purposes).* More serious is the effect of the sun, since it

* Clothing would trap a cooler layer of air next to the skin and reduce the conduction since air is an excellent insulator. Clothing would also reduce heating by convection and, if light-colored, cut the heating by direct sunlight by as much as 60 percent. This leaves us with three good reasons why the Arabs who have to live in the desert wear long, flowing robes and did not develop the bikini.

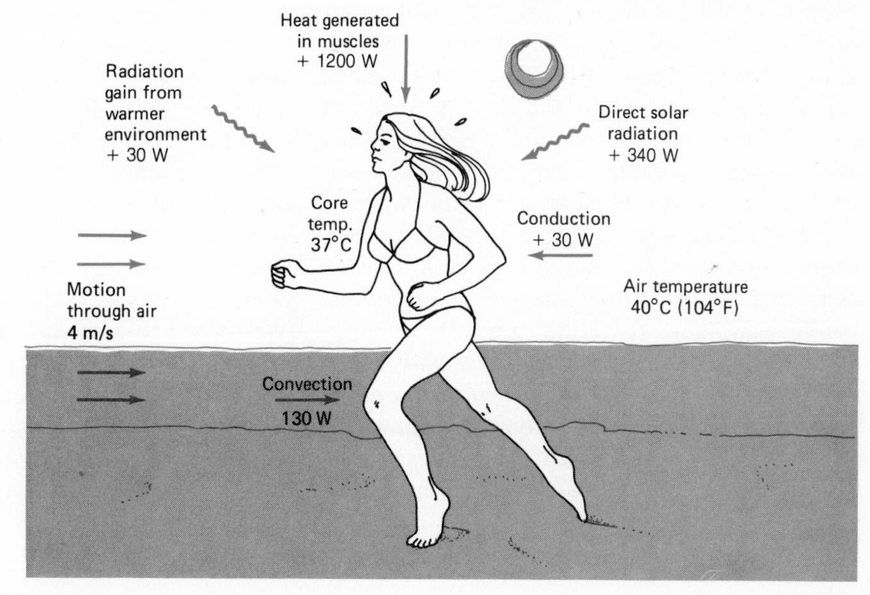

FIGURE 1 The student running on a beach gains heat from every normal source at a rate of over 1700 W. The only mechanism for the body to lose heat is through the evaporation of perspiration. If high humidity blocked efficient evaporation, the core temperature would have to rise. If it rose the full 3°C to the air temperature, it would be exactly like having a 40°C (104°F) fever, resulting in delirium, collapse, and death. With low humidity, the body handles the problem nicely.

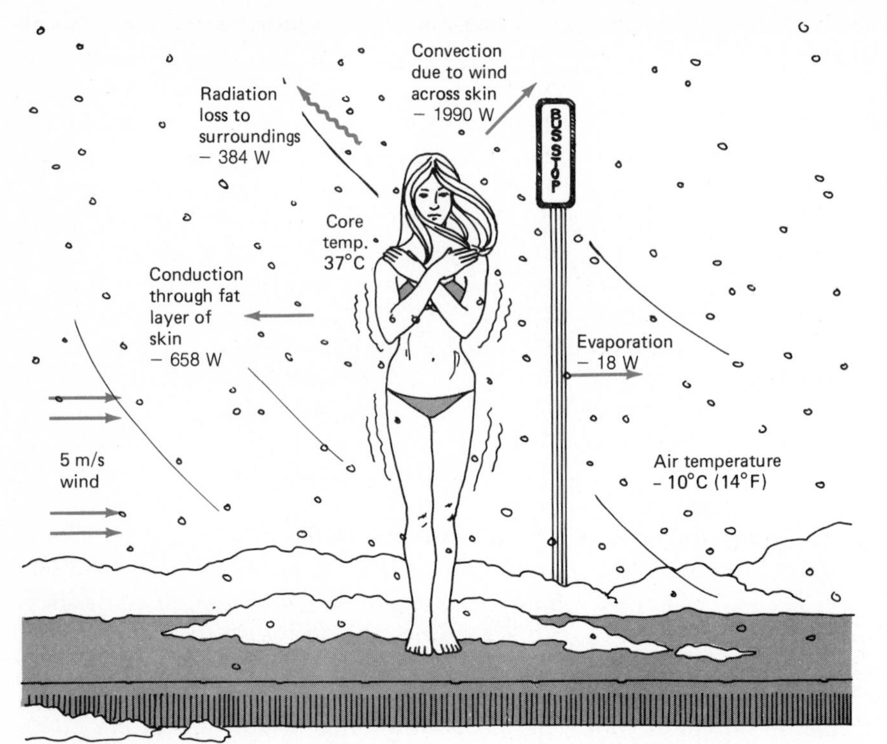

FIGURE 2
Why not wait for a bus in the winter in a bathing suit. Even violent exercise could not balance the 3000-W rate of heat loss. Suitable clothing could cut conduction and convection to the point where shivering or jogging could develop enough heat to maintain the body core temperature.

may deliver energy at a rate of 340 W to the half of the body directly exposed. Radiation from the hotter surroundings contributes another 30 W. Furthermore, to make the problem serious, we will have our student run along the beach at 4 m/s (9 mi/h). This introduces two problems for the body. Air moving across all of the body promotes the gain of heat by convection. Cooler air next to the skin is moved aside and heat is added at the rate of 130 W. More seriously, the strenuous exercise generates heat in the muscles at the rate of 1200 W. Heat is being delivered to our student at a rate of over 1600 W, or at a higher rate than the average electric toaster develops heat. Furthermore, conduction, convection, and radiation are all working against the student. How does the body cope with this problem? The heavy-duty cooling system in human beings depends upon the evaporation of water from the skin surface. Sweat glands under the skin pour moisture onto the surface to evaporate and carry away heat. Unlike boiling, in which the particles of water convert to a vapor in wholesale lots forming bubbles, in evaporation particles attain enough energy to vaporize individually. However, the heat required per gram of water converted to vapor is the same in evaporation as in boiling. From the human body temperature of 37°C, each gram of water will need 63 cal for the equivalent of temperature increase to the boiling point and another 540 cal to change to a vapor. Each gram of water requires 603 cal to evaporate, or each kilogram requires 603 kcal. Since 1600 W is about 1400 kcal/h our student will have to perspire away approximately 2.3 kg of water each hour to maintain body temperature at 37°C.* This is a great deal of water (about $2\frac{1}{2}$ qt or 5 lb of water per hour) but the human body can handle it. Heat-adapted individuals can perspire up to 4 kg/h for short amounts of time. Indeed, people are so well adapted to

* The body keeps its interior or core temperature at 37°C, but the temperature of the skin and limbs can vary widely, as you know if you have ever had cold feet in the winter. The skin temperature in the present case would be close to 40°C.

heat they can work continuously while a dog (not equipped with sweat glands) will have to lie in the shade and pant furiously to evaporate enough water from its mouth and throat tissues to cool its body. The most dramatic test of our cooling systems were experiments in which people have survived temperatures of 124°C (257°F) for a time long enough to cook a steak. That's some cooling system you carry around!

Bicycling, 290 W/m^2

Walking, 190 W/m^2

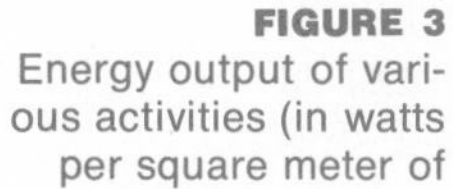

FIGURE 3
Energy output of various activities (in watts per square meter of body area). Additional activities:

Sleeping, 40 W/m^2
Sitting, 60 W/m^2
Standing, 70 W/m^2
Shivering, 290 W/m^2
Running, 700 W/m^2

Light physical work, 80–150 W/m^2

Lying awake, 50 W/m^2

dense, the air rises, much as a bubble would underwater. Thus, the hot air rising off the metal carries heat away from the metal for delivery elsewhere. You are aware of the consequences of this process even if you have never thought through the process itself. You can place a finger 1 cm below or to the side of a lighted match without harm. You know perfectly well you do not place your finger 1 cm above a lighted match for long. The convection currents from the match deliver too much of the heat from the burning for you to escape injury in this position. This example makes one additional point. The conduction of heat through air is very poor. The air in one place may be 1000°C and only 1 cm away it may be at room temperature (20°C). Most materials conduct heat from place to place much better than this. If one spot on a metal plate were glowing at 1000°C, you would know better than to touch the plate 1 cm away from the spot. Metals are excellent *conductors* of heat, while gases are very poor conductors. Materials which do not conduct heat well are called *insulators,* and air is one of the best insulators. Almost every strategy for keeping your house and body warm in winter depends on the use of air as an insulator. Houses have double walls to trap air between as insulation. Such large air spaces still have problems with convection currents set up between the warm inner wall and the cold outer wall. To break up this convection we may put in fiberglass insulation. Glass is a fairly poor conductor, but the value of the fiberglass is in the thousands of small pockets of air trapped between the fibers. Convection may be broken up even more by using a foamed insulation made up of billions of bubbles of gas trapped within plastic walls, much like the foam cups used in coffee machines. In all of these cases, the insulator retarding the transfer of heat is the gas. The same strategy is used in winter clothing. Bulky clothes made of twisted, interwoven fibers trap pockets of air to serve as insulation. This strategy completely fails when you fall in a lake and come out with trapped pockets of water instead of air.

Temperature vs. Heat

The fact that heat and temperature are really quite separate is best illustrated by an experiment. Let us take a block of ice from a freezer at $-20°C$ and place it in a container constructed as shown in Fig. 8-2. The container is surrounded by thick insulating walls to prevent heat exchange except from below. An expandable space for gases is allowed

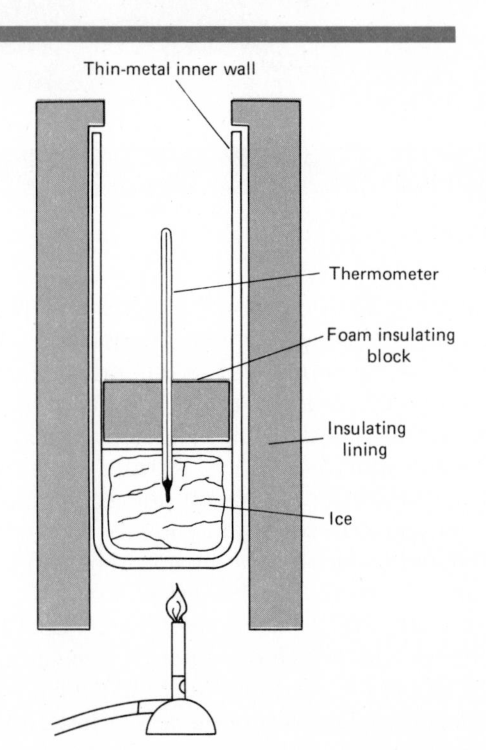

FIGURE 8-2 The chamber for heating a block of ice is designed to minimize the loss of heat and to retain the material being heated. The fitted plastic foam plug above the ice may be raised by steam pressure with very little force, preventing the loss of steam. (To provide for complete vaporization much more expansion would be needed for the block of ice shown.)

over the top of our sealed container to prevent any matter from entering or escaping. The object of all of these preparations is quite simple. We will place a burner under our container and heat the water. Once set, the burner is not changed or adjusted. It will deliver heat to the system at a constant rate ($\Delta H/\Delta t$ = constant). We will record the temperature of the water vs. time. The time, however, is a measure of the heat added to the system, since heat is being added at a constant rate.

Once we have begun we note that the temperature of the ice begins to rise. As heat is added at a steady rate, the temperature also increases at a constant rate. Soon drops of liquid water begin to form around the ice. The temperature at which the first drops appear is 0°C. As the amount of liquid water increases we notice a curious fact—the temperature remains at 0°C. It remains at this temperature for a much longer time than was required to warm the ice from −20 to 0°C (about 8 times as long, in fact). We notice at this time that the last little specks of ice are rapidly disappearing. At this time, the temperature suddenly begins to rise again. It rises evenly with time, although not so rapidly as the temperature of the ice rose at the beginning. The temperature rises smoothly for somewhat longer than it remained at 0°C. Then, as the first bubbles of boiling begin, the temperature increase stops at 100°C. As the water boils vigorously, the temperature remains at 100°C. The temperature stays at 100°C for much longer than the entire time it took to reach this temperature from the start. All this time the enclosure above our container is getting larger with an increased volume of steam. When the last of the liquid water boils away, the temperature of the steam in the enclosure again begins to rise. After noticing that this rise in temperature is constant with time, we are ready to turn off the burner and examine our data. This data is graphed in Fig. 8-3.

The graph is a conclusive demonstration that temperature is not a direct measure of the heat present in a system. For a majority of the time graphed we were adding heat to the system without any corresponding change in its temperature. The heat went somewhere, but not into an increase in temperature. At best, temperature is a measure of one place where the heat added to a system may go. But, there must then be other places where the heat absorbed by the system goes which are not recorded by a temperature increase.* There are three zones on the graph where at least part of the heat is going into increasing the temperature. These are: below 0°C where only ice was present; from 0 to 100°C when only liquid water was present, above 100°C when no liquid was present, only steam. Two zones on the graph indicate that none of the heat added was acting to increase the temperature: at 0°C while the ice was melting, and at 100°C as the liquid water was boiling. In both of these zones the water was changing its condition or *state.* The three normally encoun-

* This curve is accompanied by great personal frustration on the part of the author. More times than are comfortable he has pointed out that once water is boiling on a stove you might as well turn down the heat since the temperature of the boiling water and thus the rate of cooking will not increase with the vigor of your attack. The water merely boils away faster. This advice is always accepted as valid and ignored. The result is that after repeated exposure it is actually possible to acquire a taste for burned broccoli.

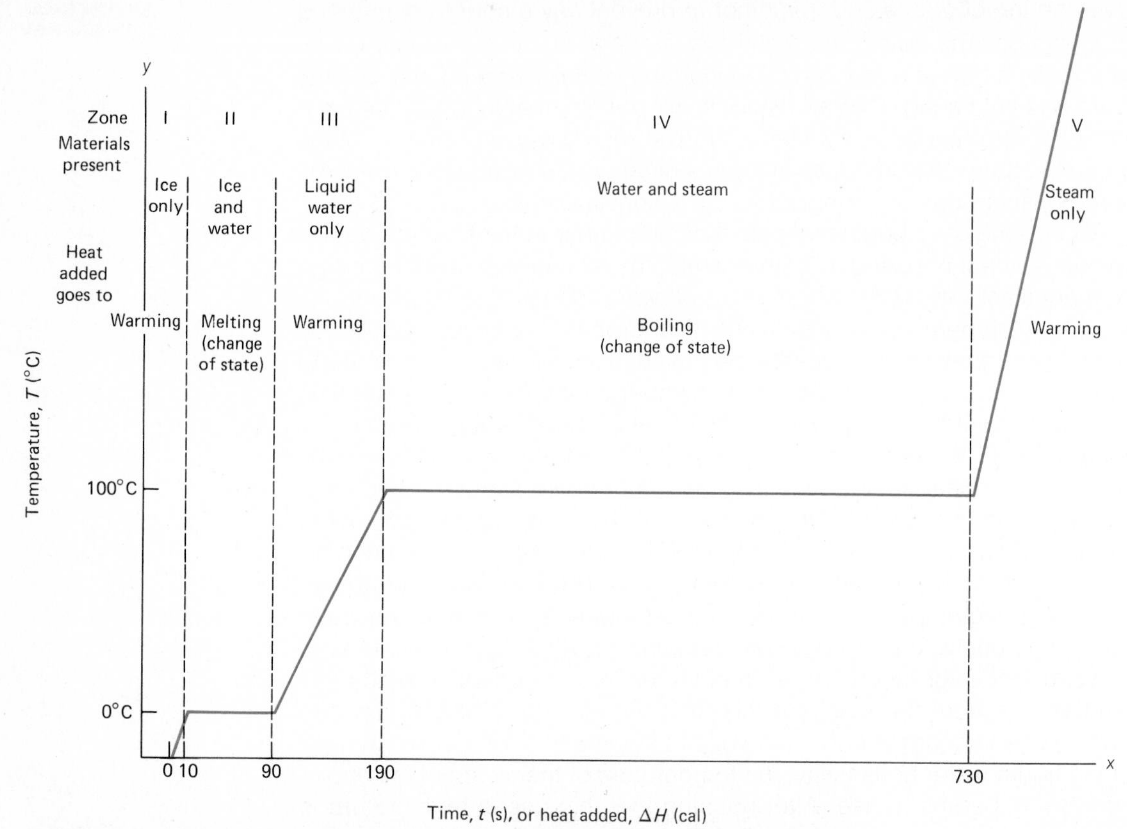

FIGURE 8-3
A graph representing the heating curve for water from ice at −20°C to steam at well above 100°C. Since heat is added at a constant rate, the x axis is a measure either of time or of heat added. While each segment of the curve is a straight line, there are plateaus where heat is still being added but there is no corresponding change in the temperature.

tered states of matter are solid, liquid, and gas. Thus, the five zones of the graph represent at least two types of activity, *warming up* and *change of state.* During each change of state, the temperature remained constant until the change was complete.

A Unit for Heat

It is necessary to be able to discuss the amount of heat transferred. The heating curve for water allows us to define a unit. Liquid water between 0 and 100°C appears to increase in temperature at a constant rate as heat is added at a constant rate. In this range temperature increase is proportional to heat added. Since water is both common and important as a material, it is convenient to define a unit of heat from its behavior in this range. *We may define one kilocalorie (kcal) as the amount of heat which must be added to one kilogram of liquid water to make its temperature increase by one degree Celsius.* Thus, if we had been working with 1.0 kg of water, we could determine the heat required for each part of our graph. To warm the 1.0 kg of liquid water from 0 to 100°C would involve a temperature change of 100°C (ΔT = 100.°C). If it requires 1.0 kcal to raise the temperature of 1 kg of water by 1.0°C, we must have added 100. kcal to raise its temperature by 100.°C.* It took 100 s on the graph to add these 100 kcal; thus, our burner was adding heat at the rate of 1.0 kcal/s. Since the burner was not adjusted during the entire data-taking process, it should have been delivering heat at this rate before and after the time when only liquid water was present. Thus, in the time between $t = 20$ s and $t = 100$ s on the graph, some 80 kcal was given to the melting ice in the system without any temperature increase. This heat apparently went into changing the solid ice into liquid water. This amount of 80 kcal/kg is called the latent *heat of fusion* (melting) of ice. It requires a comparatively large amount of heat per kilogram to melt ice.

Using similar reasoning we can see that it took 540 s to boil away our kilogram of water. The latent *heat of vaporization* of water is thus 540 kcal/kg. Any time you want to boil away 1 kg of water (about 1 qt) already at the boiling temperature, you would have to add 540 kcal of heat. This process works in reverse. Any time 1 kg of steam condenses into liquid water, it releases 540 kcal of heat for each kilogram which liquefies. This is a great deal of heat storage. It is about one-twelfth as much as you would obtain from burning 1 kg of coal. This large release of heat from condensing water provides the energy for driving hurricanes and thunderstorms, for example.

We may summarize these changes by noting that heat apparently goes either (1) into increasing the temperature of a substance *or* (2) into changing the state of the substance:

1 When heating, the amount of temperature increase we obtain depends upon three properties: what material we are heating; how much material we are heating; how much heat we add. The relationship of these items is

$$\Delta H = cm \, \Delta T$$

where c is the *specific heat,* a number describing how much heat is required to raise the temperature of one kilogram of the particular substance one degree Celsius. This specific heat of liquid water is 1.0 kcal/kg · °C. A simple problem may clarify the use of these ideas.

A volume of alcohol ($c = 0.58$ kcal/kg · °C) at a temperature of 40°C is poured into 1.0 kg of water at 10.0°C. If the final mixture has a temperature of 15°C, what amount of alcohol was added?

* A few comments on units are necessary. If one kilocalorie is the heat required to warm one kilogram of liquid water by one degree Celsius, it follows that one *calorie* is the heat required to warm one *gram* of water by the same amount. This indeed defines the scientific calorie. Unfortunately, a third of the American population is on a diet at any one time and has become familiar with the term "Calorie." The dieter's Calorie and the scientific calorie are not the same. The dieter's Calorie is really the scientific kilocalorie. Apparently, it would be too painful for dieters to read that there are about 1½ million calories in a small pizza, or 40,000 calories in an unbuttered slice of toast. (Actually, it might wake them up and improve their performance on the diet.) Thus, nutritionists report food values in kilocalories but call them Calories. Proper use can cut the confusion built into all of this. The easiest treatment is to use the kilocalorie as the unit, which we will do in this book, and there is no confusion. Whenever the scientific calorie is used it should never be capitalized, even if it means juggling a sentence so that it is not the first word. The dieter's Calorie, on the other hand, should *always* be capitalized, even in abbreviation (Cal).

Caloric intake from fast foods

College students are generally avid frequenters of fast-food chains. Below are listed the ratings of minimeals at 10 of these chains from a study completed in 1977 for *Family Circle* magazine. While it is frequently considered sophisticated to run down America's fast-food tastes, please note that the combinations listed provide a respectable portion of the recommended amounts of proteins and calories listed at the bottom.

Caloric Intake from Ten Top Fast-Food Chains

FOOD CHAIN	ITEM	PROTEIN, g	CALORIES (kcal)
A & W	Super papa burger	19	448
	Small fries	3	249
	Root beer float	3	200
		25	897
Burger Chef	Super chef	23	423
	Small fries	4	285
	Large choc. shake	9	361
		36	1069
Burger King	Whopper	29	563
	Small fries	2	218
	Large choc. shake	7	407
		38	1188
Dairy Queen	Super Brazier	43	732
	Small fries	3	239
	Large choc. shake	10	376
		56	1347
Hardee's	Delux Huskee	32	635
	Small fries	4	283
	Large choc. shake	10	328
		46	1246
Jack-in-the-Box	Jumbo Jack	28	558
	Small fries	2	226
	Large choc. shake	13	540
		43	1324
Kentucky Fried Chicken	2-pc Box Munch w/roll	32	451
	Coleslaw	2	216
	Coke	0	57
		34	724
McDonald's	Big Mac	27	543
	Small fries	4	269
	Large choc. shake	9	314
		40	1126
Pizza Hut	½ small Pizza Supreme (thin)	42	799
	Coke	0	97
		42	896
Roy Rogers	Roast beef sandwich	22	292
	Small fries	3	189
	Choc. shake	14	413
		39	894

Source: "McDonald's vs. Kentucky Fried: Rating the Fast Food Chains." Reprinted from the Aug. 23, 1977 issue of Family Circle Magazine © 1977 THE FAMILY CIRCLE, INC.

Recommended Daily Allowances

	PROTEIN, g		CALORIES (kcal)	
	FEMALE	MALE	FEMALE	MALE
Adult over 18	46	56	2000	2700
15–18 years	48	54	2100	3000
11–14 years	44	44	2400	2800
7–10 years	36	36	2400	2400

Source: National Academy of Sciences, 1974.

A We may calculate the heat gained by the water.

$$\Delta H = cm\,\Delta T$$
$$= (1.0\ \text{kcal/kg}\cdot{}^\circ\text{C})(1.0\ \text{kg})(5.0^\circ\text{C})$$
$$= 5.0\ \text{kcal}$$

B The heat gained by the water came from the alcohol: the alcohol lost 5.0 kcal. The mass of alcohol is obtained from the same formula.

$$m = \frac{\Delta H}{c\,\Delta T}$$
$$= \frac{-5.0\ \text{kcal}}{(0.58\ \text{kcal/kg}\cdot{}^\circ\text{C})(-25^\circ\text{C})}$$
$$= 0.34\ \text{kg or } 340\ \text{g}$$

(The minus signs indicate that the alcohol lost heat and went down in temperature, respectively.)

2 When heat which is added is going into a change of state, the amount of heat needed depends upon the mass of the substance and on the latent heat (fusion or vaporization) for the particular change of state.

$$\Delta H = mL$$

☼ A "Good" Bad Theory for Heat

One of the earliest theories of heat which tried to explain the observed phenomena of heat was the *caloric theory.* Science abounds with incorrect ideas, as do all fields of human endeavor. In science, however, the growth of empirical knowledge eventually weeds out inadequate explanations. Theories that are wrong just disappear, leaving only occassional traces of historic interest (for example, the word "calorie" in this case). To be successful a theory first must explain what is known about its area of concern. A theory that stops at this point is sterile. A good theory must also suggest previously unsuspected relationships and predict the results of experiments yet untried. The caloric theory was one of the most successful of the wrong ideas of science, and it is worth a brief discussion in passing.

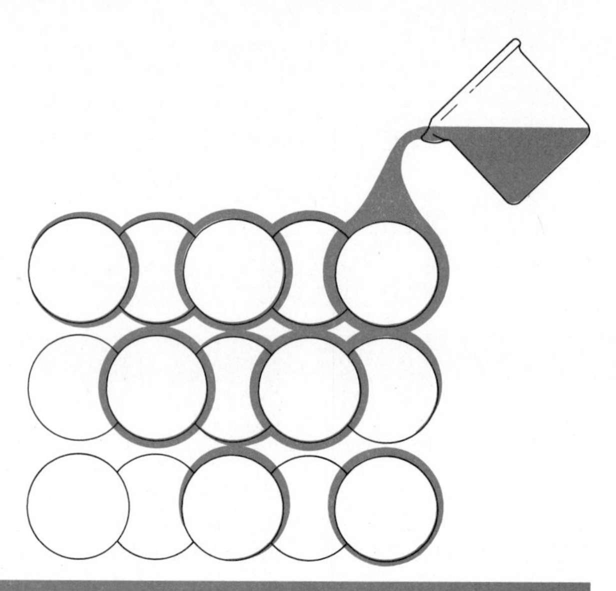

FIGURE 8-4
According to the caloric theory, heat is a fluid called caloric. When caloric is added to matter, its self-repulsion causes it to spread out and coat the particles of the matter as is indicated here.

The caloric theory stated that heat was a material called *caloric.* Some properties of caloric are simple to state (Fig. 8-4).

1 Caloric was a fluid that coated the particles of matter.

2 Caloric was self-repulsive; that is, all caloric repelled all other caloric.

3 Caloric could flow from object to object of its own accord, but it could also be transferred forcibly by friction.

This list is enough to examine its implications for matter. As an object gets hotter, the coating of caloric around its particles would get larger. If we bring a hotter object into contact with a cooler object, what would we expect? Since caloric is self-repulsive, the greater concentration of caloric in the hot object will repel caloric across the boundary to the cooler object. Indeed, caloric will flow from the hotter object to the cooler one until the concentration of caloric around the particles of both substances is the same. Experimentally we do observe that heat always flows from the hotter object to the cooler one. Such a flow of heat stops only when the temperatures of the two objects become the same. Score one for the caloric theory.

Consider the magnified view of the solid shown in Fig. 8-4. The darkish coating on the particles is their quota of caloric. One simple fact about solids is that they generally hold together and preserve their size and shape. From this we may infer that there is some type of attractive force between the adjacent particles of the solid (Fig. 8-5A). If no such attractive force were present, the particles would come apart, and a block of wood, for example, would collapse to a pile of wood powder (much finer than sawdust). Thus, whatever its nature, an attractive force must exist between particles in the solid. Now, however, we have some caloric coating the particles of the solid (Fig. 8-5B). Since the caloric is self-repulsive, the caloric on one particle is repelling the caloric on the adjacent particle. Unlike the attractive force between the particles themselves, the force due to the caloric is attempting to pry apart the particles. As we add more caloric to this system (by heating it, of course), the caloric builds up on the particles and lowers the effective forces holding the particles together. Conclusion: Heating should cause an expansion of materials as the particles are forced farther and farther apart. Furthermore, at some point the repulsive force of the caloric should equal the attractive force between the particles (Fig. 8-5C). At this point the material no longer would be able to hold its shape; it would liquefy. This is how materials behave; score two for the caloric theory.

Consider a more involved situation. *Evaporation* is the change of liquid to a vapor below the boiling temperature. If a saucer of water is left sitting in a room, the water will seem to disappear slowly and eventually be completely gone. Unlike the wholesale conversion of liquid to gas which occurs during boiling, evaporation is a slower process occurring only at the surface of the liquid (how could you prove that?). In evaporation single particles at the surface convert from a liquid to a vapor. Let us examine this prospect from the standpoint of the caloric theory.

A *liquid* has definite volume but sloshes around to take the shape of its container. Apparently the particles in a liquid are held together by forces, but they are not held strongly enough to support a definite structure like in the solid. The addition of enough caloric should be able to pry apart the particles completely and produce a gas, which is what happens in boiling. But what is the situation in evaporation? Picture the situation for a particle at the surface of the liquid as shown in Fig. 8-6. The caloric surrounding this particle is being repelled by the caloric surrounding all of the other particles. Repulsions from the sides cancel out, but the repulsions from below the particle are not balanced by any repulsions from above since this particle is at the surface. The only reason the particle is not repelled off the surface to become independent is that there are attractive forces for the other particles also. These attractions are not strong enough to hold our particle in one place in the liquid, but they do prevent it from breaking free from the mass of particles. Remember, however, that caloric is not only transferred by its own repulsion, it also may be transferred by friction. Imagine for a moment that our particle during its wanderings in the liquid has momentarily acquired more than its fair share of caloric. The repulsive forces of all of the sur-

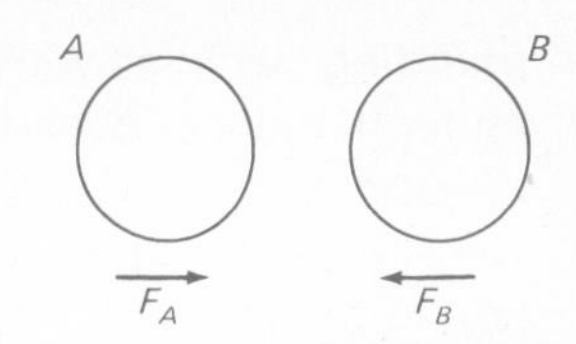

FIGURE 8-5A
In solid matter, adjacent particles must have an attractive force for each other.

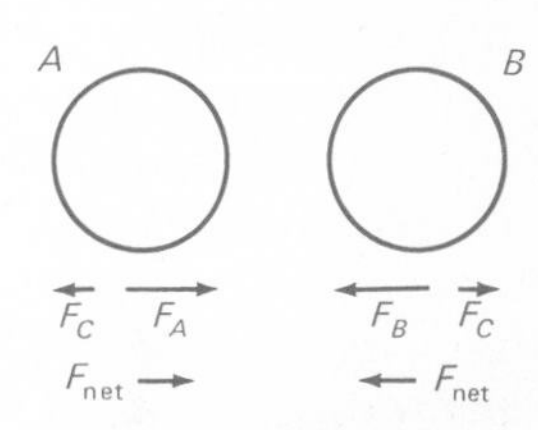

FIGURE 8-5B
When a coating of caloric is added to each particle, the repulsion of the caloric must lower the net force that binds the particles to each other.

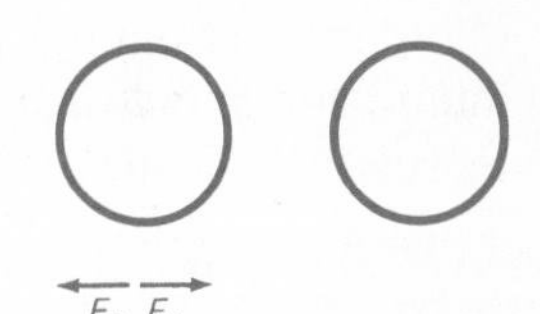

FIGURE 8-5C
With the further addition of caloric, the repulsive force grows to the size of the attractive force. There is now no net force left to hold the particles locked to each other.

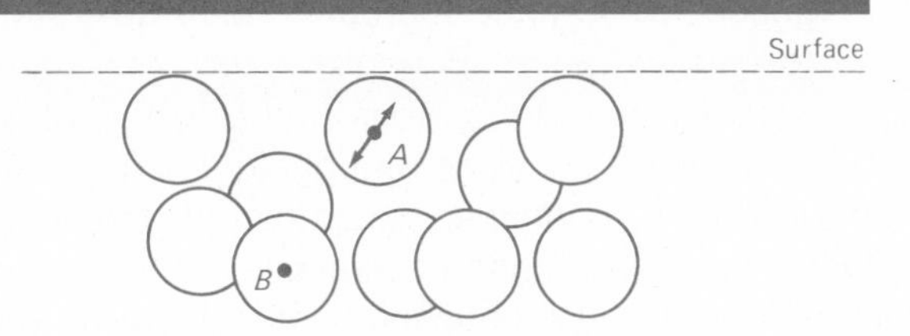

FIGURE 8-6
Particle *A* interacts with all of the particles in the liquid that surround it. For example, particle *B* attracts particle *A* at the same time that the caloric surrounding it repels the caloric on *A*.

rounding particles will now be strengthened. If our particle has acquired enough caloric over the average, this increased repulsion may be enough to overcome the attractive forces and push our particle off the surface into the space above the liquid. Our particle has made the transition from the liquid into a vapor where it is independent. If we examine this speculation, we may conclude that only particles with more than the average amount of caloric (hotter than normal) may be lost from the liquid during evaporation. Thus, the particles that are left behind must have less than their fair share of caloric. This implies that the particles left behind are the cooler ones. We may conclude that evaporation must result in the remaining liquid becoming cooler. Experiments will reveal that this conclusion is correct. Score three for the caloric theory.

The caloric theory is, of course, completely incorrect. It provided many correct predictions and even correct numerical predictions, but it eventually failed as a theory. Heat simply is not a fluid, but theories generally die hard. Since "something is better than nothing," caloric did not fade until a replacement was on the scene. One great value of the caloric theory was to force scientists to consider what was happening between the particles of a system. As you have seen in this brief section, interpretation within the caloric theory forces you to picture what must be happening at the microscopic level of matter. The development of this view of matter is very important in science. Much of this early quantitative work on heat was done by the Scotsman Joseph Black (1728–1799).

The caloric ideas of heat emerge from 40 to 20 years before the modern atomic theory was produced by Dalton in 1803. Nobel laureate physicist Richard Feynman characterized the idea that all of matter is composed of minute particles as the most important idea to be communicated to our descendants if we could pass along only a single idea in science. Caloric is associated with the development of this way of thinking about matter in science.

☼ A Boring Experiment—Rumford's Cannon Works

The earliest serious objections to the caloric theory of heat were raised by Count Rumford. Rumford experimented widely with heat, but his most famous work was done while supervising the production of cannons at Munich in 1798. To make a cannon, a solid lump of metal in the shape of the cannon was cast from molten metal. When this had cooled, a channel or shaft was bored down the length of the cannon with a drill. Gunpowder was placed at the end of this shaft followed by a cannonball. The expansion of the powder when ignited would force the cannonball down this shaft and out of the gun toward its target. Rumford noted that sharp drills generated less heat in boring a cannon than dull drills. A vat of water placed around the drill to cool it could be brought to a boil and remain boiling so long as the drilling continued, even though the water boiled away was constantly replaced. Rumford reasoned that even if caloric could be transferred by friction, sooner or later the drill and cannon would have to run out of caloric to transfer to the water. Since this did not occur, Rumford concluded that heat was not a fluid nor was it of fixed quantity. Rather, he noted that the amount of heat produced with a dull drill that was doing little boring was proportional to the energy of motion put into the process. From this he argued that heat was a form of motion. Rumford's idea was not generally accepted (few of his ideas were), and it took almost a half-century for thinking in science to swing away from caloric. Appropriately, this story moved from Munich to an independent brewer in Manchester, England, by the name of James Prescott Joule. However, before we continue to follow the ideas about the nature of heat, we will discuss the nature of gases.

Gas Laws and a Lowest Temperature

The earliest regularity in the nature of gases was noted by Robert Boyle (1627–1691). He discovered that as the pressure on an enclosed gas increases the volume of the gas lowers, or $p \propto 1/V$. Gases, however, are also very sensitive to temperature changes. This was noted by Galileo when he suggested the principle of the gas thermometer. Let us examine the temperature behavior of a gas.

If we take an inflated balloon and pack it in ice, we will find that the balloon becomes smaller. Experiments with uninflated balloons will indicate that this effect is due not to the rubber of the balloon but to the behavior of the trapped gas. The volume of this gas depends upon the temperature (Fig. 8-7). Furthermore, the increase in volume is proportional to the temperature increase ($\Delta V \propto \Delta T$). That is, if we increase the temperature by 10°C and the volume increases by 3 cm³, then another 10°C increase in temperature will produce another increase in volume of 3 cm³. (Note that temperature, as commonly measured, and volume are *not* proportional, only the changes in both. If you go from 10 to 20°C, you do *not* double the volume of the gas.) If the volume of a gas lessens as we lower the temperature, how much would we have to lower the temperature to have the volume shrink to zero? This question is valid even if it is difficult to picture how the matter of the gas could ever take up no space at all. After all, in the example above, if the volume decreases by 3 cm³ every time we lower the temperature by 10°C, there are only so many 3-cm³ drops we can go through to get to zero. Consider a similar problem. A man with a weight of 1000 N (224 lb) decides to go on a diet. After 1 week of dieting the man weighs 990 N. We may now ask the question, "Losing weight at this rate, at what time will the man reach zero weight?" The answer is quite easy. If he starts at 1000 N and loses 10 N/week, he will have lost all of his weight in 100 weeks, or about 2 years. Of course, we know that we cannot really do this experimentally. At some point the man involved would get uncooperative and die, thus ending the weight-loss process. We can still project how long the loss of all the weight would take if no additional complications arose.

FIGURE 8-7 The temperature behavior of a gas is indicated by a balloon at various temperatures. If the temperature is very low, the balloon is no longer held up since the gas has shrunk so much. Except for this final case, the various conditions represent almost constant pressure. Why?

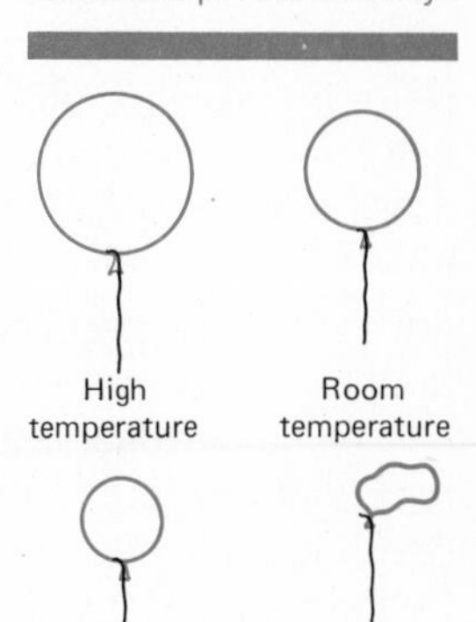

The situation is similar with gases and temperature. As we lower the temperature the volume of the gas is reduced. We may ask the question,

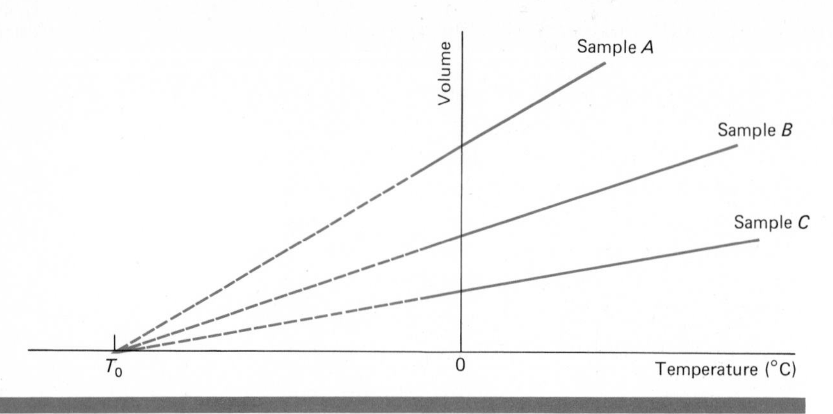

FIGURE 8-8
Graph of volume vs. temperature for three different samples of gases kept at constant pressure. The solid portion of each curve represents the region where the behavior of the gases was measured. Extrapolation to the left implies the same temperature at which each gas would shrink to zero volume. The value of T_0, the absolute zero of temperature, is −273°C.

"By how much must we reduce the temperature for the volume to shrink to zero?" We know perfectly well that nature will step in, keeping us from putting this question to a full experimental test. Long before we reach zero volume the gas will turn into a liquid, and maybe even a solid as we lower the temperature. Yet, based on the trend of the experimentally observable part of the behavior of the gas, we can project an answer. Mathematically, such a projection of results beyond the region of information is called *extrapolation.* We may extrapolate the behavior of gases to the point where they disappear. If we do this for differing volumes of several different gases, we obtain the results of Fig. 8-8. Each of these samples predicts the attainment of zero volume at the same temperature. This temperature is approximately 273° below zero on the Celsius scale. We have chosen to examine the contraction of a gas with cooling to arrive at this value, but there are many physical relationships that would indicate that this temperature is special. A temperature of −273°C represents as cold as you can get; it represents *the complete absence of heat.* Since this is the lowest temperature possible, it is reasonable to call this point the "true" zero of temperature. Thus, −273°C is called the *absolute zero* of temperature, and we may define a new temperature scale based on this new zero point. The *Kelvin scale* of temperature (K) uses degrees of the same size as degrees Celsius but begins at absolute zero. Thus, temperature changes recorded in either system are of the same size. To go from 10 to 15°C is a temperature change of 5°C, and it is also a temperature increase of 5 K (from 283 to 288 K). To obtain a temperature in kelvins, you must add 273 to the reading in degrees Celsius (K = °C + 273). If we go from 10 to 20 K, we have exactly doubled the temperature. In going from 10 to 20°C we really increase the temperature by only about 3.5 percent, even though the numbers make it look as if we double the temperature.

A major advantage we gain by converting to an absolute scale of temperatures is in expressing relationships between variables depending on temperature. From this point onward if we use the variable T for temperature it will be understood that the temperature is in kelvins.

Let us return to gases. If we double the absolute temperature of the gas in our balloon, its volume doubles. In general, we may conclude that the volume of a gas at constant pressure *is* proportional to the absolute temperature,

$$V \propto T$$

We have also seen that, if the temperature remains constant, the volume of an enclosed gas is inversely proportional to the pressure, or

$$V \propto \frac{1}{p}$$

We may combine these two expressions as

$$V \propto \frac{T}{p} \qquad \text{or} \qquad pV \propto T$$

By inserting a constant k to adjust for the various units of measurement, we may write the last expression as an equation:

$$pV = kT$$

This is called the *ideal-gas law.* It describes the behavior of most gases (ideal gases) over a wide range of pressures and temperatures.

This is an entirely empirical result that emerged from study of the temperature and pressure behavior of gases. The formula does express the relationships between these variables but does not come out of any deep understanding of the nature of gases, nor was it predicted from fundamental ideas such as the laws of motion. The link to some theoretical ideas was unclear. Heat clearly can be changed into motion, as in an automobile or a steam engine. Likewise, motion may be (and ultimately usually is) converted to heat, as when a car skids to a stop warming both tires and pavement. What is the relationship of the laws of motion to heat in general and gases in particular?

Heat and Motion

Count Rumford forwarded the view that heat was some form of motion, and that the heat which was generated was proportional to the mechanical energy lost from a system. This viewpoint was eventually resumed two generations later by several individuals, notably by James Prescott Joule (1818–1889). Experiments convinced Joule that potential energy or kinetic energy lost from a system would be converted into heat. Furthermore, he believed this conversion into heat would occur by increasing the motions of the particles making up matter. To convince others of these ideas it was necessary to demonstrate that when a given amount of mechanical energy was expended a given amount of heat was generated. In other words, 1.0 J of potential or kinetic energy, or of direct work, would become a fixed number of kilocalories of heat.* This conver-

* Joule, of course, did not use "joules" as his unit of work or energy, any more than Newton used the newton as his unit of force, or Watt used watts for power. Units derive their present names from the individuals prominently associated with the original research but usually generations after their deaths. Thus, for example, Anders Celsius died in 1744, but his name was not officially attached to the temperature scale he developed until the 1950s.

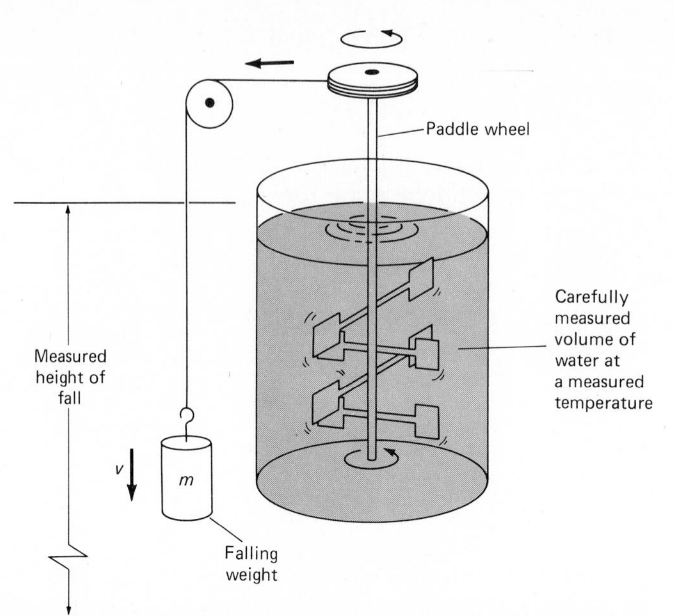

FIGURE 8-9 Joule measured the amount of heat generated from a given amount of potential energy in an exhaustive variety of ways. His most noted apparatus consisted of a falling weight driving a paddlewheel. All the potential energy was converted into stirring up the water. He concluded that it took some 4200 J to produce 1 kcal in the water. Why is it essential to have the weight fall at a slow, constant speed?

sion factor between heat and work or energy is called the *mechanical equivalent of heat.* Joule's most frequently cited experimental determination of this constant involved apparatus such as that shown in Fig. 8-9. As the weight fell at constant speed, it turned the paddles which stirred up the water. Thus, the lost potential energy of the weight, mgh, was transferred into the random swirling and mixing motions of the water. Careful measurements indicated that to deliver 1.0 kcal of heat to the water, about 4200 J of potential energy had to be expended (actually, 1.000 kcal = 4184 J or 1.00 J = 2.39×10^{-4} kcal). An equivalence of this type implies that kilocalories and joules are really different measures of the same type of variable. It is an admission that heat is a form of energy, and can be converted into increased height of masses or their increased speed, and vice versa. This point of view is correct.

Example

A blob of wet clay with a mass of 1.0 kg falls from a balcony to a concrete patio 6.0 m below. If all of the heat generated is retained by the clay, by how much will its temperature rise (the specific heat of the clay is 0.80 kcal/kg · °C)?

The potential energy lost by the clay is

$$E_p = mgh$$

$$= (1.0\ \text{kg})(9.8\ \text{m/s}^2)(6.0\ \text{m})$$

$$= 59\ \text{J}$$

We may convert this to kilocalories by using the relationship between heat and mechanical energy

$$E = 59\text{ J} = (59\text{ J})(1/4200\text{ kcal/J})$$
$$= (59\text{ J})(2.4 \times 10^{-4}\text{ kcal/J})$$
$$= 0.014\text{ kcal}$$

If this amount of heat is released in the clay, its temperature increase may be calculated using the relationship

$$\Delta H = cm\,\Delta T$$

$$\Delta T = \frac{\Delta H}{cm}$$

$$= \frac{0.014\text{ kcal}}{(0.80\text{ kcal/kg}\cdot{}^{\circ}\text{C})(1.0\text{ kg})}$$

$$= 0.018^{\circ}\text{C}$$

Consider a harder problem: A man with a mass of 70 kg climbs a staircase in a tall building. How high may he climb on the energy contained in a single slice of white bread?

If we burn a slice of bread very carefully in the laboratory, we will find that it releases about 80 kcal of heat energy. Your body does not "burn" food in the same manner but in a slow series of chemical steps that accomplish the same energy release less spectacularly. However, only about one-fifth of this energy is available for doing outside work. (The rest is released as heat inside the body.) Thus, our man can do $\frac{1}{5} \times 80$ kcal $= 16$ kcal of work. The work he is doing is to raise his weight through some height, or $W = mgh$. Each kilocalorie available is the equivalent of 4200 J, so the man may do 16 kcal $\times$ 4200 J/kcal $= 6.7 \times 10^4$ J of work. Thus, $W = mgh$ or $h = W/mg$.

$$h = \frac{6.7 \times 10^4\text{ J}}{(70\text{ kg})(9.8\text{ m/s}^2)}$$

$$= \frac{98\text{ kg}\cdot\text{m}^2/\text{s}^2}{\text{kg}\cdot\text{m/s}^2} = 98\text{ m}$$

This is the height to which the man could climb the staircase on the energy from a single slice of white bread. It is roughly 25 to 30 stories in a building, or more than halfway up the Washington Monument. (This also points out why exercise alone is not an effective way to diet. It is much easier not to eat a lettuce sandwich in the first place than to climb the Washington Monument as penance.)

How are we to interpret this conversion of mechanical energy into heat? What is happening inside matter as we heat it? Where exactly is the transformed mechanical energy going?

During the time between the American Revolution and the American Civil War, the idea of heat as a form of motion recurs frequently. If a bowling ball is rolling forward, it possesses kinetic energy. The ball hits a wall and stops. Where does the kinetic energy go? The answer given is

FIGURE 8-10
The water going over the lower falls of the Yellowstone drops through the height of a 30-story building. Joule believed that its potential energy was converted to heat and that the water at the bottom should be warmer than that at the top. He spent his honeymoon (some of it) measuring water temperatures at several European waterfalls. (*Photograph by the author*.)

that the energy goes into heat. The ball and the wall will both become measurably warmer. To consider a second case, a hiker descends a long steep trail. As the hiker descends, potential energy is being "lost." Where is this potential energy going? If the trail is steep enough, the hiker can slide down it with feet scraping and rolling through the dust and gravel. If you have done this for any length of time, you know that the shoes and feet rise in temperature. Thus, we again say that the potential energy was converted into heat. But, what does this mean? What exactly do we imply when we say that a substance has gained heat?

Joule (and others) maintained that the heat in the two above examples is just motion of another type. For example, water falls over a high waterfall (Fig. 8-10). All of the way down the water increases in speed as it falls. By the time the water reaches the bottom, its potential energy has been converted into kinetic energy. Every water particle shares considerable speed in the downward direction. When the water hits the rocks at the bottom of the falls, what happens to this motion? According to Joule, this motion is retained by the water, but its direction is shifted. Instead of all of the water sharing a common downward motion each water particle is deflected in some arbitrary direction depending upon how it hits. Thus the motion of the water is not necessarily slowed down. The water particles can be moving at the same speed as before they hit, but the direction of these motions has been *randomized.* The separate particles are now moving in every direction as the water seethes, foams, and churns at the bottom of the falls. Physically, the property of the water that has changed at the bottom of the waterfall is its temperature. The temperature has increased, just as the temperature of our hiker's feet had increased earlier. This type of consideration (and much detailed experimental measurement) leads to the conclusion that *temperature is a measure of the internal random kinetic energy of the particles of a system.* Temperature is a measure of the average kinetic energies of the particles actually making up an object. Of course, the type of motion represented by this kinetic energy would vary with the state of matter. In solids, these motions would not take the particles from one place to another in the material. (Why not?) The motion would be a rapid "jiggling" or vibration in place straining against the bonds that held the particle in place. In a fluid this random kinetic energy would allow the actual motion of particles from place to place in the liquid as they bounced and slithered against nearby particles. This seems to agree with experience, since a drop of dye placed in one spot of a container of water gradually spreads out throughout the entire water volume. This process, called *diffusion,* depends upon temperature and is slower as the liquid is cooler. Diffusion is a very important process for living things. Our knowledge of it is consistent with a random mixing of particles due to their eternal jiggling and motion, provided the average kinetic energy of these particles is directly proportional to the absolute temperature ($\frac{1}{2}mv^2 \propto T$).

While all of the above seems reasonably consistent with the evidence, a perceptive question may be brought up. What about those plateaus on the heating curve of water where heat is being added to the material but no temperature rise occurs? If we interpret temperature as a measure of the average random kinetic energy of particles, where does the energy we are adding go when we add heat without obtaining a temperature increase? What happens to the heat needed to melt ice, for example? Immediately after melting, the water is at the same temperature as the ice was just before melting. This implies that the energy "pumped into" the water did not speed up the jiggle rate of the particles and thus increase their kinetic energy. Where did the energy go? If we imagine a block of ice as a mass of small particles, what can we infer about a change of state? Since ice is a solid, it maintains its size and shape. It could not do this if the particles were free to roam around within the material. From this we may infer that forces must exist which "lock" each particle to its position. The freedom of particles to move throughout the volume of a liquid, as shown by diffusion, implies that these forces have been overcome.

Now, picture two particles with an attractive force holding them together. This force is *not* like a bond of glue between two wooden blocks holding them rigidly in place. The interpretation of temperature as a measure of kinetic energy *requires* that the particles in a solid be allowed some motion. We can heat up solids; thus we must infer we can increase the random motions of the particles even though this motion does not move them from place to place within the solid. The particles must have the ability to move back and forth in place, or vibrate somehow. Since this vibration does not cause the solid to fall apart in normal situations, we must presume that the forces holding the particles to each other operate over a range of some distance. Clearly, this distance is not great as human measures go. If we snap a board with a karate chop and then push the pieces closely back together, the crack does not "heal itself" and give us back an intact board. The closest that we can shove the pieces together is apparently not close enough for the forces between particles to resume their former strength of attraction. Thus, the range of the forces attaching particles to each other must extend across a range of some size on the scale of the particles but not great on a human scale. What would be required to overcome this force that binds one particle to the next? Clearly, we would have to apply a second force to pry the particles apart. Since the attractive force operates across a range of distance, however, we would have to exert our "prying-apart" force across enough distance to effectively move the particles out of the range of the attractive force. But, *if we exert a force through a distance, we have done work.* To break the binding forces of the particles in a solid we must do work on the particles. This is where the heat must go as we melt the ice—it does work on the particles. Yet, if the energy or the work is not to disappear, where is it after we have pried the particles apart? If we do work on a bowling ball to speed it up, that work is stored in the kinetic energy of the ball. If we do work in lifting the ball, it is stored in the increased gravitational potential energy of the ball. In either of these cases, ideally we can get the work back from the ball. Where does the work done on the ice particles to make liquid water go? Can we get this work back, too? The answer to the second question is easy—of course we get the work back. When ice melts we must put in 80 kcal of heat for each kilogram of ice we want to melt. When liquid water freezes, it releases 80 kcal of heat per kilogram of ice formed, whether we want it or not.*

Thus, when conditions are suitable, the particles can release this work again as they re-form the bonds that once held them together. Two separated particles with an attractive force between them possess potential energy equal to the work required to separate them. Just as a bowling ball releases potential energy to other forms of energy as it falls to the earth, our particles may do the same as they relock into position to form a solid. The heat required to change the state of the ice has gone into the *internal potential energy* of the water particles. Thus, as we warm up a substance, we increase its internal kinetic energy and when we add heat to change its state, we increase its internal potential energy. The sum of these internal energies is called the *thermal energy* of the system. Heat is thus related to the change in the internal energy of substances.

* Lake Superior is 580 km long and averages 150 km wide. To form a surface crust of ice on Lake Superior only 10 cm thick, nature must carry away the heat from the freezing process somehow. The lake has a surface of about 9×10^{10} m² and would release 6×10^{15} kcal in freezing to a depth of 10 cm. This is the same energy as that released by 600,000 atomic bombs of the size that destroyed Hiroshima in World War II. What would the cold Canadian air be like for the residents of the lower Great Lakes if this heat was not added to the air before it got to them? Let them ask the residents of western Minnesota. Actually, the ice on Lake Superior can reach 15 times the thickness used in this example, thus releasing 15 times as much energy.

Heat is best thought of as energy in the process of being transferred to or from thermal energy.

This way of interpreting matter as composed of large numbers of particles in various states of motion is called the *kinetic theory of matter.* We will examine the kinetic theory and its implications further for the case of gases.

Heat Engines

What if as a gas pushes against a wall of its enclosure, the wall is moved? Suppose that in the previous consideration of the gas pressure on the area A, the wall is pushed back by the gas. Now, the length of the enclosure is changed by an amount ΔL (Fig. 8-11). The force exerted by the gas particles has acted through a distance ΔL. The gas has done work on the wall—it has exerted a force through a distance. The work done is

$$W = Fd = F\,\Delta L$$

We may express this differently by multiplying by the area of the wall over itself (A/A), or 1:

$$W = F\,\Delta L = \frac{F}{A}(A\,\Delta L)$$

But, F/A is the pressure p. Likewise, the area of the end wall, A, times the change in length, ΔL, is the change in the volume of the box, ΔV. Thus the work that the gas does in expanding its enclosure is the pressure times the change in volume,

$$W = p\,\Delta V$$

This work comes from the internal energy of the gas and thus is related to a drop in the temperature of the gas. Thus, some of the heat content of the gas may be changed into the motion of the wall. Just as motion may be converted into heat, we see that heat may be converted into motion. These ideas were and are of extreme importance to our civilization,

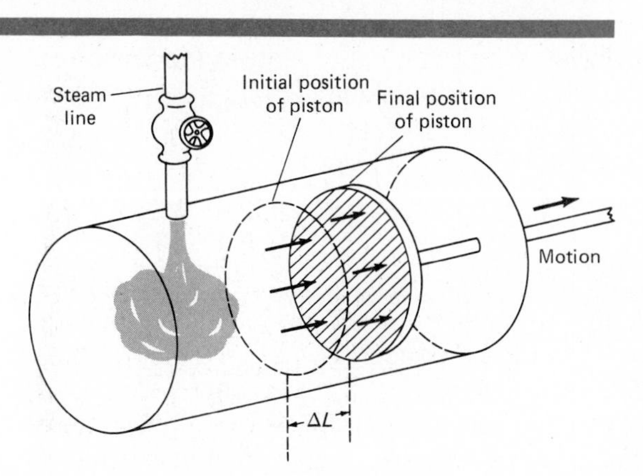

FIGURE 8-11 The steam allowed into the cylinder exerts pressure against all of the walls. Only the shaded area of the piston is movable, and this is moved some distance ΔL. The steam is exerting a force through a distance, or doing work. All that is necessary to make this a practical heat engine is some way to let the steam out and a method of getting the piston back in position for the next push.

FIGURE 8-12 The steam engine was the primary machine to power the Industrial Revolution. As the steam locomotive it was responsible for the rapid opening of the interior of the American continent and a great expansion in travel. The great power in a steam engine is indicated by noting that the engine itself is the small box in front of the large drive wheels of the locomotive. The vast bulk of the locomotive is an accessory to produce the steam. (*Photograph courtesy of David Williams.*)

although they were applied before they were completely understood. To accomplish work, early human beings relied on conversion of food to energy within their muscles or the muscles of animals. With the discovery of simple machines, tasks requiring larger forces could be accomplished, although the total work was limited to the amount which could be converted within muscles, since machines do not increase the work done. Humans next learned to convert the kinetic energy of winds and falling water into useful work through windmills, sailboats, and water wheels. Work could now be performed at rates beyond that of simple muscle power, but still not anywhere or anytime. Powerful machines were always linked to available falling water. Winds were too fickle to permit large installations except in sailing ships where there was no alternative. The development of the heat engine changed this. For the first time an engine could be made as powerful as was desired and portable (within reason). The steam engine, as the first of the heat engines, spurred the move to an industrial economy (Fig. 8-12). Humankind had a dependable source of power output that could lead to the large-scale production of almost everything. The transition to large-scale machines powered first by water power and then by heat engines marks the beginning of the Industrial Revolution.

Heat Engines and Society

The first working steam engine on record was another product of ancient Greek learning. Hero of Alexandria produced a whirling globe propelled by jets of steam. The device was more of a toy than a useful engine, and we have no evidence of practical applications for heat engines in the ancient world. Even simple, useful engines require a reasonably high industrial and technological ability of a civilization. Parts must be made to fit closely and seams must fasten tightly. While the ancient world possessed the experience to do careful work in metals, much of this declined with the fall of the Roman Empire. The centuries following the Renaissance brought a tremendous expansion in the applied science of engineering. Experience with wind and water power led to the ability to design and construct devices to handle large forces. It is impossible to imagine a village blacksmith pounding out the precision parts for the interior of your automobile engine. This requires engineers for designing and machinists for the execution of the task. The engineering tradition becomes powerful (notably in England) in the mid-1700s. This development was essential for the machine age of the Industrial Revolution. The development of science and of engineering were parallel but separate until well into the 1800s. Frequently engineers, as they made new devices, outran the scientific explanation of phenomena.

We may conveniently date the entry of the practical steam engine onto the world scene by a patent issued to James Watt (1736–1819) in 1769. Watt did not invent the steam engine; indeed, working engines were around before Watt was born. The best known of these was of a type designed and developed by the blacksmith Thomas Newcomen, who died 7 years before Watt's birth. The Newcomen engines were large and developed a great deal of power as the steam pushed the movable piston. However, they were slow-acting, requiring a long time between strokes. They were not built to careful standards and were inefficient. Even so, the Newcomen engine did its job. By burning some of the coal mined the engine could pump water out of deep coal mines more economically than humans or horses could do the same job. Watt took a job as an instrument maker at the University of Glasgow. His work with small models of the Newcomen engine convinced him that it could be greatly improved. Once a Newcomen engine had pushed the moving piston its full distance, the entire machine had to be cooled for the next stroke. Watt introduced a separate condenser or chamber where the "used" steam could be cooled and liquefied again. This allowed the engine itself to remain hot during operation. Later, the use of steam in a second chamber on the opposite side of the piston allowed the piston to be pushed back to position rather than have it slowly return as the steam condensed back to a liquid. These changes allowed the steam engine to operate much more rapidly and thus deliver more power for the same size engine. To sell his engines, Watt had to be able to describe how much work they could accomplish in a given time—the engine's power. He chose to describe this in terms of how many horses an engine could replace. He developed the unit called the *horsepower* for this purpose.*

* One horsepower (1.0 hp) = 550 ft · lb/s. That is, it is the equivalent of a 100-lb force pulled through a distance of 5.5 ft each second. 1.0 hp = 746 J/s = 746 watts (W). Actually, this is lower than most modern measurements of the continuous power output of horses, and well below the bursts of power a horse is capable of. Either Watt used anemic horses to help his sales out or the state of horsedom has improved since Revolutionary War times. Industrial expansion and the associated knowledge explosion has produced larger, stronger, and healthier people in developed countries since then. Maybe this works for horses as well.

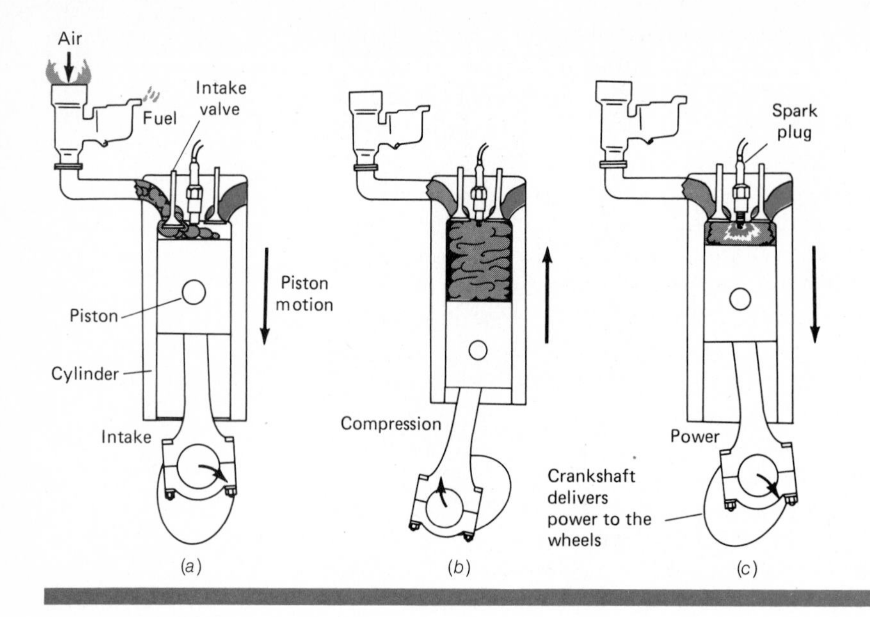

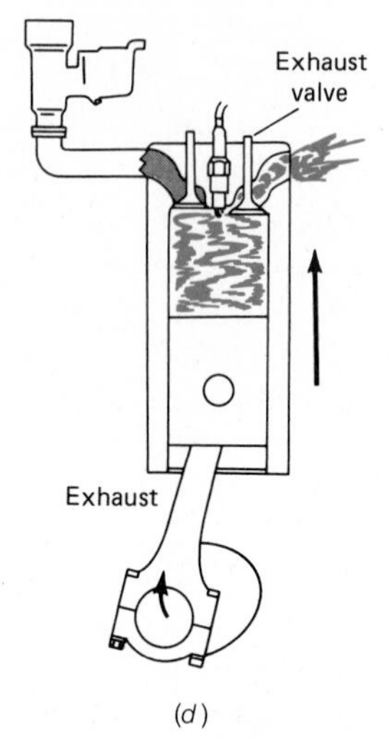

FIGURE 8-13
The four-stroke cycle engine operates in a repeating sequence of four piston motions. In the *intake stroke* (*a*) the piston moves downward drawing a fuel vapor and air mixture into the cylinder. The intake valve closes and the piston proceeds upward in the *compression stroke* (*b*). This creates a very hot, high-pressure mixture of fuel and air excellent for rapid combustion. At this point the spark plug fires, igniting the mixture and raising the gas to high temperature and pressure. This pressure forces the piston downward in the *power stroke* (*c*), which drives the car forward. Following the burning, the exhaust valve opens and the upward motion of the piston expels the burned gases from the cylinder in the *exhaust stroke* (*d*). The piston is now at the top of the cylinder ready to begin this cycle over again with (*a*).

This unit is still used by at least one conservative nation to rate electric motors and automobile engines.

Rapid-acting machines handling large forces require very well made parts. Experience with engines contributed to the rise of precision metal-working machines, most notably the metal lathe just after 1800. Precision parts allowed more powerful, faster, and more efficient engines. Thus the steam engine improved rapidly and became available for the industrialization and increased commerce of America as well as Europe. The role of the steam locomotive and railroads in shaping the nation was profound. The engineering and invention traditions grew more rapidly in America than did the older scientific tradition. We early acquired the reputation as a nation of the mechanically inclined. Population expanded from 4 million at the time of the American Revolution, primarily engaged in agriculture, to 35 million at the time of the Civil War, mostly engaged in manufacturing, some four generations later. Unrestrained by traditions and prior settlement patterns, America arrived with the age of the expanded possibilities of the engine. This timing should not be overlooked when trying to assess the development of American institutions and society.

Before the American Civil War the first theoretical paper on an internal-combustion engine was published. Just 106 years after Watt's patent, Otto patented in 1875 the four-stroke cycle, an internal-combustion engine (Fig. 8-13). Experimental gasoline automobiles followed within a generation. The second round of heat engines were on the scene ready to revolutionize the human life-style. Slowly, we were to acquire a personal mobility undreamed of in the past. Indeed, another generation of heat-engine development would even propel us through the air itself. Ultimately as important was the attack on the efficiency of the transformation of the heat from the fuel into useful work. As early as 1824,

FIGURE 8-14
The steam turbine develops a twisting force as high-pressure steam is blasted against the turbine blades shown. Wheels of rotating blades are alternated with stationary blades which direct the blast of steam. The steam enters at perhaps 15 atmospheres (atm) of pressure and expands and cools as it moves past a dozen or more sets of turbine blades. Note the controlled curvature of the blades. (*Photograph courtesy General Electric Company.*)

the French army engineer Sadi Carnot (1796–1832) had demonstrated that the highest efficiency to be expected of a heat engine was related to the difference between its temperature of operation and its exhaust temperature (or its condenser temperature for steam engines where the water is reused). In general,

$$\text{Percent efficiency} = \frac{T_{op} - T_{ex}}{T_{op}} \times 100$$

For example, if steam is fed to an engine at a temperature of 327°C (500 K) and exhausted near room temperature at 300 K, we may calculate the engine's maximum theoretical efficiency as

$$\text{Percent efficiency} = \frac{500\text{ K} - 300\text{ K}}{500\text{ K}} \times 100$$

$$= \frac{200}{500} \times 100 = 40\%$$

While real engines do not attain their theoretical efficiencies because of many types of energy loss, this calculation represents their best possible efficiency. Clearly, the higher the operating temperature compared with the exhaust temperature, the greater the efficiency can be. The most efficient type of steam engine is the steam turbine, which was developed widely during the 1880s and 1890s (Fig. 8-14). In this engine, hot, high-pressure steam is channeled against the numerous fanlike blades of the turbine, causing it to spin. Steam turbines remain the world's standard engine in the generation of electricity because of their relatively high efficiency. Even in nuclear power plants the steam turbine is the engine used. The nuclear reactor replaces only the boiler where the steam is produced (Fig. 8-15).

FIGURE 8-15
A modern mine-mouth electric generating station produces electricity at the source of the fuel. Coal from the mines is handled at the left rear of the plant with conveyor belts. The tallest buildings, two of which look like open scaffolding, are large, vertical steam boilers. The relatively low, flat building in front of them (really six stories high) is the generator room where steam turbines turn the generators to make electricity. The three cone-shaped structures at the right are hyperbolic cooling towers to condense the used steam from the turbines. The tallest stack for combustion gases is 366 m (1200 ft) high, taller than the tallest office buildings on earth. *(Photograph courtesy of the Pennsylvania Electric Company.)*

The Laws of Thermodynamics

At many times the empirical knowledge of engineers in producing engines outran the theoretical understanding of scientists. Changes in engine design were made because they worked (sometimes). As the mechanics of engines were being developed, it became clear that what we have been calling "heat" really represents energy in the process of being transferred. The sum of the internal potential energy and random kinetic energies of particles is properly called *thermal energy* rather than heat. The study of the flow of heat and its relationships became known as *thermodynamics.* Let ΔH be the heat flow into a body (and be positive if heat is entering or negative when leaving). Then, let W be the work done (and be positive if it is done *by* the body, or negative if done *on* the body). We can now summarize the relationship if we let ΔU represent the change in the internal energy of the body. The internal energy we have discussed so far is the thermal energy of the body. The relationship then becomes

$$\Delta U = \Delta H + W$$

Let us examine some situations. If heat is added with no work being done outside the body, the internal energy must increase. Should the body perform work on an outside object with no heat being transferred, the internal energy must drop. When heat flow and work are occurring at the same time, the internal energy may remain constant, but it usually does not. Examined in detail, this simple expression is widely applicable and has a great deal of power to it. It is as important in a theoretical sense as any relationship we have discussed. This may be called the *first law of thermodynamics.* Actually, it is the formal statement of the *law of conservation of energy.* The junior high school version of this idea is usually the statement that "energy can never be created or destroyed." You see it here more formally and with a great deal more subtlety. Work done by a body is either balanced by heat flow into the body or by a drop in its internal energy, or both. This statement is quite general, and it covers situations which we have not even begun to consider yet.

A second major idea emerges from the general study of energy changes. If we would have run a motion picture for a scientist of the mid-1800s, he or she would have been amazed and amused (although not startled) by this demonstration. If we had run the entire motion picture backward through the projector, we also might have disturbed our scientist. For example, people just do not jump out of pools feet first into the air to make perfect still landings on diving boards. Nor does the water jump from all directions into the hole vacated by the diver to leave a perfectly still surface to the water. *Why not?* Energy is conserved going forward or going backward in time. The laws of motion and the law of gravitation work perfectly well for the reverse situation. In fact, there was nothing in the known rules by which the physical world appeared to be operating which suggested that the reverse process was not a perfectly possible and acceptable event. Yet, the reverse events just do not happen. A watery drink does not rob water molecules of their kinetic energies to speed up cola molecules. The ice cubes do not slowly form in the drink, nor do the random motions "gang up" to eject a stream of cola upward and into a waiting bottle. But why not? What fundamental regulation of nature is allowing the process that we observe only as we move forward in time? Why shouldn't a chick respond to some innate urge by absorbing moisture from the air to get soggy and then crawl into a partial eggshell, pulling the remaining pieces into place to produce an intact egg? For that matter, why shouldn't the hen, like a biological vacuum cleaner, now absorb the egg, chemically break it down in its body, and produce little corn kernels which it places randomly about on the ground with its beak. We know that the reverse process is never observed in nature, but why not? The scientist viewing the film might well realize that *these time-reversed processes do not violate a single, known scientific law of the mid-1800s. The fact that the reverse process does not occur must mean that a fundamental statement about nature had been overlooked.*

Consideration of normal actions and motions leads to some insights about the general direction which energy processes take. Yesterday, you got up out of bed, got dressed, and proceeded to breakfast and the events of the day. Where is the energy of each of these actions now? The alarm clock used energy to make the sound that awakened you. Most of the energy went to increase the temperature of the bell and clapper of the alarm. The sound itself eventually was absorbed by the walls and furnishings and wound up in speeding up the motions of their particles. As you got up and moved around, you moved the air out of your path. The increased motion of the air, initially in a direction to swirl out of your way, rapidly became an increase in the random motion of the air molecules. The stored energy of your blood sugars and body fats wound up as heat or muscular contraction doing work on your environment. All of this is likely to have become random internal kinetic energy by this time, except perhaps for the potential energy of the comb you lifted from the floor to a dresser where it still remains. Where did all of the energies of last summer's vacation trip to the mountains go? Where is the energy now that was in the gasoline that you used, in the food that you consumed? Most of it has wound up in the eventual repository of energy transformations. It went to increase the random kinetic energies of the particles of your environment. Almost all energy winds up as increased thermal energy of the environment. Is there a pattern to all of this?

Yes, a pattern exists, but it is not easy to state completely correctly nor to express simply in an equation. We will try anyway. Energy tends to flow from a more highly ordered state to a more random state. *The general tendency of every energy transformation in a closed system is to an increased state of disorder.* A highly ordered structure such as a bridge or a building tends in time to fall apart and become a disorganized heap of rubble. You may well ask, "But how does the bridge or building get built in the first place? Is not the taking of wood and steel and stone and putting it together into the structure an increase in order?" The answer to this is simple—no! The structure is an increased ordering of its parts, but it represents an overall decrease in order for the entire system. Order in one part is bought by increased disorder in other parts. For example, a steel beam is a much more orderly arrangement of molecules than is a heap of iron ore. But to produce that beam, how much coal and gasoline and food must we take from a state of higher order to disorder? The carbon dioxide scattered throughout the atmosphere of the earth is much less ordered than a seam of nearly pure carbon under the ground. The overall effect of this transition, as of all transitions, is to increase disorder. The high degree of structure and order of the human body, for example, is purchased at the price of disordering the foods we must continually process. The measure of the disorder of a system used in science is called *entropy.* Like most things about which we can eventually make intelligent and verifiable statements, entropy is rigorously mathematically defined in science. We may clearly define the direction in which the motion picture must be running for our universe by stating that the entropy of the universe is increasing. We may take this to be a verbal statement of the *second law of thermodynamics.*

These ideas have many interesting implications. For example, if energy is neither created nor destroyed—merely transformed from one form to another—how can we possibly have an "energy crisis"? There must be as much energy today as there was at the time of Aristotle, and we will have just as much in another 100 years. While this is true, the energy crisis is yet very real. An automobile uses the energy stored in the very ordered structure of gasoline molecules. When the gasoline is burned, a minor amount of the energy is transformed into motion and most winds up as temperature increase in the environment. When the car brakes to a stop, all of the energy may have gone into heating unless we have stopped on top of a hill. The order of the gasoline molecules has been degraded. They have been broken up and the pieces scattered like yesterday's Tinkertoy crane in a nursery school. Entropy has increased. Can we put the gasoline back together again from the pieces? Unlike the king's soldiers and men, we could reassemble the gasoline molecules. But, if we did that, entropy would have to increase again, or at best stay even. Thus, for every molecule we would reassemble we would have to break apart another of at least the same complexity. In practice we do not even begin to break even in such a process. Of what practical use is it to make 1 liter of gasoline from the air when it takes 3 liters of gasoline to run the machinery to do so? Highly ordered, concentrated sources of energy are used to run our civilization. The end product energywise is random, diffuse, disordered, nonconcentrated energy in the form of increased environmental temperature.*

If all of this sounds as if it foretells of looming disaster for the human race, cheer up. Many options for the useful transformation of energy exist. The current problem is not so much a scientific one as a problem of politics, economics, and engineering. If the human race cannot muster enough internal will to manage its energy supply two generations ahead without crisis after crisis, we deserve what we will receive in the short run. It is clear that the fossil fuels that expanded the Industrial Revolution into a full machine age are a transitional energy source in the history of the earth. How long they will last is best left to committees to debate. We can be certain that such fuels will become more expensive. It is also reasonably clear that they have bought us the time to find energy options and will buy enough time in the future to develop those options if we use them wisely.

* This does not imply that the earth must get warmer through time. Heat is continually lost by the earth to space by radiation. This loss is balanced by the heat gained from the sun. Human generation of heat is not on a scale comparable to the natural process. An imperceptible temperature increase of the earth can increase the radiation to space enough to handle the human contribution to the process. More serious is the potential effect of adding combustion products to the atmosphere, especially carbon dioxide, which tend to interfere with this process of radiation itself. While the heat added by human activity is negligible on the scale of the earth, the carbon dioxide may not be. The jury is still out on this one.

Summary

Heat is transferred by *conduction, convection,* and *radiation.* Conduction always takes place from a hotter object to a cooler one. A poor conductor of heat is called an *insulator.* Gases are the best insulators.

The unit of heat is the *kilocalorie,* defined as the amount of heat required to raise the temperature of one kilogram of water one degree Celsius. When an object is heated, the increased heat may show up as an increase in temperature or as a change of state [fusion (melting) or vaporization]. Heat changes associated with temperature changes are also proportional to the mass of the substance and its specific heat ($\Delta H = cm \, \Delta T$). Heat changes related to changes of state depend upon the mass and the heat of fusion or vaporization of the substance ($\Delta H = mL$).

The temperature behavior of gases suggests that there is a lowest temperature at which gas volume shrinks to zero. This *absolute zero of temperature* is −273°C, and gives rise to an absolute scale of temperature in kelvins (K), measuring upward from this temperature in Celsius degrees (K = °C + 273).

Joule convincingly demonstrated that heat was related to motion. He determined that mechanical energy (of motion or position) could be converted to heat at the rate of 4184 J = 1.0 kcal. This led to the interpretation of temperature as a measure of the random, internal kinetic energy of the particles of a system ($T \propto \frac{1}{2}mv^2$). Similarly, heat producing a change of state is going into the increased internal potential energy of the particles of the system.

A high-pressure gas may do work against the walls of its enclosure ($W = p \, \Delta V$). This work comes from either heat added or a drop in the internal energy of the gas (ΔU). This is summarized in the *first law of thermodynamics* ($\Delta U = \Delta H + W$). The fact that some changes are observed in nature while the opposite changes are not is summarized in the *second law of thermodynamics.* This states that natural changes proceed in the direction of greater randomness or disorder, as measured by a variable named *entropy.*

Any device that produces useful work from heat is called a *heat engine.* Heat engines led to the Industrial Revolution and fundamental changes in the human life-style.

Questions

1 As a heating system comes on, there is often a great deal of "thumping and clicking" noises. Why?

2 As you drive on a concrete highway, there is often a monotonous "thump, thumping" as your tires cross tarred seams in the concrete. Why isn't the concrete made in a smooth, continuous ribbon without seams?

3 If you pick up a large piece of metal, it feels cold to the touch, while wood or plastic at the same temperature does not. Why?

4 In general, the denser a material the poorer an insulator it is. Why?

5 A ceramic pot placed on a stove top becomes as hot as the food it contains, but in a microwave oven the food gets hot but not the pot. Why?

6 Experiment with two similar pots to cook the same vegetable. Bring both to a boil at the same time. Turn down the heat on one so that boiling is barely evident and boil the other fast and furiously. Turn them both off at the same time and sample the products. What do you expect? Why? What really happened?

7 How does a pressure cooker manage to cook foods so rapidly?

8 If you ever try to cook a pasta, such as spaghetti, in the high mountains, it can come out as a horrible, pasty mess. Why?

9 A thermos flask has double-glass walls with the air evacuated between them and the inside surfaces silvered with a light metal coating. Explain how this retards all three forms of heat transfer.

10 Why does high humidity contribute greatly to discomfort on a hot day?

11 Why is a wind-chill index a better measure of discomfort on a cold winter day than the temperature? Some Scandinavians are in the habit of sunbathing in winter when the temperature may be well below freezing. How

is it possible for the body to take this? Why is this different from people who belong to "Polar Bear Clubs" and take brief dips in frozen lakes and rivers in midwinter?

12 Why is the ceiling of a house the most important area to insulate in order to prevent heat loss through the walls?

13 Glass conducts heat about 10 times as rapidly as wood of equal area and thickness. Is this the main reason storm windows cut down on heat losses?

14 A metal such as steel or aluminum conducts heat at least 1000 times as rapidly as wood. Why is it that most storm windows have aluminum frames? Would wood frames be a better retainer of heat?

Problems

1 At what temperature do the Celsius and Fahrenheit scales give the same reading? At what temperature on the Fahrenheit scale do we reach absolute zero?

2 How much heat does a 35-g ice cube absorb in melting? How many degrees Celsius would the loss of this much heat lower the temperature of 236 g ($\frac{1}{2}$ pint) of water?

3 The upper Yosemite Falls in Yosemite National Park is the highest free-falling falls in the United States, spanning 436 m (1430 ft) in a single leap (the middle and lower falls below add another 303 m, making the total drop the second highest in the world). Presuming that no heat is lost to evaporation or other places during the free fall, about how much warmer should 1 kg of water be at the bottom of the upper falls than at the top?

4 Lake Erie is the smallest of the Great Lakes in volume, having only one-twentieth the water of Lake Superior, for example. The volume of Lake Erie is 483 km^3 (116 mi^3), or 4.83×10^{11} m^3. Since 1 m^3 of water has a mass of 10^3 kg, the mass of Lake Erie is 4.83×10^{14} kg. A 1-megaton hydrogen bomb releases 10^{12} kcal of energy. How many such bombs would have to be set off to raise the temperature of Lake Erie by 10°C?

5 The caloric theory of heat is called a good theory in the text in spite of the fact that every part of it is incorrect. What makes a theory good?

6 If 210 g of ice water at 0°C is poured into a glass at room temperature (20°C), the water and the glass will transfer heat until they are at the same temperature. If the glass has a mass of 600 g, what will the final temperature be? (The specific heat of the glass is 0.15 kcal/kg · °C.)

7 How much heat is required to take 1.0 liters (1.0 kg) of tap water at 10°C and completely boil it away?

8 The table below lists the energy content of some fuels. Comparing the first three entries, what advantage do animals gain by storing excess foods in the form of fats?

Average Energy Content of Fuels

FUELS	kcal/kg
Human:	
Proteins	4,000
Carbohydrates	4,000
Fats	8,900
Other:	
Garbage	2,900
Wood	3,300
Coal	6,900
Methyl alcohol	6,700*
Hydrogen gas	4,800†
Natural gas	13,000†
Gasoline	10,500*

* 1 barrel (42 gal) is about 150 kg (1 gal ≈ 3.5 kg).
† 1 kg of a gas is about 320 ft^3 (9 m^3).

9 The average adult human need for energy is about 2000 kcal/day (range 1400 to 3600 kcal/day). If an average human is over-

weight by 11 lb and cuts his or her diet to 1000 kcal/day, all other things being kept the same, how long should it take to lose the additional weight? (Recall, 1 kg = 2.2 lb.)

10 Which provides cheaper energy, potatoes at 7 cents per pound (a carbohydrate) or gasoline at $1 per gallon?

11 Automobiles can burn pure alcohol with very little adjustment. To have the convenience of the automobile unchanged, what change in the mass of fuel carried would be necessary? (The volume change would be less than the mass change would imply; why?)

12 You might wish to compare energy prices for your home. Gas bills quote rates per thousand cubic feet, which is about 3 kg of gas. Coal rates are quoted per English ton (2000 lb) rather than metric ton (2200 lb), or 1000 kg, as we use in this text. Crude oil prices are per 42-gal barrel, while refined products are priced per gallon. Electricity is priced by the kilowatthour (kWh); 1 kWh is 3.6×10^6 J (you really could have figured that out) or 860 kcal. Wood is priced by the cord, which is 128 ft^3, or about 1.5 metric tons. Compare the costs per kilocalorie of energy delivered at your home for coal, oil, gas, electricity, and wood (*Note:* The quality of some of these fuels, particularly wood, varies widely from the numbers given in the table.)

13 Above the earth's atmosphere, in space, 1 m^2 of a perfect absorber pointed directly at the sun will gain 0.319 kcal/s (or about 1300 W). The rate of receiving energy from the sun at the surface depends upon time of day and weather. At the author's university, the December average is 1800 kcal/$m^2 \cdot$ day. Solar collectors have an efficiency of about one-third and thus can capture an average of 600 kcal/$m^2 \cdot$ day in December. Suppose that we wish to install solar hot-water heating for the water used in the dormitories. A very low water-use rate is 50 gal per person per day (check your own campus figures), of which at least 20 gal (80 kg) will be heated (one 4- to 5-min shower per day at normal water pressure and delivery).

(*a*) The water temperature as pumped on campus is 10°C, but the water is used at 50°C. How many kilocalories will we have to provide per person per day?

(*b*) How many square meters of solar collector will this require per student?

(*c*) Presume that we will need to have at least double this to store up hot water for stretches of bad winter weather. Now, how many square meters of collector area will we need for each 5000 students in dormitories?

(*d*) The area of a football field is about 4000 m^2. How many football fields of collectors would this be?

(*e*) The cost of commercial solar collectors is about $200 per square meter. Disregarding the cost of pumping and storage systems, what would the cost be for a 5000-student system?

(*f*) Suppose this calculation were for a campus receiving 5 times as much December sunshine, and it used yet undiscovered solar collectors with 100 percent efficiency and costing half as much as the current cost. What would the cost for a 5000-student system of collectors drop to?

(*g*) A ton of coal contains about 10^7 kcal of heat energy, almost all of which actually warms the water in a boiler. If coal can be delivered on campus for $35 per ton, what does it cost to supply the hot water for 5000 students for a day?

(*h*) If student usage averages 200 days/year, what is the total coal cost for 5000 students for the year?

(*i*) If the solar system in part (*e*) has a useful life span of 20 years, which is cheaper, coal or solar heating? How about the system in part (*f*)?

14 The photographs following the insert "Keeping your cool" list the energy-expenditure rate for a wide variety of activities. Use these numbers to estimate how many Calories you burned off today. Most adults have a body area between 1.5 and 2.0 m^2. Use the dimensions of the student in the insert to estimate yours or ask your instructor (the formula for calculating body area from height and mass is in the instructor's manual). The most useful conversion of units is 1.0 Wh = 0.86 Calories.

CHAPTER NINE

Waves and the Transfer of Energy

Introduction Picture yourself on the bridge of a modern oil tanker. The sea is rough and stormy with your bow crashing through waves up to 10 m high. (See Fig. 9-1.) Suddenly in front of the ship a large wave rises from the water and comes speeding toward the bow. Your first glimpse makes it appear like a wall a few kilometers long standing across your path. As it approaches, its true size becomes apparent. It reaches up perhaps to the height of a 20-story building (60 m), Worse yet, before it arrives the bow stops crashing through the "small" waves you have been buffeted with and appears to be hanging out over nothing. Then the bow plunges and the stern raises until the entire stern is lifted clear of the water. The ship begins the roller-coaster ride into the "hole in the sea" in front of the wall of water. Faster and faster it slides down the slope of water until the bow is pushed clean under the water at the bottom. At that moment the towering wall of water collapses on the length of the ship with the downward crushing force of millions of tons of water. Buried deeper below the water than its own height, deck plates crushed and seams opened, the ship continues its downward plunge, never to rise again.

This sounds like the opening of some movie. Can it happen? Ships that survived a meeting with such a "rogue wave" have reported the encounter. Joshua Slocum outlived such a wave in his single-handed voyage around the world. So did the 132,000-ton oil tanker *Wilstar* in 1974, although it lost its bow and sustained

FIGURE 9-1
Waves are carriers of energy. They demonstrate this by their ability to exert forces through distances. These waves on the North Atlantic easily move the largest of ships about. (*Photograph courtesy of the United States Coast Guard.*)

$1.2-million damage from a single wave. The giant liner *Queen Mary* came within 5° of being flipped over sideways by such a wave in 1942. Other stories abound in the literature of the sea.

Anything that can physically lift the largest ships built (which weigh more when loaded than the tallest buildings on earth and are about the same size) certainly possesses the ability to exert forces through distances. Waves are carriers of energy, whether they are water-surface waves, earthquake waves, or more routine sorts such as sound waves or wiggles in a clothesline.

Energy may be transported from place to place by several mechanisms. We have considered heat transfer and translational kinetic energy. It is time to develop the ideas of another form of energy transportation, *wave motion.* In this chapter we will examine the properties of waves in general and then study sound as a specific example.

What Is a Wave?

Consider a water surface. To disturb the flat surface of water we may either push the water away from a section to form a pit or else "pile" the water up in a hump. To do either of these, we would have to exert forces through distances. Thus, we would have done work. The water surface, however, will not allow the formation of either permanent mounds or permanent pits. If, for example, a pit is formed, the water next to the depression immediately will begin to move to eliminate the pit. While this happens we find that the depression is not just filled leaving an undisturbed surface. Instead, a wave moves out across the surface in all directions from the place originally disturbed as zone after zone of water reacts to the disturbance of its neighboring zone. Throw a rock into a still pond to see this effect. Small corks placed in the water will bob up and down as the waves thus created pass. However, the corks, and the water they are suspended on, do not move off with the wave. Thus, we may conclude that it is *the disturbance of the water* which moves from place to place, not the water particles themselves. The water particles act much like the bits of cork. They bob up and down and sway sideways, but after the wave has passed they are much in the same place as they started.

As a second example, consider a wave shaken into a clothesline. Each part of the rope returns to its original position after the wave passes. The rope does not move in the direction that the wave is traveling at all. Only the disturbance itself moves to be at different places on the rope at different times. Thus, we may define a *wave* as a *disturbance moving in some medium* (such as the water surface or the rope of a clothesline).

The motion of a wave disturbance is interesting. If we throw a rock into a pond, the waves that result form expanding circles centered on the place where the rock disturbed the water. Because the disturbance originates in the very limited area where the rock disturbed the water, we would call this central position the *point source* of the disturbance. Sources of disturbances need not be point sources. If you jump onto a floating log, the increased weight causes the log to disturb the water along the entire length of its sides. The log here is a *linear source* of waves, not a point source. If you watch the resulting wave for a short distance, you can observe that the wave resembles a straight-line ripple advancing away from the side of the log. Such a *linear wave front* has a different appearance from the *circular wave front* produced by the rock disturbing the water. In the circularity of the wave front from a point disturbance in water we have a clue to wave behavior. Careful measurement from a photograph of the ripples in a pond taken from directly over the disturbance show that the ripples are indeed *circular,* not merely "roundish." This must mean that the wave moving away from the disturbance moved equal distances in all directions by the time of the photograph. The speed of the wave in all directions along the water must be the same. A sequence of photographs reveals that the speed of the wave is constant. This result is so general that we may state it as a basic principle of wave motion. *In a uniform medium the speed of a wave is the same in all directions and is constant.*

Varieties of Waves

All types of wave have a great deal in common. The study of the properties of one type of wave—sound waves, for example—can reveal a great deal about the behavior of a different type of wave, such as water-surface waves. Before we explore the commonalities of waves, let us examine their differences.

The first difference we might already have noticed is in the geometry of the various mediums carrying the waves. A wave on a clothesline is confined to the length of the clothesline itself. The wave cannot spread out or go off in various directions; it must move back and forth along the rope. Meanwhile the wave on water is confined to a surface. The wave may move in any direction along the surface but it cannot move off the surface upward or downward. Sound waves in air are quite different in that they may move in any possible direction from the source of the wave. If a firecracker is set off in the middle of a football field, you can hear the sound 50 m away in any direction if only air separates you from the site of the explosion. This is also true if you are riding a hang glider over the field as the firecracker goes off since the sound spreads in all directions. To sum up these three cases, the motion of some waves is constrained along a straight line and is essentially *one-dimensional.* Other waves move across a *two-dimensional surface,* while others may move freely in all *three dimensions of space.* This difference is important if we begin to make measurements with waves. The spread, or lack of spread, may be expected to influence actual measures of the wave as it covers distance. For example, imagine a clothesline stretched in a vacuum chamber. Suppose we have a nearly frictionless rope, one where no work is lost in the stretching and relaxing of the fibers of the rope. (The work done by the wave in stretching the rope is restored to the wave as the rope springs back.) In such a case, as the wave moves along there is no place for the wave to lose energy, and we might assume that the wave will remain substantially unchanged with distance. On the other hand, we know that sound loses loudness the farther we move from the source, and it is logical to associate loudness with energy.* Does this drop in loudness with increased distance imply that the air is not a frictionless carrier of sound waves? Not necessarily! Sound waves are able to spread out in all three dimensions, and this fact changes the situation.

Compare the case of the clothesline with the situation shown in Fig. 9-2 for water waves. Energy is given to the water by disturbing the surface at point C. A short time later the wave has moved from point C in all directions along the surface a distance of R. The disturbance is now spread out along a circle of radius R. If the ability of the wave to lift floating weights at this position is the same as it was back near point C, we would have created energy since many more floating weights can now be fitted along the expanded wave. The energy must be distributed along the entire length of the circular wave front. Since this wave front is now much larger than it was earlier, each centimeter of the wave front must have less energy. The length of the wave front along which the energy is distributed is the circumference of a circle of radius R, or $L_1 = 2\pi R$. Consider the situation sometime later when the wave has moved twice as far from its source, or $2R$. The energy is now spread out over a

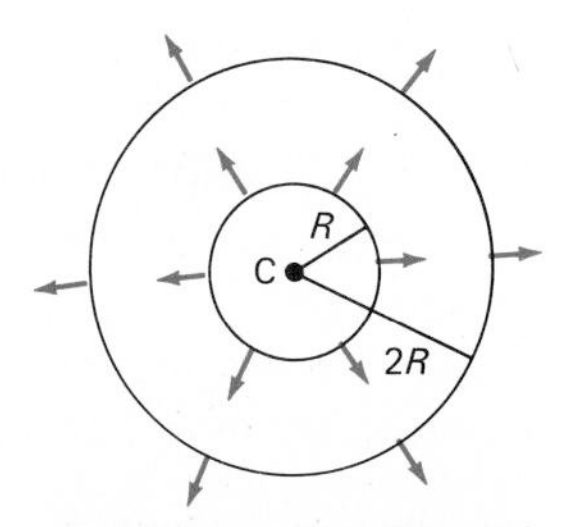

FIGURE 9-2
As a water-surface wave travels twice as far from its source, its energy is distributed along the circumference of a circle with twice the radius.

* After all, you must exert a force through a distance to blow out an eardrum, and you approach this effect as you crank up the loudness on your stereo. Actually, it takes an intensity of about 10^{-4} W/cm² to threaten your eardrums. To realize the enormous range the ear can function across, consider that this is 1 billion times the intensity of the faintest sound the ear is capable of detecting.

much larger circle, the total circumference of which is $2\pi(2R)$ or $L_2 = 4\pi R$. This is exactly twice the perimeter of the original circle, so we would expect each centimeter of wave front to have only one-half as much energy even if no energy was lost. We may define the energy content per centimeter of wave front for this wave as the *intensity* I of the wave. We expect that if water is nearly frictionless for surface waves, $I \propto 1/R$. That is, if we triple the distance, the intensity should drop to one-third, and so on. This is quite different from the case of a wave moving frictionlessly on a one-dimensional medium where the intensity does not vary with distance. In the case of surface waves we *expect* that circular waves will become weaker in intensity as we move farther from the point which is their source. Of course, if the wave does not originate at a point, the situation could be quite different. If we would generate a wave by pushing the entire east coast of Florida against the Atlantic Ocean in an earthquake, the wave would originate along a line. Such a *linear wave front* would not be spreading out to become larger and larger, and we would not expect its intensity to fall with distance in the same way that a *circular wave front* from a point source does. Worse yet, if we pushed all of the shores of the Gulf of Mexico toward its center at once, we could produce a curved wave front which would be *converging,* or drawing together, on the center of the Gulf. In this case the size of the wave front would be shrinking in time. Its entire energy would be concentrated along less and less length of the wave front. When this wave converged on the center, the intensity would be awesome even if the initial wave were tiny.*

The variation of intensity with distance for three-dimensional waves is different from that of either one- or two-dimensional waves. The wave front which moves out from a local disturbance like our firecracker is *spherical* in shape as it moves in all possible directions away from the disturbance. Thus, the energy of the wave is concentrated on the surface of an expanding sphere. The formula for the area of a spherical surface is $A = 4\pi R^2$. From this formula we can see that if we double the travel distance of the wave front ($2R$), the energy will be spread out over 4 times the surface. We can come to this same conclusion by examining Fig. 9-3. As the wave has spread to a distance R from the source, a certain amount of energy is concentrated on the area of the wave front labeled A. When the wave spreads to twice the distance, the energy concentrated

* One technical definition of a *mountain* is a mass of material with limited summit area and a local elevation contrast of at least 1000 m. If we began a 1-cm wave along all Gulf shores and this wave arrived at some central point approximately together, we would far exceed the requirements for a mountain during the period the wave converged on the center. If nothing else interfered (which would *not* be the case), the surface of the sea at the center during the buildup of the wave would be streaking skyward at several times the acceleration due to gravity. A sailor on a ship would be pinned to the deck by many times the weight of the body as the ship was rocketed skyward by the growing wave underneath it. This would be easier to live through than the collapse that would follow. Fortunately many things happen before this extreme. The largest waves encountered by oil-drilling platforms in the Gulf have been under 10 stories high—*thus far.* Converging earthquake waves of this sort have *only rarely* killed more than 100,000 people or washed more than 80 km (50 mi) ashore! I'm sure that will be a comforting thought to residents of New York, Boston, Philadelphia, the entire states of New Jersey, Delaware, etc.. There is no compelling reason why a modest giant wave (or large hurricane *surge*) should not be capable of washing completely across Long Island, for example. Fortunately, the North Atlantic Ocean is a reasonably tranquil earthquake area, and its contour does not encourage waves to converge.

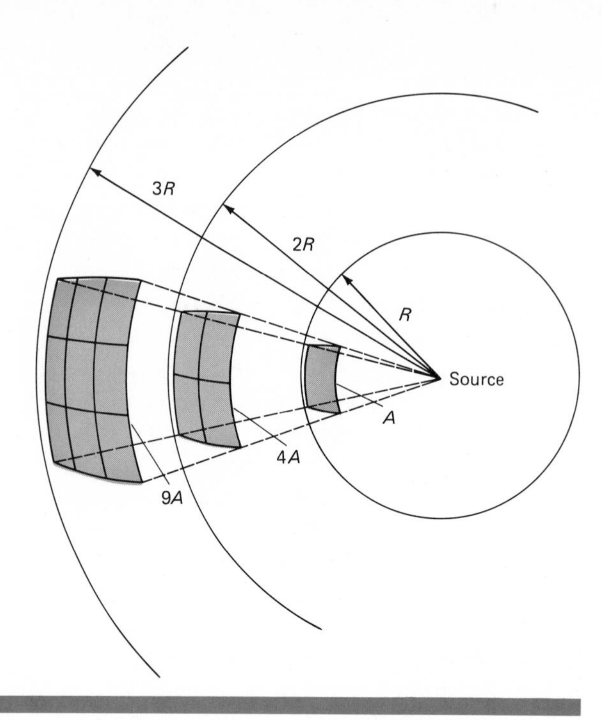

FIGURE 9-3
Waves moving in three-dimensional space distribute their energy over the face of an expanding sphere. When the wave has traveled twice as far from the source, the energy is spread over 4 times as much area; 3 times as far from the source the energy is spread over 9 times as much area.

on area A must now spread out to cover 4 times as much area. Thus, the intensity of spherical waves decreases rapidly with distance. If no energy is "lost," we find that $I \propto 1/D^2$, so that 10 times as far from the source the intensity is only one one-hundredth as great. In words, intensity varies inversely with the square of the distance. Such an inverse-square law for intensity and distance is required for anything that spreads out in space with none of it getting "used up" and none added. This form of relationship is required for anything that is "conserved" as it moves out evenly into three dimensions.

While we are considering wave differences, we might note that there is a fundamental difference in the way the energy of a wave is passed from place to place. This transfer of the wave is called *propagation.* Various materials are different in the way in which they wiggle while conducting the wave. As a wave approaches a small segment of rope in a clothesline, the segment occupies its undisturbed or *rest position.* As the wave arrives, the rope next to the segment we are considering exerts a force on the segment and pulls or *displaces* it from its rest position. In the case of a wave generated by shaking the end of the rope up and then down, the initial force causes an upward displacement of the segment. The segment is pulled upward and away from its rest position by the end of the rope carrying the advancing wave. In turn, the segment pulls upward on the next segment of the rope on its other side. If we would mark the segment of rope and watch it as the wave disturbance moved past we would see it rise and then fall. For a wave moving right to left the motion of each segment of rope is up and down. The direction of *vibration,* or wiggling, of each rope segment is perpendicular to the direction of propagation of the wave. This motion during vibration defines a *transverse wave.* The actual particles carrying such a wave do not move closer to or farther from the source as the wave passes, but vibrate in a direction perpendicular to the motion of the wave disturbance.

Sound waves cannot be transverse waves since sound passes through gases like air. In a gas, should one molecule move up and down, there is no reason for a molecule to the side to do the same. Since the molecules of a gas are not physically attached to each other, only contact between the molecules will transfer energy. Transverse waves cannot propagate through a material unless there is some attachment between the particles of the material so that a displacement of one will pull the next in line out of position also. For example, consider a gymnasium floor crowded with people. If the first row of persons begins to jump up and down, there is no reason why the next row must do so. To send a transverse wave through the people we would have to tie them tightly together. Then when the first row jumped upward they would pull the second row upward too. The disturbance could propagate across the floor as row after row of people were jerked upward by the preceding row.

We *can* propagate a different type of wave across a gymnasium floor crowded with unconnected people. All we would have to do is have the first row of individuals, on some signal, charge straight into the next row over. As they collided with the second row, these people would recoil into the third row, and so on as the wave passed across the floor. The wave that passed across the floor would appear from high above to be a zone of increased squeezing together. The actual motion of an individual person would be to move in the direction of the wave and then to recoil backward in the opposite direction. Waves of this type are referred to as *longitudinal waves.* In a longitudinal wave the vibration of the particles is in the direction the wave is moving and back in the opposite direction. Zones where the particles are squeezed together are called *compressions,* and zones of recoil where the particles are spread apart greater than the normal separation are called *rarefactions.* Longitudinal waves are the only types of waves that will propagate through gases. Thus, sound must be an example of this type of wave.

In a transverse wave, particle motion is *perpendicular* to the direction of wave propagation (Fig. 9-4). In longitudinal waves, particle motion is back and forth in a direction *parallel* to the direction of propagation. The motion of water-surface waves tends to be a combination of these two types. Individual water particles move both back and forth parallel to the wave direction and up and down perpendicular to the direction of propagation. These two motions occur at the same time and take equally long to complete one vibration, so that if we watched a single water particle it would move in an oval (Fig. 9-5). The size of the oval motion becomes smaller as we go deeper beneath the surface of the water. A relatively short distance down, no water motion is perceptible. This is why a submarine can comfortably ride out a hurricane in a submerged state while surface vessels battle for their lives a few hundred meters above. Water-surface waves are thus combination, or *composite, waves.*

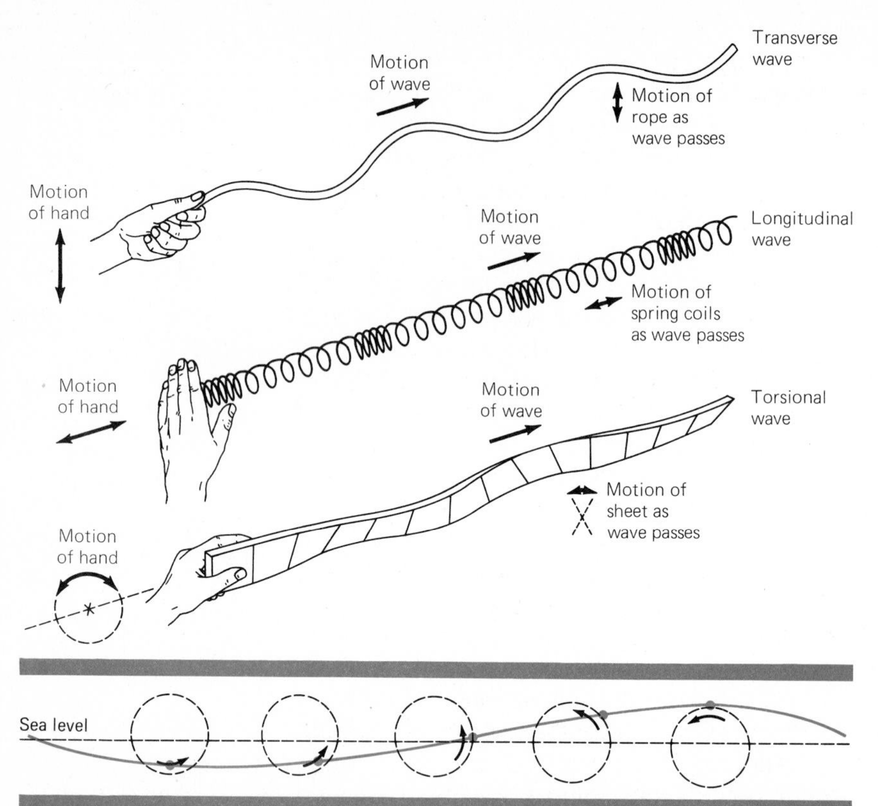

FIGURE 9-4 Types of waves are defined in terms of the motion of the material carrying the wave compared to the motion of the wave disturbance itself. The motion of the material may be perpendicular to (transverse waves), parallel to (longitudinal waves), or a rotation about (torsional waves) the direction of the wave.

FIGURE 9-5 The actual motion of a water particle in a surface wave is a circular loop. The completion of one full cycle leaves the water back where it started. The diagram may represent five equally spaced particles seen at the same time. It may equally well represent a single water particle at five different moments spaced equally in time.

However, a third type of wave is possible in which the motion of the particles of the medium carrying the wave is something different. This is a *torsional wave,* in which the motion of the particles is to rotate first clockwise and then counterclockwise around an axis parallel to the direction of motion of the wave. If you hold a large sheet of plywood vertically, you can send such waves along its length by twisting it one way and then the other. Such waves of twist or torsion are common in solid materials.

Wave Behavior

Imagine that we shake a wave into a clothesline securely fastened on its far end to a wall. To create the wave we sharply displace the end of the rope upward and then return it to its rest position. The zone of upward, or positive, displacement races along the clothesline away from our hand. Measurement from a timed series of photographs would confirm that the speed of the wave is constant as it moved along the clothesline. As the disturbance propagates along the rope, it retains its shape; that is, the distance from the beginning to the end of the disturbance remains a constant length as the wave moves. Furthermore, the relative height of each part of the displacement is constant. A portion of the wave with twice the positive displacement of another portion will stay at twice the displacement. Since the rope really is not a frictionless medium for the passage of the wave, we may note that the height, or positive displacement, of all parts of the wave is decreasing with time. We may infer that the height is associated with the energy of the wave and is decreasing as the wave does work, which it does not recover, on the rope and the air.

As the wave encounters the end which is attached to the wall, it bounces or *reflects* from the wall. *Reflection* is a general wave property, and you should not find it surprising. The *way* in which this wave reflects is more likely to be a phenomenon you have not observed before, although you probably have seen it. As our wave reflects, it changes its direction portion by portion so that the portion nearest to the wall as the wave approaches becomes the portion farthest from the wall as the wave recedes. Another way of saying this is that the leading edge of the wave stays the leading edge, even after reflection. Thus, reflection first causes a right-to-left inversion of the wave. This is not surprising. More unexpected may be the fact that the wave also undergoes an up-and-down inversion during this reflection. Upward or positive displacement of the wave becomes downward or negative displacement upon reflection. Thus, upon reflection our wave has undergone both an up-and-down and a right-to-left inversion as shown in Fig. 9-6. This type of behavior is characteristic of waves that encounter an end to the medium that is fixed and not free to vibrate. However, *fixed* or *closed-end reflection* is not the only type of reflection possible for a wave.

Instead of attaching the rope to the wall suppose we connected the end to the wall by a long, thin fishing line. The end of our rope is now free to vibrate and is not fixed or closed. Such an end to the medium is called an *open end.* Open-end reflection is different from closed-end reflection. Observation of the reflection from an open end reveals that in the reflected wave the displacement is not inverted, it is *erect,* or right side up. While the lead edge of the wave is the first to reflect and thus leads the

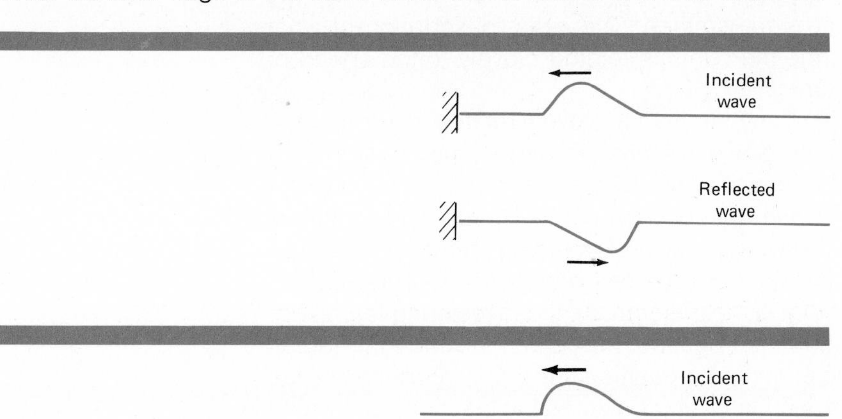

FIGURE 9-6 In a fixed or closed-end reflection, right and left on the wave are exchanged and the wave is inverted up and down.

FIGURE 9-7 In open-end reflection the wave strikes an end of the medium free to vibrate. The reflected wave has been exchanged from right to left, but up and down remain the same.

return wave back toward the hand, upward displacement remains upward displacement. Right and left have been inverted for the return wave but not up and down (Fig. 9-7).

All types of waves display both open-end and closed-end reflection. The key to the type of reflection displayed is whether or not the end of the medium carrying the wave is free to vibrate in the manner needed for wave propagation. This has nothing to do with the seeming impenetrability of the material at the end of the medium.

Change of Medium and Speed

Waves may encounter an end of the medium beyond which there is either an essentially immovable object or else nothing at all. In either case, the energy and the wave reflect. The first case is illustrated by the rope securely tied to a wall. It would be an unusual wall if a clothesline wiggle could produce a visible up-and-down motion of the wall. The second case might be illustrated by a person moving the handle of a fishing rod. The wave propagates down the shaft of the rod and reflects. The far end of the rod is not connected to anything and thus the vibration cannot be passed on or *transmitted.* Frequently, however, the end of one medium is the beginning of another. For example, if you live in a dormitory, you are perfectly aware that sound from next door can be transmitted into your room through the walls. This occurs even though the original sound wave in air encountered an end of the medium in the wall of the room. The wall simply serves as another medium for the transmission of sound. The division between one medium carrying a wave and another medium also capable of carrying the wave is called a *boundary. At the boundary between two different media we find that some of the energy is reflected back into the original medium and some of the energy is transmitted into the second medium.* This behavior is similar to the passage of light through a window pane. Some of the light is reflected by the glass even though most is transmitted. The exact description of what happens as waves change mediums is related closely to the speed of the wave in each medium.

The speed of a wave in any medium is closely related to two factors. The more massive the material we are trying to wiggle, the greater its inertia and the slower its response to our forces. Thus, as the *density* of a medium increases, the wave speed is reduced. The second factor influencing speed has to do with the size of the forces tending to restore displaced portions of the medium to their rest position. This second factor is called the *elasticity* of the material. As these restoring forces become larger, wave speed is increased. Thus, steel, in which the forces tending to snap a particle back into position are great, conducts waves rapidly, while a rope made of taffy would not. As the elasticity of a material increases, restoring forces become greater in the material. The speed of a wave depends upon both the elasticity and the density of the medium. As the wave moves from one medium to another, the speed of the wave usually changes.

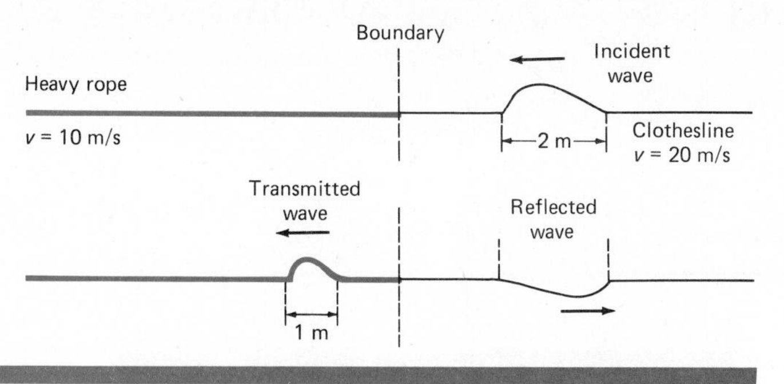

FIGURE 9-8 As a wave encounters the boundary to a new medium, it is partly reflected and partly transmitted into the new medium. When the new medium presents a slower wave velocity, as shown, the physical length of the transmitted wave is shortened. Note that the wave reflected from the boundary to a slower medium is inverted.

We can apply this idea if we picture our clothesline attached to a heavy rope used in a gymnasium for climbing. The fibers of the large rope are not necessarily much different from those in the clothesline in elasticity. However, the thick climbing rope is much more massive. The density/length ratio in kilograms of rope per meter of length may be 20 or more times as large for the climbing rope as for the clothesline. We can thus expect that waves will propagate much more slowly in the large rope. For simplicity let us assume that waves will travel half as fast in the large rope as in the clothesline. Suppose that we shake a zone of positive displacement into the clothesline in $\frac{1}{10}$ s. This wave now moves down the clothesline and encounters the boundary (Fig. 9-8). The upward tug of the clothesline sets the larger rope into upward motion and energy is being transmitted. Several things may be noted. The greater resistance to vibration implied by the inertia of the large rope makes it act like a partially fixed or closed end for the purposes of reflection. The portion of the energy that reflects from the boundary back into the clothesline will be inverted vertically as in closed-end reflection. The transmitted energy will not cause an inverted wave. The clothesline is pulling upward on the rope, and upward displacement will result. *Transmitted waves are never inverted; they are always erect.* One change that does occur is in the length of the wave as it passes the boundary. The wave shaken into the clothesline took $\frac{1}{10}$ s from beginning to end. Thus, $\frac{1}{10}$ s after the clothesline began disturbing the end of the rope, the tail of the wave will arrive and the disturbance will cease. No matter what happens, the tail of the disturbance will be $\frac{1}{10}$ s behind its head or start. Since the rope transmits waves only half as fast as the clothesline, the disturbance will be only half as long in the rope. For example, suppose that waves travel at a speed of 20 m/s in the clothesline. If a disturbance represents $\frac{1}{10}$ s, its tail will be 2 m behind its head on the line. In the $\frac{1}{10}$ s it took to generate the wave, the head had the time to move 2 m down the clothesline:

$$\begin{aligned} d &= vt \\ &= (20 \text{ m/s})(\tfrac{1}{10} \text{ s}) = 2 \text{ m} \end{aligned}$$

On the other hand, the large rope would have a wave speed of 10 m/s. If the tail of the wave is $\frac{1}{10}$ s behind the head in the rope, it would be 1 m behind it. Transmitted and reflected pulses are shown in Fig. 9-8.

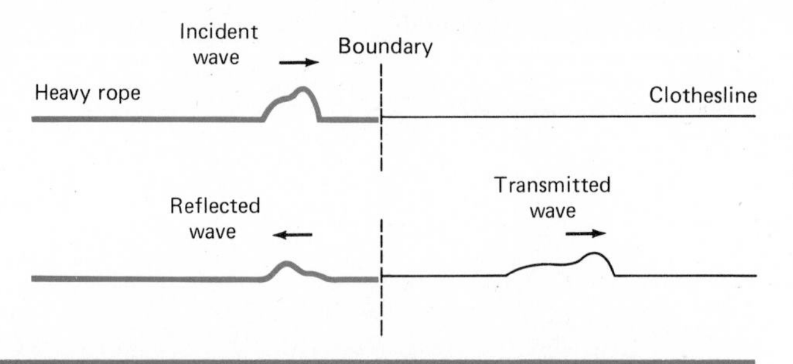

FIGURE 9-9
A wave incident onto a boundary with a faster medium also partly reflects at the boundary. In this case, the reflection is erect, not inverted. The transmitted wave is always erect, regardless of the direction at which it crosses the boundary.

A pulse in the rope approaching the boundary with the clothesline will behave similarly (Fig. 9-9). The main difference is that the reflection from a boundary with a faster medium will be erect, not inverted. The transmitted wave is never inverted, and here it will race away from the boundary at twice the speed of the reflected wave and be twice as long.

Boundaries and Speed—Refraction

Thus far we have considered the change of speed at a boundary for waves moving along a straight line only. Surface waves and three-dimensional waves also encounter a change of speed at the boundary to a different medium. With these types of waves there also may be an additional effect. If a wave front on a surface, for example, strikes the boundary to a new medium at some angle other than perpendicularly, the wave will undergo a change in its direction of motion. Let us consider this phenomenon in parts. Figure 9-10 illustrates a linear wave front in medium *A* approaching the boundary to medium *B* in which the wave speed is only one-half of the speed in medium *A*. This wave-front motion is perpendic-

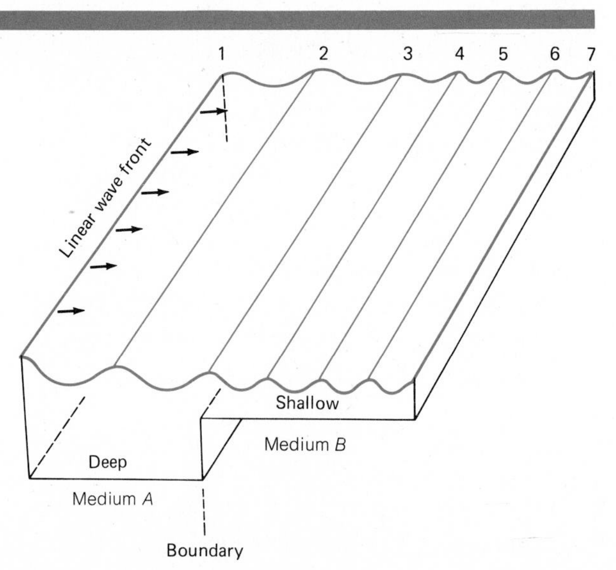

FIGURE 9-10
The boundary between the deeper water and the shallows constitutes a change in medium for the linear wave front. The motion of the waves is perpendicular to its wave front. Numbered positions represent equally spaced locations of the wave front in time. An abrupt change in speed occurs at the boundary.

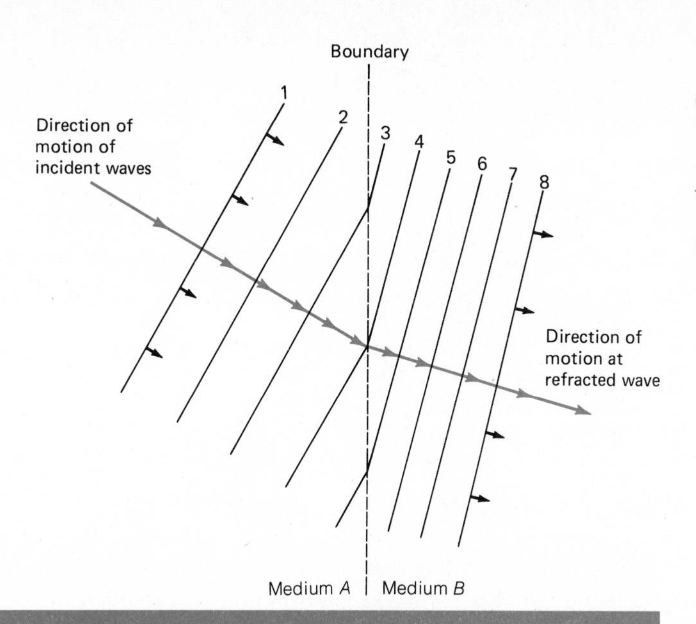

FIGURE 9-11
Wave refraction is a change in the direction of the wave at the boundary to a medium with a different speed of propagation. The numbered lines represent successive positions of a single wave front at equally spaced intervals of time.

ular to the boundary. All parts of the wave front will encounter the boundary at the same time. Thus, all parts of the wave front will be slowed down at the same time, and the motion in medium *B* will be in the same direction as it was in medium *A*. Now consider the same stiuation but with the wave front approaching the boundary at a considerable angle (Fig. 9-11). The left-hand side of this wave front will encounter the boundary first and will be the first to be slowed down in the new medium. For some time the right side of the wave front will move at the speed of medium *A*, while the left side is moving at only one-half this speed. The effect is to bend the entire wave front around to a new orientation. Since the motion of a wave is always perpendicular to its wave front, the direction of the motion of the wave has been changed. We may picture this action by imagining a line of children holding hands and running across a paved playground at a constant speed. The line encounters the end to the paved area at an angle, in the same manner as our wave front. The children on the left end of the line plunge off the paving into knee-deep mud and ooze. This slows their speed considerably, and the left end of the line is slowed down. Meanwhile, the right end of the line continues running merrily along. The entire line would wheel around to a new direction of motion in the mud as child after child hits the mud. Indeed, if the mud were waist-deep and no motion could be made in it, the line would wind up stuck in the mud at the edge of the paving aimed perpendicularly into the mud at the boundary.

This bending of waves at a boundary between mediums with different speeds of propagation is called *refraction*. To describe refraction, we consider the direction of the motion of the wave front as indicated in Fig. 9-11 by the colored pathway.

Figure 9-12 reproduces this pathway for a part of the wave front. The angle labeled θ_A is called the *angle of incidence*. It is the angle from the

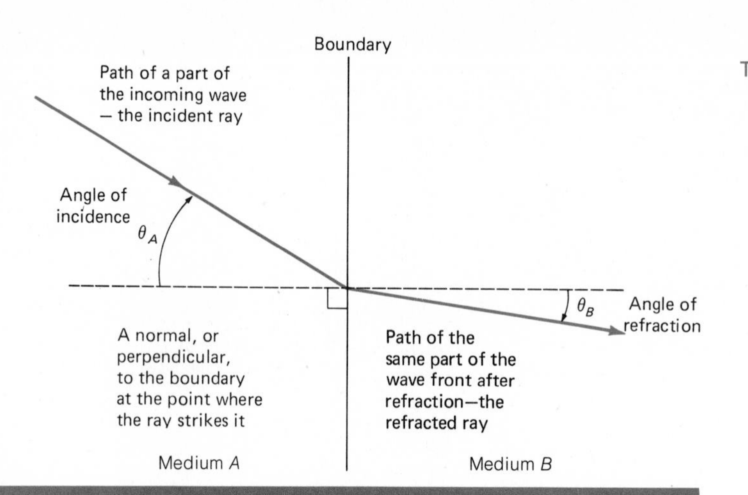

FIGURE 9-12 The terminology used to describe refraction is illustrated for the case shown in Fig. 9-11.

perpendicular to the boundary at which the wave approaches the surface. The angle labeled θ_B is the *angle of refraction,* or the angle from the perpendicular at which the wave leaves the surface. Clearly, if the speed of propagation in medium A, or v_A, is greater than the speed in medium B, v_B, then θ_A will be greater than θ_B. As a wave front is slowed down, it must bend, so that the wave plunges more into the new medium and its motion is more nearly perpendicular to the boundary. If a wave hits the boundary perpendicularly, no refraction is observed since all parts of the wave are slowed down at the same time. Thus, only the component of the motion parallel to the boundary may be observed. The relationship between angles and speeds involves the trigonometric relationship called the *sine* of the angle (abbreviated sin) and is

$$\frac{\sin \theta_A}{\sin \theta_B} = \frac{v_B}{v_A} = n$$

where n is a constant for any two given mediums and is called the *index of refraction for the boundary.*

All types of waves show refraction. The effects of refraction are quite noticeable at a seashore. When was the last time you noticed waves running parallel to the beach rather than coming inward from the open water? As water gets shallower the water wave encounters increased bottom friction and the speed is reduced. Thus, as water shallows uniformly it is a continually changing medium which refracts waves more and more. The result is that waves usually encounter the beach relatively close to the perpendicular. It is unusual to have waves arrive at the shore at a large angle if the bottom has been gradually rising for a large distance. A second effect can be observed at beaches. As the water gets shallower the speed of the wave decreases and the waves are crowded together as the wavelength lessens. The *amplitude,* or wave height, however, does not decrease. Thus the slopes of the wave become steeper. The small volume of water in the shallows in front of the wave cannot supply enough water to maintain this tall, steep shape. The lead edge of

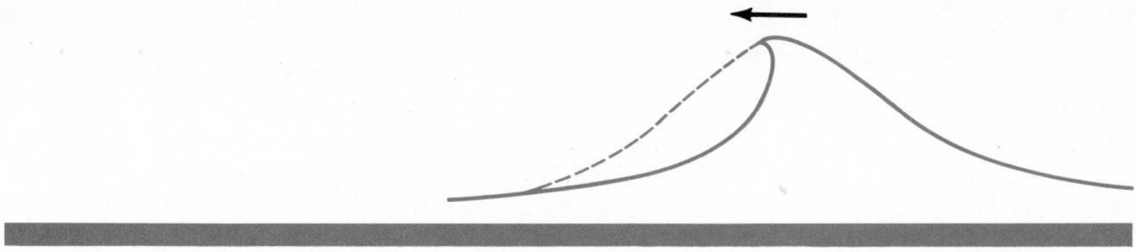

FIGURE 9-13
The beaker results as a steep wave reaches shallow water. There is not enough water in front of the wave to flow into it to maintain its shape as shown by the dashed line. The motion of the water at the crest is still a circular loop. This top water will curl over the steep forward face and actually leave an oval tunnel of air below the crest as the wave breaks. Note that the motion of the water in the steep forward face is upward. This provides the propelling force for surfers.

the wave is supplied with less water than necessary and the result is that the peak of the wave breaks. The circular motion of the water particles in propagating the wave carries the water at the crest out over the front of the wave (Fig. 9-13). Unlike normal wave motion, which preserves the energy of the wave quite well, the tumbling water of a breaker converts the potential energy of the wave into random swirling motions and eventually into a temperature increase. Breaking waves display their energy more obviously than waves encountered elsewhere.

When Waves Meet—Interference

Now that we have seen a single wave broken up into a reflected part and a transmitted part, it is reasonable to inquire about what happens when two waves meet each other. A fundamental fact about waves is that *two or more waves may be in the same place at the same time.* What happens as this occurs? Picture the situation shown in Fig. 9-14A. Two wave pulses each with positive displacement are approaching an undisturbed segment of the medium. If the medium is our familiar clothesline, we may picture what is happening. The right-hand pulse will exert an upward force on the segment. The left-hand pulse will do the same. With double the displacing force from a single pulse, the segment will accelerate upward at twice the rate. We may expect that it will reach twice the height it would have reached for a single wave before the restoring forces pull it back toward its rest position. The situation is quite different if the approaching waves have opposite displacements (Fig. 9-14B). In

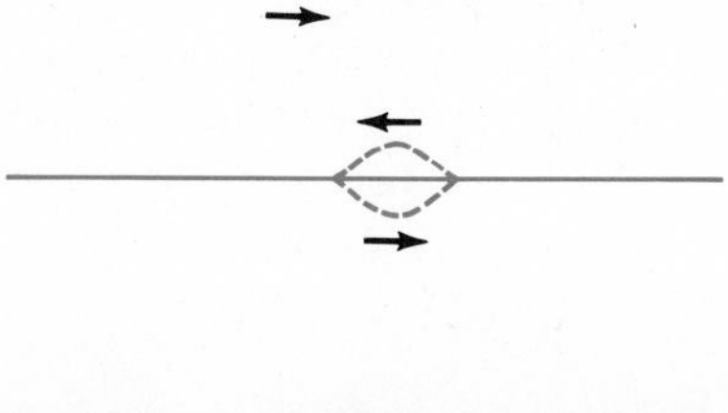

FIGURE 9-14A
When two similar pulses move toward each other in a medium, the disturbance that results represents their sum. It has the width of either pulse but twice the height.

FIGURE 9-14B
When two similar but oppositely oriented pulses cross each other, at a single instant they may add up to zero disturbance of the medium. Upward displacement of one cancels downward displacement of the other for each point on the medium at that instant.

this case one wave is pulling the segment upward while the other pulls downward. If the sizes of the displacements are exactly the same but in opposite directions, the forces will exactly balance and the segment will not be displaced from its rest positions. Such an undisturbed point occurring while waves cross is called a *node*.

This behavior of a medium under the action of two or more waves may be summarized by the principle of *superposition*.

> ***The disturbance at any point on a medium is the algebraic sum of the individual disturbances at that point at that time.***

Consider two repeating waves *A* and *B* as shown in Fig. 9-15. Suppose these two waves are exactly over each other, lined up as shown on the same clothesline. What would be the appearance of the clothesline at this time? According to the principle of superposition, the disturbance of any part of the clothesline simply would be the sum of the contribution from wave *A* plus the contribution from wave *B*. Look at the situation along the vertical line labeled 1 in Fig. 9-15. At this point wave *A* would raise the clothesline 1 unit from the dashed rest position shown. At the same time, wave *B* would create a positive displacement of 1 unit along this segment of the clothesline. The result is the sum of these two contributions, or a positive displacement of 2 units as shown on the sum at the bottom. Along line 2, wave *A* contributes a negative displacement of 1 unit, while *B* would have the clothesline at rest at this point with zero displacement. The result is a downward displacement of 1 unit as shown. Along line 3 the waves make equal contributions of 1 unit each, but in opposite directions. As a result, the medium at that point would appear in the rest position. We could continue to reconstruct the appearance of the clothesline point by point to produce the summation wave shown at the bottom. The actual appearance of the sum may be quite different

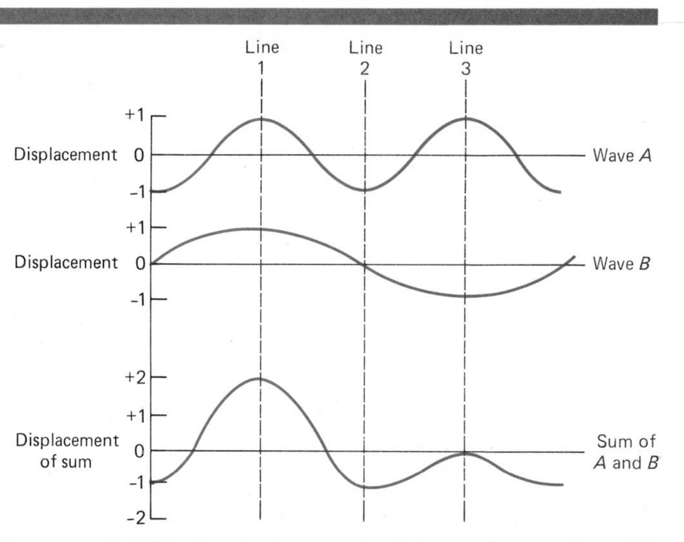

FIGURE 9-15
When two waves cross, the disturbance of the medium is the algebraic sum of the two disturbances for any point—the principle of superposition. Thus, along line 1, the displacement of wave *A* is +1. The displacement of wave *B* is also +1 at this position. These will add to a displacement of the medium from the test position of +2. For each vertical line, the wave at the bottom represents the algebraic sum of *A* and *B*.

from any of the waves comprising it and may change in quite complex ways in time. Whenever two waves are at the same place at the same time, we say that they interfere with each other. Such wave *interference* may have the two waves in agreement to produce a larger sum than either wave individually, as along line 1. This is called *constructive interference.* The two waves may also disagree to produce less of a disturbance as along line 3 in *destructive interference.*

Superposition and the resulting interference is the usual culprit in producing giant open-ocean rogue waves as mentioned at the start of this chapter.* A series of waves arriving in a single zone of ocean from storm centers perhaps thousands of kilometers away will cross each other and the local waves. The resulting shape of the ocean surface at any moment is the result of the superposition of these sets of waves. On rare occasions the timing and motions of these separate wave trains will be correct for a complete constructive interference to occur. The result is a wave of unusual height (or depth) clearly larger than any of the normal waves being encountered and usually quite unexpected. If the wave trains are arriving from nearly the same direction, the appearance might be a wall of water. If the wave trains are crossing each other at large angles, a single mound of water of tremendous height in the middle of the choppy sea might result. Such chance waves are not enjoyable encounters for sailors, but they do not last very long unless the separate wave trains are moving in almost the same direction. Oddly enough, while this type of situation might seem to have little practical importance except to sailors, the reverse is true. It was the study of this type of phenomenon with three-dimensional waves that led to the techniques to determine the structure of complex molecules. Our knowledge of the means the body uses to pass characteristics from parent to child begins here. So too does our understanding of how the blood manages to carry oxygen. This type of application across widely different fields of investigation is one of the powers of a scientific principle.

Repeating or Periodic Waves

Up to this point we have been considering the properties of single waves or pulses, such as the shock wave from an explosion. A highly important and interesting situation occurs when the wave disturbance repeats itself exactly in time. Such repeating waves are called *periodic waves* and are encountered very frequently in nature (Fig. 9-16). Periodic waves have two properties not associated with single disturbances. Periodic

* Superposition is not the cause of tidal waves, or *tsunamis,* however. These result when a low, fast wave produced by an earthquake encounters bottom friction and slows greatly near a shore. A wave a few meters high and a few hundred kilometers long can be generated by shaking the ocean floor. Such a wave may move in the open ocean at 1000 km/h. When such a wave is slowed to under 100 km/h by encountering shallows, its crest grows to great height. This is the type of wave responsible for most deaths when it hits a populated shore; it has nothing to do with superposition. Tsunami Warning System has existed since 1948 with stations all over the Pacific and headquarters in Hawaii. This has allowed several hours of warning for threatened coastlines. One 1960 tsunami tore up 800 km of the Chilean coast. A 1946 quake in Alaska produced waves some six stories tall on Hawaiian beaches, resulting in 159 deaths. Tsunamis are almost undetectable by ships on the open ocean as they race past.

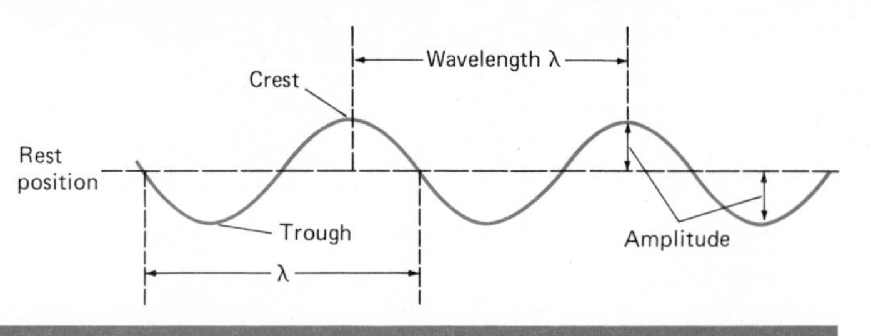

FIGURE 9-16 Terms associated with a periodic, or repeating, wave. Such a wave need not be so simple as this wave. It is periodic so long as it exactly repeats its shape in time and distance.

waves repeat themselves in time and in distance. The time required for a wave disturbance to repeat itself is called its *period* T. The length in which the wave repeats itself is called its *wavelength* λ. The position of maximum upward displacement is called the wave's *crest,* while the minimum displacement position is called the wave's *trough.* The distance from crest to crest is a convenient choice for measuring the wavelength. However, the distance from any portion of one wave to the exact repetition of that portion in the next wave may be used to measure wavelength. Wavelength, period, and the speed of the wave through the medium are related through the definition of speed, $v = \Delta d/\Delta t$. The time required to generate one full wave is the period T, and in this time the leading edge of the wave moves exactly one wavelength. Thus, $v = \lambda/T$. While this expression is correct, it is inconvenient for waves that repeat rapidly in time. For such waves, *the frequency f,* or $1/T$, is more useful. Replacing $1/T$ by f in the expression for speed, we obtain the most useful basic relationship for periodic waves:

$$v = \lambda f$$

Thus the speed of the wave is just the product of how long each wave is (λ) times the number of waves produced each second (f).

Example 1

A motor is used to disturb the end of a clothesline. If it generates a wave every $\frac{1}{10}$ s and each wave is 75 cm long, what is the speed of the waves along the clothesline?

$$T = \tfrac{1}{10}\ \text{s} \qquad f = \frac{1}{T} = 10/\text{s}$$

$$\begin{aligned} v &= \lambda f \\ &= (75\ \text{cm})(10/\text{s}) \\ &= (0.75\ \text{m})(10/\text{s}) = 7.5\ \text{m/s} \end{aligned}$$

Thus the speed of the waves along the clothesline is 7.5 m/s.

Example 2

If we wished to produce waves 30 cm long on the clothesline of the previous example, at what rate would the motor have to disturb the end of the rope?

$$v = \lambda f$$

$$f = \frac{v}{\lambda}$$

$$= \frac{7.5 \text{ m/s}}{0.30 \text{ m}} = 25/\text{s}$$

Thus the motor would have to disturb the line through 25 complete wiggle cycles each second to generate waves of the required length.

A periodic wave may have as complicated a shape as imaginable so long as it repeats its pattern exactly in time. For simplicity the most uniform type of wave is normally discussed. This is the *sinusoidal wave* of the type shown in Fig. 9-16. Such a wave is generated by a motor rotating at constant speed as it disturbs the end of a rope. An important mathematical theorem developed by Fourier indicates that no matter how complex the shape of a periodic wave, it can be broken into a series of such sinusoidal waves occurring at the same time. Many waves are a combination of several periodic disturbances occurring at the same time with different frequencies. For example, if we had the periodic waves shown in Fig. 9-15 occurring on the same medium, they would add to produce the periodic wave shown at the bottom of the drawing. This wave is very similar to the pattern of several spoken sounds or musical tones.

Because of the fact that periodic waves repeat the pattern of disturbance precisely in time, these waves produce some special effects. For example, if a burst of periodic waves is shaken into a clothesline, the train of waves will be seen to move along the clothesline quite normally. Such a moving train of waves is called a *traveling wave*. However, when a periodic wave reflects from some end of the medium and returns across the stream of incoming waves, a different pattern results. This situation is illustrated in Fig. 9-17A. The incident wave train shown at the top and the reflected wave train shown below it are moving across each other. At the instant "frozen" in the drawing these two add up to the disturbance shown at the bottom. This sum is also a periodic wave, and this represents the actual appearance of the medium at this instant. Unlike the traveling wave, however, this pattern does *not* move along the medium at the speed of the wave. (Logically, this is quite reasonable. After all, which way should the pattern move—to the left with the incident waves or to the right with the reflected ones?) We can see the change that will occur in time by moving to Fig. 9-17B. In this diagram made a short time later, the incident waves have moved a small amount to the left and the reflected waves have moved an identical amount to the right. The sum of these two disturbances is quite similar to the sum in Fig. 9-17A. There are differences between these two sums however. The crests, while in the same positions on the rope, are less high and the troughs are less deep. The positions of zero disturbance of the rope, the nodes, also have not moved with the waves. Picturing the incident wave

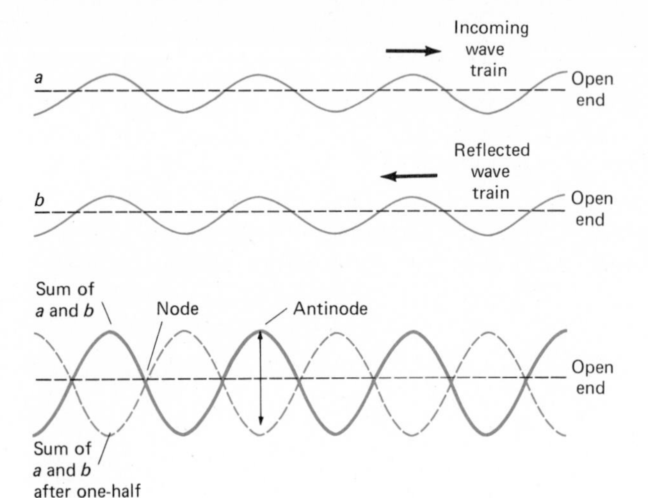

FIGURE 9-17A
When a periodic wave and its reflection cross, a standing wave results. This has fixed points of no disturbance, called nodes, between zones where the disturbance is alternately upward and downward.

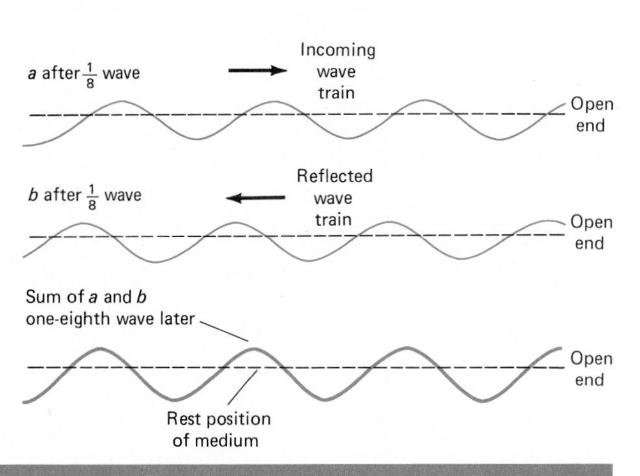

FIGURE 9-17B
If the incoming wave *a* and its reflection *b* are allowed to travel $\frac{1}{8}$ of the wavelength from the positions originally pictured, their crests and troughs no longer line up. However, the positions of crests, troughs, and nodes in their sum is the same as in the original drawing. The sum has neither rightward nor leftward motion—it is "standing" in the same position as the original sum, although reduced in amplitude.

form sliding to the left while the reflection moves to the right should convince you that these two periodic waves will *always* add to zero at the positions of the nodes. Meanwhile, between the nodes, at the *antinodes,* the two periodic waves will slip into and out of agreement with each other as the wave trains cross. This pattern of fixed points of zero disturbance with zones alternating between crest and trough in between is called a *standing wave. Whenever two periodic waves of the same frequency cross over each other, a pattern of standing waves results.* For reflected wave trains, if the barrier at which reflection occurs is a closed end, the end of the medium must be a node since it is not free to vibrate. Other nodes will be spaced every half-wavelength back along the medium. Should the place at which reflection occurs be an open end, it will be an antinode since it is free to vibrate. The first node in this case will be one-quarter wavelength from the end with other nodes every half-wavelength beyond that.

Producing Sound

The standard laboratory source for periodic sound waves has been the tuning fork. More recently, electronic signal generators have taken over in laboratories. A tuning fork is a simple mechanical device, and because of its simplicity its vibration is extremely regular. A good tuning fork produces an almost pure sinusoidal wave. When a tuning fork is struck with a block of hard rubber, the prongs (or tines) of the fork vibrate back and forth. In motion, both prongs move outward together and then snap back toward the midline of the fork. A vibrating fork produces very little sound by itself, and it continues vibrating for a long time. If the base of a vibrating fork is touched to a desk or a steel chalkboard, it sets a much larger mass into vibration against the air and sound becomes louder. The fact that the prongs of the fork are moving during vibration is easy to demonstrate by thrusting the ends of the fork into a beaker of water. In generating sound the air must be disturbed somehow. The fork does this by first pushing sharply outward against the air and then snapping back away. During the forward push, the air in front of the advancing prong is "squeezed together" or compressed. This causes an actual increase in the local air pressure within this zone. When the prong moves away from the air, it leaves an area of reduced pressure into which air molecules move. This alternation of zones in increased pressure, or *compressions,* and zones of reduced pressure, or *rarefactions,* moves outward from the fork. You hear these variations in pressure as sound if they fall within the frequency range sensed by the ear (Fig. 9-18). This range is usually listed as from 20 to 20,000 waves per second, or 20 to 20,000 Hz. One cycle per second equals one hertz (1 Hz). Below 20 Hz the body "feels" the sound rather than hears it. You probably have sensed the deep bass vibrations of a good music system by a flutter in your stomach or a vibration of the floor against your feet. The upper limit of hearing is dependent upon the particular ear and age. The frequency

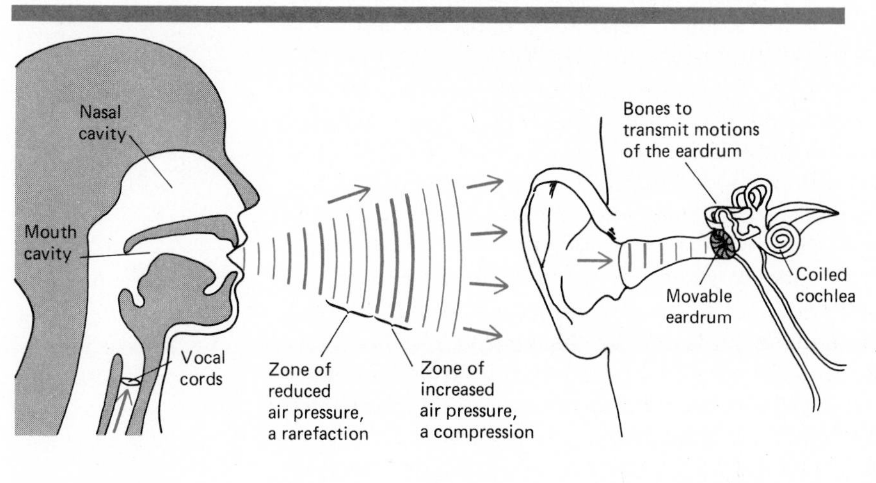

FIGURE 9-18 Flapping of the vocal cords against the airstream from the lungs alternately creates zones of compression and rarefaction in the air. These are modified by the position of the mouth, lips, and so on, and emerge to the outside as speech. Each compression enters the ear canal to push against the eardrum and depress it. Pressure on the inside of the eardrum is kept at normal by a canal to the mouth cavity. The pressure of this air causes the eardrum to pop outward into the rarefactions. Oscillations of the eardrum are transferred by tiny bones to the coiled cochlea, which is lined with fine hairs and nerve endings. Each portion of the cochlea resonates to a different frequency of disturbance. The position (frequency) and size of this disturbance is relayed to the brain by the nerve endings, enabling you to hear.

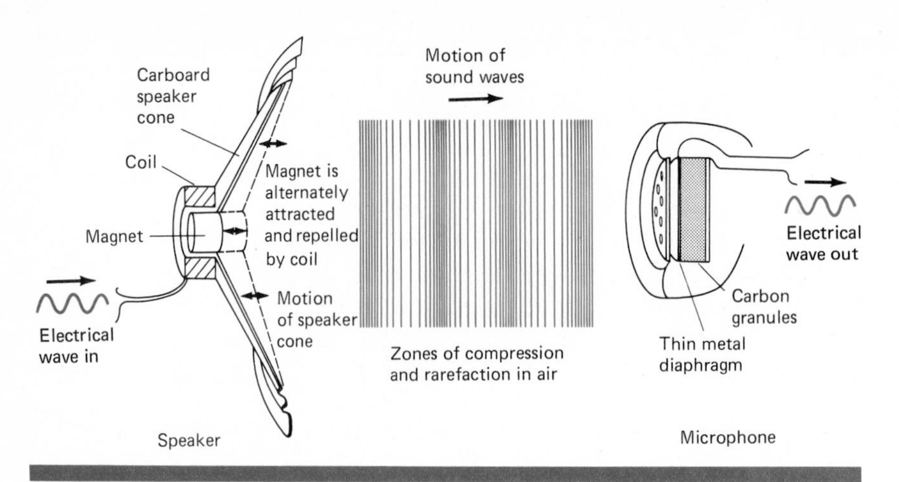

FIGURE 9-19 Electric variations corresponding to the sound are fed to coil at the rear of the speaker. This makes the coil an electromagnet alternately attracting and repelling a permanent magnet attached to the speaker cone. When repelled forward, the cone is pushed sharply against the air, only to be pulled back as the electric signal reverses. This alternately creates zones of compression and rarefaction which propagate through the air. One of the simplest microphones is found in telephones. Each compression pushes a thin metal diaphragm against an enclosure of carbon granules compacting them. The compaction increases the conductivity of the granules, allowing more electric current to flow. When the diaphragm bulges outward into the rarefaction, the granules are squeezed less tightly, causing less current to flow across them. The result is an electric variation matching the original input to the speaker.

used need not be much beyond 20,000 Hz to eliminate all but the best human ears; however, as is well known, animals can hear much higher frequencies. Sound frequencies above human hearing are called *ultrasonic.* Dogs can be taught to respond to ultrasonic whistles, and bats generate and hear sounds at frequencies of up to 100,000 Hz.

The ear detects pressure variations in the air, determines their intensities and frequencies, and relays this information to the brain. The ear can detect pressure variations almost down to the random bumping of the molecules of the air against the eardrum. However, the process of hearing actually occurs in the brain and is even more remarkable. The brain can suppress a great deal of background sound and "tune" to a particular signal. In this manner, information is actually exchanged in a crowded junior high school cafeteria and on some dance floors. Similarly, mothers can develop an uncanny ability to detect in the middle of a party an infant's cry five rooms away. (The study of what the brain hears as opposed to the sounds detected by the ear is called *psychoacoustics,* and its experimental studies make fascinating reading.)

The speakers in your phonograph produce sound in a different manner from a tuning fork (Fig. 9-19). The speaker is essentially a cone of a quality cardboard near the apex of which is a wire coil with its ends attached to the phonograph amplifier. Small juggles in the grooves of the record cause the phonograph needle to wiggle. These wiggles are changed into electric waves and are sent to the amplifier, where they are greatly magnified. These electric waves are fed to the coil on the speakers. When the crest of an electric wave passes through the coil, it becomes a small electromagnet and is attracted rearward into a large permanent magnet at the rear of the speaker cone. As the trough of the wave arrives, the magnetism is reversed and the coil and cone are repelled forward. When the cone is pulled back, it generates a rarefaction in the air in front of it; and when it is pushed forward, it generates a compression. Thus, each variation in the grooves is translated into a sound wave in air: The larger the variations in the grooves, the louder the sound; and the closer the variations, the higher the frequency of the sound. In a stereo system, the up-and-down motion of the needle generates one electric signal, while the sideways motion generates another. These signals are kept separate and fed to the right and left speaker sets without mixing.

Waves on Strings

If we consider a vibrating string, such as that of a violin or guitar, the first fact that we note is that the string maintains its vibration for some time after the original disturbance. This requires that a standing wave be produced on the string. Since both ends of such a string are fastened down, they cannot vibrate and must be nodes (Fig. 9-20). The requirements for standing waves demand that for a given length L and tension of string, only certain frequencies can form standing waves. We have seen that for a closed or fixed end of the medium, a standing wave must have its first node one-half wavelength down the medium from the end. But, for a vibrating string the position of a second node is preset by fastening down the other end of the string. Thus if the two ends are closer to each other than one-half the wavelength of a particular vibration, the wave cannot form a standing wave. The longest wave (which has the lowest frequency) that can form a standing wave on a string fixed at both ends has a wavelength of exactly twice the length of the string. In this situation, each end is a node and the center of the string is an antinode and vibrates between the positions shown in Fig. 9-20. This vibration is called the *fundamental mode of vibration* of the string, and the frequency of this vibration is called simply the *fundamental* of the string. It represents the basic musical note being played. The string, however, may vibrate in many other ways. The only restriction on its vibration is that both ends be nodes of a standing wave on the string. Since standing waves on mediums with fixed ends require nodes every one-half wavelength along the medium, the length of our string could represent one-half wavelength for the fundamental vibration. The length could also represent two-halves of a wavelength, or three-halves of a wavelength, or any whole-number–halves of a wavelength and still allow standing waves to form. Each of these *overtones* beyond the fundamental represents a shorter wave with a higher frequency. The set of possible frequencies of vibration of a string is called a series of *harmonics*. The fundamental is called the first harmonic, the next lowest frequency is the second harmonic, and so on.

This suggests a partial explanation of a question which may have occurred to you. Why is it that if a violin, a guitar, a piano, and a banjo all play the same musical note, they still sound quite different and can be distinguished by their sound alone? What gives a distinct musical quality to an instrument? While all these instruments could generate a tone of 243 Hz for the fundamental, they would vary in the relative loudness that the series of overtones produces. Overtones are encouraged or suppressed by the structure of the instrument and the way in which it is

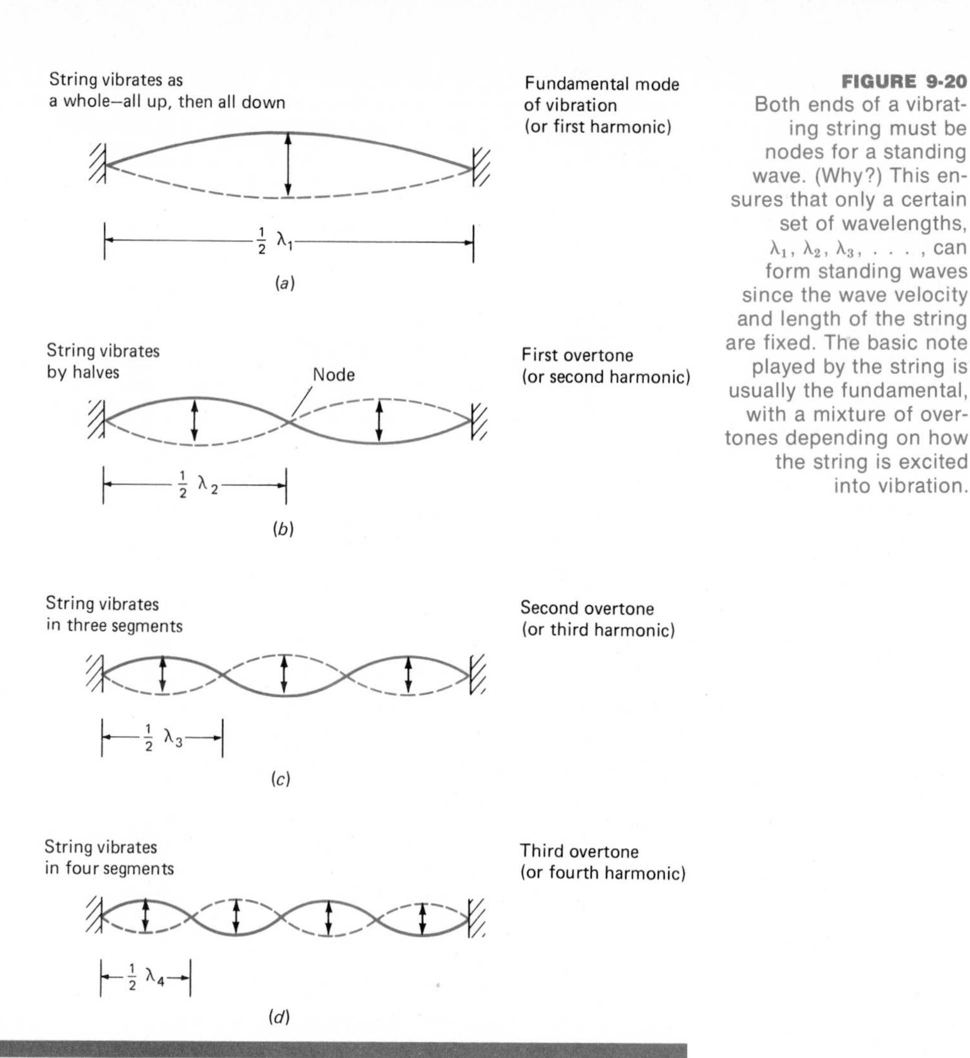

FIGURE 9-20 Both ends of a vibrating string must be nodes for a standing wave. (Why?) This ensures that only a certain set of wavelengths, $\lambda_1, \lambda_2, \lambda_3, \ldots$, can form standing waves since the wave velocity and length of the string are fixed. The basic note played by the string is usually the fundamental, with a mixture of overtones depending on how the string is excited into vibration.

played. For example, if a violin would be bowed in the exact center of the string, all the even-numbered harmonics would be absent. A violin bow is coarse horsehair which catches and holds the string while scraping across it. When the forces pulling the string back get large enough, the string is snapped back toward its rest position. The bowing of a string requires it to be in motion at the position bowed; it *cannot* be a node. If we look at even-numbered harmonics we see that they require a node in the middle of the string. If the bow is drawn across the midpoint of the string, odd-numbered harmonics are encouraged and even-numbered harmonics are suppressed. The relative strength of the various harmonics has a great deal to do with musical quality. With the tuning fork we produce a beautifully regular wave almost entirely free of harmonics if properly struck. Yet, no one has started up an all-tuning-fork orchestra. The tone of a tuning fork is monotonous and lacks the musical subtleties we demand of instruments.

Waves in Pipes

Another system frequently used for producing sound is that of a tube or pipe of air. Blowing across the open mouth of a soda pop bottle generates a sound. So do tubas, trombones, and pipe organs by much the same means. Unlike a vibrating string, the ends of a pipe full of air may be either open or closed ends. To produce a standing wave in an air column with one end open and one end closed is somewhat different from producing a wave on strings. The closed end must be a node, but an open end must be an antinode. The simplest situation in which this can occur is if the pipe is one-quarter wavelength long, or $L = \frac{1}{4}\lambda_1$ (Fig. 9-21*a*), or if one node is between the ends as in Fig. 9-21*b*, with $L = \frac{3}{4}\lambda_2$. Successive possibilities have L representing any odd number of quarters of a wavelength. We may arrive at this result by consideration of another effect, which occurs whenever the input of a wave to a system is precisely timed to the oscillation of the wave already in the system. If a wave is reflecting back and forth and we add a small contribution making sure that our crests always add to crests and our troughs to troughs, we will find that the energy of the wave will increase in time. This timed addition of energy to a system to provide constructive interference achieves what is called *resonance.* A mechanical example of resonance is provided by picturing a child on a swing. If we provide a small push, the swing will begin to oscillate back and forth. What happens with a series of small pushes depends upon our timing. If we give pushes at random times determined by a random number table, as many pushes will retard the motion of the swing as will increase it. Not only is the child likely to suffer whiplash injury from the jerky ride, but he or she is likely to think us pretty stupid for not mastering such a simple case of timing. A series of small pushes administered exactly in time to the natural

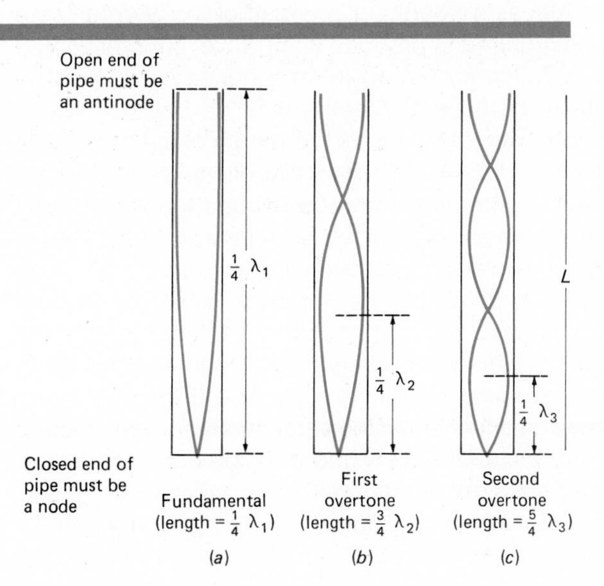

FIGURE 9-21
To produce a standing wave in a tube with one closed and one open end, we must have a node at the former and an antinode at the latter. (Why?) For a given length of tube, there is a set of wavelengths (λ_1, λ_2, . . .) that will form a standing wave. Some horns actually may not play the fundamental, merely a mix of the overtones. The brain "hears" the missing fundamental anyway.

swinging frequency has a very different effect. As we add our energy exactly "in tune," the size, or amplitude, of the swing grows bigger and bigger on each swing. Natural sources of friction usually take energy from the system too fast for us to get to looping-the-loop with the swing, but the amplitude becomes large.

Resonance is a very important effect in nature. For musical instruments it is essential, but for most mechanical systems we try hard to avoid resonance. It is undesirable to have your picture window resonate to the vibrations of a heavy truck. The last thing we want in an automobile is to have the springs resonate to regularly spaced bumps in the road so that the up-and-down bouncing increases in time. This leads to the complete loss of control of the car. Resonance is a powerfully useful phenomenon in the right situation. It allows very small, timed inputs to become very large disturbances. This is useful in a musical instrument and essential in a transistor radio. It also allows you to be heard. If we cut out the vocal cords and had them vibrating on a ring stand, the volume of the sound would be so low it would be inaudible at a few feet. Your voice attains its full volume by setting into resonance the air in the various tubes and cavities of the body. This is why you sound quite different with some cavities blocked, as when you have a head cold.

Let us examine resonance for a tube open at one end and closed at the other. If a source of sound waves, such as a tuning fork, is placed over the open end of the tube, we may follow the disturbance it generates. A crest (compression) produced by the fork proceeds down the tube (Fig. 9-22). At the closed end the crest inverts to become a trough and comes back toward the fork. Now, if the returning trough finds the fork producing a trough at the instant it returns, the energy of the fork may add to the energy reflecting in the tube in constructive interference, and we will have resonance. Thus, our part of the wave, which left as a crest, must come back to the fork as a trough, or one-half of the wiggle cycle later. In one-half the time required to produce a full wave, our crest could have moved one-half wavelength. Thus, for resonance to occur, the round-trip distance down the tube and back must be one-half wavelength. This means the one-way distance, or the length of the tube, must be one-quarter wavelength, or $L = \frac{1}{4}\lambda_1$. This is the same result we obtained for the resonance in the tube from a consideration of standing waves at the start of this section. Of course, we could produce a resonance no matter which trough we came back on—it would not have to be the first trough after we left. If we returned one-and-a-half full wiggle cycles after we left the fork, we could attain resonance. In this case, $L = \frac{3}{4}\lambda_2$ by the above reasoning, which is the same result that we obtained in considering the second resonance by standing waves (Fig. 9-23).

Open-end pipes may be analyzed in the same manner. Here, a crest reflecting at the open end remains a crest and must be back one full vibration after it left. Thus the length of the total trip is one wavelength, or the tube must be one-half wavelength long, $L = \frac{1}{2}\lambda_1$. Consideration of standing waves produces the same result. Open ends must be antinodes since reflections are not inverted.

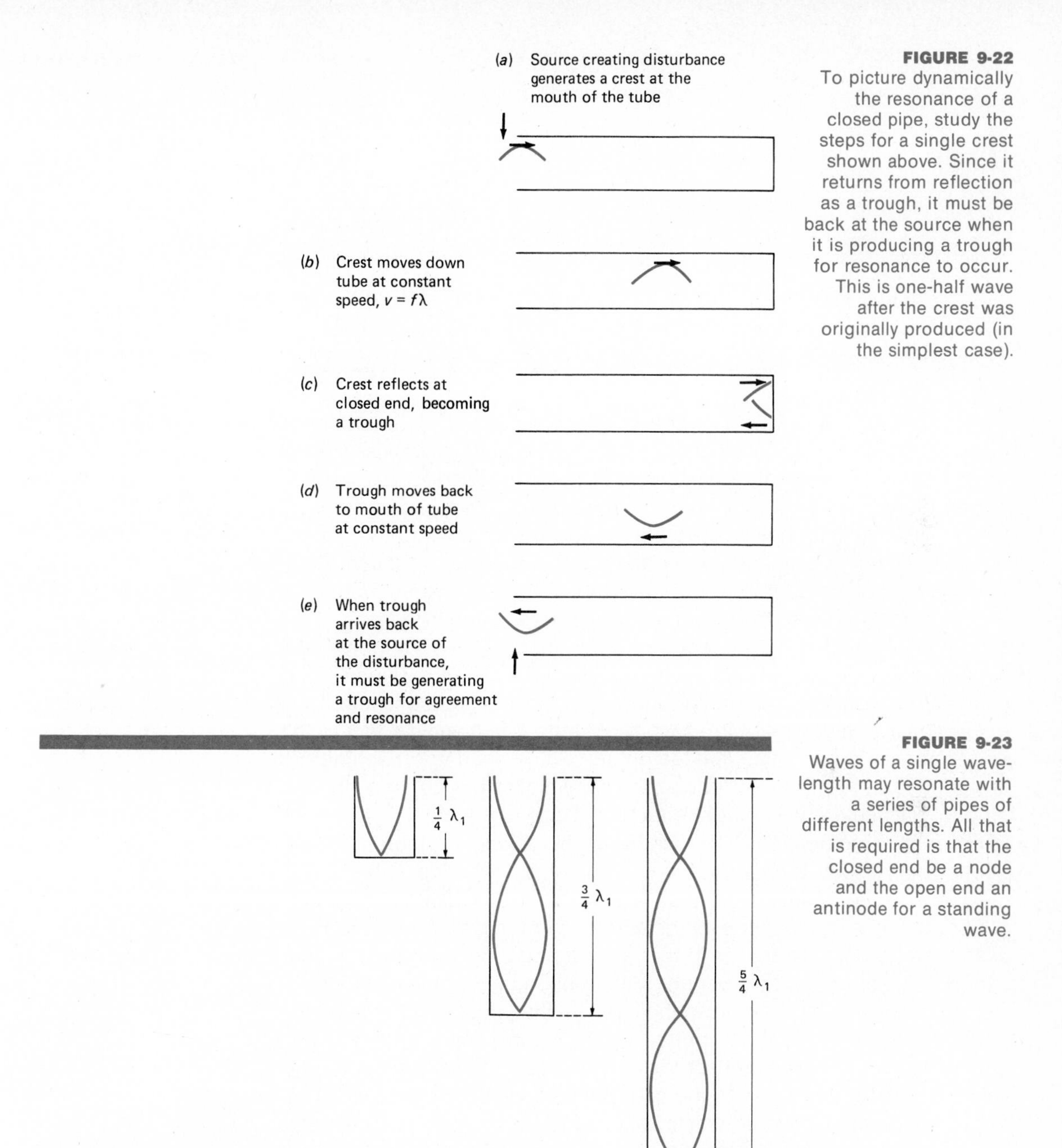
(a) Source creating disturbance generates a crest at the mouth of the tube
(b) Crest moves down tube at constant speed, $v = f\lambda$
(c) Crest reflects at closed end, becoming a trough
(d) Trough moves back to mouth of tube at constant speed
(e) When trough arrives back at the source of the disturbance, it must be generating a trough for agreement and resonance
$\frac{1}{4}\lambda_1$
$\frac{3}{4}\lambda_1$
$\frac{5}{4}\lambda_1$

FIGURE 9-22 To picture dynamically the resonance of a closed pipe, study the steps for a single crest shown above. Since it returns from reflection as a trough, it must be back at the source when it is producing a trough for resonance to occur. This is one-half wave after the crest was originally produced (in the simplest case).

FIGURE 9-23 Waves of a single wavelength may resonate with a series of pipes of different lengths. All that is required is that the closed end be a node and the open end an antinode for a standing wave.

Waves from Moving Sources

The last set of wave effects we will consider for sound are those caused by the motion of the source of the sound. When a source moves as it generates sound, each wave is produced at a particular point. The wave will

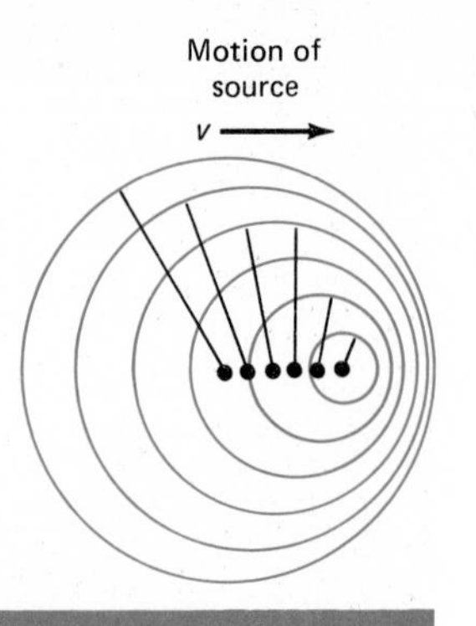

FIGURE 9-24 Waves from a moving source are not concentric as they move outward since the source is at a different position as it generates each. The waves tend to "bunch together" (increase in frequency) in the direction of the motion and "spread farther apart" (decrease in frequency) for a stationary observer in the rearward direction.

still spread out in an expanding sphere from the point of its generation at constant speed. However, these spheres of expanding waves will not have a common center; they will not be *concentric*. Each sphere centers on the place the source was at the moment that it was generated (Fig. 9-24). The result is that successive crests are closer together in the direction of the motion, and farther apart in the opposite direction. A listener in front of the source would have more waves arrive each second than were being given off by the source. As a result, the frequency of the source seems higher than it really is. If the source is receding from the listener, the waves arrive at a lower frequency than the frequency of the source. This apparent shift in frequency of a moving source is known as a *doppler shift.* It can be quite obvious at times. The next time you watch an automobile race, notice that the engine whine of an approaching car is shifted to a lower-pitched tone as the car whizzes by you, or by the microphone if you are watching the race on TV or in a movie. A movie affords another comparison. The whine to the driver is different in frequency than it sounds on approach or as he or she recedes. The driver is moving with the source and hears the unshifted frequency being produced, which is between the extremes heard in the pits. A less pleasant example of a doppler shift is in the radar that police use to clock your speed. The set produces a very high frequency beam of rays which reflect from anything substantial in their path. If the target is still, the reflections will arrive back at the police car with the same frequency with which they left. As waves reflect from an approaching car after one crest reflects, the car advances to a closer position where the next crest reflects. Successive reflected crests are crowded on the heels of the crests before them and arrive back at the detector with shorter intervals between arrivals. By measuring the shifted frequency of the returning waves, your speed of approach may be measured quite accurately, as many motorists have found out. (In such a case your speed v is just the frequency of the radar set times the shift in wavelength or $v = f\,\Delta\lambda$.)

When the motion of a source through the air becomes quite large, another interesting effect occurs. If you were moving through the air at the speed of sound, the wave disturbances you produced could not outdistance you in the forward direction. Your speed would match that of the

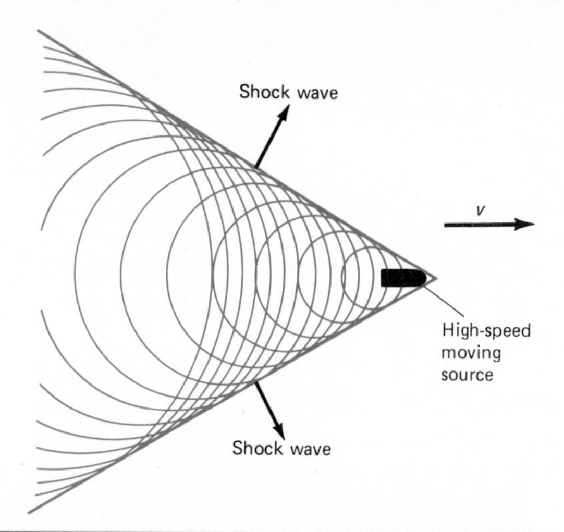

FIGURE 9-25
A source moving through a medium faster than its waves propagate in that medium produces a shock wave. Expanding wave fronts generated by the source at successive positions merge with each other in an expanding cone with the source at its apex. This is the origin of the "sonic boom" of a supersonic jet.

forward racing wave, and successive disturbances would be piled on top of each other, creating a very large wave disturbance (Fig. 9-25). If the motion of the source of waves is greater than the speed of the waves in the medium, this effect may be recognized more easily. Note that the waves in Fig. 9-25 add along a triangular wedge to produce a much larger disturbance than any one wave would produce. In air, such a triangular wave front becomes a cone of intense compression and rarefaction and is the wave responsible for the *sonic boom* associated with supersonic aircraft. Such a shock wave can do considerable damage on the ground because of its strong and rapid pressure variation. You have seen this type of wedge-shaped shock wave representing a line where small disturbances constructively interfere. The wave produced by the bow of a boat moving faster than surface waves on water represents this type of disturbance.

Summary

In this chapter we have examined the properties of waves, with particular attention to sound waves. Most properties of waves are quite general and shared by all varieties of waves. A wave represents a disturbance of some material called the *medium* of propagation. This disturbance moves from place to place in the medium with time. Three types of waves may be distinguished by the motion of the particles of the medium to propagate the wave. In *longitudinal waves,* particles vibrate parallel to the motion of the wave to transmit the motion. This is the only type of wave which may travel through a gas, and sound is the most common example. *Transverse waves* are transmitted from place to place by a particle motion which is perpendicular to the direction of motion of the wave. *Torsional waves* are propagated by a rotational motion of particles around the axis of wave motion. Waves *reflect* at the end of the medium or at boundaries to other mediums. Such reflections may invert the wave or not depending on whether the end is fixed and thus not free to vibrate or is open and capable of vibrating. Waves may represent single disturbances or repeating disturbances of the medium. A wave that repeats in time is called a *periodic wave* and is characterized by a given repeat time and repeat length called the *period* and the *wavelength*. *Frequency* (which is the inverse of—or 1 over—the period) is related to the wavelength through the relationship $v = \lambda f$, where v is the speed of the wave. This speed is constant if the medium is uniform in its properties.

Waves encountering a boundary are partly reflected and partly transmitted. The transmitted wave adjusts in wavelength to the speed of the new medium, and may change direction in the process of *refraction*. When two waves meet, the size of the disturbance is determined by addition according to the principle of *superposition*. Such addition may result in a larger wave by *constructive interference* or a smaller wave by *destructive interference*. Wave trains of the same frequency that cross produce zones of each type of interference. On a one-dimensional medium such crossing of wave trains generates points of zero disturbance called *nodes*, which are fixed in position and between which are zones of constructive interference. Such a pattern is called a *standing wave*. Standing waves are particularly important in the generation of sound by musical instruments, and we examined the cases of vibrating strings and waves in open and closed pipes. Waves from sources in motion undergo a *doppler shift*, which is an apparent shift in the frequency for a stationary detector, increasing the frequency if the source is approaching. If the source is moving at or beyond the speed of waves in the medium, a zone of constructive interference is generated by the waves which, in the case of sound waves, is the cause of sonic booms.

Questions

1 What evidence do you have that the speed of all frequencies of sound is the same, from high notes to low notes?

2 The speed of sound in air (room temperature) is about 340 m/s. In water it is 1500 m/s, and in steel it is almost 6000 m/s. What can you conclude from these speeds?

3 Would the reflection of a sound wave from a water surface be erect or inverted (see Question 2)?

4 If a sound in air with a wavelength of 1.7 m entered a body of water, what would be the wavelength of the wave in the water? What would be the frequency of this sound in air? In the water?

5 As a sound wave entered water at a small angle to the perpendicular to the surface, would the wave front bend toward the perpendicular to plunge downward into the water or bend upward more parallel to the surface?

6 If waves having a wavelength of 4 m reflect from an open end to the medium, at what distances from the end will nodes develop? What could be the distances for an open end?

7 Transverse earthquake waves do not pass through the core of the earth, while longitudinal ones do. What might you conclude from this?

8 How would you determine whether the wall of a bathtub is an open or closed end for water-surface waves?

Problems

1 A 5-m-long sailboat is anchored for fishing on a windy day. The skipper notices that as one wave crest is just leaving the stern another is arriving at the bow. If 15 waves are counted arriving in 10 s, what is the speed of the water waves?

2 A motor is used to disturb the end of a

clothesline. If it generates a wave every $\frac{1}{10}$ s and each wave is 75 cm long, what is the speed of the wave along the clothesline?

3 If we wished to produce waves 30 cm long on the same clothesline as in Prob. 2, at what rate would the motor have to disturb the clothesline?

4 Many cassette recorders have a volume unit (VU) meter built in to measure sound intensity. Place a speaker in an open window pointing to the outdoors. Measure the sound intensity at 1 m, 2 m, 3 m, . . . with the VU scale. Translate the readings into watts per square centimeter using the diagram. Plot these intensity readings vs. $1/d^2$ and determine whether a straight line results.

5 Examine the strings of a piano. The keys of the piano play notes with frequencies ranging from 27.5 Hz to almost 4200 Hz, giving it the largest range of a standard musical instrument. What is changed about the strings to change the frequency? Since the speed of sound in air is about 340 m/s, what is the range of wavelengths produced by the piano?

6 The outer ear can be thought of as a tube with an end closed by the eardrum. If the ear is about 2.5 cm deep to the eardrum, around what frequency should human hearing be the sharpest? (*Hint:* What frequency should be in reasonance with a closed-end tube this size?)

7 Orchestras like to play with more "brilliance" than their competitors. Thus, since many instruments are tuned by ear, musical scales have increased in frequency through time. Between Handel's *Messiah* and "The Stars and Stripes Forever" of John Philip Sousa the frequency for the same note had increased by 10 percent. This creates problems for an instrument designed and built for a specific scale. For example, consider the flute as an open-end pipe (a simplification). The lowest note the flute can play is middle C with a frequency of 261.63 Hz on the current accepted international scale (A = 440). How long would a flute have to be to play Handel's middle C of 250 Hz? To play Sousa's middle C of 274 Hz?

8 Draw a diagram like Fig. 9-22 for an open-end pipe. What are the conditions for setting up a standing wave at each end? Extend this to a drawing such as Fig. 9-23 for a pipe with two open ends. Why should the sound energy reflect from an open end?

CHAPTER TEN

Other Sources of Forces—Electricity and Magnetism

Introduction Why is it that a clothesline snaps back into place after a wave passes? What allows the chair you are sitting on to counteract the downward pull of gravity and support your weight in its position? In both cases, up to now, we have answered that there must be a force. While this logically follows from the laws of motion, it is not very informative. What kind of force? How does it act? What is its cause? Up to now we are armed with the ability to describe only one type of force—the attraction of mass for mass in gravitation. There must be at least one other type of force around. First of all, gravitation is too weak a force to be of any considerable effect until masses are huge. As you enter a room the furniture does not jump toward you in a mad display of attraction—not even one loose picture or book. If a book did come flying toward us as we entered a room, our first reaction (after ducking) would be to find who had thrown it. In other words, we would immediately ascribe this behavior of the book to something beyond an attraction for the matter in us. The second reason gravitation will not explain everyday forces is that it is always an attractive force—lots of pull but no push! How could we speed a bowling ball down the alley away from us with a purely attractive force? The everyday interactions of matter require a force that can be negligible across even small distances but can become large enough at a moment's notice to prevent us from falling through our chair. We need a force that will allow the molecules of a rope to be stretched out of place by a passing wave and then pull them rapidly back into position. The search for such a force will take us through the field of electricity and magnetism and on to a consideration of the microscopic structure of matter. In the electric force we will find the agent responsible for most of the everyday interactions of matter.

Both electrical effects and magnetic effects were known to the ancient Greeks, who gave us both these titles. By the beginning of the modern age in science so much folklore had grown up around magnetism that it was difficult to isolate the real from the fanciful. One man who did was William Gilbert (1546–1603), court physician to Queen Elizabeth I of England, and part-time natural philosopher. In a splendid example of the new experimental philosophy of science, he threw out everything about magnetism except what he could confirm by direct experiment. His book *De Magnete* (*Concerning Magnets*) is the beginning of the modern treatment of the subject.

We will consider magnetism and electricity side by side since from the beginning they appear to have much similarity. Ultimately, we will find that this similarity is real and that both represent different aspects of the same phenomenon.

Natural Magnetism

Magnets have always been a source of joy and fascination for humans willing to play with them and observe their unusual behavior. This was true when the Greeks named the effect after magnetic iron ore, or *lodestones,* found in the province of *Magnesia* (in modern Turkey). It is true in our own day when small magnets are packaged in "executive tranquilizer kits" for adults to fidget with. The behavior of a magnet is unlike that of most normal matter we handle daily. Magnets are capable of exerting definite, perceptible forces on some types of other matter across distances and with no physical contact. The closest experience we have with this type of phenomenon is in the attraction of matter for matter in gravitation. However, the differences between magnetism and gravitation are many, and some are immediately apparent. First of all, magnetism is much stronger than gravitation. There is no perceptible attractive force between a paper clip hanging from a string and a 200-g lump of clay placed 5 cm to the side of the clip. When a 200-g magnet replaces the clay, the paper clip is immediately pulled toward the magnet and no longer hangs straight down on its string. If the paper clip is placed on the desk top, the downward pull of all the matter in the entire earth cannot equal the upward pull of the tiny magnet which lifts the clip off the desk. A second major difference from gravitation is that magnetic effects may be "concentrated" or "diluted" by the surrounding matter. Your weight remains much the same whether you are standing on a continent, sailing across an ocean, or flying through the air on a passenger jet. No type of matter seems to concentrate gravitational effects or to shield us from them. This is not true of magnets as a simple experiment will show us. If we place a magnet at a reasonable distance to the side of the sus-

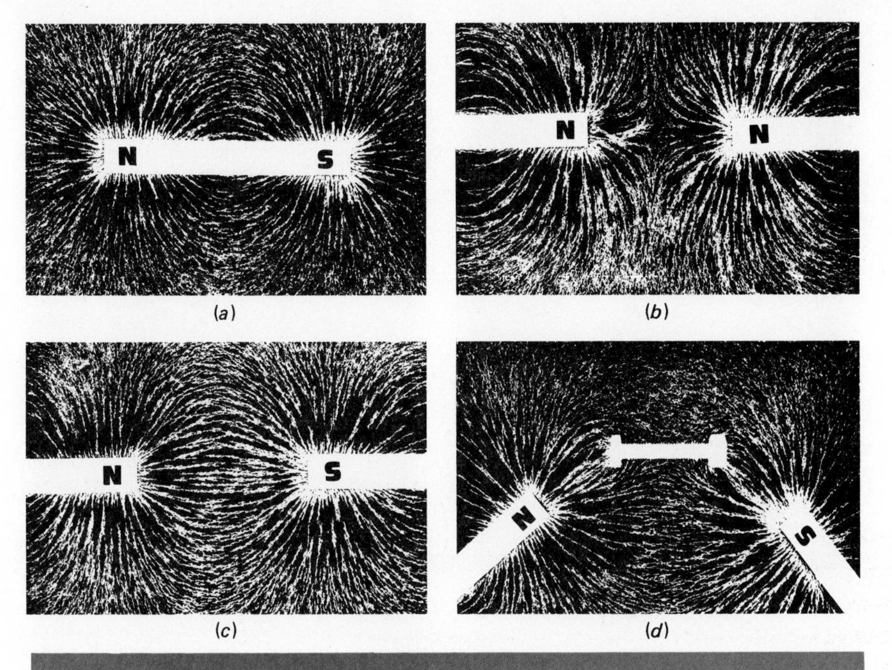

FIGURE 10-1 Patterns of lines of iron filings around magnets. Lines may be thought of as originating on an N pole and ending on an S pole. The lines spread and narrow but do not cross each other. This is most striking with the N poles (*b*). Note the influence of the ferromagnetic bolt in "gathering up" the lines when placed between poles (*d*). (*Photographs by the author.*)

From his studies with magnets, Faraday concluded that a magnet alters the space surrounding it. He called this alteration the *magnetic field.* The visual representation of the magnetic field is the pattern of lines of force. It is not useful for us to define units in which to measure magnetic effects at this point. However, the pictorial method of representing a field does define both the size and the direction of the magnetic force for any point around a magnet. Figure 10-2 represents the conventional drawing of a magnetic field. If we were to place a small compass at any point on the drawing, we could predict in which direction the compass would point and how strongly it would be pulled in that direction. The direction of the force is shown by the lines of force. Since these lines never cross, the compass would point parallel to the lines in its vicinity. We define the direction of the lines as running from the N pole of the magnet to the S pole. Thus, a compass at point *A* on Fig. 10-2 would align with its N pole (the darkened side of the compass needle) pointing straight down. At point *B*, the needle would point to the right on the diagram, and so on. Figure 10-2 shows several compasses at various places near a magnet. Are the directions of pointing consistent with the magnetic field of the magnet?

We could measure the strength of the force by which a compass is pulled into line with the field. For example, a small spring attached to the compass needle would reveal the size of the force by its stretch. The field drawing of Fig. 10-2 also provides enough information to obtain the relative size of the magnetic force for various positions. If we could place a little square of paper in the diagram perpendicular to the lines, the size of the force would be proportional to the number of lines passing through

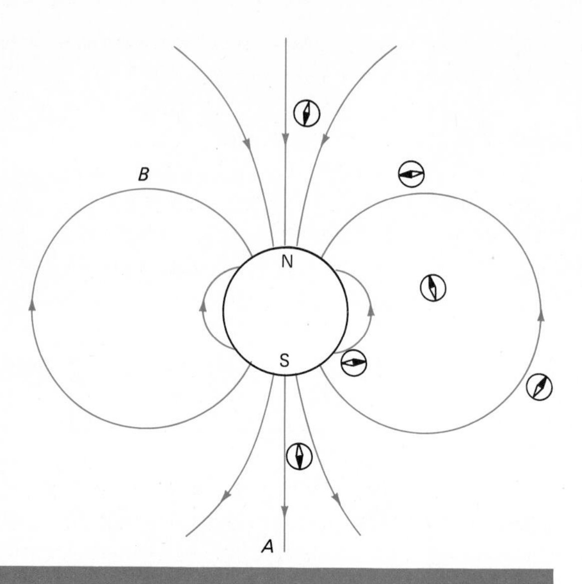

FIGURE 10-2
A conventional representation of the magnetic field around the earth. The object of the lines is to define the size and direction of the magnetic force at any point in the space represented. The direction of the force is parallel to the lines of the field in any region, and the size increases as the lines pack more tightly in space. The small circles show the direction a small compass would point in six different locations on the drawing.

the paper at each position. The more tightly packed the lines are in space, the greater the size of the magnetic force on our compass.

This use of compasses in defining a magnetic field requires some consideration for the most common magnetic field in our experience. Compasses all over the earth align themselves on an approximate north-south line. This is evidence for an overall magnetic field of the earth itself. Gilbert investigated the data on the direction of pointing

Michael Faraday

"Being Bad at Mathematics Could Win a Nobel Prize in Physics."

The most renowned British scientist of his day was Sir Humphrey Davy. When Davy was asked what his most important discovery was, he answered without hesitation, "Michael Faraday." Actually it is difficult to decide who did the finding, Faraday or Davy. Son of a blacksmith, Faraday had very little formal education. He began work at the age of 12. He was working as an apprentice to a bookbinder at the age of 19 when a customer gave him a ticket to a series of lectures by Davy at the Royal Institution. Faraday apparently had been reading the books as well as binding them for he had developed an interest in science. Yet, the lectures were his first formal association with the science of his day and a turning point in his career. Within 2 years, in 1813 and at the age of 21, Faraday managed to obtain the position of laboratory assistant at the Royal Institution. At the age when Newton was reaching his peak years of scientific achievement, Faraday began his formal work in science. Within a year he was traveling with Davy through Europe and assisting in the research as they moved from place to place. Davy's main work consisted of using the battery, discovered by Volta in

1800, to break down various materials and discover new elements. Among the elements discovered by Davy in this manner were sodium, potassium, and calcium. Faraday thus began his work in science with the union of electricity and chemistry. His discoveries were many and he grew in scientific prominence, but his work was leading deeper into electrcity.

Faraday's problem was that electricity, and especially the relationships of electricity and magnetism, are quite complex. The normal scientific method to handle a complex idea with accuracy and ease is via mathematics. Once the translation to mathematical statements is made, progress becomes much easier. Faraday, however, lacked experience with the language of mathematics. His education was in simple reading, writing, and elementary arithmetic. Denied the use of advanced mathematics, Faraday had to be able to describe and understand the forces exerted across distances between currents and magnets. How does one magnet manage to exert a force on another? Faraday constructed elaborate mental models to simulate this action. He imagined space as full of invisible gears and levers, tubes, and wheels. Eventually he described phenomena in terms of fields of force with invisible lines of force running through space. Many of the well-trained scientists of his day regarded Faraday highly for his discoveries but did not take his mental models seriously.

During his lifetime, Faraday's crowning achievement was the discovery of how to produce an electric current from the motion of a magnet—electromagnetic induction. This discovery is the basis of the electrical civilization in which we live, and without it you might be reading these words by the light of an oil lamp. Faraday lectured widely on his discoveries at the Royal Institution and his lectures were always quite popular. One story, probably not true, had a newly elected politician ask after a lecture of what use electromagnetic induction was. Faraday's reply was, "Well sir, in 15 years you will be taxing it." The time table was off, but a 1-cent tax per kilowatthour of electricity produced today by this discovery would bring in revenues of $22 billion per year *for the United States alone!* This is roughly 10 times the entire national budget for medical research—not a bad dividend for an elementary school education.

The final chapter of this story comes after Faraday's death. His successor as director of the Royal Institution was an upper-class gentleman with an excellent education—James Clerk Maxwell. Faraday had been an experimental genius; Maxwell was a mathematical genius with a long string of important work. Maxwell was criticized for spending his time in translating Faraday's ideas of lines of force and fields into mathematics. Many regarded these as mental crutches Faraday had used to guide his work and nothing more. Maxwell succeeded in compressing all of the knowledge of electrical and magnetic phenomena into six short equations. Symmetry obliged him to add one term to an equation and then inspect the result. The result was a revolution in science! Solved simultaneously, Maxwell's equations (they are always known by this name) produced a mathematical form immediately recognized by a physicist as being the equation of a *wave.* Electric and magnetic fields can have ripples race through them. The equation predicted the speed of these ripples as precisely the speed of light. The second great theoretical system of physics was born, that of *electromagnetism.* Tragically, Maxwell died of cancer at 48 in 1879, just 9 years before Hertz demonstrated the vision of Faraday and Maxwell by producing the first radio waves in the laboratory. By the time the Nobel Prizes began in 1901 all three of these men had died. It would have been interesting to have had one of the first of these prizes awarded to an individual with the equivalent of only a fifth-grade education.

compasses—both the north-south point and the *dip,* or angle of point from the horizontal, for the compass. A compass near the equator displays little dip, but nearer the poles the dip becomes quite large. Gilbert concluded that the earth's magnetic behavior was similar to that which would be produced by a gigantic bar magnet, some 650 km long, inside the earth. He made a large model of the earth with a lodestone in the center to show that the direction of pointing of a compass along the surface as well as the dip would simulate the pattern for the earth. We know today that this picture is too simple. For one thing, permanent magnets weaken with heating and the earth's interior is too hot for a bar magnet to retain its magnetism. More startling is evidence frozen into rocks that the earth's magnetic field can completely "flip" in orientation, with the N pole and S pole reversing in direction. As lavas cool, magnetic minerals tend to be lined up with the earth's magnetic field. When the rock solidifies, the direction of magnetization of the minerals preserves a record of the direction of the earth's field at that time. The rock record indicates that the earth's magnetic field reverses itself periodically at intervals of about 20,000 to 30,000 years. Such a *geomagnetic reversal* is currently overdue. Magnetic north has been in approximately the same position for about 50,000 years, one of the longest spans without a reversal on record. These reversals are important in the study of the earth and we will return to them later. Flipping over a bar magnet the size of Florida in the interior of the earth is not a reasonable mechanism for such a reversal. Before we can consider the earth's magnetic field further, we will need another mechanism for producing a magnetic field.

Electricity

Electricity and magnetism have many properties in common. Even the origins of the study of electricity are similar to those of magnetism. Evergreen trees seal wounds of their trunks with a thick yellowish substance which is very sticky. If these globs are allowed to harden and crystallize over decades or centuries, the semiprecious stone called amber is produced. The Greeks noted that when amber was rubbed with wool it acquired the ability to attract small scraps of wool. The Greek word for amber was *electrum,* from which the study of this attraction was called electricity.* Actually, the effect is not very hard to produce in our day since we have access to many amberlike substances. The plastic shaft of a ball-point pen is an adequate substitute for amber, as are most combs. If you run the shaft of the pen across your hair several times, you will find that it acquires the ability to attract and lift dust specks, loose hairs, and flakes of dandruff off your desk top. You might also notice its ability to attract the hairs on your forearm at a distance of a centimeter or

* Aristotle ascribes the first study of magnetism to Thales of Miletus, and Greek tradition credited Thales with the discovery of electricity. This is not likely, but in any event both studies were common enough by the fourth century B.C. for Socrates to criticize the "attractors of iron and rubbers of amber" for their waste of time in these pursuits. Living in an age where we are likely to use electricity a dozen ways before we leave the house in the morning, we are likely to disagree with Socrates.

two, or to attract the edge of a loose sheet of paper toward it. One of the best combinations for electrical effects is to rub hard black rubber (called *ebonite,* and found in many combs) with a fur (dry, undoctored human hair is good; cat hair is better). The combination of a rubber rod and fur has become fairly standard in laboratories. To display an electrical attraction, some light yet visible chunk of matter is desirable. Many early experimenters used bits of cork or the dried inner stem, the *pith,* of large grasses. The easiest material to obtain for this purpose today are bits of a Styrofoam cup.

If a small ball of light material is suspended on a silk thread, it displays an attraction for a rubber rod that has been rubbed with fur by being pulled away from the vertical and toward the rod. Some materials would not display this power of attraction after being rubbed with fur; for example, no metal would. Materials such as glass, sulfur, and ivory would display various amounts of attraction. Eventually some ideas as to what was happening began to grow. The rubber rod is said to be *charged* when it is rubbed with fur. If a charged rubber rod is suspended free to move and another charged rubber rod is placed nearby, a small repulsion may be observed. This effect is easier to see if the small ball of cork or pith is used. At first, the ball is attracted to the charged rod. But, if the ball is allowed to touch the rod, it is then repelled by the rod. Investigators concluded that something was transferred to the ball when it touched the rod. This was the same something that was placed on the rod by rubbing it with fur. This "something" was called *electric charge.* If the silk thread holding the ball was replaced by a fine wire, the repulsion for the charged rod was quickly lost. This suggested the conclusion that some materials, such as metals, allowed charge to move away through them, while others, such as the silk, did not. Materials which allowed the free movement of charge from place to place were called *conductors* of electric charge. Materials that seemed to confine charge to wherever it was placed were called *insulators.* The requirements for producing a charge on a material become clearer. *A charge is generated whenever two different insulators are rubbed across each other.*

One complication arises almost immediately. If a charged rod is touched to a ball suspended on an insulating thread, such as silk, the ball will then be repelled by the rubber rod. However, the ball will now be attracted to a glass rod that has been rubbed with silk. This process works in reverse. If the uncharged ball is touched with the charged glass rod, it will then be repelled by the glass. Now, however, the ball will be attracted to the charged rubber rod. Different investigators interpreted these facts differently. The conclusion which ultimately proved to be correct was that there are two types of electric charge. The type on a rubber rod that has been rubbed with fur is called *negative charge,* and indicated by a minus sign (−). The type on the charged glass rod is the opposite of this and is called *positive charge* (+). We may conclude from the above behavior that: *Charges of the same type repel each other, but opposite types of charges attract each other.*

A second conclusion is suggested by the behavior of a neutral, or

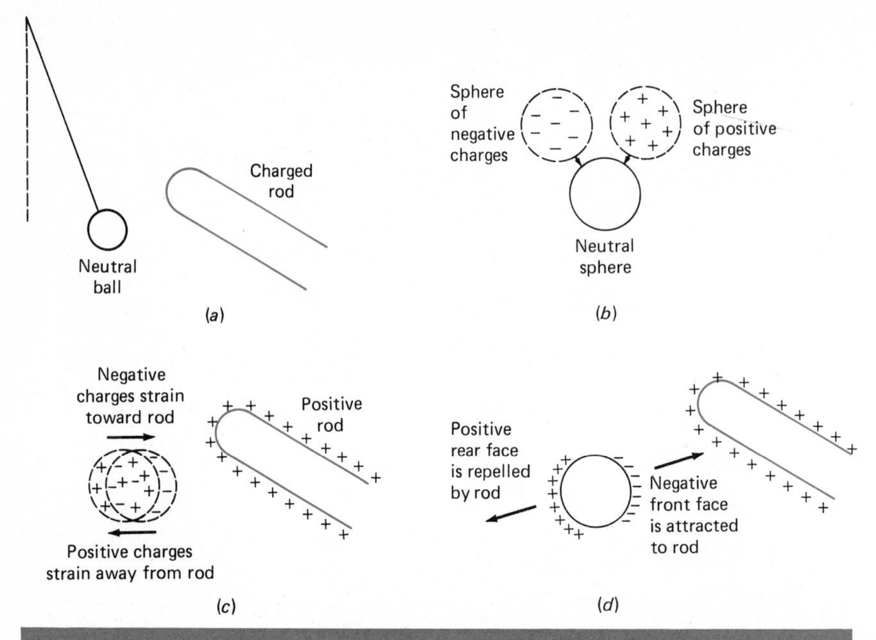

FIGURE 10-3 Attraction of a neutral object by a charged rod. (*a*) Neutral objects may be attracted by a rod with either a positive or negative charge. (*b*) Neutral objects contain both positive and negative charges in equal amounts. (*c*) In the presence of a charged rod the two types of charges on the sphere would react differently. One type would be drawn toward the rod and one type would be pushed away. Even a tiny shift would leave the near edge with an opposite charge from the rod. (*d*) The opposite charge on the near side of the sphere is closer than the similar charge on the back of the sphere. Hence, the attractive force for the rod is larger, and the sphere is attracted. A small charge separation of a neutral object induced by outside charges is called an *electric polarization.*

uncharged, ball toward charged rods. The neutral ball is attracted either toward a positive rod or toward a negative rod. One interpretation of this is that the neutral ball really contains positive and negative charge in the same amount. When a charged rod is brought close, the opposite charge on the neutral ball is attracted more than the same charge is repelled for some reason. This is correct, but it is not the only way to interpret the experimental facts (Fig. 10-3). For some time several perfectly reasonable theories about electricity competed for acceptance.*

The Electroscope

A sensitive device for detecting the presence of an unbalanced electric charge is the *electroscope* (Fig. 10-4). An electroscope consists of an electrically isolated metal system. One end of the system is accessible to the experimenter. The other end is usually sealed and has some delicate arrangement for the detection of unbalanced charge. The earliest such systems used a very fine sheet of gold leaf attached at its top to a metal plate. Gold is the most malleable substance known—that is, it can be beaten into a thinner sheet than can any other material. A sheet of gold leaf is much lighter than small balls of cork or pith, and thus can

* During the period of the American Revolution it was fashionable at parties to perform science demonstrations as part of the entertainment. No area of science was more popular for this than electricity. Most of science then was conducted by true amateurs. To Americans an "amateur" is thought to be anyone who does not get paid for the task and it sometimes implies that the person is not very skilled (amateur athletes being a possible exception). The real meaning of the term is "nonprofessional expert." Copernicus and Gilbert were amateurs, as were almost all scientists until well into the nineteenth century. The number of new scientists beginning a professional career each year today exceeds the total number that lived up to 1900.

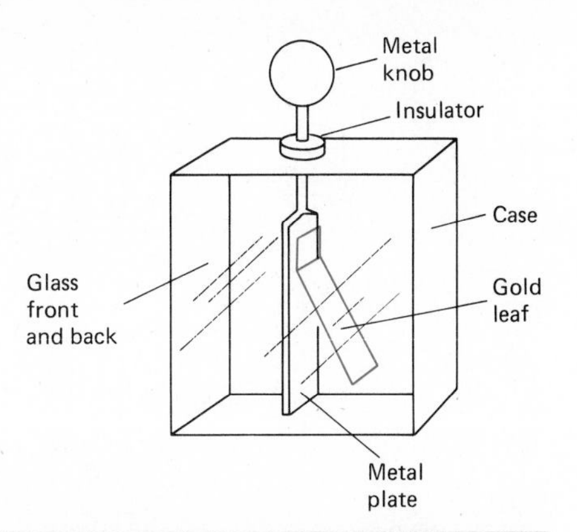

FIGURE 10-4 The electroscope isolates a metal plate connected to a knob from outside contact by using an insulator. Charges placed on the knob may move freely to the plate but not off the electroscope. Even a small charge will cause the light metal leaf to be repelled from the plate with which it shares its charges.

record the presence of smaller forces. If a charged rod is brought near the knob of an electroscope, the gold leaf at the bottom deflects, or stands away, from the plate to which it is attached. This indicates that there are unbalanced charges at the bottom of the electroscope and the gold is being repelled from the shaft. If the rod is withdrawn, the leaf collapses to its rest position once again. The interpretation of this behavior is reasonably easy. Neutral matter contains both positive and negative charges in the same amount. If a negative rod, for example, is brought near the top of the electroscope, then the positive charge in the material of the electroscope will be attracted. Similarly, the negative charges in the electroscopic material will be repelled. But the central shaft of the electroscope is made of metal, a conductor. In a conductor, charge is free to move. Thus, positive charge may move toward the top of the electroscope, and/or negative charge is repelled away. However, the shaft of the electroscope is electrically isolated. This means that only good insulators such as air and glass are allowed to touch the shaft. Thus, the shuffling charges in the electroscope cannot leave; they may only rearrange. As they do this, the lower plate and the attached gold leaf both acquire some additional negative charge repelled from the upper part of the electroscope. The unbalanced negative charges in the plate and those on the leaf repel each other. Since the leaf is very light, this force is enough to push away the gold leaf and have it stand away at some angle from the plate. The greater this angle is, the larger the force of repulsion must be. Thus, the electroscope allows for easy measurement of the relative size of electric forces.

If the electroscopic plate is actually touched by the charge rod, the gold leaf is deflected and remains so when the rod is withdrawn. Apparently charge is physically transferred from the rod to the electroscope by contact. Charging by direct contact with another charged object is called *charging by conduction.* Belief that charges are actually transferred is strengthened by noting that on occasion there will be a crackle as the rod approaches the plate. Following this the electroscope will be charged just as if it had been touched by the rod. The rod has discharged to the electroscope with a small spark that jumped across the

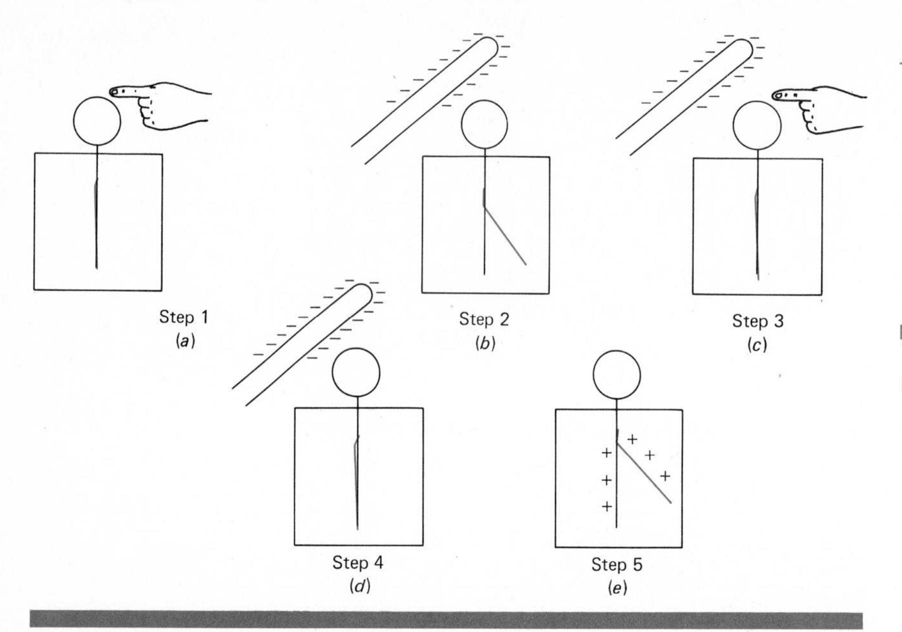

FIGURE 10-5
The steps in charging by induction are shown. (*a*) The electroscope is touched momentarily to ensure that it is electrically neutral. (*b*) A charged rod (with either sign of charge) is brought near the knob but not touched to it. As the rod approaches, the leaf is deflected. (*c*) With the charged rod still in position, the knob of the electroscope is touched with the finger. The leaf is observed to collapse to the undeflected position. (*d*) While the charged rod is kept in position, the finger is withdrawn. No change in the electroscope is observed. (*e*) The charged rod is now removed. The leaf of the electroscope deflects, indicating the presence of a charge. Testing will indicate that the charge is of the opposite sign from that on the rod.

gap between these objects.* Charge really does move. In both of these cases the charge on the electroscope is of the same sign as the charge on the rod. (How would you prove that experimentally?)

To restore a charged electroscope to neutrality, you only have to touch the plate momentarily with a finger. While you will feel nothing electrically, the fact that the electroscope discharges suggests that human beings are conductors of electricity (at least at the voltages of rods and furs). This provides another method of charging an electroscope (Fig. 10-5). If a charged rubber rod is brought near the plate, but not touched to it, the leaf will deflect. While holding the rod in this position the plate is touched momentarily with a finger. As this is done, the leaf collapses. Now, after the finger is removed, the rod is taken away. As it is, the leaf once again will be deflected, indicating that the electroscope is charged. Interestingly, while the rubber rod is negatively charged, the charge that resides on the electroscope will be positive. This process is called *charging by induction.* Can you explain what is happening to charges at each stage in the illustrations of Fig. 10-5?†

* Although we have not defined the volt yet, this is an appropriate place to note that it requires about 10,000 volts (V) to drive a spark across 1 cm of dry air. On a cold, dry day a rubber rod charged with fur can discharge to an electroscope at a distance of perhaps 2 cm—and farther yet if in place of a rounded plate we use a more pointed metal end on the electroscope. Thus, the rod must be operating at 20,000 V or more in these simple experiments. This has always tempted the author to place a rubber rod and a fur in a neat little box on the laboratory wall. On the glass front would be bold letters in red declaring, "DANGER: 20,000 VOLTS." Unfortunately, if this were tried, some student surely would become petrified with fright. Rubbing the rod with the fur is *exactly* the same process as running a black rubber comb through your hair, producing the same results electrically. If you really want to see some electrical effects, take off a dacron or orlon sweater or shirt on a dry night in the dark. The sparking can be incredible, and, like the rod and fur, harmless.

† The only charges which actually move in this and in all normal electrical effects are the negative charges. While positive charges are present in equal amount in neutral matter, they are confined to their positions and do not wander about in response to electric forces. Of course, if some of the negative charges leave an area of neutral matter, it will then have a net positive charge.

Forces between Charges

If a neutral metal sphere is suspended on an insulating stand, it may be charged by conduction. If a second, identical sphere is touched to the first momentarily and then withdrawn, it is found that the two spheres share the charge evenly, each retaining half of the original charge. This fact allows us to measure the forces between charges of known sizes. A small sphere is given a charge of q_1, and a second small sphere is given a charge q_2. The spheres are separated by a distance of R. The electric force between the spheres will be proportional to the size of q_1 and proportional to the size of the charge q_2. The force will also be inversely proportional to the square of the distance between the spheres. Mathematically,

$$F = K\frac{q_1q_2}{R^2}$$

where K is some constant to adjust for the size of units in which we will chose to measure charges. Doubling of either charge doubles the force, while doubling both charges makes the force 4 times as large. If the distance between charges is doubled, the force drops to one-fourth of its previous value. This relationship is known as Coulomb's law, after Charles Coulomb (1736–1806), its discoverer.

We immediately encounter a difficulty, one that occurred historically. How do you define a unit in which to measure charge? A unit for length or mass is no great problem. Once a meter or a kilogram has been decided upon, one very careful standard is produced. From such a standard, copies are made and provided for laboratories (or butcher's shops) all over the world. Defining a unit of time was more difficult. In that case scientists took a repeating event, the year, and subdivided it down to obtain the second. Today we tend to define the second by a more precise event;* we count the vibrations of cesium atoms (much as quartz watches count the vibrations of a quartz crystal). How do you "put up" a standard charge for comparison so that everyone will agree on the size of a given charge? Charge flows, and no matter how nicely sealed an electroscope is, a charge placed on it slowly "leaks" away. The charge produced on a rod by rubbing depends upon the humidity, the pressure, the type of fur, and much more. So how do we produce and keep a standard of charge?

Today, we have found out that there is a charge of the smallest size that we have observed so far in nature. The smallest "piece" of negative charge is the same size as the smallest positive charge. These charges are very tiny; thus there must be many of them to a practical working unit. The standard, everyday unit for measuring charge is called the *coulomb* (C). Like meters, kilograms, and seconds, the coulomb is a basic unit of measure for the metric system. In fact, it will be our *last* basic measure. All future units will be defined in terms of these four basic measures

* This is more precise because the earth is slowing down all the time. In 1972 we had to insert an extra second in all the clocks on earth to adjust for the slowdown since 1900, and another second was inserted on January 1, 1980.

(meters, kilograms, seconds, coulombs), much as the joule of energy was defined as 1 kg · m^2/s^2.

We will define a coulomb as a charge equal to 6.24×10^{18} of the tiny fundamental charges carried by particles such as the electron. Historically, the first good measure of the coulomb comes from the study of the electroplating of metals, but our definition is simpler unless you really had to measure out 1 C of charge. If we use this definition of the coulomb of charge, the value of the constant K in Coulomb's law is 9×10^9 N · m^2/C^2. Having defined a unit of charge we may see how Coulomb's law operates by working a problem.

Example

A small sphere with a charge of +0.02 C is placed 75 cm from a sphere with a charge of −0.01 C. What is the force between the spheres?

$$F = K\frac{q_1q_2}{R^2}$$

$$= (9 \times 10^9\ \text{N} \cdot \text{m}^2/\text{C}^2)\,\frac{(+0.02\ \text{C})(-0.01\ \text{C})}{(0.75\ \text{m})^2}$$

$$= -3 \times 10^6\ \text{N}$$

Thus, the spheres would experience an attraction of about 300 tons of force!*

The moral of this is that an unbalanced charge as large as this is not normally encountered. The unbalanced charges left on a rubber rod by rubbing fur are a very tiny fraction of a coulomb. Furthermore, electric forces are enormous compared to gravitational forces. Suppose it were possible to grind up 1 kg of rubber rod and separate the positive charges and the negative charges. If we fastened the jar containing the positives here on earth, we could then ship a similar jar to the moon with the negative charges. There is no experimental method known which would detect the gravitational attraction of these masses for each other over even a few meters of distance. Could we detect the electrical attraction at the distance of the moon? The answer is definitely yes. Even at the distance of the moon the force pulling the negative charges back would be over 100 tons! Electric forces are on a completely different scale than gravitational forces.

Electric and magnetic forces can be either attractive or repulsive. The form of the two force laws is identical. In both cases the force law becomes very difficult to apply except in cases of simple geometry. For example, if a small sphere with a charge of $+q$ is placed near a large metallic sphere with a charge of $-q$, what will the force between the spheres be? Even though we know the distance from the center of one sphere to the center of the other, we will find this a difficult question to

* The metric ton is the weight of 1000 kg of mass or 9800 N. It is about 10 percent larger than the English ton of 2000 lb (or 8896 N).

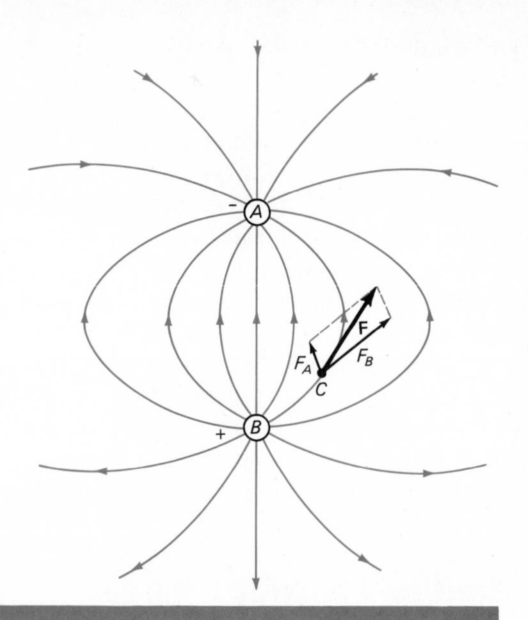

FIGURE 10-6
The charges on spheres A and B both exert a force on the small positive charge at C. These forces are shown as is there sum. The electric-field lines drawn summarize the total force for any position in space around A and B. The direction of F is parallel to the lines and its size (length in the drawing) is proportional to how densely packed the lines are in space at the position of F.

answer. The reason for this is simple. Charges can move in the metal. The presence of the positive charge attracts negative charges toward the small sphere. Thus, the negative charges will not be uniformly distributed on the surface of the large sphere but will cluster toward the area nearest the positive sphere. This rearrangement of charges will continue until the zone of negative charge repels other negative charges back along the sphere as strongly as the positive sphere attracts them. The distribution of charges on the surface of conductors thus tends to be both changeable and nonuniform. Only if charged objects are very far apart compared to their size is Coulomb's law easy to use directly.

As in the case of magnetic forces, it is most convenient to consider forces in terms of a field. The presence of an electric charge requires that an electric force would exist for a small charge placed anywhere in the space near the charge. The total force at any point is the sum of the forces contributed by all electric charges in the universe. We define the force acting per unit of charge at any point as the *electric field, E,* at that point.

$$\vec{E} = \frac{\vec{F}}{q}$$

Just as the magnetic forces at various places around a magnet can be illustrated by using iron filings, electric fields may be displayed. The commonest way to do this is with small seeds floating on liquid surrounding a charged object. Consider a particularly simple case. Figure 10-6 shows the electric field around two small charges of equal size but opposite in sign. The drawing summarizes the direction and the size of electric forces for any position in the vicinity of these charges. For example, consider a small positive charge placed at point C in the diagram. This charge would experience a force of attraction for the charge at *A*, which is labeled F_A. It would also experience a larger force of repulsion for the positive charge at *B* since it is closer. This force is labeled F_B. The relative sizes of these forces are determined from Coulomb's law. The total force on the charge at point *C* would be the sum of F_A and F_B and is labeled *F* in the diagram. Note that *F* points in the direction indicated by the drawing of the electric field (parallel to the lines of force). Repeating this drawing procedure for different positions would confirm that the resulting force from both charges is always in the direction indicated by the electric-field drawing. Furthermore, such repetition would indicate that the size of the force varied with the nearness of the lines of force to each other. Thus, the electric-field drawing is a convenient way to summarize the forces that would be experienced by a charge in the vicinity of spheres *A* and *B*. For more complex distributions of charge, electric-field considerations may be the only convenient interpretation.

The Work Required to Transfer Charges

Consider an area in which the electric field is not zero. For example, imagine a large insulating wall uniformly charged with negative charges. At some distance from the wall forces will be exerted on charges, attracting positive charges toward the wall and repelling negative charges away from the wall. If we were to take a sphere with a net negative charge from this position and move it to the wall, we would have to exert a force to overcome this repulsive force. Furthermore, we would have to exert this force through a distance to move the sphere to the wall. *It requires work to move this charge against the electric field.* Conversely, if we released a sphere of negative charge near the wall, the charges on the wall might exert enough force to move the sphere from our hand and away from the wall. *Work is done on objects moved by the electric field.* For charges to change position either with or against the electric field requires work to be done. A negative charge close to a large body of negative charges can do work during its movement away from these charges. It possesses potential energy in the same way a compressed spring or a bowling ball above the earth's surface possesses energy. It is very useful to define this potential energy carefully. Suppose that a sphere containing exactly one coulomb of positive charge is moved from point *A* to point *B*. Suppose further that we began with zero kinetic energy (at rest) and ended the motion the same way. If exactly one joule of work had to be done to accomplish this, we would define the *electric potential difference* between *A* and *B* as exactly one *volt* (V). *The volt is a measure of electric potential-energy difference between positions, and is the work required per unit charge to transfer the charge between these positions.* Mathematically, if *V* represents voltage,

$$V = \frac{W}{q} \qquad \text{or, in units,} \qquad \text{volts} = \frac{\text{joules}}{\text{coulombs}}$$

Suppose we had a charge of -0.5 C and walked it to the negatively charged wall described. If we found that it took 100 J of work to do so, the potential difference between where we started and the wall would be

$$V = \frac{100\ \text{J}}{0.5\ \text{C}} = 200\ \text{V}$$

Similarly, when we read on an automobile battery that it is a 12-V device, this means that 12 J of work must be done to transfer 1 C of charge against the electric field from one terminal of the battery to the other (which is what we do when we charge the battery). Alternately, it also means that 12 J of work will be done by each coulomb of charge from the battery when they move with the electric field from one terminal to the other (which is what we do when we use the battery to start the car). Thus, charging a battery is a bit like storing a set of bowling balls on a high shelf. Using the battery is a bit like allowing the balls to fall to the floor doing work on the way down.

Electricity in Motion

Up to this point we have been considering *static electricity*. Except for a little initial motion and shuffling, the charges just stay in place, attracting and repelling. While a great deal of knowledge about electricity can come from this study, the real utility of electricity in society comes from moving, or *dynamic or current electricity*. The change in emphasis came about from an accident. Experiments with charges had accidentally led to a way to store large charges in a device called a *Leyden jar*. If a charged rubber rod is touched to an isolated metal sphere, some negative charge will leave the rod and move onto the sphere. If the rod is recharged, still more charge may be delivered to the sphere. This process quickly ends, however, as the sphere is soon as negatively charged as the rod is, and as many charges are transferred from the sphere to the rod during contact as will move the other way. The Leyden jar is a glass jar coated on the inside and outside with a metal foil layer (Fig. 10-7). The inner metal layer is connected electrically to a knob at the top of the jar. The outside metal layer is *grounded;* that is, it is given a connection by a conductor to the earth itself. (A wire to the nearest metal water pipe that runs into the ground does well for this purpose.) As negative charges from a rubber rod are delivered to the knob of the jar, they move to line the inner metal foil. The repulsion of this negative charge causes negative charges in the outer foil to move out the ground wire to the earth. Thus the inner and outer metal walls acquire opposite charges. When the recharged rubber rod is brought to the knob again, the negative charges on it are repelled by the negative charges lining the interior of the jar, but they are attracted by the positive charges of the outer wall. These forces are almost equal. Since the repulsion of other negative charges on the rubber rod is greater than the sum of these forces from

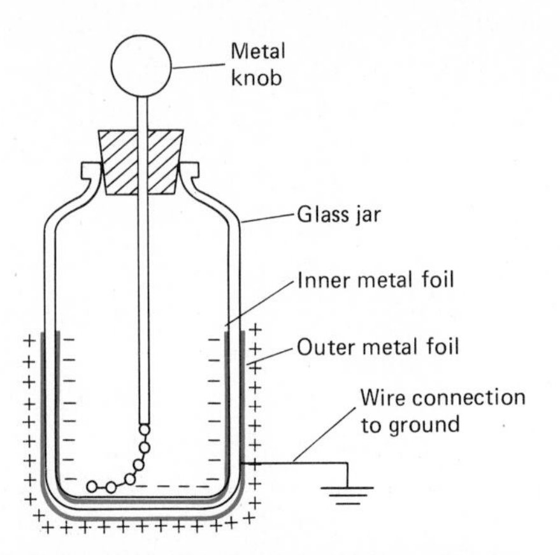

FIGURE 10-7
The Leyden jar was the earliest device for the storage of large amounts of separated electric charges.

the jar, more negatives leave the rod and move into the jar. As the total negative charge in the jar increases, so does the positive charge on the outside of the jar. Thus, the rubber rod can continue delivering charge to the jar until a very large charge has been stored. (Modern devices storing charge in similar ways are called *capacitors* or *condensers,* and you use one every time you dial a station on a radio.) Connecting the knob of the jar and the outside foil causes the jar to discharge. Large Leyden jars can store a potentially dangerous amount of charge. Several early experimenters quit this line of investigation after reviving from an accidental discharge of the jar through their body. However, many machines were produced which would generate and store a large charge. People were (and still are) fascinated by the crackling arcs that were made visible during the discharge. Demonstrators lined up in a ring as many as a hundred people holding hands and delivered a jolt to each as the machine was discharged through the ring.

Working with a small static discharge, Luigi Galvani (1737–1798) noted that the discharge caused the muscles of a frog to contract. By itself this was not surprising. Many experimenters had picked themselves up off all sides of their laboratories wherever their own muscles had thrown them during a discharge; why should frogs be different? Galvani's point was that the muscles of thoroughly dead frogs could be made to contract. In fact, the leg muscles did not even have to be attached to the rest of the body to produce the contraction. This ability was lost if the muscle was dried out or was out of the body for several hours. The contractions also became weaker as they were repeated. Supposedly, one day he hung some freshly killed frogs from an iron hook on his balcony to drip dry. He was startled to see that when the wind blew the frog's legs into contact with the brass balcony railing the legs jerked into contraction. Following this up, he determined that if a wire was connected to the spinal nerves of the frog and another to the foot, the leg could be made to contract when the wires were touched. However, this occurred *only* if the two wires were of different metals. Galvani was convinced that he was onto an important discovery, and he asserted that current electrical phenomena required a life force. He was wrong, but he had made a key discovery.

Alessandra Volta (1745–1827) showed that the key was in the two dissimilar metals. Using stacks of metal discs and blotter paper soaked in conducting solutions, he built what he called a *battery,* which could produce a continuous flow of electricity or *current* along a wire. This discovery represents the beginning of the study of electric currents.

The Flow of Charges

If a plate of one metal and a plate of a second metal are both partly immersed in a conducting solution, we have the basis for an *electrochemical cell* (Fig. 10-8). Zinc and copper are good choices for metals, and a weak acid solution is a good liquid. The plates should not touch; they must merely be in the same bath of solution. To flow, electricity must have somewhere to go. A complete pathway must be provided from one side of a cell to the other for electricity to flow steadily. This complete pathway is called an *electric circuit.* If the exposed end of the zinc plate is connected to the copper plate by a wire (of any metal), a flow of charge along the wire will take place. We could prove this today by placing a small light bulb in the wire and observing it light up. Galvani happened to use a frog's leg as his detector for the flow of electricity, but at today's prices light bulbs are cheaper and are usable for a longer time.*

The action of the cell is interesting. Imagine that we make one cell by using a small drinking glass, half full of acid solution and small strips of zinc and copper. Then, let us make a second cell by using a school swimming pool of acid and dropping an automobile-sized clump of copper at one end and a similarly sized clump of zinc at the other. If we

* The energy for the contraction of the muscle comes from stored chemicals called *ATP molecules* and not from the electricity. The bulb, meanwhile, uses the energy of the moving electricity itself to light up. This implies that the number of contractions of the muscle will be limited and the electric signal to contract will stop producing a response from the muscle. The bulb life, however, is limited only by the time it can operate before the hot wire in it breaks.

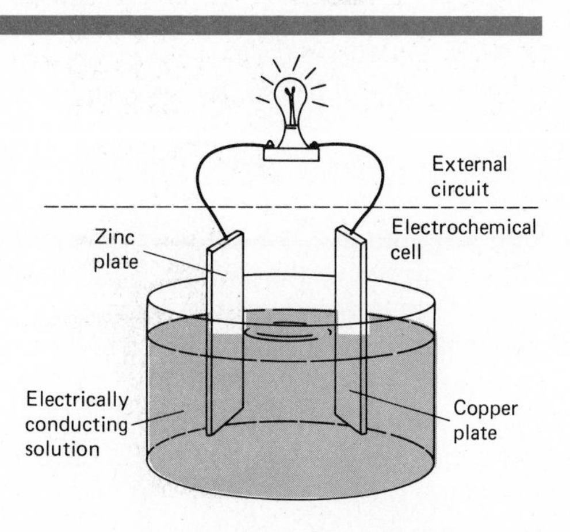

FIGURE 10-8 An electric current may be produced from the chemical potential energy of two different metals. The electrochemical cell develops an electric potential difference, or voltage, between the plates. If the plates are connected by a conducting path, a current will flow.

connected the two metals with the wire containing our light bulb, you might be forgiven for guessing that the bulb would be made brighter if connected to the swimming pool cell. You would be wrong. Bulb brightness does not depend upon the size of the cell at all. The voltage of a single cell depends only upon the materials from which it is made, not on its size. There is a difference, of course, but not in the brightness. The swimming pool cell might well keep the bulb lit for the balance of your life. The small cell, however, will light the bulb for a much shorter time until we observe the zinc plate dissolve and flake apart and electrical action cease. While voltage does not depend upon size, the total energy that may be delivered by a cell during its operating life span does.

Each cell produces a flow of electric charges from one metal, through the bulb, to the other metal. Such a flow of electric charge is called an *electric current* I. *Current is the rate of flow of charge.* $I = \Delta q/\Delta t$ or, if the flow is constant, $I = q/t$. The unit in which current is measured is the *ampere* (A). One ampere is a rate of flow of one coulomb of charge each second.

Example

The potential difference (voltage) between the zinc plate and the copper plate of a cell is 1.5 V. If a current of 2.0 A is measured flowing through the light bulb connected between the plates, at what rate is work being done in the light bulb? A volt, you will recall, is the work done in transferring charge in joules per coulomb, or voltage $V = W/q$. The current is the rate of flow, $I = q/t$. In this circuit we have 2.0 A of current, or 2.0 C of charge passing each second through the bulb. But, from the voltage we know that each coulomb must do 1.5 J of work to move between the plates. Thus,

$$IV = (2.0\ \text{C/s})(1.5\ \text{J/C})$$

$$= 3.0\ \text{J/s} = 3.0\ \text{W}$$

Electric potential energy is converted into heat and light in the bulb at the rate of 3.0 J/s or 3.0 W. The electric power of this circuit is 3.0 W.

The method of obtaining this result is quite general. By the nature of their definitions, the product of current times potential difference is the power P of a circuit, or

$$P = IV \qquad \text{or, in units,} \qquad (\text{C/s})(\text{J/C}) = \text{J/s} = \text{W}$$

If we want our bulb to burn brighter, we do not need a larger cell; we need a higher voltage. Increasing the voltage will also increase the current flowing through the bulb, thus the power (brightness) increases even more rapidly than the voltage. It was Volta who demonstrated how to connect cells to increase the voltage in a device he called a *battery*. If two cells are connected side by side (Fig. 10-9A), with the zinc plates attached electrically and the copper plates connected, we have merely increased the effective sizes of the plates and extended the operating life

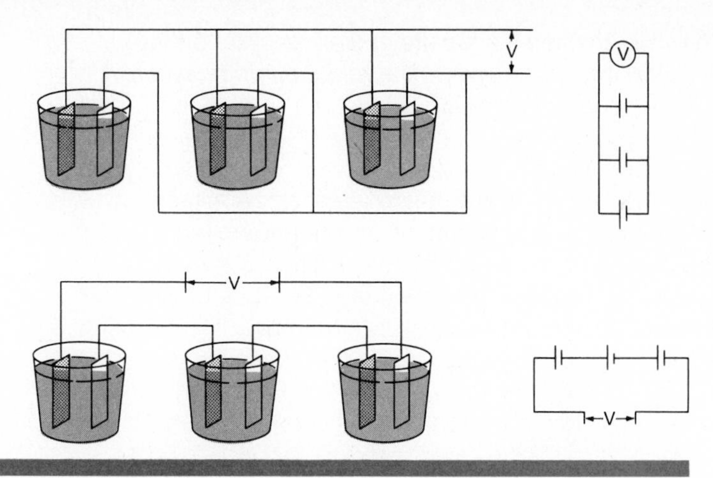

FIGURE 10-9A When cells are connected in parallel, similar plates are attached by electric conductors. The potential difference or voltage developing between the wires connecting the zinc plates and that connecting the copper plates is the same as that of a single cell using these metals. The electrical symbolism for this situation is shown at the right.

FIGURE 10-9B When cells are connected in series, each plate is connected to one of the opposite type in the next cells. Charges passing through this arrangement have their potential energy increased in each cell they pass through. The voltage measured across the end wires of this battery of cells would be 3 times the voltage of a single cell.

of the cells. We have taken a step toward the swimming-pool cell. We call this connecting the cells *in parallel.* We may use an analogy of bowling balls in a gravitational field to picture the action of the cell. Height difference in the gravitational case is analagous to a voltage difference in the electrical case. If the bowling balls represent charges, the cell represents an elevator which raises the bowling balls to a shelf and thus increases their potential energy. Each elevator can raise the balls through a fixed height, which in the electrical case is the voltage of the cell. If we place two such elevators side by side, in parallel, the number of bowling balls which may be lifted could be increased. However, the potential-energy increase of a single bowling ball would not be changed since either elevator would deposit the balls at the same height. The work that could be done by a single bowling ball in falling from the shelf would not change.

Clearly, if we wish to increase the potential energy of the bowling balls to a very high value, the elevators should not be used side by side, but in sequence, or *in series.* Now, a particular bowling ball will be raised by each elevator in turn and will end the process much higher than if it were raised by a single elevator only. To increase the voltage of cells it is necessary to connect them *in series* so that charges will gain potential energy from each cell. Such a connection of cells is shown in Fig. 10-9B. If the voltage of a single cell is 1.5 V, we will find that the voltage measured across all three cells will now be 4.5 V. In general, for cells A, B, and C:

$$\text{In parallel: } V_{\text{total}} = V_A = V_B = V_C$$

$$\text{In series: } V_{\text{total}} = V_A + V_B + V_C$$

We now need to be able to predict for a particular battery voltage and a given device, such as a light bulb, how much electricity will flow. In general, as the voltage increases, the current flowing through a particular device increases in direct proportion ($I \propto V$). How the current varied with the specific device was studied extensively by Georg

Simon Ohm (1787–1854). He devised a measure of the electrical behavior of devices which is called the *resistance R. The relationship between voltage, current, and resistance is called Ohm's law.*

$$V = IR$$

where V is the voltage in volts, I is the current in amperes, and R is the resistance in *ohms.* An ohm is the equivalent of a volt/ampere and is usually indicated by the capital Greek letter omega Ω. Thus, if a device had a resistance of 10 Ω, it would require a potential difference of 10 V applied across the device to cause a current of 1 A to flow.

Example

A toaster has current of 11 A flow through it when operating on a 110-V line.

(*a*) What is the resistance of the toaster?

$$R = \frac{V}{I}$$

$$= \frac{110\ \text{V}}{11\ \text{A}} = 10\ \Omega$$

(*b*) How much charge flows through the toaster in 1 min of operation?

$$I = \frac{q}{t}$$

or $$q = IT$$

$$= (11\ \text{A})(60\ \text{s})$$

$$= (11\ \text{C/s})(60\ \text{s}) = 660\ \text{C}$$

(*c*) What is the wattage of the toaster?

$$P = IV$$

$$= (11\ \text{A})(110\ \text{V})$$

$$= 1200\ \text{J/s} = 1200\ \text{W}$$

(*d*) If electricity costs 6 cents per kilowatthour (kWh), how much does it cost for electricity to operate the toaster for 10 min?

$$P = \frac{W}{t}$$

or $$W = Pt$$

$$= (1200\ \text{W})(600\ \text{s}) = 7.2 \times 10^5\ \text{W} \cdot \text{s}$$

$$= (1.2\ \text{kW})(1/6\ \text{h}) = 0.20\ \text{kWh}$$

$$\text{Cost} = (0.20\ \text{kWh})(6\ \text{cents/kWh}) = 1.2\ \text{cents}$$

Note from this example that the electrical item you pay for is work or energy. The kilowatthour is a unit of work or energy.

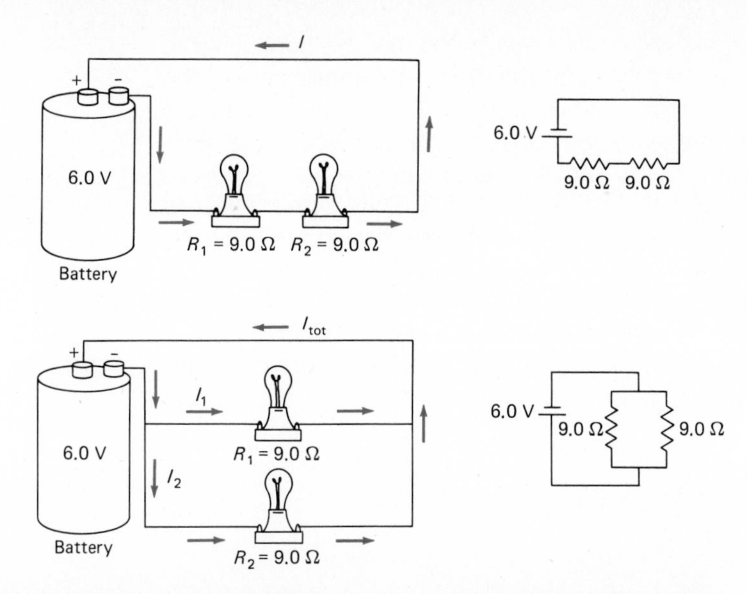

FIGURE 10-10A When resistances are connected in series, charges emerging from the battery must flow through each one in turn. Some of the work of these charges must be done in each of the resistances.

FIGURE 10-10B When resistances are connected in parallel, the path for the flow of charges branches, so that a given charge will flow through one or the other of the resistances but not both. The higher the resistance of a branch, the fewer charges will flow through it and the lower the current.

Ohm's law is useful in that it may be applied between any two points of a circuit, not just across an entire circuit. For example, suppose that we connect two identical light bulbs in series with a 6.0-V battery as shown in Fig. 10-10A. If each light bulb has a resistance of 9.0 Ω, how much current will flow?* The measured potential difference of 6.0 V across the terminals of the battery means that each coulomb of charge will do 6.0 J of work as it moves from the one terminal to the other. The work required for the charges to move through a small length of normal wire is negligible. Thus, these 6.0 J of work per coulomb of charge will be done in the two light bulbs. Each coulomb of charge must pass through each light bulb as it moves through the circuit. Since the bulbs offer equal resistance to the flow of electricity, equal amounts of work will be done in each bulb. Thus the measurement of the potential drop across one of the bulbs would read 3.0 V. Applying Ohm's law,

$$I = \frac{V}{R}$$

$$= \frac{3.0\ \text{V}}{9.0\ \Omega} = 0.33\ \text{A}$$

If the current through the first bulb is 0.33 A, the current through the second bulb must be the same. For resistances in series, the current through any resistance is the same as the current through any other resistance, and may be called the *total current* (since all of the current must pass through each resistance).

* Light bulbs are used here for clarity since they are an electric device we all relate to constantly. Actually, the light bulb is about the most complex device that could be used in this context. The resistance of a wire depends upon temperature, and a light bulb is a hot-wire device. As a result, light bulbs are almost the only common example of a nonlinear resistance, where R_{tot} would not be exactly $R_1 + R_2$. We will ignore this special behavior and treat light bulbs as if they have normal resistance.

There is a second way in which these bulbs may be connected to the battery, which is electrically quite different. The connection of the bulbs in parallel to each other is shown in Fig. 10-10B. This situation is different in that we now present alternate paths through which the electricity may flow. Charges arriving from the battery at the junction of wires leading to the two bulbs will flow either through the upper bulb or through the bottom bulb. No single charge will flow through both bulbs. Likewise, no charge may do work in both bulbs. Every coulomb of charge which flows through the bottom bulb will do its full 6.0 J of work in that bulb since this is the only (substantial) resistance which that coulomb of charge will encounter in the circuit. Thus, the potential drop measured across the bottom bulb will be 6.0 V, and the current through that bulb will be

$$I = \frac{V}{R}$$

$$= \frac{6.0\ \text{V}}{9.0\ \Omega} = 0.67\ \text{A}$$

This is *not* the total current flowing in the circuit, however, since it completely neglects any contribution of the upper bulb. If the upper bulb were removed, this would be the total current, since it does not depend upon the presence or absence of the other bulb. This lack of dependence upon other resistances in the circuit is an important feature of resistances wired in parallel. Since the upper bulb has the same resistance as the lower bulb, a similar calculation would indicate that the same current is flowing independently through that bulb. The total current would thus be 1.3 A, half flowing through each bulb. If we use this total current with the voltage of the battery, we may calculate the *total resistance* of the circuit.

$$R_{\text{tot}} = \frac{V_{\text{tot}}}{I_{\text{tot}}}$$

$$= \frac{6.0\ \text{V}}{1.3\ \text{A}} = 4.5\ \Omega$$

The total resistance is sometimes called the *equivalent resistance* of the circuit since the entire circuit could be replaced by a resistance of this size and not change the power requirement placed on the battery. Note that the total resistance of the parallel circuit is less than that of any of the resistors. Adding resistors in parallel always increases the current flowing and thus lowers total resistance. We may calculate the total or equivalent resistance for parallel resistors quite easily. Consider the circuit shown in Fig. 10-10B. Here, I_1 is the current flowing through resistor R_1, and I_2 is the current through R_2. The total current is just the sum of I_1 and I_2, or

$$I_{\text{tot}} = I_1 + I_2$$

By Ohm's law, $I_{\text{tot}} = V/R_{\text{tot}}$, where V is the voltage of the battery. Similarly, $I_1 = V_1/R_1$ and $I_2 = V_2/R_2$ by Ohm's law applied to each resistor separately. Substituting these expressions for the currents, we obtain

$$\frac{V}{R_{tot}} = \frac{V_1}{R_1} + \frac{V_2}{R_2}$$

However, essentially all of the work done by the current flowing around the circuit and through R_1 is done in R_1. In other words, the voltage or potential difference measured across the ends of R_1 will be V, the voltage of the battery. Likewise, V_2 will be V, the voltage of the source (battery). Since all of the voltages are really the same value, they will divide out to leave us with

$$\frac{1}{R_{tot}} = \frac{1}{R_1} + \frac{1}{R_2}$$

This is the general expression for calculating the total resistance of resistors connected in parallel.

Example

Resistors of 2 Ω and of 6 Ω are connected in parallel to a 1.5-V flashlight cell. What total current will flow from the cell?

$$\frac{1}{R_{tot}} = \frac{1}{R_1} + \frac{1}{R_2}$$

$$= \frac{1}{2\ \Omega} + \frac{1}{6\ \Omega} = \frac{3}{6\ \Omega} + \frac{1}{6\ \Omega} = \frac{4}{6\ \Omega}$$

If $\frac{1}{R_{tot}} = \frac{4}{6\ \Omega}$

then, by inverting both sides,

$$\frac{R_{tot}}{1} = \frac{6\ \Omega}{4} = 1.5\ \Omega$$

If the total resistance is 1.5 Ω, we obtain

$$I_{tot} = \frac{V}{R_{tot}}$$

$$= \frac{1.5\ \text{V}}{1.5\ \Omega} = 1.0\ \text{A}$$

which is the total current flowing from the cell.

The Interaction between Electricity and Magnetism

While what we have covered thus far forms the basis for much of the use of electricity in our own day, it is not enough to have produced an electrical age. The reason is simple. Electricity was fine as a subject of parlor games and incredibily interesting for serious experimentation, but it would be simply too costly to use for any serious, sustained purpose. So far in our discussion, the battery offers the only source of a potential difference to produce charges flowing through circuits. Suppose your house operated on batteries. How many hours a day could you afford to use electric lighting, your TV set, or any of the dozens of electric appliances you are accustomed to? How much zinc and copper—or other conductors—would have to be produced continuously to run our society on batteries? The mass use of electric power depends upon a simple and inexpensive way of changing plentiful forms of energy into electricity. The chemical potential energy of pure metals is neither cheap nor plentiful. Like most discoveries which change the nature of life on earth, these came from the simple curiosity of humans trying to understand how nature operates. The search was not for a way to open up an electrical age, but to clear up a nagging problem. The properties of electricity and magnetism were so similar that there had to be a connection between the two sets of phenomena.

The first experimental connection between electricity and magnetism was discovered accidentally by Hans Oersted (1777–1851), supposedly during a lecture. He noted that when a current was passed through a wire running over the needle of a magnetic compass, the needle experienced a force twisting it out of its normal position. The needle was not aligned with the wire by this force, nor was it attracted or repelled by the current. Instead it was twisted so that the axis of the needle was perpendicular to the wire. *An electric current could exert a force on a magnet.* More interestingly, magnetic forces were exerted by a current even in the absence of a magnet. If a wire is run vertically through a sheet of paper on which iron filings are sprinkled, when a current is sent through the wire the filings will tend to line up, recording the presence of a magnetic field. *Electric currents generate magnetic fields!* This field does not point toward the wire or away from it. Rather, it runs circularly

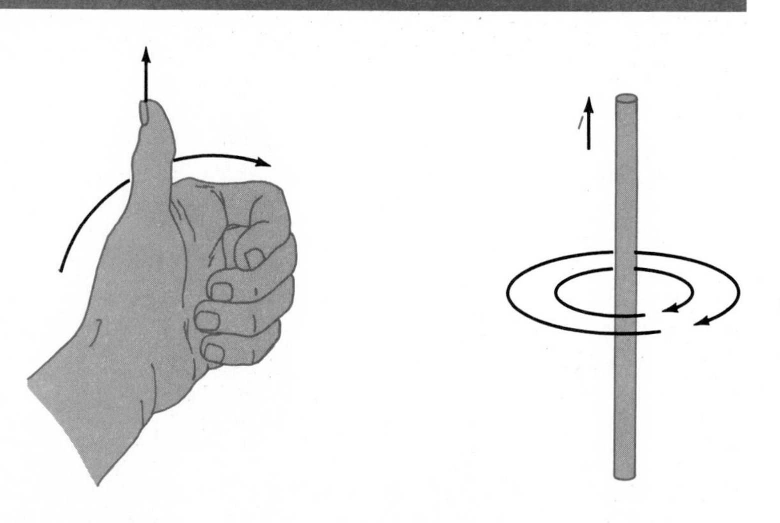

FIGURE 10-11
A current of negative charges flowing up a wire establishes a magnetic field surrounding the wire. If the thumb of the left hand points in the direction of the current, the magnetic field will curl in the direction indicated by the fingers. What do the circular lines indicating the magnetic field tell you?

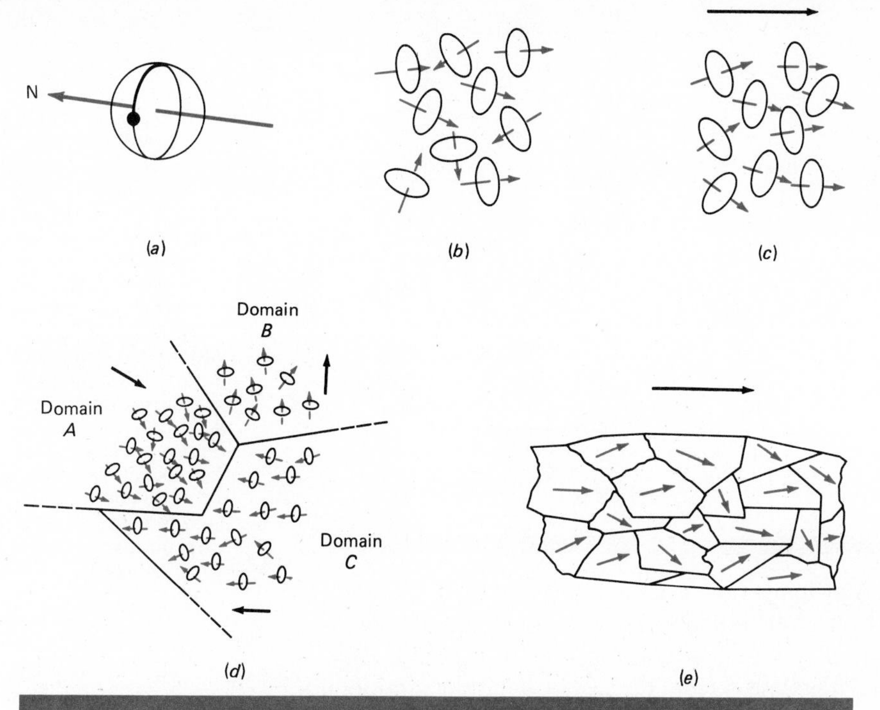

FIGURE 10-12 All magnetism, even of permanent magnets, results from the motion of electric charges. (*a*) A charged particle in a looped orbit constitutes a current flowing in a loop. The negative charge shown would set up a tiny magnetic field, the orientation of which is indicated by the arrow. (*b*) Such magnetic fields of the particles of matter would have no effect if each "atomic magnet" had a random orientation in space. The field for a large number of atoms would cancel out. (*c*) If the magnetic orientation of the atomic magnets would line up, they would add to an effect which could be measurable outside of the matter itself. A large group of such lined-up atomic magnets is called a magnetic domain. (*d*) In a ferromagnetic material each magnetic domain has an independent direction of magnetization. Normally, these cancel to produce no external effect. (*e*) In the presence of an external magnetic field the direction of magnetization of domains in ferromagnetic materials can rotate to agree essentially in direction with field. If the domains can be "frozen" in this orientation, a permanent magnet results.

around the wire, with each line of force ending back on itself. If we allow the thumb of the left hand to point in the direction of the flow of the negative charges in the wire, the fingers of that hand will wrap around in the direction in which the N pole of a magnet is pulled (Fig. 10-11). If two wires run side by side near each other, they will attract each other if current flows through them in opposite directions, or repel each other if the currents are parallel. If a single wire is looped around into a coil rather than allowed to run straight, the magnetic fields generated by the current in each loop will reinforce each other down the center of the coil and far outside the coils. Thus, a small current can be used to produce a relatively strong magnetic field by running it through many loops which are closely spaced in a coil. For small currents, the magnetic field may be strengthened dramatically by placing a ferromagnetic material, such as an iron bar, in the center of the coil. The resulting magnetic field is identical to that produced by a conventional bar magnet, with the advantage that it can be turned on or off, or adjusted in strength, by controlling the electric current through the wire. Today we believe that *all magnetic effects, including those of permanent magnets, result from the motions of charges* (Fig. 10-12).

The ability of a coil of wire to act like a bar magnet provides us with a method of converting electric energy into motion in the *electric motor.* If a current is sent through a coil of wire as shown in Fig. 10-13, it generates a magnetic field with one end of the coil acting like the N pole of a bar magnet and the opposite end acting like a S pole. If this is done while the coil is between the magnetic poles as shown, the upper end of

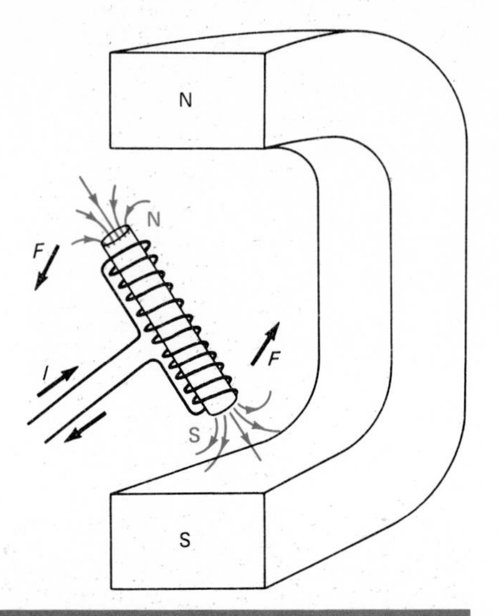

FIGURE 10-13 Current is passed through a coil of wire wound around a ferromagnetic bar. The magnetic field set up by each loop of wire reinforces down the center of the coil to make the upper end of the bar an N pole and the lower end an S pole. These poles will be immediately repelled by the poles of the large external magnet around the coil. If the coil is provided with a shaft to rotate on and we change the direction of the current every half-turn, the coil will continue to spin. We have produced a simple electric motor.

the coil will be repelled by the N pole above it and attracted to the lower S pole. Likewise the S pole of the coil will be repelled by the pole nearest to it. If the coil is mounted so that it is free to rotate, the set of forces on the coil will tend to twist it around so that its S pole will point straight upward. Suppose that we allow the coil to rotate to this new position, but just as the S pole of the coil approaches the straight-up position, we reverse the direction in which electricity is flowing through the coil. This will cause the coil to establish a new magnetic field which has the reverse polarity, or the opposite pole directions, as before. The upper end of the coil will cease being a S pole and will become a N pole. Instead of being attracted to the N pole above it, the upper end of the coil will experience a renewed repulsive force. Since the same reversal would have occurred at the lower end of the coil, this end too will be repelled by the magnetic pole near it. The coil will continue to rotate, propelled by these new repulsive forces. Of course, we will never allow the coil to attain the magnetic poles it is attracted toward. Each time the coil nears the desired poles, we will reverse the direction of the electricity and renew the repulsive forces which keep the coil rotating. Of course, the upper and lower magnetic poles do not have to be permanent magnets; instead they can be electromagnets generated by the flow of electric currents. Thus they may have a much stronger magnetic field than permanent magnets will sustain. A direct-current motor of the type we have described can generate a great deal of twisting force (torque). If you doubt this you should consider the fact that almost all railroad locomotives in use today are moved by such motors. Such locomotives are called "diesels," but electric motors actually move the train. The diesel engine is used only to generate the electricity for these motors. Diesel engines as well as automobile engines share the problem that they only deliver torques with great efficiency when they are rotating at a high speed. At low speeds of rotation (low rpm) such engines are likely to stall if presented with a heavy load to move. To solve this problem, automobiles use gears in a transmission. These allow the engine to rotate rapidly while the rear wheels move slowly during start-up. Other gears allow the rotation rates of engine and wheels to be more nearly the same for other rates of motion. Direct-current motors do not require gears. They can effectively deliver a torque at any rate of turning, from barely moving to high-speed rotation.

Producing a Current with a Magnetic Field

A common situation in science occurred during the historical development of electromagnetism, in which individuals working separately made the same discovery at about the same time. The individuals were Michael Faraday in London and Joseph Henry (1797–1878), the first director of the Smithsonian Institution in Washington, D.C. Faraday published his results well ahead of Henry; he wrote about the 7 years of experimentation it took to arrive at the discovery of *electromagnetic induction.* If an electric current can produce a magnetic field, why should not a magnet be capable of producing an electric current? It was; but not by the methods which were tried early in the search. We should note that a static or stationary electric charge does not produce a magnetic field. Only if the charge moves does it produce a magnetic field, which swirls around the line of motion of the charge. In the same way, a stationary magnet does not produce the electric field in its surroundings. Yet, an electric field, or a force acting on charges, is necessary to move the charges which will then become a current. A nonzero electric field is produced only if the magnet moves relative to the charges. More precisely, an electric field is generated in a volume of space by a magnet only when the magnet approaches or moves away. Let us examine this relationship more slowly.

Imagine that a magnetic field runs through the space above your desk such that the N pole of a compass would point straight down into the desk. Further imagine a tiny sphere of positive charge floating in position 10 cm above the center of your desk in this field. There is no interaction between the magnetic field and the charge since both are stationary. We could have the magnetic field increase or decrease its strength as though a magnet were approaching or receding, but it is easier to keep the field constant and imagine the charge as moving. If the charge moves from above the center of the desk directly toward you, the charge would experience a force due to its motion relative to the magnetic field (Fig. 10-14). In direction this force is the strangest one we have yet encountered. The force is not in the direction of the charge's motion, nor is it opposed to it. Neither is the force in the direction of the magnetic field or opposed to the field. We find that the force induced on a charge moving relative to a magnetic field is *always* perpendicular to the direction of the field, and it is *always* perpendicular to the motion of the charge through the field. In our example, the positive charge would

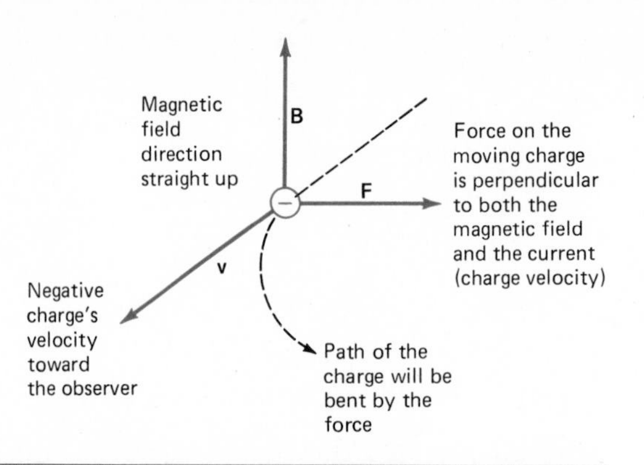

FIGURE 10-14 A charged particle moving across a magnetic field experiences a force perpendicular to both its motion and the magnetic field. If the magnetic field points straight up as shown, a negative charge moving out of the page would experience a force deflecting it to the right. A positive charge would be deflected to the left.

experience a force to your right as it moved toward you. Now replace the sphere of positive charge with a wire that runs from left to right above the center of your desk and in the magnetic field. The wire contains equal amounts of positive and negative charges, as does all neutral matter. If you pull the wire toward you, it will move through the magnetic field. Every positive charge in the wire will experience a force pulling it to the right along the wire. Every negative charge will experience a force pulling it to the left. Thus, the right end of the wire becomes positive and the left end becomes negative as charges respond to these forces. A potential difference is generated between the two ends of the wire, and if they are connected to each other, a current flows. When the motion of the wire relative to the magnetic field stops, the forces on the charges stop and the current drops to zero.

Faraday pointed out that if a current is to flow, the ends of the wire must be connected to form a loop. He determined that the important factor in generating this current was the total amount of magnetic field passing through the area determined by this loop. He called the number of magnetic lines of force passing through an area perpendicular to the lines the *magnetic flux.* Only if there is a change in this flux through a circuit, such as a loop of wire, will a current be induced. For example, if a bar magnet is thrust through a loop of wire connected to an electric meter, the meter will indicate that a current is flowing as the magnet approaches the loop, or as it leaves the loop, but not while it is still. To picture this, imagine a uniform magnetic field from ceiling to floor across the entire area of your desk. If a square loop of wire is pulled toward you through this field, no current will flow (Fig. 10-15A). The negative charges in the front wire of the loop are cutting the magnetic field and will experience a force tending to move them to your left. However, the negative charges in the back wire of the loop are cutting the field at the same rate, and they will experience the same force to the left. The front part of the loop experiences a force which tends to move negative charges in a clockwise direction around the loop, but those in the rear wire tend to move counterclockwise with the same force. Faraday pointed out that a current will result only if the amount of magnetic field through the area defined by the loop changes. For example, if the for-

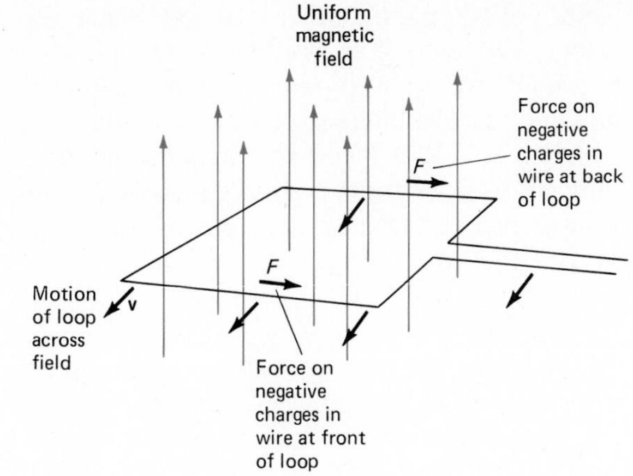

Uniform magnetic field

Force on negative charges in wire at back of loop

Motion of loop across field

Force on negative charges in wire at front of loop

FIGURE 10-15A If a loop of wire is pulled across a magnetic field, individual charges in the wire cutting the field will experience a force. However, the forces on the front of the loop will be in the opposite sense of rotation from the forces on the charges at the back of the loop, and no flow of charges is produced.

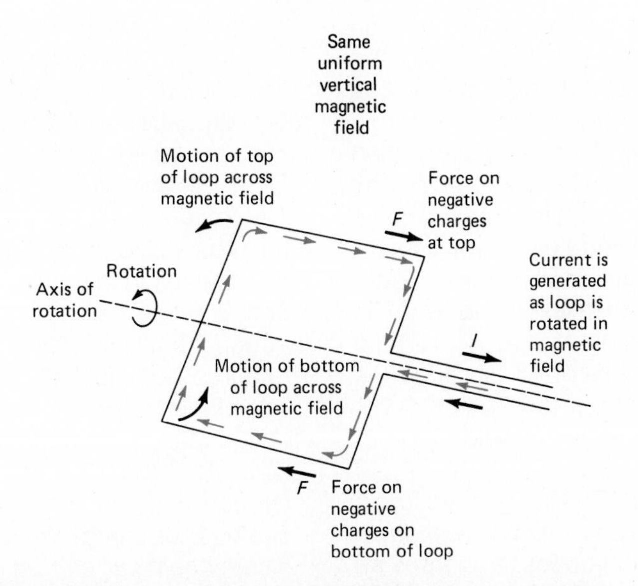

FIGURE 10-15B If the same loop is rotated in the same magnetic field, the forces on charges on opposite ends of the loop will be in the same sense along the wire (here clockwise), and a flow of the mobile negative charges can occur. This it the principle of the electric generator.

ward wire cut through the field faster than the rear wire, this would produce a current. One way to change the flux through the loop would be to rotate the loop around a horizontal axis (Fig. 10-15B). When the loop is in a vertical plane, it is parallel to the magnetic field, and no magnetic lines of force pass through its area. Hence the flux is zero. However, when the loop is rotated 90°, the flux is a maximum; as the loop is rotated, the flux changes from zero to maximum. Thus, if a loop of wire is rotated in a magnetic field, a current may be produced. As a loop of wire is rotated about an axis perpendicular to a magnetic field, an electric potential difference is generated between the ends of the loop which will produce a current if the ends are joined. This is the principle of the electric generator. To produce the electricity for our homes, all that is necessary is to rotate coils of wire so that the magnetic flux through the coil continually changes. The voltage produced depends upon the rate of change of flux and the number of complete loops in the coil. In this way we can take the motion of falling water or of expanding steam, and use it to rotate a coil of wire in a magnetic field to generate electricity. This principle has been of fundamental importance in changing the nature of human life-styles in an electrical age. With electricity we have the ability to transport any needed amount of power from place to place with great convenience. Very few home workshops would have a power drill if each required a small steam engine to power it, or a water wheel.

Summary

Magnetism and *electricity* are the names given to phenomena involving the ability of some materials to exert measurable forces on other objects at a distance. Objects called *magnets,* if free to rotate, align themselves on earth with one side always facing approximately north. This end of the magnet is called a *north-seeking pole,* or simply N pole. The opposite side of the magnet is called an S pole. All magnets contain both N poles and S poles. Like poles repel each other, and opposite poles attract. Magnets exert a force on all magnetic substances in the space around them. The pattern of this force is called the *magnetic field* of the magnet.

Electrical effects are ascribed to a property called electric *charge,* or q. The unit of charge is the coulomb (C). Charges are found to be of two types, *positive* and *negative.* Like charges repel, and opposite charges attract. Normal, electrically neutral, matter appears to have both types of charge in equal amounts. Materials called *insulators* confine charges to a particular place, while *conductors* allow charges to move through them in response to forces. Frictional contact between dissimilar insulators can transfer charges from one to another and produce unbalanced charges. The type of charge residing on a rubber rod rubbed with fur is defined as negative charge. The presence of an unbalanced charge may be detected by a sensitive device to measure net charge such as the electroscope. The force between charges is found to follow *Coulomb's law,* $F = Kq_1q_2/R_2$, although complex geometry will make this law difficult to apply. As with magnetism we may define the electric forces in the space surrounding charges in terms of an *electric field,* or E. This field is the force acting per coulomb of charge placed at that point, $E = F/q$, and has a direction as well as size for every point in space. Fields (electric or magnetic) are usually illustrated as a series of nonintersecting lines, called *lines of force,* which describe the force at any point. The force is in a direction parallel to the lines and has a size proportional to the concentration of lines in that area of space. The motion of charges with or against an electric field requires that work is done. We define the work required per coulomb to transfer charge from one place to another as the *electric potential difference,* or *voltage,* between these positions. The volt (V) is the unit of potential difference and represents a joule of work per coulomb of charge transferred, $V = W/q$. An *electrochemical cell,* or, if more than one is used, a *battery,* produces a constant potential difference between its terminals for its useful life. This constant potential difference allows a continuous flow of charge to occur from one terminal of the battery to the other during its useful life. Such a flow of charge is called a *current,* designated I. A flow of one coulomb per second is defined as a current of *one ampere* (A), $I = q/C$. The actual charges to move in such a current are found to be the negative charges. To allow such a current to flow, a complete pathway for the motion of charge from one terminal of the battery to the other must exist. Such a pathway is called a *circuit.* A current flowing in a circuit does an amount of work in the circuit depending on its size and on the potential difference of the battery. The rate at which work is done, or the power P, is the product of current times voltage, $P = IV$. The amount of current which will flow in a circuit depends not only on the potential difference of the battery but also on the nature of the circuit. The property of the circuit restricting the flow of charge is called resistance R, which is measured in ohms (Ω). The relationship of potential difference, current, and resistance is $V = IR$, which is called *Ohm's law*. This relationship applies between any two points as well as for the entire circuit. Resistances in a circuit may be arranged sequentially so that charges must flow through each resistance one after another. Such an arrangement is called a *series* circuit. In general, the total resistance of such an arrangement is just the sum of the individual resistances. Resistances may also be connected *in parallel*, so that some part of the current flows through each, but no charge may flow through each resistance. The total resistance R_{tot} in such a case is found from the formula $1/R_{tot} = 1/R_1 + 1/R_2$ for resistances R_1 and R_2.

Flowing charge sets up a magnetic field in the space surrounding the charge. The field surrounding a vertical wire in which negative charges are flowing upward circles the wire in a clockwise direction, with each line of force ending on itself. If the wire is looped into a coil, the field produced by individual segments of the current will reinforce each other down the center of the coil, resulting in a magnetic field similar to that of a bar magnet. The magnetic fields produced by currents exert attractions and repulsions similar to those of permanent magnets. These forces can be used to make a current-carrying loop move in a fixed magnetic field, which is the principle of the *electric motor*. A charge moving perpendicular to a magnetic field experiences a force perpendicular to both its motion and the direction of the magnetic field. If the amount of magnetic field (the *magnetic flux*) through a coil changes, a current will flow in the coil. This is the principle of *electromagnetic induction* upon which most production of electric currents in our society depends.

Questions

1 Liquid oxygen is weakly attracted to a strong magnet. What type of material does this make it?

2 Since the N poles of magnets all over the earth are attracted in the direction of the magnetic North Pole of the earth, what type of magnetic pole must be located in that vicinity?

3 The pole pieces of very large electromagnets often taper down to a very fine gap. Why?

4 Do the small magnets used for attaching notes to refrigerator doors attract each other, repel each other, or what?

5 What could you say about the magnetic field in a volume where the magnetic lines of force were all parallel to each other?

6 Why should the magnetic field of the earth remain constant in direction for tens of thousands of years and then, in a relatively short time, completely reverse itself in direction?

7 Why cannot the N pole and the S pole of a magnet be separated by just snapping the magnet in half to produce one pole of each type?

8 Which of these combinations would generate a static electrical charge on the objects; which would not, and why not?

(*a*) A boy running a stick along an unpainted picket fence.
(*b*) A cat scratching its back on a metal grate.
(*c*) A person sliding across a plastic seat cover.

(*d*) Leather shoes scraping on a nylon carpet.
(*e*) The windows of a spacecraft plunging through air.

9 Can you think of a way to show that it really is the negative charges in matter that move to show electrical effects, while positive charges remain in place?

10 An electroscope is seen to be charged by its leaf standing away from the plate. How could you experimentally determine the sign of the charge with a fur and a rubber rod?

11 Explain how charging by induction leaves a charge of opposite sign from the charged rod used. Use a set of diagrams to show the action of charges at each step.

12 Solid objects maintain their size and shape. Why don't they just fall apart and sag into blobs under the action of the force of gravity?

13 If there was an electric field pointing down the axis of a long, straight metal wire, what would happen to the charges in the wire? What does this tell you about the electric field in metals under most conditions?

14 We have a 1.0-Ω resistor. Suppose we connect a 1,000,000-Ω resistor in parallel with it. Without doing any mathematics, what can you predict about the equivalent resistance of these two resistors connected in parallel compared to the original 1.0-Ω resistance?

15 How do we produce most of the electricity used in our homes?

16 If I took an ordinary metal coat hanger and held it in a horizontal plane, what would happen in the coat hanger as I dropped a small magnet through the center of it? How could I make this effect more pronounced?

17 How is an electromagnet made? Describe how it works.

18 What really causes an electric motor to turn as electricity is supplied to it?

Problems

1 Two small spheres, each with a charge of q, are separated by a distance R. If the force of repulsion between the spheres is F, what would the force become if we did the following:

(*a*) Doubled R?
(*b*) Doubled both charges?
(*c*) Cut R in half and doubled one of the charges?
(*d*) Tripled each charge and R?

2 (*a*) In a hydrogen atom, a negatively charged electron is 5×10^{-11} m from a proton with the same size charge, but positive. What is the attractive force of the proton on the electron? [*Note:* If 6.24×10^{16} electrons make up 1.0 C of negative charge, the charge on each electron is $1/(6.24 \times 10^{18}$ C).]

(*b*) If the mass of the electron is found to be 9.11×10^{-31} kg, what acceleration should the electron experience from this force?
(*c*) The acceleration due to gravity is 9.8 m/s²; this amount of acceleration is usually called 1 g. How many g's of acceleration does the electron experience?

3 A sphere with a positive charge of 0.01 C experiences a force of 50 N pulling it to the north. What is the electric field at the position of the sphere? If this force was constant and pulled the sphere 10 m to a wall, what would be the electric potential difference between the original position of the sphere and the wall?

4 A flashlight used two 1.5-V cells in series to light the bulb.

(*a*) At what voltage does the bulb operate?
(*b*) If the bulb has a resistance of 0.5 Ω, what current will flow when the flashlight is switched on?
(*c*) At what rate is work being done in the circuit? (What is the electric power?)

5 An automobile has a 12-V battery. When the key is turned, 150 A flows through the starter motor, which turns over the automobile engine. What is the power (wattage) of the starter motor? Since 776 W is the equivalent of 1 horsepower (hp), what is the horsepower of the starter motor?

6 What is the resistance of a 9.0-Ω and a 3.0-Ω resistance if they are connected in series? In parallel?

CHAPTER ELEVEN

Light and Electromagnetic Waves

Introduction All the primates are "eye-centered"—that is, they rely on sight for the bulk of the information they obtain about the environment. At night some monkeys retreat to the trees to hide through a fitful rest deprived of their sharpest sense. Some children sleep with the covers over their heads from a similar motivation. Yet, for all its importance to humans, the nature of light was a subject of controversy well into this century. One ancient theory proposed that light was continually generated by objects under the action of the sun. This idea was not impossible. A fluorescent wall poster works this way. A lamp gives off ultraviolet rays which the eye cannot see. When these rays strike the fluorescent pigments in the painting, they radiate visible light. However, the idea that all objects are the source of the light we see when we look at them is unrealistic. When we stare down into a still pool of water, what conspiracy of nature would be required to have the water radiate a picture of ourselves back at us? It is much more reasonable to regard some objects—the sun, a flame, a firefly—as true sources of light, and other objects as simply reflecting the light from these sources.

Normally, light moves in straight lines. We have considerable confidence that objects are in the direction in which we see them. Catching a baseball would be a shortcut to the hospital if the ball we see in front of us could really be approaching from any direction and the light was bent out of position. Considering this behavior, some investigators believed light to be a stream of fine particles. This view presents several problems. Light travels extremely fast. Lightning bolts or the flashes of cannon fire may be seen well before their sound may be heard. The assumption that light was *instantaneous* (took no time at all to travel) yielded very consistent values for the speed of sound for early investigators. Sound requires 3 s to move 1 km (5 s/mi), and light was clearly much faster than this. Thus, if light was composed of a stream of particles, they must have very little mass. A sand grain moving at even 1 percent the speed of light has momentum equal to a bowling ball streaking toward the pins. The eye intercepts millions of bits of information about the environment each second. If each bit came in with a particle only one-millionth the mass of a sand grain, the net momentum transfer would be like being hit in the eyeball with a bowling ball each second. Thus, if light is made up of a stream of particles, they must have unimaginably small masses.

Another promising school of thought believed light to be a wave form. Waves transfer energy and can carry information, like sound waves, but they need not transfer mass. A wave theory of light also poses problems. First of all, if light is a wave, what is wiggling? What would be the medium for light? For every type of wave we have studied, the speed of the wave depends upon the elasticity and the density in the medium carrying the wave. To account for the very high speed of light, the medium would have to possess very high elasticity and almost no density. The forces tending to snap the medium back into place would have to be much larger than in steel, for example. Furthermore, if this medium possessed even as much mass per cubic centimeter as air, it would be difficult to see how these forces could overcome the inertia fast enough to shake the medium back and forth at the speed light passes. Thus, what we need for a medium is something which is almost nothing but is held together tightly. As silly as this sounds, the proposed medium for light was given a name, the *ether.* Further problems arose in the early 1600s with experiments with gases. Toricelli, a pupil of Galileo, invented the *barometer* and used it to measure the air pressure, or the weight of the atmosphere pushing on the device. Air pressure drops rapidly with altitude. At sea level, the weight of the atmosphere will support a column of mercury in the barometer some 760 mm high. This implies a pressure of about 10^5 N/m^2, or slightly less than 15 lb/in^2. At the altitude of the city of Denver, the pressure drops to 85 percent of this. At an altitude of a few kilometers, half of the air is below you; 100 km or so, the air is effectively all behind you. Thus, space is largely empty. While this clarifies why planets can move about the sun frictionlessly, it presents problems. Particles might encounter no difficulty in streaming across space, but a wave needs something to wiggle.

The interplay between wave and particle interpretations of light has been complex. Light is of such great importance to us that we will examine its properties for their own sake. However, keep in mind that the great historical argument over the nature of light was never far behind the facts on the observed behavior of light.

Light and Materials

Light coming from a small source, such as a candle or a small light bulb, spreads out in all directions from the source. Measurements disclose that if we move twice as far away from the source, it appears only one-fourth as bright. Thus, the intensity of light from a small source drops off in proportion to the inverse square of the distance ($I \propto 1/d^2$). If we place at a considerable distance from the source a screen with a small opening, or *aperture,* some light moves through the aperture. Provided the source is indeed far from the aperture, this light will continue in its motion directly away from the source and not expand to become perceptibly wider as it moves short distances from the aperture. If the aperture is wide enough so that the bundle of light coming through has noticeable width, it is called a *beam* of light. If the aperture is narrow enough that the width of the bundle of light passing through is negligible; it is called a *ray* or a pencil of light. A transparent-glass light bulb with a small, hot filament is good for casting sharp shadows and isolating a beam (Fig. 11-1A). A common frosted-glass light bulb is a poor choice as a light source for isolating a ray of light. Such a source must be very far from the aperture for it to appear small enough to produce a crisp ray. Indeed, the reason the bulb is frosted is so that the entire surface of the bulb scattering light will soften shadows and reduce the contrast of the area it illuminates (Fig. 11-1B). When we discuss a ray, we are considering a thin bundle of light retaining negligible thickness while it moves.

When light strikes the surface of an object, any of several things may happen. If the object does not allow light through, it is said to be *opaque.* Opaque objects absorb some of the light, and scatter or reflect the rest. A mirror, for example, absorbs a small percentage of the light and reflects the rest. The printed page of a book, on the other hand, absorbs most of the light falling on the black printed letters but scatters most of the light falling on the white page between the letters. If we hold the page up and view the light through it, we notice that a considerable amount of light also passes through the page. An object that allows such "scrambled" transmission of light through it is called *translucent.* Paper is partially translucent but not so clearly as the frosted glass of a light bulb. Some materials also allow light to pass through them rather freely

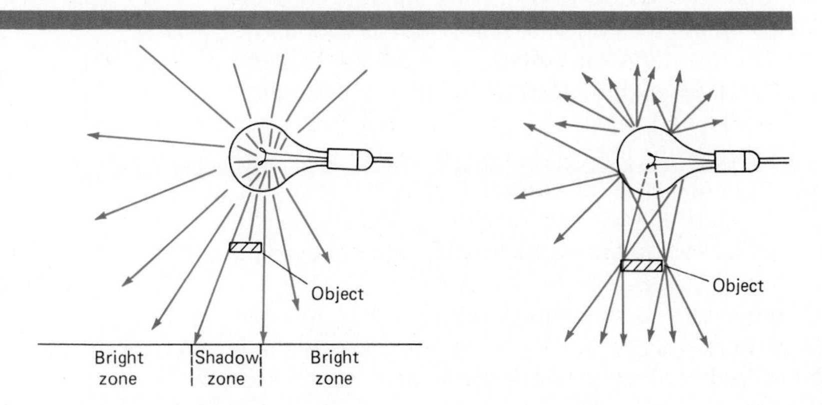

FIGURE 11-1A A transparent-glass light bulb radiates light directly from the white-hot filament at the center. When an object blocks light from this filament, its shadow zone receives almost no light. This great contrast promotes eyestrain.

FIGURE 11-1B The inner workings of a frosted-glass bulb are the same, but each part of the bulb's surface scatters light in all directions. Rays coming from different parts of the bulb partially "fill in" shadow zones, eliminating harsh contrasts.

without scrambling or randomizing the direction of its travel. Such materials are called *transparent,* and the air must be such a material or you would have great difficulty reading this. Thus we have the following set of behaviors of materials to light. The light may be:

1 *Absorbed,* as with the dark letters on this page

2 *Scattered* or "bounced" in all directions, as with the white parts of the page

3 *Reflected,* or bounced in a particular single direction, depending on the direction at which it reaches the surface

4 *Transmitted with scattering,* or randomization of direction, as with translucent substances

5 *Transmitted without scattering,* as with transparent substances

Usually, each substance will behave in more than one of these ways. The shiniest mirrors absorb some light and the blackest of paints reflect some, but we characterize materials by their dominant behavior.

Most objects scatter light in all directions. Consider the tip of your nose. It is illuminated by light falling on it from some source—the room lights or the sun. The skin then scatters the light in almost all directions so that no matter where someone is in front of you, the tip of your nose sends some light in the direction of his or her eye. While all objects reflect some light, not all of them scatter it in this manner. If you are standing in front of a vertical mirror, some of the light leaving your nose strikes each part of the face of the mirror. However, if you want to examine the tip of your nose in the mirror, there is only one place on the surface of the mirror where you must look to see the reflection of the light from your nose. While light from your nose is bouncing off all parts of the mirror, only that reflecting in one place leaves the mirror at the appropriate angle to reach your eye. The reflection of light follows the same patterns as the reflection of waves discussed in Chap. 9. The point at which a ray strikes the mirror is called the *point of incidence* of the ray. From this point we may construct a perpendicular or *normal* to the surface of the mirror. The angle between the ray coming into the surface, the *incident ray,* and the normal is called the *angle of incidence,* θ_I. The angle between the reflected ray and the normal is called the *angle of reflection,* θ_R. The law of reflection of waves simply asserts that the angle of incidence equals the angle of reflection, or $\theta_I = \theta_R$. Here, we may determine that this relationship also works for light. (The same relationship also holds for particles. If you are passing a basketball on one bounce to another player, you aim for the point on the floor midway between the two of you. This is because $\theta_I = \theta_R$, for particles, as anyone who has spent much time at a pool table will cheerfully confirm for you.)

The image in a plane mirror is realistic enough to fool you into believing you are seeing objects directly. This is because the relationship between rays is preserved upon reflection. Arriving at a plane mirror

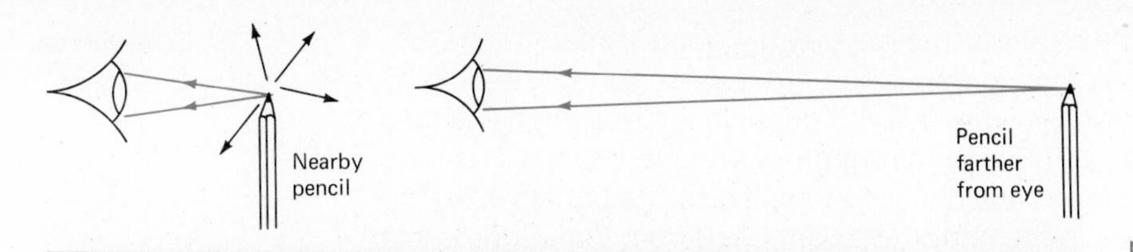

FIGURE 11-2A *Seeing a real object.* The tip of the pencil scatters light in all directions. The eye intercepts a spreading cone of such rays. This process occurs for every point on the visible surface of the pencil—the tip is used only for illustration.

FIGURE 11-2B The cone of light rays intercepted by the eye from a nearby object has a greater angular spread than the cone of light rays intercepted from a faraway object. The rate of spreading gives information about the distance to the object.

light rays traveling parallel to each other are still parallel after reflection. Rays that are drawing closer, or *converging,* still converge after reflection, and at the same rate. Likewise, rays that reach the mirror spreading apart from each other, or *diverging,* diverge at the same rate after reflection. This last situation is particularly important for the image formation in a plane mirror in terms of the method by which the eye forms its own image. Consider the information coming to the eye from some nearby object, say the tip of a pencil (Fig. 11-2A). Light enters the front surface of the eye. The light striking at the dark inner circle of the eye, or *pupil,* passes into the interior for analysis. The pupil gets larger as the eye requires more light and smaller as it requires less light in bright situations. Even at its smallest, the pupil is a disk, not a spot, and thus allows not just a single ray but many rays to enter the eye from our pencil tip. Light scatters from the tip of the pencil in all directions, and separate rays diverge as they move farther from the pencil tip. The pupil intercepts a cone-shaped bundle of these diverging rays from the pencil tip. The rate at which this bundle of rays is spreading apart carries to the eye information about the distance of the pencil tip from the eye. The farther away the object the more nearly parallel the separate rays from that object are as they enter the eye (Fig. 11-2B). The reflection of light in a plane mirror preserves the rate of spread of light rays. Thus the eye could correctly record the distance the bundle of rays must have traveled to reach the eye. Not aware of the fact that the actual path of the rays has been "folded" by the mirror, the observer perceives the pencil to be at the position of the image (Fig. 11-3). This may be demonstrated by taking a

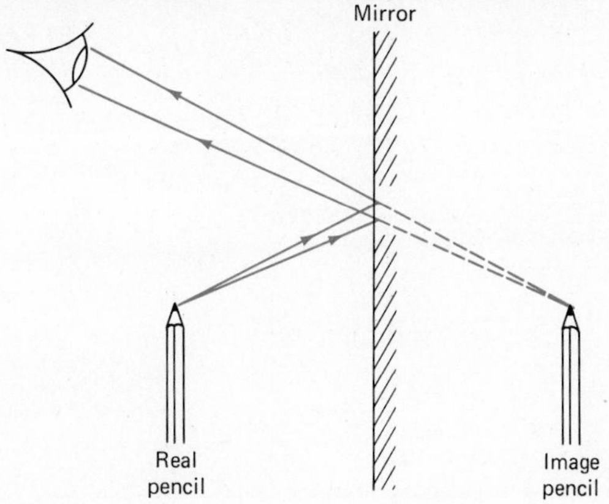

Mirror

Real pencil

Image pencil

FIGURE 11-3 *Seeing an image in a mirror.* Rays scattered from the tip of the pencil strike the mirror and are reflected. The reflection does not change the rate at which they are spreading apart, or diverging. The eye intercepts a spreading cone of rays, which behave exactly as if they were coming from the tip of the image pencil behind the mirror. Such an image is called virtual since the light does not really pass through the point marked by the tip of the image pencil.

picture of yourself in a mirror with a reasonably good camera. The camera adjusts for the different spread of light from near and far objects by an adjustment of *focus*. If you are 3 m in front of the mirror when you take the picture, the result will be in sharp focus when the camera is set for 6 m. This distance is the actual distance traveled by the light to enter the camera (3 m from you to the mirror and 3 m back from the mirror to the camera).

Images in plane mirrors trigger great fun with babies, as most adults know. Animals with strong territorial instincts often get enraged by the intruder seen in a mirror. Some of these are frightened half to death when "the other" animal comes on more ferociously every time they try to scare their opponent off. A few of these will actually leave the battleground in possession of their image and some will accept domination with their image the victor. (Could this be the opposite of a poor self-image?)

Reflections in Curved Mirrors

The image observed in a plane mirror is called a *virtual image*. (See Fig. 11-4.) The eye perceives the light as coming from a place in space through which the light does not actually pass. The image appears behind the plane mirror, but the mirror is opaque and light is really confined to the space in front of the mirror. Another type of image may be produced, although not by plane mirrors. This is the *real image*. In certain situations the light coming from one part of an object and scattering in many directions may really be reassembled at a particular point in space to form a *real image*. You are just as familiar with real images as you are with virtual ones. As you watch a movie projected on a screen,

FIGURE 11-4 Light reflecting from any very smooth surface reflects in an orderly fashion and forms an image. If the person at the left were not shown, how could you tell which part of the photograph was real and which was an image? Is the image real or virtual? (*Photograph by the author.*)

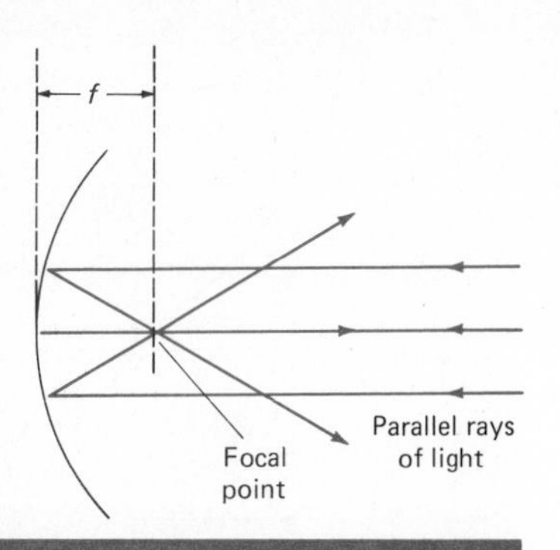

FIGURE 11-5 Parallel rays of light striking a concave mirror are reflected to converge to a point called the focal point.

you are observing a real image. The light spreading out from each part of the small film is reassembled at a particular position on the screen. This reassembly of light occurs at a particular position in space. A mirror may produce a real image, but to do so the surface of the mirror must be curved inward at the center, or *concave,* not flat. A concave spherical mirror reflects light at every point by the law of reflection. If only a tiny part of the surface were observed it would appear much like a plane mirror. A fly walking on the surface of our mirror would record about the same image of itself as if walking on a normal mirror. However, the surface of the mirror is curved, and each portion of the mirror would treat the same ray differently since it would strike the surface at a different angle. If parallel rays of light strike a concave mirror, they *converge* upon reflection (Fig. 11-5). Except for the rays striking near the edges of our mirror, the rays are brought together almost to a point.

The point where parallel rays of light are brought to a focus is called the *focal point* of the mirror. The shortest distance between the surface of the mirror and the focal point is called the *focal length* of the mirror. This distance may be measured along the ray that passes through the focal point on its way to the mirror, strikes it perpendicularly, and is reflected back through the focal point. The line of this ray is called the *central axis* for the rays. Mirrors are usually described in terms of their focal length (as are lenses). The focal length is half of the radius of the sphere from which a spherical mirror is cut. Thus, the focal point is halfway between the *center of curvature* of the mirror and its surface. If rays of light come to the mirror diverging from some object in front of the mirror and not parallel to each other, these too may be brought to a focus, but not at the focal point. Diverging rays are brought to a focus farther from the mirror than the focal point. Consider the rays from the object shown in Fig. 11-6. Rays from the tip of the object strike all portions of the mirror. These rays are reflected according to the law of reflection as shown and converged to a point marking the tip of the image. The rays from each point on the object form a separate point of the image upon reflection.

We may define the distance of the object from the mirror as the *object distance,* D_O. Likewise, we may define the *image distance,* D_I, as the distance of the image from the mirror. If we measure the position of the image for various positions of the object, we obtain data similar to that

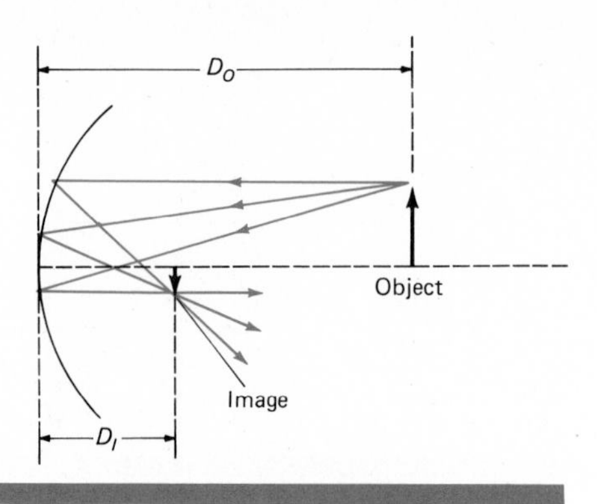

D_O
Object
Image
D_I

FIGURE 11-6 Diverging rays of light from an object striking the concave mirror may also be converged upon reflection. In general, the image distance D_I will be larger than the focal length f for the convergence of parallel rays.

graphed in Fig. 11-7. As the object is placed very far from the mirror, the image approaches the focal point. Why? When the object is drawn closer to the mirror, the image moves out from the focal point more and more rapidly. Finally, as the object is moved to the focal point, rays of light from it emerge from the mirror parallel to each other. The graph of Fig. 11-7 is similar to one encountered in Chap. 2. It is the graph of a hyperbola, but it is not in the standard form. The standard form of such a curve would have the two limbs of the curve approach the x axis and the y axis, respectively, as they raced off to infinity. In such a form, the equation of the curve is quite simple—namely, the x-variable value times the y-variable value equals a constant, or $xy = c$. This curve has each limb approach the numerical value of the focal length f rather than the axis.

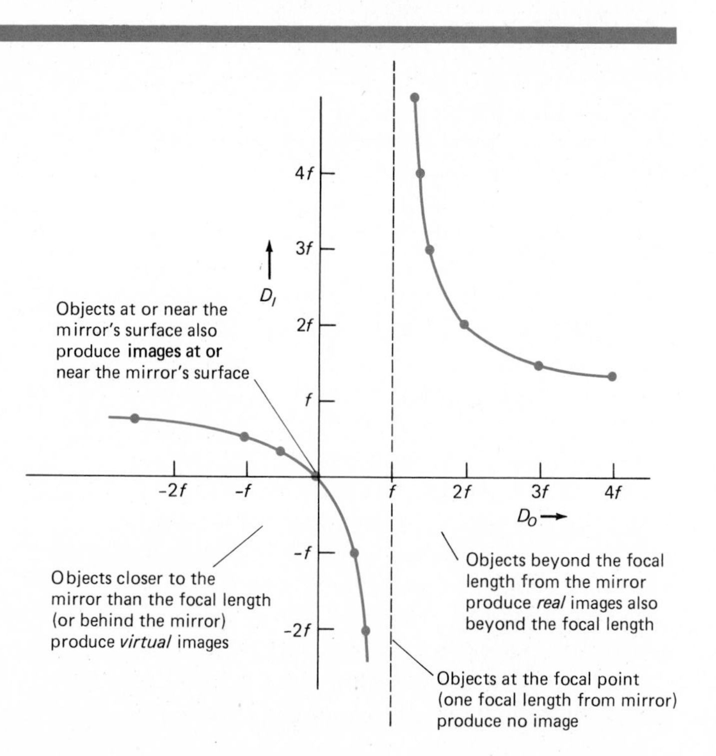

4f
3f
2f
f
-f
-2f
D_I
-2f
-f
f
2f
3f
4f
D_O
Objects at or near the mirror's surface also produce images at or near the mirror's surface
Objects closer to the mirror than the focal length (or behind the mirror) produce *virtual* images
Objects beyond the focal length from the mirror produce *real* images also beyond the focal length
Objects at the focal point (one focal length from mirror) produce no image

FIGURE 11-7 Graph of the distance of the object from the mirror (D_O) vs. the distance of the image from the mirror (D_I) for a spherical mirror. Distances are in focal lengths, and cases using the back (convex) surface of the mirror are included.

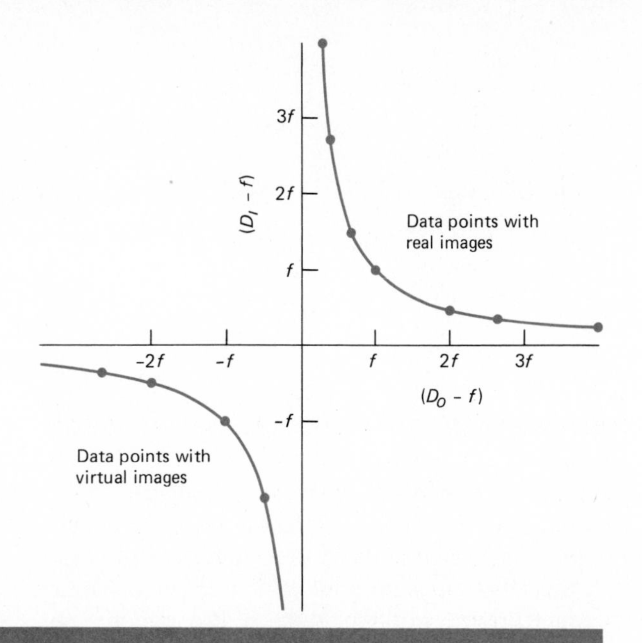

FIGURE 11-8
The graph of Fig. 11-7 may be placed in the standard position on the graph paper by subtracting the focal length from all values of D_O and D_I. The graph of these new variables $D_O - f$ and $D_I - f$ has the form of a simple inverse relationship discussed in Chap. 2.

Subtracting f from every value of D_O moves the curve over to the y axis. Subtracting f from values of D_I drops the curve to the x axis and puts the hyperbola in standard position, as shown in Fig. 11-8. The equation for the curve is now of the form $xy = c$, where our x variable is $D_O - f$ and our y variable is $D_I - f$. Furthermore, inspection of the numerical value of the constant c indicates that it has the value f^2. We thus obtain for $xy = c$, the expression

$$(D_O - f)(D_I - f) = f^2$$

Expanding the quantities in parentheses and dividing by the common denominator yields the final equation for the relationship

$$\frac{1}{D_O} + \frac{1}{D_I} = \frac{1}{f}$$

This is known as Gauss' equation of the *law of optics;* and it describes the relationship between object distance, image distance, and focal length of the mirror.

Actually, Gauss' equation describes the relationship beyond the range in which we tested it. Our data ended when we had moved the object toward the mirror as far as the focal point. There is no physical barrier to prevent us from moving the object still closer to the mirror. Will there be an image with the object inside the focal length? You may recall that we have already observed that a fly sitting on the mirror would observe an ordinary mirror image of itself. Closer than the focal point the concave mirror produces a virtual image similar to a plane mirror but with one notable difference. The image in the concave mirror will be enlarged in size from the original object. To understand this process it is useful to study the ray diagram in Fig. 11-9. To locate the image of a

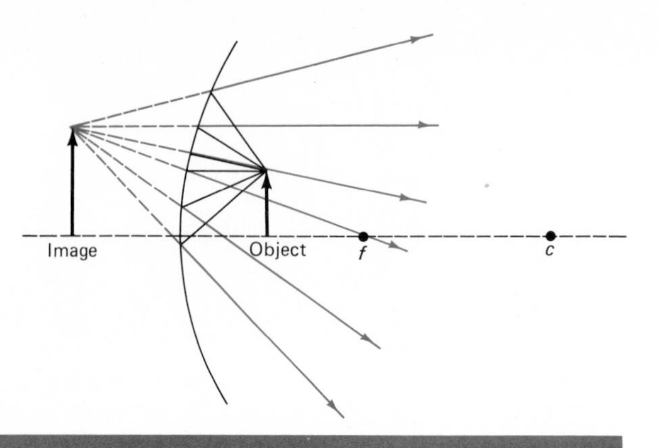

FIGURE 11-9 The formation of a virtual image in a concave mirror by an object closer to the mirror than the focal point. Wherever an eye to the right intercepts the diverging rays after their reflection from the mirror, they will appear to have originated at the virtual image point to the left of the mirror.

point on the object it is necessary to draw at least two rays of light. While an infinite number of rays is possible, the two that are the easiest to construct both pass through the focal point of the mirror. A ray of light from the object point moving toward the mirror parallel to the central axis must pass through the focal point upon reflection. (Why?) The second ray passes through the focal point on its way to the mirror. This ray reflects to move away from the mirror parallel to the central axis. This is because light rays are *reversible.* That is, the geometry of reflection works no matter which ray is the incident ray or the reflected ray. If we direct a ray into the mirror parallel to the axis, it reflects through the focal point. Therefore, by the principle of reversibility, if we direct a ray through the focal point at the mirror, it will come out parallel to the central axis. Actually, in the ray drawings you have seen thus far, the only way of knowing which direction the light was moving is by the little arrows drawn on the rays. Every one of these rays work perfectly well in reverse. Indeed, this reversibility of light is so complete that object positions and image positions are reversible. If an object at *A* sends out rays to form an image at *B*, then an object at *B* would send out rays to form an image at point *A*. We can observe this reversibility in the graph of Fig. 11-7. If we graphed this data with D_O on the *y* axis and D_I on the *x* axis, the graph would be identical to the original graph. Object and image are interchangeable in the relationship because light rays are reversible.

Numerous practical uses are made of curved mirrors. The enlarged virtual image just discussed is the basis for shaving mirrors or beauty mirrors which give the observer an enlarged image of the face. Light from a source placed at the focal point of a concave mirror will emerge after reflection as a parallel beam of light. This effect is used in innumerable light sources from flashlights through searchlights. Dual-purpose headlights for cars operate on this principle, but they present a particular problem. A hot-wire filament placed at the focal point sends out a strong beam directly forward from the headlight with little spread. This tends to be brilliant enough to dazzle an oncoming driver caught in the beam. Thus, a second filament is located above and slightly closer to the reflector than the focal point. This "low beam" sends out light in a spreading pattern mostly downward and sideways from the forward direction. For these uses concave mirrors must be corrected from a spherical shape for complete efficiency at their edges. Sir Isaac Newton demonstrated that perfect focusing of a mirror required that its cross section was not a circle but a parabola. (Alhazen of Basra had produced and used parabolic mirrors in the eleventh century.) With this shape the edges of the mirror are "tipped" more toward the focal point and edge distortions of the image do not occur. Newton constructed a parabolic mirror to make the first reflecting telescope. Today, all of the really large optical telescopes are *reflectors,* using parabolic mirrors, rather than *refractors* using large lenses. Large lenses weigh so much that they bend themselves out of shape when pointed to different parts of the sky. Unlike mirrors, the light must pass through the lens; thus the lens cannot be braced from behind to support its great weight in a large telescope.

Refraction

The general topic of refraction was treated in Chap. 9. As the incident ray of light strikes the boundary from one transparent medium to another, it generally undergoes a change in the direction of its travel. This change of direction, or bending of light, at a boundary is called *refraction.* As light moves from one material to another that is optically more dense—for example, from air into water—the ray is bent so that it plunges into the water at a steeper angle that it was traveling in the air. If we construct a perpendicular, or normal, to the surface at the point where the ray hits the surface we may define two angles. The angle in the first material between the incident ray and the normal is defined as the *angle of incidence,* θ_1. The angle in the second material between the refracted ray and the extension of the normal into that material is called the *angle of refraction,* θ_2 (Fig. 11-10). The larger of these two angles will be in the

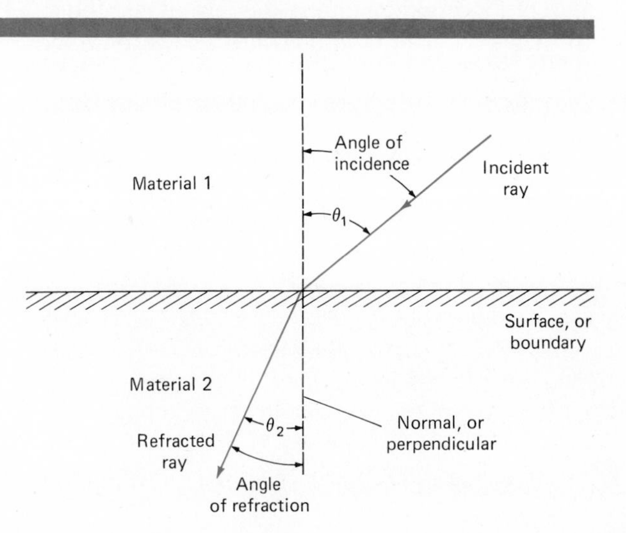

FIGURE 11-10 A definition of the terminology used in describing the refraction of light.

material which is optically less dense, and the smaller angle will be in the denser material. The relationship between these two angles is more subtle than most we have examined. Although the exact relationship had been sought for centuries, it was first stated by Willebrord Snell (1591–1626) in Holland in 1621. Snell's law of refraction is

$$\frac{\sin \theta_1}{\sin \theta_2} = n$$

where the subscript 1 refers to material 1, the subscript 2 refers to material 2, and n is a constant.

The *sine* of an angle (abbreviated sin) is the ratio of the length of the side opposite that angle to the hypotenuse (or side across from the right angle) in a right triangle. For example, many right triangles exist with one angle of 30°. Some of these are big, some are *small. What do all of them have in common? Answer–in every right triangle in which* one of the angles is 30°, the side across from that angle is just one-half as long as the side across from the right angle. If the longest side, or *hypotenuse,* is exactly 1 km long, the side across from the 30° angle is 0.50 km long. We summarize this by writing sin 30° = 0.50. While the size of any one side of a triangle depends upon how big the triangle is, the ratio of the lengths of the sides depends only on the angles involved. Thus, sin 30° = 0.50 is true for all triangles, big or small. We will not use the idea of the sine of an angle beyond its definition, but the idea is reasonable enough and of very great utility in both mathematics and science.

The n in the expression refers to a number which may be assigned to each boundary as a result of observation and measurement. It is called the *index of refraction,* and is a measure of what was referred to earlier as relative optical density. The index of refraction of a vacuum is set at exactly 1 unit. On this scale, air has an index of refraction of just about 1.00 (actually, 1.0050 at sea level). Pure water has an index of refraction of 1.33 relative to air. Thus, as light passes from air into water, the sine of the angle in air will always be $\frac{4}{3}$ as large (1.33 times as large) as the sine of the angle in water. Some of the implications of this are illustrated in Fig. 11-11, where several rays from a light source are shown entering

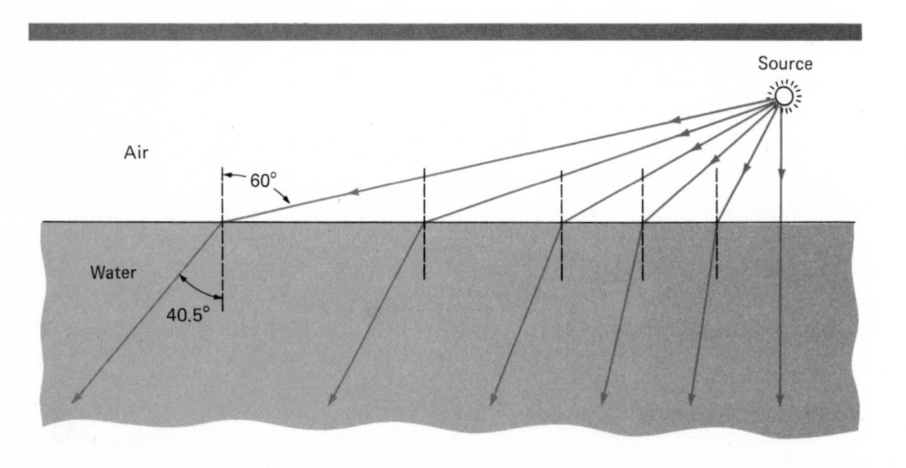

FIGURE 11-11 The refraction of various rays of light entering water from the air. The ray at the right strikes the surface perpendicularly and is not bent. All other rays undergo a change in direction at the surface of the water.

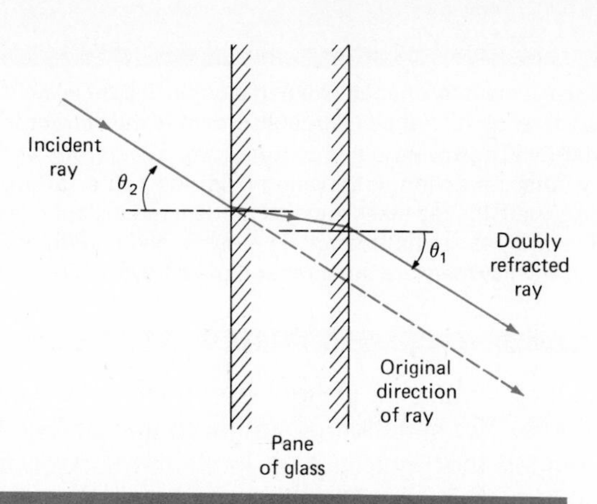

FIGURE 11-12
Light passing through a pane of glass is refracted at both the front surface and at the back surface of the glass. If the two surfaces are parallel, the light emerges traveling in the same direction as originally but slightly displaced to the side.

water at various angles of incidence. A light ray passing perpendicularly from the air to the water shows no refraction at all. This is always true of light rays striking a surface perpendicularly; they do not bend. For all other angles, the bending is apparent. In general, the greater the angle of incidence, the greater the angle of refraction. Of course, all of these light rays are reversible. That is, they could just as easily have started in the water and encountered the surface at the angle shown. In that case they would have been refracted as indicated and then moved in the direction toward the labeled source.

The bending of light passing through a boundary between two materials presents some interesting situations. A straight rod placed in water and viewed from the side will appear bent where it enters the water. A fish underwater scanning the air above the water for a flying meal must correct for refraction if it is to make a successful jump. Since the light bends at the surface, the prey will be farther away horizontally and lower than observed underwater. A similar situation occurs for a person spearfishing in shallows or from a boat. The refraction of light at the surface will always make the fish appear farther horizontally and nearer the surface than it really is. If the spear is thrown at the image of the fish, it will generally pass above the fish, as all aboriginal spearfishers have learned. To strike the fish it is necessary to aim nearer to your feet than the direction of the image.

One of the difficulties in measuring refraction is that light will bend at each boundary between different materials. Thus, if light is passed through a block of material, two refractions will occur before the light emerges into the air to be measured: one when it enters the block and another when it leaves the block. This is the reason why refraction is not apparent when we see an image through a well-made pane of glass (Fig. 11-12). Refraction at the front surface and back surface of the pane leave the ray traveling in its original direction, although slightly displaced sideways. To measure refraction, one standard strategy is to ensure the the light will strike one surface of the block perpendicularly so no refraction occurs at that surface. Any refraction that is observed thus must have occurred at the other surface of the block.

One way of doing this is to use a material shaped like half a circle. If light is incident on the exact center of the flat face of such a semicircular block, it will refract into the block. The center of the flat face, however, is the center of the circle from which the block is cut. Thus, the ray, after refraction, is traveling from the center of the circle toward its circumference. Any path the ray takes must be along a radius, but a radius of a circle is always perpendicular to its circumference. Therefore, no bending occurs as the ray leaves the block, and we may measure the direction of its travel while it is inside the block.

In Fig. 11-13 we see several rays of light approaching a water-air surface from a light source. The situation is similar to that of Fig. 11-11 except that we have placed the light source under the surface of the water, and the rays approach the surface in the water. As refraction occurs the ray in the air will bend away from the normal so that the angle of refraction in air is larger than the angle of incidence in the water. As the size of the angle of incidence increases, the refracted ray bends over to lie closer and closer to the water surface. At an angle of 48.5° the refracted ray runs exactly along the water surface. This angle is called the *critical angle* for water at a surface with air. If the angle of incidence is increased beyond the critical angle, the refracted ray would have to be bent around back under the water surface. This does not happen. Instead the underside of the water surface acts like a mirror, and the light is reflected back into the water from the surface by the law of reflection. This process is called *total internal reflection,* and it implies that if the light strikes the surface at too large an angle of incidence the light is trapped and cannot leave the water. Every time light passes from a material with a higher index of refraction to one with a lower index of refraction, there is a critical angle beyond which the light is trapped in the optically denser material. Reflection by this means is *total;* that is, none of the light escapes through the surface and all is reflected. For this reason, quality optical systems use glass prisms rather than mirrors to redirect the path of light where weight or thickness is no problem. Similarly, light can be "piped" along a fiber-optic bundle of glass cylinders which are thin enough to be as flexible as a cable. The light reflects back and forth along the walls of each cylinder, but always at an angle greater than the critical angle, and thus is confined to the glass until it reaches the end of the cylinders. Devices such as this are used in communications and in

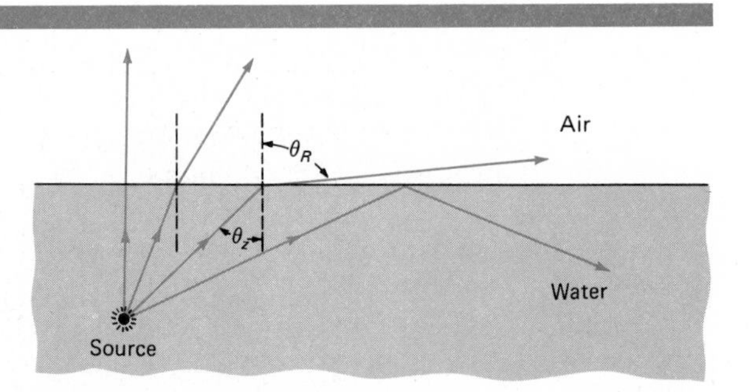

FIGURE 11-13 Light rays emerging from water into air bend so that the angle of refraction in air, θ_R, is larger than the angle of incidence in water, θ_I. At $\theta_I = 48.5°$ the refracted ray just skims the surface of the water, and beyond this angle reflection rather than refraction occurs.

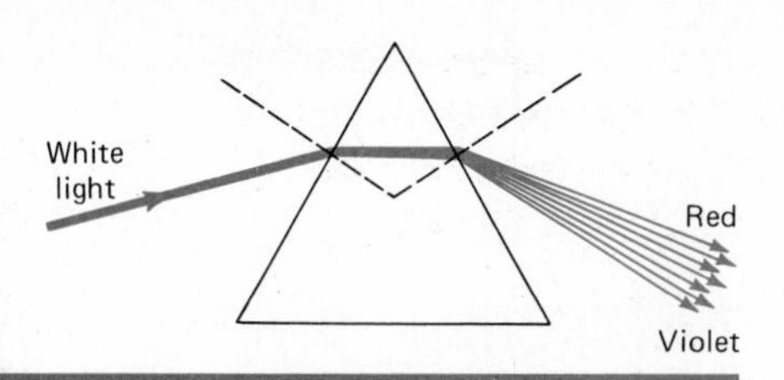

FIGURE 11-14 Each color of light bends through a slightly different angle during refraction in the phenomenon called dispersion. The result of using a prism, as shown, to reinforce this effect is a continuous smear of separated colors from white light called the spectrum.

medicine. For example, it is possible to examine the lining of the stomach for ulcers by having the patient swallow one end of a fiber-optic cable.

Since the bending of light at a surface depends upon the difference in the index of refraction of the two materials involved, diamond, with the largest index of refraction ($n = 2.4$), causes the sharpest bending with a given material of any substance. Thus, diamond has the smallest critical angle with air of any substance. Light must strike the surface of the diamond within 24° of the perpendicular in order to escape into the air. If not, the light is trapped inside the diamond. This effect combines with another to produce the sparkle of diamonds. The second effect is that as light is refracted at a surface each color is bent through a slightly different angle. This phenomenon is known as *dispersion.* Newton is noted for investigating dispersion by passing light through a triangular prism to form a *spectrum* of colors (Fig. 11-14). This phenomenon was not new with Newton, but he was the first to infer the correct conclusions. Once a particular color was isolated from the spectrum, no amount of reflection or refraction could make it yield white light again. White light was obtained only if the color was remixed with the other colors of the spectrum. From this Newton concluded that white light was a mixture of all of the colors of the spectrum, and our eyes perceive this mix as white. Earlier investigators had believed that colors were somehow produced on the spot from white light by the process of refraction, which is not the case. As a ray of white light enters a diamond, the ray is strongly refracted into the interior. Each color is bent through a slightly different angle by dispersion and moves along a separate path through the diamond. However, it is relatively difficult for light to emerge from a diamond since it must strike the surface nearly perpendicularly. Thus, the light bounces from face to face inside the diamond until it strikes at an appropriate angle to emerge. All of the time the light is bouncing back and forth inside the diamond, each color, which was refracted at a slightly different angle than any other color, is drawing apart or diverging from its neighboring colors of the spectrum. As a result, when the light finally manages to emerge it does so as flashes of pure colors separated from the other colors of the spectrum. Thus, as a diamond is rotated under white lights you observe flashes of red, sparkles of yellow, shafts of green, and so on, rather than the mixed white light that entered the diamond. This sparkle is both beautiful and fascinating. A brilliant-cut diamond has 58 separate faces cut an angles calculated to increase this effect. Because of its high index of refraction, a diamond produces this color separation and sparkle better than any other transparent substance. However, some synthetic crystals today come close to diamond in optical behavior if not in price.

Optical Uses of Refraction

Refraction is employed in many devices in everyday use, primarily for image formation. If we send a set of parallel rays of light through a flat pane of glass, the rays emerge parallel. If we curve the front surface of the glass, each of the parallel rays will encounter the surface at a different angle of incidence and will be refracted through a different angle. Such a piece of glass with a curved surface that produces differential refraction of light is called a *lens*. The commonest form of simple lens has one surface, or both surfaces, curved as a segment of a large sphere and is thicker at the center than at the edges. Any lens thicker in the center than at the edges is called a *convex* lens and will act to converge rays of light. The lens illustrated in Fig. 11-15 has both front and back surfaces curved to bulge outward at the center and is called a *double convex lens*. The central ray strikes both the front surface and the rear surface of the lens on the perpendicular and thus does not show bending. The remaining rays strike neither surface perpendicularly and are visibly refracted at each surface. All of the rays are converged to the focal point. The distance between the focal point and the optical center of the lens for parallel rays of light is the focal length f of the lens.

Similar to concave mirrors, convex lenses form both real and virtual images. If an object is placed closer to the lens than its focal length, an enlarged virtual image will be seen through the lens. The commonest use of this principle is in the magnifying glass (Fig. 11-16). Even though the double convex lens of a magnifying glass will act to converge light rays, rays from a point on an object near the lens are diverging too rapidly with distance to be physically brought together by the lens.

Consider the rays from the antenna of the insect in Fig. 11-16. After being refracted the rays will never cross; thus no real image will be formed. However, as these rays reach the eye, they come from a direction different from the true direction of the antenna from the eye. The rays also are spreading apart at a slower rate with distance, indicating that they have come from farther back than they really have. The information arriving at the eye is consistent with the antenna being at the position of the image antenna shown. Similarly, rays from the tip of the stinger ap-

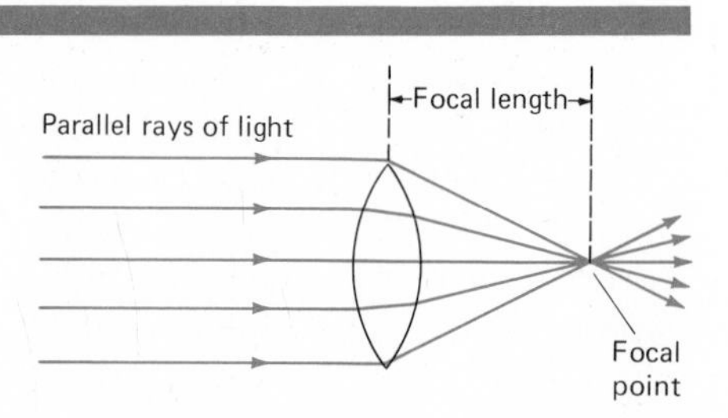

FIGURE 11-15 A convex lens bends parallel rays of light so that they converge to a single point called the focal point.

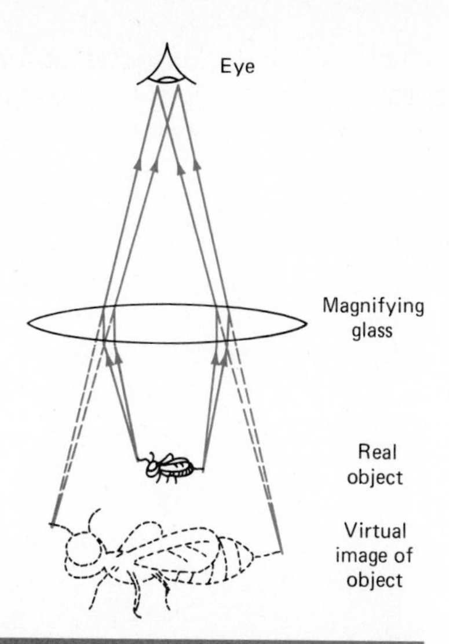

FIGURE 11-16 The magnifying glass produces an enlarged virtual image of a real object. Every point on the object sends light rays to all parts of the lens. Only rays from each end of the object which will enter the eye are shown.

pear to be coming from the image point of the stinger which is indicated. Each point on the object is sending out rays, some bundle of which may be refracted to the eye. The entire image appears farther away than the real insect, and much larger.

Other than eyeglasses, which always produce virtual images of the outside world for the eye, most lenses are used to produce real images. With cameras we want the lens to reconstruct the light into a real image on the film so it can provide a record. Likewise, projectors form real images on screens. To form a real image with a convex lens, such as a magnifying glass, the object must be placed farther from the lens than its focal length (Fig. 11-17). Measuring object distance from the lens, D_O, and image distance D_I, we may graph these variables. If we do this, we obtain the same graph and the same mathematical relationship for lenses as we found for spherical mirrors:

$$\frac{1}{D_O} + \frac{1}{D_I} = \frac{1}{f}$$

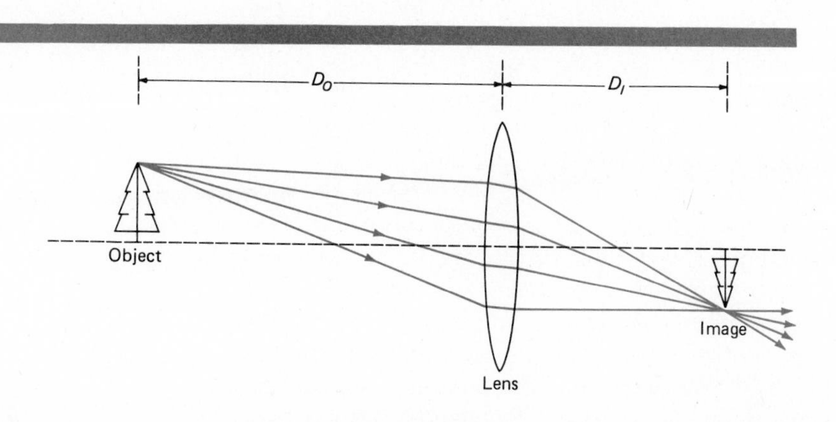

FIGURE 11-17 Convex lenses are often used to form real images, as in cameras, projectors, and in the eye.

This relationship allows us to predict the position of the image for any given position of the object. A camera, for example, allows us to vary the distance between the lens and the image on the film over a small range. This adjustment of D_I is necessary because the various objects which we may wish to take a picture of will be at various distances from the camera, thus causing D_O to vary.

The actual optics involved in a camera lens are much more complicated than this simple discussion implies. For example, one complication is caused by dispersion—the separation of colors in white light upon refraction. Since red light is refracted the least of any visible color, a lens will converge red least strongly. Thus, for a simple lens, an image in red light would be formed farther from the lens than an image in some other color. The focal length of a simple lens is slightly different for each color of light, with f for red light being the largest and f for violet light being the smallest. Thus, if a camera is focused to give a sharp image in red light, the violet light from the same object will be out of focus and blurred. This departure from a perfect image, or *aberration,* is called chromatic aberration.

Spectra

The dispersion of light does more than create problems for photographers. In this phenomenon several other bits of information about the behavior of nature were found. In 1800 the astronomer William Herschel separated a light beam of the sun with a prism as had Newton. Herschel's objective was to determine what portion of the spectrum was associated with the heat of the sun. By inserting a thermometer in various portions of the spectrum he could measure the heating caused by separate parts of the spectrum. He found that most of the heating effect of the sun was associated with rays bent even less than red light by the prism. These rays were invisible to the eye, but behaved like light in a laboratory. They could be refracted, reflected, and focused much like light, but they could not be seen. Since this invisible form of light was below red in the spectrum it was called *infrared* radiation. Independently, other invisible rays were discovered in sunlight which were bent even more than violet light was. These were called *ultraviolet* radiation. It slowly became clear that human eyesight was limited to a small band of radiations, which we call *visible light,* out of a much larger spectral range of radiations. What was not appreciated for some time was how tiny a portion of the range of similar radiations visible light actually represented.

At about the time that radiations beyond the range of visible light were being discovered, another curious phenomenon was noticed in the dispersed spectrum of sunlight. Using a thin beam of sunlight and examining the projected spectrum carefully, scientists noted that some colors were missing. Instead of a bright, continuous smear of light with one color blending smoothly into the next, dark lines were observed at certain portions of the spectrum where a particular shade of red, for example, might be missing. Actually, there are thousands of such dark

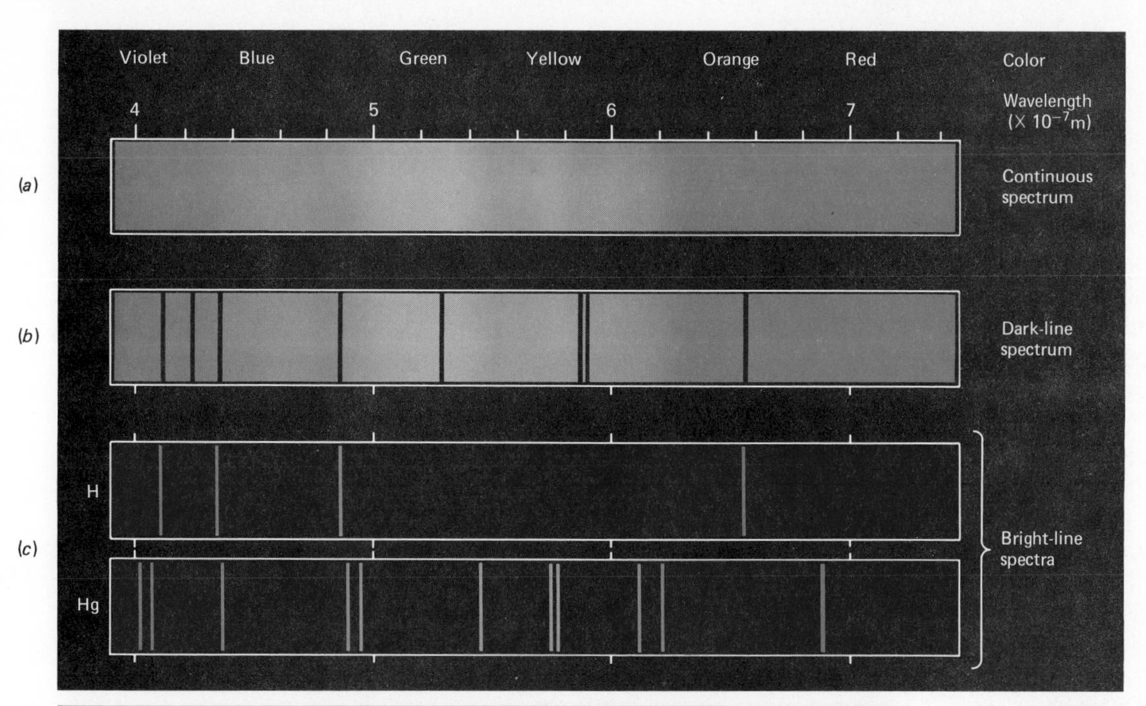

FIGURE 11-18
Luminous materials produce three types of spectrum. (*a*) In the continuous spectrum, all colors of light are present. (*b*) The dark-line spectrum resembles the continuous spectrum with certain specific colors missing. (*c*) The bright-line spectrum represents only certain, specific colors. The spectra illustrated are for hydrogen (H) and mercury (Hg).

lines in the sun's spectrum, but it requires good instruments and very careful experimental work to detect them. This led to the study of the light given off by a wide variety of light sources. Eventually, spectra were determined to be of three types (Fig. 11-18):

1 *Continuous spectra,* representing all colors of visible light, are given off by solids, liquids, and high-pressure gases, which when heated radiate light owing to their high temperature (*incandescence*). A good example is the white-hot tungsten wire in an ordinary light bulb.

2 *Bright-line spectra* (or *emission spectra*) occur when a low-pressure gas is excited, either electrically or by heating. In this case we obtain only certain very specific colors of light instead of all colors. For example, mercury-vapor streetlights cause a great deal of light, but persons or objects standing under them have an odd appearance. Lips look dark blue, and certain colors of clothing appear black, or blue, but not the same as under white light. This is because the lamp does not give off all colors of light. It emits the bright-line spectrum of mercury, which consists of several widely separated colors ranging from an orange through two shades of violet. If an object does not reflect any of these colors, it will appear black. Other examples of bright-line spectra are in the light given off by neon-type advertising signs (only the red-orange signs of this type use neon gas), and the yellow sodium-vapor lamps used to light highways, particularly in high-fog areas.

3 *Dark-line spectra,* (or *absorption spectra*), such as the spectrum of

the sun, represent all colors of light but with certain colors missing. Actually, thousands of "missing colors" may appear as dark lines in the detailed spectrum of a star. These lines were studied extensively by Fraunhofer in 1814 and are often called *Fraunhofer lines* when referring to the sun's spectrum. They represent colors "subtracted out of" the continuous light of a source. For example, sodium vapor, if excited, glows with two very close shades of yellow dominating its glow. If white light is passed through a chamber of cool sodium vapor, the sodium will absorb light of exactly these colors, causing them to appear less bright in the white light emerging from the chamber. Such dark lines in the sun's spectrum indicate light absorbed from the continuous glow of the sun's surface by the halo of low-density gases surrounding the sun.

Of the three types of spectrum, the bright-line spectra were the most curious. Why should a material emit only certain colors of light? Furthermore, the patterns of light emitted were found to be an absolute "fingerprint" of the type of matter emitting it. All mercury samples emit one pattern of colors, and no other type of matter emits this pattern. The dark lines in the sun's spectrum corresponding to the sodium pattern of bright lines was evidence that the sun's outer atmosphere contained sodium, since no other material emits or absorbs that pattern of lines. As experience with spectra grew, the pattern of light given off by a vaporized sample of matter became the most precise means of identifying the types of matter present until well into the twentieth century. Thus *spectral analysis* became an important technique for investigating materials. Even though spectral analysis became one of the most precise fields of experimental science, for over a century there was no idea as to why matter should behave this way.

Imagine an analogy. You are given the problem of identifying the nationality of a person in the next room. You may not talk to or even see the person. You enter the dark room and grope around on the floor until you find feet. At that point you take a large wooden mallet and hit the toes sharply. You then emerge from the room triumphantly and announce that the person is French. When asked how you know, you announce, "Simple; the cry of pain was shrieked on the note of high C. Now the English will shriek in B or in A over high C; Italians in F sharp or one of six other notes; Navahos in middle C or E; but only the French will shriek in high C!" Bright-line spectra exhibited even stranger properties; at high pressure even mercury or sodium will provide a continuous spectrum. To extend the analogy, imagine the Vienna Boy's Choir that can sing any song when squeezed together, but separately each boy has a range of only the same three notes and no others. If spectra, especially bright-line spectra, now seem a little strange and intriguing to you, imagine what they represented to workers in the field.

The first hint of a pattern in spectra did not emerge until 1885, but for all their mystery, spectra were useful. Aristotle had asserted that the heavens were made of special materials unlike those of the earth. Spectra indicated that the sun had sodium, magnesium, calcium, iron, oxygen, hydrogen, and other materials identical to the materials of the earth in their optical behavior (Fig. 11-19). One material that was

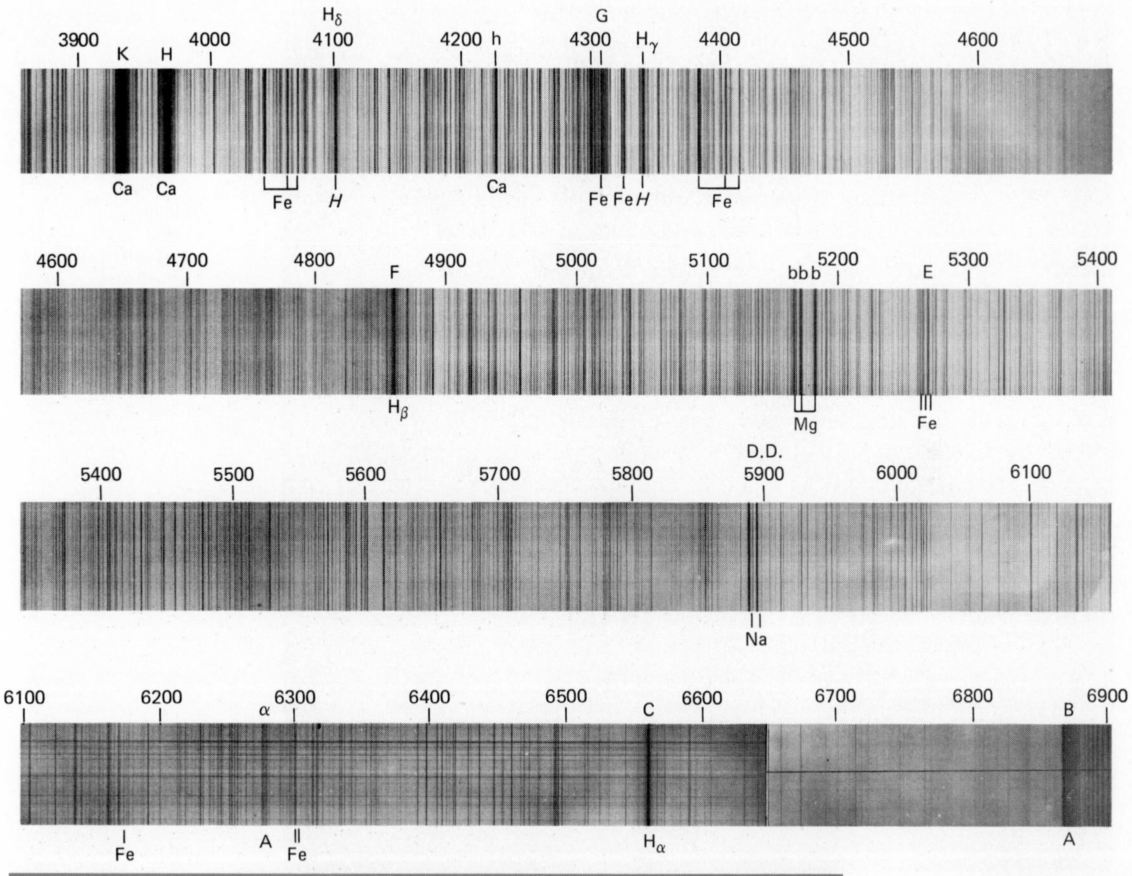

FIGURE 11-19
The spectrum of the sun showing dark lines representing "missing" colors (Fraunhofer lines). While the lines appear very dark, actually they are just greatly reduced in brightness. The four strips shown join to cover the same range of wavelengths illustrated in Fig. 11-18. Numbers are in 10^{-10} m. Letters below each strip identify the elements associated with several prominent lines. *(Photograph courtesy of the Hale Observatories.)*

unknown on earth was represented in the dark lines of the solar spectrum; it was called *helium,* after Helios, Greek god of the sun. Helium was discovered in 1895, and even little children know about it now because of its use in balloons. Thus spectra indicated that the sun is made of the same types of material found on earth, as are the stars which also send us light for analysis. However, the interpretation of what was happening in matter to produce spectra had to wait for many other discoveries.

The Nature of Light

Thus far, we have discussed three effects (reflection, refraction, and dispersion), which could be successfully accounted for by individuals believing light to be a wave and by other individuals believing light to be a particle. We have also discussed spectra, which in the nineteenth century could not be explained by any theory of the nature of light and its production. The wave-vs.-particle argument concerning light reached an early peak toward the end of the seventeenth century with Sir Isaac Newton and Christian Huygens, the leading advocates of the opposing

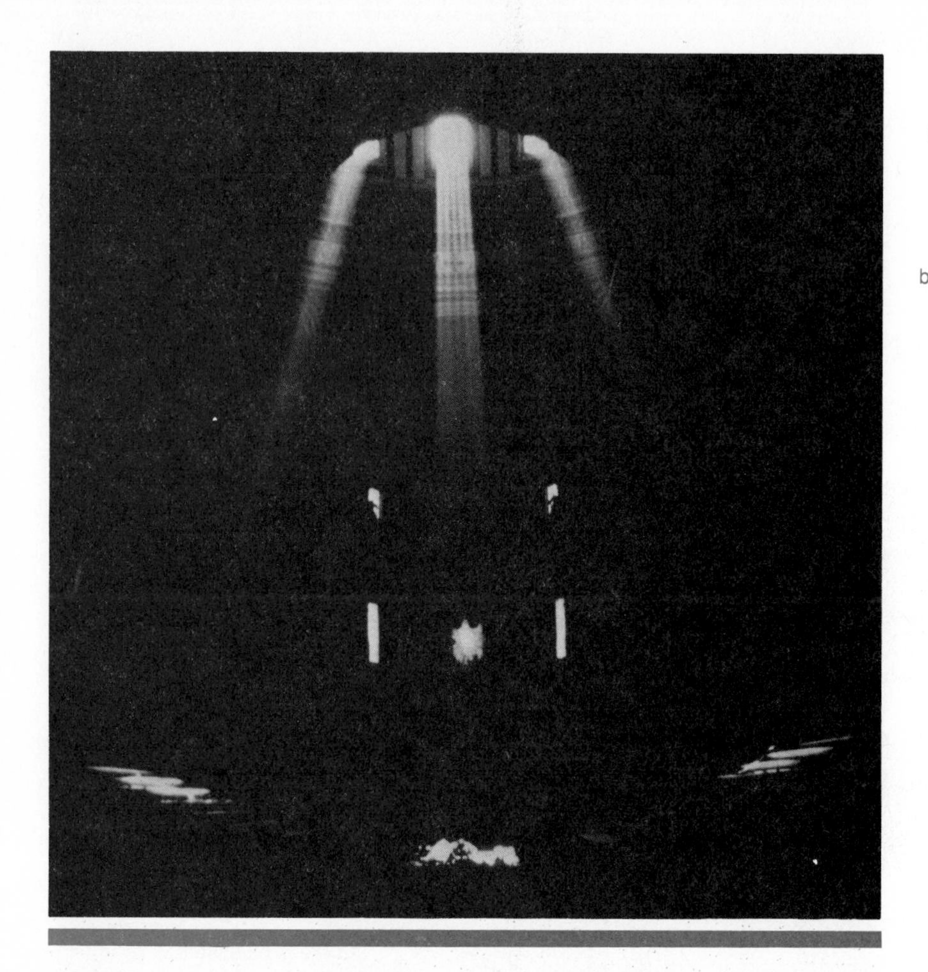

FIGURE 11-20 The everyday observation of light suggests that it travels in straight lines, as seen here in St. Peters Basilica in Rome. If the sun is very far away, shouldn't the beams be parallel? Why do they appear to diverge? (*Photograph by the author.*)

views. Newton leaned toward a particle, or "corpuscular," nature for light for many reasons. For example, he argued that if light was a wave, it should bend around the edges of objects and fill in the zone of shadow. Huygens had countered that by showing that if the wavelength of light were very small, the bending at edges would be extremely minor (Fig. 11-20). Nevertheless, the debate went back and forth between the two schools of thought, and both lacked a critical experiment to demonstrate their case convincingly. During the eighteenth century the corpuscular view became the majority opinion in science, but the wave theory retained some advocates.

How do we decide between rival theories, both of which can adequately handle the known facts? How did the civilization resolve the difference in the views of Ptolemy and Copernicus? In that case it required better measurement to see which set of ideas predicted the facts more closely. There was one area in which a wave theory and a particle theory made quite different predictions. This concerned the speed of light in materials other than air. Both wave and particle theories can account for refraction but on very different grounds. Well before Newton, Descartes had pointed out that the refraction of light entering glass from air, for example, could be explained if a force acted at the surface of the glass pulling the particles into the new medium. This force would act perpendicular to the surface of the glass and have the effect of speeding up the light so that it would travel more rapidly in the glass than in air. As the light left the glass, this force would tend to pull the particles back toward the glass, and thus slow them down as they reentered the air. The main implication of this was that if light was composed of particles, the speed of light should vary with the index of refraction. In glass with a n of 1.5, light should travel 1.5 times as fast as in air. This is certainly a definite prediction for experimental determination. Furthermore, the prediction of a wave theory is exactly the opposite. We have seen that the refraction of waves depends on a change in the speed of waves. But, to bend toward the normal in entering a new medium, as light does as it enters glass from air, this new medium must have waves move at a slower speed. In fact, wave theory predicts that the ratio of the speed in air to the speed in glass is the index of refraction. Thus, if we have a glass with $n = 1.5$, we would expect the light to travel 1/1.5 or $\frac{2}{3}$ as fast in the glass. Now, these predictions are very different. The speed of light in air (vacuum) is about 300,000 km/s. Water has an index of refraction of 1.33. Wave theory thus predicts the speed of light in water to be 225,000 km/s, or $\frac{3}{4}$ of its speed in air. Particle theory, meanwhile, predicts the speed of light in water to be 400,000 km/s, or $\frac{4}{3}$ its speed in air. The difference is very large, and the experimental result would certainly lead us to reject one, or both, theories. Unfortunately, by 1800 the means of measuring the speed of light in a laboratory were not yet available. Direct measurement had to wait another half-century.

The beginning of the nineteenth century did see a shift in view toward the wave theory of light. Several wave effects simply are not duplicated by the actions of particles. In particular, two waves may add together at a single position to produce no disturbance in the phenomenon called *interference.* In 1801, Thomas Young (1773–1829) demonstrated that two beams of light may be added together to produce darkness! The largest experimental problem for Young was to obtain two sources of light which were giving off the same light at the same time. (In wave terminology this is called having two sources *in phase.*) For example, it would be difficult to show wave effects with sound using two speakers if one were playing Beethoven's Fifth while the other were playing the Peanut, Butter, and Jelly rock. There would be no agreement as to what kinds of waves were being given off and when. There would be nodes, or points of total cancellation of sound, but these would be very limited in time and in position. No fixed pattern of destructive interference could be established. Young, working with candles, had a similar problem. Two candles would not give off light in phase. We may appreciate this problem today by picturing the candle as made of atoms, which are the radiators of the light. As the atoms gain heat, they move faster and collide. In collisions as well as in chemical reaction, bursts of energy are radiated as light. Billions of atoms in a seething mass of activity are radiating light individually—this one a burst of red, that one yellow, and so on. There is no way to have agreement in wavelengths and timing between two candles any more than there is between the two speakers above. What Young did was to split up the light from a single candle so that it acted as two separate light sources. He did this by using a set of fine slits.

The behavior of light with a slit is interesting. As a slit is made smaller and smaller, the light passing through is constricted to a finer and finer beam—*up to a point!* After a certain slit width is reached, any further narrowing of the slit results in a widening of the beam. This is caused by a phenomenon known as *diffraction.* A narrow slit, therefore, instead of passing a fine ray of light, emits an expanding arc of light which will spread sideways to cover considerable area. Young's experiments used a single candle illuminating both slits. Whatever light was arriving at one slit also would be arriving at the other slit at the same time. Thus, the two slits acted as if they were two sources vibrating together, or in phase. Beyond the slits the light would spread out so that the light from one slit would cross over the light from the other slit. The pattern observed on the screen consisted of a bright central zone directly opposite the midpoint between the slits. From the bright central zone outward in either direction were alternating zones of darkness and brightness representing zones of destructive and constructive interference. This interference pattern could be explained in terms of the different distances light must travel from each of the slits to a particular position on the pattern.

Figure 11-21 shows the apparatus with sets of concentric circular segments representing waves moving from two slits. If each arc represents an outward-moving wave crest, the intersection of two arcs would represent positions where a crest from slit *A* and a crest from slit *B* are superimposed. This would represent constructive interference, or, for light, a position of brightness. Similarly, where the arc from one source falls midway between two adjacent arcs from the other source, we would find a crest falling on a trough for destructive interference. This pattern has an entire series of zones where at each moment in time the contributions of the two slits will cancel each other to produce no disturbance. Such *nodal lines* are indicated by dashed lines in Fig. 11-21. Whether a particular spot on the screen will be bright or dark depends upon its relative distance to the two slits. We have stated that the two slits are in phase. This means that at the exact instant a crest leaves one slit, a crest will also leave the other slit. If the respective distances from the two slits to a point on the screen are the same, these two crests must arrive at that point at the same time since they travel at identical speeds. Thus, point *C*, which is equidistant from the two slits, will always be a point of constructive interference or brightness.

As we proceed downward along the screen in Fig. 11-21, we eventually arrive at a position of destructive interference, *P*. The distance from the upper slit, *A*, to *P* is just one-half wavelength farther than the distance from *B* to *P*. Thus, a crest leaving *B* arrives at *P* one-half of a wave before the crest that left *A* at the same time. The delayed crests from *A* thus fall

Light source
Double slit
Alternating bright and dark zones
Screen

(a)

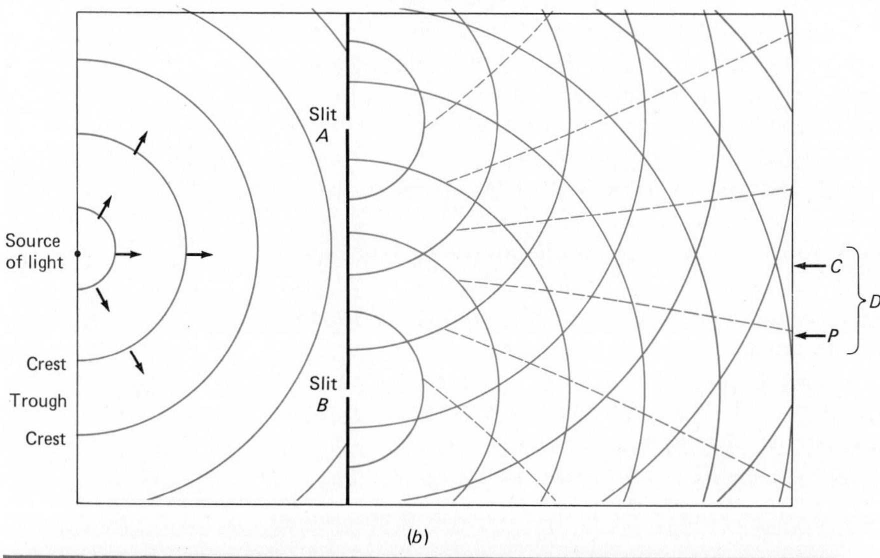

(b)

FIGURE 11-21 In Young's experiment, each line slit acted as a source of light. Wave fronts bending from these slits crossed to generate an interference pattern of bright and dark zones.

on the troughs from slit B, producing cancellation, and P is a node, or position of darkness. Mathematically, the following relationship can be shown to be approximately correct:

$$\frac{\text{Difference in path lengths } (\tfrac{1}{2}\lambda)}{\text{Distance between slits } (d)} = \frac{\text{distance of node from central spot } (D)}{\text{distance of screen from slits } (L)}$$

This implies that the wavelength of the light used is

$$\lambda = \frac{2dD}{L}$$

From this Young could obtain a measurement of the wavelength of light from three relatively easily measured distances. Red light has a wavelength of about 7×10^{-7} m, while violet light has a wavelength about half as large, or 3.5×10^{-7} m, and this constitutes the entire range of wavelengths visible to the human eye. (Another way of looking at these numbers is to say that waves of red light, with the largest wavelength, would require about 10,000 waves to reach from one line to the next on a sheet of notebook paper, while twice as many waves of violet light would be required to reach the same distance.)

Interference is a pure wave effect. It may be predicted from a knowledge of wave properties and it occurs exactly as predicted. On the other hand, there is no particle interpretation of interference. Two bowling balls simply will not pass through each other at a point, disappearing totally as they cross. Yet even though the experimental evidence was compelling, there was not a stampede to embrace the wave theory of light among the scientists of the day. Resistance to the wave theory collapsed slowly, as it usually does in such cases. (Several individuals from the time of Galileo onward have speculated that a fundamental change in a basic viewpoint of science requires that a new generation of scientists grow up with knowledge of both sides of the story and not just the bias of a single view.) Some individuals came to accept the wave theory when Fresnel (1799–1827) demonstrated the interference of light without using diffraction to produce crossing beams from the same source in 1815.

Young also correctly interpreted another effect of light in terms of waves—the phenomenon of *polarization.* This is an effect that has been observed since the sixteenth century with certain crystals. It can best be described today with the aid of a pair of old Polaroid sunglasses. If you look at the lenses of the sunglasses, you will note that they reduce the intensity of the light passing through them. That is not unusual since all sunglasses are designed to do exactly that. However, if you look at some surface with a bright glaring reflection through the Polaroid glasses, you will note that it strongly reduces, or even eliminates, the reflected glare. While watching the area of the glare, if you rotate the sunglasses to various angles you will find that at some angles the glare comes through the lenses and at others it is stopped. With the two lenses of the sunglasses separate, you may investigate this dependence on angle for the transmission of light. Placing one lens behind the other in the normal, upright position, the second lens seems to have little additional effect. Holding the first lens steady and rotating the second around the line of sight, you will notice the amount of light passing through both lenses begin to drop. When the second lens has been rotated through 90°, almost all of the light is stopped. In this position we say that the polarizers are *crossed.* Two polarizers thus act like a type of valve for light, allowing more or less through depending on their orientation. Light reflecting from a surface is polarized so that a single polarizing screen can allow more or less through depending on the screen's orientation.

Young's interpretation of this phenomenon was quite simple but was very difficult to arrive at. He concluded that light was a transverse wave and that a polarizing screen allowed waves to pass only if they were vibrating in one plane. Picture an analogy. A rope is tied to a tree and you may shake transverse waves into the free end. The waves you shake into the rope can vibrate up and down, or sideways, or at any angle between

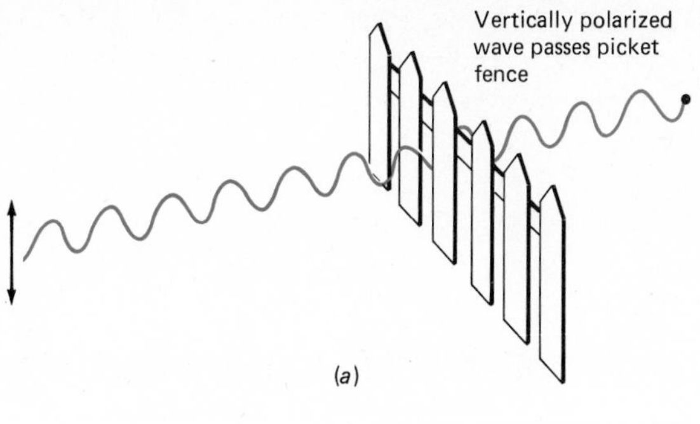
Vertically polarized
wave passes picket
fence
(a)

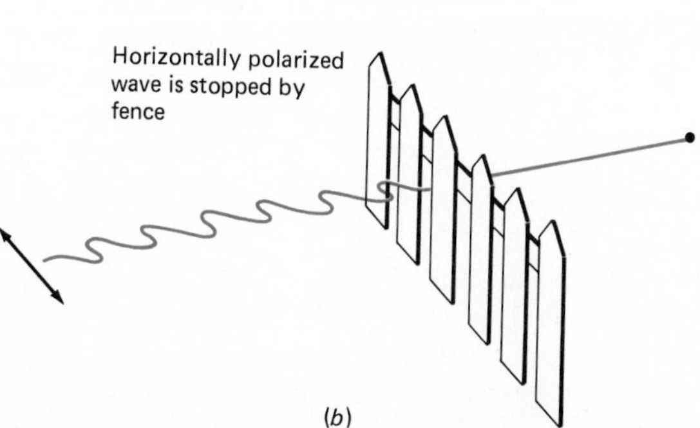
Horizontally polarized
wave is stopped by
fence
(b)

FIGURE 11-22 The polarization of transverse waves selects a plane of oscillation of the wave. (*a*) The picket fence allows the up-down wiggles of a vertically polarized wave to pass through it. (*b*) The slats of the picket fense stop the sideways wiggles of a horizontally polarized wave on the clothesline.

these extremes. If the rope is now threaded between the slats of a picket fence you may still shake in waves vibrating in any plane perpendicular to the direction of travel. However, only the up-and-down vibrations will be able to move through the fence as the rope moves up and down between the slats. Sideways motion of the rope is stopped by the fence. A polarizer of light allows through only that light which is vibrating in one plane (Fig. 11-22). Vibrations not in that plane are not transmitted but are absorbed. When two polarizers are *aligned,* any light passing the first polarizer is vibrating in the plane also passed by the second polarizer, so it seems to have little effect (Fig. 11-22*a*). When the polarizers are crossed, however, all the light passing through the first polarizer is vibrating in exactly the plane stopped by the second polarizer, and none is transmitted (Fig. 11-22*b*). Polarization is an effect observed only with *transverse waves.* (The effect has nothing to do with poles, magnetic or electric. The name comes from some speculations of Newton as he was considering the effect in terms of particles of light.) It was difficult for Young to come up with the explanation for polarization because from the very first speculations about light waves, everyone had thought that any waves of light would have to be longitudinal waves. After all, only longitudinal waves were known to pass through liquids and gases. Transverse waves require materials rigidly hooked together so that a sideways wiggle can move along the medium. The fact that light is demonstrated to be a transverse wave by polarization would seem to imply that the medium carrying the waves must be a solid! The acceptance of a wave theory for light had enough problems without requiring this. Yet, the implication was that whatever was carrying light was a solid, more rigidly

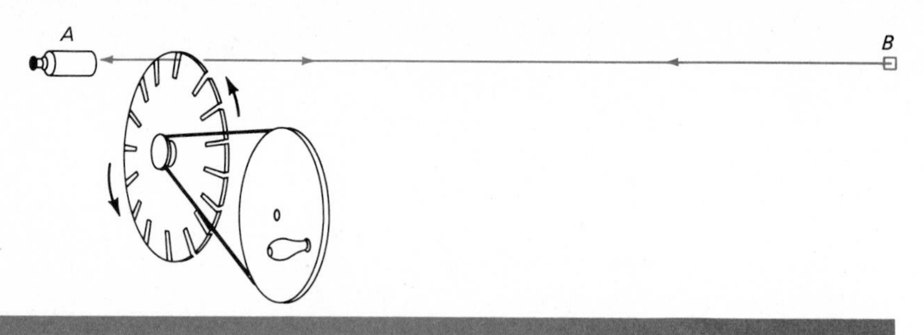

FIGURE 11-23 Fizeau produced the first laboratory method for determining the speed of light by the use of a rapidly rotating toothed wheel. As a notch in the wheel passed the light source and telescope at *A*, a burst of light emerged to move toward the distant mirror *B*. When this pulse reflected at *B* and returned, it was blocked from entering the telescope by the rotation of a tooth between notches in the wheel. Only if the wheel were rotated fast enough to the correct speed would another notch be moved to the precise position to allow the burst of light to be seen in the telescope.

bound together than diamond, filling all space and having no mass! It is actually not difficult to see why there was considerable resistance to a wave theory of light.

Resistance to a wave theory collapsed in 1850 after Fizeau (1819–1896) developed a method to measure the speed of light in the laboratory (Fig. 11-23). His coworker Foucault (1819–1868), using improved apparatus, demonstrated that light travels much slower in water than it does in air. This confirmed Huygens' prediction of light as waves and rejected Descartes' prediction of light as particles. The wave theory of light won wide acceptance even though it still presented serious problems.

Electromagnetic Waves

Just as the wave nature of light was gaining widespread acceptance, a new contribution came from a different field of investigation in 1864. This was the first publication of James Clerk Maxwell (1831–1879) on the electromagnetic theory of light (his major work was published 8 years later). To summarize this theory we must pick up the thread of the work on electricity and magnetism from the last chapter. Maxwell was an extremely gifted Scottish aristocrat who had taken on himself the task of translating Faraday's idea of a "field" into solid mathematical form. Maxwell tried to rewrite the accumulated knowledge about electrical and magnetic effects into general mathematical statements. He eventually succeeded in compressing this knowledge into four statements (four equations). To have one equation make complete sense physically, he was obliged to add a term to the experimentally derived form of another equation. When he was done, his four equations could fit nicely on the back of a postage stamp and be clearly legible, yet they held the second great theoretical revolution of physics. Faraday had shown that moving a magnet past a conductor would generate a current; that is, a changing magnetic field would make an electric field. Maxwell's modification asserted that a changing electric field could produce a magnetic field. Having completed an extremely difficult mathematical task, Maxwell checked his equations for their consistency. Almost immediately, two of the equations when solved simultaneously led to a very interesting mathematical result. They described a phenomenon moving from place to place in time—a wave! While electric fields are generated by charges and magnetic fields are generated by moving charges, this wave was not tied to charge once produced, but moved at constant speed away from its origin. The linkage between electric and magnetic fields tied them together and allowed them to "break off" and travel through space.

Picture the effect this way: Imagine a charge at a point. The charge produces an electric field in the space around it. If we could shrink our charge down to nothingness, the electric field would shrink with it and eventually disappear. Now imagine that instead of shrinking the charge, we very rapidly ripped the charge away and out from our field of view. The electric field would try to collapse to zero. But, going from some value down to nothing is changing, and a changing electric field produces a magnetic field. Once produced, our magnetic field tries to shrink to zero also, but a changing magnetic field produces an electric field. We are back to square one! The isolated electric and magnetic fields are doomed to move off into space shuffling energy back and forth like two children playing catch with a hot potato. The fields will be locked, symmetrically exchanging energy until they encounter a charge to deliver the energy back to (and be absorbed). This type of wave is called an *electromagnetic wave.*

The breakaway of an electromagnetic wave from charge can best be illustrated in the case of a *dipole,* or two charges of opposite sign. Figure 11-24*a* shows two such charges and the electric field between them. For simplicity, only the electric field on one side of the two charges is shown (the same thing is happening in all directions around the line of the charges). The charges are now sharply accelerated toward each other (Fig. 11-24*b*). Portions of the field near the charges react rapidly to the changing position, but portions farther out collapse with some delay and remain substantially unchanged. By the third position (Fig. 11-24*c*) the charges are almost passing each other and are moving at great speed. At the instant the charges are on top of each other, the field should be zero, since there is then no isolated charge and the effects should cancel. However, the electric field is not collapsing as rapidly as the charges are approaching each other, and a "leftover," or residual, field will be present in space at the moment the charges are together. Figure 11-24*d* shows the charges as having overshot the central position, and they are now moving apart. The positive charge is now below the negative charge on the paper, and a new electric field is being generated pointing in the opposite direction from the original field. The action of the

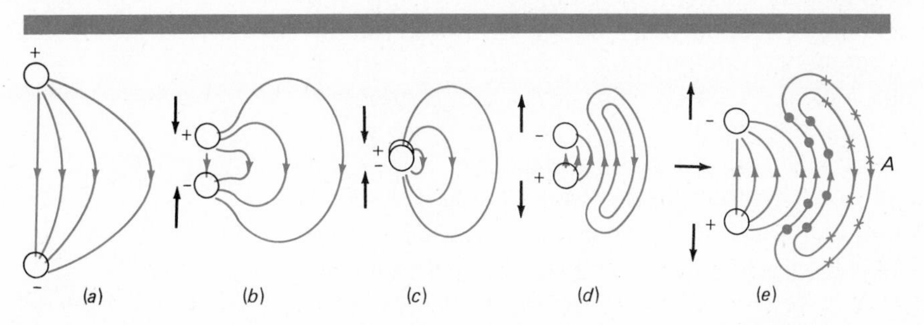

FIGURE 11-24 Stages in the production of an electromagnetic wave by two charges accelerating past each other.

charges crossing by each other had "snipped loose" the loop of the old field which was collapsing. This loop now moves out as an electromagnetic wave. Figure 11-24*e* shows only the detached electromagnetic wave but indicates the direction of the magnetic field associated with the electric field shown. Crosses indicate the magnetic field is pointed into the page, and solid circles indicate that the magnetic field comes perpendicularly out of the page. The wave is moving off to the right at constant speed.

Imagine yourself an observer at the point labeled *A*. As the first part of the wave passes, the electric field is pointed downward, indicating that a positive electric charge in your hand would experience a downward force. As the back portion of the wave passes, the same charge would experience a force that was upward. This upward-and-downward motion of the charge in your hand is at right angles to the direction of the wave's motion past you. Thus, you would perceive it to be a transverse wave from its effect on your charged particle. This closed loop, free from the oscillating charges which were its origin, will move outward. While preserving this form, the electric and magnetic fields will actually oscillate back and forth in their directions (as indicated by the arrows or the crosses and dots) as they exchange energy back and forth.

While the illustration of two oscillating charges produces the easiest situation to visualize the production of an electromagnetic wave, it is not the only way they are produced. The implication of Maxwell's equations is that *electromagnetic waves are produced whenever charges are accelerated.* The greater the acceleration, the tighter the closed loop of the electromagnetic wave produced or, in other words, the higher the frequency of the wave.

Two constants in the wave equation Maxwell obtained usually represent the elasticity and density of the medium. These yield the speed of the wave. In this case, they were the fundamental electric and magnetic constants of empty space. Maxwell used these to calculate the speed of the predicted electromagnetic waves and obtained about 300,000 km/s (later measurements and calculations by this method yielded 299,781 km/s). Fizeau's measurement of the speed of light in 1849 had been 313,300 km/s, and Foucault's later improvement of the method had yielded 298,000 km/s. *Maxwell concluded that light was an electromagnetic wave.* Actually, light represents only a tiny part of the range of electromagnetic radiations that are known today (Fig. 11-25).

Maxwell's work was intensely mathematical, highly theoretical, and largely unappreciated. He died of cancer at the age of 48 in 1870 without having seen a single experimental confirmation of his greatest theoretical work. Ten years after Maxwell's death Hertz was able to generate electromagnetic waves in a spark gap and coil and have them detected by a completely separate spark gap placed at a distance. This is the principle by which radio and TV broadcasting works today, but Hertz was too busy investigating the behavior of nature to be interested in the commercial possibilities of his invention. The century after Maxwell saw

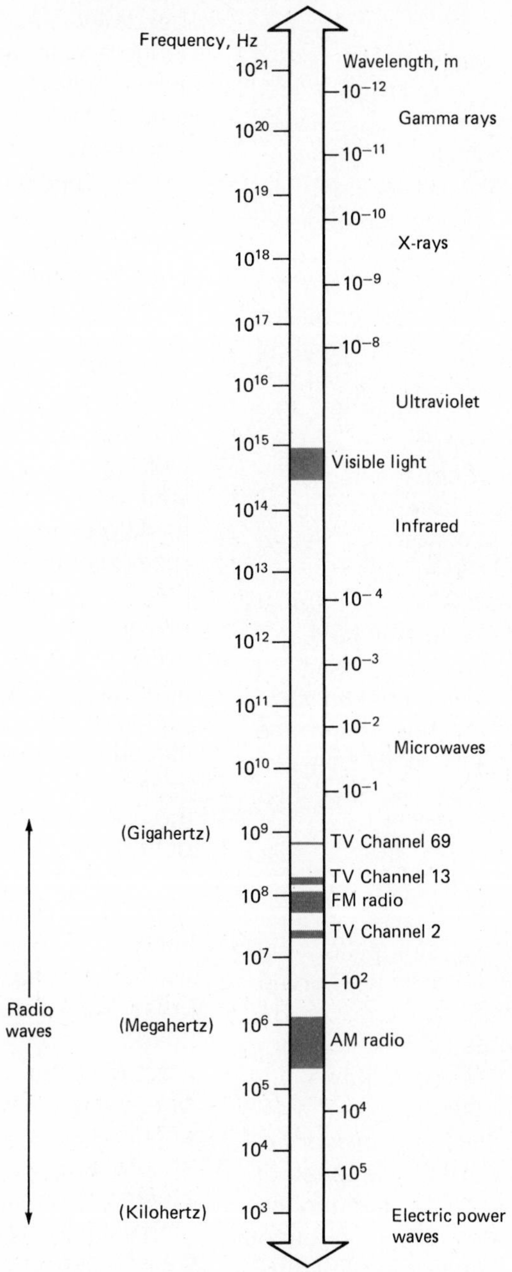

Frequency, Hz

Wavelength, m

10^{21}

10^{-12}

Gamma rays

10^{20}

10^{-11}

10^{19}

10^{-10}

X-rays

10^{18}

10^{-9}

10^{17}

10^{-8}

10^{16}

Ultraviolet

10^{15}

Visible light

10^{14}

Infrared

10^{13}

10^{-4}

10^{12}

10^{-3}

10^{11}

10^{-2}

Microwaves

10^{10}

10^{-1}

(Gigahertz) 10^{9}

TV Channel 69

TV Channel 13

10^{8}

FM radio

TV Channel 2

10^{7}

10^{2}

Radio waves

(Megahertz) 10^{6}

AM radio

10^{5}

10^{4}

10^{4}

10^{5}

(Kilohertz) 10^{3}

Electric power waves

FIGURE 11-25 The electromagnetic spectrum includes all wavelengths which are currently measurable. Notice that visible light makes up a tiny portion of the range.

the expansion of our knowledge from the original forms—visible light, infrared, and ultraviolet light—to longer waves (microwaves, radar, TV, radio) and shorter waves (x-rays, gamma rays). All of these forms are similar except for their varying wavelengths and frequencies.

Summary

Light is given off in all directions by luminous objects. Nonluminous objects become visible as they scatter, reflect, or transmit with distortion the light falling on them. Mirrors reflect light in the same manner waves reflect from a smooth surface; namely, the *angle of incidence* equals the *angle of reflection,* where both angles are measured from the perpendicular (normal) to the surface at the point where the ray strikes. Parallel rays of light striking a mirror thus remain parallel after reflection. Curved mirrors alter this and can either converge or diverge the rays upon reflection. The distance from the mirror at which parallel rays of light are converged or their extensions appear to converge is called the *focal length* of the mirror. All mirrors have the ability to create images of the original source of light. Such an image may be *real* if the light is in fact reconstructed in space at the position of the image. If the light only appears to come from the image position, the image is *virtual.* The relationship between image distance, object distance, and their focal length is given by

$$\frac{1}{D_I} + \frac{1}{D_O} = \frac{1}{f}$$

The identical relationship holds for the images formed by lenses. Lenses are curved transparent materials that converge or diverge light by refraction as studied in Chap. 10. The ability of materials to bend light at the surface between them depends upon a constant for each material called the *index of refraction.* This number is actually the speed of light in a vacuum divided by its speed in the material. Different colors of light bend through slightly different angles in refraction in a phenomenon known as *dispersion.* Red bends the least and violet the most upon entering a material with a higher index of refraction. Study of this effect convinced Newton that white light was a mixture of all colors, which could be separated by a glass prism into the *spectrum.* This separation provided a means to analyze the light given off by various sources. It led to the discovery of rays bent less than visible light (called *infrared*) and other rays bent more (*ultraviolet*). Some sources of light do not produce all visible colors, or a continuous spectrum. Sources such as hot, low-density gases emit only certain very specific colors (*wavelengths*) and nothing else in *bright-line* spectra. Other sources, such as stars, have a spectrum with all colors except for certain "missing lines" or greatly reduced wavelengths. Young showed that light demonstrates the wave property of *interference.* This also gave him the ability to measure the wavelength range of visible light (about 3.5 to 7×10^{-7} m). Young interpreted light to be a transverse wave from the phenomenon of *polarization.* Single polarizers will let a fixed amount of light pass through. A second polarizer placed in the path of the light will either transmit the light from the first polarizer or stop it completely, depending upon its orientation. These observations posed several problems. If light was a wave, what was wiggling? Furthermore, transverse waves had been observed only in solids. Maxwell, working on the theory of electricity and magnetism, derived a form predicting that whenever electric charges are accelerated a wave should result. This wave should represent a transverse vibration of electric and magnetic fields and should detach itself from the accelerated charge which was its origin and move off in space at 3×10^8 m/s, the speed of light.

Questions

1 If light is made of particles, what suggests that their mass is very, very small?

2 What were the presumed properties of the ether, and what physical evidence suggested each property?

3 What problem in interpreting light as a wave was introduced with the barometer to measure air pressure?

4 Why are normal light bulbs frosted rather than clear?

5 Where do you have to stand to see your reflection in a plane mirror?

6 How does the reflection from a plane mirror differ from the light bounced by a white painted wall?

7 A billiard ball and the cue ball are both on the midline of the table and are 50 cm apart. You wish to bounce the cue ball from the cushion to hit the ball. Describe where you aim and why.

8 Do the highways going into your town converge or diverge?

9 Like the eye, a good camera will have an iris. What does the iris do? What other means does a camera use to control the light falling on the film? Why does the camera need this while the eye does not?

10 The right and left sides of a human face tend to be similar but not identical. Hold up a large picture of yourself next to a mirror and compare how you look to others with how you see yourself (which is which?). Is there a difference?

11 Why was Newton unsatisfied with the image produced by a spherical concave mirror, and why did he use a parabolic mirror instead?

12 Why are all of the extremely large optical telescopes produced today reflectors using a mirror rather than refractors using lenses?

13 What kind of a mirror is the spherical antishoplifting mirror found mounted in some stores?

14 How can you tell if one transparent material is optically denser than another? Light traveling from peanut oil to kerosene bends away from the normal. Which substance is optically denser (has the higher index of refraction)?

15 How does the idea of reversibility of light rays unite searchlights and reflecting telescopes?

16 Why may extremely clear and smooth-flowing mountain streams be deceptively deep?

17 As the density of a gas increases, generally its index of refraction increases also. The atmosphere of Venus at its surface is extremely dense. Explain how this might prevent light reflected at the surface from escaping from Venus. (Hint: To picture this, imagine the atmosphere as a set of shells of decreasing index of refraction as they get farther from the surface.) What color of light should this influence the least?

18 As the light of stars comes to our eyes at night, should there be some bending of the light as it passes through the atmosphere of the earth? If so, should the amount of this bending be the same in all directions or not?

19 Describe what happens to the various colors of light when white light strikes an apple.

20 Describe what happens to the various colors of light when the white light from a taillight bulb on a car strikes the plastic filter that leads to the outside.

21 Chlorophyll is the green pigment in plants that produces food from raw materials and sunlight. From your answers to the last two questions on color behavior, in what single color of light could you grow a plant and be ensured that it will die from the inability of chlorophyll to do its work?

22 How can you tell whether an image is real or virtual?

23 What is fluorescence? If we examine the spectrum of a fluorescent light, we see an essentially continuous spectrum, but the spectral lines of the element mercury are brighter than

the rest of this background. How can you explain this?

24 What effects of light are pure wave effects with no particle interpretation? Does the fact that light displays these effects prove that it is a wave?

25 Lasers produce light that is monochromatic and coherent—that is, all the same wavelength and all at crest at a particular time and place. Would the use of a laser as a light source make Young's experiment easier or harder to perform? Why?

26 What wiggles to make a light wave? How are they produced?

Problems

1 The standard lens of a 35-mm camera has a focal length of 50 mm. How far is the center of the lens from the film when taking a picture of a distant scene ($D_O = \infty$)? How far is the lens from the film when taking a close-up portrait of a person 1.0 m in front of the lens?

2 The human eyeball is about 2.0 cm in radius. What is the focal length of its lens when focusing on a very distant object? When focusing on a printed page 20 cm distant?

3 A light ray strikes the surface of a plate of glass at an angle of 60° to the normal. If the ray travels at 30° from the normal inside the glass, what is the index of refraction of the glass? (Hint: sin 30° = 0.500; sin 60° = 0.866.) What is the speed of light inside the glass? (c_{vacuum} = 300,000 km/s.)

4 What is the speed of light in a diamond (n = 2.42)?

5 The critical angle for a ray of light striking the inside surface of a crystal in air is 30°. What is the index of refraction of the crystal? (*Hint:* sin 90° = 1.00.)

6 (*a*) A beam of a particular color of red light falls on two fine slits separated by 0.01 mm. On a screen 1 m beyond the slits, a pattern of light dots and dark spots is formed with the center of the first dark zone 3.0 cm from the bright central spot. What is the wavelength of the red light being used?
(*b*) How fast would an oscillating charged particle have to vibrate to generate a wave with this wavelength?

CHAPTER TWELVE

The Beginnings of Chemistry

Introduction The variety of types and forms of matter as we look around us is bewildering. It seems impossible to account for this diversity through any simple insights into the nature of matter. For millennia investigators of matter were reduced to listing and subdividing the properties of various types of matter. Yet, there had to be much more to matter than was immediately apparent. Mix flour, water, and yeast; throw in a few ingredients for flavor; knead and let it sit for a while; bake. The bread you produce does not resemble the white powder you started with, nor the colorless liquid. Nor does the loaf resemble the plant, or the seeds of the plant, that the process began with. Matter is highly transformable. Perhaps the best example of this is one that grew up with early civilization. It is difficult for a human to tear animals to shreds or to fell trees for houses with bare hands. The earliest evidence of humans worldwide are the discarded stones shaped for such purposes. Even greater control of the environment became possible through the use of fire. Stone tools and fire allow the development of a simple agriculture and technology, but more is needed. Anthropologists have long recognized the various stages of development by the names assigned to epochs of human history. The Paleolithic Era means literally the "Old Stone Age." Time produced a refinement in the technique of working stone to produce tools that lead to a new stone age (Neolithic). From this point a giant leap leads to the ages of metal: copper, bronze, iron, then steel.

Metals are strangely different from the normal materials of the earth. We might think of brightness and a shiny surface as the distinction, but native (naturally occurring) lumps of copper can appear very stonelike. The wonder of the metal is that it can be shaped and it is not brittle. Hammering may dent and deform a metal, but it does not shatter into a shower of fragments. Copper was the first working metal. Although copper initially was found in native lumps of metal, technologists of prehistory learned to roast the metal from its colorful ores. When copper is mixed with molten tin, an interesting result forms. Examination does not reveal silver flecks of tin and reddish brown pieces of copper when the mass is cooled. A new material has been formed called *bronze.* Although it is a metal, bronze resembles neither copper nor tin, and is much stronger structurally than either. Such a substance is called an *alloy.*

The supreme achievement of the ancient world in *metallurgy* was yet to come. The rarest and most precious metal known was to become the heavy working metal of the civilization. This was iron. Iron does not occur naturally on earth as a metal. It was known only through fragments which fell from the skies in meteorites. A blazing charcoal campfire built on a rich seam of iron ore might barely be capable of producing some lumps of metallic iron. It is usually assumed that some person kicking around the ashes of the previous night's big roast became the Andrew Carnegie of the ancient world in this manner. The origin of iron production dates to the same period as the origins of Greek science, less than 1000 years before Christ.

Iron is plentiful on earth, but never in the form of the metal. To obtain the metal, powdered, reddish iron-bearing rock is mixed with burning black charcoal and provided with a strong flow of air. In the white-hot inferno that results, molten-iron droplets are formed which trickle down to form a pool at the bottom of the furnace. The complexity of matter is again apparent since none of the materials used provide any hint of containing the metal. Finding order in the combinations and recombinations of matter is a much more difficult problem than determining the regularity in the fall of a bowling ball.

The Greeks had several ideas on the nature of matter, as might be expected. The first idea of note was the four-element theory discussed in Chap. 1. In earth, air, fire, and water the Greeks believed to have isolated the basic essences which composed matter. This view implies that matter might be *mutable;* that is, that one form might be changed, or *transmuted,* into another. For example, if some fire could be bound to the structure of limestone, coal might result. Later ages were to become obsessed with the transmutation of cheaper metals into gold. This quest provided the support for centuries of *alchemists,* who were responsible for the accumulation of considerable experience with the properties and changes of matter.

The second Greek idea of importance was suggested by Democritus of Miletus. He held that there had to be a smallest, indivisible part of each type of matter. He maintained, largely on philosophic grounds, that a lump of iron, for example, could be subdivided only up to a certain point at which you had the smallest possible piece of iron. After this no further division would be possible. He called this smallest piece *an atom,* and his theory was called the *atomic theory*. The ideas of Democritus were not based on any type of physical evidence, and were largely ignored. Yet, the idea of a smallest particle of matter was revived and dropped many times as chemistry struggled to be born from the magic of alchemy.

Alchemy reached its height with the most famous of its practitioners, Paracelsus (1493–1541). At the same time Copernicus was working toward a new theory of the heavens, Paracelsus was discarding the largely wrong ideas of the ancients for a largely wrong system of his own. Like Copernicus, he was to create the crack in the ancient sphere of thought which would lead to its collapse. A move in the correct direction occurred at the time of Newton, principally in the work of Robert Boyle (1627–1691). Boyle believed matter was composed of particles in constant motion. This led him to his law relating the pressure and volume of an enclosed gas. He also tried to explain the process of dissolving one substance into another in terms of sizes and shapes of atoms. After a generation of fruitful discoveries, the English school following Boyle faded, and the emergence of a science of chemistry had to wait for another century. Meanwhile, a German school in the tradition of Paracelsus developed a theory of chemical combination that would dominate the study of matter until after the first firm beginnings of chemistry were established. Known to us as the *phlogiston theory,* the system of Joachim Becher (1635–1682) was a three-essence theory in place of the ancient four elements. Metals in this theory were a combination of a mineral substance called a *calyx* and a fire essence called *phlogiston.* The smelting of iron from its ore was interpreted as the addition of phlogiston to the ore to form the metal. To produce the ore back again, it would be necessary to drive off the phlogiston. This happened slowly in the rusting of iron. Fires were the visible emission of phlogiston from the burning substance. One objection which was raised was that ores weighed more than the metals that were obtained from them. However, it was counterargued that phlogiston, like the Greek element air, had a "positive buoyancy" and sought to rise. Thus, in combination it contributed negative weight and made substances lighter.

The second half of the eighteenth century was a time for revolutions—American, French, and chemical. Black discovered the gas we now call carbon dioxide and showed that it was quite different from air in its properties. Cavendish isolated "inflammable air," our hydrogen, from the action of acids on metals. Priestley and Scheele separately isolated "dephlogisticated air," in which substances burned more brightly and vigorously than in air. Cavendish showed that when "dephlogisticated air," our oxygen, was burned with "inflammable air," our hydrogen, the only product formed was ordinary water. Conservative opinion was shaken, since if water can be made by burning one gas in another, how can it be an element? Two things characterized the work of these men and distinguished it from much of the earlier efforts. The first was the careful and detailed measurement of all materials entering into or coming from the reactions of materials. Each was a meticulous worker (particularly Black and Cavendish). Secondly, and probably as a consequence of their careful experiences in the laboratory, each came to assume that the amount of matter involved in any reaction did not change. The weights of the materials entering into the reaction (the *reactants*) was exactly equal to the weight of all of the substances produced (the *products*). Hiding here was the first great principle of matter, *the law of conservation of matter.* The demonstration of this idea along with the overthrow of the phlogiston theory was the work of Antoine Laurent Lavoisier (1743–1794), who is often regarded as the father of modern chemistry. Black, Cavendish, Priestley, and Scheele all agreed in varying de-

The loss of a clear head

Antoine Laurent Lavoisier and a New Science

(*The Metropolitan Museum of Art, Purchase, Mr. and Mrs. Charles Wrightsman Gift, 1977.*)

Antoine Laurent Lavoisier was born rich, was an adored child, and received the best education money could buy. Similar circumstances have produced self-indulgent, useless people who have done not one whit for the sea of less privileged humanity about them, nor has this type been noted for using the wealth and leisure of their position to further human knowledge. While Cavendish was an exception, Lavoisier was *the* exception, *par excellence!* Lavoisier's scientific life was devoted to the attempt to make order of the field we now call chemistry. To this he devoted most of his time and, *not* incidentally, some of his fortune (science *is expensive* at the research level and always has been). By the age of 23 he had received a gold medal from King Louis XIV for his research. Four years later he achieved the highest recognition of his peers by being made a full member of the French Academy of Sciences while still extremely young.

Several investigators had asserted that pure water, if boiled long and hard enough, could be reduced to a small amount of a powdery material, or calyx. Lavoisier studied this intensely. The difference he brought to the study was a precise weighing of everything entering into the process. He was able to show that the weight of the residue from boiling was equal to the very small weight loss of the glass vessel in which the boiling of water took place. In other words, the calyx remaining after boiling represented the amount of glass dissolved by the water before it was driven off. With this study (1770) Lavoisier introduced into chemistry a quantitative precision which it had lacked.

Originally, Lavoisier had intended to introduce into chemistry a consistent system of naming compounds. Previously, each chem-

ist had chosen names for new materials produced with no semblance of any system. In time, it became apparent to Lavoisier that to produce a really consistent system of the naming of substances (nomenclature) he would have to repeat, carefully, the entire body of known chemical reactions, which he proceeded to do.

The spirit of quantitativeness and precise measurement of materials demands of chemists that they, sooner or later, accept the law of conservation of matter. The amount of matter involved in a chemical reaction, no matter how complex, does not change. Lavoisier carried this belief to extremes. At one point he sealed a human assistant in an airtight enclosure for several days measuring the weights of all materials going in to sustain life and all waste products expelled. This emphasis on precise weighing (mass determination) promoted a science of chemistry.

Lavoisier is most noted for his oxygen theory of combustion. The theory then in vogue asserted that burnable substances gave off a substance called phlogiston as they burned. Lavoisier's consideration of the work of Cavendish and of Priestley convinced him that this was all wrong. Burning substances combined with part of the air in burning. The one-fifth of the air that Priestley called "dephlogisticated air" Lavoisier recognized as a new element, which he called oxygen. He asserted that burning was a chemical combination of the fuel with oxygen, and he measured the weights which would back up this idea. He "burned" (today we would say "oxidized") materials in a sealed container, measuring the weights of reactants and products. From this he could show the consistency of the oxygen theory of combustion.

In 1789 Lavoisier published what is now recognized as the first modern textbook of chemistry. However, he had made two fatal errors. The first error was in the investment of his fortune. Then as now, the collection of taxes was a cheerless business resented by the populace—at best, a necessary evil. French kings sold at a discount the right to collect taxes to nobles and business executives. Lavoisier invested in the right to be a tax collector. Equally disastrous was the fact that he was a seeker after truth. As such, he was obliged to point out the errors of a journalist, named Marat, who fancied himself a scientist. Neither of these two errors would have been a major problem except that Lavoisier lived at the time of the French Revolution. This was a bad time to be a tax collector, and even worse time to be an enemy of a powerful radical journalist. Actually, Lavoisier favored the revolution and the overthrow of a corrupt aristocracy. His father had paid a small fortune to buy him the first rank of nobility, but Antoine refused to be "de Lavoisier." In a letter to Franklin he advocated an American-style republic and regreted that Franklin was not in France to help guide them away from some of the excesses of revolution. During the Reign of Terror, Lavoisier was sent to the guillotine in spite of the protests of French scholars. The court's judgment was "the revolution has no need of scientists." Human understanding was not avenged when Marat was stabbed in his bathtub shortly thereafter, nor when most of the condemning radicals were led to "Mme. Guillotine." The loss of the greatest mind in chemistry of his age at the peak of his productivity cannot be estimated.

The painting of Marie and Antoine Lavoisier is by the renowned artist Jacques Louis David. Marie was the daughter of a rich merchant who was under great pressure to marry her to a penniless, middle-aged count to secure her entrance to the royal court. Instead, he arranged her marriage to his business associate Lavoisier. Marie was intelligent and artistically gifted. She studied painting with David and also worked in the laboratory with her husband, keeping records and producing many of the technical drawings which illustrated his works.

grees with the phlogiston ideas, as can be seen from some of the names they assigned their discoveries. All were to outlive the tragic figure of Lavoisier, yet they could not abandon phologiston. Lavoisier seems almost alone in his refusal to try to formulate a grand, overlying scheme for all of matter. He confined himself to measuring and interpreting the experimental facts, and led science, once again, away from building sand castles with premature speculations and overextension. Among other things, Lavoisier gave us the modern idea of an element, and with him begins the modern system of naming chemical materials.

Elements

An *element,* in modern usage, is a material which cannot be broken down into any simpler substance by normal chemical means. Iron is an element because there is no process which will break it into simpler materials. Iron heated in an enclosure of oxygen will burn fiercely, once hot enough, and will produce a red powder. This red powder is *not* a simpler substance than the iron since it weighs more. This indicates that something was added to the iron to produce the red powder. In fact, the added weight of the powder is exactly the same as that of the oxygen used up in the reaction. Thus, we infer that the powder is not an element, but is instead a chemical combination of oxygen and iron. Much experience indicates that *all* substances obtained from iron are heavier than the iron used in making them and thus represent something added to the iron. On the basis of this experience we would be forced to conclude, temporarily at least, that iron was an element.

In contrast to the identification of an element, a material such as bronze would fool no one. The tin and copper in bronze may be separated in a variety of ways, and if bronze is burned in oxygen, a mixture of powders is obtained. One is identical in its properties to the powder produced when copper is burned in oxygen and the other is the product of burning tin in oxygen. Thus, although both copper and tin would eventually prove to be elements, bronze is not.

The composition of bronze offers an important distinction from the red powder formed in burning iron. Different batches of bronze will have different proportions of copper and tin in them, much as the chocolate cake baked by aunt Harriet has more chocolate in it than the one baked by aunt Mabel. Bronze is a *mixture,* and different batches may vary in the ingredient proportions yet still be bronze. The red iron rust obtained by burning iron in oxygen is not this way at all; it is not a mixture. Every pure sample of the red powder, whether made in a laboratory from pure elements or scraped from a rusting warship on a beach in the Philippine Islands will yield the same result when chemically separated, or *analyzed.* Every kilogram of the pure red powder yields 699.4 g of iron and 300.6 g of oxygen. This invariable proportion of the elements making up a substance identifies it as a *compound.* Unlike the mixture of a chocolate cake, a compound has very definite physical and chemical properties.

Cakes may be lighter or darker, sweeter or less sweet, moister or dryer, and still be chocolate cakes. Not so with red iron oxide. Physically, it has a definite color, hardness, boiling point, melting point, and even a definite type of crystal if it occurs in pure large chunks. Chemically, the compound is equally definite in what it will and will not react with. Perhaps the most important compound is water, which Cavendish measured to require 8 times the weight of oxygen as of hydrogen to produce. (Even though the oxygen outweighs the hydrogen 8/1 in water, hydrogen is such a low-density, "light" gas that twice the volume of hydrogen as oxzygen is required.) If less hydrogen is added than 1 kg for every 8 kg of oxygen, some of the oxygen will not be used up in the reaction. This type of behavior led to the *law of definite proportions: Elements combining chemically to form a compound do so in fixed proportions by weight.*

Lavoisier recognized normal burning as being due to a chemical combination with oxygen. The chemical terminology he introduced reflected this. Compounds of another element with oxygen he called *oxides,* and he would write the reaction for forming water as:

$$\underset{\text{(1 unit of weight)}}{\text{Hydrogen}} + \underset{\text{(8 units of weight)}}{\text{oxygen}} = \underset{\text{(9 units of weight)}}{\text{hydrogen oxide (water)}}$$

The vigorous burning of materials in oxygen compared to air could be attributed to the fact that only a small part of air is oxygen. Four-fifths of the air is another elemental gas named *nitrogen,* which generally does not participate in most reactions and serves to dilute the oxygen and slow down burning. Similarly, we might write:

$$\underset{\text{(699.4 units of mass)}}{\text{Iron}} + \underset{\text{(300.6 units of mass)}}{\text{oxygen}} = \underset{\text{(1000 units of mass)}}{\text{iron oxide (rust)}}$$

TABLE 12-1
Elements Known at the Time of American Civil War

Known to the ancients (10)
Carbon, copper, gold, iron, lead, mercury, silver, sulfur, tin, zinc
Medieval additions (1300–1500) (3)
Arsenic, antimony, bismuth
Century following Boyle (1660–1770) (5)
Phosphorus (1669), platinum (1735), cobalt (1735), nickel (1751), hydrogen (1766)
1770s (5)
Fluorine, nitrogen, oxygen, manganese, chlorine
1780s (6)
Molybdenum, tellurium, tungsten, titanium, uranium, zirconium
1790s (4)
Strontium, yttrium, chromium, beryllium
1800s (10)
Niobium, tantalum, palladium, cerium, iridium, osmium, sodium, potassium, calcium, barium
1810s (4)
Iodine, cadmium, lithium, selenium
1820s (4)
Silicon, aluminum, bromine, thorium
1830s (3)
Magnesium, vanadium, lanthanum
1840s (2)
Ruthenium, terbium
1850s (0)

Lavoisier recognized over 20 substances as elements, many known since ancient times. In addition to the gases hydrogen, nitrogen, and oxygen were carbon, sulfur, phosphorus, and a long list of metals including iron, copper, tin, lead, gold, silver, and mercury. The century following Boyle saw four new elements discovered. The 80 years from the first publications of Lavoisier added 38 new elements identified for a total of 55 known elements by 1850 (Table 21-1).

Multiple Proportions and the Atomic Theory

One of the first serious complications of a simple theory of chemical combinations is that there can be more than one way in which the same elements may combine to form a compound. For example, carbon and oxygen combine to form the gas discovered by Black now called *carbon dioxide.* These same two elements also combine to form the gas *carbon monoxide.* Each of these two substances is a perfectly definite compound with its own set of properties, but the first is always 72.7% oxygen by weight and the second is only 57.1% oxygen by weight. In both cases, the remainder of the weight is made up of carbon only. Thus, there are at least two different ways in which carbon and oxygen may unite to form compounds. There was even more variety in the combinations of oxygen with nitrogen, where several gases were known. In each of these the ratio of nitrogen to oxygen is definite and does not vary from sample to sample. Yet, the gases are quite distinct, ranging from a dense brownish, poisonous vapor through the substance known as laughing gas.

The resolution of the definite proportions observed in compounds came through the revival of an idea which had been suggested many times before but without solid experimental reasons, the idea of *atoms.* The English schoolteacher John Dalton (1766–1844) proposed the *modern atomic theory* in 1808, which contained several parts:

1 The smallest part of an element is an *atom* of that element.

2 Atoms of different elements are different, especially in their weights and in their abilities to combine with other atoms.

3 In a chemical reaction no atoms are ever gained or lost, they are merely rearranged into new combinations.

4 In a reaction, two (or more) different types of atoms are combined in simple ratios to obtain the smallest unit of a compound. (Dalton called this smallest piece of a compound a "compound atom." Today we call it a *molecule.*)

Dalton could account for two elements forming more than one compound by assuming that the ratio of the atoms of each in the molecules varied. For example, if you made a molecule containing one hydrogen atom for each oxygen atom, it would have a certain proportion of hydrogen by weight. If another molecule was made having two hydrogen atoms for each oxygen atom, it would have approximately twice the proportion of hydrogen by weight. Thus, Dalton asserted the *law of multiple proportions* (1804). *When atoms join to form more than one compound, the proportion of any type of atom in the several compounds will be in a simple whole number ratio by weight.* In the case of the two gases containing carbon and oxygen, carbon dioxide is about 27% carbon by weight and 73% oxygen, while carbon monoxide is 43% carbon and 57% oxygen. In the first case, we have a carbon/oxygen ratio of 27/73, or 0.37. In the second case the ratio is 43/73, or 0.75. The ratio of 0.75/0.37 is about 2/1, or carbon monoxide has just twice as much carbon for its oxygen as does carbon dioxide. The ratio 2/1 is a simple whole number ratio by weight as required by the law of multiple proportions. Thus, the simplest explanation of the two compounds would be to assume that carbon monoxide had one carbon atom for each oxygen atom and that carbon dioxide had one carbon atom for *two* oxygen atoms. This could be expressed in the method developed shortly after Dalton's work by assigning one or two letters to each element as an *atomic symbol.* Thus, C stands for an atom of carbon and O for an atom of oxygen. The symbol 2C would mean two separate atoms of carbon while the symbol O_2 would indicate two *joined* or *bonded* atoms of oxygen. Thus we have

CO	CO_2
Carbon monoxide	Carbon dioxide

as the simplest representations of the molecules of these two gases.

As more work was done, the usefulness of Dalton's atomic theory grew as a way of interpreting experimental findings. The most reasonable approach is to take the lightest substance known, hydrogen, and assign its atoms a weight of 1 unit. By carefully studying the weight of hydrogen that will combine with other types of atoms (carbon, oxygen, and nitrogen, for example) and the weights of these atoms that will combine with each other, the weights of these atoms can be worked out relative to hydrogen. This would yield a weight of 12 for carbon, 14 for nitrogen, and 16 times the hydrogen atom for oxygen. Using these weights it is possible to determine the simplest possible molecules for various substances. For example, water contains 8 times as much oxygen as hydrogen by weight. But, if each oxygen atom is 16 times as heavy as a hydrogen atom, water must have two hydrogen atoms for each oxygen atom. We could then write the formula for the molecule of water as H_2O, indicating two atoms of hydrogen and one atom of oxygen locked together chemically to form the smallest piece of a new substance, water. While this formula is the simplest possible combination in order to have the oxygen in the molecule weigh 8 times the hydrogen, it is not the only possibility. Any formula with two atoms of hydrogen for each atom of oxygen would produce the 1/8 weight ratio observed experimentally. Thus, H_4O_2 is a possible formula, as is $H_{10}O_5$ and so on. To pin down the correct formula, we must also determine the weight of the smallest particle of water. Since this molecule of water can be shown to have a weight 18 times that of hydrogen, we know the formula is H_2O, where each of the two hydrogen atoms weighs 1 unit, the oxygen atom weighs 16 units, and the total package thus weighs 18 units. We may also work out several other formulas

CH_4	NH_3	N_2O
Methane (Natural gas for stoves)	Ammonia (Sharp odor, poisonous)	Nitrous oxide (Laughing gas)
NO	NO_2	HCN
Nitric oxide (Colorless, poisonous)	Nitrogen dioxide (Brown, mildly poisonous)	Hydrogen cyanide (Very poisonous, gas-chamber gas)

Since all of these substances are gases, it is relatively easy to determine the relative weights of the molecules because of the following two related discoveries.

Joseph Gay-Lussac (1778–1850) had noted in 1808 that gases combine with each other in simple ratios of volumes. For example, exactly 2 liters of hydrogen gas will combine with a single liter of oxygen gas to form just 2 liters of water vapor if the temperature is kept constant. Similarly, 3 volumes of hydrogen gas will combine with 1 volume of nitrogen gas to form exactly 2 volumes of ammonia. Gay-Lussac was a meticulous worker in the laboratory, and his error was so small as to leave no doubt to the conclusion. Yet, it was not immediately clear why these strict ratios by volume should be the case. The answer was supplied by Amadeo Avogadro (1776–1856) in 1811 (although it was mostly ignored for over 40 years until revived by his compatriot Cannizzaro). Avogadro hypothesised that *equal volumes of any two gases at the same temperature and pressure contain equal numbers of molecules.* With this he had to combine the speculation that the smallest part of some elements were not single atoms but interlocked groups of atoms. For example, the molecule of hydrogen gas was not a single atom, H, but two interlocked atoms, H_2. Similarly, some other elements (primarily gases) were not found singly but joined, such as N_2, O_2, and so on. Applied to the formation of ammonia, for example, this accounted for the strict combining ratios observed by Gay-Lussac.

Avogadro's hypothesis points out that whatever volume of nitrogen gas we chose to use, we would have to have just 3 times that volume of hydrogen gas to provide three atoms of hydrogen for each atom of nitrogen to make the ammonia. To mentally picture the same reaction in terms of the atoms involved, we may represent the hydrogen atom by a small sphere, and the hydrogen gas molecule by two such interlocked spheres. Nitrogen atoms may be represented by larger spheres as shown in Fig. 12-1. The reaction thus occurs as shown where three hydrogen molecules ($3H_2$) plus a nitrogen molecule (N_2) rearrange to form two ammonia molecules ($2NH_3$). Rather than drawing this type of reaction each time it occurs, we may write it symbolically. The expression, "hydrogen gas plus nitrogen gas yields ammonia," gives us the basics of the reaction but tells us nothing about how much of each of the reactants is required, nor how much ammonia is produced. Instead we may introduce the *balanced chemical equation:*

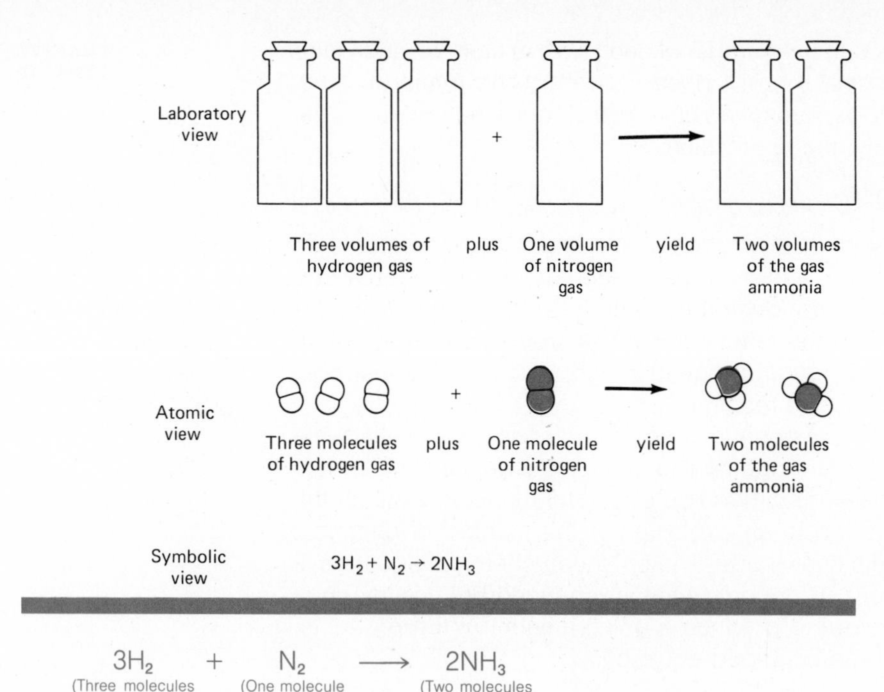

FIGURE 12-1 Three ways of representing the same reaction. From the even numbers of volumes determined in the laboratory, Avogadro reasoned that the same volumes of any gases (at the same temperature and pressure) must contain the same number of molecules. This idea would work only if elemental gases, such as hydrogen and nitrogen, came in molecules made of two joined atoms each—giving us the atomic level view. This is summarized symbolically in the equation.

$$\underset{\text{(Three molecules of hydrogen gas)}}{3H_2} + \underset{\text{(One molecule of nitrogen gas)}}{N_2} \longrightarrow \underset{\text{(Two molecules of ammonia)}}{2NH_3}$$

An equation is *balanced* if, first of all, the formula for each type of molecule represented is correct (H_2, N_2, and NH_3 in this case). Secondly, the *coefficients* (or numbers in front of each type of molecule telling you how many of them participate in the reaction) must be adjusted correctly so that the same number of atoms of each type occur on each side of the equation. Here we start with three molecules of hydrogen gas, each of which has two atoms of hydrogen, for a total of six atoms of hydrogen beginning the reaction. We end with two molecules of ammonia, each of which has three atoms of hydrogen, for a total of six atoms of hydrogen after the reaction occurs. Thus, the number of hydrogen atoms balances. Likewise, we start with two atoms of nitrogen locked into a single molecule of nitrogen gas, and end with two atoms of nitrogen, one in each of our two ammonia molecules. The nitrogen atoms also tally, and we have neither gained nor lost any atoms during the reaction. We start with six hydrogen atoms and two nitrogen atoms and we end with the same. We have merely indicated on paper what nature does in any chemical reaction: All of the parts (atoms) are preserved, but their linkages or *bondings* have been rearranged.

By contrast, the expression

$$H + N \rightarrow NH_3 \tag{1}$$

is *not* satisfactory because it does not reflect the reality that in nature 3 times as many hydrogen atoms as nitrogen atoms are required to produce the ammonia molecule, nor are the numbers of atoms balanced on both sides of the equation. Even the expression

$$3H + N \rightarrow NH_3 \tag{2}$$

is *not* correct, even though the number of each type of atom on each side of the equation balances, since it does not reflect the reality that hydrogen atoms and nitrogen atoms do not come singly in nature but as the pairs H_2 and N_2. Finally, the expression

$$6H_2 + 2N_2 \rightarrow 4NH_3 \qquad (3)$$

while it *is* balanced, with 12 hydrogen atoms and 4 nitrogen atoms on each side of the equation, and *does* have each molecular formula correct (H_2, N_2, and NH_3), is *not desirable* since it is *not the simplest possible correct statement* of the reaction. Here, all coefficients may be divided by two to yield our original equation, which is the simplest possible correct statement of the reaction.

The balanced chemical equation does more for us than just list what happens in terms of the rearrangement of parts. It answers the vital question, "How much?" The widespread use of careful measurements in the times of Lavoisier had led to an explosion in the number of chemical facts added to scientific knowledge. The balanced chemical equation is in this spirit since it tells us not only what is produced but now much of the reactants are required to produce a given amount of the products. Once again, consider the balanced equation

$$3H_2 + N_2 \rightarrow 2NH_3$$

A hydrogen atom has a mass of 1 unit, and the hydrogen molecule, H_2, therefore has a mass of 2 units. If we take $3H_2$ we will thus be using a total of 6 units of mass of hydrogen. Nitrogen atoms have a mass of 14, so the nitrogen molecule, N_2, has a mass of 28 units. The balanced equation informs us that for every 6 mass units of hydrogen used, we will need to add 28 units of nitrogen gas to produce 34 units of ammonia. If we add more nitrogen than this, it will be unused and mixed with the ammonia produced. If we add less nitrogen than the 28 units, some of the hydrogen will not be able to find nitrogen atoms to combine with, and the ammonia produced will be mixed with the excess hydrogen.

Normally, the units worked in are the mass units of grams or kilograms. If we take a number of kilograms equal to the atomic mass of an element (for example, 12 kg of carbon), we are taking 1 *kilogram atom* of the substance. For compounds, the basic unit is the mass of the molecule. For example, an ammonia molecule, NH_3, has a mass of 17—14 mass units for the nitrogen atom and 1 mass unit for each of the three hydrogen atoms. If we take 17 g of ammonia, we are taking the mass of the molecule in grams. This amount is referred to as a *gram molecular mass* or *mole* (mol) of the substance. Similarly, 17 kg of ammonia would be a *kilomole* (kmol). Consistency would have us use kilomoles, but if you take $3H_2$ in kilomoles, you would have 6 kg of hydrogen gas and be on your way toward filling the Hindenburg Zeppelin. Kilograms of gases represent much more than is commonly encountered in the laboratory, where the gram is a very useful working unit for most measurements.

One mole of any substance contains the same number of molecules as one mole of any other substance. Thus, 17 g of ammonia has the same number of molecules as 28 g of nitrogen gas. Since these are both gases, they also occupy the same volume at the same temperature and pressure. At *standard conditions* (SC; or *standard temperature and pressure,* STP) of 0°C and 1 atm of pressure (760 mmHg barometric pressure, or 1.013×10^5 N/m^2, or 14.7 lb/in^2 of pressure) *1 mol of any gas occupies a volume of 22.4 liters* or 0.0224 m^3. (A liter is 1000 cm^3 and is close to a quart. Therefore 22.4 liters is about $5\frac{1}{2}$ gal.) The number of molecules in a mole of any substance is called the *Avogadro number,* although it was not determined until long after his death. One mole of any substance contains about 602,000,000,000,000,000,000,000 molecules. Thus, the Avogadro number is 6.02×10^{23} molecules per mole. This implies that $5\frac{1}{2}$ gal of air (1 mol) contains about 100 trillion molecules for each man, woman, and child on earth—give or take a few million.

The Search for Pattern

As much careful laboratory work led to the ability to determine how many atoms of one type combined with how many of another type to form a molecule, patterns emerged. By now, you have probably noted that metals tend to combine with nonmetals to form compounds. If one atom of a metal combined with one atom of chlorine to form a molecule, then two atoms of the metal would combine with each oxygen atom to form an oxide. Similarly, three atoms of that metal would combine with each atom of nitrogen. For example, consider the compounds of sodium:

$NaCl$	Na_2O	Na_3N
Sodium chloride	Sodium oxide	Sodium nitride
(1/1)	(2/1)	(3/1)

The same pattern of numbers may be seen with hydrogen:

HCl	H_2O	H_3N
Hydrogen chloride (hydrochloric acid when dissolved in water)	Hydrogen oxide (water)	Hydrogen nitride (ammonia)

In contrast, atoms of metals combining 1/1 with oxygen atoms require two atoms of chlorine apiece when forming chlorides.

$MgCl_2$	MgO	Mg_3N_2
Magnesium chloride	Magnesium oxide	Magnesium nitride

These compounds are all called *binary compounds* because each contains only two types of atoms. According to Lavoisier's system of naming compounds, the *-ide* ending identifies them as binary.

By the mid-nineteenth century this led to describing the combining tendency of elements in terms of a number called the element's *valence.* The valence was a number, usually between 1 and 4. Metals were given positive valences while nonmetals were assigned negative valences (Table 12-2). The formula for a neutral molecule was reached when the sum of all the valences of the separate atoms in the molecule totaled to zero. For example, sodium atoms have a valence of +1 and oxygen

TABLE 12-2

Some Common Valences of Elements

Valence	Elements
+1	Na (sodium), K (potassium), H (hydrogen), Ag (silver)
+2	Mg (magnesium), Ca (calcium), Fe (iron), Cu (copper), Zn (zinc), Pb (lead)
+3	Fe (iron), Al (aluminum)
−1	Cl (chlorine), F (fluorine)
−2	O (oxygen), S (sulfur)
−3	N (nitrogen), P (phosphorus)
−4	C (carbon), Si (silicon)

atoms have a valence of −2. One sodium atom linked to one oxygen atom, NaO, should *not* be a stable molecule since the valences of the separate atoms do not add up to zero, but to −1: $(+1) + (-2) = -1$. In order to have a neutral molecule, two sodium atoms, with valence +1 each, would be needed for each oxygen atom. Thus the correct formula for sodium oxide is Na_2O. Similarly, if the valence of the metal iron, Fe, is +3, we can determine the formula for iron oxide. Only the combination Fe_2O_3 or a multiple of it would produce a neutral molecule between iron atoms with valence +3 and oxygen atoms of valence −2. Two iron atoms with valence +3 each gives a total of +6. Three oxygen atoms, each having a valence of −2, would total −6. The sum of +6 and −6 is zero, so this combination, Fe_2O_3, should produce a neutral stable molecule—and it does.

Two complications immediately arise with the idea of valence. The first is that several elements display more than one valence in their various compounds. For example, carbon forms both carbon dioxide, CO_2, and carbon monoxide, CO. If we presume that the oxygen has a combining tendency or valence of −2 in both of these molecules, the carbon must have a valence of +4 in the first case and +2 in the second case. This problem of multiple valence is not serious enough to be allowed to interfere with what is otherwise a simple scheme for determining the correct formula for many molecules, and we will ignore it for the moment. The second complication is really something of a simplification in disguise. There are certain groups of atoms which engage in reactions as though they were a single metal or a single nonmetal having a single valence. During the bulk of their reactions, these groups, or *radicals,* stay firmly bonded to each other; yet they need some outside partner to form a neutral molecule. An example of such a group is the carbonate radical, CO_3, with a valence of −2 (Table 12-3). This is the same valence that oxygen has in its various oxides. Thus, a carbonate radical usually may be substituted for an oxygen atom in a molecule, and it will lead to the formula for a real molecule. For example:

In H_2O (Water) replace O with CO_3 and get H_2CO_3 (Carbonic acid, the "bite" in soda pop)

TABLE 12-3
Some Common Radicals and their Valences

FORMULA	VALENCE	NAME
NH_4	+1	Ammonium
NO_3	−1	Nitrate
HCO_3	−1	Bicarbonate
$H_2C_3O_2$	−1	Acetate
OH	−1	Hydroxide
ClO_3	−2	Chlorate
CO_3	−2	Carbonate
SO_4	−2	Sulfate

or in Na_2O (Sodium oxide) replace O with CO_3 and get Na_2CO_3 (Washing soda)

While the idea of valence is useful to help us arrive at the correct formula for a wide variety of molecules, it is an empirical scheme derived from laboratory experience. Valence considerations will correctly lead us to the formula for sodium chloride, NaCl—not Na_2Cl or Na_3Cl or $NaCl_2$, but NaCl. But then, while we know that one atom of sodium, no more nor less, will link or bond with each atom of chlorine, we are stopped. We still have no idea as to how a soft, silvery, violently reactive metal such as sodium and the yellow-green gas chlorine, actually used as a poison gas in World War I, can combine together at all. Furthermore, it is incredible that when they do combine they produce the white crystals of common table salt. Far from its poisonous and violent parents, salt is actually necessary in small amounts for life. At this point our knowledge of the mechanism of atomic bonding remains so rudimentary we could not even predict the commonest behavior of salt—that it dissolves in water. Clearly, while we have come a long way in understanding the interactions of matter, nature has many more subtleties waiting for us if we can only figure out where to begin the search.

Relative Chemical Activities

The laws of conservation of mass, definite proportions and multiple proportions, the atomic theory, Avogadro's hypothesis, and the definition of the mole establish chemistry as a quantitative science. Once whatever is happening in a reaction is worked out, this set of relations can handle the question of "How much?" However, by themselves, these relationships do not predict what would happen in a reaction. Why, for example, do two hydrogen atoms combine with one oxygen atom to form a water molecule? Why not 1/1, or 3/1? We are left in somewhat the same position we were in after Galileo. He had shown us the power of number to describe what was happening with falling objects, but he did not show us what this implied about the nature of nature. In the simpler field of motion, it required Newton to summarize the apparent behavior of all natural phenomena. The more complex area of chemistry required over a century of continual discoveries to move from "how much" to "what." Before we expand into the theory of chemical combination, let us acquire some of the working factual knowledge of how matter is observed to react.

The point at which to begin, as pointed out by Lavoisier, is with the simple reactions of the elements. The elements may be divided on the basis of their physical properties into *metals* and *nonmetals.* Metals tend to have shiny surfaces, be *malleable* (able to be pounded out into sheets), be *ductile* (able to be pulled or drawn out into wires), and to be good conductors of electricity. Nonmetals tend not to have these properties, although individual elements may form shiny-surfaced crystals (for example, iodine). A few elements are difficult to classify. For example, carbon is not shiny, malleable, or ductile, but it *is* a good conductor of electricity. In Lavoisier's day as today the largest group of known elements were the metals. If we examine the chemical behavior of the metals, we find that there are some similarities beyond physical appearance. In an atmosphere of pure oxygen most metals will burn brightly to form powdery products. Some metals do this rapidly and with a great amount of heat and light being given off. For example, a photographic flashbulb is nothing more than a wire of aluminum metal sealed in a glass bulb containing pure oxygen. A small spark is sufficient to start the rapid burning of the aluminum, which is over in a fraction of a second. Analysis of the white powder produced indicates that it contains two atoms of aluminum for every three atoms of oxygen. Under Lavoisier's scheme for naming compounds, this product is called *aluminum oxide,* and the formula for its molecule is written as Al_2O_3. The chemical equation for the reaction is:

$$\underset{\text{Aluminum metal}}{4Al} + \underset{\text{Oxygen gas}}{3O_2} \longrightarrow \underset{\text{Aluminum oxide (white)}}{2Al_2O_3}$$

The reaction of copper with oxygen is not nearly so spectacular. The metal must be heated to a high temperature before it burns with a dull glow.

$$\underset{\text{Copper metal}}{2Cu} + \underset{\text{Oxygen gas}}{O_2} \longrightarrow \underset{\text{Copper oxide (black)}}{2CuO}$$

As the formula indicates, the common oxide of copper has one oxygen atom and one copper atom in each molecule. Mercury is even more sluggish than copper in combining with oxygen. This combination occurs only when the mercury is heated in the presence of oxygen, and the reaction stops if the heating is stopped.

$$\underset{\text{Mercury metal}}{2Hg} + \underset{\text{Oxygen gas}}{O_2} \longrightarrow \underset{\text{Mercury oxide (red)}}{2HgO}$$

This particular reaction is of considerable interest because it is easily *reversible,* that is, the red oxide may be easily induced to break apart into mercury metal and oxygen gas again. While heating causes the elements to combine and form the red oxide, intense heating causes the oxide to separate into the two elements. We may indicate this as:

$$2Hg + O_2 \rightleftharpoons 2HgO$$

difference between fizzing, sputtering, and exploding in the reactions of the three. If the metals are dropped in water, they react vigorously, liberating hydrogen and forming powerful *alkalis* or *bases.* For example:

$$2K + 2HOH \rightarrow 2KOH + H_2\uparrow$$

$$2Na + 2HOH \rightarrow 2NaOH + H_2\uparrow$$

(If you have not recognized it, HOH is our old friend water written another way.) The main difference between KOH, NaOH, and LiOH is one of degree. If three people of similar size and shape were dropped into vats containing concentrated solutions of these three hydroxides, the one in the KOH would dissolve first. However, the remaining two people would not be far behind. Lye, or NaOH, is the active ingredient in drain cleaners since it dissolves fats, oils, hair, and so on. Meanwhile, KOH is of use to biologists studying insect structure. When an insect is dropped in a concentrated solution of potassium hydroxide, all of its soft tissues are dissolved away leaving a beautiful transparent exoskeleton for study. Thus, you will probably concede that there is a family resemblance in the behavior of these elements. Actually, several such groups of elements with related chemical behavior were known in the nineteenth century. The important resemblance defining a family is *chemical* behavior and not physical appearance.

A sizable step in determining order in the properties of the various elements was provided by Dmitri Mendeleev (1834–1907). Mendeleev arranged the known elements in the order of increasing atomic weight. In this list he noted that elements with similar chemical properties repeated at relatively regular intervals along the list. He set out to form the elements in a two-dimensional array, or *table,* rather than a one-dimensional list. Mendeleev listed the elements in order of increasing atomic weight until he came to an element with a strong chemical resemblance to one already listed. When this occurred, he began a new line under the first. (Mendeleev's table actually ran up and down while today we always run them sideways as described.) Let us follow the procedure he used. The lightest known elements were hydrogen, lithium, beryllium, boron, nitrogen, oxygen, fluorine, and so on. We may list these by symbol as

H Li Be B N O F

The next lightest element known was sodium. As we have seen, sodium has a strong chemical resemblance to lithium. Mendeleev therefore broke off his first line at fluorine and placed sodium under lithium in his table. The next element, magnesium, bears a chemical resemblance to beryllium. The element after that, aluminum, is related to boron. In fact, eight elements in a row fall neatly under a close relative. The next element is potassium, which we know must fit under lithium and sodium. The element after this is calcium, which indeed is a close relative of magnesium. For the moment we will stop at this point and consider the table (Table 12-5). [We have placed hydrogen above the group of alkali

TABLE 12-5
Abbreviated Periodic Table

	FAMILY OR GROUP						
	I	II	III	IV	V	VI	VII
Row or period							
1	H						
2	Li	Be	B	C	N	O	F
3	Na	Mg	Al	Si	P	S	Cl
4	K	Ca					
Normal valence	+1	+2	+3	±4	−3	−2	−1

metals in the table by default. While hydrogen is not an active metal like the rest (and hydrogen hydroxide is not likely to dissolve you), hydrogen does have the same combining power, or valence, as these metals.] It is valence that is likely to convince us that this listing by atomic weight in a table is indeed hinting at a pattern for elements. Each column of the table constitutes a separate family of elements, and represents elements with a different normal valence from every other column. From this much alone we can state Mendeleev's periodic law of the elements:

> *The properties of the elements repeat themselves periodically with increasing atomic weight.*

From this law, each row in the table is called a *period,* while each column is called a chemical *family* or group of elements, identified by Roman numerals. Even as short a chart as drawn in Table 12-5 indicates a great deal of chemical logic. For example, as we proceed to the left and downward on the chart, metals become more and more active, with potassium (K) the most active of those listed. Conversely, as we proceed upward and toward the right we reach more and more active nonmetals, with fluorine (F) being the most active of all. Unfortunately for keeping this ordering by chemical properties simple, there are several times this number of elements left to put into the chart. The very next element in Mendeleev's list after calcium, where we stopped, is the metal erbium, which is not like the aluminum family at all. In fact, a whole series of metals now occur which are more like each other than any family on the chart. Mendeleev noted that some way down the list of elements arranged by increasing weight of the atoms, the pattern repeated. The set of arsenic, selenium, and bromine, in order, fitted neatly into the nitrogen, oxygen, and fluorine families according to their chemical behavior. Furthermore, 19 elements farther down the list was another set of three elements that had to fall into these same three families. To fit both of these periods in, blanks or gaps had to be generated in the chart where an element should have been but none were known. This was the test of Mendeleev's scheme of organizing the elements. At least three gaps in the chart predicted the existence of yet unknown elements. Based on the family resemblances made apparent by his chart, he could predict the chemical behavior, atomic weight, melting point, boiling point, and even (to a limited extent) where to look for the new elements. Table 12-6 indicates various physical properties for the set of alkali metals (group I in Table 12-5). In almost all cases, the physical properties of the elements in a family change smoothly as the atoms become more massive. Mendeleev published detailed predictions of the physical and chemical properties of three new elements which proved in time to be extremely accurate.

A further confirmation of Mendeleev's scheme for ordering the elements came 30 years after his original publication with the discovery of an entirely new family of elements. In the eighteenth century Cavendish had noted that most of the air is nitrogen, while about one-fifth of it is oxygen. When both of these gases are removed by a reaction, a tiny bubble of gas remains which makes up less than 1 percent of the air. William Ramsay (1852–1916) analyzed this bubble with a spectroscope in 1894 and found the lines of an entirely new element which he named *argon*. Within the next 5 years he isolated from the air three more gases present in very much smaller amounts, neon, krypton, and xenon. Finally, in a gas given off by some rocks in tiny quantities he isolated a fifth element called helium. This set of five elemental gases was definitely a family since their chemical behavior was identical—they did nothing! Under no circumstances could chemists at the turn of the century make them combine with any other substance. These elements form the family of the *inert gases*. In fact atoms of these elements do not even display an attraction for their own kind. Atom for atom, xenon gas is heavier than silver, yet it is a gas at room temperature. Atoms of other heavy elements, even relatively inert ones such as gold, still have enough attraction for each other to lock together into a solid at high temperatures. The inert gases remain gases to very low temperatures, and the lightest, helium, remains a liquid and refuses to solidify even at absolute zero.

The inert gases convinced any remaining skeptics that Mendeleev's scheme for arranging the elements was really showing some important principle in nature. Each of these gases has an atomic mass which places it next to an alkali metal in the table. We may conveniently end each period of the table with one of the inert gases. When we do so, we have produced the completed periodic table as is found inside the back cover of the text. Period VIII is that of the inert gases—look for it. You should also note the string of elements in the middle of the table labeled

TABLE 12-6

Properties of Alkali Metals

METAL	ATOMIC MASS	VOLUME OF 1 MOL, CM^3	DENSITY	MELTING POINT, °C	BOILING POINT, °C	IONIZATION POTENTIAL*
Li	6.94	13	0.53	186	1336	5.4
Na	23.0	24	0.97	97.5	880	5.1
K	39.1	46	0.86	62.3	760	4.3
Rb	85.5	56	1.87	28.5	700	4.2
Cs	133	71	1.87	28.5	670	3.9

* Voltage required to make the vapor of the element conduct electricity.

Source: M. J. Sienko and R. A. Plane, "Chemistry," 5th ed., McGraw-Hill, New York, 1976.

the *transition elements.* This table is identical to our original table (Table 12-5) for the first three periods except for the addition of an inert gas at the end of each period. However, a large gap has been introduced in that original table between the families of group II and group III to allow the introduction of these transition elements. Actually, another even wider gap is necessary in periods 6 and 7 for the elements below the table called the *rare earths.* (These are not displayed in their proper position since the table would become too long and thin to fit on a page and still be legible.)

Clearly, the various elements are not a miscellaneous collection of substances with random properties. There is too much order in this arrangement to doubt that some underlying factors control the properties of each type of atom—but what? Why should the chemical properties of atoms repeat in a pattern with increasing mass? What must atoms be like to show such relatedness? We will consider these questions in more detail in the next chapter.

Summary

While a promising start was made toward a modern science of matter and its combinations around 1660, the chemical revolution did not take place for another 100 years. Once started, the new chemistry was characterized by precision measurement. This gave rise to the belief that the amount of matter was not changed during a chemical reaction, merely transformed. This is the *law of conservation of matter.* Lavoisier introduced a modern system for the naming of chemical substances, stated the modern definition of an element, and wrote the first textbook in chemistry. To Lavoisier an *element* was any substance that could not be broken into simpler (lighter) parts by ordinary chemical means. *Compounds* were substances with two or more elements joined together in a particular way. In the *oxygen theory of combustion,* Lavoisier asserted that ordinary burning is just the chemical combination of atoms in the fuel which is burning with oxygen atoms from the air. The compounds that result, like all compounds, were characterized by having *definite proportions* of each element composing them by weight. Dalton, noting that two elements often combine to form two or more distinct compounds, stated the *law of multiple proportions.* This asserted that in such cases the weights of the elements involved in the set of compounds were in a simple ratio of whole numbers from compound to compound. This led Dalton to his *atomic theory* (1809), which asserted that the smallest piece of an element was a particle of fixed size called an *atom.* The smallest piece of a compound was a joined set of atoms called a *molecule.* With this idea Dalton could account for the law of multiple proportions in terms of the combining power of atoms. The *Avogadro hypothesis* stated that equal volumes of gases at the same temperature and pressure contained the same number of separate particles. To work correctly, this hypothesis required that many elemental gases exist not as atoms but as molecules of two atoms joined together. The relative weights of the atoms, originally based on the lightest (hydrogen) being 1 unit, were called *atomic weights.* The *molecular weight* is equal to the sum of the weights of the atoms in a single molecule. A molecular weight in grams is called a *gram molecular weight* or a *mole* of the substance. It became possible to write *chemical equations* to indicate the rearrangement of atoms occurring during a reaction. Such an equation is *balanced* if no atoms are gained or lost during the reaction and all molecular formulas are correctly written. *Valence* is a number assigned to atoms to indicate their combining tendency with other atoms. Metals are assigned a positive valence and nonmetals a negative valence. Metals may be arranged in a series of relative chemical activity called a *replacement series* or *activity series* since more-active metals will replace less-active metals from compounds in solution. Only metals which are of limited chemical activity are found free in nature. Moderately active metals, such as iron, are *reduced* to metals from their ores using the element carbon. Very-active metals are reduced from their ores by electricity. This *electrolysis* was used extensively just after 1800 to isolate many active metal elements.

Mendeleev arranged the elements in a table. He found that the chemical properties of the elements repeated periodically with increasing atomic weight. In his table, atomic weight increases as each row is read left to right. Each time a series of elements occurred which repeated the properties of an earlier series, he arranged them in a new row, or *period,* under the previous elements. In this way he produced a table where each column represented a *family* of elements with similar chemical properties. From blanks in the table, Mendeleev was able to predict the existence of several unknown elements. By the study of where these blanks occurred in the family, he was able to predict most of the properties of these elements. When these elements were discovered, most of the predictions were verified. One entirely unknown family of elements was discovered after Mendeleev's work. This was the set of chemically inert gases, which perfectly fit a new column in Mendeleev's *periodic table.* Experience convinced many investigators that this grouping of the elements reflected something of fundamental importance and was not a chance arrangement. However, no one was quite certain why this pattern existed or what it meant.

Questions

1 Why are hydrogen gas and oxygen gas simpler substances than water?

2 Why isn't apple pie a compound? Why isn't concrete a compound?

3 Pure silica sand is silicon dioxide, SiO_2. When you look at the average beach or the sand in a normal sandbox, how do you know it is not pure silica sand?

4 Steel is an alloy of iron and carbon. Why is it not a compound of these elements?

5 Avogadro is an example of a person whose work was not appreciated in his own time. Do you recall any other examples?

6 What advantages do we get with balanced chemical equations over simple word equations or unbalanced chemical equations?

7 Mendeleev's table lists oxygen and sulfur in the same family. Oxygen is a colorless, odorless gas while sulfur is a yellow, powdery solid. These two react with each other to produce poisonous "brimstone fumes," or sulfur dioxide, SO_2. How can they be placed in the same family?

8 Why did the discovery of the inert gases provide strong support for Mendeleev's scheme for arranging the elements?

9 A Martian working in the ordering department of an assembly shop notices that every time an order is filled for bolts one is also filled for nuts. Furthermore, for every 6 tons of bolts 1 ton of nuts is needed. This ratio holds even for small orders, such as 120 kg of bolts and 20 kg of nuts. Our Martian has no idea what these items are, but several conclusions can be made nonetheless. What are they?

10 If all of a sudden the Martian in Question

9 was asked to order 1800 kg of bolts and 600 kg of nuts for a new product, what would the Martian conclude?

11 Why does the law of multiple proportions argue for the existence of a smallest piece of an element?

12 What would be three different strategies for obtaining the metal nickel from the ore nickel oxide?

13 What would be the name of the following binary compounds?

PbO	Na_3N	HCl
FeF	K_2O	Al_2O_3
$HgCl_2$	H_3N	Mg_2C

14 Name the following compounds.

$NaOH$	K_2CO_3	$Mg(NO_3)_2$
	NH_3Cl	H_2SO_4

Problems

1 What would be the correct formula for the following binary compounds?

zinc oxide	carbon chloride (dry-cleaning fluid)
aluminum chloride	iron nitride
sodium fluoride (fluorite)	mercuric oxide
calcium chloride (drying agent)	hydrogen oxide (cleaning agent)

2 Write the formula for the following compounds:

sodium sulfide	iron carbonate
hydrogen sulfate	zinc hydroxide
calcium nitrate	hydrogen hydroxide

3 Write a chemical equation for adding lithium metal to water.

4 Limestone is calcium carbonate. When it is roasted, as in the making of cement, it breaks down into calcium oxide (lime) and carbon dioxide. Write a balanced chemical equation for the roasting of limestone.

5 What is the molecular weight of the following compounds?

H_2O	$NaCl$	Na_2O	HF
CaF_2	MgO	FeO_3	

6 What percentage of a ton (1000 kg) of salt (sodium chloride) is made up of chlorine?

7 What percentage of the weight of water is hydrogen?

8 Hydrogen and oxygen also unite to produce the unstable compound hydrogen peroxide, H_2O_2. What percentage of the weight of hydrogen peroxide is hydrogen?

9 What is the approximate ratio of the percentages in Probs. 7 and 8? Does this agree with the law of multiple proportions?

10 How many grams of hydrogen peroxide, H_2O_2, would be required to have 1 mol?

11 Hydrogen gas and chlorine gas react to produce the gas hydrogen chloride. Write the balanced chemical equation for this reaction. How many liters of hydrogen gas would be required to react with 2.0 liters of chlorine gas? How many liters of hydrogen chloride would result?

12 What would the volume of 8.0 g of oxygen gas be at STP?

13 The standard ore of aluminum is bauxite, Al_2O_3, which is separated electrically into aluminum and oxygen gas.

(*a*) How much aluminum can be obtained from 1 ton (1000 kg) of its ore?
(*b*) What volume of oxygen can be obtained (STP) from the same ton of ore?

14 According to the Avogadro hypothesis, a fixed volume of any gas (or combination of gases) contains a fixed number of molecules (for example, 22.4 liters of gas contain 1 mol or 6×10^{23} molecules). The commonest molecule in air is that of nitrogen gas, N_2, with a

molecular weight of 28. As we raise the humidity, we add more and more molecules of water vapor to the air. Should this make the air more dense or less dense? Why?

15 Uranium has an atomic weight of 238. How many atoms are there in 1 g of uranium (about the mass of a dime)?

16 The Hindenburg Zeppelin was the largest airship ever built, with a length of 245 m and a volume of 2×10^5 m^3. Less well known is that the United States built several large airships. Largest of these was the U.S.S. *Akron* (length = 239 m), launched in 1931 and designed as an aircraft carrier. The *Akron* could launch and recover its five fighter planes while it was in flight. One option was available to the American airship denied the Hindenburg. The United States did (and does) control the world supply of helium gas, and we could use this in place of the inflammable hydrogen in the Hindenburg. Presume both of these airships had the same volume. What would be the relative lifting power of the two filled with these different gases? (Remember, the air is essentially N_2 with a molecular weight of 28.)

CHAPTER THIRTEEN

The Bonding of Atoms

Introduction Dalton, in formulating the atomic theory, pictured atoms as far from tiny, featureless spheres. As a working model for what might be happening he pictured atoms as having some form of "hooks" which might allow them to attach to each other. Such hooks could account for the combining tendency of atoms. For example, if each hydrogen atom had one hook, and each oxygen atom two hooks, the structure of water made sense. A hydrogen atom would attach to one hook of an oxygen atom, leaving the other one free to attach to a second hydrogen atom (Fig. 13-1). When this was done, all of the "hooking," or combining power, was used up, and the stable molecule H_2O resulted. Furthermore, such a scheme might indicate why isolated hydrogen atoms would form hydrogen molecules, H_2, rather than remain separate atoms. Still further, such a model could explain why hydrogen gas, H_2, and oxygen gas, O_2, could be mixed with no reaction occurring until a small spark caused them to combine explosively. The spark could provide the small amount of energy, called the *activation energy*, necessary to break H_2 and O_2 molecules into separate atoms. Once these hooks were freed from attachment to atoms of their own kind they would then be free to attach to the other kind of atom. This view of the activation energy is correct even though the concept of atomic hooks is not. Still, imagining each atom as though it had a number of hooks equal to its valence provided a rough explanation for a large number of reactions. While it helps human thought

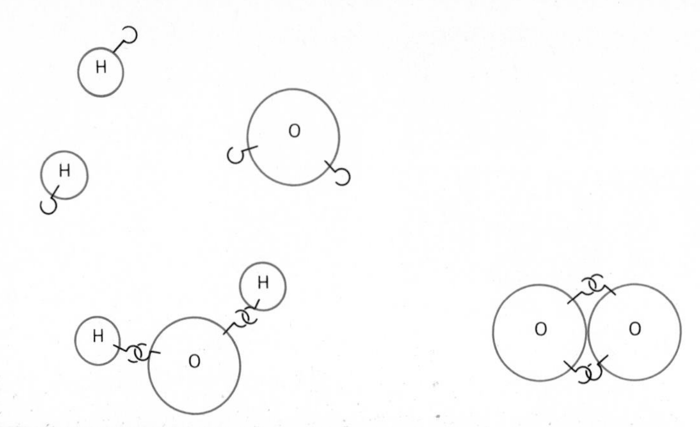

FIGURE 13-1
Some of the combining tendencies of atoms could be understood by imagining them to have various numbers of hooks.

greatly to have some kind of mental picture (or model) of what is happening, there were serious problems with the theory of the physical existence of hooks on atoms. One problem is in trying to explain the chemical preference of metals for nonmetals. Metals will mix with other metals readily, but generally they will not combine chemically with each other. One way out of the problem might be to suppose the connection between atoms to be "plugs and sockets." However, if we imagine metals to have plugs and nonmetals to have sockets, for example, we still have problems. This would not explain combinations between two nonmetals, such as nitrogen dioxide, NO_2. It also would not explain why two oxygen atoms in the absence of suitable metals for partners will combine with each other to form the much less active oxygen molecule, O_2.

A final problem with hooks or plugs and sockets as models for atomic interactions is that elements may have more than one valence or combining power. For example, how are we to interpret the two oxides of carbon, CO and CO_2? If carbon has the ability to "hook to" or *bond* four separate atoms, as displayed in methane gas, CH_4, and other compounds, how can one carbon bond to one oxygen and use up all of its connectors? The problem is even more severe with the oxides of iron, FeO, Fe_2O_3, and Fe_3O_4; or the oxides of nitrogen, N_2O, NO, NO_2, and so on. Some simple type of mechanical connector on atoms just will not explain the variability in combining power which some atoms demonstrate.

What type of mechanism allows one atom to attach to another? What clues do we have as to the nature of chemical bonding? The first important clue goes back to Cavendish in the late eighteenth century. He passed electricity through water and produced two colorless, odorless gases. At the negative wire, or *cathode,* the evolved gas burned with a pale blue flame in air. This was, of course, Cavendish's discovery of hydrogen. Likewise, at the positive wire, or *anode,* he discovered the gas later called oxygen. The point to be made here is that while water is extremely stable chemically, it could be separated electrically. As has been pointed out, Sir Humphrey Davy used the same method to separate several very active metals from their compounds, metals so active they could be separated in no other manner. If compounds may be separated by electricity, is it not reasonable to guess that they might be held together electrically? Faraday extensively investigated the process of *electrolysis,* as this separation by electricity is called. He came to several conclusions:

1 Not all materials could be *electrolyzed* or separated into elements by electricity. For example, most materials from living systems (which the chemist Berzelius called *organic compounds*) could not be separated at any reasonable voltage.

2 To be separated the molecules of a substance had to be rendered mobile somehow. That is, the substance either had to be dissolved in a liquid or heated until it liquefied. Even then, however, electrolysis was not guaranteed. For example, pure sugar dissolved in pure water cannot be electrolyzed.

3 If a substance could be electrolyzed, there was a minimal voltage that had to be supplied before the separation would begin. The more chemically active the metals and nonmetals involved in the compound were, the higher the voltage required to achieve separation. Thus, lead chloride, $PbCl_2$, could be separated more easily than sodium chloride, NaCl, since sodium is a more active metal than lead. Generally, however, the voltages required were not very high and were easily within the range of the batteries available.

4 The amount of an element produced was directly proportional to the total electric charge passed through the material. If a current of 1 amp flows through a pool of molten salt for 1 h, it produces 0.86 g of sodium metal at the negative cathode. It would require about 27 h of electrolysis to produce 1 mol of sodium at this rate. Expressed differently, the total charge required to liberate 1 mol of sodium would be 96,500 C. Whatever combination of time and current was used, whenever the total of 96,500 C was reached, 1 mol of sodium metal had been liberated at the cathode. During the same time, exactly 35.5 g of chlorine gas would bubble free at the positive anode. This amount of electricity, 96,500 C of charge (called 1 *faraday* of charge now), is the amount required to liberate 1 mol of any substance *provided it has a valence of* $+1$ *or* -1*!* The same amount of charge will liberate only $\frac{1}{2}$ mol of an element with a valence of $+2$, or $\frac{1}{3}$ mol of an element with a valence of $+3$, and so on.

Clearly, one conclusion we may draw from all of this is that there is a very definite, quantitative link between electricity and at least some chemical bonding; but what is it?

A Piece Provides a Puzzle

The next step in understanding the atom occurred in the most unlikely place imaginable for the study of the properties of matter—a vacuum. In 1879 William Crookes (1832–1919) reported on studies of the passage

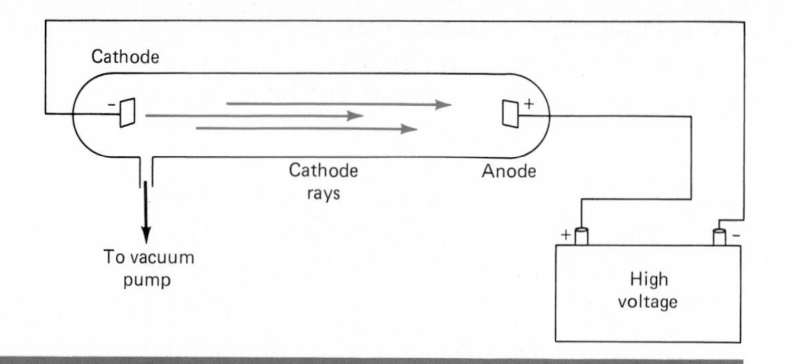

FIGURE 13-2
As the air is pumped from the glass tube, streams of particles, called cathode rays, streak across the tube from the negative cathode to the positive anode.

of electricity through a vacuum. He was using a sealed glass tube with a metal electrode at each end from which the air was evacuated (Fig. 13-2). At very high vacuum the electricity would still pass through the tube, provided a high voltage was applied across the electrodes. This transfer of charge was associated with a beam or ray which streaked continuously across the tube if the potential difference was maintained. The ray was repelled by a negatively charged plate and attracted by a positive one. Thus, it was concluded that the ray carried negative electric charge and streamed from the negative plate, or cathode, to the positive plate. From this, the rays were known as *cathode rays,* and the tube was called a *cathode-ray tube.* A ray implies a massless emanation such as a beam of light. Cathode rays could turn a paddle wheel suspended in the tube; thus the rays had mass and could transfer momentum. These rays were really a stream of very fast moving particles.

J. J. Thompson (1856–1940) became convinced of the particle nature of the cathode ray, and he named the particle the electron. While the particles were too small for Thompson to be able to measure their mass directly, he measured their charge/mass ratio, or q/m. The most highly charged particle yet measured had been the "electrified hydrogen atom" or *hydrogen ion* obtained by passing hydrogen gas through an electric arc. He found that electrons carried almost 2000 times as much charge for their mass as did these hydrogen ions. For some time the suspicion had been growing that electric charge came in small "chunks" or "pieces," just as matter had a smallest piece, referred to as an *elementary charge.* If this view was correct, the electron and the hydrogen ion might be expected to have the same size electric charge, namely, one elementary charge each. *But,* if that was true, the electron could have only about one two-thousandths the mass of the lightest atom known. The electron would clearly be smaller than the atom (in mass). It would be a *subatomic particle.*

Furthermore, where did the electrons come from? Electrons streaked from the cathode regardless of whether the cathode was lead, copper, aluminum, carbon, or any other electric conducting material. When the high voltage was turned off, these subatomic particles had to be residing in the system somewhere. Since electrons could be produced from any type of electric conducting matter, it seemed likely that the electron was a normal constituent of matter. In other words, *electrons were normally found in atoms;* they were tiny pieces of atoms. If this view was correct (and it was), *the atom must have some sort of internal structure.*

The suspicion grew rapidly that in watching the electrons streaking across a cathode-ray tube, we are seeing the electric current in a wire minus the wire.* Furthermore, the electrons which transit the tube come from the internal structure of the atoms making up the electric conducting system itself. Again, this view proved to be correct. Electrons are the charged particles that move from place to place during the course of all normal electrical phenomena. For example, an electric current is a sustained flow of electrons along a conductor, and the charging of an electroscope is just a reshuffling of the distribution of electrons. Furthermore, one or more electrons are found in *all* atoms. They are, therefore, one of the basic constituents of matter. Thompson went a long way in demonstrating many of these points from 1897 onward. It is interesting to note that this work was begun at a time when a large number of scientists were yet unconvinced of the physical reality of atoms. Resistance to the idea of atoms had been shrinking toward the end of the nineteenth century, and largely ceased within a generation of the discovery of the electron. This was not the result of a single key experiment. Rather, it was the growing weight of experimental results in widely diverse fields, which could be interpreted successfully only in terms of atoms, that finally convinced the skeptics.

A Model for the Atom

The experimental work which led to an understanding of the atom belongs to the early twentieth century. This development will be presented later in the text. The model presented here is quite probably an old acquaintance of yours from junior high school or even earlier. By your stage in learning, most individuals picture an atom as almost a miniature solar system with orbiting electrons swirling around a central blob. The most important thing to remember about any simple picture of the atom is that it is just that—a simple picture. At worst, such a picture is just plain wrong, and at best, it is not completely right. Having said this, we will list the features of a simplified Rutherford-Bohr model of the atom which *does* have great utility in interpreting chemical bonding. The features of the model are listed here without any experimental evidence.

* Actually, the electron stream is invisible. When this stream hits the glass, or a tiny amount of gas molecules in the tube, or a chemically coated target, these things give off light or *fluoresce.* It is by this mechanism that the beam is rendered visible. In case you have not recognized an old friend yet, American statistics suggest you have stared intently at a cathode-ray tube for over 20,000 h of your life by now. The largest modern use of the cathode-ray tube of 100 years ago is as a television picture tube. When a fine beam of electrons streaking across the tube strike the chemicals, called *phosphors,* coating the inside of the screen, the phosphors glow. A more intense beam of electrons produces a brighter glow; and if the beam is cut off, no glow results. The beam is bent at the back of the tube to strike different parts of the screen by two sets of electrically charged plates. By varying the position and intensity of the beam very rapidly, the screen glows in a pattern recognizable as a picture. Thus, a research apparatus used to unlock the secrets of atomic structure a century or less ago evolved to powerfully modify the nature of our society today.

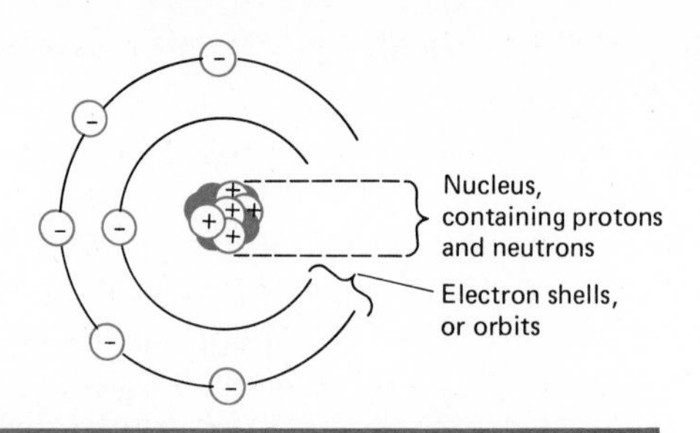

FIGURE 13-3
Representation of the parts of an atom using a simplified Rutherford-Bohr model. Protons and electrons are labeled with their positive and negative signs, respectively, while neutrons are shaded. If the electron shells were drawn to scale, all of the particles together would be smaller than the tiniest dot which could be visible on the paper.

1 Atoms are composed of three basic subatomic particles called *electrons, protons,* and *neutrons.*

2 The protons and neutrons account for almost all of the mass of the atom and are located in a tiny cluster in the center of the atom called the *nucleus* (Fig. 13-3).

3 Protons carry an elementary charge of + 1 and have a mass of 1 unit. The number of protons in the nucleus of an atom is called the *atomic number.* All atoms with the same atomic number are atoms of the same element. For example, all atoms with one proton are atoms of hydrogen, while all atoms with any other number of protons are *not* atoms of hydrogen.

4 Neutrons have a mass only slightly greater than protons, and they carry no electric charge. The sum of the number of protons plus neutrons in the nucleus of an atom is called the *atomic mass number* and is quite close to the mass of the atom.

5 Different atoms of the same element must have the same number of protons, but they may have different numbers of neutrons. Such atoms differing only in the number of neutrons are called *isotopes* of the element. The atomic weight of a large sample of an element represents the average mass of atoms in the mixture of the various isotopes of that element. Thus the element copper, which has two isotopes with atomic mass numbers of 63 and 65, has an experimentally determined atomic weight of 63.54.

6 The vast bulk of the volume of the atom outside its nucleus contains nothing but the electrons of the atom. Electrons have a mass only 1/1836 of the proton mass, and may therefore be regarded as negligible in normal circumstances. Electrons possess an elementary charge of − 1. This is exactly the same size as the charge of the proton, but it is of opposite sign. To be electrically neutral, an atom must possess the same number of electrons as it has protons.

7 Electrons are not randomly distributed throughout the volume of the atom but occur in *shells* or *orbits.* Each electron shell varies in the number of electrons it may contain. The innermost shell may hold up to two electrons. The second shell may hold eight electrons, the third eighteen electrons, and so on, with each shell having increased ability to

hold electrons. *Except,* whichever shell is the outermost shell will never contain more than eight electrons.

8 Since the nucleus resides in the center of the atom surrounded by one or more shells of electrons, it is unaffected by the normal interactions of matter. All touching of matter against matter as well as all chemical reactions are interactions of the electrons in the outermost shells of atoms with each other. Thus, chemical properties are determined primarily by the outermost shell of electrons called the *valence shell.*

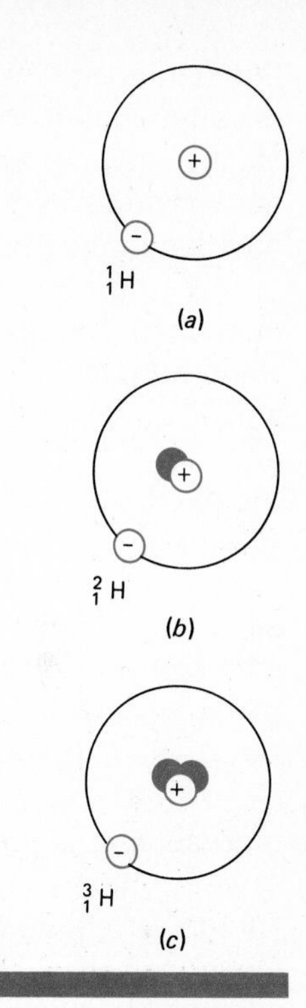

FIGURE 13-4 The three isotopes of hydrogen—(*a*) normal hydrogen, (*b*) heavy hydrogen or deuterium, and (*c*) tritium.

To develop a picture of an atom from this model we will start with hydrogen, the simplest atom. Hydrogen has the lowest possible atomic number of 1. Thus, each hydrogen atom must have exactly one proton in its nucleus in order to be hydrogen. The most common hydrogen atom has an atomic mass number of 1 also. This informs us that this hydrogen atom has only one particle in its nucleus, which must be its one proton. Since the atom is electrically neutral, it must contain one negative electron to balance the positive charge of its single proton. We now have all of the necessary parts for the simplest hydrogen atom—one proton, no neutrons, one electron. We may express this symbolically as ^1_1H, where the upper number is the mass number and the lower number is the atomic number. This is drawn in Fig. 13-4*a*, with the proton indicated in the nucleus at the center of the atom by a plus sign. The single electron is shown in the first electron shell out from the nucleus and is indicated by a minus sign. This is the simplest possible atom that can exist by the rules of the atomic model. However, it is entirely possible to have a more complex atom which still remains hydrogen. About 1 in 7000 atoms of hydrogen have an atomic mass number of 2 rather than 1. Since one, and only one, of these particles must be a proton in order to have an atom of hydrogen, the second particle must be a neutron. Figure 13-4*b* shows a representation of this atom of *heavy hydrogen* (^2_1H) or *deuterium* (^2_1D), as it is frequently called. (The separate letter D is frequently used for deuterium even though it is a species of hydrogen atom.) Whatever normal hydrogen atoms will do chemically, deuterium atoms will also do. The only distinction between these two isotopes of hydrogen is that the deuterium atom has twice the mass. The increased mass of deuterium produces a difference in the rate at which it performs the same functions as normal hydrogen. This is a bit like identical twins who happen to weigh 150 and 300 lb, respectively, but with *no difference in size, shape, or appearance.* To all simple indicators they might be indistinguishable. The difference, however, would rapidly appear if the twins competed in the high hurdles or the pole vault. Similarly, ^1_1H atoms react chemically about 10 times as rapidly as ^2_1D atoms. Likewise, compounds prepared from pure deuterium vary somewhat in their physical properties from the same compound containing only normal hydrogen atoms. Water made from deuterium is called deuterium oxide, or *heavy water,* and is usually written as D_2O rather than H_2O. Heavy water boils about a $1\frac{1}{2}$°C above normal water and freezes at a temperature almost 4°C higher. There is a third isotope of hydrogen with two neutrons in its nucleus. This atom, ^3_1H, is called *tritium* (Fig. 13-4*c*). It is very rare since the nucleus containing two neutrons is *unstable* and comes apart in time. If every person on earth represented a hydrogen atom, we would need 100 million earths to have one person representing a tritium atom.

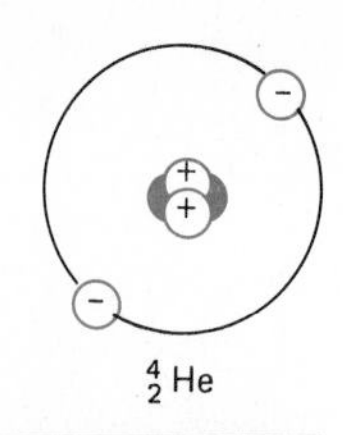

FIGURE 13-5
The helium atom, a particularly stable arrangement of particles.

The next heaviest atom after the set of hydrogen isotopes is the helium atom. The common form of helium has an atomic mass number of 4, while all helium atoms have an atomic number of 2. This atom may be written as ^4_2He. This implies that the atom has two protons and two neutrons in its nucleus. To be electrically neutral, the atom must also have two electrons, both of which fit into the first shell, as is drawn in Fig. 13-5. This particular combination of parts is incredibly stable, and helium is, of course, an inert gas. Helium refuses to combine chemically with any other type of atom under any circumstances at any time. Furthermore, helium atoms do not even display the courtesy, common among atoms, of being weakly attracted to their own kind. At temperatures near absolute zero helium atoms remain independent of their neighbors.

With the atom of lithium, ^7_3Li, we encounter the first complication in representing atoms. The lithium nucleus presents no problems, having seven particles, three of which are protons and the other four neutrons. The complication is that lithium atoms have three electrons, but the first electron shell will hold only two electrons. Thus, with the lithium atom we must begin filling the second electron shell as shown in Fig. 13-6*a*. If you glance at the periodic table, you will note that lithium is the atom which begins the second row (the second period) of the table. To determine whether this is coincidence we may jump out of sequence and draw the atom of sodium, which begins the third period (Fig. 13-6*b*). Checking the periodic table, we find that sodium has an atomic number of 11 and an atomic weight of 22.99. Since the atomic weight represents the average weight of a sodium atom, we may guess that the commonest isotope of sodium has an atomic mass number of 23. Thus, the sodium atom we will draw is $^{23}_{11}Na$. Again, the nucleus presents no new problems. It contains 23 particles, 11 of which must be protons. Thus, the nucleus consists of 11 protons and 12 neutrons. If there are 11 positively charged protons in the nucleus, there must be 11 negatively charged electrons in the elec-

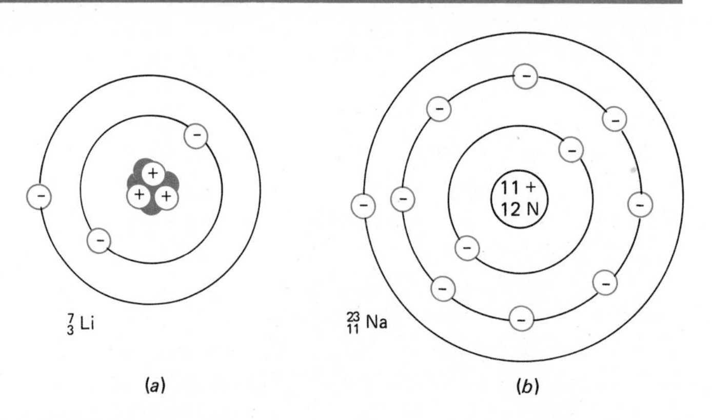

FIGURE 13-6
The two lightest alkali metals: (*a*) lithium (Li), and (*b*) sodium (Na). No attempt is made to draw separately the 23 particles in the sodium nucleus.

tron shells to produce a neutral atom. Two of these electrons will fit in the first electron shell and eight in the second shell, making a total of ten. This leaves us with one additional electron left over from our eleven. Thus, with sodium it is necessary to begin to fill the third electron shell, just as with sodium we begin the third row of the periodic table. Indeed, we have found that the first row, or period, of the periodic table contains all the elements with only one shell of electrons. Furthermore, the second period of the table contains all elements with electrons in two shells. With sodium we begin the third period and require a third electron shell. This third period will contain all atoms with three shells containing electrons. In general, each atom is listed in the same period as the number of shells of electrons found in that atom; e.g., all atoms in the sixth period will have six shells of electrons.

Toward a Model of Chemical Activity

Using our atomic model, placing the atoms in periods makes sense based on the *configuration,* or arrangement, of electrons in shells. What in the structure of the atom makes a chemical family into a family? What property makes the chemical behaviors of different atoms be quite alike? We have already drawn the structure of the atoms of two members of a family—lithium and sodium, both alkali metals. Let us draw a third member of the family of alkali metals and determine whether there is any common structure. The third alkali metal is potassium, $^{39}_{19}K$, with 19 protons, 20 neutrons, and 19 electrons in its structure. Of the 19 electrons, 2 will fill the first shell and an additional 8 will fill the second shell. By the rules of our atomic model, the third electron shell may hold up to 18 electrons, *but not if it is the outermost shell of electrons.* Whichever shell is outermost may never contain more than eight electrons regardless of its total capacity for electrons. With potassium we are left with nine electrons after filling the innermost two shells. Placing all nine of these in the third shell violates the rules, and we are required to place only eight in the third shell. We must begin to fill the fourth electron shell by placing the only remaining electron there. Rather than draw all of this detail, Fig. 13-7 indicates the electrons in the outermost shell only. Potassium is seen as the fourth atom in the first column. This process also produces the commonality of structure between lithium, sodium, and potassium we are seeking. Examination of the drawings of the three atoms reveals that the only common point between the three is that they all have just a single electron in their outermost shell, which could hold eight electrons. Since this outermost shell is physically the outermost part of the atoms, this is the only part of the atom that normally comes into contact with the outside world. Thus, lithium, sodium, and potassium are extremely similar in the part of their structure exposed to contact with other atoms. All have just a single electron in their valence shell, which could contain up to eight electrons.

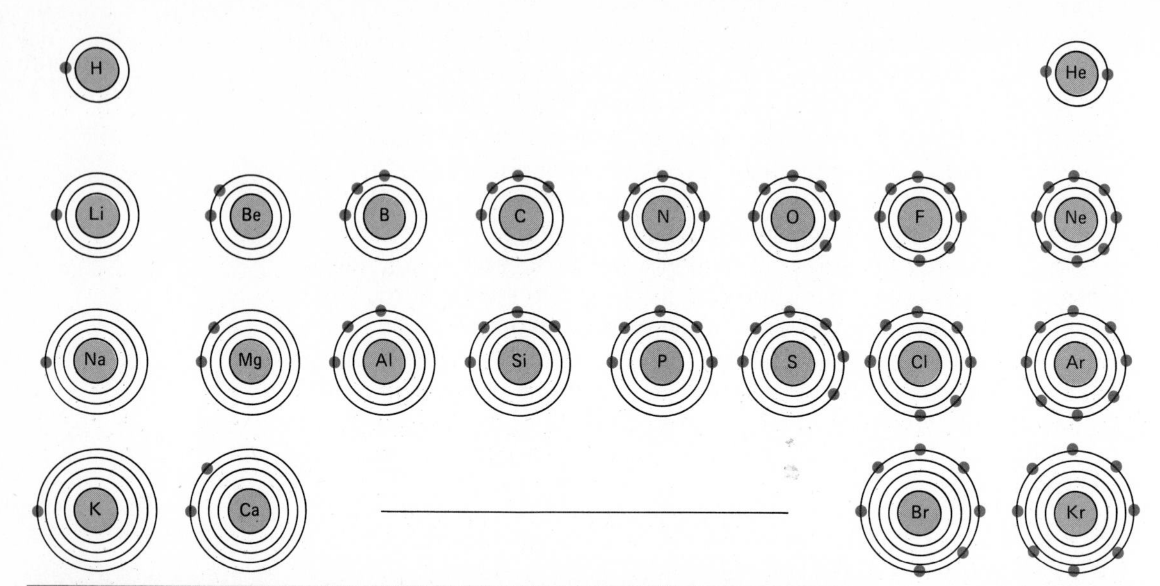

The single electron in the outermost shell accounts for the similarities between hydrogen and the alkali metals, even though hydrogen is quite different in its reactions. Hydrogen, however, has its single electron in a shell that can hold only two electrons, and under its valence shell hydrogen has nothing but its nucleus. The alkali metals have their single valence electron in a shell that can accommodate eight electrons, and under the valence shell each atom displays a complete shell of electrons with no vacancies.

FIGURE 13-7
Simplified diagram of the atoms arranged in the order of the periodic table. Only the electrons in the valence shell are drawn in.

Chemical Reactions and Energy

An extremely important observation about chemical reactions has not been mentioned up to this point. *All chemical reactions that proceed spontaneously release energy.* This is usually in the form of thermal energy or increased kinetic energy of the molecules, but vigorous reactions give off light in a flame as well. Slow reactions, such as the rusting of iron, give off energy so slowly as to be imperceptible without delicate instruments. Reactions liberating energy are called *exothermic reactions.* The opposite type of reaction, called an *endothermic reaction,* absorbs energy. All chemical reactions involve changes in energy. If the reaction occurs spontaneously, energy is always given off and the products are in a lower energy state than the reactants were. This is very much like the spontaneous motion of a bowling ball. If the ball moves by itself, it will always move downhill to a lower potential energy state. The lower the ball becomes compared to its surroundings, the less likely it is to be disturbed, or, the more stable it is.

Chemical reactions are very similar. To separate hydrogen and oxygen from water requires energy. In fact, a completely balanced equation would list the amount of energy required since it is as much a part of the reaction as the atoms are. The complete reaction might be written as

$$2H_2O + 136 \text{ kcal} \rightarrow 2H_2 + O_2$$

If the 136 kcal of energy is not provided, the water will most certainly *not* split up into the separate elements. In this sense, the energy is one of the reactants. Likewise, it should not surprise you that the amount of heat produced when 2 mol of hydrogen gas is burned in oxygen gas is exactly 136 kcal:

$$2H_2 + O_2 \rightarrow 2H_2O + 136 \text{ kcal}$$

In this reaction, a tiny amount of energy must be supplied to start the reaction, but this is returned plus an additional 136 kcal/mol of oxygen gas consumed once the reaction is started. By analogy, a bowling ball sitting in a small depression on the side of a mountain might require a small amount of energy to get up over the lip; however, the ball returns this energy and much more on its way down the mountainside. The amount of energy involved in a reaction is called the *heat of reaction,* ΔH. Highly endothermic reactions produce materials as products which are quite dangerous because of the large amount of energy they will return if the reaction reverses. Such compounds are *unstable,* and the addition of a small amount of energy can cause their explosive breakdown. Conversely, compounds formed in exothermic reactions are *stable.* The more heat liberated in the reaction the more stable is the product formed. The formation of rust from iron and oxygen liberates kilocalories of energy as more rust is formed. This is liberated rapidly if the reactants are heated enough to begin burning, or it is liberated slowly if the reaction is allowed to proceed slowly on the moistened fenders of your automobile. The heat of reaction is great enough to ensure that rust is quite stable and difficult to restore to iron metal and oxygen gas. However, when formed from its elements, the oxide of aluminum has a heat of reaction of almost 400 kcal/mol and represents a more stable compound than rust. Thus, as we have seen, aluminum metal may displace iron from its oxide and form the even more stable aluminum oxide. *All spontaneous chemical reactions move toward more stable, lower energy states and release energy in the process.*

Having formed this conclusion, we may now return to the question of the bonding of atoms. Which atoms possess a structure that represents the most stable, lowest energy states? Clearly, the most stable atoms are those of the inert gases, which are so stable they do not react. This lack of reaction must imply that there is no lower energy state available to these gases—they represent the bottom of an energy pit. If we examine the structure of the inert gases helium, ${}^{4}_{2}He$; neon, ${}^{20}_{10}Ne$; argon, ${}^{40}_{18}A$; and krypton, ${}^{84}_{36}Kr$; we see that there *is* a similarity in structure as shown in the last column of Fig. 13-7. Can you identify it? Each inert gas has a full, or completed, valence electron shell. The complete outer shells of the inert gases are the likely cause of their stability. Indeed, the difference one electron can make in chemical behavior is extreme. The fluorine atom is the most chemically active of all nonmetals. Comparing this atom with the neon atom in Fig. 13-7, we see that the only difference in its electron structure is that fluorine lacks one electron to complete its valence shell. If we examine the drawing of the active metal sodium, we

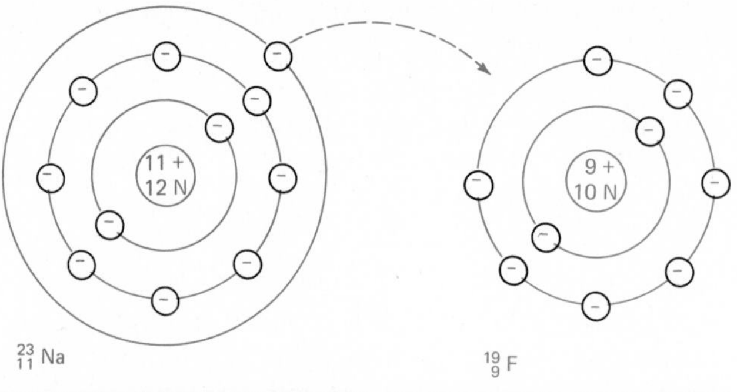

FIGURE 13-8
In electrovalent or ionic bonding, the valence electron or electrons of a metal atom (here, Na) are physically transferred to the vacancies in the valence shell of a nonmetal (here, F). After the transfer, the outermost shell of each atom involved must have its full complement of electrons, as do the inert gases.

note that its electron structure is also identical to that of neon except for the single electron in its valence shell. If fluorine were capable of gaining an electron, it would have the electron configuration of neon. Likewise, if the sodium atom could lose its single valence electron, it would also resemble the neon electron structure. What would be more reasonable than to imagine that both atoms do exactly this?

When a sodium atom is near a fluorine atom, a *physical transfer of an electron occurs.* The sodium atom *loses* or gives up its single valence electron to leave it with the same electron structure that neon has (Fig. 13-8). The electron that is lost by the sodium atom is gained by the fluorine atom to fill the single vacancy in its outside shell. Of course, if a neutral atom gains or loses electrons, it is no longer neutral; it is electrically charged. Such an electrically charged particle that results from the transfer of electrons is called an *ion.* The sodium atom loses an electron to become the *sodium ion* with a charge of +1. Similarly, the fluorine atom began with nine positive protons and nine negative electrons. When this atom gains a tenth electron it becomes the *fluoride ion* with a charge of −1. While sodium atoms are nasty, violently active objects to handle, sodium ions are actually necessary in your diet for the proper functioning of your nervous system. Likewise, while fluorine is a killer gas that maimed a century of chemists trying to produce and contain it, fluoride ions are actually good for your teeth and are added to drinking water. Thus, the properties of the ions are not at all the same as those of the atoms they come from. Furthermore, if the sodium ion has a positive charge and the fluoride ion has a negative charge, they will electrically attract each other. This attraction allows the product, NaF, to be held together in a definite structure.

The type of bonding between atoms accomplished by the gaining of electrons from one atom by another atom is called *ionic bonding.* The compounds that result from such bonding are called *ionic compounds,* and the ions that form them are held together by their electrical attractions for each other. Since the forces bonding ionic compounds are entirely electric in nature, ionic compounds may be separated electrically. In our sample compound, the sodium ion (written as Na^+ to distinguish it from the sodium atom) and the fluoride ion, F^-, are held together by the electrical attraction of their opposite charges. If the ions are rendered mobile by either dissolving the NaF or melting it, the sodium ions will migrate toward a negatively charged cathode. The battery supplies an excess of electrons to the cathode. If the voltage is great enough, the repulsions of these electrons for each other combined with the attraction of the sodium ion causes an electron to leave the cathode, join the ion, and produce the neutral atom once again. Meanwhile, at the anode the battery has caused a deficiency of electrons to build up, leaving it positively charged. When this deficiency is great enough, it exerts a greater attraction for the excess electron of the fluoride ion than the ion does itself. Thus the anode pulls the electron away from the ion leaving a fluorine atom. This atom quickly finds a companion and becomes fluorine gas, F_2. This all may be summarized in a set of chemical equations:

Forming the compound: $2Na + F_2 \rightarrow 2NaF$

Melting the compound: $NaF \rightarrow Na^+ + F^-$

Electrical separation:

At the cathode $Na^+ + e^- \rightarrow Na^0$

At the anode $F^- \rightarrow F^0 + e^-$

The electrical separation does not occur until the battery voltage is great enough to supply as much energy in the last two steps as the elements released in their formation.

This view of ionic bonding leads us to the ability to redefine some terms. A *nonmetal* in this view is any type of atom which tends to acquire additional electrons in a chemical reaction. A *metal* is any type of atom in which the outer electron or electrons are weakly held to the atom and tend to be lost in a reaction. This explains why metals are good conductors of electricity. In a cluster of metal atoms the valence electrons are so weakly held to a particular atom that they drift freely from one atom to the next. While these valence electrons are continually in the process of drifting through the structure, a "frozen picture" of the situation may be visualized (Fig. 13-9). The metal atom minus its valence electrons is referred to as the *kernel* of the atom, and is represented by the large sphere. Electrons drift in the spaces between the kernels while the electrons and kernels exert electric forces that hold the structure together. This view allows us to picture the origin of several properties of metals. For example, if a voltage is applied across the ends of the wire, the electrons experience a force (as do the kernels in the opposite direction, but

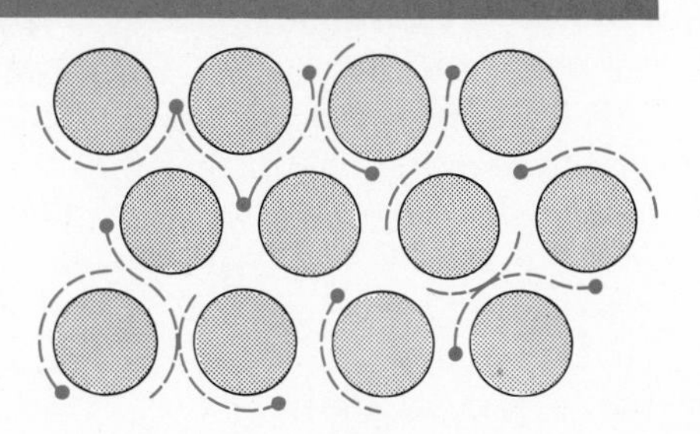

FIGURE 13-9 In a metal, the outer shells of adjacent atoms merge so that the valence electrons may drift from atom to atom. The inner "kernel" of the metal atoms are shaded, and only valence electrons are indicated as separate particles.

they are too massive to respond). The electrons respond to this force and their randomly directed drift becomes drift in the direction of the force acting on them. As electrons drift off the end of the wire into the positive terminal of the battery, the negative terminal is supplying replacement electrons at the other end. Actually, because of the tiny mass of the electron and the small resistance to their motion by the kernels, this drift is very rapid—a respectable percentage of the speed of light. Some metals, such as the Nichrome heating wires in a toaster, do offer more resistance to this flow of electrons. When this occurs, the electrons proceed in a series of jumps, losing their kinetic energy at each kernel they stop at. This lost kinetic energy is transformed into the heat that makes your toast. This model of metallic structure also allows a ready interpretation of the ability of metals to deform without breaking. The pieces in the structure, electrons and kernels, are not rigidly held to each other and will adjust their positions in response to outside forces applied to the metal.*

Ionic bonding nicely accounts for the ratio of atoms in which a metal and a nonmetal will combine. For example, the magnesium atom has two electrons in its valence shell. To revert to the electron structure of an inert gas it must lose both of these. The fluorine atom, as we have seen, has a single vacancy in its valence shell and will accept only one electron. Therefore, one magnesium atom will combine with two fluorine atoms, giving one electron to each (Fig. 13-10*a*). This results in the ionic com-

* Perhaps the most interesting point about the ionic theory of bonding is that most of its details were worked out before there was any evidence for a structure to the atom. The ionic theory was published in 1884 by Svante Arrhenius (1859–1927) before Thompson found the first subatomic particle and named it the electron. Arrhenius asserted that atoms could acquire positive and negative charges, and that the resulting ions (as he named them) would be held together electrically. He went on to a complete description of the process of dissolving ionic compounds and the electrical separation of the basic elements—all without an idea of atomic structure.

While it was a remarkable achievement in our view, it almost cost Arrhenius his degree. He was 25 years old and working on his Ph.D. The ionic theory was his research for the doctorate. Many of his professors were very unconvinced that atoms could acquire charges. In fact, in 1884 there were still holdouts who could respectably argue against the existence of atoms at all. Grudgingly, he was allowed to graduate. Less than a generation later, in 1903, he was awarded the third Nobel Prize granted in chemistry for this work, so the story had a happy ending.

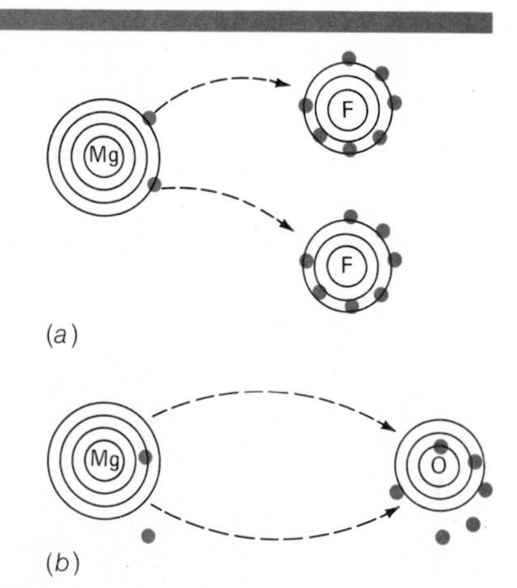

FIGURE 13-10 Magnesium atoms have two valence electrons to lose in ionic bonding. (*a*) Since fluorine atoms have only one vacancy to fill, it takes two fluorine atoms per magnesium atom to form a compound. (*b*) Because oxygen atoms have two vacancies in their valence shell, they will combine with magnesium atoms 1/1.

pound MgF_2. Meanwhile, if the same magnesium atom were to combine with oxygen, we find they would combine one atom to one atom. This is reasonable since the oxygen atom has two vacancies in its outer shell, while the magnesium atom has two electrons to donate (Fig. 13-10*b*). Examination of the valence numbers we used previously to determine molecular formulas again agrees with this model of ionic bonding. We may see that the valence listed is just the charge which will reside on the ion after bonding. Oxygen with a valence of -2 tends to acquire two electrons when combining, and the oxide ion has a charge of -2.

Covalent Bonding

Unfortunately, ionic bonding does not explain all of the cases of atomic combinations we have become familiar with. For example, by what mechanism do two oxygen atoms join together to produce the oxygen gas molecule, O_2? Surely two atoms with an identical tendency to acquire electrons will not gain or lose electrons in interacting with each other. Any mechanism for the physical transfer of electrons from one oxygen atom to another would leave one of the two atoms not resembling an inert gas structure at all. Furthermore, how would ionic bonding account for the fact that two elements may combine in more than one way, as with CO and CO_2? Another mechanism for atomic bonding is necessary to explain such cases.

A problem is most apparent if we examine the structure of the carbon atom. Carbon is element number six and has an atomic mass number of 12. Thus, the typical carbon atom has six of each type of particle—protons, neutrons, and electrons. Figure 13-11 shows the representation of the carbon atom by the Rutherford-Bohr model. Note that carbon has four electrons in its outside shell. If carbon were mixed with a strong metal having a tendency to donate electrons, we might expect the carbon atom to acquire four electrons to complete its outside shell. Imagine the situation after the carbon atom has acquired *three* of these electrons. The ion still has six positively charged protons, but now it has nine negative electrons for a total charge of -3. Being all negatively charged, electrons repel each other. The fourth excess electron for the carbon atom will have to approach and join the structure of the carbon. A great deal of work will have to be done "pushing" the electron into place against this electrical repulsion. As the charge becomes greater on the ion, at some point more energy is required to make it accept another electron than is liberated as the electron joins the structure. In the analogy of a bowling ball on the earth's surface, more work is done rolling the ball up to the edge than is released when the ball rolls down into the pit on the other side. As a result, there is no energy advantage in acquiring the electron, and a C^{4-} ion is unlikely unless the energy is supplied externally.

FIGURE 13-11 Representation of the carbon atom.

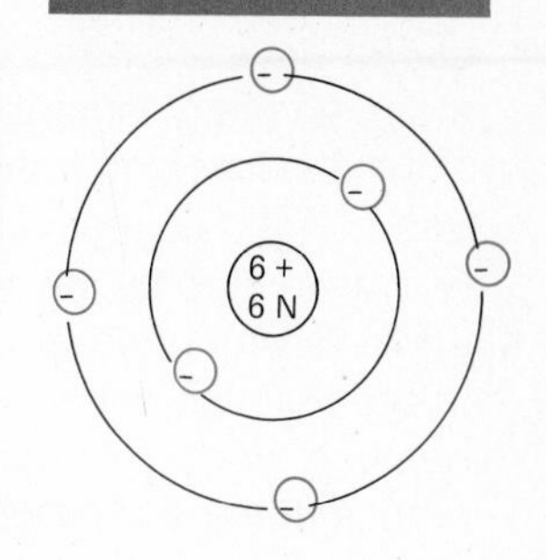

Of course, it is reasonable to imagine the carbon atom proceeding in the other direction in a chemical reaction. With four electrons in its valence shell, carbon could lose these four to attain an inert gas structure—it could wipe out the second shell entirely. However, this is not an easier problem than gaining an additional four electrons. Picture the ion that has lost three of these valence electrons and has one more to lose. The ion now has its original six protons but has only three electrons including the one we wish to lose. The ion has an unbalanced charge of +3, which is attracting the negative electron. A considerable force will have to be exerted across a distance to pluck the last valence electron away from the carbon. The energy which must be put into the freeing of this election is greater than the stability gained by completely removing the valence shell of electrons. Thus it is difficult to see how carbon atoms could form bonds by either gaining or losing electrons. Carbon atoms actually bond by a different scheme called *covalent bonding.*

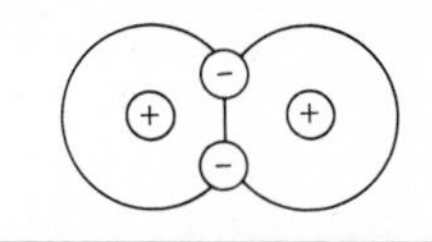

FIGURE 13-12 Covalently bonded hydrogen atoms in the hydrogen molecule, H_2. Neither electron "belongs to" either atom—they are shared equally.

A similar situation exists with the hydrogen atom, which has one electron in a shell which could hold two. Losing this electron would leave just the positively charged proton of the hydrogen nucleus behind. If this were to happen, the atom would essentially disappear since the nucleus takes up a negligible volume of the atom. However, gaining an electron seems just as improbable since the ion would then have twice as much negative charge as positive. In the hydrogen molecule H_2 we can picture the situation as close to the case of two perfectly matched bullies. Each has a single lollipop in one hand and wants one for the other hand. Since they are evenly matched, it is difficult to see how one will wrestle the lollipop away from the other. If one does succeed, however, this ties up both hands while leaving the opponent with both hands free for the next phase of the struggle. It is easiest to picture this match as winding up with both bullies having one hand on each lollipop, pushing and pulling to wrest each from the other. In doing this, interestingly enough, each has attained his original goal of having one lollipop in each hand, although without complete possession. Minus the personalities, covalent bonding is quite similar. As in all bonding, the "struggle" is to attain the lowest possible, most stable energy state. This is attained by the sharing of electrons in the case of the hydrogen gas molecule. Effectively, each atom throws one electron into the pot and both atoms share the pot. Figure 13-12 pictures the hydrogen gas molecule. Neither electron "belongs to" either nucleus any more. (If you need a mechanical picture, you may visualize the electrons as doing figure eights or large loops about both nuclei. While not particularly correct, such a mental model will serve you well.)

Bonding between atoms achieved by the sharing of electrons is called covalent bonding. Electrons are shared only in pairs, with each atom normally contributing one of the electrons to each pair to be shared. Covalent bonds are quite common. Molecules of elemental gases with two atoms joined are all formed by covalent bonding. Thus, H_2 was formed by the sharing of a pair of electrons. The fluorine gas molecule is produced in a similar fashion (Fig. 13-13). When the electrons in the valence shell are counted, both electrons in the shared pair are counted for each fluorine atom. Thus each fluorine atom in F_2 would have eight electrons in its valence shell. Each atom has a complete outer shell with the same electron configuration as the inert gas neon.

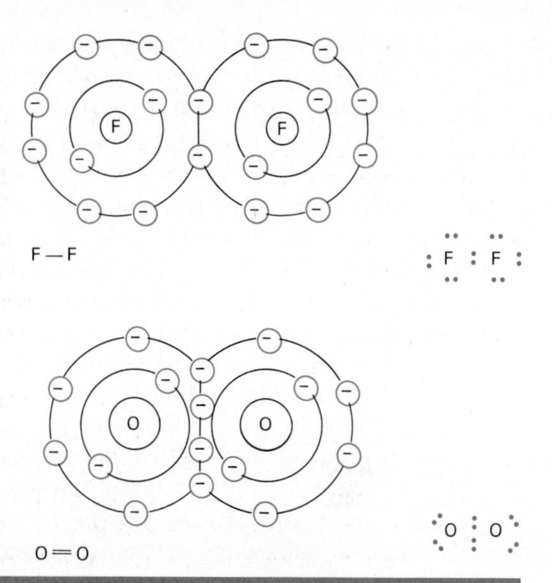

FIGURE 13-13 Covalent bonding in the F_2 and O_2 molecules is indicated by a complete drawing of the electrons, electron-dot diagrams, and by dashes representing a single shared pair of electrons. Note that the bond between oxygen atoms is a double covalent bond, requiring the sharing of two sets of four in all electrons.

Oxygen atoms also form molecules with two atoms each, or *diatomic* molecules. The oxygen atom has six electrons in its valence shell, and thus has two vacancies which must be filled to produce a stable molecule. The sharing of one pair of electrons between oxygen atoms will not be enough since this still leaves a vacancy in each atom's outer shell. Oxygen atoms must share *two pairs* of electrons, for four common electrons in all, to provide each atom with a completed valence shell (Fig. 13-13). When two pairs of electrons are shared between atoms, the resulting bond is known as a *double covalent bond,* or *double bond* for short. Actually, *triple bonds* between atoms are also common, as in the nitrogen molecule, N_2. In this case, three pairs of electrons are shared with each atom, contributing one electron to each shared pair.

One of the most important of all covalent molecules is the water molecule, H_2O. The oxygen atom has two vacancies in its outermost shell. It may fill each of these by sharing a pair of electrons with a hydrogen atom

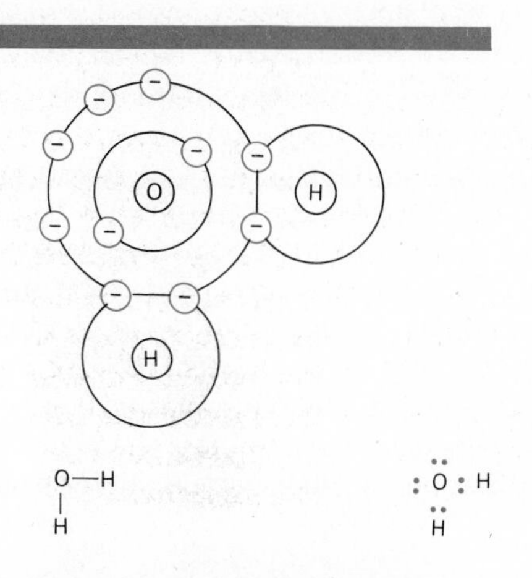

FIGURE 13-14 Structure of the covalently bonded water molecule indicated by three separate schemes.

as illustrated in Fig. 13-14. While drawing such molecules helps us picture the arrangement of parts in the molecule that is formed, drawing is time-consuming. A faster scheme of drawing results when we realize that only the valence shell of electrons is involved in the formation of these molecules. We can represent the atom beneath the valence shell by its atomic symbol, Na for sodium, or N for nitrogen, for example. To this we may add dots to represent the electrons in the valence shell as in Figs. 13-13 and 13-14. This *electron-dot model* of the atom allows us to work out covalent bonding without representing much of the unnecessary detail. An even simpler representation of covalent binding is available for simple cases. If we let a single dash represent a pair of shared electrons between those atoms, we may indicate normal covalent bonds rapidly. For example, H_2 becomes H—H, or O_2 becomes O═O. These representations contain all of the information presented in the more detailed drawings. We will use the dash between atomic symbols frequently to represent a covalent bond for the balance of the book. However, this indicates only normal chemical bonds with ease. Unusual cases of bonding are best represented by electron-dot drawings.

The Process of Dissolving

When a purely ionic compound is placed in water, it *dissolves*. That is, the electrical attractions holding positive ions and negative ions together are somehow overcome, and the ions separate. For example, if we watch this process as a salt crystal dissolves, we see the crystal shrink in size and disappear. Our taste will reveal that the salt winds up intimately mixed with the water it dissolved into, since, in time, all parts of the water taste salty. Furthermore, there is a limit to the ability of water to dissolve the salt. At a given temperature a certain amount of water will dissolve a certain amount of salt. Any excess salt remains undissolved. What does this imply?

Water is the foremost material for the pulling apart of ionic compounds. It will dissolve a greater range of materials than any other common substance. Yet, water itself is not ionic. Chemically pure water will not conduct a current or be separated by electricity into hydrogen and oxygen. [Truthfully, a better statement is that water is overwhelmingly nonionic. Experimental measurements reveal that about 1 water molecule in every 10 million (10^7) breaks apart into H^+ and OH^-!] Yet, the fact remains that compounds held together by the electrical attractions of their parts do come apart in water. Thus, water is an electrically neutral covalent compound that retains the ability to "pluck apart" molecules which are held together electrically. It does this because, while water is an electrically neutral molecule, not all zones of the molecule are neutral. Simply stated, the oxygen atom in the water molecule is an "electron hog" and keeps more than its quota of the set of electrons shared with the hydrogen atoms. Although the molecule is neutral, the oxygen atom is somewhat negative while each hydrogen atom is partly denuded of its cover of electrons. Thus the molecule has a negative zone and two positive zones. All covalent bonding occurs at strictly regulated bond angles. Normally, when oxygen bonds to two atoms, these bonds are not on opposite sides of the oxygen atom but are at right angles to each other (see Fig. 13-14). In water we thus might expect the two hydrogens to be attached with a 90° separation as seen from the oxygen nucleus. This is not the case. In monopolizing more than its fair share of the pair of shared electrons, the oxygen atom makes each hydrogen atom somewhat positive. The hydrogen atoms thus repel each other, and the bond angle between them stretches from the normal 90° to over 104°.

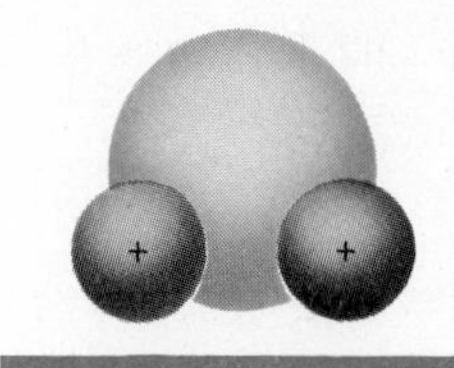

FIGURE 13-15 The actual structure of the water molecule leaves both attached hydrogen atoms partially denuded of their electrons and thus positively charged. The opposite zone on the oxygen atom has more than its fair amount of electron coverage and is thus negatively charged.

A neutral molecule with positive and negative zones is called a *polar molecule.* Water is the commonest example of such a molecule, and its polarity explains many of its unusual properties. For example, based on molecular weight, water should be a gas at normal temperatures. Water has a molecular weight of 18 and is a liquid at room temperature. Yet, oxygen gas has a molecular weight of 32 and is a gas, as is N_2 with a weight of 28, and even CO_2 at 44. Normal molecules with a weight under 60 or so are gases at room temperature. The motion required at this temperature is great enough to "jiggle" them free of their kind and have them act as independent particles. (Recall that temperature is a measure of the average kinetic energy of the particles of a system, $\frac{1}{2}m\bar{v}^2$.) Given the mass of a water molecule, and its average speed of motion at room temperature, other molecules can break loose of any attractions for their companions, but not water! Water molecules have an incredible tendency to stick to each other due to their polarity. The positive portions of one water molecule are attracted to the negative portion of another. In this manner water remains a liquid at normal temperatures. The tendency for forming this type of polar molecule is characteristic of hydrogen atoms. These atoms are comparatively so small that they are frequently denuded of an equal share of the electrons in a covalent bond (Fig. 13-15). The hydrogen compounds of nonmetals usually have higher melting and boiling points than molecules without hydrogen but with similar molecular weights. Thus, ammonia, NH_3, behaves a bit like water in its unusual physical properties. (For this reason science-fiction writers can speculate about life forms based on NH_3 rather than H_2O as a solvent at a different range of temperatures and pressures than is found on earth.) It too is polar with significant attractions between molecules due to its hydrogen atoms. On the other hand, methane, CH_4, behaves normally and indicates no unusual attractions between molecules. This indicates that the electrons in the covalent bonds are shared evenly between the carbon atom and each hydrogen atom. The attraction of some hydrides, such as water, for molecules of their own kind is really another form of bonding, called *hydrogen bonding.* Hydrogen bonds are caused by the tendency for attached hydrogen atoms to be partially positively charged and attract negatively charged particles. Ionic or covalent bonds customarily have energies on the order of 100 kcal/mol of bonds. (For example, 1 mol of NaCl would represent 1 mol of ionic bonds, while 1 mol of $AlCl_3$ would represent 3 mol of bonds.) Hydrogen bonds represent an energy of only about 5 kcal/mol of bonds. Thus, hy-

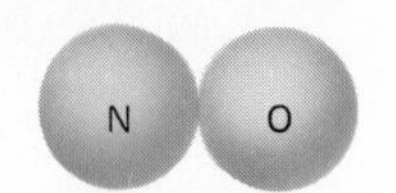

Unbonded:
N ↔ O distance =
2.9×10^{-10} m

Covalent bond:
N ↔ O distance =
1.3×10^{-10} m

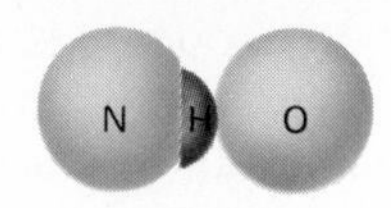

N - H bonded,
nonpolar H:
N ↔ O distance =
3.2×10^{-10} m

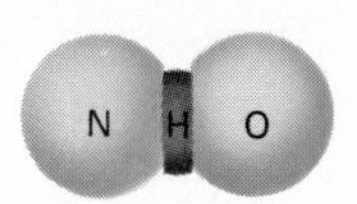

N - H bond, H - O
Hydrogen bond:
N ↔ O distance =
2.6×10^{-10} m

FIGURE 13-16 Bonding pulls atoms closer together. The strength of the covalent and polar bonds are shown for the case of a nitrogen atom and a hydrogen atom. Notice that a hydrogen atom bonded to the nitrogen and forming a hydrogen bond to the oxygen atom reduces the separation of the nitrogen nucleus and oxygen nucleus to *less than their closest unbonded approach* without a hydrogen atom between them.

drogen bonds are much more easily broken than other types. However, they do supply the additional "stickiness" of water molecules that makes water unusual in its properties. Hydrogen bonds are also of fundamental importance in determining the shape and activity of many large, biologically important molecules.

Perhaps the best way to consider hydrogen bonding is in terms of its effect on the distances between atoms. The closest approach of a nitrogen atom and an oxygen atom without bonding has their nuclei some 2.9×10^{-10} m apart. If the two atoms are covalently bonded to each other, the distance between the nuclei is about 1.3×10^{-10} m. Thus, as bonding increases the nuclei pull closer together (Fig. 13-16). If a neutral hydrogen atom is bonded to nitrogen and squeezed between it and an oxygen atom, the distance is increased to 3.2×10^{-10} m. However, if a hydrogen atom bonded to a nitrogen atom and partly denuded of its electron is placed between the nitrogen and the oxygen, it is capable of a hydrogen bond with the oxygen atom. In this case, the nitrogen-oxygen distance shrinks to some 2.6×10^{-10} m.

Acids and Bases

Two groups of compounds that have been singled out for special interest since ancient times have been the *acids* and the *bases* (or *alkalis*). The acids have a sour taste, and this property gave them their name. In their most concentrated forms, the acids corrode or dissolve metals. Thus, medieval alchemists knew that a lump of zinc dropped into oil of vitriol (sulfuric acid) would fizz, gurgle, and eventually be completely dissolved. This action of a concentrated acid is quite impressive, and it marked the acids as a special group. While some combinations of acids are so powerful they will dissolve a lump of pure gold, others are so mild that we include them in our diet. Vinegar and orange juice are both acidic; indeed, this is what gives them their characteristic tang. Lavoisier thought he had found the common property of acids when he discovered oxygen in several of them. He named oxygen, which means "acid-former," based on this information. Unfortunately, he was wrong, and many of the most-used acids contain no oxygen. The element he named hydrogen ("water-former") is the one contained in common acids, and it is in the behavior of water that we may find our clues to acid behavior.

Water molecules ionize very weakly. The reaction

$$H_2O \rightleftharpoons H^+ + OH^-$$

does occur. This reaction, which is called the *dissociation* of water, is reversible, but the bulk of the matter remains on the left side of the equation as water. Only 1 atom in 10 million splits up into the hydrogen ion (H^+) and the hydroxide ion (OH^-). When this point is reached, the ions come together to form water molecules as fast as other molecules split up to become ions. We say that we have reached *equilibrium,* or a balance in the rates at which the reaction proceeds in each direction. For pure water we will reach equilibrium when we have about 1 mol of water molecules remaining, and 10^{-7} mol of H^+ and of OH^-, respectively. Equilibrium is characterized by a number, or constant, which expresses how much of a reaction is in products compared to how much is in reactants. We read $[H_2O]$ as "the concentration of H_2O," in moles per liter. In general, for the reaction

$$A + B \rightleftharpoons C + D$$

we could determine the concentration of the final mixture since

$$\frac{[\text{Products}]}{[\text{Reactants}]} = \frac{[C][D]}{[A][B]} = k$$

where k is a constant characteristic of the reaction called the *equilibrium constant.* If we measure the concentration of hydrogen ions, $[H^+]$, we obtain $[H^+] = 10^{-7}$, since we have 10^{-7} mol of hydrogen ions in our solution. Likewise, $[OH^-] = 10^{-7}$. The product of these concentrations for the dissociation of water is a constant:

$$k = \frac{[H^+][OH^-]}{[H_2O]}$$

$$= \frac{(10^{-7})(10^{-7})}{1} = 10^{-14}$$

Now, the important thing about such a constant is that it is *constant* for a particular reaction. Anything that is done to upset the concentration of one of the substances involved in a reaction will cause the other concentrations to shuffle to keep the value of the constant the same. (This is called *Le Chatelier's principle.*) If a substance is dumped into the solution that tended to liberate hydrogen ions, for example, $[H^+]$ would increase. With more hydrogen ions about, more hydroxide ions would encounter them and unite to form water molecules. Thus, $[OH-]$ would decrease as these ions were removed from the solution. If we added H^+ to the solution until $[H^+] = 10^{-6}$, we would find that enough OH^- had been taken back into water molecules so that $[OH^-] = 10^{-8}$, and the value of the equilibrium constant would be unchanged. We would also find that our solution was now a weak acid. As the value of $[H^+]$ rises, the value of $[OH^-]$ drops accordingly *and* the solution becomes more strongly acidic. If we got to the point where one-tenth of the solution was made of hydrogen ions, $[H^+] = 10^{-1}$, we would find $[OH^-] = 10^{-13}$, or only 1 in every 10 trillion particles would be a hydroxide ion. We would also find this to be an extremely powerful and corrosive acid. As $[H^+]$ increases, the solution becomes more powerfully acidic.

TABLE 13-1

The pH Scale and Some Important Values

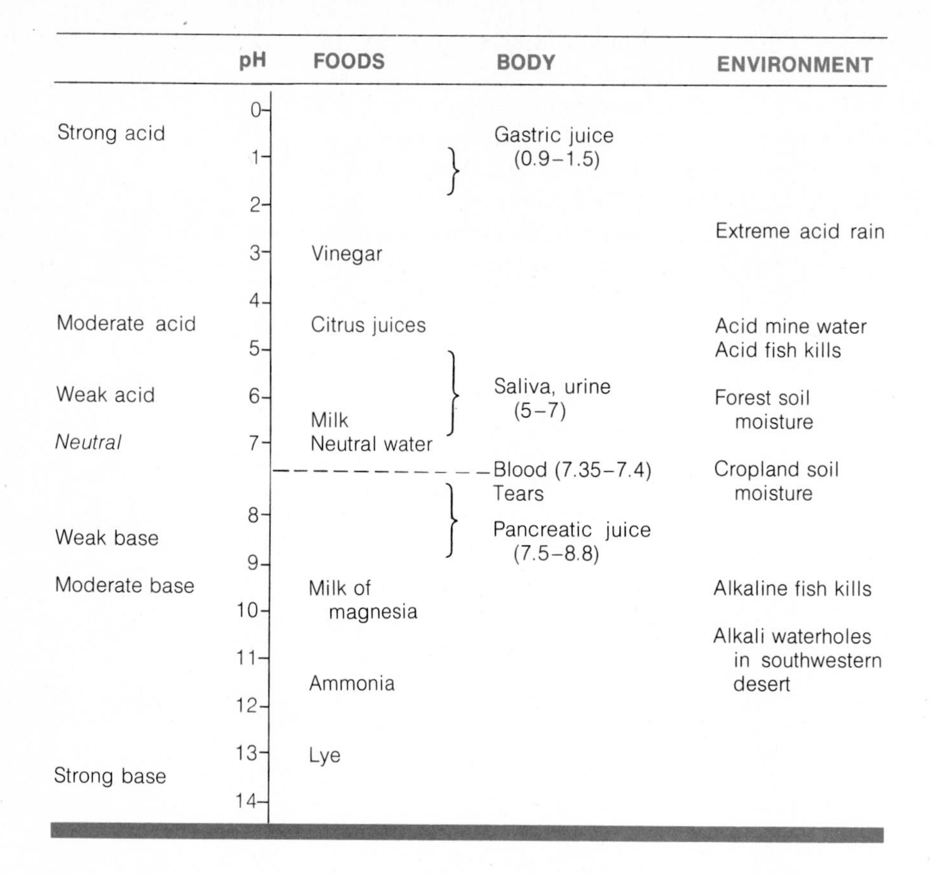

	pH	FOODS	BODY	ENVIRONMENT
	0			
Strong acid	1		Gastric juice (0.9–1.5)	
	2			
				Extreme acid rain
	3	Vinegar		
	4			
Moderate acid		Citrus juices		Acid mine water
	5			Acid fish kills
Weak acid	6		Saliva, urine (5–7)	Forest soil moisture
		Milk		
Neutral	7	Neutral water		
			Blood (7.35–7.4)	Cropland soil moisture
			Tears	
	8			
Weak base			Pancreatic juice (7.5–8.8)	
	9			
Moderate base		Milk of magnesia		Alkaline fish kills
	10			
	11			Alkali waterholes in southwestern desert
		Ammonia		
	12			
	13	Lye		
Strong base				
	14			

This process works in the opposite direction. As a material liberating hydroxide radicals is added to a solution, the concentration of the hydroxide ions increases. As this happens, hydrogen ions from dissociated water molecules can locate and combine with hydroxide radicals more easily, and $[H^+]$ declines. As this happens, the solution becomes *basic* or *alkaline.* A powerful base, such as lye, is every bit as extreme in its chemical reactivity as a strong acid.

When a base is mixed with an acid, *neutralization* takes place, and the products are water and a *salt.* For example, the reaction

$$HCl + NaOH \rightarrow NaCl + HOH$$

produces common table salt. In general, the product of the neutralization of any acid with a base is called a salt. Thus, the reaction of milk of magnesia with stomach acid would produce the salt magnesium chloride.

One of the most important properties of a solution is whether it is *acidic, basic,* or *neutral.* This is particularly true of the solutions involved with living things. For example, most trees grow best if the soil is somewhat acidic. Prolonged tree growth alters the soil in this direction. Grasses, however, thrive on basic soils. Thus, land freshly cleared of forest might be a poor area for growing corn (which is a grass). Soil alkalinity may be controlled by spreading a basic material on the ground, the commonest of which is lime, CaO. In water this dissolves to produce $Ca(OH)_2$, which neutralizes soil acids and makes the groundwater slightly aklaline. Since the acid-base balance of many fluids is very important, it is necessary to have both a simple test for it and a way of describing it. Acidity is described by a number called *pH*. This is just the numerical value of the power describing the concentration of hydrogen ions. For example, if $[H^+] = 10^{-7}$, the pH would be 7 and the solution would be called *neutral* since it would have equal numbers of H^+ and OH^- in it. If $[H^+] = 10^{-1}$, the pH would be 1 and solution would be very highly acidic. Should the pH be greater than 7, there are more hydroxide ions than hydrogen ions and the solution is *basic*. Table 13-1 indicates pH values of some common substances. Many chemical substances change form as a particular pH is reached. If the change in form involves a color change, the substance may be used as an *indicator* for determining when that level has been reached. Filter paper soaked in a variety of such chemicals can indicate by its color when dipped in solution what the pH level is. This is a great convenience when testing many samples. Indicators are also used in precision measurement of pH in the process of *titration*. In this procedure a solution of known pH is slowly dripped into the sample until the anticipated color change occurs. From the volume of the known pH solution needed to neutralize the sample its original pH may be calculated.

Reaction Rates and Catalysts

In theory all reactions are reversible and thus eventually achieve equilibrium. However, many reactions produce materials which escape, such as gases, or insoluble solids. Thus, we are much more likely to view reactions as beginning in one state (reactants) and ending in another (products). Reactions take time to go from where we start to where we want to end, or to their equilibrium position if the reaction is strongly reversible. The rate at which a reaction takes place can be altered by various conditions. For example, foods will break down much faster with boiling than if they are just thrown into a pot of water and allowed to sit. As a general rule of thumb, the rate of a chemical reaction doubles for every 10°C temperature increase. Water at boiling is 80°C above room temperature, so eight doublings of the reaction rate should occur. This implies that whatever reaction occurs with water should occur over 250 times as fast at boiling. We can boost this even higher by using a pressure cooker in which water will boil at a temperature greater than 100°C. We can understand this temperature behavior if we picture what is happening between molecules. For two molecules, say *A* and *B*, to react they will first of all have to encounter each other. Secondly, they may have to encounter each other in the correct orientation or alignment. If *A* is a large molecule and *B* will react with only one end of *A*, it does no good for *B* to bump into the end of *A* it does not react with. Thirdly, the molecules may have to bump with enough energy to break existing bonds and form new ones. Recall that H_2 and O_2 may be mixed at room temperature with no reaction occurring. When these molecules collide at this temperature, they do not have enough energy to break their existing bonds to enable them to react with each other. When given enough energy to do this, they react explosively. Increasing the temperature of materials speeds up the motion of molecules. This results in more collisions and higher-energy collisions and thus increases the reaction rate.

An entirely separate mechanism for increasing the reaction rate is of great practical importance. (It is especially important to the chemistry of living organisms where it is not practical to raise the temperature.) This is the use of *catalysts.* A catalyst is a substance that changes the rate at which a reaction occurs but is not itself used up or changed by the reaction. An analogy might be that of a teacher in a junior high school study hall. With the teacher present, the rate at which studying occurs is vastly greater than the rate with the teacher absent, even though the teacher need not do anything in particular to secure this result except be present. The action of chemical catalysts is not so magical as this, although their effect on the reaction rate may make them seem so. Some catalysts work by forming a temporary product with one of the reactants that is then much more likely to react with the other reactant. Such activity may serve to lower the "energy hump" the reactants must cross in order to react. Other catalysts serve as surfaces on which the materials remain pinned until a reaction can take place. This can be quite important if the molecule must be aligned in a particular way for the reaction to occur. However they perform their role, catalysts are of extreme importance to chemistry. What catalysts do *not* do is to change the nature of an equilibrium. No more product will be produced using a catalyst than would eventually have been produced without one under the same conditions. The catalyst may shorten the time to reach the equilibrium by millions of times, and thus may appear to be causing something to happen which was not apparent before the catalyst was added.

Summary

Since many compounds can be separated electrically in the process of *electrolysis,* it is logical that some form of electrical attraction holds these molecules together. The electrolyte must be dissolved or molten for electrolysis to work. Then, metals collect at the negative plate, or *cathode,* and nonmetal elements form at the positive plate, or *anode.*

High-voltage batteries connected across a vacuumized tube provide a flow of charged particles across the space inside the tube. These were originally called *cathode rays,* but were renamed *electrons.* Electrons have mass and are negatively charged. All matter contains electrons, which implies that atoms have some form of internal structure.

Determination of the structure of the atom was the most important problem studied during the early part of this century. The Rutherford-Bohr model of the atom that emerged has the following points. Atoms are composed of three basic types of particles. *Protons* and *neutrons* are found in a tiny cluster at the center of the atom called the *nucleus.* Protons have a mass of 1 unit and an electric charge of positive 1 (+1). Neutrons have essentially the same mass but carry no net electric charge. The number of protons in the nucleus is called the *atomic number,* and this determines which element the atom represents. The sum of protons and neutrons in the nucleus is called the *atomic mass number* and is close to the atomic weight of the element. Different atoms of the same element with different numbers of neutrons are called *isotopes of the element. The bulk of the volume of the atom is occupied by electrons which have negligible mass and a charge of* -1. The electrons are arranged in *shells* or *orbits,* each of which may hold a definite number of electrons. The shells, in order from the nucleus outward may hold 2, 8, 18, 32, . . . electrons. However, if any shell is the outermost shell of electrons, or *valence shell,* it may not contain more than eight electrons.

All naturally occurring reactions proceed with the release of energy, or are *exothermic. Endothermic* reactions, which absorb energy, proceed only if energy is supplied to the reaction. The group of atoms with the greatest stability is the group of *inert gases.* Each of these gases is characterized by an atomic structure having a *full* or *complete valence shell* of electrons with no vacancies. Chemical bonding may thus be viewed as an attempt to acquire an inert, gaslike structure by acquiring a complete valence shell of electrons.

Ionic bonding is accomplished by the physical transfer of electrons. An atom with a tendency to lose its valence electrons is called a *metal. Nonmetals* tend to acquire electrons to fill vacancies to complete their valence shells. After electron transfer, metal atoms acquire a net positive charge and nonmetals a net negative charge. Such particles no longer have the properties of their parent atoms, but are new materials referred to as *ions.* Since the two types of ions formed in a reaction have electric charges of opposite signs, they attract each other. Ionic compounds are held together by this electrical attraction and may be broken apart by superior electric forces. In a pure sample of a metal, valence electrons are free to move from atom to atom. This metallic state of matter clarifies a number of the properties of metals including their electrical conductivity.

The second form of bonding is *covalent bonding,* requiring the sharing of electrons between two atoms. Electrons are shared in pairs with one electron of each shared pair coming from each of the two atoms connected by the covalent bond. It is possible for two atoms to share two pairs of electrons in a *double bond,* or three pairs of electrons in a *triple bond.*

If the pair of shared electrons is not shared evenly between the atoms in a molecule, zones of the molecule may have a small net electric charge even though the entire molecule is neutral. Such a molecule is called a *polar molecule.* Bonds formed between charged zones on neutral molecules are called *hydrogen bonds,* and are only about one-twentieth of the strength of typical ionic or covalent bonds. Such bonds explain the ability of water molecules to dissolve a wide variety of substances.

The rate at which a reaction occurs depends upon the temperature and the concentration of each of the reactants. The rate of reaction may be increased by a *catalyst.* Catalysts do not enter into the reaction and are not used up during it.

Questions

1 Draw the atoms $^{8}_{4}Be$ and $^{29}_{14}Si$.

2 A sulfur atom has 16 electrons. How will they be arranged? Will this make sulfur a metal or a nonmetal? Why?

3 Cobalt atoms (Co) have two electrons in their outermost shell, while iodine atoms (I) have seven. What will be the correct formula for a compound of cobalt and iodine?

4 Would you expect the reaction to produce nitroglycerin to be exothermic or endothermic? Why?

5 What would the charge be on an aluminum ion? A beryllium ion? A phosphorus ion?

6 Draw the ionic molecule lithium fluoride indicating all electrons. Draw lithium oxide.

7 Draw the covalent molecule of ammonia, NH_3, indicating each electron. Draw the F_2 molecule.

8 What is the distribution of electrons in the shells of a calcium atom with an atomic number of 20?

9 Would you expect the molecule of silicon dioxide, SiO_2, to be covalent or ionic? Why?

10 Why is a strongly polar molecule required as the fluid in which life can take place?

11 Washing soda, Na_2CO_3, is the salt formed from the neutralization of the strong base, lye, with weak carbonic acid (as in soda pop). Would you expect washing soda dissolved in water to be slightly basic or slightly acidic? Why?

12 Lakes resting on rock types such as limestone, $CaCO_3$, have remained near a pH of 7 despite acid rains. Meanwhile, lakes on other types of rock, such as granite, have become so acidic that most plants and no fish can live in them. Use your answer to the last question to suggest why lakes might show this variability.

13 Cars use catalytic converters containing small spheres coated with platinum to eliminate undesirable exhaust products. What do you think these accomplish judging by the name? Lead in gas can "poison" the catalyst. What does this mean?

14 Copper will burn slowly in pure chlorine to form copper chloride. Sodium reacts violently with the chlorine, liberating a great deal of heat and light. Which material, $CuCl_2$ or NaCl, should require a higher voltage for its separation by electrolysis? Why?

15 Using dashes for a single pair of shared electrons, draw the structure of the following covalent molecules: O_2; CO_2; vinegar, $C_2H_3O_2$; acetylene, C_2H_2; hydrogen peroxide, H_2O_2; ozone, O_3.

16 Write the balanced equation for the neutralization of stomach acid (HCl) by milk of magnesia (magnesium hydroxide).

CHAPTER FOURTEEN

The Chemistry of Carbon and of Life

Introduction Science-fiction writers are fond of pointing out that all life on earth is based on the chemistry of the element carbon. Their aim is to inquire whether some distinctly different form of life is possible. For our purpose, we will note that you can divide chemistry into two convenient parts: the chemistry of carbon, and the chemistry of the hundred-and-some other elements. Of these two fields, *organic chemistry*, the chemistry of carbon, is by far the larger. It involves more known substances, occupies more investigators, and bears more importance to us as living beings than *inorganic chemistry*—the study of everything but carbon. Part of the fascination, of course, is that only carbon compounds are alive, and anyone capable of reading this has a vested interest in life. Beyond that, however, the page is a carbon compound, printed with inks that are carbon compounds, stitched with threads and glued with substances that are carbon compounds, bound with covers that are more carbon compounds. Furthermore, you are sitting on a carbon compound and wearing carbon compounds unless you are naked and standing, in which case both hair and skin are more carbon compounds. Meanwhile, your digestive system works on, for example, a lettuce leaf and a french fry, both of which are carbon compounds, to rearrange them into simpler carbon compounds your body may use. The chemistry of carbon has variation and subtlety denied to the other elements. This is largely because carbon is the premier covalent atom, having four electrons in its

TABLE 14-1
Some Variations on a Single Carbon Atom

FORMULA	STRUCTURE	NAME AND USE
CH_4	$H-\overset{H}{\underset{H}{C}}-H$	*Methane*—Natural gas, not particularly poisonous; suffocates only when it crowds out the oxygen molecules you need.
CH_3Cl	$H-\overset{H}{\underset{H}{C}}-Cl$	*Chloromethane* or *methyl chloride*—Highly volatile; used to spray and freeze skin for minor surgery.
CH_2Cl_2	$H-\overset{H}{\underset{Cl}{C}}-Cl$	*Dichloromethane* or *phosgene gas*—A poisonous gas used in World War I; corrodes the lungs, producing slow-healing sores.
$CHCl_3$	$H-\overset{Cl}{\underset{Cl}{C}}-Cl$	*Trichloromethane or chloroform*—This was one of the original hospital anesthetics, now rarely used for this purpose.
CCl_4	$Cl-\overset{Cl}{\underset{Cl}{C}}-Cl$	*Carbon tetrachloride*—Dry-cleaning fluid and fire extinguisher.
CF_2Cl_2	$F-\overset{F}{\underset{Cl}{C}}-Cl$	*Dichlorodifluoromethane*—Freon gas, used as a coolant in air-conditioning systems and refrigerators, formerly the propellant in spray cans.
CO_2	$O=C=O$	*Carbon dioxide*—An old friend; your body monitors its level in the blood to control your breathing rate and heartbeat.
CH_2O	$\begin{matrix}H\\ & C=O\\ H\end{matrix}$	*Methyl aldehyde* or *formaldehyde*—Common embalming fluid when dissolved in water; original base of the plastics industry.
CH_3OH	$H-\overset{H}{\underset{H}{C}}=O-H$	*Methyl alcohol* or *wood alcohol*—Not the drinkable alcohol; causes blindness if you drink a small amount.
HCN	$H-C\equiv N$	*Hydrogen cyanide*—Highly lethal gas used in gas chamber executions.
HCOOH	$H-\overset{O-H}{C}=O$	*Methanoic acid* or *formic acid*—Found in ant bites; base material for formica; one of the strongest organic acids.
H_2CO_3	$O=\overset{O-H}{\underset{O-H}{C}}$	*Carbonic acid*—The fizzer in soda pop; falls apart into carbon dioxide and water as the pop goes flat.

outer shell. Carbon's nearest elemental relative, silicon, also has four electrons in its outer shell, but it is a much larger atom with fourteen electrons in all. This larger size alone prevents the silicon atom from performing some of the "atomic gymnastics" which carbon atoms do.

Energy forces the carbon atom to be almost purely covalent and share electrons whenever possible. Carbon is simply the most versatile atom in combining with other atoms—most notably other carbon atoms. These bonds tend to be strong and durable, requiring a relatively large amount of energy to separate them. (Recall that a diamond is a case of pure carbon-carbon bonding.) When an ionic substance such as salt is dissolved in water, the solution becomes electrically conductive, indicating that the weak electrical attractions of neutral water molecules have pulled the compound apart into separate ions. When a carbon compound, such as sugar or alcohol, is dissolved in water the solution is very rarely conductive, indicating that the water has not pried the molecular structure apart but only separated adjacent molecules from each other.

To appreciate the diversity of carbon compounds, consider just a few of the common molecules that can be built with the base of a single carbon atom. Table 14-1 lists a few of these, where each dash indicates a single covalent bond between two atoms. While the structures are interesting, the range of properties and uses is amazing. But remember, these are only some of the structural variations of a single carbon atom. It is not unusual to find molecules in your body with thousands and thousands of carbon atoms, each in its own place and attached in a particular way to the atoms surrounding it. Even with molecules this large, changing a single bond can change a life-supporting molecule into a useless blob of atoms. In fact, this is exactly how many of the deadliest of poisons work—they attack a small part of a large molecule essential for life and render it useless. This is the nature of carbon compounds, the most diverse and important group of compounds that exists.

Hydrocarbons

One of the commonest groups of carbon compounds is that of the *hydrocarbons.* Hydrocarbons are compounds formed of hydrogen and carbons atoms only. The simplest subgroup of these may be viewed as a chain of carbon atoms bonded by single covalent bonds with hydrogen atoms attached to all other available carbon bonds. If there are no double or triple bonds between carbon atoms, as many hydrogen atoms as possible are linked to the carbon atoms. This is the series of *saturated hydrocarbons,* or the *alkane series.* To recognize the series, they are named so that all of their names end in *-ane.* The simplest of this group is natural gas, CH_4, or methane, with a single carbon atom bonded to four hydrogen atoms. The next most complex is ethane, C_2H_6, with two carbon atoms bonded to each other and six hydrogen atoms attached to the remaining bonds. Table 14-2 illustrates some of the simplest alkanes. Note that each carbon atom is bonded to two hydrogen atoms except for the end atoms of the chain, which are each bonded to three hydrogen atoms. This ensures that each alkane will have twice as many hydrogen atoms as carbon atoms plus two more hydrogens for the ends of the chain. We may thus write the general chemical formula for the al-

TABLE 14-2
The Alkane Series of Hydrocarbons

STRUCTURE	NAME	FORMULA
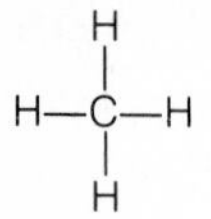 H H—C—H H	Methane	CH_4
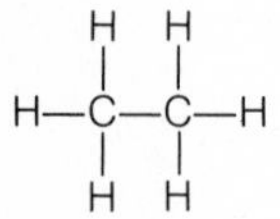 H H H—C—C—H H H	Ethane	C_2H_6
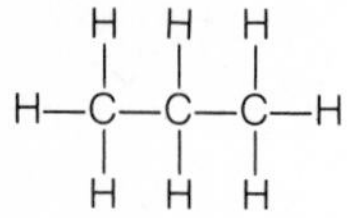 H H H H—C—C—C—H H H H	Propane	C_3H_8
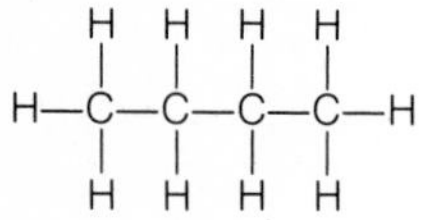 H H H H H—C—C—C—C—H H H H H	*n*-Butane	C_4H_{10}
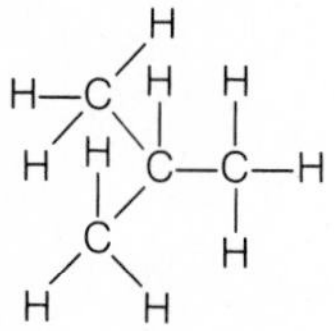 H H—C H H H H C—C—H C H H H	Isopropane	C_4H_{10}
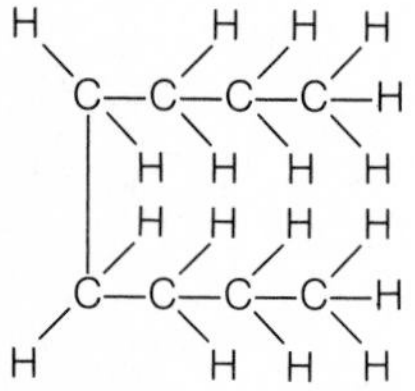 H H H H C—C—C—C—H H H H H H H H H C—C—C—C—H H H H H	*n*-Octane	C_8H_{18}
In general: $CH_3(CH_2)_nCH_3$		$C_{n+2}H_{2n+6}$

kanes as C_nH_{2n+2}. If $n = 1$, this becomes CH_4, or methane; or, if $n = 6$, this produces C_6H_{14}, which is the formula of hexane, and so on.

One complication arises from butane on. While all butane is C_4H_{10}, not all C_4H_{10} is butane. There is another way to arrange these atoms to produce a saturated hydrocarbon. Molecules with the same formula but different structures are called *isomers.* The isomer of *n*-butane has three carbons in a chain with the fourth carbon attached to the middle atom in this chain. This molecule is called *isopropane* and shares its formula with butane. To indicate this difference, molecules with an unbranched string of carbon atoms are referred to as "normal" and are written with a small "*n*" preceding their names. Since isopropane and *n*-butane are entirely separate molecules with distinct physical and chemical properties, this is an important distinction. The number of isomers goes up very rapidly with increasing numbers of carbon atoms.

Physically, the important distinction between the alkanes is that their properties vary by the size of the molecule. The shortest-chained molecules are normally gases. Propane and butane are gases at normal temperatures but may be liquefied quite easily by applying pressure. Alkanes with a few more carbon atoms are liquids, but are quite volatile—that is, they vaporize quite readily when exposed to the air. As the number of carbon atoms in the molecule increases, the volatility decreases until compounds with perhaps 20 carbons are quite sedate liquids with little tendency to vaporize. When the number of carbon atoms per molecule becomes larger still, we reach the range of solid hydrocarbons, such as asphalt or paraffin.

The most important reaction in the use of hydrocarbons is that they burn in air. Because of this property, the hydrocarbon group contains some of the most important fuels on earth. As such, they are usually found mixed together in nature and are used in mixtures. Crude oil is composed of a mixture of most of the possible liquid hydrocarbons with some dissolved solid hydrocarbons added. Because of this wide range of molecular properties, crude oil can't be used by itself; it must be separated (*fractionated*) into portions having approximately the same number of carbon atoms per molecule. Thus, the hydrocarbon range from about seven to twelve carbon atoms per molecule vaporizes well enough for use in automobiles, but not so readily as to be difficult to store and transport. A mixture of hydrocarbons in this size range is called *gasoline.* In the wintertime shorter alkanes that vaporize more easily are included in the mixture to allow your engine to start more readily. In the summertime, the mixture will include somewhat larger molecules which evaporate less readily. However, in all cases, gasoline is a complex mixture of hydrocarbons. When the very first oil wells were being drilled about 100 years ago, the most desirable portion of the crude oil was kerosene for lamps. Half a century ago gasoline became more desirable as the internal-combustion engine became more common. Long-chain hydrocarbons can be split ("cracked") into shorter chains in the gasoline range by passing them over a catalyst at high temperature and pressure. In this manner the fraction of crude-oil hydrocarbons in the gasoline range of sizes is increased. The separation of crude oil into its various fractions is accomplished by the process of *distillation.* Each fraction has a narrow range of boiling points. Heating the crude oil at the base of a fractionating tower boils away all but the longest-chain compounds. The temperature of the tower gets lower as it rises, and each fraction of the crude oil condenses back to a liquid at a different height in the tower. The typical oil refinery thus consists of fractionating towers, calalytic crackers, and storage areas.

When burned in oxygen the alkanes may be converted completely to carbon dioxide and water. Thus, the complete combustion of octane in gasoline would be

$$2C_8H_{18} + 25O_2 \rightarrow 16CO_2 \uparrow + 18H_2O \uparrow$$

If there is a deficiency of oxygen for the amount of hydrocarbon present (called a "rich" mixture in an automobile engine), carbon monoxide as well as unburned hydrocarbons will be present after combustion. Since such rich mixtures burn much better in a typical engine, past practice left these materials in the exhaust gases. Environmental concern caused this practice to be changed even though the power developed by the engine might be reduced. "Lean" mixtures, in which the amount of oxygen is appropriate for the amount of hydrocarbon, are difficult to ignite in the engine. One solution is a stratified-charge engine, with a small zone using a rich mixture to begin combustion leading to a much larger volume of a lean fuel-to-air mixture to provide most of the combustion. The commonest practice, however, is to burn off the excess materials after they leave the engine in an afterburner or catalytic unit of some sort. Most American cars are equipped with catalytic converters to promote the complete combustion of the fuel as well as reduce the oxides of nitrogen in the exhaust. Chemical reactions in the catalytic converter can raise its temperature to over 900°C (1600°F) during prolonged use. This indicates the highly reactive nature of engine exhaust gases.

Hydrocarbons generally produce nearly 4.5×10^7 J/kg of fuel. A kilogram represents about $1\frac{1}{2}$ liters of liquid fuel ($\frac{3}{8}$ gal). This is about 50 percent more energy per kilogram than coal and 3 times as much as most wood products. As the proportion of hydrogen in a fuel increases, its energy value per kilogram tends to increase. Thus, methane gas has the highest value with four hydrogen atoms per carbon atom. Methane, or natural gas, releases about 5.5×10^7 J/kg during combustion. While over 10 percent higher than a typical hydrocarbon, this number is only about one-third of the fuel value of 1 kg of pure hydrogen gas. Of course, methane, like hydrogen, *is* a gas and occupies a very large volume per kilogram, unlike a liquid or solid fuel. At normal temperatures and pressures methane requires up to 1000 times as much storage space per kilogram as does gasoline.

The alkane family, while by far the commonest, is not the only series of hydrocarbons. The set of carbon compounds with a double bond between two carbons is called the *alkenes.* The names of these compounds end in *-ene.* The commonest of these is the gas ethene. Likewise, the series of hydrocarbons with a triple bond in their structure exists called the *alkynes,* the names of which end in *-yne.* These are not so common as the earlier families. Chemicals added to an alkyne tend to break down the triple bond to a double bond and attach to two carbon atoms at this point in the chain. Another situation exists where the end of a chain coils around to attach to the other end. These are called the *cyclic hydrocarbons.* The most common carbon ring has six members. As an element, carbon tends to form planes made up of such six-sided, or hexagonal, rings. This is the common form of pure carbon called *graphite.* Geraphite makes an excellent lubricant since the locking of ring to ring in a plane is quite strong, but the bonds between adjacent planes of rings are quite weak (Fig. 14-1). Thus, the planes of atoms in graphite will easily slide over each other much as a stack of slippery, plasticized sheets of paper will.

From this discussion you might well imagine that the most important

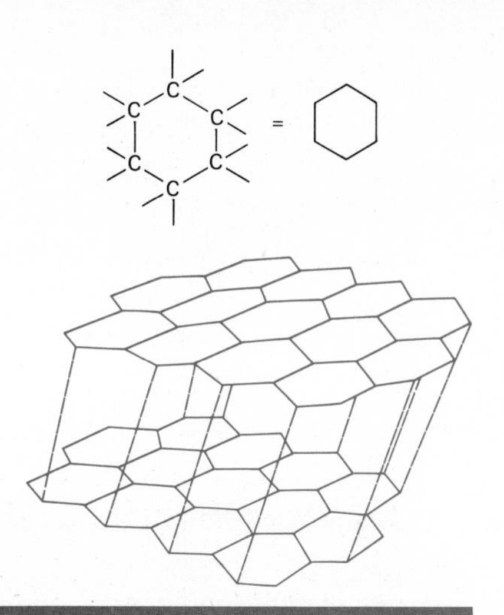

FIGURE 14-1 Six carbon atoms singly bonded in a hexagonal ring can be represented by a simple hexagon. In the structure of graphite such rings join to form sheets or planes of carbon atoms. Note that a carbon atom in the middle of such a plane has one covalent bond left to bond to either the plane above it or below it, but not both. Thus, bonding in the plane of rings is strong, but between planes it is quite weak.

cyclic hydrocarbon would be a six-member ring with two hydrogen atoms attached to each carbon atom. While cyclohexane, C_6H_{12}, is common and important as a chemical, its unsaturated relative *benzene* is the most significant hydrocarbon with a ring structure. The three series of alkanes, alkenes, and alkynes, including members having ring structures, make up the *aliphatic hydrocarbons.* Benzene and its derivatives make up an entirely separate branch called the *aromatic hydrocarbons* (because many of them have distinctive smells, frequently pleasant). The structure of benzene was a serious problem for the chemists of the last century. Benzene has the formula C_6H_6, and it can be drawn as a ring with one hydrogen atom per carbon. In the drawing, double bonds and single bonds alternate between the carbons. While easy to draw, the structure was wrong experimentally. A double bond is stronger than a single bond. If we picture carbon atoms as tiny billiard balls, joining two by a single bond produces a distance from one nucleus to the other of 1.54×10^{-10} m (Fig. 14-2). With a double bond the distance is reduced further to 1.34×10^{-10} m, and a triple bond pulls the nuclei to within 1.20×10^{-10} m, or only 77 percent of their separation with a single bond. This appears reasonable, since bonding represents an attractive force between atoms which holds them together. The stronger this force, the more closely the atoms are pulled to each other. The difficulty is that all

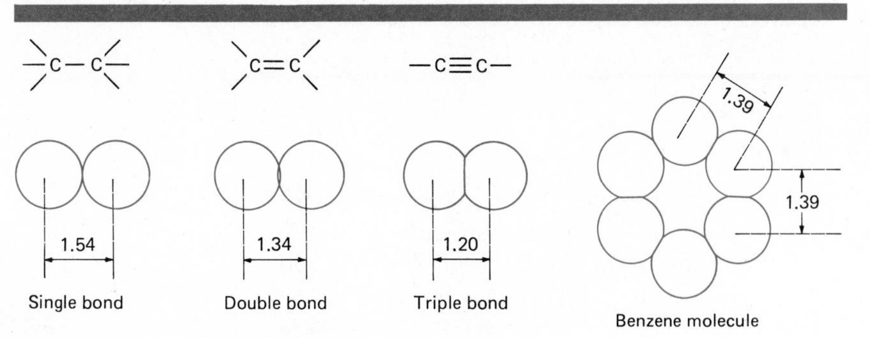

FIGURE 14-2 Increased bonding pulls atoms closer together. The distance between the nuclei of carbon atoms joined by a triple bond is less than that of double or singly bonded carbon atoms. The bonds in the benzene ring are special. All bonds are equivalently strong, but the distance between adjacent carbon atoms indicates a stronger attractive force than a single bond yet less than a double bond. All distances are times 10^{-10} m.

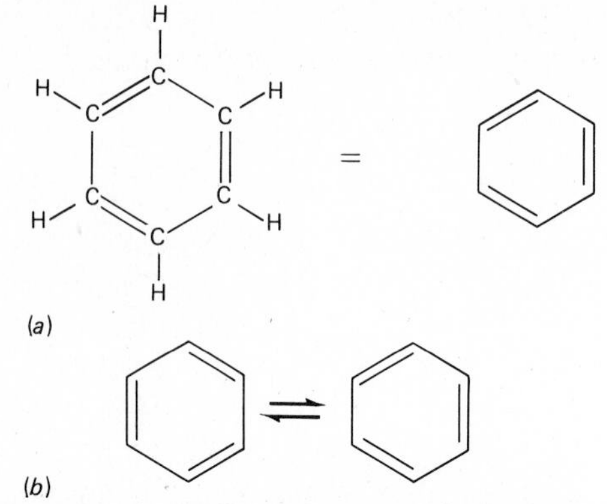

FIGURE 14-3
The benzene structure may be drawn as a six-carbon ring with alternating single and double bonds, or by the abbreviated hexagon with alternate double bonds shown. This structure is definitely *not* correct. Kekulé proposed the scheme at the bottom where the double bonds flip rapidly back and forth between carbon atoms, which he called resonance. This would give adjacent carbon atoms an effective "bond and a half."

carbon-carbon bonds in benzene are the same. All six bonds are equivalent in strength, none are stronger or weaker, and the carbon-carbon distances are a constant 1.39×10^{-10} m. This would imply that the bond is stronger than a single covalent bond but less than a double bond.

Measurements of interatomic bond lengths belong to our own century, but the unique character of benzene became apparent over 100 years ago. The individual who standardized most of the methods we have been using to represent organic molecules was Friedrich Kekulé (1829–1896). Kekulé also first recognized the ring structure in organic compounds, particularly in benzene. Chemical reactions indicated that the benzene molecule was particularly stable and did not act as though a normal double bond was present. To make all bonds equivalent Kekulé imagined the double bonds in benzene as "flipping" rapidly back and forth between carbons in a process called *resonance.* Figure 14-3*b* represents the two states of resonance in this model. Carbons connected by a single bond in one state are connected by a double bond in the other state or resonance. If the resonance was very rapid, this would amount to almost a "bond and a half" between adjacent carbons, and all bonds would be of the same strength. Kekulé announced this idea in 1865, and it turned out to be a brilliant guess, requiring science the better part of a century to understand in terms of atomic structure. While we will leave this interpretation for later, the importance of a model or picture of what was happening at the level of the molecule was essential for chemistry. The expansion of organic chemistry due to Kekulé's model was enormous.

Organic Groups

Thus far we have introduced only the beginning cast of characters for organic chemistry. Hydrocarbons are found in petroleum or may be produced from coal. They form the basis for materials as different as red paint and vanilla flavoring, nylon and heroin. To begin we will clarify the terminology of some simple groups. If one hydrogen is removed from methane, the remainder, CH_3, may be bonded to a variety of other substances. We call this group a *methyl group*. For example,

$$\underset{\text{Methane}}{CH_4} + \underset{\text{Chlorine gas}}{Cl_2} \longrightarrow \underset{\text{Hydrogen chloride}}{HCl} + \underset{\text{Methyl chloride}}{CH_3Cl}$$

Table 14-1 illustrated some of the variations possible with substitutions on a single methane molecule. Similarly, when a hydrogen is removed from an ethane molecule, the remaining group is called an *ethyl group*, as in ethyl chloride, CH_3CH_2Cl. Various covalent complexes that could attach to these groups cause the molecules that result to behave in particular ways. For example, a hydroxyl group, —OH, may be attached to the open bond of a methyl group producing CH_3OH. The set of hydrocarbons with an —OH group substituted for a hydrogen are known as *alcohols*. Methyl alcohol, or methanol, is known as wood alcohol since it may be distilled from wood (and is poisonous). Among the alcohols, the human favorite is ethyl alcohol, or ethanol. Since this may be distilled from fermented grain mashes, it is also called grain alcohol. Ethanol is the least poisonous of the alcohols—a fact noted by almost all subcultures of the human race since before written history began. Another familiar member of the group is isopropanol, or rubbing alcohol. Any chemical name ending in *-ol* is likely to be an alcohol—for example, menthol.

The addition of the hydroxyl group to simple hydrocarbons accomplishes an interesting change. It is frequently noted that "oil and water do not mix." Any bartender, however, can inform you that alcohol and water do mix, and in any proportion. No matter how long a capped dry martini or bottle of beer is left sitting in a refrigerator it does not separate into layers of water and alcohol. This is because the hydroxyl group is *polar*. As in the structure of the water molecule, the oxygen atom partly denudes the hydrogen of its electron coverage, leaving it somewhat positively charged. As a result, alcohol molecules and polar water molecules have an attraction for each other. Nonpolar hydrocarbons tend to be *hydrophobic* and prefer to clump together with molecules of their own kind in beads when placed in water. Small alcohol molecules are *hydrophilic* and are actually attracted to water. You may have noted that many items such as medicines, flavorings, and so on are frequently found in an alcohol-water mixture. The alcohols are very similar in structure to the hydrocarbons and thus mix with and dissolve a wide variety of organic compounds which water cannot dissolve. If the compound is dissolved in alcohol and the alcohol is mixed with water, we may obtain a good compromise. The medicine is in a dissolved, well-mixed state (thus one spoonful contains exactly the tenth of one sand grain you need in one dose—try to measure that out as a solid in your kitchen). Meanwhile, you are spared the necessity of taking a spoonful of a fluid with the consistency of diesel fuel.

A second organic complex which may be attached to a hydrocarbon to produce a particular chemical activity is the *acid group*, or *carboxyl group*, —COOH (Table 14-3). This consists of a carbon double-bonded

TABLE 14-3
Some Organic Groups

STRUCTURE	GROUP NAME	EXAMPLE
Hydrocarbon groups:		
$-CH_3$	Methyl	Methyl benzene—The solvent toluene
$-CH_2CH_3$	Ethyl	The gasoline antiknock compound tetraethyl lead, $Pb(CH_2CH_3)_4$
$-CH_2CH_2CH_3$	Propyl	—
CH_3 | $-CH_2-CH$ | CH_3	Isopropyl	Isopropyl alcohol—Rubbing alcohol, C_4H_9OH
Functional groups:		
$-OH$	Hydroxyl	Alcohol group, with polar hydrogen
O ‖ $-C-OH$	Carboxyl	Acid group, tends to ionize hydrogen as in CH_3CH_2COOH, vinegar
$-NH_2$	Amino	Found in all proteins
$-NO_2$	Nitro	Unstable group, as in trinitro toluene, or TNT for short

to an oxygen atom and attached to a hydroxyl group. The oxygens of this complex tend to attract the electron away from the hydrogen so strongly that it frequently comes loose as a hydrogen ion, H^+. This behavior, of course, is characteristic of acids. The simplest organic acid is formic acid, with a hydrogen atom attached to the organic acid complex, $HCOOH$. The commonest acid, however, is acetic acid, CH_3COOH. A dilute water solution of this acid is called vinegar. This is another compound that has been used by people from prehistoric times. The combinations possible with functional groups are staggering. For example, a benzene molecule with an acid group forms benzoic acid (Fig. 14-4). If a hydroxyl group is attached on the carbon of the ring next to that bearing the acid group, salicylic acid results. (Actually, this is less strong as an acid than most swimming pool water.) If an acetic acid molecule is placed near this molecule, it is possible to remove the hydroxyl group along with the hydrogen ion of the acetic acid molecule as a molecule of water. If we attach the remaining large parts, we obtain a molecule with the frightening name of acetylsalicylic acid. When this solid was first produced in 1889, it was found to have some interesting properties. It was one of the first compounds known to have the ability to reduce human fever. While the mechanism of fever is not entirely understood even today, it has always been clear that high fever in itself is dangerous to the human. If an adult runs a fever of more than a few degrees Celsius, he or she will enter a coma. If the fever continues, irreversible brain damage and death may result from the fever alone, thus saving the disease causing the fever the trouble of finishing you off. Not only did acetylsalicylic acid reduce fever, it tended to reduce swelling and, most miraculously, it reduced pain itself. *This was unheard of 100 years ago!* Through thousands of years of human history pain was simply endured.

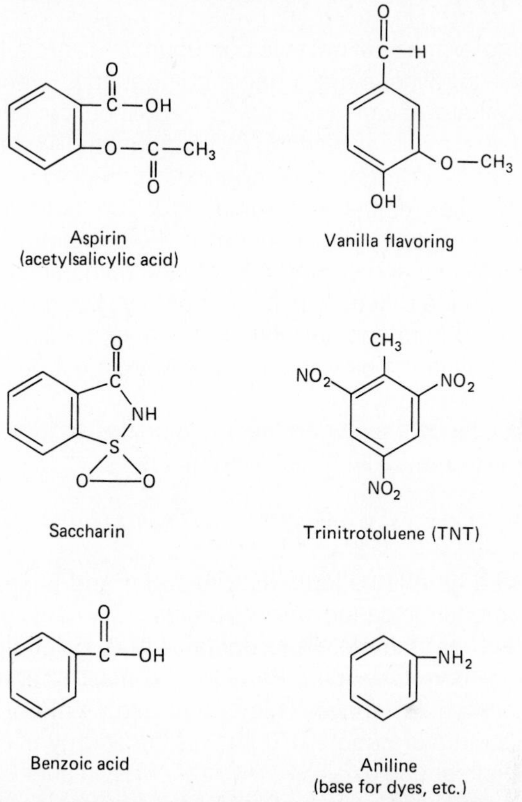

FIGURE 14-4 Thousands of common substances have one or more benzene rings in their structures.

Aspirin (acetylsalicylic acid)

Vanilla flavoring

Saccharin

Trinitrotoluene (TNT)

Benzoic acid

Aniline (base for dyes, etc.)

Treatment with large doses of ethyl alcohol consumed internally was the historical treatment for humans. This did not reudce the pain so much as it reduced the human's ability to sense anything. It is easy to picture the German chemist Dresser trembling as he patented his process for producing acetylsalicylic acid in the most profitable move of his lifetime. Of course, the patent rights of the Bayer chemical empire for *aspirin* have long since run out. Yet, aspirin remains one of the most remarkably effective of all drugs for treatment of human beings.

Even if you know the structure of the molecule you want to end up with, the real skill in organic chemistry is to work out a series of reactions to get you there from available substances. The act of chemical synthesis of particular molecules involves so much skill and experience that organic synthesis is almost an art form.

Other organic functional groups are of importance. Several of these are listed in Table 14-3, along with some familiar compounds containing these groups. Combinations of groups in a single molecule are also possible. It is easily possible to lose sight of the fact that most of these materials are the common, ordinary substances we encounter every day. Several of the structures shown in Fig. 14-4 have scientific names that only the bravest of chemists would pronounce. Yet, time and again we come upon a structure which nature has produced first.

The Substances of Life

Human beings are a complex maze of organic compounds, as are all life forms. Skin, hair, blood—all tissues are organic compounds. Even the inorganic calcium phosphate of our skeleton is deposited by microscopic factories of organic molecules working away diligently. Certainly the most interesting property of organic compounds is that some aggregates of them possess life. Almost as fascinating is the fact that all known life forms are extremely similar chemically. If we examine the chemical activity of a bacterium and compare it with the human cell, the similarities are extreme and the differences are minor. If you came upon a thinking, talking humanoid from another planet, it is a virtual certainty that you would have more in common chemically with an oak tree than with it.

The reaction which sustains the myriad life forms on the earth is that of *photosynthesis*. Expressed in its simplest form this is

$$6CO_2 + 6H_2O \xrightarrow{\text{sunlight}} C_6H_{12}O_6 + 6O_2\uparrow$$

The specific product listed here is called a *simple sugar* and is an example of the family of compounds called *carbohydrates*. This name originally comes from the fact that the molecule contains one molecule of water for each atom of carbon, $C_n(H_2O)_n$. Perhaps the most important part of the equation is not written above. This is that each mole of the product has locked in its structure about 670 kcal of the energy of sunlight. Green plants run this reaction as shown. All life forms, in effect, run this reaction in reverse to break the carbohydrate back into simpler parts and obtain as much of the 670 kcal/mol as they can to provide the necessary energy for living. The green plants run the reaction in reverse to live, too. The only fact which allows us nongreen creatures a chance at life is that the green plants are incredible misers. They produce food whenever they can, yet use only what they need. The balance of the food they produce goes into storage or to build structures to allow them to compete more successfully with their neighbors—stems, roots, leaves, flowers, and so on. Most of these structures are complex, and higher animals lack the chemical machinery to break them down to usable foods. Even grazing animals such as deer or cattle play host to thriving internal colonies of bacteria which specialize in this process. Without these bacteria the grazing animals could starve to death while continuously eating. Humans can digest and live on only more refined foodstuffs than tree bark or stems. Most human food comes from the reproductive peculiarities of plants. It only takes one seed to reproduce one plant, but nature takes no chances on the process. In their competition to survive, plants produce many more seeds than ever could possibly find space to grow on the earth. To give the seed a fighting chance at starting life, the plant packs with it a supply of foods that can be easily broken down for its use. Thus we obtain wheat, rye, oats, corn, barley, rice, and other grass seeds as food. These are the mainstay of the human diet. To this we add peas, beans, and other nongrass seed structures. Additionally, some plants pack their seeds in structures to be eaten by animals. The seeds survive the digestive system of the animal nicely and are taken to new locations and deposited with their own glop of manure as fertilizer. Thus we obtain the fruits for the animal diet. Finally, we do not have to eat plant foods directly but can allow some animal to do it for us and then eat the animal. This is a biological overview of the human food supply starting at photosynthesis. Now, let us examine the chemical overview of life.

Requirements of the Body

To live we need supplies of chemical substances to supply energy and the materials to make and remake the structures of our own bodies. We may obtain energy from three groups of compounds—*carbohydrates, fats,* and *proteins.* A small group of chemical substances provide specific chemical structures the body needs and cannot manufacture from spare parts—these are the *vitamins.* Finally, we need water and basic inorganic materials to work with which we call *minerals.* Water is the fluid in which all life takes place.

The simplest carbohydrates are the *sugars.* Sugars may taste sweet—many do, but this is not required of a sugar. The commonest group of *simple sugars* has the formula $C_6H_{12}O_6$, and there are 16 of them. While this is an example of isomerism, it is also extremely important to the body. Fed on one of these forms your body can obtain all of the energy it needs and store a surplus. If fed exclusively on another one of these, you might die of starvation. The chemical factories of the body are highly specific in the size and shapes of molecules they will deal with. These molecules are arranged in a six-member ring with an oxygen atom

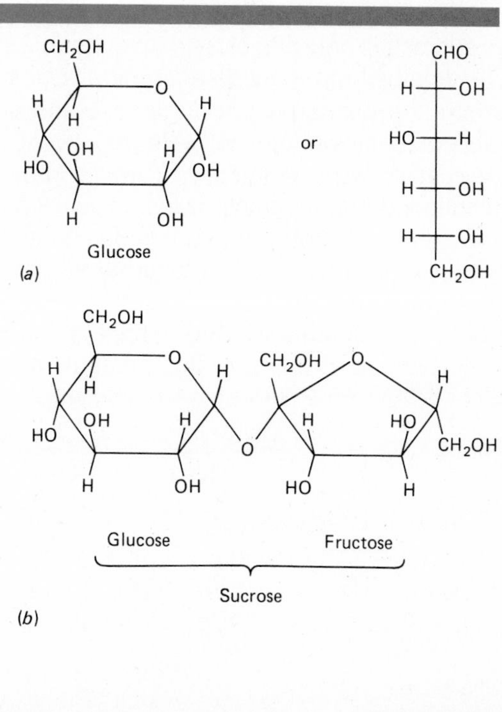

FIGURE 14-5
The single sugar glucose is the only sugar found in the human body. The actual structure is the ring shown at the left. Carbon atoms occur at each intersection of bonds and are not drawn. Glucose also may be pictured in the linear form at the right. The direction of the H's and OH's from the central spine is critical in determining which sugar is represented. A molecule of glucose joined with a molecule of fructose as shown makes the double sugar sucrose, or table sugar.

as one member of the ring, although their structure is frequently represented as a linear molecule. Figure 14-5 gives both representations for the sugar glucose. Sugar names end in *-ose,* allowing you to recognize them. Glucose is the basic fuel of your body and is supplied to all of your cells via the blood. For this reason it is also called *blood sugar.* Your cells consume between 0.5 and 0.6 kg of glucose a day. The differences between these simple sugars (*monosaccharides*) is in the direction from the ring in which the —OH groups lie. Each time we would flip a particular —OH group from down to up in the drawing, or vice versa, we would obtain a different molecule. Glucose is the only six-carbon sugar found in the body. (Two simple sugars with five carbons each, $C_5H_{10}O_5$, are also found in the body—ribose and deoxyribose.) Normally the concentration of glucose in the blood is controlled within a small range by the chemical *insulin.* Failure to produce sufficient insulin causes diabetes. It can be tested for by the presence of glucose in the urine.

The sugar we are familiar with is not a simple sugar, but a *double sugar* (*disaccharide*). Table sugar is called *sucrose* and is made by attaching a molecule of glucose to a molecule of another simple sugar, fructose. The process is called *condensation* and involves *dehydration,* or removal of a water molecule. Starting with an —OH group on each molecule, a hydrogen is removed from one and the —OH group from the other to form H_2O, leaving the two single sugars joined via an oxygen atom into a double sugar:

$$2C_6H_{12}O_6 \rightarrow C_{12}H_{22}O_{11} + H_2O$$

This process is illustrated in Fig. 14-6 with each sugar represented by a simple ring. The digestion of sugar is the reverse of this process and is called *hydrolysis.* In hydrolysis a water molecule is added to the double sugar causing it to become two simple sugars again. Just as glucose is not the only single sugar, sucrose is only one of a group of double sugars with the formula $C_{12}H_{22}O_{11}$. Another important disaccharide is *lactose,* or milk sugar. This results from the condensation of glucose and another simple sugar called *galactose.* Lactose does not taste very sweet; hence milk is not particularly sweet in taste. *Fructose,* or fruit sugar, is the sweetest of the sugars mentioned and is almost twice as sweet as table sugar. What causes the sweet taste is not well understood. Consider the structure of saccharin (Fig. 14-4), which is not a sugar at all and yet is much sweeter.

Simple sugars are, of course, soluble in water. This is essential for transport in the blood or in the sap of trees. It also presents a serious

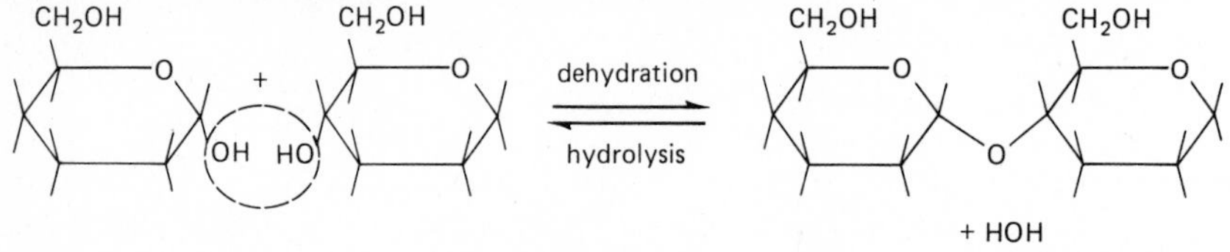

CH₂OH CH₂OH O O + OH HO dehydration hydrolysis CH₂OH CH₂OH O O O + HOH

FIGURE 14-6
Sugar molecules may easily be joined through their —OH groups. Splitting the H from the —OH group on one sugar and the entire —OH group from another (as circled by the dashed line) leaves the two molecules joined through an oxygen atom. Since the materials removed make up a water molecule, the process is one of dehydration. Because each carbon atom has an attached —OH group, the process may easily continue to form long chains of joined molecules.

problem for a plant. Consider the potato, which stores its excess food underground in a tuber. If this food is stored as a sugar, it is highly vulnerable to being dissolved and washed away by the groundwater. The process of saccharide condensation used to produce sucrose provides a solution to this problem. Sugars are amply provided with —OH groups, and the process of condensation does not need to stop when two simple sugars are joined. By attaching sugar molecule after sugar molecule in a long chain, molecules may be built which are too large to be soluble in water. A chain of this sort formed from several hundred individual glucose molecules is called *starch.* (For ways to accomplish the partial hydrolysis of a starch, consult your local cookbook under "potatoes.") Plants produce a second type of molecule by saccharide condensation of sugar molecules. This *polysaccharide* is composed of about 1500 glucose units attached in a slightly different manner, which your digestive enzymes cannot attack, and it forms the prime building fiber of plants—cellulose. Cellulose is used for the cell walls making up the roots and stems, trunks of trees, and so on. Cellulose and starches are both constructed entirely from glucose subunits and thus are excellent food sources if they can be hydrolyzed. Only certain bacteria seem to be able to accomplish the hydrolysis of cellulose with any ease. If a convenient way could be found for this hydrolysis, the trunk of a pine is potentially as nourishing as a potato.

The mechanism for obtaining energy in the body is complex and occurs in a series of steps. The following description contains many simplifications which may be objected to in detail, although the overall picture is essentially correct. If we burn a mole of glucose, we would obtain the 670 kcal stored in it by photosynthesis. Fires, internal or external, are hazardous to your health. The body must accomplish the oxidation of glucose at normal body temperature. To do this the molecule must be dismembered in a sequence of small steps, and the energy release attained when each bond is severed must be stored in some form. This is a highly specialized chemical task, yet the body needs energy when it needs it and where it needs it. Thus, there must be a temporary storage device which is easily transportable. By analogy, the manufacture of an automobile is a complex task. When you needed a new car, you could set up an assembly line in your backyard and make one. It certainly is more convenient to centralize the production somewhere and just ship the product to individuals needing cars. Glucose breakdown in the cell begins in the cytoplasm and finishes in small assembly lines located on

Shedding tears for bacteria

Chemical warfare appears to be a dirty topic in the press, which denounces it whenever possible. Actually, nature beat us to most of the really deadly substances we know of. A shot glass of the toxin of the bacterium *Clostridium botulinum* could wipe out a city if properly deployed. Bacterial toxins are part of the chemical warfare which has been going on for millennia inside human bodies. Like all vast chemical plants, the human body has a very serious security problem.

Outsiders running around turning the wrong valves can destroy the entire plant. The prime security force of the body is the group of white blood cells, leukocytes, who wander throughout the body checking credentials. To do this, they are the only type of human cell with the separate ability to move and wander between and through other tissues. The white cells verify the nature of the objects they encounter as having a right to be in the body. Should a foreign object be detected, the white cells engulf and destroy it—at least, this is the plan. However, bacteria have been at the business of combat with leukocytes for some time and have their own defenses. Since bacteria cannot move at will and the white cells can, mobility is all with the defensive forces. Bacteria produce chemicals called *toxins*, which leave the cells and saturate their surroundings. These toxins could be natural bacterial waste products, but they include some of the most virulent poisons known. In a body where billions of cells die and are replaced every day, a small colony of a few million bacteria should have little effect. That the whole body can be laid flat in bed is a testimony to the chemical potency of some of these bacterial products. White cells migrate by the millions to any site where bacteria have penetrated the outer defensive line of the skin. They also die by the millions if the chemical attack of the bacteria is sophisticated enough. The prime strategy at this point is to contain the invasion, and it may be accompanied by suicide charges of the white cells. The real weapon to combat infection is superbly effective in most cases. This is the chemical machinery of the body itself. From the beginning of the infection it starts work on a chemical solution to the particular attacker. While it sounds stupid at first reading, the human body will win if it can manage to live long enough. The classic job of the physician was not to do anything, but to tell the family whether the patient's body would have enough time to find a solution to the problem. However, the change of medicine in the middle-third of this century has been unbelievable. You cannot appreciate the difference between old medical movies and a modern clinic, or the explosion in health care costs, unless you appreciate this change. Consider this description by Dr. Lewis Thomas, President, Memorial Sloan-Kettering Cancer Center and Professor of Pathology and Medicine, Cornell University (abstracted from "Biomedical Science and Human Health: The Long Prospect," *Daedalus*, vol. 106, Summer 1977, pp. 163–171):

Accurate diagnosis became the central purpose and justification of medicine, and as the methods for diagnosis improved, accurate prognosis also became possible, so that patients and their families could be told not only the name of the illness but also, with some reliability, how it was most likely to turn out. By the time this century had begun, these were becoming generally accepted as the principal responsibilities of the physician. . . .

I was a medical student at the time of sulfanilamide and penicillin, and I remember the earliest reaction of flat disbelief concerning such things. We had given up on therapy, a century earlier. . . . [Some diseases] were fatal in 100 percent of cases, and we were convinced that the course of master diseases like these could never be changed, not in our lifetime or in any other.

Overnight, we became optimists, enthusiasts. The realization that disease could be turned around by treatment, provided that one knew enough about the underlying mechanism, was a totally new idea just forty years ago. . . .

We need reminding, now more than ever, that the capacity of medicine to deal with infectious disease was not a lucky fluke, nor was it something that happened simply as the result of the passage of time. It was the direct outcome of many years of hard work, done by imaginative and skilled scientists, none of whom had the faintest idea that penicillin and streptomycin lay somewhere in the decades ahead. It was basic science of a very high order, storing up a great mass of interesting knowledge for its own sake, creating, so to speak, a bank of information, ready for drawing on when the time for intelligent use arrived.

In time the body will build specific molecules that will destroy bacterial toxins, called *antitoxins*. With the toxins gone, leukocytes engulf and digest the bacteria. Once the body has learned to make the molecules to combat a particular disease organism, it tends to retain the ability to produce these molecules on short notice. This is why many diseases can strike a particular human being only once, and after that the human is immune from further attack. The strategy for vaccines comes immediately from this view of chemical warfare in disease. If the body is injected with a small amount of the chemical agents it would have to combat in a particular disease, it will develop its chemical defense. Thus, a vaccine can frequently be prepared from the chemical residues of a disease organism. Learning to cope with this prepares the body to fight off the organism in short order if it later appears.

Of course, anything so complex as the human body will have areas of weakness to attack and because of that must have several lines of defense. Consider the eyes! The eyeball must be free to move around in its socket and is connected to the body only through the optic nerve at the back and the muscle strands that move it around. The front of the eye cannot be richly supplied with a stream of red blood since it must be transparent for you to see through. Between the eye and its socket is a fluid layer which is open to the outside. Everything seems perfect for bacteria. The area is moist, it is warm, it is technically outside the body and largely immune to leukocyte attack. (Besides which the leukocytes are busy providing nourishment to the front of the eye, where the red cells aren't allowed. These are the moving spots you sometimes see in front of your eyes.)

How does the eye protect itself? Actually, it's quite simple. The eye produces a protein for tears which has such voracious habits it makes a piranha look like an angelfish. This is the enzyme *lysozyme*. Like all enzymes, or organic catalysts, lysozyme causes a reaction to take place without being itself used up in the process. The reaction is the disintegration of bacteria. The molecule lysozyme looks very much like a miniature set of false teeth, and acts like them, too. It has two hinged parts with an unusually deep cleft between them. On the inside of the cleft are chemically active sites which bind sugar molecules. Now, a bacterium is essentially a plant which has been forced to look elsewhere for a living since it lacks chlorophyll to make its own food. Bacterial structures are very plantlike. In particular, the cell walls of bacteria are thick structures made of cellulose, a polysaccharide chain of simple sugars. As lysozyme wanders about, when the end of a cellulose strand straggling from the bacterial wall falls in the cleft, it is bound by the active sites. As soon as this happens, the molecular structure changes and lysozyme snaps shut. This action does two things. First of all, it neatly severs a small chain of four sugar molecules from the end of the cellulose fiber. Secondly, this action once again changes the molecular structure, which causes lysozyme to release the four-sugar chain and open its "jaws" for the next "bite" out of the cellulose fiber. In very dilute concentrations lysozyme molecules work so fast on bacteria that you can picture them clicking away like a typewriter. Four-sugar unit by four-sugar unit the bacterial cell wall is dismembered. That is all there is—the bacterium is destroyed by leaking out into the tear fluid. This dilution means death to a living cell since in the diluted fluid key molecules cannot find the substances they need to keep operating. Lysozyme is an incredibly efficient way to skin a bacterium.

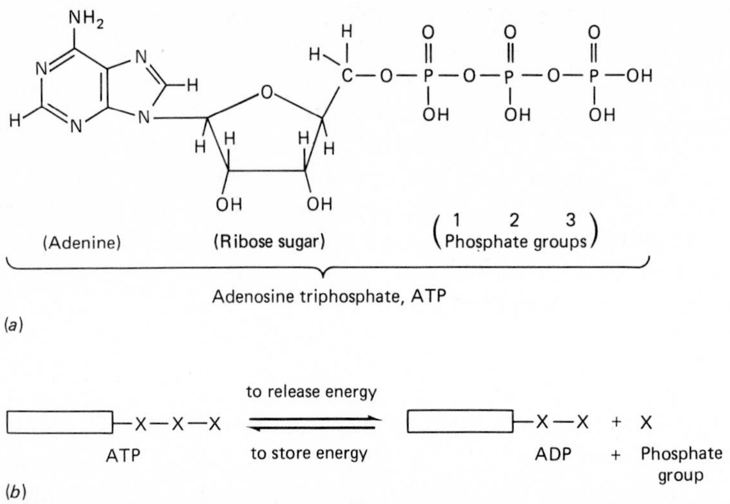

(Adenine) (Ribose sugar) (1 2 3 Phosphate groups)

Adenosine triphosphate, ATP

(a)

to release energy

ATP to store energy ADP + Phosphate group

(b)

FIGURE 14-7 The basic energy carrier of the body is the ATP molecule. This is composed of two common body molecules with three phosphate groups attached as a "tail." The energy from the breakdown of foods may be stored by attaching the third phosphate group to this tail. Detaching the last phosphate group releases a burst of energy.

structures called *mitochondria.* These have an "assembly line" of highly specialized molecules, each of which accomplishes a single task. Organic catalysts, or *enzymes,* reduce the glucose into pyruvic acid in 10 fast and easy-to-accomplish steps releasing less than a tenth of the energy of the glucose molecule. This preparatory set of steps is called *glycolysis.* The string of enzymes on a mitochondrion takes another set of 10 repeated steps to reduce the remaining energy. The end result of all this is the exact reverse of photosynthesis:

$$C_6H_{12}O_6 + 6O_2 \rightarrow 6CO_2 + 6H_2O + \text{energy}$$

The energy is not released to be wasted. Instead, it is taken at one end of the enzyme from the breaking of a glucose bond, ripples through the enzyme, and is used at the other end to put together another molecule, called *adenosine triphosphate* (ATP). ATP molecules are the carriers of instant energy packets to all parts of the cell requiring energy. Much as thousands of devices are designed to work off standard flashlight batteries, thousands of cellular processes are designed to work with bursts of ATP energy. ATP is made by attaching a third phosphate group PO_4^{3-}, onto ADP (adenosine diphosphate), which has only two (Fig. 14-7) in the reaction

$$\text{ADP} + \text{phosphate group} \xrightarrow{\text{energy}} \text{ATP}$$

The net effect of the complete breakdown of a molecule of glucose is to produce some 38 ATP molecules out of ADP molecules. These ATP molecules store some 300 kcal of the original energy of each mole of glucose. In turn, about half of this can be used in chemical activity such as thinking or tightening a muscle. Thus, the body uses between one-fourth and one-fifth of the energy of the glucose for purposeful activity. The remainder is lost as heat. In times of extreme activity energy may be needed faster than oxygen can be delivered to the cells. The body may compensate in a number of ways. One is to step up blood delivery by boosting the breathing rate and heart pumping rate. Generally, this is not adequate. Several *anaerobic* pathways to increased energy are possible which do not require oxygen. The most common pathway takes the pyruvic acid from glycolysis, converts it to acetic acid for an additional 6 kcal/mol, and simply allows this to accumulate in the cells.

Lipids or Fats

The human diet consists of a wide variety of organic substances. The duty of the digestive system is to break these down into usable molecules and then allow them to enter the body. As we have seen the end product of carbohydrate digestion is glucose. The body can chemically change a wide variety of molecules into glucose. Yet humans can live quite well on diets totally devoid of carbohydrates. The eskimos are one example of a human population with such a diet. A prime source of energy in the eskimo diet is from lipids. A *lipid* is a plant or animal material that will dissolve in oil but not in water, and the most important subgroup of lipids is the fats. Both plants and animals use fats as molecules for the storage of food, but the resulting materials are more apparent in animals. Fats, or *glycerides,* are composed of two parts. *Fatty acids* are long-chain hydrocarbons with an acid group attached to one end. These chains are usually 10 to 20 carbon atoms long. To make a fat, three fatty acids (alike or different) are attached to a molecule, usually the alcohol glycerol, using their acid groups (Fig. 14-8). The properties of the fat that results depend upon the nature of the carbon chains. If the chains have all single bonds between carbon atoms with other bonds all holding hydrogen atoms, the fat is called *saturated.* Saturated fats tend to be solids at room temperature—such as lard or butter. However, if the chains contain less hydrogen than they could with many double bonds between carbon atoms, they are called *polyunsaturated.* Such polyunsaturated fats tend to be liquids at room temperature—such as corn oil or olive oil. The main physical difference between the animal fats and the vegetable oils is one of boiling point, with the fats boiling at a higher temperature. This relates directly to one of our main uses of these compounds. If a po-

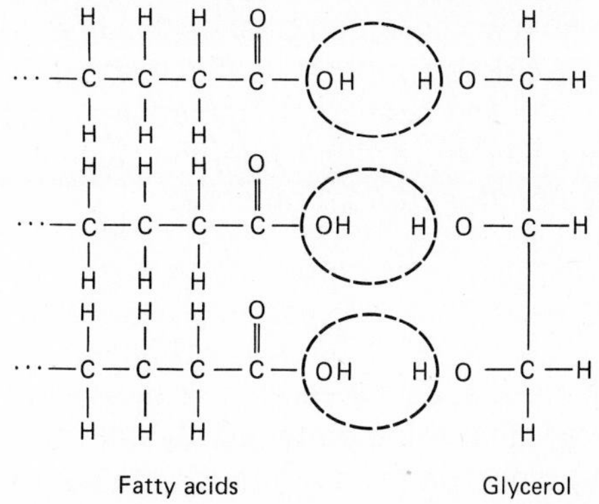

```
          H   H   H   O                    H
          |   |   |   ||                   |
    ···—C—C—C—C—(OH    H) O—C—H
          |   |   |                        |
          H   H   H                        |
          H   H   H   O                    |
          |   |   |   ||                   |
    ···—C—C—C—C—(OH    H) O—C—H
          |   |   |                        |
          H   H   H                        |
          H   H   H   O                    |
          |   |   |   ||                   |
    ···—C—C—C—C—(OH    H) O—C—H
          |   |   |                        |
          H   H   H                        H

             Fatty acids              Glycerol
```

FIGURE 14-8 A long hydrocarbon with an acid group at one end is a fatty acid. Three of these may be attached to a glycerol molecule by splitting out the water molecules shown by the dashed lines. When attached in this manner, these pieces make up a fat.

tato is dropped into boiling water, it cooks at 100°C and gets mushy in time. If the same potato is dropped into a pool of boiling oil, it cooks at a higher temperature, depending on the oil. This higher temperature produces a solid, crisp structure rather than a soft one, producing potato chips, french fries, and so on. While animal fats are desirable for frying because of their higher boiling point, vegetable oils are cheaper. The oils may be made into solid fats by breaking their double bonds and adding hydrogen atoms. In the process of *hydrogenization* vegetable oils are passed over a platinum catalyst along with hydrogen gas and emerge as a solid fat with a higher boiling point. Spry and Crisco are just two such hydrogenated vegetable oils.

Saturated fats are frequently implicated in the production of the complex fatty molecule *cholesterol* in the human body. Buildup of deposits of cholesterol inside the arteries leads to restricted blood flow and may contribute to high blood pressure and premature hardening of the arteries. Cholesterol itself is a natural substance found in all human bloodstreams and in many foods. The body is perfectly capable of making cholesterol from other molecules. More cholesterol is produced this way than is consumed in a normal diet. One of the easier molecules for the human chemistry to modify to obtain cholesterol is the fatty acid from a saturated fat. It is not clear whether cholesterol deposits in the arteries are a cause of arterial disease or a symptom of malfunction of some other mechanism of the body. The biological sciences are much more difficult areas of experimental science than the physical sciences, and this is an example. It is not possible to "turn off" the thousands of separate reactions occurring in a body to study just the one we want to. Trying to study a specific reaction in a living human body is a complex task. It is a bit like trying to hear a tenor singing an operatic aria in the opposite stands of a stadium during an exciting football game.

Fats are built up in the body in two-carbon units from acetic acid. Storage of energy in fat represents a more economical method than storage in starches. More ATP units are generated from a fat than from a starch of comparable size. Each kilogram of fat will store about 6.0 kcal of energy. A kilogram of carbohydrate stores only 4.5 kcal. Thus, fat is 33 percent more efficient for storage of energy, especially for an animal which must carry its storage facilities around with it. To obtain the energy from a fat, the body first breaks the molecule into fatty acids. Each of these goes through a repeated cycle of processing which "snips" two carbon atoms off as an acetic acid molecule during each repetition of the cycle. The enzyme chain which accomplishes this breakdown works perfectly well in reverse if ATP is supplied. In this manner the energy of excess food is stored in fats. The acetic acid molecules produced in fat breakdown enter what is called the *Krebs cycle,* the same cycle entered by acetic acid molecules produced from glucose.

Proteins

The third category of chemical substances which can provide the energy necessary to maintain life is that of the proteins. We will not discuss their use as an energy supply since proteins are much too valuable for that.

Protein molecules are the substance of life itself. Everything that is living is an assemblage of protein molecules. While proteins may be dismembered to supply energy for the body, they may also be used as the "spare-parts" bin for new life. In the time it takes you to read this chapter several thousand humans will have died of starvation on earth. Some will have died from insufficient energy supply to run their bodies. More will have died from an inadequate spare-parts supply to maintain their bodies—from protein starvation.

Proteins are made from subgroups called *amino acids.* These are molecules with an amino group ($—NH_2$) and an organic acid group (or carboxyl group, $—COOH$), *both* attached to the molecule. All important amino acids have these groups attached to the same carbon atom. The simplest amino acid is glycine, CH_2NH_2COOH (Fig. 14-9). Biologically and chemically, the most important thing that these molecules do is attach to each other. They do this by means of a *peptide bond,* in which the amino group of one amino acid attaches to the acid group of another (Fig. 14-9). Every time a peptide bond is formed, a water molecule is split out of the structure, so this is a process of dehydration. Similarly, to break a peptide bond you must pry apart the parts and add the appropriate parts of a water molecule to the places where they are missing. The digestion, or breakup of protein molecules in your stomach, is thus a process of hydrolysis, or the chemical addition of water to the structure.

Since amino acids have both amino groups and acid groups on them, they may attach to other amino acids in two ways. Nature uses this property to form long chains of amino acids in the proteins. The amino acids used to form living proteins differ only in what is attached to the carbon containing the amino group. For example, if we pluck a hydrogen from that atom in glycine and replace it with a methyl group ($—CH_3$), we obtain the amino acid *alanine* (Fig. 14-9). While the amino group and the acid group of these molecules allow them to attach to each other, it is the tails of the molecules, not involved in protein formation, which will determine the behavior of the protein once it is formed.

All human proteins are made of some 23 different amino acids.

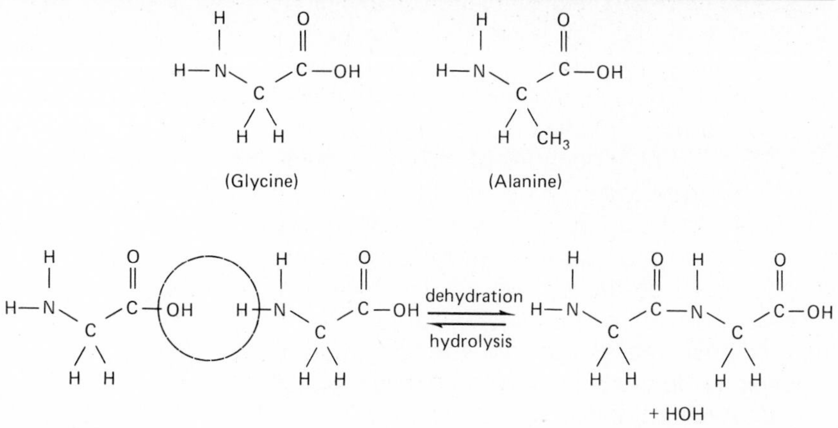

(Glycine)

(Alanine)

dehydration

hydrolysis

+ HOH

FIGURE 14-9
Amino acids all have an amino group and an acid group attached to the same carbon atom as shown in the glycine molecule. Other amino acids differ only in the side group also attached to this carbon, such as the $—CH_3$ group shown for alanine. Amino acids may be joined by a peptide bond by splitting the —OH portion of the acid group on one molecule and one of the hydrogens from the amino group of the other molecule as indicated by the dashed line.

Thus, whatever specialized protein is necessary, from muscle tissue through brain cells in a forming embryo, it can be made from 23 different parts. At this point a problem arises. The human chemical machinery is incredibly versatile at manufacturing needed molecules from miscellaneous spare parts, but it is not perfect. Eight of the twenty-three amino acids have tails which the human body cannot synthesize, and two more cannot be produced in the amounts necessary to maintain your life. Thus, it is essential that your diet contain these 10 molecules. These 10 *essential amino acids* are represented in Table 14-4. In each case the molecule is identified by the three-letter code, which is standard in discussing protein structure.

Now the problem becomes more complex. The more closely related two life forms are, the more closely their proteins resemble each other. For example, identical human twins have identical protein structure, and thus there is no rejection problem in a transplant of organs. All humans have the extreme similarity of chemical structures required for reproduction; yet, we are sufficiently different from cattle, for example, that there is no likelihood of human-cattle reproduction (not that we would want it). However, and most important, we are sufficiently related to cattle that we can live comfortably on a diet of beef. While significant differences exist in protein structure between humans and cattle, this is not important in the use of beef as a food supply. All proteins are broken down into amino acids before they enter the body anyway. The point is that beef proteins (or pork, or dog, or kangaroo proteins) are similar enough to human proteins to ensure that the essential amino acids will be present and in about the correct proportion to allow us to make human protein from them. Life as a vegetarian is more complex. If even one of the essential amino acids is missing from your diet, or present in insufficient amounts, you will suffer protein starvation. Human beings are not sufficiently similar to plants that plant proteins have just what we need in the correct amount. A diet built on one type of plant exclusively tends to be fatal to humans. The Asian living on only rice will die, as will the Mexican living on only corn. Plant diets must be built around a variety of plants to give an adequate supply of each essential amino acid or else be supplemented with animal proteins. A human may die of starvation on a million-Calorie-a-day diet. He or she may die fat, but nonetheless the culprit will be starvation—starvation for essential amino acids which the body cannot synthesize and which must be present in the diet to support life. In principle there is no reason why a single plant could not supply all that is needed in the human diet. Recently, after decades of selective breeding, such a plant has indeed been developed—a special type of corn. Normal corn proteins are deficient in the amino acid lycine and will not prevent protein starvation. A strain of high-lycine corn recently developed could have a profound impact on the protein starvation rate in underdeveloped areas if the corn is used as the dietary staple.

Amino acid chains form two basic types of protein. The first group is that of the *filamentous,* or linear, proteins. Silk or wool are examples of collections of such molecules. Although these may be formed of very

TABLE 14-4 The Ten Essential Amino Acids

STRUCTURE	NAME AND ABBREVIATION
$H-C(H)(H)-C(H)(OH)-C(NH_2)(H)-COOH$	Threonine (Thr)
$CH_3-S-C(H)(H)-C(H)(H)-C(H)(NH_2)-COOH$	Methionine (Met)
$H-C(H)(NH_2)-C(H)(H)-C(H)(H)-C(H)(H)-C(H)(NH_2)-COOH$	Lysine (Lys)
$CH_3-C(H)(CH_3)-C(H)(NH_2)-COOH$	Valine (Val)
$CH_3-C(H)(CH_3)-C(H)(H)-C(H)(NH_2)-COOH$	Leucine (Leu)
$H-C(H)(H)-C(H)(H)-C(H)(CH_3)-C(H)(NH_2)-COOH$	Isoleucine (Iso)
(ring)$-C(H)(H)-C(H)(NH_2)-COOH$	Phenylalanine (Phe)
(benzene ring fused to ring with $C=CH$, $N-H$)$-C(H)(H)-C(H)(NH_2)-COOH$	Tryptophan (Try)
$NH=C(NH_2)-N(H)-C(H)(H)-C(H)(H)-C(H)(H)-C(H)(NH_2)-COOH$	Arginine* (Arg)
$H-C=C-C(H)(H)-C(H)(NH_2)-COOH$ (ring: $N=C(H)-N(H)$)	Histidine* (His)

* Synthesized in body but not in adequate amounts.

long molecular strands, the molecules do not execute any special chemical functions. In the second class of proteins, the molecules bend and kink to assume a particular three-dimensional shape. These are the *globular* proteins, and they include most of the critically important structures of the body. For example, hemoglobin is a globular protein molecule.

Hemoglobin, a Protein Molecule

Decoding the structure of a biologically important protein molecule is a task that can absorb the working life span of a team of scientists. The attack usually involves specialists from several areas of science. Even though modern instruments have speeded this work enormously, it remains an exacting, detailed area of study. In spite of the fact that years of effort can be required before any useful results begin to emerge, morale in such working teams is usually quite high. This is because it is near miraculous that some of these structures can be determined at all. Yet, the last few decades have seen an unbroken string of successes in the area. Consider a common protein molecule—hemoglobin. Hemoglobin is the pigment in blood cells that makes red blood red. Functionally, this is the molecule that binds oxygen molecules in the lungs and delivers them to the external body, where it picks up carbon dioxide molecules to transport them to the lungs. Clearly, this is a molecule which is completely essential to human life. A *single drop* of human blood contains about 100 million red blood cells, each of which is equipped with many thousands of hemoglobin molecules. The molecule may be obtained by smashing or dissolving the red cells and separating the mixture with solvents. Much careful labor can isolate relatively pure hemoglobin. From this the molecular weight of the molecule may be measured. Water, for example, has a molecular weight of 18—16 for the oxygen atom and 1 for each of the two hydrogen atoms. Hemoglobin, by contrast, has a molecular weight of about 60,000. This would imply that a single molecule of the protein has between 10,000 and 20,000 atoms. Washing the pure substance with acids alternately breaks the molecule into four separate protein chains, each of which binds a heme structure, which gives the molecule its name. The *heme group* is a platelike structure centered on an atom of iron. The heme structure actually binds the oxygen or carbon dioxide in gas transport. (Production of this structure is the main reason iron is required in the human diet.*) Since they come apart quite easily, these four subchains are not covalently bonded to each other. The four chains are not alike, but represent two each of two different types.

Analysis from this point represents a difficult problem. Of course, we could just digest the entire protein with a brew of stomach acid. This completely hydrolyzes the peptide bonds and gives us separate amino

* It is extremely interesting that this structure has another very much like it. The heme structure is the most important functional pigment of animals and is built around an iron atom. The most important functional plant pigment is green chlorophyll, which is built around a magnesium atom. I will let you look at the two structures and decide whether you see a family resemblance between heme and your friendly neighborhood pond scum (Fig. 14-10).

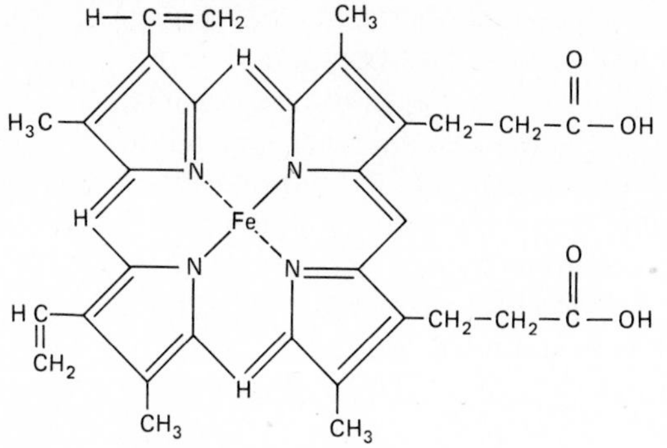

(a) Heme molecule

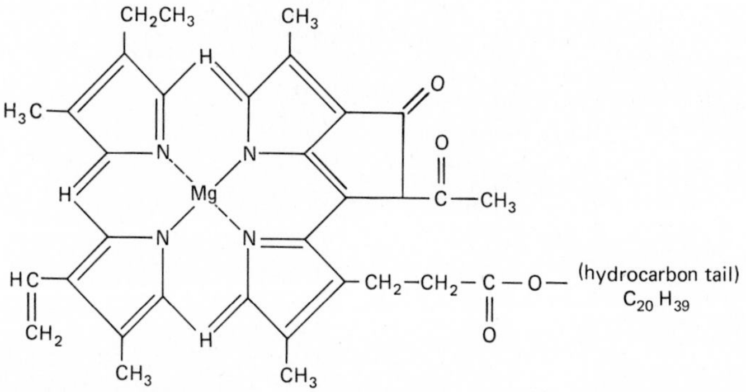

(b) Chlorophyll molecule

FIGURE 14-10 Structures of the heme molecule, the red pigment of blood, and the green chlorophyll molecule of plants.

acids, producing 604 amino acids per molecule. Tied for first place are the amino acids alanine and leucine with 72 each, and at the bottom end are two lonely cysteine molecules. Altogether 19 different amino acids are present, and we can determine how many of each goes into the molecule. What use is this? Looking at a heap of steel beams piled on the ground will not tell us what the Eifel Tower looked like before we took it apart. In proteins, as in art, the structure is more than the pieces. Connected in an arbitrary manner these amino acids would produce a reasonable imitation of used bubble gum but not a structure that performs a function essential to our lives. It is not the pieces but how they are put together that counts. A big rock is just more deadness than a little rock. A big assemblage of amino acids, if put together exactly right, is more than a little one; it *functions*. It does things small random peptide linkages cannot do. It may move oxygen from where it is plentiful to where it is scarce. If the assemblages get big enough and we have enough of them, they do more yet. They move, they react—collectively, they live! Big, nonrandom protein molecules are the essence of life.

Now, as such things go, hemoglobin is a reasonably manageable molecule in size. Once we have the parts list, what next? Next is the order of the parts—the assembly directions. If the parts of the molecule give us the chemical formula, how they are put together makes up the *primary structure.* In this we are aided by a set of chemicals that digest proteins. Some of these mindlessly attack all peptide bonds and separate amino acids into a jumble of parts. These cannot be used at this state. Other chemicals are much more selective. Some, such as the enzyme carboxypeptidase, snip apart a protein one amino acid at a time, starting at a particular end. Carboxypeptidase works from the end of the peptide chain with the free acid group. Other enzymes work their way in from the end with the free amino group. By carefully controlled digestion we can determine the sequence of amino acids by the order in which they appear during digestion. For example, hemoglobin digests in this sequence: Val-Leu-Ser-Pro-Ala-Asp-Lys-Thr-Asn-Val-, plus lots more. Thus, the amino acid on the end is valine, the next is leucine, and so on. We now know how the parts are put together. Do we have the structure of the molecule? No, it's not just the sequence of the parts but also how they are arranged in space that makes the molecule functional. We need the *secondary structure:* the spatial arrangement of the side chains and curvatures of the spine of the molecule. Even if we were told the order in which several hundred Tinkertoys were plugged together one after the other, we would not know the bends and kinks in the chains used to make this a work of art. The next step is to determine the position of each part of the molecule in space relative to all other parts, the *tertiary structure.* This is far from simple.

To determine the structure in space of a molecule, use is made of the wave property called *diffraction.* To picture how this works imagine a large pier sitting on heavy round pilings. As small waves wash across the pilings, each acts as a reflective surface and scatters some of the wave in all directions (Fig. 14-11). As these reflections move away from the pier, there are certain positions far from the pier where the reflections from each of the pilings will come together in phase and produce a disturbance. There are other positions where the reflections will arrive out of phase, crests on troughs, to produce no disturbance. For a set of pilings which are spaced irregularly the pattern of constructive and destructive interference will be very complex. Even so, this complex pattern can be read mathematically to infer the position of each piling in the pier. Now

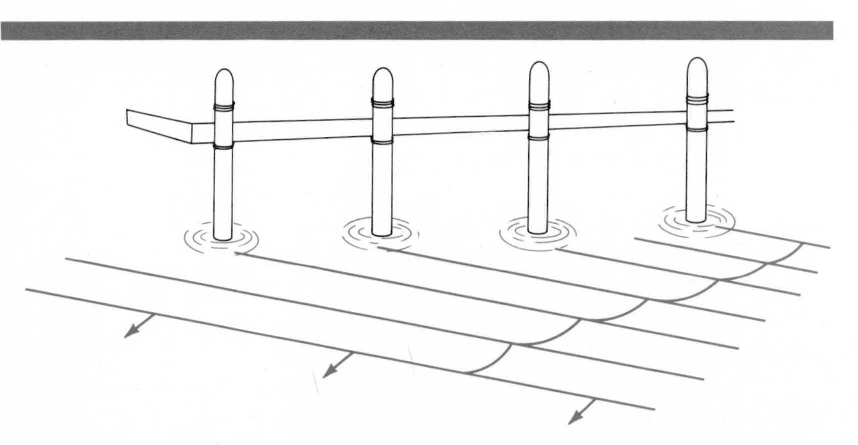

FIGURE 14-11
When a train of waves washes past the posts of a pier, each post is the source of ripples which move away in all directions. Far from the pier, these sets of ripples will add together at some spots to make a larger wave and cancel out in other places. A very detailed mathematical analysis of the way these ripples agree and disagree at a distance would allow us to calculate the size, location, and spacing of the posts without ever seeing the pier.

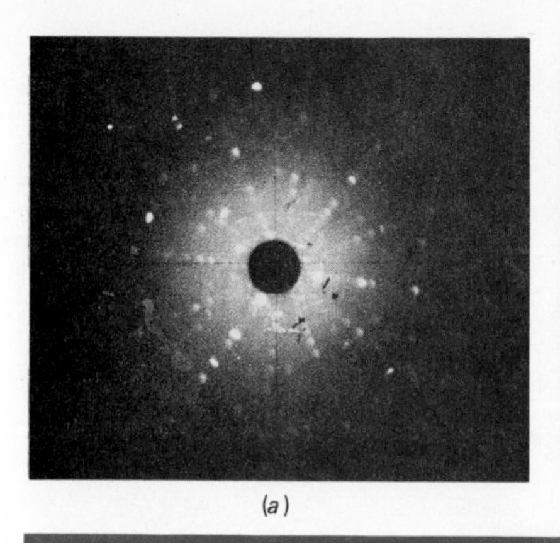

(a)

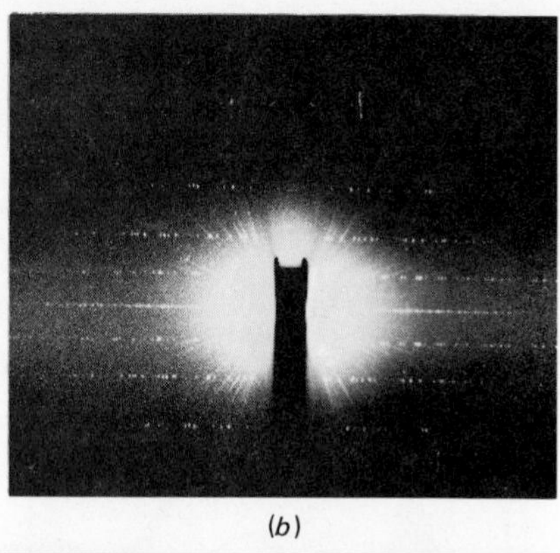

(b)

FIGURE 14-12 When a beam of x-rays is directed into a crystal, diffracted rays reinforce to produce bright spots in some locations and cancel for no effect in other positions. The information in these patterns allows us to determine the atomic spacings and placements in the crystal. (*a*) The simplest pattern is produced when a beam of x-rays directed toward the dark central opening of the film encounters the sample in front of the film. (*b*) A much more detailed pattern is produced if the sample and the photographic film are separately rotated about the beam direction. This is the technique necessary for use with most protein samples. (*Photographs courtesy of Mr. Fred Park.*)

let us look at our protein molecule. The size of waves appropriate for use with a molecule are x-rays. Generally, a wave does not detect a structure much smaller than about one-fourth of its wavelength. Thus, just as a tidal wave would not reflect from the pier pilings, light cannot be used to detect an atom, since the smallest light rays are many thousands of times longer than an atom and will not reflect. Light can reflect only from gigantic clumps of atoms. One molecule will not reflect enough x-rays for us to detect the pattern of constructive and destructive interference. Thus it is necessary to prepare a crystal of pure protein where the molecules are lined up one after the other in an even array, like a neat set of stacked boxes. Even then, many problems remain in using diffraction to determine the location of the atoms within the molecule. With persistence these can be solved and photographic film can record "hot spots" and "cold spots" of x-ray interference in the space near the sample (Fig. 14-12). For a complex molecule it may be necessary to take hundreds of such film records at precisely controlled distances from the sample. The intensity of each of these spots is measured, and a computer is used to calculate the position of the atoms causing the diffraction. Without the computer, the calculations for a complex molecule could require several human lifetimes. The result of this calculation is called an *electron-density map* of the space inside the molecule.

Now, what about our hemoglobin molecule? Well, it is a compact mass where the four heme groups are separated by a distance some 28 times the diameter of a hydrogen atom. These groups are mostly on the outside of the molecule and are spaced like the corners of a triangular-based (equilateral) pyramid. Interestingly, if the molecule is plotted when binding oxygen molecules, it shrinks by some seven hydrogen atom radii. Time measurements of this action indicate the molecule "snaps shut" into a changed structure when binding molecules. Furthermore, there seem to be "spring-trigger" sites on the molecule that cause it to open to release its cargo. The opening and closing positions on the molecule are so far apart that information must be transported along the body of the protein itself by some mechanism. Much more has been learned about this molecule, but the key ideas already have been

FIGURE 14-13
Representation of a molecule of hemoglobin. The flat plates in the structure are the heme molecules used in binding oxygen. The circle drawn in represents the approximate size of a single carbon atom to scale.

implied. Protein function depends upon its specific size; its shape; the spatial relationship of its parts; the location of special chemically active sites; the distribution of small forces along the coiled protein spine, which can be modified at an active site to cause a "ripple effect" down the protein chain. In short, the geometry of the molecule is as important as its chemistry in determining its function.

The Attack on Life's Plan

The most important development in molecular biochemistry in recent times has been the decipherment of the genetic structures by which organisms pass along a complete blueprint for life. This is ultimately as important to biology as the decipherment of hieroglyphics by Champollion was to Egyptology. For the first time we may read life's records, or blueprints. Probably the most serious problem a living system must solve is that of reproduction. A simple organism has very little problem. An amoeba can just duplicate every structure and then split into two halves, each of which will grow up to be just like the parent. However, each daughter cell will have all of the faults and flaws of its parent as well as the strong points. Survival in a changing system depends upon the ability to adapt. Today, large manufacturers of buggy whips have either diversified into new fields or they have gone into extinction—as have shipyards specializing in full-rigged sailing ships. All complex organisms (such as grass and oak trees, salamanders and people) have solved this problem by sexual reproduction. Instead of only one pattern to solve problems, we have two parents, and inherit one pattern from each. In this shuffling of traits some offspring will be extremely well equipped to solve the problems, while other traits will die out. Thus the organism adapts to changes through time. Of course, the problems are usually chemical in nature, and the reshuffling is not perfect. For example, no human being on earth can synthesize 8 of the 10 essential amino acids. This ability has presumably been lost somewhere, since most other animals and some plants can synthesize these chemicals. The human diet through time must have been sufficiently rich in these amino acids that the possession of the solution to their synthesis did not really matter for human survival, and the pattern was lost. Cattle still have it, as do dogs and kangaroos.*

Somehow, in reproduction we must transmit a complete blueprint for life to the new organism. The human egg is about $\frac{1}{10}$ mm across and can barely be seen by the human eye. Most of the egg is food, however, and this contains no information. The sperm cell of the male has about one-millionth the volume of an egg and is little more than a swimming cell nucleus. A billion sperm cells could make up a single liquid drop. Yet in the tiny volume of a single sperm cell, plus a similar volume in the egg cell, rests the complete information to construct a total, functioning human being.

Gregor Mendel (1822–1884) completed his experiments with pea plants and wrote his theory of genetic inheritance in 1868. This declared that characteristics in an offspring were determined by two separate instructions, one from each parent. Mendel called the carrier of these instructions the *gene.* Mendel's work was ignored for over 30 years until his gene became associated with structures found in the cell nucleus called *chromosomes.* Thirty more years of work definitely established that the genetic blueprint for a life form was contained in the chromosomes. Chemically, chromosomes are deceptively simple. They are not proteins, and the structure appears much less complex than a protein. A sample of pure genetic material can be separated into only six different types of molecules: the five-carbon sugar deoxyribose, phosphate groups, and four molecules called *bases.* Of these four, two are chemicals called *purines* named *adenine* and *guanine.* The other two are compounds called *pyrimidines* named *cytosine* and *thymine.* Some repetition of these six molecules can make up a complete blueprint for you and for every mechanism that allows you to live. For example, somewhere in the genetic material are the instructions for building a hemoglobin molecule, since neither the egg nor the sperm contains a sample to work from. Because it is weakly acidic, is found in the nucleus, and has the sugar deoxyribose dominating its structure, the molecules making up the chromosomes are named *deoxyribonucleic acid,* called DNA for short. Of course, not all DNA is exactly alike. It would hardly do for a grass plant and a human being to be made from the same blueprint. The place to start an investigation then is to determine what is constant or

* Another critical area of lost chemical patterns is that of the vitamins. These chemical structures must be present in our food for humans to remain in good health. One example is vitamin C or ascorbic acid. Just about anything any human would ever consider eating contains some vitamin C, so why bother keeping the instructions for making it? We didn't! However, vitamin C is easily destroyed by several things, such as lengthy exposure to the air or cooking. People living off of fresh, raw plants and animals have no problem with maintaining vitamin C intake. The other 99 percent of the human race cook most of their food and may well have some problems.

different about DNA from species to species. The proportions of sugar and phosphate turn out to be about constant. Furthermore, if we divide the weight of a separated sample by the molecular weights of each molecule, an interesting fact emerges at once. All samples of DNA contain one phosphate for each sugar present. Furthermore, one sugar-phosphate pair is present for each molecule of one of the four bases. This suggests that the basic chemical unit making up DNA has three molecules connected somehow, sugar-phosphate-base, and that this unit repeats. Since the only variability from species to species is in the proportions of the bases, these must contain the genetic code. Examination of the composition of DNA reveals one more fact. Bases must somehow occur in sets. Every time a molecule of adenine occurs, a molecule of thymine is found also. Similarly the number of molecules of guanine always matches that of cytosine. Thus, the bases must exist in pairs. However, the number of adenine molecules compared to guanine molecules varies widely from species to species. (This number also varies widely from chromosome to chromosome in a single cell. To obtain a large enough sample of DNA to work with, the chromosomes of billions of cells must be used. Thus the numbers are the average for a species.) An important clue to the DNA structure was produced by x-ray diffraction studies. The diffraction photograph produced was characteristic of a molecule tightly coiled or twisted in a shape called an *alpha* (α) *helix*. DNA is composed of immense strands twisted into a tight spiral. Furthermore, some structure repeated at intervals the thickness of a few atoms down the center of the spiral.* However, unless a structure repeats exactly in space it cannot be detected by x-ray crystallography. Any part of the structure that varies from molecule to molecule provides no information.

From this set of information, Watson and Crick proposed a model for the structure of DNA in 1953 (Fig. 14-14*a*). DNA is made of two linear chains each of which has a backbone of alternating sugar molecules (S) and phosphate groups (P). Projecting to the side from each sugar molecule is one of the four bases (A, G, T, or C). The sequence of these bases is not the same in the two chains. Every time an A occurs in chain 1, the corresponding position on chain 2 has a T, and vice versa. Likewise G-C base pairs are matched on opposite chains. The base pairs are *not* covalently bonded but are held together by hydrogen bonds (Fig. 14-14*b*). Thus, the DNA molecule is a bit like a zipper. Each side chain is tightly held together by covalent bonds, but the two strands can be opened down the middle by breaking the much weaker string of hydrogen bonds. If "unzipped," the molecule presents a variable code written in four letters, A, G, T, and C. Subsequent research has decoded the "message" contained in these bases. A series of three letters gives the code for a particular amino acid; for example, GGT would indicate glycine,

* An α helix might be loosely compared to a coiled rope. The chain that is coiling has considerable structure sticking out sideways from the chain. The α helix allows a coil with almost no vacant interior space and is an incredibly efficient way to fill space. Most important proteins have some of their structure in the α-helix form.

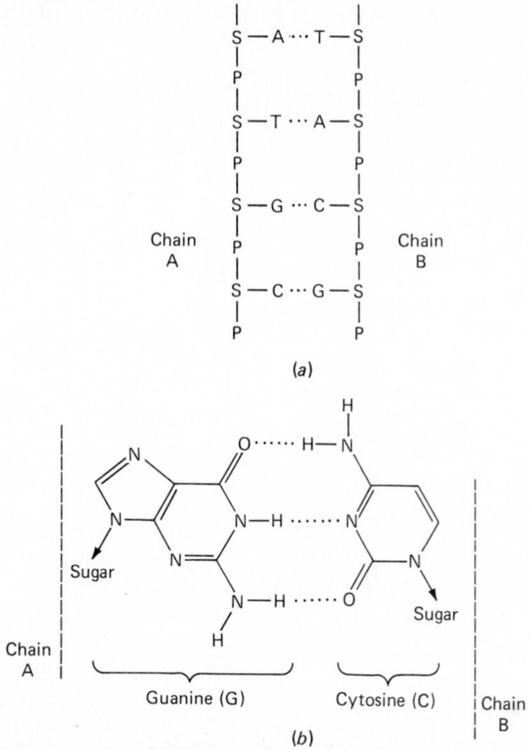

FIGURE 14-14
(a) The DNA structure is made of two chains held together only by the hydrogen bonds down the middle. The backbone of each chain alternates between sugar molecules (S) and phosphate groups (P). Projecting from this each sugar has one of four base molecules—adenine (A), thymine (T), guanine (G), or cytosine (C). The sequence of these molecules forms a code for the assembly of a protein from amino acids. (b) The hydrogen bonds between guanine and cytosine hold the two chains together. Since hydrogen bonds are weaker than the covalent bonds holding each chain together, they may be "unzipped" for "reading" the code with smaller forces than would threaten to break the chains.

while GCC would specify a molecule of alanine. There are even three-letter codes which provide "start" and "stop" marks for messages. The message on a DNA molecule is how to make a particular protein molecule. The practical implications of the determination of DNA structure and the decipherment of its code have been too great for us to estimate yet. We are living through the golden age of molecular biochemistry. There probably has been no more exciting period for researchers in this field. The application of the techniques of the physical sciences to a fundamental problem in biology yielded results beyond anyone's imagination.*

As an example, about a million people in American suffer from some form of diabetes. Most frequently the problem is the inability of their bodies to produce the molecule of insulin. Serious cases require that a supply of insulin be provided to the body daily. Since the insulin molecules of all higher animals are quite similar (although not identical), many forms of livestock can supply insulin, although the extraction of

* The degree to which the techniques of x-ray diffraction were unusual to classical biology is easy to estimate. James Watson was trained in classical biology, and his fellowship and support money from the United States were withdrawn when he moved to work with a crystallography group in England. The committee reviewing the support of students abroad did not believe that a person not trained in the mathematical techniques of x-ray physics could do useful work in molecular biology. Watson was awarded the Nobel Prize in Medicine and Physiology with Francis Crick and Maurice Wilkins (the crystallographer) for the discovery of DNA structure. This was the same year that Max Perutz and John Kendrew won the Nobel Prize in Chemistry for the determination of the structure of hemoglobin. Not until 1968 did Nirenberg, Holley, and Khorana win the Nobel Prize for deciphering the code in DNA.

large amounts of insulin from an animal body is an expensive process. The structure of human insulin was deciphered by Sanger; since then it has been possible, in theory, to make the molecule in the laboratory. In reality, this is much more difficult than extracting it from animal bodies. Knowledge of the genetic code offers another solution to the problem of insulin supply. Since the structure of the molecule is known, the genetic code for the molecule can be written. If a sample of DNA is made in the laboratory with this code, it could serve as the gene for producing insulin. If this gene could be introduced into the chromosomes of a fast-growing, easily harvested organism, this organism would synthesize insulin while it lived. In 1978 the production of such a strain of insulin-synthesizing bacteria was announced. The insulin is of no value to the bacteria whatever. This ability was introduced into the bacteria to provide for humans a convenient supply of the insulin molecule.

Equally exciting was the announcement in 1980 of the production of a strain of bacteria with the human gene to produce interferon. *Interferon* is a protein produced in the human body to attack viruses. Previously it had to be extracted in tiny amounts from collections of white blood cells, a process that was prohibitively expensive. Research has indicated that interferon is effective in combatting several forms of cancer, as well as simpler diseases such as the common cold. When it becomes widely available in the late 1980s, it could trigger a revolution in medical care comparable to the development of antibacterial drugs such as penicillin two generations ago.

Summary

The chemistry of the element carbon is called *organic chemistry* and includes more variety than the chemistry unrelated to carbon. Carbon atoms tend to form four covalent bonds, and do so readily with each other. Single, double, and even triple bonds occur between carbon atoms, but this leaves at least one bond free to join to another atom. Specific variations in the patterns of bonding produce widely different substances. *Hydrocarbons* are compounds of hydrogen and carbon atoms only, and include the major fuels, natural gas and petroleum. Burned in oxygen, hydrocarbons may be reduced to carbon dioxide and water. Straight-chain hydrocarbons are given names according to the number of carbons. For example, methane, ethane, propane, and butane represent carbon chains from one to four carbon atoms long. The names are useful since they carry over into the naming of many other substances. Distinctly different are the aromatic hydrocarbons built around the ring structure of the *benzene* molecule, C_6H_6. If a hydrogen atom on a hydrocarbon is replaced by one of several specific groups of bonded atoms, its properties may be changed. For example, the —OH, or *hydroxyl group,* may attach to produce a substance called an *alcohol.* Similar important organic groups are the acid group (or *carboxyl group*) —COOH and the *amino group* —NH_2.

Life forms are complex assemblages of organic molecules. Foods are composed of three groups. *Carbohydrates* are combined in the proportion of one molecule of water for each atom of carbon as in $C_6H_{12}O_6$, a single sugar. Two sugars may be chemically joined by *dehydration,* or removing a molecule of water, to form a double sugar, $C_{12}H_{22}O_{11}$, such as table sugar. Longer chains of sugar molecules joined similarly form starches. Digestion in the body involves reversal of this process by *hydrolysis,* or chemically adding the water molecule to produce separate sugar molecules. The body obtains energy from breaking down the sugar glucose into carbon dioxide and water and transports the energy as ATP molecules. A long hydrocarbon with an acid group at one end forms a fatty acid. The commonest *fat* has three such molecules attached to the multiple alcohol *glycerol.* Formation is by dehydration between the acid group of the fatty acid and the hydroxyl group of the alcohol. Digestion begins with hydrolysis.

The most important biological molecules are *proteins,* which are composed of chains of *amino acids.* An amino acid has both an amino group and an acid group. In a *peptide bond,* the OH of the acid group on one amino acid combines with a hydrogen from the amino group of another to join the two amino acids. This is a process of dehydration and its reversal, in protein digestion, is hydrolysis. The specific structure of large protein molecules gives them functions important for life. To determine the structure several steps are necessary. The chemical formula listing the number of atoms of each type is of limited use for a protein since they are so large. The sequence of amino acids in a protein makes up its *secondary structure.* The actual spacing of the parts in the molecule is the *tertiary structure* and allows us to interpret the activity of the molecule. The tertiary structure is determined by x-ray diffraction of a pure crystal of the protein. Critical to our understanding of the chemistry of life was the decoding of the (nonprotein) structure of DNA, the molecules that life forms transmit in the genes to store the patterns for constructing protein molecules.

Questions

1 Draw the three atomic arrangements of the hydrocarbon C_5H_{12}.

2 How high would the energy content of 1 liter of a hydrocarbon, such as gasoline, be able to raise a car with a mass of 1000 kg if burned with 25 percent efficiency in an engine?

3 Would you expect the distance between carbon nuclei to be greater or smaller in the structure of diamond or of coal? Why?

4 All of the alcohols have hydroxyl, or —OH, groups on their molecules. Why are they not classified as bases, like NaOH or $Mg(OH)_2$?

5 Originally, scientists expected to find some elaborate code mechanism to instruct long protein strands to coil into the specific globular shapes which make them biologically active. Instead it was found that the correct coiling occurred automatically because some amino acid side groups were hydrophilic and others hydrophobic. Explain what this means and how it could cause the coiling to take place.

6 The acid, or carboxyl, group, —COOH, has both a hydrogen atom and an —OH group on it. Why is it considered an acid group and never a basic group?

7 Why aren't vitamins and minerals considered foods?

8 Give some evidence that the shape of a molecule makes a great deal of difference in how the human body will treat it.

9 Why would it not be possible to live on a pure fat or pure carbohydrate diet?

10 The body uses 0.5 to 0.6 kg of glucose ($C_6H_{12}O_8$) a day for energy. If 1 mol of glucose stores 670 kcal, how much total energy does this represent? If a person is on a crash diet and all of this energy had to come from body fat, which stores 6.0 kcal/g, how much fat could be lost per day?

11 If the human body has 23 different amino acids in its proteins, why are only 10 of them considered essential in your diet?

12 A single unit of code on a DNA double strand consists of two bases, two sugar molecules, and two phosphate groups for a total molecular weight of around 700. Three of these units code to a single amino acid in a protein. A human sperm cell is little more than a mass of swimming DNA. If it has a mass of 2×10^{-8} g, how many amino acids may it carry the code for?

13 Distinguish between the primary, secondary, and tertiary structures of a protein.

14 The primary function of the stomach is to hydrolyze proteins. Express this in everyday language.

CHAPTER FIFTEEN

The Materials of Nonlife – The Matter of the Earth

Introduction Thus far in considering matter we have looked at the chemistry of relatively pure substances, such as metals and ores. We have also examined the chemistry of the element most important to us as life forms—carbon. However, the great bulk of matter is not relatively pure in form, nor is carbon a very abundant element in nature. Carbon is about as common as rubidium, strontium, or zirconium in the crust of the earth. That is to say, if we are concerned about the behavior of the bulk of matter on earth, carbon would be a footnote. Two types of atoms dominate the earth and between these two, seven of every ten atoms are accounted for. We have access only to the outermost portion of the earth, down to depth of many kilometers. In this sample, every other atom is an atom of *oxygen*, and over half of the rest are atoms of *silicon*. It is a truly remarkable fact about the earth that it is so diverse, and yet the preponderance of its matter is made up of only two types of atoms. The oxygen you are familiar with is the elemental gas in the air. The contribution of the atoms of oxygen in air to the oxygen total is trivial. The air is only one-fifth oxygen, so it serves to *lower* the oxygen percentage. Limestone, sandstone, concrete, and solid granite are all about half oxygen. The earth we know of is a collection of silicon-oxygen compounds, and the rest is such a small fraction of the total that they may be considered impurities. But if we view the incredible divergence of the earth's surface, it becomes clear that the impurities make a great deal of difference in the end.

The rocks of the Grand Canyon are different from the rocks of the high Alps. They differ in color, texture, elevation, shape, size, hardness, contour, economic worth, and history. Yet, both collections are mainly oxygen and silicon. To study the earth, therefore, it is well to begin with the two main elements, but more is also needed. Rocks are an incredibly complex jumble of molecules—made, remade, sorted, arranged, and displayed by nature. Nature is like a fine artist shading one combination into another so that attempts to define separate types are difficult. Any simple treatment of the materials of the earth is an oversimplification. Lines that appear clear on paper are not so in the field or in the laboratory.

Composition of the Earth

A listing of the chemical composition of the materials in the entire earth is not possible. Obtaining a sample from 1000 km straight down would be a more difficult task than bringing back samples from Pluto. Thus, when we discuss the earth, we mean the part we know about, the outermost part, or the *crust.* This includes the volume we live in, burrow into, or drill into. While the earth is some 12,700 km from where you are, through the center, to the surface on the other side, we know about perhaps only 100 km of this total thickness. For this material we can give a total composition by element (Table 15-1). The major impurities in the oxygen-silcon crust are aluminum, iron, calcium, sodium, magnesium, and potassium. After this, the abundance of other elements falls off rapidly. The composition of the crust is not unlike the crust of slag which forms on a pool of molten iron in a blast furnace. While it is not consistent with the data to push this idea very far, it is certainly true that the earth cannot be uniform in its composition throughout its entire depth. The gravitational effect of the earth on the moon allows us to calculate the total mass of the earth as 6×10^{24} kg. When compared to the earth's volume, this makes the earth the densest object known in the solar system. Each cubic centimeter of the earth averages some 5.5 g of mass, or 5.5 times the density of water (Fig. 15-1A). The crust has an average density of

TABLE 15-1
Composition of the Earth's Crust*

ELEMENT	PERCENTAGE BY WEIGHT
Oxygen	46.5
Silicon	28
Aluminum	8
Iron	5.5
Calcium	4
Sodium	2.5
Magnesium	2
Potassium	2
(All others)	1.5
	100

* Varying methods of averaging cause a small variation in these percentages. All are quoted to the nearest ½ percent by most schemes.

FIGURE 15-1A
On a cosmic balance it would require 5.5 spheres of water the same size as the earth to have the same mass as that of the earth. Thus, the earth is 5.5 times the density of water.

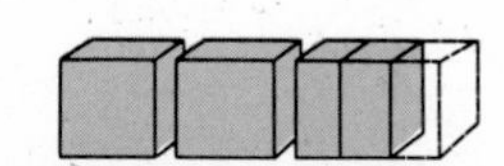

FIGURE 15-1B
The same type of cosmic balancing for the surface materials of the earth works out differently. If we sliced out a typical cubic kilometer from a continent, say a chunk of Kansas, we would find that only about 2.7 km^3 of water would be required to balance the rock. Rocks at the surface average only half the density of the earth as a whole.

about 2.7, which is much lower than the earth's average (Fig. 15-1B). To make up for the fact that the outer layers of the earth are below average in density, the deep interior must have a much greater packing of matter than the rocks we know of.

The typical rock at the earth's surface is a mixture of chemical compounds. To find some order in the crustal structure it is well to begin with *minerals.* A mineral is a chemical compound with a definite chemical composition and crystal type. The commonest minerals are compounds of oxygen and silicon. What might be unexpected is that the structure of the mineral can be more important than the chemical composition. Sand, sandstone, quartz, amythyst, quartzite, opal, flint, chert, chalcedony, and many more forms may be nearly pure forms of silicon and oxygen. Yet, no individual familiar with these types of rock would accuse them of being alike. To understand this let us examine the nature of the compound formed between these two elements.

Silicon atoms are similar to carbon atoms in the tendency to combine, although they are larger. The basic unit of a silicon-oxygen molecule has a silicon atom with four equally spaced bonds to oxygen atoms radiating out from it (Fig. 15-2). The SiO_4^{4-} structure generates a shape like a triangular-sided pyramid, or tetrahedron. This form "fits" in space quite well. In forming this tetrahedron, each oxygen atom has used up only one of its two bonds and thus each corner of the tetrahedron may bond to something else. Usually, this is another such tetrahedron so that each oxygen atom winds up as a corner atom in two such tetrahedra. If this pattern repeats in all directions, each oxygen atom belongs in two tetrahedra, and is bonded to two silicon atoms. Thus, each silicon atom "has" one-half of four separate oxygen atoms for an average of two oxygen atoms per silicon atom, or SiO_2. The formula SiO_2 is easy to write and is referred to as *silica,* the common mineral formed from silicon and oxygen atoms only. The way the basic unit repeats in space makes up much of the observed difference in the forms of silica, which is not reflected in the simple formula. In the quartz structure, six tetrahedra form

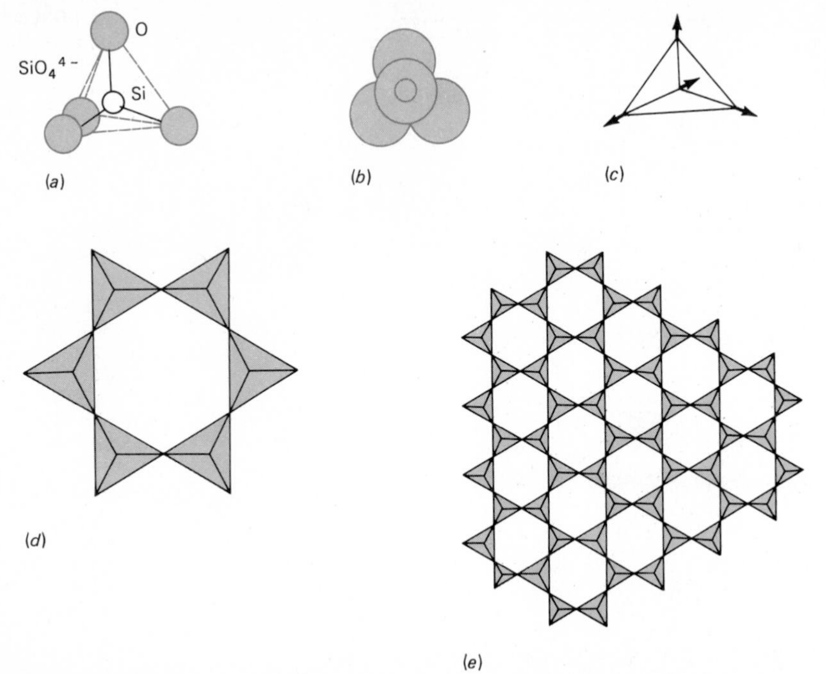

FIGURE 15-2
Silicon atoms form four bonds and oxygen atoms form two. The basic unit formed from these two types of atoms has a silicon atom at the center bonded to four oxygen atoms at the corners of a tetrahedron (*a*). Drawing (*b*) shows this structure from above with the atoms more nearly to scale. Note that each corner oxygen is free to form one additional bond. Drawing this tetrahedral structure by a triangle (*c*) allows the representation of some possible combinations. Six such tetrahedra free to form 12 outside bonds from the mineral beryl (*d*). If tetrahedra link endlessly in a plane, the structure of mica results (*e*). Many other structures are possible. Note also that the "hole" in a six-member ring could hold an atom or ion without necessarily bonding to it.

a ring which is interlocked to other rings at each side, above and below the original ring. It produces the very orderly arrangement of repeating units in space called a *crystal*. Figure 15-2 indicates some of the ways in which SiO_4 tetrahedra can be arranged in space. If some corners of the tetrahedra are not shared with another tetrahedron, it indicates that the oxygen atom in that corner is free to bond some other substance.

For example, the six-unit ring (Fig. 15-2*d*) is the basis of the mineral *beryl*, $Be_3Al_2Si_6O_{18}$. which is the principal source of the element beryillium. If we crystallize beryl in repeating units properly and slip an occasional chromium atom into the structure, we obtain the green crystal known as *emerald*. Possession of a 1-kg crystal of perfect emerald would secure your early retirement. Yet, chemically, this is the equivalent of 1 kg of fine sand (SiO_2), a few dollars worth of beryllium, and a quarter's worth of aluminum with a pinch of chromium (less than a penny's worth) for color. Clearly, as with organic chemistry, the arrangement of the parts is every bit as important as the chemical composition, if not more so. If another example is needed, recall that diamond is chemically identical to a bag of pure charcoal.

If the chemical composition of a wide variety of rocks is quite similar, what makes their physical form so very different? This is almost entirely a function of how they are formed into solid materials. Imagine that we take a huge vat and drop in the elements that make up the earth's crust in the correct proportion. We now heat this and apply pressure until we have a molten, seething brew. Exactly what comes out of the pot will depend on how we treat the material from this point. Will we allow volatile gases to escape? Will we keep it uniform and molten for long enough to allow denser minerals to sink and less dense minerals to rise to the top of the brew? Will we cool it rapidly to freeze everything in place, or allow separate crystals to form slowly and grow in the molten solution? The types of rocks generated will depend upon the physical conditions of formation. Let us examine a particular combination.

Minerals occurring in one range of compositions incorporate silicon and aluminum and are of relatively low density. These are called *sials* for short, from the elements involved (Si and Al), and they rise in our pot. Another group of minerals are built around silicon and magnesium complexes. These *simas* are denser and sink. As we gently lower the temperature, the first of many minerals in the mixture approaches the point where it will solidify. This is likely to be a calcium magnesium silicate, or a mineral called *plagioclase,* combined with a mineral built on unconnected SiO_4 tetrahedra called *olivine.* A mixture rich in these materials is the rock type known as *basalt.* Both are simas and will settle as they form, taking much of the iron, calcium, and magnesium from the molten rock mass. As we continue to slowly cool our brew, the material at the top will represent a purer and purer sialic brew of minerals. If we cool this very slowly, each mineral has time to grow crystals slowly in solution. With a long time for cooling, visible crystals of sodium or potassium aluminum silicate, called *orthoclase,* will grow. Crystals of a *mica* will form with the structure of interlaced hexagonal sheets of SiO_4 (Fig. 15-2*e*). All of these crystals will be embedded in a structure of *quartz,* the last common form to solidify as cooling continues. When we reach this point, we have produced the most commonly occurring rock in the continental masses of the earth—*granite.* A granite is a large cyrstal (long-cooling) rock composed of 60 to 70% feldspar (usually orthoclase), 10 to 20% mica, and at least 10% quartz as the material holding the mixed matrix of crystals together. Thus, granite is a complex mixture of minerals, each of which has a characteristic chemical formula and structure.

We can classify rock types formed by cooling from a molten mass according to two characteristics—the mineral mixture they represent and the relative time they had to cool (crystal size). Table 15-2 attempts to do this for the commonest categories of rocks formed from molten material, or *igneous* ("fire-born") *rocks.* The granite formed in our example was light in density as well as in color and had large crystals. If the identical mixture of minerals was spilled onto the earth's surface, or

TABLE 15-2
Classification of Igneous Rocks*

	CRYSTAL SIZE		
CRYSTAL COLOR AND DENSITY	**LARGE**		**VERY FINE**
Light minerals	Granite	Andesite	Rhyolite
Dark (heavy) minerals	Gabbro	Diorite	Basalt

* The modern tendency is to drop the class of igneous rocks for the classes of *volcanic* rocks and *plutonic* rocks (which includes most metamorphic rocks). Since this text develops no subtle distinctions in geology, the traditional scheme is used.

extruded, it would cool rapidly and not develop large crystals to form a *rhyolite.* The best example of this is probably the gigantic plain of yellow-pink rhyolite which gives Yellowstone National Park its name. Most volcanoes extrude rock brews from great depths and thus produce *basalts.*

Restructuring the Surface— The Rock Cycle

If you are standing on the natural surface of the earth, you are likely to be on *soil,* which is a complex mix of rock fragments and organic debris (chunks of stems and roots, etc.). If you dug down, the type of rock you would hit first in about half the places on land is limestone, which does not contain any silica at all. Yet, basalt underlying the ocean basins is the commonest rock in the earth's crust and granite is the commonest rock under the continents. How do these observations fit together? Once rock is exposed at the surface of the earth, it is exposed to a variety of agents that tear down and destroy its structure. To understand the transformation of the materials involved we can list the stages of breakdown and buildup of materials known as the *rock cycle* (Fig. 15-3). Beginning with molten rock in the earth's interior, or *magma,* we may proceed into the cycle. If this material cools and solidifies, an *igneous* rock results. The faster the cooling of this rock the smaller the resulting crystals. In-

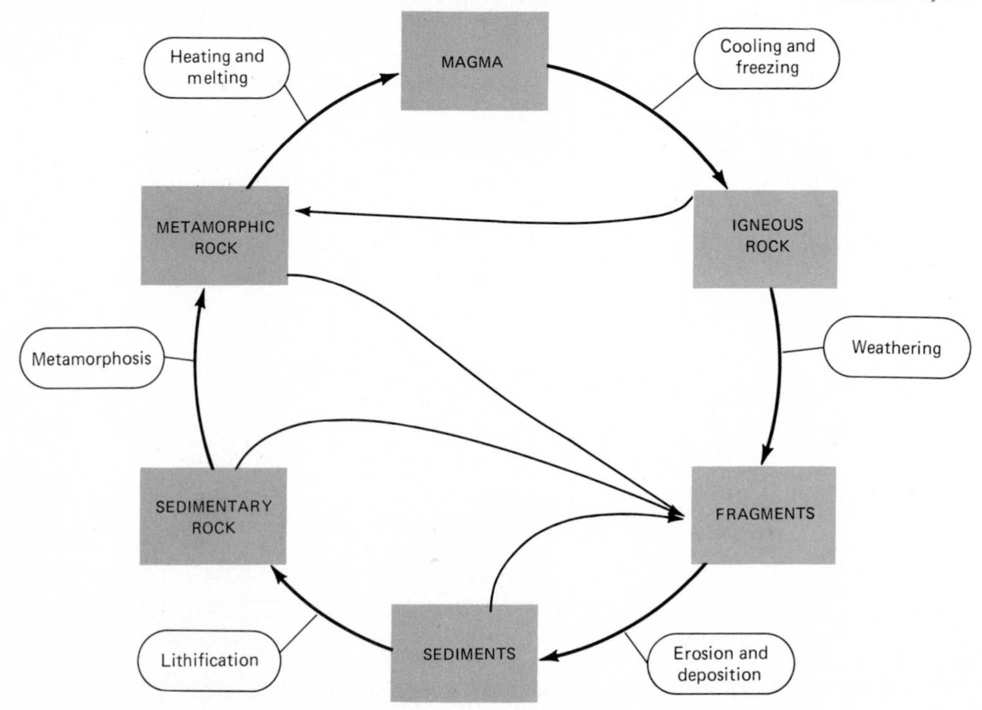

FIGURE 15-3 The rock cycle.

deed, if the cooling begins from a state where crystal formation has not started and proceeds at once to low temperatures at the surface of the earth, a volcanic glass, such as *obsidian*, may result. In this, no crystal formation has had time to take place.

Exposed to the surface, the breakdown of any type of rock begins immediately in the process of *weathering*. Weathering may be mechanical, chemical, or biological in nature. The rock may have another rock fall on it resulting in cracks, or it may be struck by lightning. These are examples of mechanical weathering. However the rock is broken down, weathering results in the production of separate chunks of rock called *fragments*, some of which are dissolved pieces carried in groundwater. Once fragmented, the rock is subject to the movement and separation of the pieces by water, wind, and other agents in the process of *erosion*. Erosion is the movement of particles of rock from their original site. One of the by-products of erosion is the *sorting* of materials of similar mechanical or chemical characteristics. Soluble materials may be dissolved and carried away by rivers to the sea. Fine chunks that are insoluble may be moved in the direction of a prevailing wind, and so on. Once materials have been separated by the process of erosion they may be deposited in heaps of similar materials. Layers of materials deposited after erosion are called *sediments*. The sandy bank of a river is made up of materials selected by erosion and deposited at a particular place because the water could no longer suspend them and carry them along. Such sediments are immediately subject to further weathering and erosion. Some sediments, however, are buried deeply by further sediments and may slowly be compressed or cemented into a hard, durable material which is a new rock type. Such a rock is called a *sedimentary rock*, and most are formed under shallow bodies of water. Of course, compression may be carried to extremes. If heat and/or pressure are continued past a certain point, the actual structure of the material may alter. Substances may be forced into new crystal types, grains may melt at point of contact with other grains, and so on. Under such extreme conditions a *metamorphic rock* results (from "metamorphosis," or "complete change"). If this process continues, the entire material may melt and we are back to magma, the beginning of the cycle.

An Example—The Future History of a Granite Tombstone

It is well to consider the rock cycle as it applies to a particular material. Imagine the block of granite that we produced earlier being exposed at the surface (Fig. 15-4). What will happen to it? Initially, it will be washed by water. Virgin granite has never been exposed to water during its formation, and several minerals in a granite may be soluble. In particular, sodium and potassium compounds in the feldspars are quite soluble, and calcium compounds somewhat less so. Thus, the sodium and potassium compounds are systematically *leached*, or dissolved from the rock

FIGURE 15-4 A block of granite exposed at the earth's surface. The dark flecks are the mineral hornblende. Bright reflections come from flat sheets of the mica and duller reflections from crystal faces of feldspar. White, glassy areas are the matrix of quartz. (*Photograph by the author.*)

to a depth of many millimeters as it is exposed to water. Calcium compounds dissolve less readily, but are attacked by the water if it is acidic. Normal water is easily rendered acidic by contact with the air. As carbon dioxide from the air dissolves in water, it forms carbonic acid, H_2CO_3, which can leach calcium. Furthermore, lightning discharges through the air produce nitrous and nitric acids, which dissolve in the rainwater. These are much more powerfully acidic than carbonic acid and serve to leach calcium compounds. Any of these acids can produce soluble calcium compounds which are washed away with the water. The sodium, potassium, and calcium are usually flushed all the way to the sea. There, the sodium and potassium usually will remain dissolved unless the sea is cut off from the ocean and evaporates. If this happens, the sodium may be deposited in layers of *halite,* or rock salt, while the potassium remains dissolved. As a last resort during the drying of a sea, potassium may be deposited. The largest example on earth is the huge Chilean deposits of Chile saltpeter, KNO_3, which is the richest source of both potassium and of nitrates on earth. (Of course, the last deposited is the first dissolved. The Chilean desert has essentially no rainfall and has not had any for thousands of years. If it had, this extremely soluble material would have dissolved and washed away before human history.)

The history of the calcium is likely to be different. The calcium or lime tends to form many insoluble materials, the commonest of which is calcium carbonate, $CaCO_3$. In a pure crystalline form called *limestone,* this is the commonest of all sedimentary rocks, being the rock nearest the surface at 48 percent of the area of the continents. To become limestone, the dissolved lime in water first must be deposited as sediments by a settling process. A great deal of this deposition is performed by life forms. Seashells take the lime from the water and change it to $CaCO_3$ to make their shells. As the organisms die their shells collect in layers on the bottom of the sea. These layers may be buried by later sediments, compressed and cemented together, or *consolidated* in the processes of *lithification* ("rock making"). The result of this process is a series of horizontal layers or *strata* of the solid rock called limestone. Once produced, the limestone may be exposed to weathering, just as the sediment would have been. In either case, the result would be to break up and disperse the material as fragments again. An alternative path for the limestone might occur if it is buried deeply, in which case great heat and pressure might cause fundamental changes in the rock. Some portions of the rock might be rendered nearly molten. This could allow for the slow regrowth of crystals from the consolidated fragments making up the limestone. Impurities could be "baked out" or rendered mobile enough to collect in separate zones. This set of possible events would change the nature of the rock in the process of *metamorphosis*. The rock produced from the metamorphosis of limestone is called *marble*. While crystals usually are not apparent in limestone and are randomly oriented when they do occur, marble is a compacted crystalline mass with the same chemical form but a highly altered structure.

Meanwhile, back at our granite, the processes of weathering and erosion have been continuing. The materials which will be produced once the readily soluble parts have been dissolved away depend greatly on the climate. The quartz content of the granite may be reduced to a fine, pure silica *sand* (Fig. 15-5). A larger volume of sands, however, re-

FIGURE 15-5 Weathering and erosion can separate and then accumulate beds of sand from a particular mineral in the rocks. This sand is really white grains of the mineral gypsum, $CaSO_4 \cdot 2H_2O$, which gives White Sands National Monument its name. The 250 mi^2 of these sediments are the largest area of their type in the world. Gypsum is formed in the evaporation of shallow seas; and if roasted to drive off the bonded water, it becomes plaster of paris. (*Photograph by the author.*)

TABLE 15-3 Sediment Types by Particles Sizes and Names

PARTICLE SIZE (mm)	PARTICLE NAME	SEDIMENT NAME	ROCK PRODUCED BY CEMENTING GRAINS
Over 256	Boulder	Gravels	Conglomerate
64–256	Cobble		
4–64	Pebble		
2–4	Granule		
1–2	Very coarse sand	Sands	Sandstones
$\frac{1}{2}$–1	Coarse sand		
$\frac{1}{4}$–$\frac{1}{2}$	Medium sand		
$\frac{1}{8}$–$\frac{1}{4}$	Fine sand		
$\frac{1}{16}$–$\frac{1}{8}$	Very fine sand		
$\frac{1}{256}$–$\frac{1}{15}$	Silt	Muds	Shale
Under $\frac{1}{256}$	Clay		

sult from the remains of the more plentiful *feldspar.* The more persistent and more acidic the waters washing the granite are, the more likely the feldspars are to break down into the extremely fine aluminum silicates called *clays.* The composition of a brown-red sand and a clay may be identical since they both can be the products of feldspar breakdown. As the chemical attack on feldspar continues, more and more silica is freed from the structure. The eventual product in cases of extreme leaching may be the pure aluminum oxide called *bauxite.* Once fragmentation has occurred, pieces are produced which are subject to erosion. Mechanical erosion generally separates or sorts fragments on the basis of particle size. Fast-moving fluids can suspend larger particles than can slow-moving fluids; and denser fluids, such as water, can suspend larger particles than can less dense mediums, such as air. Thus similar particles of our granite may wind up in sandbars, dunes, clay beds, and so on. Table 15-3 lists sediment types by particle sizes. A mixture of particles larger than a BB is called a *gravel.* Particles finer than this down to the limit of naked-eye visibility are called *sand,* and below this size *mud* (Fig. 15-6). Very fine clay particles may be suspended by water indefinitely since they acquire a net electric charge frictionally and repel each other. You can test this by shaking a fine clay-water mixture in a jar. While larger particles settle rapidly, the water still will not be clear days later. These fine clay particles will not settle until the electric charge is neutralized and they clump together into larger particles. This normally occurs when the water encounters the ions present in seawater at a river delta, and this is where such particles settle.

Erosion, Hero or Villain?

The process of erosion almost always gets a "bad press." Mud slides in California, dams silting up faster than predicted, dust storms scouring away the topsoil all are examples of erosion; and all bear negative connotations for the reader. The general public almost never hears anything

FIGURE 15-6 Sediments along the banks of the Rio Grande which have been partially eroded by later action of the river. Ripple marks are seen at the top, and bedding marks at the side showing the various layers deposited. Changes in the patterns at the side allow us to infer stages in the river action in depositing these sediments. (*Photograph by the author.*)

good about erosion. It shares the status of the dandelion. Almost the only time most people think about dandelions is when they turn up in the middle of a lawn where they are not wanted. Yet, there are few flowers so complex or beautiful as a dandelion if it is studied carefully. However, when called to the attention of the average suburbanite the dandelion is a call to instant chemical warfare. There is no cause to weep for the flower, however, since the human race will be long gone when the survival of this hardy little plant is threatened. What could do in both human and the dandelion would be to stop erosion, since it is necessary for the survival of both. Erosion, however, is an essential process in the renewal of the fertility of the earth. The companion process that must work in balance with erosion is weathering.

Plants are incredibly versatile organisms and can survive and thrive in a wide variety of situations (Fig. 15-7). Chemically, almost all of the substance of a plant is made from air and water. Even the most versatile of plants, however, need a few additional substances. These are various minerals (or chemical compounds containing specific atoms not found in the air or in pure water). For example, we saw in the last chapter that a key atom in the structure of the plant pigment chlorophyll was magnesium. Just like calcium, nitrogen, phosphorus, and several other atoms, magnesium must be present in the groundwater in trace amounts for a plant to survive. If you have a blazing wood fire, all of the molecules the tree produced from air and water molecules go up with the flames. The remaining ashes represent the mineral portion of the plant, which were dissolved in the groundwater. These minerals got into the water either

FIGURE 15-7
The sentinel pine grows directly out of small cracks in the surface of the granite Sentinel Dome in Yosemite National Park. The angle of the tree is the result of strong prevailing winds. (*Photograph by the author.*)

from the decaying bodies of other plants or animals, or from the rocky materials of the earth itself as these materials weathered.

Thus, a balance is necessary for the continued fertility of the earth's surface. If material is eroded faster than the rock surface is weathered, we have barren rock on which plants have difficulty finding a foothold. If weathering proceeds much faster than erosion, the upper materials of the earth become depleted in the minerals essential to support life. Without human intervention such as the spreading of chemical or animal fertilizers on the ground, the soil is sterile and will provide little, if any, crop growth. The nature of the soil is determined by the balance of weathering and erosion along with plant action. At one extreme we have tropical soils, known as *laterites* (latosols). The temperature, acidity, and abundant rainfall all work to dissolve away soluble minerals. As the waters soak into the soil, minerals are leached from the upper layers. Those most desirable for plant growth are usually the most soluble; these are removed first and are carried the farthest downward and away from the reach of plants. In extreme cases, normally insoluble minerals, such as iron, are dissolved and moved away from the surface. These, however, lose their solubility once away from the acidic surface zone and are sometimes deposited in a rocklike layer of hardpan perhaps a meter under the surface. The tradgedy of a lateritic soil is that it is very difficult for it to recover from an injury. As you look at a picture of a tropical rain forest, it is easy to obtain the impression of incredible fertility since plants are present everywhere in abundance. This is true only because the plants are very efficient at retaining and recycling a limited mineral supply. Roots reabsorb the minerals from decaying matter as they are

FIGURE 15-8
Probably the most famous erosional landforms in the United States are the columns of Bryce Canyon in Utah. Vertical cracking (jointing) of the rock allows water to erode structures vertically along cracks. Measurements of the material flowing in Bryce Creek after a rainfall disclosed less than 10 percent water! The remainder was dissolved and suspended solids. (*Photograph by the author.*)

washed into the soil, and only the tiny amount lost must be made up from the leached soil itself. If the vegetation covering a lateritic soil is removed over a large area, reestablishment of the plant community is extremely slow. If crops are grown on this land, they will be good crops only so long as the decaying plant matter remains in the upper soil. After one or two harvests, the soil is usually depleted, and the absence of minerals is fatal to the next crop.

Different soil types are produced where erosion remains somewhat ahead of weathering. (See Fig. 15-8). Once erosion is brought into a balance by plants anchoring the surface, growth is limited by the rate of weathering. Usually, slow weathering implies a short supply of water. Desert regions contain some of the most fertile soils on earth simply because the mineral content of the soil has not been dissolved away. All these soils need is the water of irrigation to be greatly productive. A natural compromise exists in the *prairie soils* (or brunizems) of moderately arid regions such as the American great plains. These soils are very fertile since the upper layer is richly interwoven with roots and decaying plant debris, while the slight rainfall has leached usable minerals only to

Guns or butter

Nitrogen Fixation in History

The compound known as Chile saltpeter, or potassium nitrate (KNO_3), interacts with human history. The extensive deposits of nitrates in Chile have a long political history. Nitrates are very valuable as fertilizers, but they are also an essential ingredient of most explosives. Gunpowder is a mixture of a nitrate, charcoal, and sulfur. The latter two can be obtained in many places, but nitrates are so soluble in water that their occurrence is rare. Europe has too much rainfall to have any extensive nitrate deposits above ground. Before World War I every advanced nation kept a fleet of ships which went back and forth to Chile to stock up on nitrates. In good times these were used as high-quality fertilizers. All living things require some chemically combined nitrogen for protein-building. Only one small group of bacteria can take this nitrogen directly from the air. All of the rest of life require it "fixed" into some combined form. Land cultivated in a crop such as corn or wheat declines in productivity every year until it is desertlike, since these crops deplete the fixed nitrogen in the soil. The law under the feudal system in medieval Europe required each field to remain unplanted every third year. By doing this, forage crops, known as legumes, were planted every third year. Legumes have growths on their roots where the nitrogen-fixing bacteria live, and the growth of legumes promotes nitrogen buildup in the soil. The practice of "third year fallow" preserved the fertility of the soil of medieval Europe. Other areas such as the part of North Africa around Carthage (Tunisia), or the Middle East, became deserts from overuse of the soil. In modern times crops are grown every year, but nitrates are added chemically to the soil from time to time (currently, every year in developed nations).

In World War I, Germany was cut off from the supply of nitrate in Chile. Great Britan had a larger battle fleet than the Germans, and the German fleet fought it only once. This was the Battle of Jutland. Germany won this battle, but lost the naval war here since the English retained control of the sea. This not only meant no fertilizer for soil, it also implied that no nitrates were available for explosives. In a war that was the first modern war, this was an intolerable situation. The introduction of the machine gun and artillery bombardments against an entrenched enemy used up explosives at an unpredicted rate. The storehouses of Chile saltpeter in Hamburg and elsewhere in Germany shrank at an incredible rate as firepower demands increased. The solution for Germany lay in its scientists—at this point unquestioned as the best chemists in the world. In desperation the German generals ordered the German chemist Fritz Haber to come up with a substitute for Chile saltpeter in 6 months using only common materials. His effort took a bit longer, but Haber perfected the process for the production of nitrates from common materials. The basic process produced ammonia from air and water using electricity:

$$N_2 + 3H_2 \xrightarrow{e} 2NH_3$$

The process allowed Germany to produce explosives and maintain the war for years after its natural supply of nitrates ran out. As a result, Haber was criticized for using his talents for military purposes. Even today, most histories of science ignore the Haber process as a prime development in applied science. Yet, the population of the earth today is past 4 billion.

Left to the ability of nitrogen-fixing bacteria to enrich the soil and grow crops, one-half to one-fifth of this number would be doomed to starvation. The Haber process today is the principal means for the fixation of nitrogen in soils that have known no legume for 40 or more years. The continued fertility of the soil in the developed world depends upon the ammonia (NH_3) produced by the Haber process. Many millions of needless ca-

sualities occurred during World War I. How many would die today from a crop failure across all the developed areas of the world due to soil nitrogen deficiency? While the development of Haber's process for the fixation of nitrogen from the air and water was a product of wartime necessity, it has proved to be one of the most important developments in history for the expansion of agriculture, along with the discovery of irrigation and the plow.

a manageable depth. If the removal of leached rock debris from the surface balances the downward migration of minerals, such a soil may remain fertile indefinitely.

Human civilization has always been tightly locked to soil fertility. Overuse tends to deplete the mineral reserves of almost any soil, since, in nature, the food removed by humans would have stayed with the soil, releasing its minerals for new plant growth. Only soils which are constantly renewed do not have this problem. The Nile Valley was one such area until the building of the Aswan Dam. Volcanic eruptions killing tens of thousands of people have been common throughout history. If you have ever wondered why people choose to live near a volcano, you now know the answer. Volcanic soils also are unusually fertile and highly desirable as farmlands if flat enough. Other unusual soils exist. When the great ice sheets left North America a geologically short time ago, they left a gigantic plain of debris behind. Prevailing winds piled fine rock flour from this expanse into thick layers called *loess*. A large strip of the Midwestern farm belt sits on these deposits. Similarly, some of the most fertile areas of China are on deposits of loess, which are in places several hundred meters thick.

Sedimentary Rocks and the Interpretation of Earth History

The commonest type of rock at or near the surface of the earth are the sedimentary rocks. Most sedimentary rocks are composed of a clear sequence of horizontal layers of strata. For example, Fig. 15-9 shows the upper part of the walls of the Grand Canyon. The layering is obvious as we examine the structure of the rock. The appearance is almost that of a flat blanket of rock placed upon a flat blanket of rock. The erosional cut of the Colorado River through the rock walls of the canyon reveal this sequence. The walls resemble what might be seen if we cut through a multilayered cake.

Obtaining the age for a sample or structure in the earth is a complex business. It is frequently possible to infer the relative age of two samples with ease. This is based on a principle first stated by Nicolaus Steno (1638–1696). From his study of strata Steno concluded that only a fluid could deposit a sediment; thus, there could be nothing but a fluid above it while it was being deposited. For undisturbed strata this implies that

FIGURE 15-9 Sedimentary rocks are deposited in layers or sheets. The area shown is the uppermost portion of the Grand Canyon. The layers shown are larger than might be expected since the vegetation on the various ledges are small trees. The light cliffs above the obviously layered zone are over 30 stories tall, and the cliffs above this layer are taller yet. In spite of this, the descent into the canyon has barely begun at this point. (*Photograph by the author.*)

the lower ones are the older ones. In the same manner a desk top acquires piles of paper. If these papers have not been disturbed since they were deposited, we may conclude that the lowest papers have been there the longest. Steno also observed that while the bottom of a sediment acquired the contours of the basin it filled, the top of the sediment was always very close to horizontal. Steno's third principle was that sediments do not have edges. When deposited the stratum is either bounded by some wall of the basin or else it just thins out to the vanishing point. Thus, when we look at the rock wall that represents the edge of a stratum, we must infer that much of the rock has been removed. Steno's principles allowed geologists to determine which layers were younger and which were older. Such *relative dating* requires a huge amount of work, but is a

FIGURE 15-10
The sandstones of Zion National Park in Utah show the ripple marks and bedding planes characteristic of sand dunes. In addition, the rock tends to hold old sand surfaces as it weathers and erodes. In the Washington Arch we see the curve of an old dune preserved. Incidentally, the arch could accommodate the U.S. Capitol building, dome, and all underneath it. (*Photograph by the author.*)

FIGURE 15-11
These layers of rock have been bent into a tight arch after their formation. Such *folding* of rock layers may occur across tens or hundreds of miles or over a few meters. One certain feature is that such folding occurs after the rocks have been deposited in flat layers or strata. (*Photograph courtesy of Mr. Fred Parks.*)

means by which scientists may piece together the history of the earth. However, some sedimentary strata in nature are not horizontal at all, but lie at a considerable angle or are bent into curved three-dimensional shapes. A few of these are easy to interpret. For example, the curves in the Navajo sandstone in the walls of Zion National Park shown in Fig. 15-10 give every evidence of being the curved surfaces of sand dunes frozen through time and cemented together into stone. Other sedimentary strata appear as warped sheets of rock, bent and cracked into distorted shapes (Fig. 15-11). There is no known mechanism for depositing

FIGURE 15-12 Fossil ferns from the Mazon Creek formation of Illinois from the author's collection. When found, such fossils are elongated, ellipsoidal rocks. A few sharp taps on the edges with a rock hammer cause them to split into a positive and negative cast of a fern frond. Dozens of such casts may be recovered in a single day. (*Photograph by the author.*)

sediments that could have built these materials in their current form. A second line of evidence indicates that the form of these rocks has been changed in time. Sediments on their way to becoming sedimentary rocks are a perfect "modeling clay" for recording the events of their time. Sedimentary rocks abound with the imprints of life forms called *fossils*. Some represent the imprint of the body of a dead plant or animal, which has long decayed. Others represent the remains of the body where, molecule by molecule, the dead tissue has been replaced by more durable minerals in the groundwater (Fig. 15-12). Very common remains of this type are the durable shells of some marine creatures such as crabs. Other life forms, such as clams, have shells so durable they may be preserved essentially intact. Still other fossils are merely records of the life form's presence, for example, footprints. Particularly common on the surface of some sedimentary strata is the preservation of ripple marks, indicating they were washed by a steady current as they were deposited. Each of these types of evidence allows us slowly to develop a picture of the past history of the earth. From this picture the first great unifying principles of geology emerged.

The first clear attempt to deal with the earth's past from the study of fossils was that of the naturalist Georges Cuvier (1769–1832). To account for the diversity of life forms and the distortions of the rocks in which they occurred he developed a picture of great catastrophes. According to Cuvier, great sweeping upheavals must occur from time to time, causing massive death of life forms and great deposition of materials into new sedimentary layers: the theory of *catastrophism.* This theory was used by Cuvier to explain the principles that emerged from his study of fossils, which was the most extensive and most detailed study completed to his time. In fact, Cuvier is often named as the father of *paleontology,* the study of fossil life forms (as well as the father of the science of comparative anatomy). In areas of undisturbed sedimentary rock, the first sediments to be laid down, hence the bottommost, would be the oldest. In such regions where a great thickness of sedimentary rock is exposed for study, the highest fossils are invariably the most familiar types of life forms, bearing the greatest resemblance to living species. As you proceed deeper into the sediments, you are examining older and older life forms, and these bear less and less resemblance to living species. Clearly, Cuvier had to account for the massive extinction of types of living organisms that are no longer found on earth but once were quite plentiful. With his theory of repeating catastrophes, Cuvier believed he had found the mechanism to account for the fossil record.

The opposing viewpoint to catastrophism was first clearly held by James Hutton (1726–1797). Hutton studied strata carefully and concluded that most of them must have been deposited quietly over great periods of time. He recognized that most sandstones and shales, and effectively all limestones, could have been produced only underwater. He pictured the lake beds and the floor of the sea slowly building up deposits, which ultimately could become rock in some distant future era, during his own life. Hutton pictured the processes reshaping the earth as continuous and incredibly slow on a human scale. The individual who extended Hutton's views to a unified picture of geology was born in the year of Hutton's death. Charles Lyell (1797–1875) profited by much of the work accomplished during the early decades of his life. In this period of the Industrial Revolution, people were digging up the countryside at an increased rate. Canals were being carved through the bedrock of Europe, and building materials were being excavated at a new pace. This led to several discoveries, some by chance and some by careful examination of newly exposed materials. One extremely important observation was that certain fossils were confined to particular strata in the rock. Such *index fossils* can be used to identify a particular layer of rock across a considerable distance where the rock is missing or buried deeply (Fig. 15-13). While other life forms may be found in many sequential layers, index fossils first appear and then disappear in the rocks in a sharply limited time interval. Overlap of the time span of such life forms allowed the first detailed time sequence or relative dating of the rocks to be developed.

While rocks were being dated in this way, a much more interesting, if less important, discovery was made. Much of the work of science represents long, tedious work requiring a great deal of attention to detail and considerable persistence. The tracing of some obscure shell through thousands of rock samples to determine the extent of its existence in time was of this sort. Such work on index fossils was absolutely essential to the continued development of geology and the under-

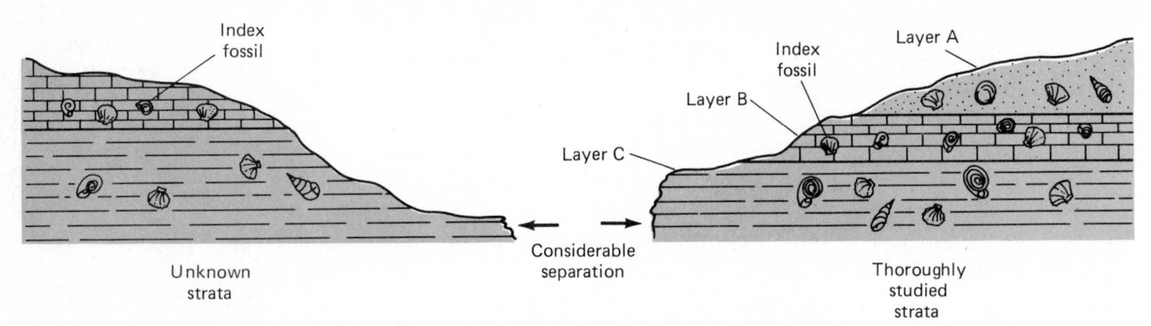

FIGURE 15-13
Index fossils are useful in establishing strata deposited at the same time over considerable distances. Detailed studies of the strata at the right establish a particular type of shell that is never found in earlier strata (C) nor in those deposited later (A). Such an index fossil can identify the period of deposition of the unknown layer at the left by its occurrence. This technique can work only with literally lifetimes of detailed study of the fossils in each stratum.

standing of earth history, yet this work was unspectacular and unlikely to capture the popular imagination. The second discovery was quite the opposite. In 1824 a paper was published describing some fossil bones found in southern England. From a collection of over a dozen bones William Buckland pieced together a carnivorous reptile over 12 m long which roamed the English countryside in days long past. From the study of the thigh bone even the great Baron Cuvier, the most famed and respected naturalist of the day, agreed to the conclusions. The existence in the past of a flesh-eating reptile the size of a one-bedroom apartment is dramatic. It captured the public imagination far more than the almost microscopic creature that was conveniently common and then died in droves to mark one point in time in the rock record. Once discovered, dinosaurs became a popular rage—during one period in the 1830s new books on geology outsold popular novels. This surge hit just as Lyell produced his great study of geology. Lyell's studies caused him to reject the idea of catastrophism. His basic conclusion was that the entire world as we know it was produced over a vast period of time by the very same forces we see at work today on its materials. This is the principle of *uniformitarianism.* Mountains today are being ground away by water, wind, and ice. Elsewhere, the land is rising to provide the mountains of a far tomorrow. At the same time, the steady, relentless trickle of particles falls like snowflakes to the bottom of the sea depositing today's uplands in the basins to form tomorrow's surface. Furthermore, Lyell introduced a systematic reconstruction of the past eras of the earth and much of today's terminology. He documented the introduction of plant and animal groups into the fossil record, their expansion to become numerous as both individuals and separate species, and the eventual decline of some groups.

Earth History

A contemporary of Newton, Bishop Ussher, worked diligently through the Biblical records to attempt to determine the age of the earth. He concluded that the earth was created in the year 4004 B.C. Newton commented that problems in methodology made the age of the earth much more uncertain than this, but the attempt was praiseworthy. Another contemporary, Sir Edmund Halley, tried for the same goal by a different method. Halley assumed that all of the salt in the oceans was originally part of the rocks of the surface. From the rate at which the earth's rivers carry salt to the sea Halley hoped to determine how long the sea had existed. No matter how conservative a set of assumptions Halley made, he was forced to conclude that the earth was much, much older than Ussher had estimated. This tends to be the pattern that has repeated itself since the seventeenth century. Each new method of dating the earth by some indirect method has produced a greater age. Consider that the typical rate of producing new sediments is on the order of centimeters per century. Then consider that the thickest beds of sedimentary rock are on the order of tens of kilometers in thickness. What sort of number would you arrive at?

Modern methods of dating rock by means of the radioactive decay of minerals in the rock have provided the ability to determine the age of samples with great consistency. These methods indicate that the oldest rocks yet measured are over $3\frac{1}{2}$ billion years old. Thus, the crust of the earth has been around for at least this long. This is a huge period of time for human beings to imagine. If this length of time was shrunk to a single year, all of human written history would have occurred in the last 10 min of that year. Only very exceptional human beings would live as long as 1 s on this scale. Geologic time is divided today in much the same pattern originally set out by Lyell and his contemporaries (Table 15-4). The most recent rocks represent the *Cenozoic Era,* or age of "recent life." Fossils are dominated by life forms that are quite familiar even though

TABLE 15-4 Divisions of Geologic Time

MILLIONS OF YEARS BEFORE PRESENT	ERA	PERIOD	EPOCH
	Cenozoic	Quaternary	Recent Pleistocene
2		Tertiary	Pliocene Miocene Oligocene Eocene Paleocene
65	Mesozoic	Cretaceous Jurassic Triassic	
225	Paleozoic	Permian Carboniferous Pennsylvanian Mississippian Devonian Silurian Ordovician Cambrian	
600	Precambrian		

FIGURE 15-14 The trilobite was the dominant lifeform on earth for a longer time span than there have been vertebrate animals. Each segment along the body is arched in three lobes, giving the animal its name. Most types were a few inches long, although some species were well over a foot in length. (*Photograph by the author.*)

many are extinct—mammals, birds, insects, seed plants, and so on. Earlier is the *Mesozoic Era,* or age of "middle life," and older yet the *Paleozoic Era,* or age of "ancient life." The Mesozoic Era was the time when reptiles, especially the dinosaurs, were dominant on land, and a form of shellfish, the *ammonities,* were plentiful in the seas. Both of these forms disappeared entirely at the beginning of the Cenozoic Era, serving as a convenient marker in the rocks to divide time. The Paleozoic Era lasted longer than the later two eras combined and had a variety of life forms dominant at various times. One very successful form that developed considerable diversity in the Paleozoic Era was the *trilobite* (Fig. 15-14). At one point these creatures were the dominant life form on earth. Their extinction marks the end of the Paleozoic and the beginning of Mesozoic time. Trilobites go back in the fossil record to the dawn of recorded life in the rocks. The oldest rocks found by Lyell were in Wales, and he called these Cambrian, after the Latin name for Wales (Cambria). Thus, the *Cambrian Period* is the oldest portion of the Paleozoic Era. Rocks formed earlier than these yielded no trace of life to the geologists of Lyell's day, and these were collectively called *Precambrian.* We know today that the Precambrian Era represents about 85 percent of the entire history of the earth. The largest exposure of Precambrian rocks at the surface of North America is across much of central and eastern Canada where higher (thus younger) rocks have been scoured off by the grinding of great ice sheets. These rocks of the *Canadian Shield* formation document a vast span of rock-building on earth in which no life forms left a record of their existence.

Each era may be subdivided into periods and each period into epochs of time as is shown in Table 15-4. Some of these periods and epochs marked particularly interesting or important times in the history of parts of the earth. For example, the *Carboniferous Period* of the Paleozoic Era is of great importance to us today. Worldwide, the rocks of this period are associated with the great deposits of coal. While these rocks record the first abundant animal forms on land, the amphibians and the insects, their importance is in the explosion of plant life. The expansion of plant forms capable of living in air produced a volume of vegetation new on earth. Over the years the bodies of dead plants accumulated in thick layers at the bottoms of marshes and shallow seas. The pressure of sediments later deposited over these plant remains slowly compressed these forms and worked them into the rock we call *coal*. The rock beds of the Carboniferous Period are the greatest accumulation of fossil fuels on earth. The United States alone has known deposits of 1.6 million million tons of coal (1.6×10^{12} tons), or about one-fifth of the known world supply. This represents 93 percent of our fossil fuels in the United States. If this coal were mined at the rate of 1 billion tons a year (we should reach this in the 1980s), the coal represents a 1000-year supply, one of the great resources of the nation.

The study of the locations of water-deposited rocks from a particular period allows us to determine the ancient shoreline of the continent. If we do this for North America we are forced to conclude that the shape of the continent above sea level has been very changeable through time (Fig. 15-15). From Cambrian time through Pennsylvanian time much of the central United States was a shallow sea. We may infer this from the marine life forms deposited in the rocks in these areas during these times. In the Carboniferous Period alone as much as 6 km of sediments which became rock were deposited across parts of Arkansas. In fact, only from the Tertiary Period onward has most of the United States been above sea level. One reason for this may be found in the history of the earth in relatively modern times. The Pleistocene Epoch was the time of great continental glaciers. These masses of ice advanced across the land over huge areas. In places the ice of the glaciers was up to 5 km

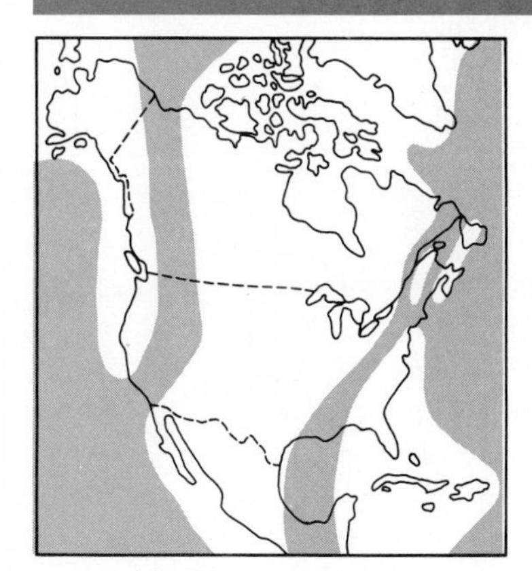

(*a*) 600 million years ago

(*b*) 450 million years ago

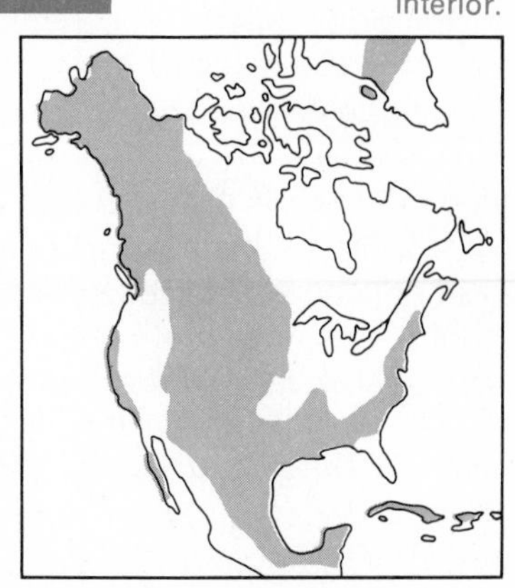

(*c*) 100 million years ago

FIGURE 15-15 The portion of the North American continent above water has varied considerably through time. The past shape of the continent may be inferred from sea-bottom deposits of various ages now found on the dry continental interior.

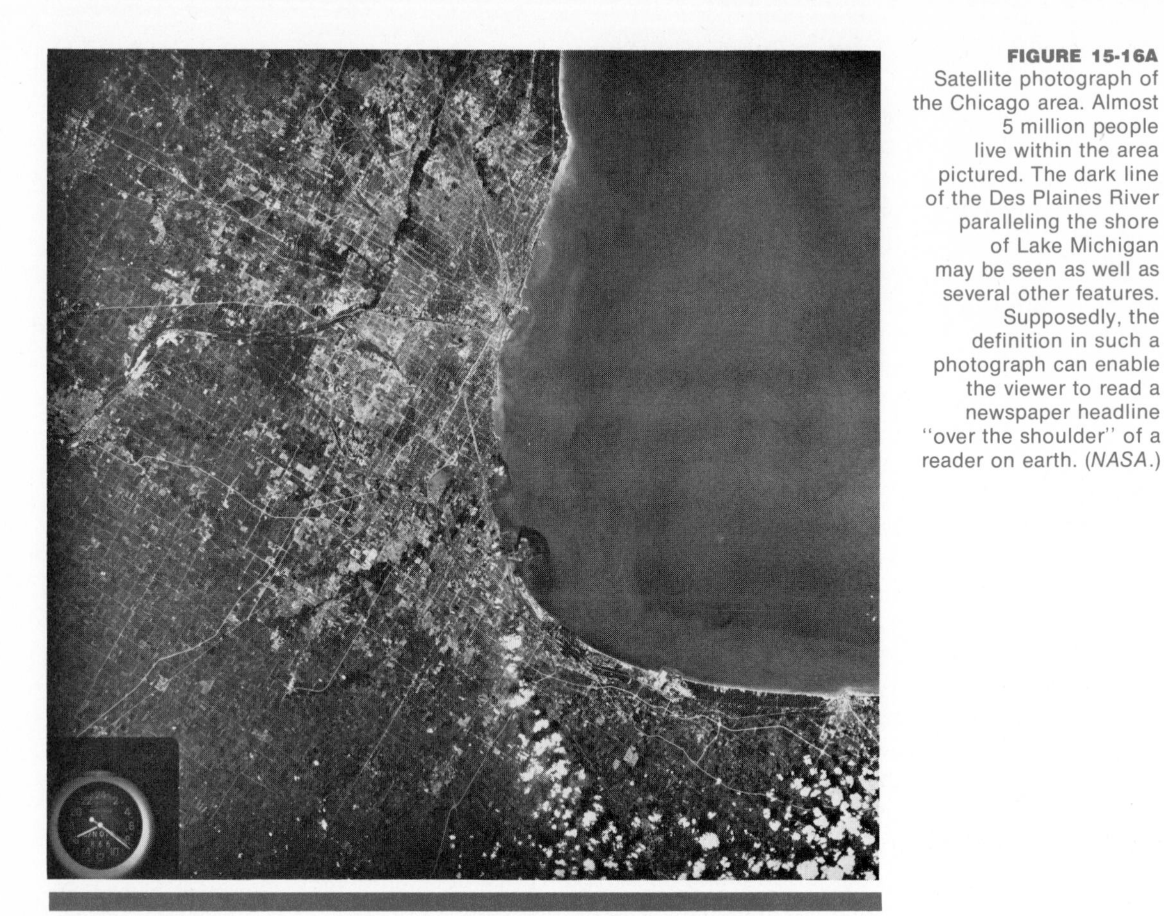

FIGURE 15-16A Satellite photograph of the Chicago area. Almost 5 million people live within the area pictured. The dark line of the Des Plaines River paralleling the shore of Lake Michigan may be seen as well as several other features. Supposedly, the definition in such a photograph can enable the viewer to read a newspaper headline "over the shoulder" of a reader on earth. (*NASA*.)

thick. One effect of the glaciers was to lower the level of the seas so that a greater area of the continental platforms was dry land. Even today there is some argument as to whether we are really out of the "ice age" of the Pleistocene Epoch. The ice caps of the earth trap enough water to raise the level of the sea by almost 100 m. If this ice melted, no part of the state of Florida would be above water except for the penthouse level of some Miami Beach hotels. Historically, this would be a much more normal situation than the present contour of sea level across the continent.

Evidence for Pleistocene glaciation abounds over the northern United States. The average depth to undisturbed bedrock at Chicago is over 50 m. It is likely that many individuals living in Chicago on the debris of the great ice sheets, and not vacationing elsewhere, have never encountered the undisturbed bedrock of the planet. On top of the bedrock is a mass of chopped and ground rock deposited by the glaciers, called *till,* and lake sediments from a larger Lake Michigan (Figs. 15-16A and 15-16B). Till is different from most sediments in that it is not sorted by size. The glacier moves particles of all sizes equally well. In this respect, a glacier is like an immense conveyor belt. To understand this we need to examine the actions of a glacier. Any time a layer of ice exceeds a thickness of about 100 m, the bottom of the ice is subjected to

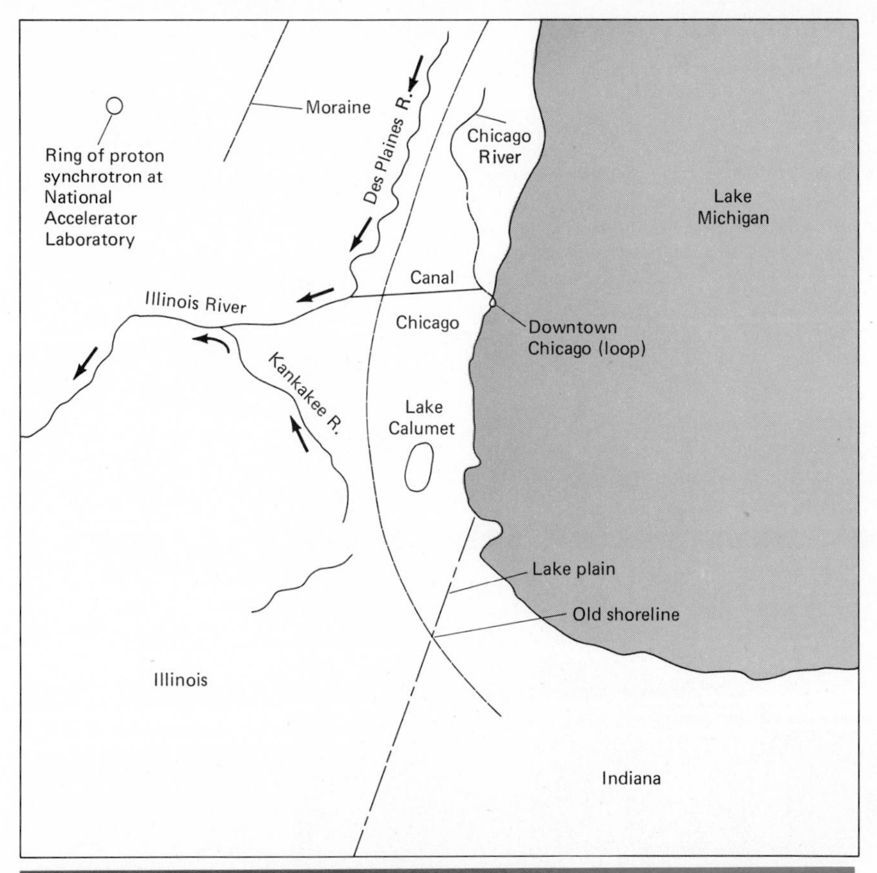

FIGURE 15-16B Even though the body of Lake Michigan dominates the drainage, only the tiny Chicago River drained naturally into the lake (for most of this century locks backed it up and flushed it, and its pollution, through the canal shown to the Illinois River). The Kankakee River and the Des Plains River drain the inland area and do not empty into Lake Michigan, but drain away from the lake into the Illinois River. Much of Chicago sits on the lake plain and terraces of old Lake Michigan before it drained to its present level below Lake Calumet to the Illinois Channel.

great pressures by the weight of the materials above. Ice under this pressure becomes *plastic* and can flow while remaining solid. If ice of this thickness is on sloping land, it will begin to flow downhill. At the same time, this plastic ice is injected into every crack in the rock floor beneath it. When this happens, even momentary relief of the pressure causes a powerful expansion of the ice. Freezing water expands to form ice. This expansion generates forces well beyond the ability of rock to withstand. Rock slabs pried loose from the bottom rock under the glacier are "plucked" from the structure, suspended in the ice, and moved with the advancing ice mass (Fig. 15-17). To the casual eye, a large ice mass may appear stationary, but glaciers are always advancing, *even when they seem to retreat.* If the advance of the ice is faster than the melting at the forward edge, the glacier advances to cover more area. If the melting exceeds the advance, the glacier seems to retreat. In either case, the glacier moves rock fragments forward. Should the melting of a glacier at its forward edge equal its advance rate, it will appear to remain stationary in time. However, the rock debris carried by the glacier is moved to the front edge and dumped there as the ice melts (Fig. 15-18). Should the advance rate and the retreat rate caused by melting balance each other, a mound of rock material is deposited called a terminal *moraine.*

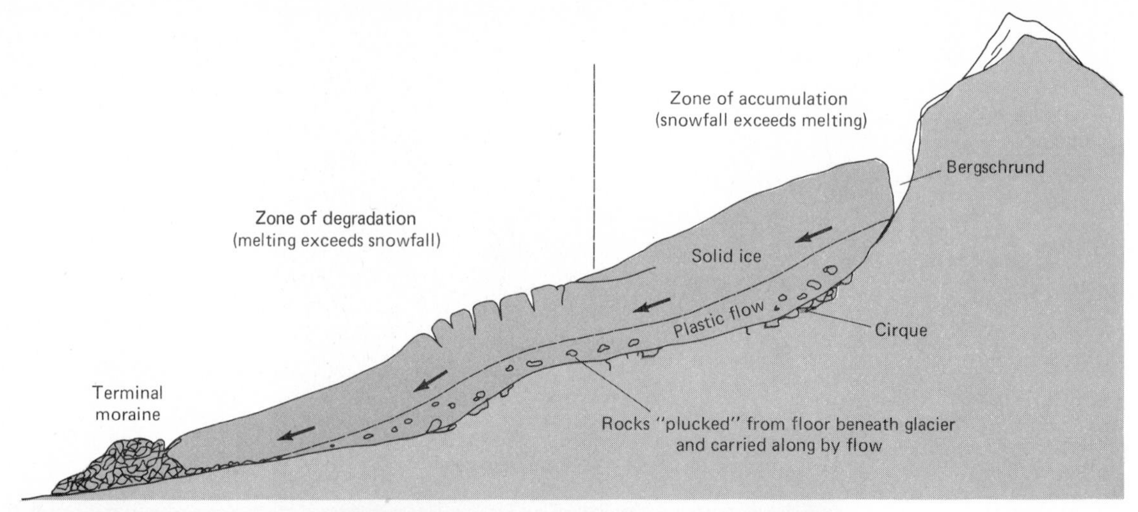

FIGURE 15-17
A slice through a mountain glacier is similar in many ways to one through a great continental ice sheet. Both have zones of growth and of excess melting, and both always advance by plastic flow. However, while ice sheets are miles thick and can ride over much of the terrain, mountain glaciers are channeled by existing valleys.

FIGURE 15-18
The side edge of a glacier in Alaska and its lateral moraine, or piles of rock debris. Melting has been active at the lower end of the glacier and enough ice has been removed so that its surface is covered with rock fragments which were suspended in the ice. The cleft drops over 15 m to a river flowing along the side of the ice, draining its meltwaters. This glacier continues almost 50 km (30 mi) in the direction of the background to its origin high in the mountains. (*Photograph by the author.*)

For a great ice sheet such as that coming down from the Hudson Bay area of Canada and carving out the basins of the Great Lakes on the way, these terminal moraines can be huge. To such an ice sheet it makes almost no difference whether the rock fragment is the size of a dust grain or of the Washington Monument. The exception is that the larger fragment is dragged at the base of the glacier while rock flour, the fine fragments, may rise in the plastic ice mass. Thus, the huge fragments drag on the rock surface under the ice and create gouge marks in the bedrock under-

FIGURE 15-19
Mountain glaciers are effective at cutting downward into rock, but are less effective in cutting sideways. Upon their retreat they leave steep-sided, flat-bottomed valleys with a characteristic U-shaped profile such as these in Alaska. Compare this with the V-shaped cut formed by water erosion as shown in Fig. 8-10. (*Photograph by the author.*)

FIGURE 15-20
A large, thick glacier can cut downward through rock more rapidly than a smaller one. After the glaciers melt, the valley of the smaller glacier is often left hanging high on the valley wall of the larger glacier. The most frequently viewed American example of such a *hanging valley* is at the entrance to Yosemite Valley in California. Here, Bridalviel Creek spills in a falls from its valley to the Yosemite valley below. Note the U-shaped profile of the valley. (*Photograph by the author.*)

neath as they plow along the surface. Such gouge marks are apparent wherever the glacier advanced and then retreated too fast to leave much debris.

Deciphering the advances and retreats of a glacier is a particularly subtle form of geology requiring great patience and a huge volume of data. The basic problem is that each new advance obliterates the record of past activity by moving away all of the structures created by earlier activity. All of the Great Lakes represent "puddles" of meltwater trapped

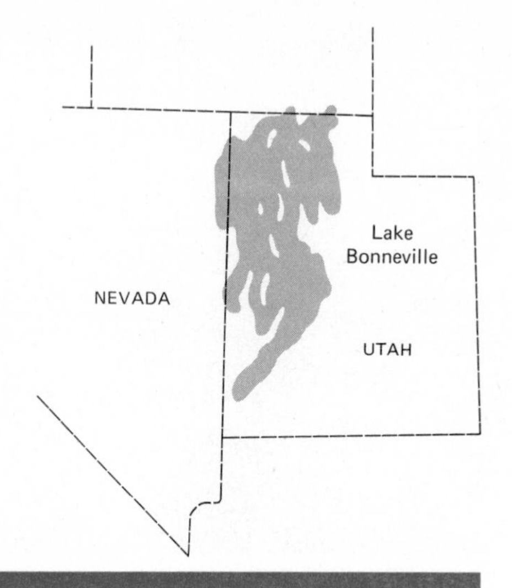

FIGURE 15-21
At its greatest extent, glacial Lake Bonneville occupied most of western Utah and was on the scale of the Great Lakes. Numerous other large glacial lakes filled the basins of the western United States at the melting of the ice. The remnant of Lake Bonneville remains in the Great Salt Lake.

from the disappearing ice sheets. In geologic terms these puddles are as temporary as the puddles in your backyard after a rain. They are momentarily trapped behind a few resistant rock structures that withstood the ice sheets and the immense piles of debris left around the edges of these sheets. Several of the great glacial lakes of the United States have already drained. At its greatest extent, Lake Bonneville in the western United States covered parts of Utah, Nevada, and Idaho (Fig. 15-21). This lake had a surface area of almost 52,000 km², which is larger than Lake Ontario and Lake Erie *combined* (see Table 15-5). It was deeper than any of the Great Lakes except Lake Superior. The lake's undoing was its extreme elevation above sea level of 1580 m (almost exactly 1 mi) and its dry surroundings. As glacial meltwater caused the lake level to rise, it eventually found the lowest spot in the surrounding rock structures that penned it in. Waters spilling through this channel had a great descent to sea level and they rushed downhill. This action quickly cut an exit for the lake. Actually, even then much of Lake Bonneville simply dried up. Today, the largest remnant is the Great Salt Lake of Utah with no exit for its waters. In a moist climate the lake would grow until it cut a permanent exit channel. In the Great Salt Lake water evaporates as rapidly as it is provided to the lake. The great Bonneville salt flats surrounding the lake attest to its once gigantic size, as do various lake terraces on the mountains surrounding its valley.

The future of the Great Lakes cannot be quite the same as the fate of

TABLE 15-5
Comparative Statistics of Three Lakes

	DEPTH, m	SURFACE AREA, km²	ELEVATION OF SURFACE ABOVE SEA LEVEL, m
Lake Superior	406	82,000	183
Lake Bonneville	320	51,800	1580
Lake Erie	64	12,900	174

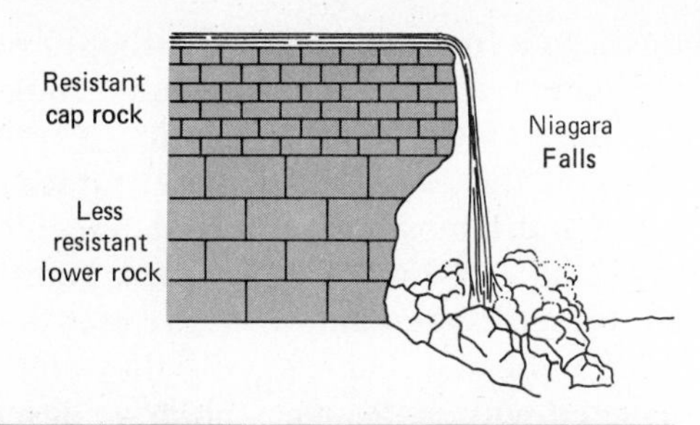

FIGURE 15-22 Niagara Falls is being cut back by erosion of the support from under the stratum of highly resistant Niagaran limestone-dolomite on which the Niagara River flows. The cap rock loses support, cracks off in large blocks, and drops to the base of the falls in a random pattern but always moving the falls upstream.

Lake Bonneville. As Table 15-5 shows, Lake Superior has more depth below sea level than the elevation of the present surface above sea level. Even if Lake Superior were opened to the ocean, a great inland bay of some sort would replace the deeper parts of the lake. For the moment, the upper Great Lakes have met their match in the ridge of limestone supporting Niagara Falls. This ridge wears away slowly by undercutting rather than by surface erosion from above (Fig. 15-22). The falls are being worn back at a rate of slightly more than 1 m/year. They have been worn back about 10.5 km thus far. From this you should be able to calculate the age of the upper Great Lakes; can you? Eventually, the falls will cut back all the way to Lake Erie, and the upper lakes will undergo increased drainage and a change in shoreline.

A second effect is working to drain the Great Lakes eventually. We may picture all of North America as a rigid slab of rock. At one time, all of Canada was covered by several kilometers of ice. The weight of this ice depressed the northern part of North America. Now that the ice is gone, the northern end is "bobbing up" to its original height again. As we raise the northern end we tend to tilt the entire slab of rock. The Great Lakes are like shallow puddles on a sheet of plywood that is being tilted. As Canada bobs up in reaction to the lost weight of the ice, the Great Lakes tend to spill toward the south. The level of the lake rises at the southern

Carbon-14 dating

"When Organisms Die They Stop Eating"

One of the best modern examples of a chain of inferences is the one that led to an absolute dating system. Cosmic rays bombard the earth from space, and while you read this several fragments from their impact high in the atmosphere will pass through your body. By the 1930s a great deal was known about cosmic rays (although more was not known), and balloons were sent to high altitudes to measure these rays and their products. In 1932 a new particle, the *positron* or positive electron, was discovered in the tracks of atomic fragments shattered by a cosmic ray at high altitude. This particle had been predicted by theory but never before observed. Another predicted particle was found the same year on the ground when Chadwick demonstrated the existence of the neutron. Almost immediately the workers with cosmic rays prepared to measure the neutrons produced in the high atmosphere by cosmic-ray impacts on atoms of air. The first flight late in the 1930s did indeed find a steady level of free neutrons in the high atmosphere.

We may now begin our chain of infer-

ences. What would happen to a free neutron high in the atmosphere? Some would escape to space, of course, but the majority would end up striking the atoms of the air. Four-fifths of the air is nitrogen and almost all of the remainder is oxygen. Thus, 4 out of 5 times the neutron would hit a nitrogen atom first. Studies on the ground indicated that oxygen atoms seldom interacted with neutrons, but nitrogen atoms did. Almost all nitrogen is the isotope $^{14}_{7}N$. The reaction that occurs changes N 14 into an atom of carbon and an atom of hydrogen as follows:

$$^{14}_{7}N + ^{1}_{0}n \longrightarrow ^{14}_{6}C + ^{1}_{1}H$$

The carbon produced is not the common isotope C 12 but the rare isotope C 14, or carbon 14. Now, carbon 14 is *radioactive*. That is, its nucleus will eventually change by the emission of a particle in a process called *radioactive decay*. The details of this are discussed in Chap. 18 and need not concern us here. The important point to remember is that the rate of radioactive decay is a constant which is not influenced by the chemical or physical environment of the radioactive atom. What will happen to this carbon atom suddenly produced high in the atmosphere? The answer is easy. It should rapidly combine with the oxygen atoms present to become part of a carbon dioxide molecule. What should happen next? This depends upon two rates, the rate at which carbon 14 decays and the rate at which carbon dioxide high in the atmosphere mixes with that lower. Rough measurements suggested that the mixing is much faster than the decay process. What does this imply? Again simple—a certain percentage of the carbon dioxide of the air at the surface should contain radioactive carbon 14. If the rate at which cosmic rays bombard the earth has been constant through time, then the percentage of C 14 in the lower air should be constant also. So what? Aha! We are now near the point. All green plants use the carbon dioxide of the air for their supply of carbon. From this they manufacture all the organic molecules which make up their bodies. Furthermore, all animal food ultimately traces back to these plants. Thus, a fixed percentage of the carbon of all plant and animal bodies should be radioactive C 14.

Enter the Nobel Prize! When organisms die they stop eating! This is exactly the key point as described by Willard Libby (1908–1980) to a group of high school physics students. While alive, carbon is exchanged with the environment by eating, elimination, and so on. Once the carbon is trapped in the dead body, the carbon 14 decays to nitrogen 14 and is *not replaced*. Thus, the carbon 14 percentage in each gram of body carbon declines as time goes on, and the amount present becomes a measure of how long the body has been dead.

These ideas occurred to Libby just before World War II, and remained totally untested until after the war. In fact, after the war, Libby's colleagues at the University of Chicago were amazed that he would want to devote his time and talent to such an unexciting project as the precision measurement of the rate of decay (half-life) of an obscure carbon isotope. Carbon 14 decays to one-half its original amount in about 5600 years. Once this had been established, it was necessary to test the decline of carbon 14 with time. One of Libby's collaborators crawled through the sewers of Baltimore collecting fresh methane (CH_4) from decaying sewage. An old, old sample presented no problems. Natural gas from deep underground has been there for millions of years, and certainly long enough for all of its carbon 14 to have decayed. These measurements agreed perfectly with Libby's theory, and he moved on to ancient materials of known age. The results provided the first consistent mechanism for determining the age of carbon-bearing samples of the recent past. Willard Libby was awarded the Nobel Prize in Chemistry in 1960 for this fundamental advance.

FIGURE 15-23 The great ice sheets advanced from two centers of accumulation in Labrador and west of Hudson Bay. They made five separate advances into the United States, each farther west than the previous one. Curiously, a large area of western Wisconsin was left untouched by the separate sheets moving around it. Note that the edges of the various sheets have defined most of the major river channels by their margins.

end of Lake Michigan, for example, and becomes shallower at the northern end.

Field investigations have established a series of advances and retreats of the ice sheets of the Pleistocene Epoch in the United States (Fig. 15-23). Ice spread down from centers of accumulation in Canada in a series of separate advances separated by interglacial periods. Several of the rivers of the northern United States established their present channels when they served to drain away the huge volumes of meltwater from the glaciers. The Mississippi River at the Illinois-Wisconsin border flows on a flat valley almost 2 km wide between rock bluffs. Drilling in the sediments of this valley indicate that it is a water-carved trench cut into the bedrock to depths of over 300 m. This places the base of the valley well below present sea level. Carving this valley into solid rock would require a volume of fast-flowing water of gigantic proportions. Similarly, the Great Lakes were scooped out by the glaciers so that their bottoms at places are lower than sea level. The basic shape of the western Great Lakes was largely dictated by a rock structure called the Michigan Dome. The ridge of rock that had the most success in withstanding the ice sheets carving the Great Lakes was the Niagaran limestone. This is the rock formation that Niagara Falls falls over. The Green Bay Peninsula in Wisconsin and Georgian Bay in Canada represent this rock standing where it refused to be overwhelmed by the glacier. At Green Bay this ridge divided the main force of the ice sheet into the lobes that carved the bay and Lake Michigan. South of this point a convenient accident occurred. The last advance of the ice in this area slid across a living forest burying the trees and packing them in an airtight container of rock debris. The entire forest lay there, pushed over in the direction of the ice advance. When Willard Libby developed the radioactive carbon 14 dating process just after World War II, wood from these trees were among the early samples dated. These timed the last advancing gasp of the ice age in the central United States as having occurred about 10,000 years ago. This is only a stone's throw as geologic time goes. In fact, the interglacial periods between ice advances during the Pleistocene Epoch frequently lasted far longer.

Summary

The bulk of the crust of the earth is made up of oxygen (48%) and silicon (23%). The commonest unit of these elements is the SiO_4^{4-} tetrahedron. The density of the entire earth is twice that of its crustal rocks, so it is almost certain that the materials of the deep interior are different from the materials of the crust. A *rock* is a piece of the earth's crust, and is usually a mixture of materials. Rocks are made of *minerals,* which are pure substances with definite physical properties and a single chemical composition. Most dense of the common *igneous rocks,* or rocks cooled from a liquid state, is *basalt*—a finely crystalline, dark rock. The opposite extreme in igneous rocks is the light-colored, low-density rock *granite.* Between basalt and granite a substantial majority of all igneous rocks are accounted for. Once exposed at the surface a rock is subjected to *weathering,* or the breakdown of the rock into smaller pieces. These pieces may then be moved from the original site of the rock in the process of *erosion.* These and subsequent steps in the possible history of the rock are described in the *rock cycle.* If weathering proceeds faster than erosion, soluble minerals are *leached,* or dissolved from the upper layers of the soils. The balance between weathering and erosion as modified by the presence of plants determines the *soil type* of the surface.

Once weathering has reduced a rock to fragments, erosion may transport and then deposit these fragments in *sediments.* One agent of erosion, glaciers, moves all sizes and types of fragments equally well and produces *unsorted* sediments. Other agents of erosion, such as wind or running water, sort sediments by the ease with which they may be transported. Once deposited, sediments may be turned into a new type of rock called a *sedimentary rock* in a process called *lithification.* If this is accomplished by the application of heat and pressure, the process may proceed beyond the sedimentary rock stage, to produce a *metamorphic rock.*

Sedimentary rocks are particularly important in determining the past history of the earth. During their deposition as sediments these rocks frequently trapped the bodies of dead plants and animals which were turned into *fossils,* or rock impressions of life forms of the past. Deciphering the fossil record led to two interpretations around 1700. *Catastrophism* maintained that the earth was subjected to recurring periods of great catastrophes. Opposing this is the currently held belief that the processes we see working in the modern world have changed very little in time and have produced the rock record as we find it. This is the theory of *uniformitarianism.*

Geologic time is divided into four *eras.* The *Precambrian Era* represents the time before the appearance of any complex life form in the rocks and it accounts for 85 percent of the rock record in time. The *Paleozoic Era* came next and was marked by several life forms found fossilized, notably the family of forms called *trilobites* (now extinct). The disappearance of the trilobite is the marker to begin the *Mesozoic Era.* Much of this era was dominated by the large reptiles called *dinosaurs,* and their extinction marks the end of the era and the beginning of the *Cenozoic Era,* which is the last great division of geologic time. Each era is divided into recognizable *periods* characterized by a particular set of life forms. Each separate time within a period can be identified by particular varieties of life which appeared in the rocks, thrived, and became extinct in a geologically brief span. Such fossil types may be used as *index fossils* to associate rocks in time across a large physical separation on earth. Several periods in time are of particular interest. One of these is the *Carboniferous Period* of the Paleozoic Era since this was the time when most of our fossil fuels were originally deposited as sediments of plant material. More recently, the last epoch, or *Pleistocene Epoch,* of the Quaternary Period of the Cenozoic Era is of interest as our recent geologic past. This epoch was one dominated by four separate advances of massive continental ice sheets, separated by long interglacial periods. This ice age was responsible for much of the surface appearance of the northern United States as well as Canada.

Questions

1 A rock represents large, visible crystals of orthoclase feldspar imbedded in a mix of quartz with no crystalline structure visible even at the microscopic level. How could such a rock have formed?

2 A rock represents visible crystals of quartz imbedded in a mix of orthoclase feldspar with no crystalline structure visible even at the microscopic level. How could such a rock have formed?

3 A mixture of vegetable oil and water is whipped in a blender and blast-frozen. How will its appearance differ from a similar blended mixture allowed to freeze slowly over a 2-day period?

4 Where would you expect basalts to be more common, near the surface of continents or deep beneath them? Why?

5 How could you go about changing a granite to a rhyolite in the laboratory?

6 What process in the rock cycle are you performing if you:

- (*a*) Dig some coal from a cliff face and carry it home?
- (*b*) Throw your garbage into a ravine daily?
- (*c*) Pack a snowball so tightly it becomes an iceball?
- (*d*) Spread rock salt on an icy sidewalk?
- (*e*) Sculpt a block of marble into a statue?
- (*f*) Mix and pour a batch of concrete?

7 The strata shown in Fig. 15-9 are over 1 mi above sea level yet they contain fossils of ocean-dwelling marine creatures. How could you account for this?

8 How would Cuvier account for the abrupt disappearance of trilobite fossils from rocks all over the earth of the same age? How would Lyell account for this?

9 Deposits are currently being formed in the China Sea at the rate of 2 cm/century. How long would it take to form a layer of shale such as that shown in Fig. 15-9, which is 150 m thick if the sediments are compacted 20 percent during lithification?

10 The fineness with which we divide geologic time increases as we approach the recent past. Why?

11 Which is more likely to be preserved and wind up as a fossil, a squirrel or a shark? Why? Which do you suspect we have more of, fossil fish or fossil birds?

12 Most agents of erosion sort particles by size. Give a few examples of this. What agent of erosion does not sort particles?

13 When glaciers melt they leave large mounds of debris in moraines at their sides and front. They also leave the following forms.

Give a probable reason for the existence of each.

(*a*) Snakelike mounds of sorted and rounded gravels stretching for miles and not associated with the sides or ends of the glacier—called *eskers.*
(*b*) Teardrop-shaped hills blunt in the direction from which the glacier came and tapered in its direction of motion—called *drumlins.*
(*c*) Pits or depressions, sometimes large enough for a house—called *sinks.*
(*d*) Cone-shaped mounds of moderately sorted, fairly small debris frequently along a moraine—called *kames* (this one is hard; think).

14 All lakes are relatively young, temporary features on the earth's face. Why?

15 Can you name some of the major river courses which Fig. 15-23 suggests might have been determined by the positions of glaciers?

16 What are the two mechanisms by which a waterfall may be produced?

17 Every gram of carbon in a human body contains 6.48×10^{-10} g of radioactive carbon 14. If half of this decays to other atoms in 5600 years, how much would decay in a single year (average)? In a single day?

CHAPTER SIXTEEN

Shaping the Crust

Introduction One of the problems in interpreting the earth is that of scale. A human is a bit like an ant running around on an open page of *The New York Times*. We perceive pattern but are unlikely to obtain a good overview because we are continually in contact with a maze of detail. If we were approaching the earth from space for the first time, our perceptions might be quite different. First of all, we would note that the earth is divided rather clearly into landmasses and large bodies of water. This division is very clear-cut even from space. Very few areas exist on earth where a jumble of various-size blocks of land are mixed with seas and inlets of all sizes. More common is the situation which occurs between New York and Europe. A relatively crisp coastline borders a giant expanse of water with few, or no, small landmasses for an expanse the size of a continent. None of the world's oceans have a landmass the size of the British Isles out in the middle somewhere. Continents and oceans are clearly divided, and groups of large islands are associated with the edges of continents and not with the open oceans. Is this some accident of flooding or is it a basic part of the structure of the earth?

If we could drop the level of the sea by about 500 m, the picture would become clearer than it is if we just look at the current surface. Many "puddles" that are an accident of flooding would disappear. For example, all of Hudson Bay would be a dry, flat plain, as would be the entire Baltic Sea and much of the North Sea. None of these has an average depth over 100 m. The British

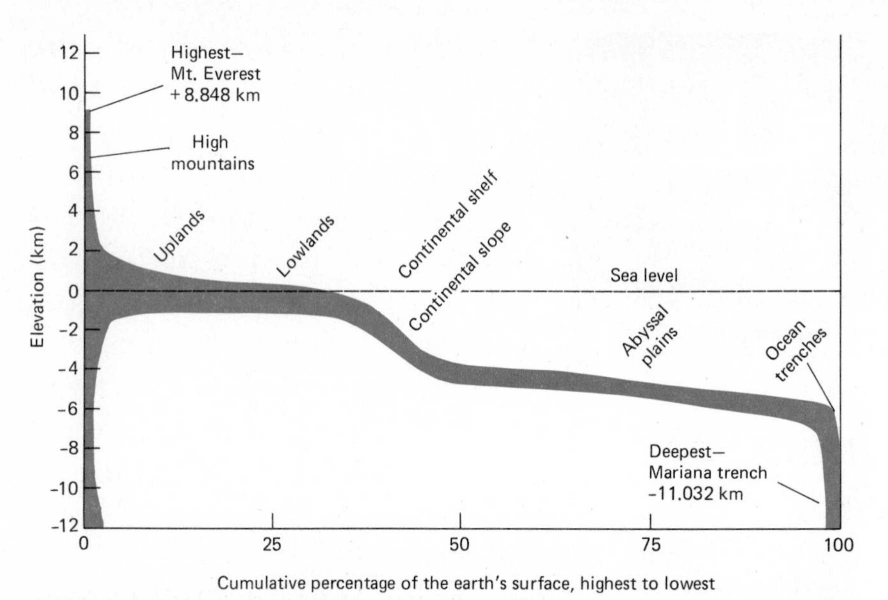

FIGURE 16-1 This graph shows the percentage of the earth's surface at any elevation. The earth is really a two-surface planet—continental surfaces and abyssal ocean plains. About 29 percent of the surface is dry land and fewer than 3 percent is in the form of mountain heights.

Isles would be isles no longer. However, no new continents would appear. In fact, the basic outline of most continents would be easily recognizable. Furthermore, flying low over the earth we would note an even more pronounced change. Unlike the gently sloping plains that dip gradually into the sea today, most continental margins would end in cliffs. In places the cliffs would be almost vertical, but steeply angled plunges would be more common. Almost totally absent would be broad plains at nearly the level of the sea. Continents would be plateaus sticking up out of the sea. In fact, if we could remove the oceans entirely, the most fundamental fact about the surface structure of the earth would become clearer. The earth would have broad plains at a low level from which huge, broad blocks rose miles into the air. *The earth is essentially a two-surface planet.* The earth's overall appearance might be described as that of a roughly peeled onion. As the onion is peeled broad chunks can be pulled from one layer, leaving a two-surface globe. Hills, mountains, canyons, and other minor irregularities are small in comparison to the regularity of two surfaces. Continents are gigantic tabular blocks sticking up out of the oceanic plain. Figure 16-1 shows the percentage distribution of the land surface of the earth by elevation and confirms the "peeled-onion" model of the earth's surface. *The difference between continental platforms and ocean basins is basic and fundamental, and other features are minor by comparison.* In this chapter we will examine the earth's deep interior and processes occurring to produce the surface we know.

The Earth's Deep Interior

Almost all of our knowledge of the deep earth comes from the study of earthquake waves. Like all waves, these refract and reflect at boundaries of mediums and travel at constant speed in uniform materials. Earthquake waves represent a physical motion of the rock of the earth itself. As such, their detection is relatively simple by the principle of inertia. If a large mass is freely suspended (on springs, for example), when the earth itself shakes and moves the mass will tend to remain stationary. If a pen is attached to the large mass and a recorder is attached to the earth, the relative motion of the two will produce a record of the shakes and wiggles of the recorder. Such a device is called a *seismograph*. A complete record usually involves three separate masses at one place and the traces they produce on the recorder. One instrument records up-and-down motion of the earth, one north-south motion, and the third the east-west motion. From the three records the actual movement of the earth is determined.

Earthquakes result from the motion of one block of the earth across another along a break or crack between the blocks called a *fault*. As the blocks move, friction between their surfaces locks the two blocks together at the fault. The continued motion increases the forces tending to produce motion along the fault. These forces bend the edges of the block which are locked to the face of the other block. At some point the force at the fault is great enough to overcome the friction and slide one block across the other. The distorted rock now snaps rapidly back to its normal shape by moving some distance along the fault. In a single large earthquake the actual motion along the fault may be 10 m or more.

While faults are common in the rock structure of the entire earth, most are minor and many have long since lost the motion of blocks which leads to earthquakes. The best-known fault in the United States is the *San Andreas fault* in California. This fault runs along the eastern edge of Baja California and through the state of California for two-thirds of its length until moving into the Pacific Ocean. Motion on this fault was responsibile for the great San Francisco earthquake of 1906. (The fault runs past San Francisco at the ocean front parallel to the Golden Gate Bridge.) That earthquake was the most catastrophic one yet suffered in the United States. The action along the San Andreas fault is to move a thin strip of western California to the northwest relative to the rest of the state. (This strip contains many of the largest population centers in the state.) The total displacement of this strip across the rest of the state may be measured by matching structures to the right and left of the fault. To date, about 300 km of motion has occurred on the fault. This is roughly comparable to sliding Chicago to the position of Detroit, or moving New York City to Washington, D.C.

When an earthquake occurs several types of waves are produced. The fasting moving waves are the primary, or *P*, waves; later come secondary, or *S*, waves. The designation *P* is frequently remembered as standing for "push" or "pressure" since the *P* waves are longitudinal. The *S* waves are often called "shake" or "shear" waves since they are transverse. Since the *P* and *S* waves travel at different speeds, the time difference between their arrival at a seismograph is a measure of the distance of the earthquake from the instrument. The combined records of a

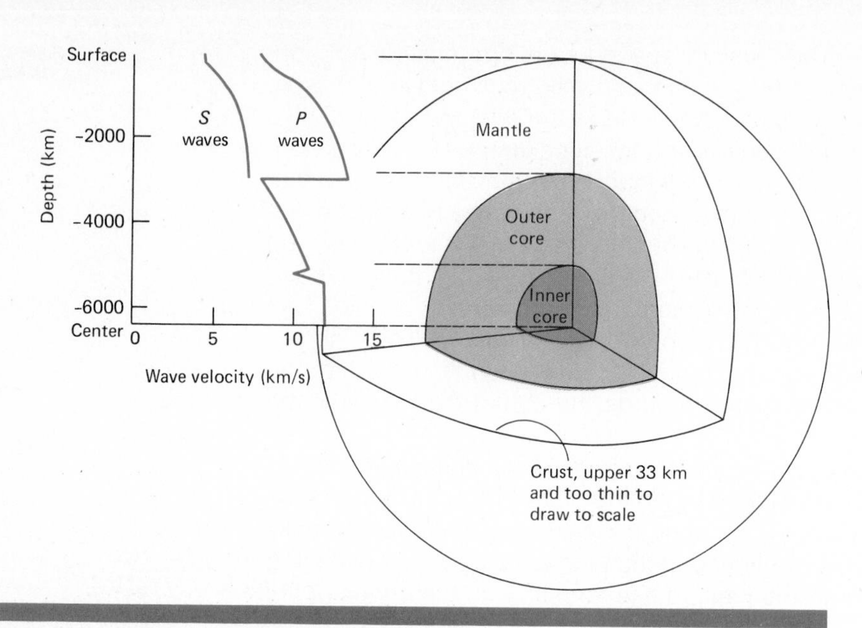

FIGURE 16-2
Variations in the velocity of earthquake waves with depth allow us to divide the interior into several regions. Transverse *S* waves increase in velocity with depth until they are stopped entirely at 2900 km. This leads us to believe the outer core is liquid. Note that the crust of the earth is too thin to represent on this scale.

series of stations thus can pinpoint the location of the earthquake as well as its depth. Earthquakes are largely confined to well-identified belts on the earth. The largest of these is a ring around the shores of the Pacific Ocean. The second most active zone runs through the Mediterranean Sea over to northern India. Active zones also include lines running down the middle of ocean basins.

The earliest guesses about the earth's interior were based on observations from deep mines. For every 30 m one descends into the earth, the temperature increases about 1° C. If this rate were maintained beyond the depths attained by humans, the temperature would reach about 1000° C at a depth of 3 km. This implies that at the relatively shallow depths of 10 to 20 km almost all types of known rock would reach their melting point. From this, some individuals concluded that only a small fraction of a percent of the total mass of the earth formed a solid outer crust, underneath which the bulk of the earth was molten. Modern evidence suggests that this is wrong, although we retain the word *crust* to describe the outer layer of the earth. The study of earthquake waves reveals the depth of the crust. When a wave encounters an abrupt change of the medium in which it is traveling, part of its energy is reflected and part is refracted into the new medium. Such an abrupt change in medium is called a *discontinuity*. In 1909 the Croatian geologist (Croatia is now part of Yugoslavia) Mohorovičić discovered a distinct discontinuity from his study of earthquake records. Later studies showed this boundary exists all over the world at depths averaging 33 km under the continents and about 5 km under the denser floors of the ocean basins. This is called the *Mohorovičić discontinuity* (or *moho* to save considerable time) and is the boundary between the crust and the next layer of the earth, the *mantle*. It was originally thought that the moho might be the transition to a zone of liquid rock (or at least plastic rocks that flow under pressure). Two separate types of evidence indicate that this view cannot be correct. First of all, *S* waves have no difficulty in moving through the material of the mantle. Transverse waves, however, cannot move through a liquid, and *S* waves are transverse. Second, the moho does not mark the limit of deep earthquakes, some of which occur more than 600 km deep (one-tenth the distance to the center). Some rigidity of the rocks must be preserved to at least this depth.

At a depth of 2900 km another abrupt discontinuity is observed in the record of earthquake waves. The speed of *P* waves, which had been rising slowly with distance, drops abruptly to about half of its previous value and *S* waves are stopped completely. This is exactly the behavior to be expected if a boundary to a liquid layer were encountered. Below 2900 km we have encountered the *core* of the earth. A further transition at 5100 km below the surface indicates that an *inner core* may be a solid with a radius of some 1200 km surrounded by a liquid *outer core* 2200 km thick (Fig. 16-2).*

Ups and Downs— Isostasy and Geosynclines

The first clear ideas about the causes of elevation differences of the surface emerged from a study of the area with the greatest elevation contrast on the land surface of the earth. In the 1850s British surveyors in India conducted an extensive survey. The Ganges River basin is a plain that runs nearly east-west across northern India (Fig. 16-3). From its

* A 1 megaton hydrogen bomb releases 10^{15} J or 2.4×10^{14} cal of energy. This is about the same as a moderately large earthquake. The advantage of the bomb is that exactly where and exactly when it was detonated is known. The *S* and *P* waves from American bomb tests of the 1940s and 1950s were recorded by a worldwide network of seismographs. These records have contributed a great deal to our knowledge of the earth's deep interior.

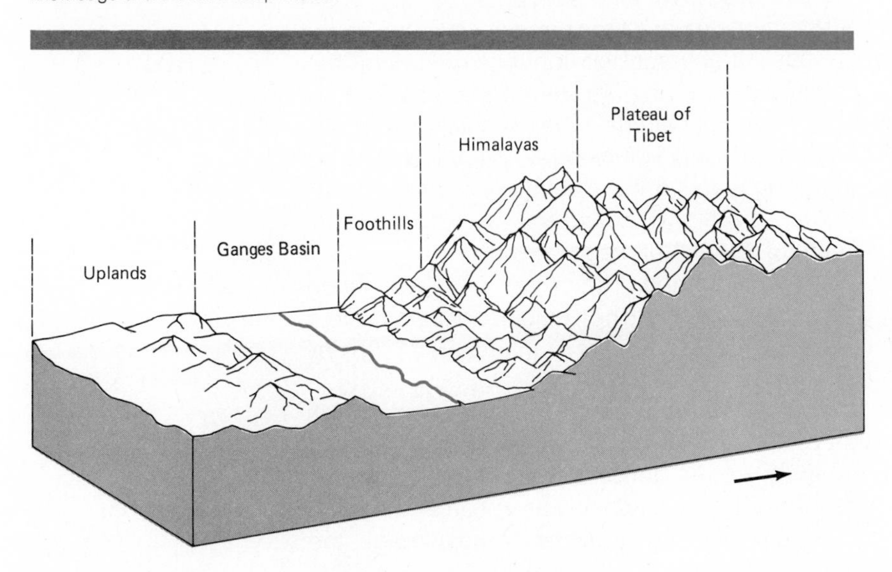

FIGURE 16-3
A block drawing across northern India indicates the Ganges Basin near sea level next to the highest mountain mass on earth. The rock walls rise almost 9 km (6 mi) above the lowlands in a horizontal distance of a few hundred kilometers.

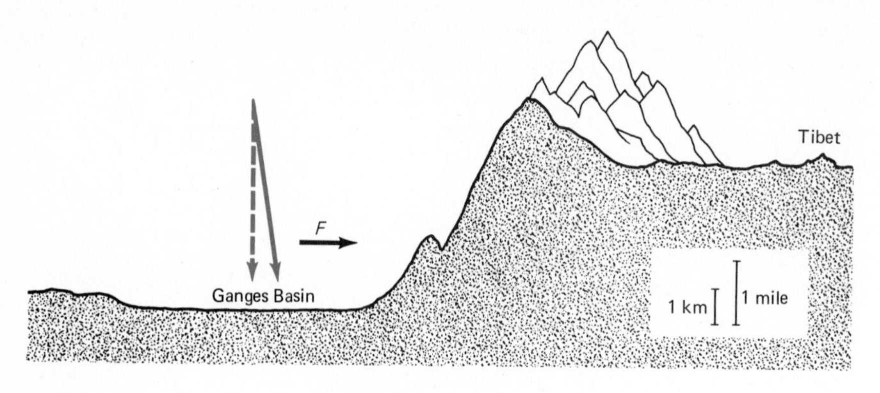

FIGURE 16-4A The huge mass of rock in the Himalayan uplift might be expected to exert enough gravitational attraction to deflect a plumb bob from a true vertical. In fact, this does not occur.

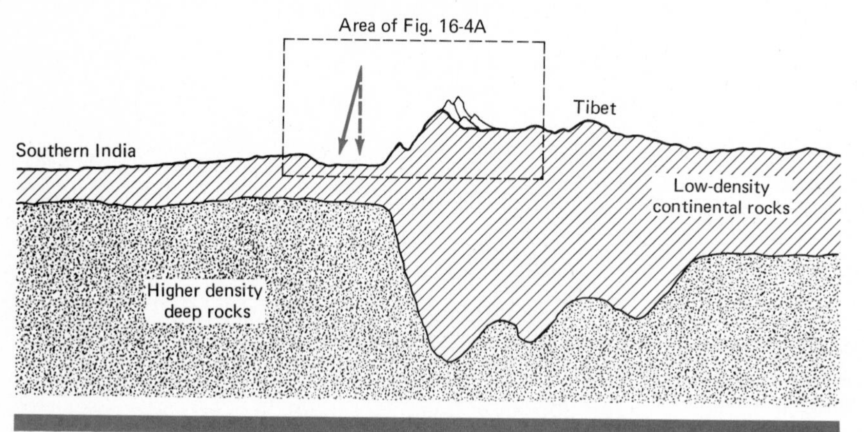

FIGURE 16-4B Airy suggested that the Himalayas had a gigantic "root" of low-density rocks projecting below them into the higher-density rocks found at great depth. The more dense rock under southern India then could exert a greater pull, offsetting the effect of the mountains.

northern edge rises the great Himalayan mountain chain with many of the tallest peaks on earth. Farther north, beyond the Himalayas, lies the plateau of Tibet some 3 km above sea level. Standing in the Ganges Basin you might expect that the enormous mass of rock above your level to the north would influence the local direction of straight down (Fig. 16-4A). This relatively close matter should exert a northward pull on objects not balanced by a corresponding mass of material from the south. Astronomical measurements determined "true up," or the direction precisely away from the center of the earth. "Local up," or the direction exactly opposite that of a plumb bob hanging on a string, could also be measured. (In practice this required a careful survey running some 600 km.) The two measures disagreed but not as expected. The plumb bob was pulled much less in the direction of the Himalayas than calculations perdicted! One explanation of this was suggested almost immediately (1855) by astronomer royal Airy (Fig. 16-4B).

It has long been known that the earth as a whole had a very high density, some 5.5 times that of water. The rocks of the earth which we encounter every day are generally not as dense as the earth as a whole. Typical rocks from the continents are less than half as dense as the earth itself. But if the surface rocks are only half as dense as the average, the

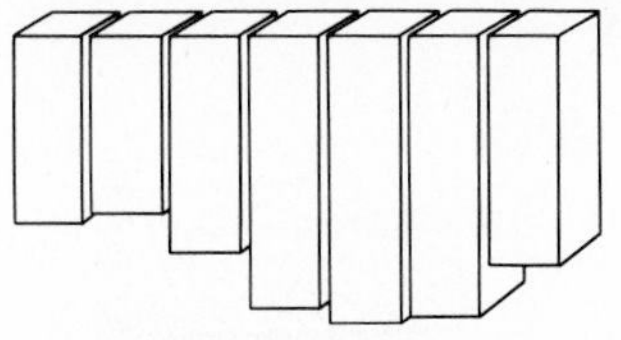

FIGURE 16-5A
Square wooden blocks are cut into random lengths from wood with a density six-tenths that of water.

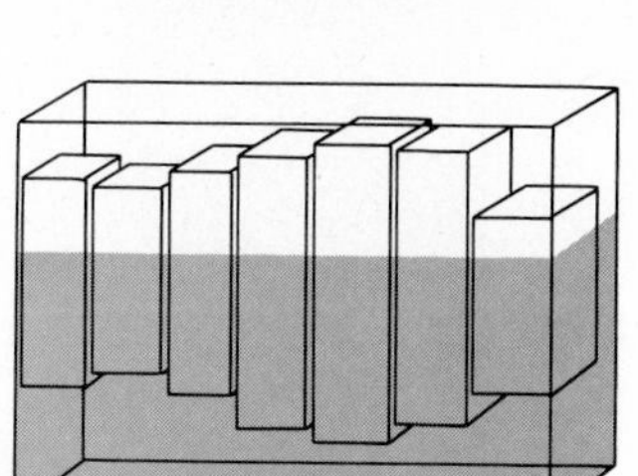

FIGURE 16-5B
In a tank of water, each block will float with four-tenths of its volume above the water. Blocks floating higher than their neighbors must also project deeper into the water.

interior must contain materials with much higher density to make up the average value. Airy calculated that the deflection which was observed could be accounted for if the Himalayas had a huge "root" of low-density, continental rocks projecting downward into regions of much denser rock. If this were so, the mass of this mountain root would be much lower than the mass of the higher-density rock under central India. Thus, the plumb bob would not be attracted so strongly toward the Himalayas as expected. This view led to the idea of continents as masses of lighter rock "floating" on denser materials underneath.

Let us explore further the implications of a density difference in rocks. Picture the following situation: Cut some 10-cm² beams of a fairly dense wood into random lengths as pictured in Fig. 16-5A. The density of this wood is about six-tenths the density of water, so the wood will float with about four-tenths of its volume above water. What will happen if we drop these blocks vertically into the tank of water at the same time, so that none can tip over? Each block will sink into the water until six-tenths of its length is below the water surface and four-tenths is above it (Fig. 16-5B). Thus the tops of the longer blocks will float higher above the water level than the tops of the shorter blocks. To float high each block must have a longer "root" going down into the water. Another way to express this situation is to say that each 10-cm² column in the tank—air from the top of the tank plus wooden block plus water under the block to the bottom of the tank—must have the same total weight. If there is more block, there must be less air above it and water below it. Applied to the earth, this suggests that if we would slice out columns of the earth, say 1 by 1 km, down well into the dense lower rock, all such columns would be expected to have the same weight. The only way that light rocks could project up above the surface in mountains would be if they also extended deeper than average in a root. This idea that all parts of the earth's surface are freely floating on denser materials underneath and that greater height also must imply greater depth of low-density rocks is called the *principle of isostasy* (Fig. 16-6).

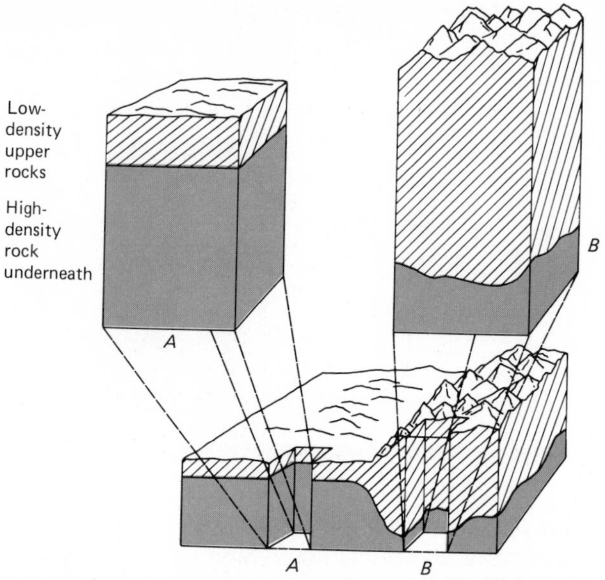

FIGURE 16-6
The principle of isostasy asserts that all large blocks of the earth with the same cross-sectional area cut to a depth well below continental rocks will have the same weight. While column *A* is shorter, more of it is made of high-density rock and thus it has the same weight as column *B*. The increased elevation of column *B* implies that its lower-density rocks also extend to greater depth. In effect each portion of the crust "floats" on the denser rock underneath.

Isostasy is used to interpret many phenomena. For example, very minor earthquakes occur in the upper Mississippi River Basin from time to time. These are associated with the action of the river removing by erosion millions of tons of material from the surface of several states. In reaction to the removal of this low-density material, the land surface rises by the principle of isostasy. An even greater motion is the rising of the northern portion of North America and Eurasia. With the gigantic weight of the great ice sheets removed, the continents are in the process of "bobbing up," much as a log bobs up in the water after a person jumps off. Thus, Scandinavia is becoming larger each year as it emerges from the sea. Changes in the shoreline of the north end of the Baltic Sea indicate it is rising about 1 cm/year, and has been rising for the past 7000 years.

Even with the principle of isostasy to interpret elevation differences we are left with a problem. Erosion constantly acts to wear down the higher elevations and deposit the materials in depressions. In the long history of the earth why have not the low-density rocks of the continents been eroded away and deposited in the ocean depths? Without active processes to build new mountains, nature has had more than enough time to produce a relatively flat, uniform surface. The evidence indicates that in places, erosion has piled up sediments *over 30 km thick* (19 mi thick) from the debris of long-gone mountain ranges and uplands. In the absence of mountain building the continents should be worn down to at least the level of the oceans. Yet mountains today thrust their peaks over 9 km above the level of the sea. Furthermore, with the exception of volcanoes, the great mountains and ranges of the earth are made up largely of sedimentary rocks brimming with fossils of marine creatures. In the highest peaks on earth you can look at the frozen imprints of the shells of animals that live only in shallow seas. Mountain walls expose layer after parallel layer of rock that was deposited in the tranquillity of a sea undisturbed for eons of time. Mountains must be lifted from the depths by some process; but how? One clue comes from the detailed study of the sedimentary layers themselves. A particular layer of sedimentary rock has a thickness at any one point and an extent in each direction horizontally. As you proceed in a particular direction sooner or later the layer would become thinner and then disappear entirely. Sedimentation takes place in particular areas only. The sediments being created during our life span in the Gulf of Mexico are not being formed in Tennessee. Some future geologist may be able to infer the outline of the Gulf of Mexico by the extent of today's sediments.*

Deposits of sedimentary rocks sometimes occur in beds covering a very large area, representing several large states in extent. Examination

* Actually, the Gulf of Mexico is a poor example, and is used here for familiarity. The Gulf of Mexico represents a real chunk missing out of the platform of North America and averages $1\frac{1}{2}$ km, or over 1 mi, deep. Deposits of sediments in the central Gulf occur in waters too deep for creatures typically found as fossils in the sedimentary rocks of the mountains. Relative "puddles" such as Hudson Bay or the Baltic Sea are less than one-twentieth as deep.

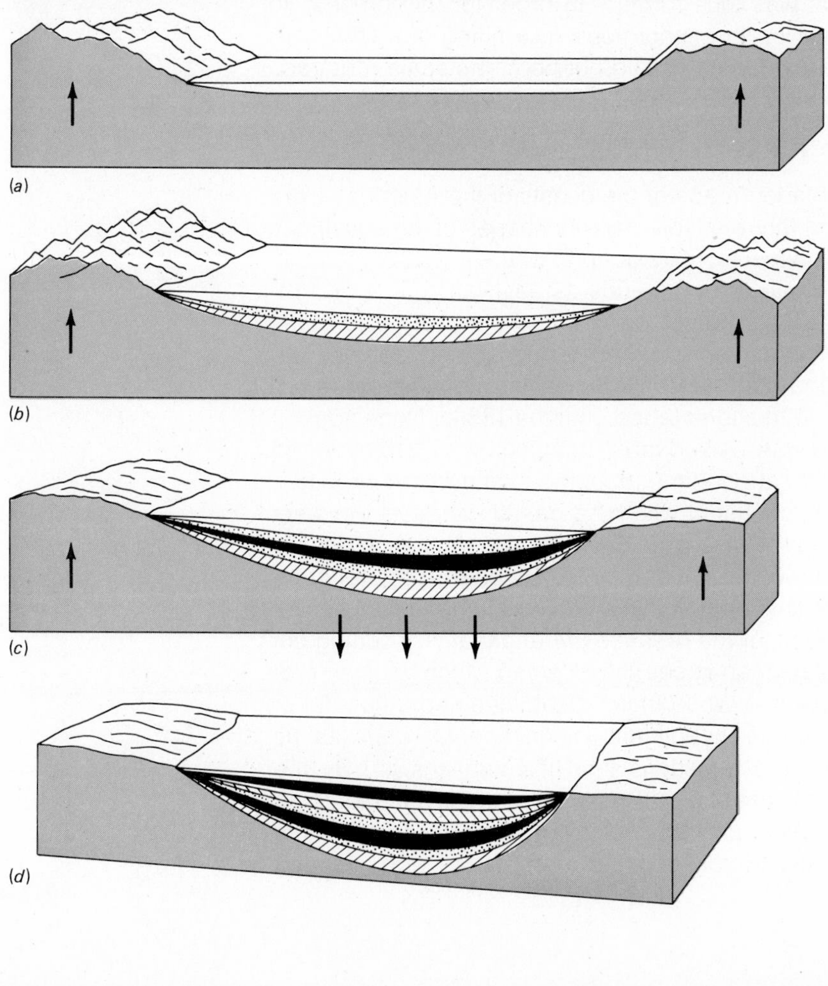

FIGURE 16-7 The development of a geosyncline begins with a shallow sea surrounded by rising uplands (*a*). As the uplands erode they deposit their sediments in the sea, while continued rising presents yet more upland for erosion (*b*). Eventually the thickness of the sediments creates pressures converting the lowest layers into rock. The weight of this rock added to the basin causes it to move downward (*c*). Eventually the uplift may end, leaving beds of rock many miles thick underlying the sea. Generally, these beds are thickest in the middle and thin out identically toward the edges.

of the variation in thickness of these beds discloses that many layers share the same extent and run out in about the same positions in each direction. Furthermore, these beds get thicker toward the same positions. Sedimentary strata are frequently thickest in the very mountain ranges where they are also highest. These characteristics led to the description of the large structure called a *geosyncline*. Geosynclines represent enormous beds of sedimentary rock originally deposited in shallow seas. The proposed formation of a geosyncline is illustrated in Fig. 16-7. A lifting of the surface creates uplands on the margin of a depressed part of the continent which is a shallow sea (Fig. 16-7*a*). For a long period of time erosion wears away these uplands and mountain ranges about as fast as they rise. The eroded materials are carried out to the shallow sea and deposited as sediments (Fig. 16-7*b*). As the lifting of the uplands continues, the weight of the sediments deposited in the sea causes it to sink. This sinking is balanced by the arrival of new layers of sediment (Fig. 16-7*c*). Eventually, the lifting of the uplands ceases and erosion wears them down to lowlands. By this time the deposited sediments have reached very great thickness in the formation of the *geosyncline*. Later events could raise this structure to reveal the great thickness of sedimentary rock that was thus formed. In order for this process to work, a geosyncline must be truly enormous. The filling of a small depression, such as the Great Lakes, would not be of the scale required.

An Even Larger Pattern Emerges

Shortly after the first extended maps of the coasts of the Americas were available, scholars noted the relatively parallel shapes of the coastlines of the old and the new continents. For example, where Africa bulges outward toward the west, South America ends its eastward bulge and retreats to central America. By the seventeenth century the suggestion was made that at one time the continents may have been joined together and somehow broke apart and separated. However, the fit, although good, was only approximate and this idea faded until the 1850s. Once again it submerged until the idea was revived most ambitiously in 1910 by Alfred Wegener. He asserted that at one time all of the continents were joined together in a single supercontinent which he called *Pangaea*. He went on to conclude that during the Mesozoic Era Pangaea cracked apart and the pieces began to drift away from each other. Although Wegener compiled extensive evidence to support his views, it was far from conclusive, lacking a reasonable cause for the breakup of Pangaea. (Try cutting out the continents on an outline map to see the extent to which you can produce "jigsaw-puzzle"-type fit.) While these ideas attracted a handful of ardent supporters, Wegener's hypothesis of *continental drift* (as he named it) raised many more questions than it could answer. By the 1930s the proposition of continental drift was under attack and fading. In another generation the idea was reduced to a paragraph or so in the more careful geology texts, which devoted more space to explaining why it could not work.

However, great changes were occurring in the instruments available to geologists. In particular, the submarine geology of the ocean basins had come of age. World War II produced the instruments which would allow a thorough mapping of the ocean floor. The last 20 years have produced more knowledge of the deep oceans than all of the work done previously. Sonar sets send out a sound pulse which travels to the ocean floor. At the floor some of the sound reflects and travels back to the ship. By measuring the time required to obtain this "echo," the depth of the water may be calculated. An interesting fact emerged about the earth's great oceans from this data. The oceans have high rises or *ridges* running down their centers, which only occasionally are high enough to emerge through the surface as islands. Furthermore, these ridges form an unbroken, interconnecting set between all of the earth's oceans, much like the seams on a baseball. Lastly, and particularly in the Atlantic Ocean, the ridge runs with considerable pattern midway between the continents on one side of the ocean and the continents on the other side. Actually, the ridge is not an unbroken line running down the middle of the ocean. Rather, it is a series of straight segments separated by sharp breaks where the line "jumps" to the east or west in a direction almost perpendicular to the line of the ridge (Fig. 16-8). The lines marking these jumps are called *transform faults* and are the location of most midoceanic earthquakes. There appears to be too much pattern in all of this for it to be coincidence. The data were revealing something basic about the structure of the earth's surface, but what? One final hint produced the modern revolution in our interpretation of the earth's structure. This hint came from magnetism.

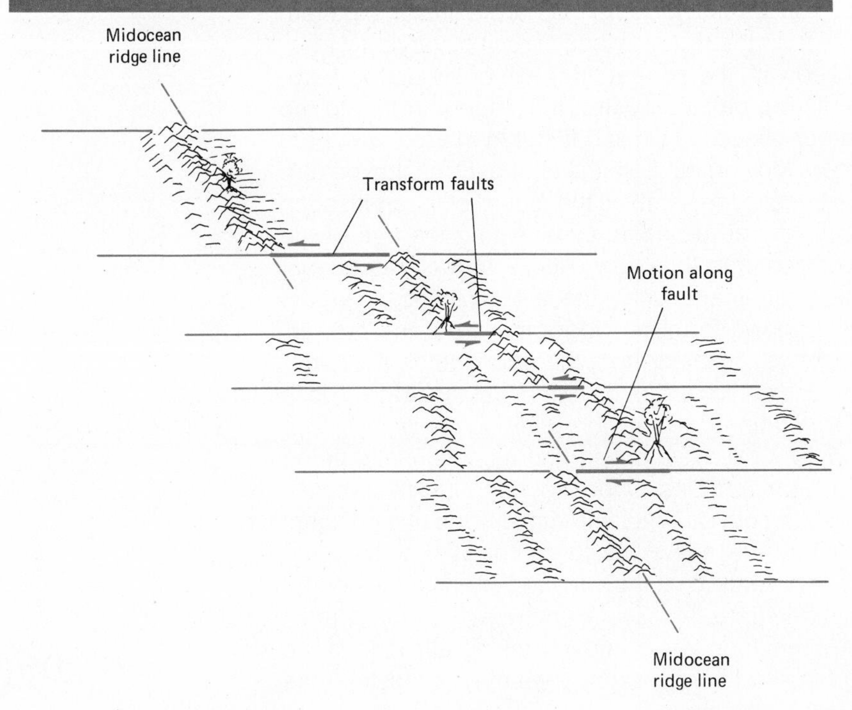

FIGURE 16-8
The midoceanic ridge proceeds in straight segments with "jumps" from segment to segment. The lines of these jumps are transform faults and are the location of most midoceanic earthquakes.

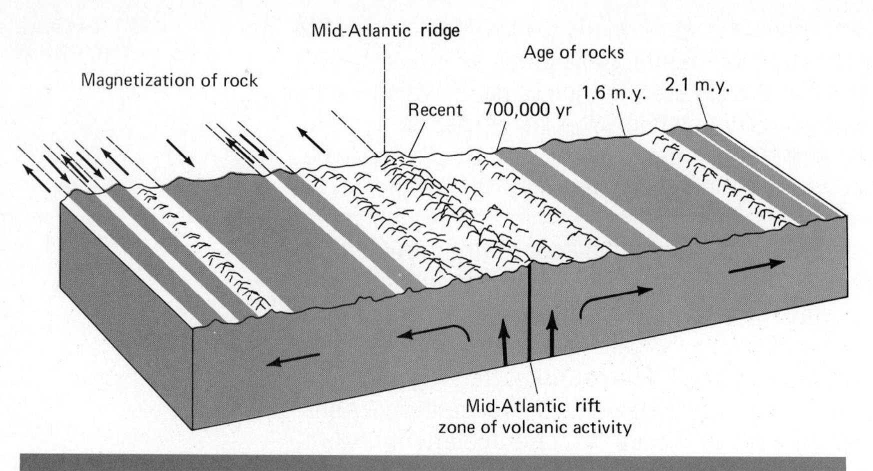

FIGURE 16-9 The direction of magnetization for zones of rock on each side of the midoceanic ridge is indicated (clear = normal; dark = reversed). The pattern of magnetization is a mirror image on either side of the ridge and reveals the direction of the earth's magnetic field when the rock solidified. Dating of the rocks confirms that new ocean floor is continually produced at the midoceanic ridge and spreads out in both directions. Distance from the ridge is a direct measure of the age of the rock.

In the 1950s very sensitive devices to detect the strength and direction of magnetization of rocks became available. Such *magnetometers* could be towed behind a ship and continuously record the magnetization of the rocks of the ocean floor. As explained in Chap. 10, magnetic crystals in molten rock line up with the earth's magnetic field. When the rock cools, the strength and direction of the magnetization of the earth's magnetic field is recorded in the rock. Samples of lava flows that can be dated indicate that the earth's entire magnetic field reverses direction every few hundred thousand years with some 19 reversals in the last 4 million years (reversals of very short time duration might easily be overlooked in the rocks at the earth's surface). A magnetometer towed perpendicularly across a midoceanic ridge reveals an interesting pattern (Fig. 16-9). Moving away from the ridge, bands of rock of varying widths are alternately magnetized with the current direction of the earth's magnetic field and against it. The *pattern* west of the ridge and that to the east of the ridge are mirror images. This is difficult to interpret until another regularlity is noticed. Moving out from the central ridge, the pattern of magnetization vs. distance is the same as the record of magnetization vs. time for the recent history of the earth. Determining the age of the rocks of the ocean floor confirms this view. Rocks at the midoceanic ridge itself are of recent origin, made within the last few thousand years. The farther the sample is taken from the ridge, the older the rock is. Graphing the age of rocks vs. distance from the ridge yields a straight line with a slope that is constant for any one segment of the ridge. These data have one clear interpretation: *The sea floor is being produced at the ridges and is spreading outward.* New ocean floor is continually being produced along the midoceanic ridges. Once produced, the floor moves out from its place of origin in both directions from the ridge at a constant rate. The constancy of this rate allows us to calculate how long the process has continued. For example, at the rate measured for the central Atlantic Ocean we may calculate that South America and Africa must have been joined some 60 million years ago. By taking all parts of the oceans and measuring the sea-floor spreading, we may reconstruct the positions of all the continents at that time. By following the sea-floor movement backward in time, an extraordinary fit between the continents is obtained. This fit is not the fit of the sea-level profiles of the continents, but of their true edges at the continental slope into the ocean abyss. The continents do resemble the pieces of a jigsaw puzzle. Continents are rafts of low-density material carried by the spreading denser crust of the ocean floor.

One question arises almost at once. If the ocean floor is growing in the Atlantic (and elsewhere), where is the surface shrinking? Since the earth is not becoming larger in time, for every 1 m of Atlantic Ocean produced there must be 1 m less of something else. The Atlantic coasts of both Africa and South America appear to be very stable through time (otherwise the fit of their edges would not have been preserved). The evidence suggests that South America is overriding the floor of the Pacific Ocean. The floor of the Pacific Ocean is spreading also. Both ocean floors are moving toward South America. However, the profile of the two coasts is quite different. On the Pacific Coast, a deep ocean trench parallels the coast, while just inland the Andes mountain range forms the highest mountains of South America. A first guess might be that as the continent smashes into the Pacific Ocean floor the rocks are crushed, cracked, and thrown upward in a gigantic "bow wave." (This was Wegener's guess at the beginning of the century.) However, a detailed consideration of the stress of rock will not allow this interpretation. If two huge masses of rock collide with each other (say that North America would smash into Asia), what happens depends very much on the relative speed of the masses of rock. At low speeds, cracking and crumbling could indeed occur. However, after a critical speed of about 5 cm/year such processes cannot absorb the momentum of the collision quickly enough. Instead, the masses must behave as intact sheets and slide over or under one another somehow. Just such a process is suggested by the pattern of earthquakes on the west coast of South America (Fig. 16-10). Just offshore, earthquakes are near the surface. At the coastal area the quake centers are at moderate depth. Inland, most earthquake centers are located at great depth. If the earthquakes are taken to represent the slippage of one block of the earth across another, they trace a

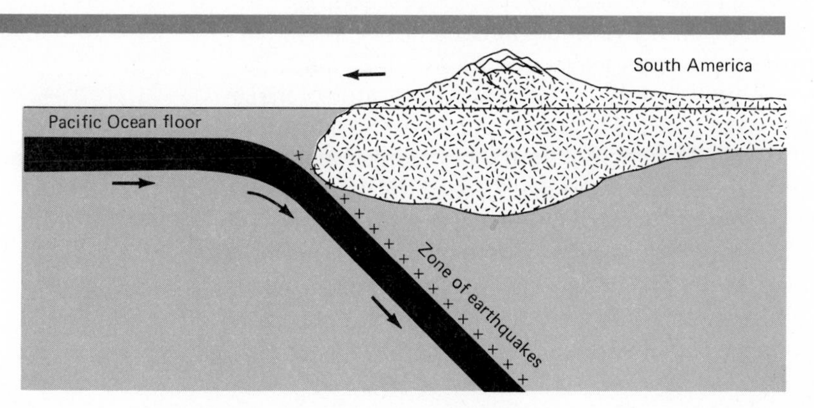

FIGURE 16-10
Records of the depth of earthquakes off the west coast of South America support the idea that the floor of the Pacific Ocean is being pulled under the continent to be destroyed in the depths. Earthquakes offshore are shallow and get progressively deeper as they proceed inland. Earthquakes one-third across the continent are some of the deepest ones recorded (about 900 km).

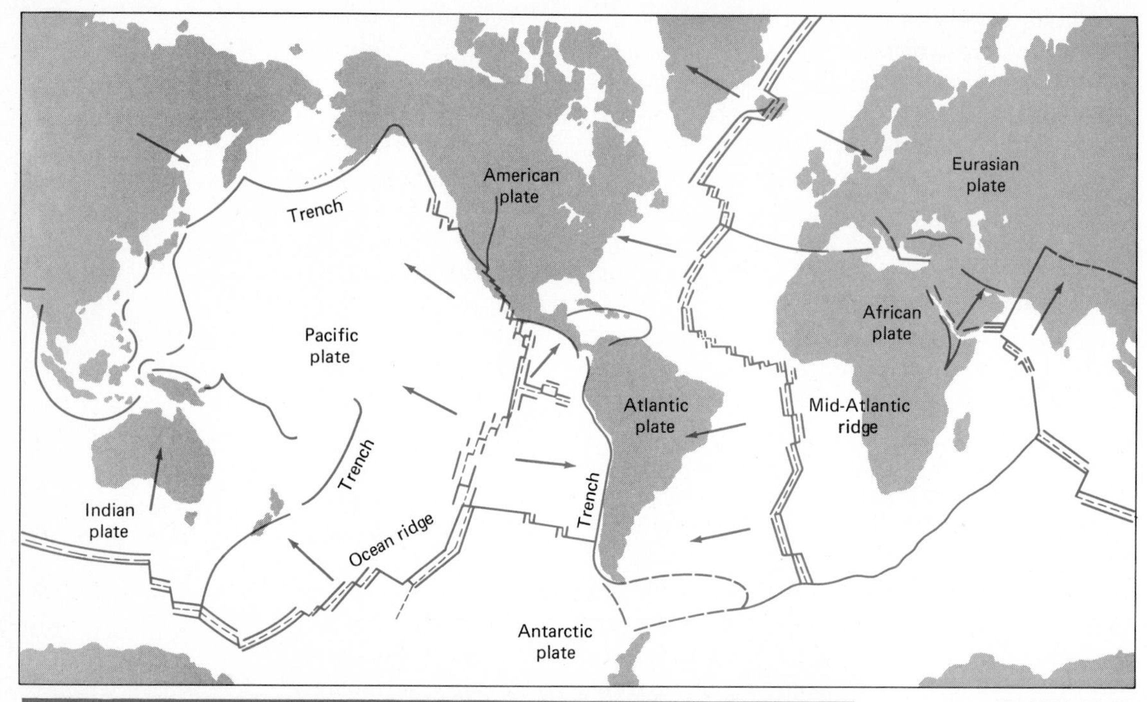

FIGURE 16-11 Boundaries of the major plates of the earth are indicated. New surface is generated for each plate at the midoceanic ridges. The surface then moves out in the direction indicated, eventually to be consumed at a subduction zone. Continents may be quietly "rafting" on the plates. Note that the western United States has overridden a midoceanic ridge, and part of the landmass is being pulled off by the Pacific plate motion.

surface where the floor of the Pacific Ocean is sliding under South America and into the deeper earth. A zone where a plate of the earth's surface is compressed and deflected downward (to destruction) is called a *subduction zone*. Thus, ocean floor is produced at midoceanic ridges and is destroyed in subduction zones marked by ocean trenches. Throughout all this the continents ride on the surface as passive rafts of floating material—indeed, they are not required for the process to occur.

Figure 16-11 shows a map of the earth with oceanic ridges and subduction zones indicated. This series of boundaries separates the earth's surface into a number of gigantic slabs or *plates* moving relative to each other. The study of the origin, motion, and destruction of these plates forms the new field of *plate tectonics*. This revolutionary new view emerged in the mid-1960s after a decade of separate discoveries.

There are six major plates and several small ones covering the surface of the earth (Fig. 16-11). Geologists now believe that major reforming of the crust of the earth is largely confined to the edges of these plates. Worldwide, the motions of these plates relative to each other measures from over 1 cm/year to almost 20 cm/year. North America is rafting westward at 3 cm/year on the surface of its plate. This has already created problems for the continent, since the western edge has overridden the boundary to the next plate. The rift along the San Andreas fault marks the line where our continent has pushed over the plate boundary. A slice of the far west including Baja California and the coast north to San Francisco is now riding on the Pacific plate and moving north-northwest at 5.6 cm/year.

Independent Evidence for Continental Drift

Ultimately, the best evidence for continental drift is the actual measurement of the motions involved. A worldwide series of measurements has now confirmed the actual drift and determined the rates of motion for various plates. Other, less-direct evidences supporting this idea also exist, however. The first of these is structural. We have discussed the structure of a geosyncline which accumulates beds of sediments many kilometers thick. There is no way to have a small geosyncline, since a small basin is simply not large enough to cause the local depression of the crust of the earth as it fills with sediments. Yet, there are several places on the earth which do not extend far horizontally, where rocks are piled to the thickness and conformation expected only in geosynclines. Invariably, these "tiny geosynclines" end on a continental margin with one side open to the ocean abyss. If the continents are "moved" back to the position where they were joined according to the theory of continental drift, this problem resolves itself. Geosynclinal areas of Brazil, for example, join a much larger similar area in Africa. Studies of the sediments of some of these regions have demonstrated that rock sequences in the geosynclines match each other over the span of an ocean. Thus, these thick beds of deposits happened to be sheared apart as landmasses were torn on a line running through them to form separate continents.

The second type of indirect evidence for continental drift has to do with the fossil record. Why should a particular type of dinosaur be found fossilized in South America and India, but not in the rest of Asia or North America? For different periods of geologic time, separate continents share the bulk of their life forms; yet at other times the types of life on these landmasses are distinct from each other. How could a land-based dinosaur migrate from South America to India without passing through Asia on the way? Linkages based on the fossil record of the continents strongly argue that land bridges existed between various continents at various times in the past. During the Permian Period all of the continents were apparently joined in the supercontinent of Pangaea (Fig. 16-12). Later, this was separated into a northern supercontinent called *Laurasia* (North America, Eurasia) and a southern supercontinent called *Gondwanaland* (South America, Africa, India, Australia, and Antarctica). A shallow ocean separated these two about on the line of the Mediterranean Sea and was called the *Tethys Ocean.* This allowed the northern and southern supercontinents to develop separate plant and animal forms after a period of great similarity in the fossil record. The breakup of the supercontinents began back in the Triassic Period, with Gondwanaland fragmenting more severely. Measurements of the Indian subcontinent indicate it has been moving northward since at least the Tertiary Period. The continuing collision of this subcontinent with Asia is generating the greatest deformations occurring on earth today in the Hima-

The Grand Canyon

Erosion is responsible for many landforms, large and small. Probably none of these is so spectacular as the Grand Canyon of the Colorado River in Arizona. Crossing the relatively level Colorado Plateau, Spanish explorers suddenly found the ground drop away in a series of cliffs and ledges as much as 1 mi deep. Unable to descend, they walked back the way that they had come. Even today, over 100 road miles must be traveled to go from the South Rim to a point directly across from it on the North Rim of the canyon. It is still possible to approach within feet of the rim with no warning that one of the wonders of nature is at hand.

Geologically, the rocks of the Grand Canyon open up a record of life forms spanning 1 billion years of earth history. The Grand Canyon series of Proterozoic rocks record some hints of the earliest life forms known, while isolated hills of Mesozoic strata back from the rim take us to the age of dinosaurs. There are gaps in the record. Apparently the top of the Redwall limestone was dry land during the Pennsylvanian portion of the Carboniferous Period, only to be submerged and resume deposition in the Permian Period. Such an interruption in the rock record is called an *unconformity.* A lengthier time span is represented by the missing periods right after the Cambrian. Whether rocks were deposited and then scoured off again before the deposits present today began is not known. However, any vertical movement during the entire Paleozoic Era must have been gentle since the separate strata are parallel. Below these an even greater thickness of rock (over 2 mi thick) was deposited, tilted, and eroded as the base on which the Paleozoic rocks currently lie.

FIGURE 1
The scale of the Grand Canyon is difficult to imagine. The cliffs of the Redwall Limestone far below on the right average 55 stories in height, or the height of the Washington Monument. Note the people at the upper left. (*Photograph by the author.*)

The structure of the Grand Canyon is fascinating for a separate reason. In most places, water spilled near the rim will drain away from the canyon. Cape Royal right on the North Rim is the highest point in any direction for tens of miles. If the Colorado Plateau represents a gentle upward blistering of the surface, the river has chosen to cut the canyon through the very top of the upward bubble. The rock strata dip gently away from the canyon on all sides. This could have occurred only if the rising of the plateau was so slow that the river could cut downward as fast as the rocks rose. A river meeting this structure already formed would detour around the blister rather than slice through a mile of rock to form a channel down the middle.

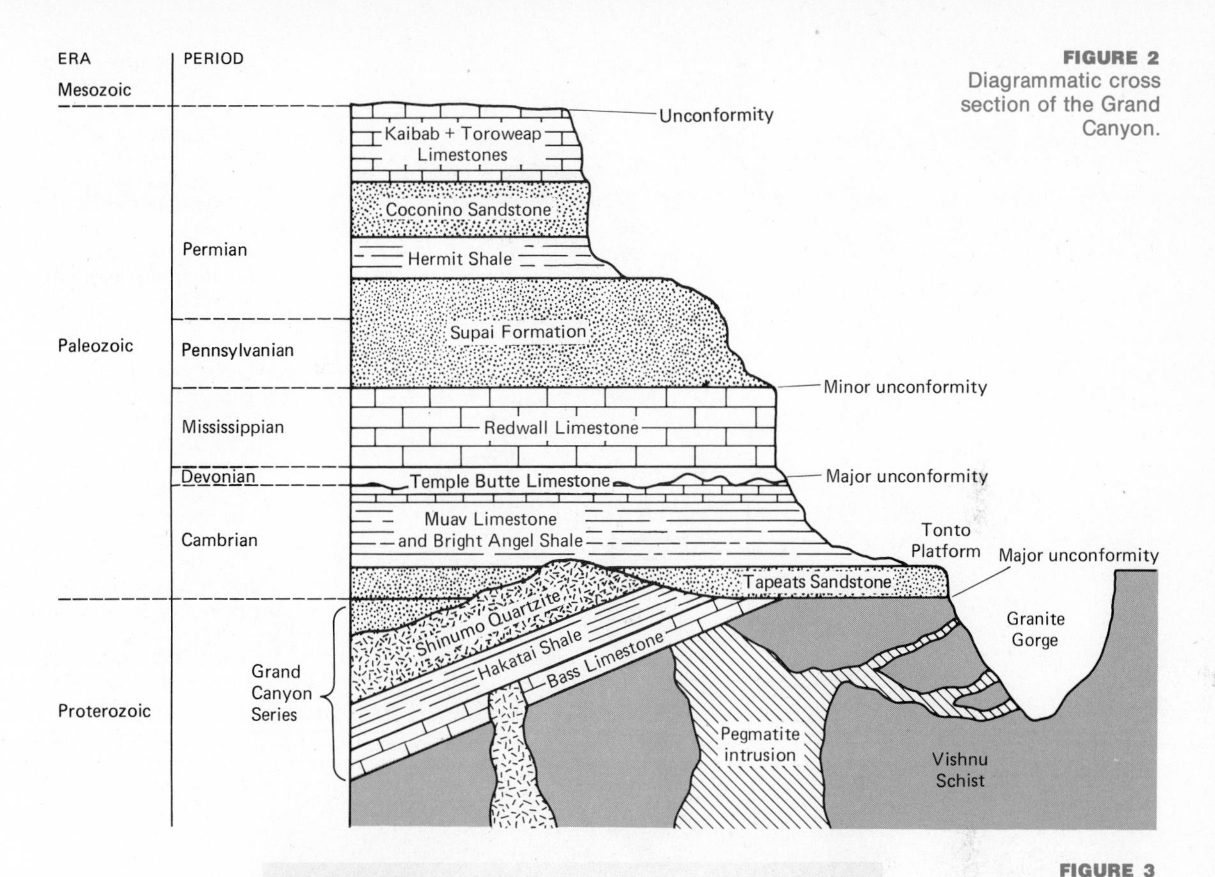

FIGURE 2
Diagrammatic cross section of the Grand Canyon.

FIGURE 3
Motion along the Bright Angel fault has raised the right-hand strata about 60 m (190 ft) compared to the same strata on the left. Although access to the canyon is difficult, erosion along the fault allows a trail to snake downward along its face. (*Photograph by the author.*)

FIGURE 16-12 The closure of the Atlantic Ocean as occurred during the Permian Period. Continents have been moved backward in the direction of their plate's motion. Dark zones are sections of overlap of two continents, while white spaces at edges indicate "missing" continental fragments. The fit is to the edges of the continental shelves rather than to the sea-level contour of the landmasses. The fit is phenomenally good, although Central America and the isolated continental blocks making up some large Caribbean islands (e.g., Cuba, Puerto Rico) have not been reassembled.

layas and associated ranges. Meanwhile a great crack has been opening up in Africa, called the great rift valley, as the eastern part is shearing from the rest of the continent. Back in our own hemisphere, the land bridge between the Americas seems to be a temporary accident in an extremely broken-up area of continental fragments.

Smaller Structural Features of the Earth

The development of the ideas of and evidence for plate tectonics is one of the most exciting developments in science in the second half of the twentieth century. *Tectonics* itself is the study of the structural deformation of the earth. While plate tectonics interprets the grand structural patterns of the crust, smaller structures are more easily appreciated. The earth's surface is constantly being worn down by weathering and erosion. It is also being built up by patterns of vertical movement of rock. Although most of these activities are associated with the larger picture of plate tectonics, we will discuss them separately.

The most direct way by which the surface of the earth may be built up is by the action of molten rock, or *igneous activity.* If at the surface, this activity is called *igneous extrusion.* The most obvious form of igneous extrusion is a volcano. Perhaps more important is the flow of liquid rock from a seam or rift as it occurs at the midoceanic ridges. There are numerous examples of this activity on land as well as underwater. When magma comes to the surface it is called *lava,* and the western United States preserves several large *lava flows.* As mentioned before, the Yellowstone Plateau is an extensive plain of yellow rhyolite filling a broad valley between mountain ranges. Rhyolite is a microcrystalline cousin of granite, having the same mineral composition but having cooled too rapidly to allow the crystals found in a granite to have formed. More frequently lava flows are formed from the dark rock basalt. One of the largest of such flows in the United States is the Snake River plain of southern Idaho (Fig. 16-13). Lava issuing from cracks at the southern edge of this plain flooded an extensive area about equal in size to most of New England to depths of over 100 m. Here and there in the plain isolated hilltops stick up out of the lava as reminders of a curved surface.

Better known than lava flows are the world's *volcanoes.* With a volcano the extrusion of igneous rock is confined to a single location. The name comes from the Romans, who imagined them as the vents of the underground forges of the god of the underworld, Vulcan. Actually, volcanoes vary widely in their characteristics. On the one hand, eruptions can be so benign that major activity in the Hawaiian Islands becomes a tourist attraction. Sightseers from all over the world book flights to watch Kilauea spouting fountains of molten rock, and try to get as close as possible to film the spectacular events. Yet, on the other hand, massive eruptions rank in the popular folklore as some of the most awesome and

FIGURE 16-13
The bleak lava landscape of the Snake River plain is typical of many large western areas flooded by lavas to depths of hundreds of meters. (*Photograph by the author.*)

powerful natural events known—and rightly so. The mightiest eruptions have produced statistics no mere hydrogen bomb could equal. The most violent in modern times, the eruption of Krakatoa in 1883, killed 300,000 people with tidal waves, was heard 3000 mi away, and produced shock waves in the ocean recorded all over Europe—half a world away. Additionally, volumes of fine dust were blown high into the stratosphere in such quantity that the next year was called the "year without a summer" because of the lack of sunlight. No one but scientists ever try to get closer to a volcano of the Krakatoa-type when they show signs of impending eruption; quite the opposite.

This extreme difference in the reaction of people to volcanoes is not irrational. It is based on the fact that there are three different types of volcanoes, which differ greatly in destructive potential. The main difference is one of dissolved gases in the rocks feeding a volcano. The smallest type of volcano is called a *cinder cone.* Cinder cones tap a pocket of magma underground rich in *volatiles,* or materials which are easily vaporized. As a result, the events which follow the development of a crack to the surface are easily understood. They are much the same as what happens when a bottle of warm soda pop is shaken violently and then uncapped. Like the magma, the soda pop is rich in dissolved gases. These are kept dissolved by the high pressure of their container. When that pressure is relieved, the gases form bubbles and rush upward and outward carrying the surrounding materials with them. The lava from a cinder cone is blasted upward into the air where it cools rapidly and solidifies. The resulting cinders, or *pumice,* is not so much a solid rock as a series of gas bubbles surrounded by solidified rock walls. The materials of a cinder cone are ejected straight up, solidified, and fall right back down. This dictates the characteristics of a cinder cone. It is a mound of loose debris stacked up at the steepest angle gravity will allow, the so-called *angle of repose.* Cinder cones are never very tall, since as they reach a critical height, the weight of the mound begins to crush the materials at the bottom. A typical maximum height is about 150 m. The eruption of a cinder cone is violent, noisy, spectacular, and not particularly threatening unless you are sitting near the vent. Cinder cones have very brief life spans; when the pressure of the magma pool is relieved, they become extinct.

The extreme opposite of the cinder cone is a volcano characterized by very little dissolved volatiles in its magma. This is the typical Hawaiian volcano called a *lava* or *shield volcano.* The magma of such a volcano quietly cools underground unless there is some immense pressure in the rock to be relieved. All of the Hawaiian chain of volcanoes has been generated by the sea floor traveling over such a pressure point. Only the southeastern tip of the chain now sits over the vent and is active. Completely unlike the cinder cone, this type of volcano is immense. Measured from its base on the ocean floor to its summit, Mauna Kea (on the island of Hawaii—the tallest of the chain) is over 1 mi higher than Mount Everest. However, height is the least remarkable attribute of a shield volcano. Almost all of the cliffs and sheer walls of the Hawaiian Islands are the result of erosion, not volcanic activity. Shield volcanoes never maintain a slope as great as even 10°! A typical eruption of a shield volcano proceeds as follows. First of all, the ground tremors and rumbling indicating mass movement underground begins. After this, the lava sheet sealing the main vent may melt and crack, revealing molten rock underneath. One area in a capped plain 1 km across may spout molten rock. Sometimes density differences will cause a rising column of magma to shoot into the air as a fountain, splashing on the nearby rock. As all this activity is going on, the magma underground is probing the flanks of the volcano for a weak spot at low elevation. If found, such a flank vent cracks open and a flow of hot, fluidy lava issues forth gurgling its way to the sea. Frequently, these vents are underwater to begin with. Once the necessary volume of lava has been vented, the volcano settles down to a troubled sleep. One of the safest places to be in such a minor eruption is on the rim of the volcano crater. Except in the most determined of situations the volcano will not have the potency to push molten rock up and over the rim of its crater. When an eruption with this determination occurs, run! Mauna Loa is the great active Hawaiian volcano, and Kilauea is a lesser vent on its side. Kilauea is the most consistently active volcano on earth. It takes a back seat only on the rare occasions when big brother acts up. In the absence of gas pressure, these volcanoes simply (literally) do not have the steam to be unpredictable. In their ceaseless activity throughout history, however, the volume of lava expelled to the earth's surface is incredible. Much less of the Hawaiian Islands shows up above sea level than the proverbial tip of the iceberg. If these islands were removed, the level of the oceans around the world would drop by about 200 m. It would be possible to walk from England to Russia without leaving dry land and without passing through any existing country.

The truly dangerous type of volcano combines both of these extreme forms and is called a *composite cone.* The composite cone begins its eruptions like a cinder cone and ends them like a shield volcano. After raining down cinders and ashes, molten lava gushes from the vent and cements the cinders together. Thus, composite cones have the steepness of the cinder cone with some of the great size of the shield volcanoes. Almost all of the famous volcanoes of the world are composites. Needless to say, composite volcanoes are dangerous and the prime type of spectacular volcano. This is because they have both the pressure buildup of volatiles and the molten rock to reseal the bottle after an eruption. Numerous composites around the world have literally "blown their top" as did Krakatoa in its famous eruption. The prime American example occurred when Mount Mazama blew off the upper mile of its cone to leave a pit now filled with the third deepest lake in North America. The walls of Crater Lake in Oregon drop down to the water surface anywhere from 50 stories to twice the height of the tallest building on earth. Yet, the water plunges to record depths beyond this drop. Klamath Indian legends might still recall this prehistoric eruption, which probably dwarfed Krakatoa's performance. An eruption of this type seems to be a

America becomes aware of its volcanoes

The Cascade Range stretches from Lassen Peak and Mt. Shasta in Northern California to Mt. Baker astride the Canadian border. The range includes over a dozen major composite volcanic cones which inhabitants regarded as part of the scenic backdrop of the area and little else. The only eruptive activity of the chain in this century had been some steaming, venting, and mudflows from the southernmost member, Lassen Peak, ending at the time of the First World War. This was sufficient to give the peak and its environs the status of a National Park. The long slide down its northern snow-slope was more than adequate compensation for the long hike to its summit for thousands of tourists. This pleasant record for Continental American volcanoes closed on the morning of May 18, 1980.

The San Andreas fault marks where North America is riding over the boundary of the great Pacific plate (see text). This fault ends at the boundary to a minor plate called the Gorda plate, running roughly from the California-Oregon border to slightly north of the Canadian border. The midocean ridge producing the Gorda plate is still hundreds of miles offshore in the Pacific Ocean. Where North America and the Gorda plate move toward each other in opposite directions the Cascade Range is born. As the continent depresses and overrides the advancing ocean floor, a subcontinental zone of large magma pools is generated. Some of these work their way to the surface by melting channels in the overlying rocks to form the Cascade peaks. A 1978 report by the United States Geological Survey nominated Mt. St. Helens, near the Washington-Oregon border, as a likely candidate for future activity. Although its last eruption (1854) had occurred well before the American Civil War, Mt. St. Helens has been the most explosively eruptive member of the Cascades since slightly after the pyramid age of ancient Egypt (about 5000 years). Before that, the mountain had displayed about 4000 years of inactivity. When is a volcano extinct?

On March 20, 1980, the first of several serious earth tremors was recorded under the mountain. Activity came and went for some two months. During this time an ominous bulging of the mountain occurred as if it were subjected to great forces from underneath. Then, on May 18 the northern flank of the mountain gave way in a great avalanche triggered by an earthquake. This abrupt release of pressure allowed water in contact with the hot rock below to flash into steam, splintering the top of the mountain and uncapping the magma. The resulting eruption, as the gases in the magma suddenly had the pressure of their rock cap removed, was hundreds of times greater in explosive power (several megatons) than the bomb destroying Hiroshima (10 kilotons). Within 10 km all vegetation was scoured clear of the landscape. Some 21 km from the peak, a truck was flipped over and its plastic parts melted in the blast. Many cubic kilometers of the top of the mountain were completely blown away. Thousand-ton boulders were thrown for miles and fine rock dust scattered across a dozen states. The original avalanche poured down with such vigor that it crossed a valley, climbed over a 200-m ridge on the opposite side, and continued on its way. Walls of water set up in nearby lakes flooded valley walls over 100 m off the lake surface. Pulses of these waves surged down river channels, tearing apart settlements tens of kilometers from the eruption.

Since this eruption, Mt. Hood has shown some signs of activity. In the preceding decade both Mt. Shasta and Mt. Baker have shown minor activity. The Cascades as a volcanic range are very much alive. The settlement pattern of the Pacific Northwest has largely discounted the range as a significant threat due to their long quiet period. Today, Mt. St. Helens remains the likely candidate for major explosive eruptions based on its actions of the last few thousand years. However, any of the Cascades could come to life with similar suddenness. The basic cause for the volcanoes still exists and remains active in the motion of crustal plates. Future eruptions along the range are inevitable. Recent history suggests that the earth averages an eruption on the scale of Mt. St. Helens every 10 years or so. By good fortune, none of these were in the Cascades during the settlement of the Pacific Northwest. How long such good fortune can continue is anybody's guess.

constant possibility with composite volcanoes thought to be extinct. After Vesuvius blew away much of its cone, burying Pompeii and Herculaneum, it was active for 1000 years. Then, it entered a period of 500 years of rest before it again resumed activity in the seventeenth century. The volcanoes of the Cascade Range of the American west coast haven't been on record for half the quiet period of Vesuvius. The possibility always exists that Mount Rainier will make Seattle a footnote in American history much as Pompeii has become in Roman history.

Igneous Intrusion

Liquid rock does not have to come to the earth's surface to form features apparent in the rock today. Even volcanoes leave remnants of themselves when they are long gone. The molten rock that solidifies in the vent of a volcano slowly cools under great pressure, which makes the vent much more resistant to weathering than the surrounding materials. The pressure withstood by such vents is indicated by the fact that they are the only known bedrock source of diamonds. South African diamond mines dig straight down such huge vents. If molten rock does not come to the surface to form the features we see, the process is called *igneous intrusion* rather than extrusion. Volcanic vents represent igneous intrusions later exposed by erosion. Perhaps most famous is Devil's Tower in Wyoming. One of the most impressive of these is Ship Rock in New Mexico, which rises like a skyscraper out of the flat desert about it. Radiating from its central spires in several directions are long ridges of similar materials called *dikes*. A dike results when magma is injected into a vertical crack in the rock structure and cools there. The dikes of Ship Rock are some of the most dramatic in the world, running like 10-story walls across the desert and blocking passages on the desert floor. After erosion removed the original rocks that contained them, the dike stands erect, making it obvious how it got its name. Similar injection of magma not across rock layers but between them where they join is called a *sill*. This type of horizontal igneous sheet may erode much more slowly than surrounding rock. As such, sills support numerous waterfalls and rapids. Thus, the origin of this name is also apparent.

Volcanic necks, dikes, and sills all tend to be reasonably small features on a geologic scale. The action of molten rock under the surface makes much larger structures. The central United States is basically flat when viewed on a large scale. The exception you might note to the latter is a distinct "bump" or "blister" out in the great plains where it does not seem to belong. This describes the Black Hills of South Dakota. The Black Hills represent the injection of a large amount of magma under the surface rocks of the earth, causing the surface to bulge upward in a gigantic blister (Fig. 16-14). This structure is called a *batholith*. A similar, but smaller, structure occurs when magma is injected between layers of rock with enough force to buckle the upper covering rocks in a structure called a *laccolith*. Either of these structures bends and

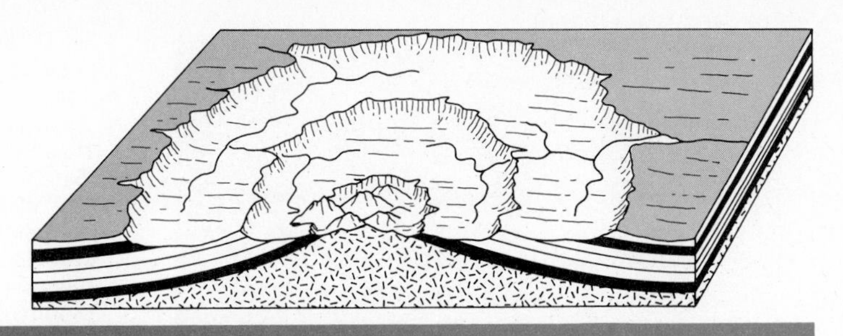

FIGURE 16-14 Block diagram showing the general structure of the Black Hills of South Dakota. Sedimentary rocks have completely eroded off the central part of the dome, exposing the granites into which Mount Rushmore's figures were carved. More resistant strata maintain elliptical ridges about the center. The broad valley between ridges is so well formed it is known as "the racetrack" even though a single lap is over 100 km.

stretches the rocks above them. Rocks may be incredibly resistant to crushing, but they have limited strength to resist breaking when being stretched. A structure such as the Black Hills horribly fragments the rocks covering it. This makes these rocks subject to accelerated weathering and erosion. As a result, when we look at Mount Rushmore in the center of the Black Hills, we do not see the covering layers of the batholith. Instead, we see the granite formed from the magma injected to produce the entire structure in the first place. All of the cap rocks, along with much of the granite, has been carried away. The carving of the surface of Mount Rushmore only aided a process that has roots much longer in time than the existence of the human race.

Folding and Faulting

Not all of the vertical building of the earth's surface is a result of the activity of liquid rock. Much of it results from the deforming of the surface due to forces transmitted through the rock in the twin processes of *folding* and *faulting.* If motion does not occur along a crack in a rock, it is called a *joint* in the rock. If motion does occur along the crack, it is called a *fault* (Fig. 16-15). At the simplest level, one block may rise and the other fall to create a ridge. Such a fault generated what is perhaps America's most photogenic mountain range—the Grand Teton Mountains of Wyoming. The mountains rise in a dissected wall from the valley floor with no foothills to soften the contrast in elevation (Fig. 16-16). Motion along a fault may also be horizontal, producing a *transcurrent fault,* such as the San Andreas fault and the transform faults of the ocean floor. Faults rarely occur singly. If a block between two parallel faults drops or sinks relative to its sides, the structure is called a *graben.* The background in Fig. 1-3 shows one wall of the Nile graben. The opposite process may occur where the detached block rises between the faults. This structure is called a *horst.* Both of these structures are common in the great structural system east of the Sierra Nevada of California. This system is a classic case of a large structure determined primarily by motions along faults. Imagine a row of books standing on a desk top. The lines between separate books represent faults. If the stack is pushed sideways, it slips so that each book slides against its neighbors to produce a slanted stack. Make the books of various thicknesses and dif-

FIGURE 16-15 A crack in the rocks with no motion along it is called a fault, and the types are shown. A transcurrent fault where crust is not conserved, as at midoceanic ridges, is called a transform fault.

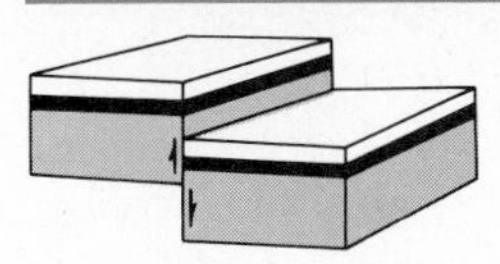

(*a*) Normal fault

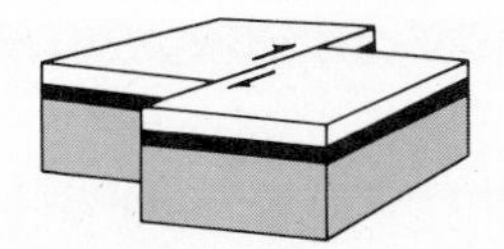

(*b*) Transcurrent fault

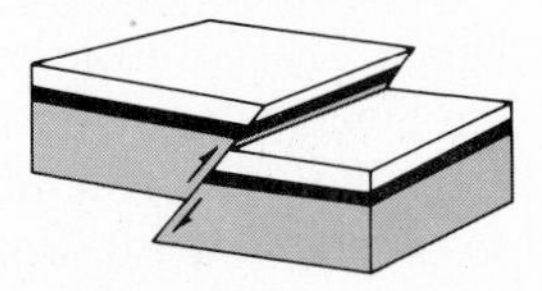

(*c*) Thrust fault

(a)

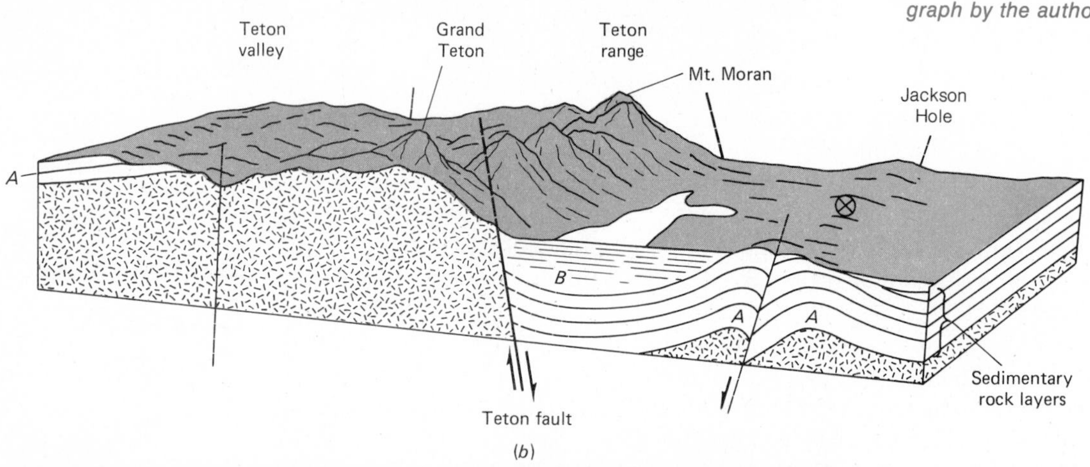

(b)

FIGURE 16-16 A fault-block mountain range (*a*) shows considerably more detail than the block of Fig. 16-15*a*. The Teton Range of Wyoming was generated by the uplift of a block of Precambrian granite along the Teton fault. In diagram (*b*), layer *A* represents the oldest sediment, of Cambrian age. Almost a mile of sediments, including *A*, have been eroded from the top of the Teton block. There is no way of estimating the amount of granite removed. Even so, the Grand Teton comes within 15 m of being the tallest mountain in Wyoming. The area (*B*) in front of the range has been filled with eroded materials. ⊗ marks the position where the photograph was taken. (*Photograph by the author*.)

ferent heights (so that there may be a separate up or down sliding) and you have a rough model of the structure of California from the Pacific Coast through the Sierras into the state of Nevada (Fig. 16-17). The best-known block with the downward slippage of a graben now holds Death Valley the lowest area on the continent. The lowest spot on earth, the Dead Sea, also rests on top of a sunken graben.

One additional type of fault is worth mentioning. This occurs when the crack in the rocks is at a very shallow angle or *dip* to the horizontal. Forces on the rocks can then cause one block to ride up and over the sur-

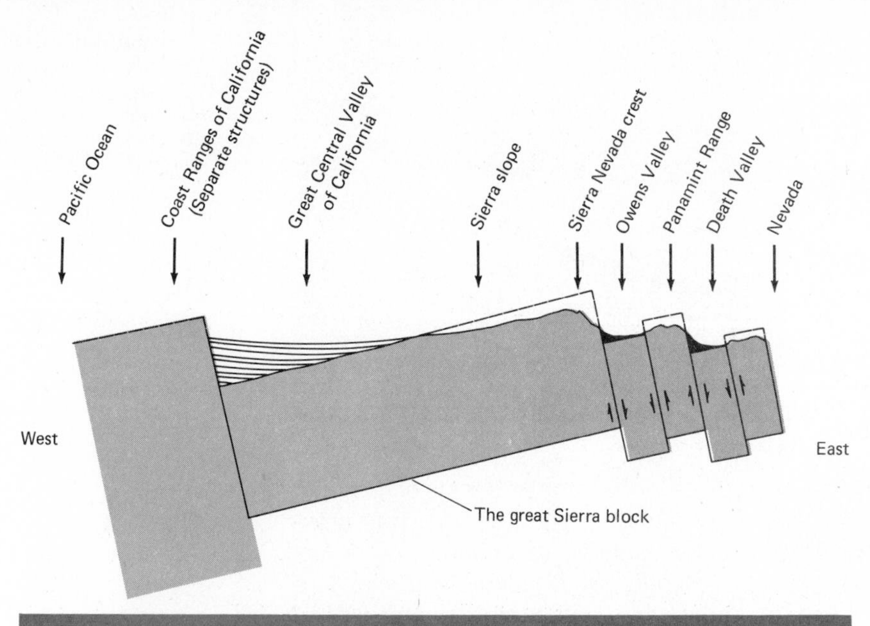

FIGURE 16-17
A highly idealized section from the Pacific Ocean through California and into Nevada. The great Sierra block dominates much of the structure of California. Sediments in the San Joaquin and Sacramento Valleys bury the western edge of the block to a depth of over 5 km. The highest point on the Sierra crest is at Mount Whitney, highest mountain in the contiguous United States. From it you can look to the sunken block of Death Valley, lowest spot in North America. The western slope of the Sierra block rises gently from the valleys of California, while the eastern edge plunges steeply into Owens Valley.

face of the other in the structure known as an *overthrust.* Motion along an overthrust is frequently dramatic, reaching tens of kilometers. A prime example is in the Glacier National Park area of Montana. Here, one isolated mountain remains sitting on the great plains detached from the distant wall of similar rocks miles away. This would not be unusual except the Chief Mountain represents a series of older sediments sitting on top of a younger base. This is the type of unexpected phenomenon which the tourist would not notice but which receives great attention from geologists. Actually, the entire front wall of the Glacier Park area has been thrust out over the plains in the Lewis overthrust (Fig. 16-18).

FIGURE 16-18
The entire surface of Glacier National Park, shown here, has been moved out over the plains of Montana in the Lewis overthrust. This zone of thrust faulting is some 500 km long and represents a movement of the block with the mountains from 20 to 50 km eastward. Note the glacially produced U-shaped profile of this valley. In the distance cirques may be seen. (*Photograph by the author.*)

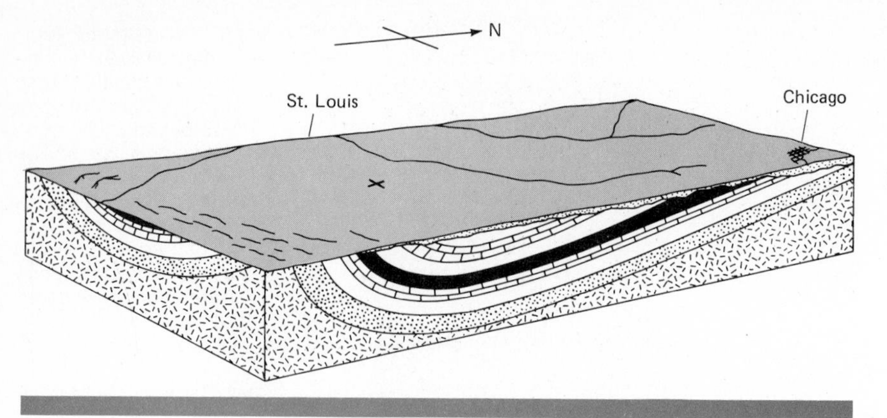

FIGURE 16-19 The Illinois Basin, a much larger structure than a single range, affects the human uses of the area. The *x* on the block marks the approximate area where each stratum attains its greatest depth.

Rocks do not have to break to be deformed. Large sheets of rock can be quite flexible when placed under considerable pressure. This pressure can then warp or *fold* the rocks in a variety of ways. Imagine a loose carpet laid over a hardwood floor. If a child with a running start makes a fast stop on the carpet, the forces generated will stretch the carpet behind the child and cause the carpet in front to ripple upward in a series of waves. The behavior of sheets of rock may be similar. The main difference between the rug and the crust of the earth is that rocks are normally hard, relatively incompressible, and brittle. Laboratory studies suggest that rocks under suitable pressure for long periods of time can be quite plastic, as is the ice under a glacier. The folded features which are observed in a wrinkled carpet may also be observed in the rocks. Two large-scale features are the *dome* and the *basin.* Michigan Dome was described in the last chapter in connection with the formation of the Great Lakes. Features of this scale are indicated by a gentle dip in the rock layers in all directions from a central area. A companion structure would be the Illinois Basin (Fig. 16-19). This is a bowl-shaped formation filled with layers of sedimentary rock. These layers do not thin out greatly with distance from the center of the basin. We may thus infer that the basin was produced by folding after the rock layers were deposited. Each stratum curves uphill in every direction from the center of the basin, and this structure determines the pattern of coal mining in the state of Illinois. The coal-rich beds, indicated in black in the diagram, rise gently to the surface in the northern portion of the state, ideal for strip-mining. At the southern end of the state the same strata plunge steeply into the ground and require underground mining. Oil and gas is generally found where these beds attain considerable depth. More importantly, the structure of the rocks determines the availability of well water in the state. Water-bearing strata, or *aquifers,* follow the general trend of the structure. The lowest of these comes to the surface under Lake Superior and is thus well supplied with water at its northern end. The movement of water through porous rock is very slow, probably requiring over 1000 years to trickle under Wisconsin to wells in Illinois. The problem arises in the population growth and industrial expansion of northern Illinois (Chicago itself uses water from Lake Michigan). Increased pumping in the northern part of the state has lessened the deep well water available in the central portion of the state. There, wells have had to be deepened or drilled into other aquifers to maintain a water supply. All over the world, underground structure strongly influences the pattern of water availability. Development must take these patterns into account or serious shortages can develop. (As another example, Long Island in New York depends upon surface seepage in the central area of the island to provide the water for the wells sustaining its suburban population. Development of this area threatens the water supply. Storm sewers, paved parking lots, homes, roads—all remove water from the collecting area for seepage.)

Domes and basins tend to be large features which can be inferred only by collecting data over a large area. However, the processes of folding work from this scale down to small folds in the rock less than 1 m across. The most common forms are a simple upward ripple called an *anticline* and a similar downward fold called a *syncline* (Fig. 16-20). Between these is the *monocline,* which can be thought of as half of either of the other two forms. A reasonably isolated example of a large anticlinal fold is that of the Bighorn Mountains of Wyoming. Approaching from the east or the west rock layers tip upward toward the crest of the range. Nearing the top on either of the roads through the range, the traveler comes upon a parklike upland which is reasonably flat where the cap rocks have been worn away. From this area, the innermost part of the fold rises to the summits of the range. The western United States has many mountain ranges formed from a single anticlinal fold. It is in the east, however, that folding attains its greatest prominence in shaping the landscape. The Appalachian mountain chain is a series of ridges and valleys formed by rippling the rocks off the eastern coast. This appears a

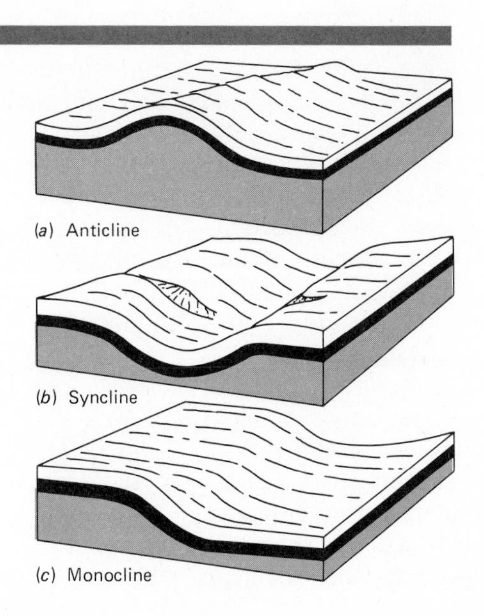

FIGURE 16-20 The types of folding of rock structures are shown.

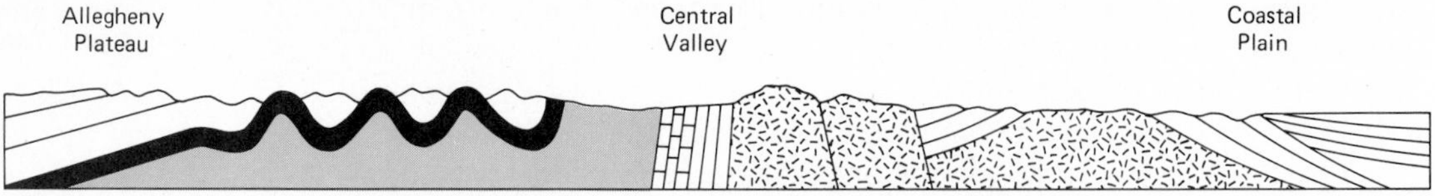

Allegheny Plateau

Central Valley

Coastal Plain

FIGURE 16-21 The structure of the Appalachian region is very complex, as shown by this simplified slice from Pittsburgh, on the Allegheny Plateau, to Washington, D.C., near the coast. Strata east and west of the central valley are not similar. Could you work out a sequence of steps that would account for such a structure?

bit like the rippled carpet after a fast stop, with an alternation of anticline following syncline across the area (Fig. 16-21). One very major difference exists which is really remarkable at first glance. The Appalachians are a much older range than those in the western part of the country, thus weathering and erosion have been working longer. The rocks at the top of an anticline are inherently weakened because they have been placed in *tension*. Large-scale jointing may be expected, and the increased elevation also aids in the destruction of these rocks. Synclines are different. During formation of a syncline the rocks at the bottom are put under great *compression*. This seals joints in the rocks and generally hardens them and makes them less susceptible to attack by weathering. The Appalachians have been exposed long enough for the implications of these differences to have reached their logical conclusions. Ridges in the Appalachians frequently are held up by rocks that were at the bottom of synclines in the original structure. Likewise valley areas are frequently the deeply eroded summits of anticlinal ridges. In general, the Appalachians represent an old system of mountains probably formed when North America collided with neighboring continents to form Pangaea. The rock structure is not so obvious on the ground partly because of the geologic age and partly because the dense vegetation does not allow the layering in the rock to be seen easily.

Summary

The earth represents a two-surface planet with high plateaus of light rock, called *continents,* floating on a base of denser rocks making up the *ocean basins*. The height at which a large rock mass will stand above the surface is determined by the principle of *isostasy*. This asserts that each large column of rock, measured to a constant depth well below the crust, must have the same weight. Large-scale deposition of sediments occurs mainly in structures called *geosynclines.* These are large enough to depress the floor of the shallow sea they represent as sediments are deposited. The evidence of earthquakes indicates that the earth is not uniform throughout its thickness. The outermost layer is the *crust,* which ends abruptly at the moho (Mohorovičić discontinuity), which averages 33 km deep under the continents and one-sixth of this under the ocean floor. Beneath this is the area called the *mantle*. The upper mantle retains some rigidity, since earthquakes occur in it to depths of 600 km. Earthquakes result from the abrupt movement of one mass of rock past another and would not occur in a plastic or liquid material. Earthquakes generate both a rapid longitudinal wave called a *P* wave and a slower transverse wave called an *S* wave. At 2900 km of depth is the *core* of the earth, which is probably liquid. Deeper still, at 5100 km or some 600 km from the center of the earth, earthquake waves indicate a change to the *inner core*.

The pattern of continental platforms on

earth is not static in time, but all are in motion relative to each other in what is termed *continental drift.* The crust is not a single, solid piece, but is composed of moving *plates.* This implies that somewhere crust is being destroyed and at other places it is being produced. New crust is continually being formed at *midoceanic ridges* and being destroyed at *oceanic trenches* or *subduction zones.* Most of the dramatic geologic activity on earth is associated with the margin of these crustal plates. The geologic movements of the West Coast along the San Andreas fault occur where North America has overridden an oceanic ridge, producing new crust. A portion of the continent has thus been deposited on a new plate and is moving off in a different direction at a new speed.

At a smaller level, the surface may be modified by a number of processes. *Igneous extrusion* is the spilling of lava onto the surface of the crust. This occurs in *lava flows* and in association with *volcanoes.* Volcanoes differ mainly in the amount of volatile substances dissolved in their magma, which gives rise to their separate characteristics, from the huge but tranquil *lava* or *shield volcanoes,* to the small but violent *cinder cones.* Between these extremes is the group containing most famous volcanoes, the *composite cones.* Composites may be steep-sided and have violent eruptions like cinder cones, but they may grow to great size approaching the shield volcanoes. *Igneous intrusion* injects liquid rock into or under the crustal rocks without reaching the surface. This can produce blistering of the surface in *batholiths* or smaller *laccoliths.* Magma may be injected between strata to form *sills* or into joints across strata to form *dikes.* The surface is also rearranged by the bending and breaking of crustal rocks in the processes of *folding* and *faulting,* respectively. *Normal faults* involve vertical motion of the separated blocks, while *transcurrent faults* involve horizontal motion. If no motion occurs along a crack in the rock structure, it is called a *joint.* A similar form to the normal fault is a very low-angle fault along which one block is pushed to override the other. This is the *thrust fault,* frequently involving motions in the tens of kilometers. A similar form to the transcurrent fault is that in which the horizontal motion results in the creation and/or destruction of crust. This is called a *transform fault* and occurs mainly in the plate boundaries near midoceanic ridges. Simple folding consists of a single upward arch called an *anticline,* a single downward arch called a *syncline,* a half-arch formed called a *monocline,* and combined forms. A complex combination of folds is most apparent in the Appalachian Mountains. A larger bending of the rocks includes the broad swelling of the *dome* or the *basin.* From the center of a dome, rock strata dip downhill in all directions, unlike the anticlinal structure which has a linear, ridgelike trend.

CHAPTER SEVENTEEN

The Atomic Nucleus

Introduction In the last several chapters we have jumped a half-century ahead of the development of science to present two of the new fields emerging in the second half of the twentieth century. Effectively, our story was interrupted at Arrhenius, Maxwell, Mendeleev, and Faraday. The fundamental feeling in the late nineteenth century was one of supreme confidence and satisfaction with science. True, a few nagging problems remained here and there, but the guidance of Newton's ideas, with some help from Maxwell's work, would eventually lick them—wouldn't it? It was a time when leaders who should have known better described the future of physics as a mopping-up operation. Albert Michelson, who was the first American to win a Nobel Prize in science (1907), said, "It seems probable that most of the grand underlying principles have been firmly established . . . the future truths of physical science are to be looked for in the sixth place of decimals." From our viewpoint, this was about as timely a remark as that of Queen Marie Antoinette to the nobleman complaining that the peasants had no bread. Her remark, "Let them eat cake!" survived much better than the queen. Even philosophers succumbed to the image of a vast "Newtonian world machine." If we are given the position and momentum of every particle in the universe at one instant, does this not imply that we can work out past and future? What is the future but the consequences of the present? If momentum is conserved and energy is conserved, can we not work out what must happen in an interaction of matter? What is life but a complex train of interactions of matter? Is thought itself anything but a pattern of molecular rearrangements in the neurons of the brain? Then is not everything predetermined by the condition of the molecules in the past? The philosophy of mechanistic determinism weighed on the human spirit. Small wonder that the runaway bestseller of the period was the masterful translation of the poems of an obscure Arabic astronomer by Edward Fitzgerald. *The Rubaiyat of Omar Khayyam* appeared in 38 editions in the United States alone by the end of the century and most editions were reprinted over and over. The fatalism of *The Rubaiyat* struck a responsive chord. The "We have studied the universe and we have won" of the scientists translated philosophically to "We are powerless, will-less cogs in the machine of the universe." *The Rubaiyat* makes the point with chilling clarity:

The Moving Finger writes; and, having writ
Moves on: not all your Piety and Wit
    Shall lure it back to cancel half a Line,
Nor all your Tears wash out a word of it.

[5th Edition; quatrain #71, 1889]

But, be of good cheer, it all was an illusion. Science was passing through its sophomore year. The classical period of science was drawing to a close and ahead lay revolution and turmoil from which we emerged a bit wiser and a lot less naïve about the transparency of nature. Like Bilbo Baggins, science had slain the dragon and brought home a few trinkets in the process. Little did it realize what complexity had been generated for the next round of Baggins. The seeds for a profound change in science had been planted in the last decade of the century. Some of the problems which seemed solvable but which would require a reworked science were:

1 The orbit of Mercury will not conform to the predictions of Newtonian mechanics.

2 Why should incandescent gases radiate only certain, very specific colors of light in producing their spectrum?

3 The speed of light is a constant independent of the motion of the earth through space.

4 Maxwell's equations along with Newtonian mechanics seem to imply that the energy radiated by an object hotter than absolute zero should be infinite, which is absurd.

5 What properties of the elements permit them to be arranged into a periodic table?

Actually, this list could go on for some time, but by now you have the idea that the science of the late nineteenth century had some serious problems. Science at all times has serious problems, but these were fundamental problems, requiring changes in the basic way in which we view the natural world. In all, a new world for science was in the offing, but one in which the uses of science were to dominate its quest for knowledge in the public mind (and life-style).

A New Form of Radiation

Little trinkets are sold today which will glow after the room lights are turned out. These are particularly popular for children's rooms. The phenomenon is know as *phosphorescence.* The material absorbs light while the room lights are on or while the sun is shining and then slowly radiates this energy back in its own characteristic glow. This effect was very popular in the late nineteenth century. Proper Victorian homes used wallpaper with designs in phosphorescent materials. The cool glow of the phosphors after the sun went down was very stylish. (Unfortunately, many of the phosphors used were also very poisonous chemicals. This was found out after the Victorian mansions became tenements and babies began eating the crumbling wallpaper.) The ability of certain forms of matter to absorb and then reemit radiation was intimately tied to a series of changes in science. Cathode-ray tubes produced a stream of negative particles which J. J. Thompson called "electrons." The electron stream in such a tube is invisible in a very high vacuum. If a small amount of air is left in the tube, the air will fluoresce under electron impacts, rendering the stream of particles visible. A superior way to render the beam visible was to implant a screen in the tube coated with a phosphor which would glow brightly when struck by electrons. We still do this on the inside surface of TV picture tubes to produce the picture from the electron impacts inside the tube.

Wilhelm Roentgen (1845–1923) was experimenting with his cathode-ray tube when he noticed a nearby fluorescent screen glowing. The screen was not inside the tube, nor was it in the path of any electron stream. The screen fluoresced whenever the cathode-ray tube was turned on in its vicinity. A foil of aluminum reduced the glow only slightly, while heavy metal sheets, such as lead, stopped the glow entirely. The most fascinating effect occurred when Roentgen inserted his hand between tube and screen. He saw on the screen a shadowgram of his hand in which the bone structure under his skin showed clearly. Undoubtedly, some very penetrating form of radiation was being given off by the cathode-ray tube. These radiations were unlike any kind known and Roentgen named them *x-rays.* Within a few months he had discovered most of their important properties. The *x-ray* was a massless, chargeless ray, like light. Also like light, the x-ray would expose photographic film wherever they struck it. Unlike light, the x-ray would expose the film right through all of the normal light-tight film containers used to store film. This ability to penetrate normal matter was the most unique property of the x-ray, which was to find rapid application in medicine. Roentgen's paper of 1895 created an immediate stir.* However, as unexpected as this discovery was, it was not unreasonable, and did fit into what was known about nature. The x-ray would soon come to be viewed as electromagnetic radiation of a much shorter wavelength than light, shorter even than invisible ultraviolet rays. However, only months away from the announcement of the discovery of x-rays was another discovery which would not be absorbed so easily by classical science.

Rays from Rocks

Henri Becquerel (1852–1908) was a professor of chemistry at the Sorbonne (the University of Paris). He decided to determine whether Roentgen's x-rays were given off by phosphorescent materials. Each day he would place one or more rock samples from the large collection at the Sorbonne on a windowsill where they could sit in bright sunlight all day. At the end of the day he would place the exposed rock on top of a sheet of photographic film. At this point one of the more famous accidents of science occurred. It was a cloudy day! Becquerel dropped the rock he was going to expose into a desk drawer on top of the wrapped film he was to use later, and forgot all about them. Much later, he opened the drawer to rediscover the sample which had never been tested still sitting on the wrapped film. He placed the sample in line for its turn at the windowsill and was about to throw away the film. Becquerel decided to develop the film to see how well this batch of film was lasting.† To his surprise, he found on the film a picture of the rock, even though the film had never been unwrapped from its light-tight container while in the drawer! Here was a totally new and unexpected discovery. He was forced to conclude that the rock, called *pitchblende,* was giving off a ray that penetrated the dark paper wrapping of the film. A chunk of freshly mined pitchblende which had never been exposed to strong light would show the same effect. Realizing that he had stumbled onto a totally new process, Becquerel gave it a name. This spontaneous emission of penetrating rays from matter was called *radioactivity.*

Pierre and Marie Curie began to study radioactivity. Pattern emerged rapidly. Only those rocks which contained one or both of two

* Not the least of the public excitement over x-rays was the indignation of proper Victorian ladies that they were about to lose the privacy of their homes and clothing. This type of speculation reached ridiculous extremes before it became clear in the public mind just what x-rays will and will not do. As soon as you have seen your first medical x-ray it becomes clear that the pornographic market is not likely to get much competition from this source. The only carryover from this naïve, hysterical picture of the implications of x-rays is in the Superman comics.

† When you look at a Matthew Brady photograph of the American Civil War, you should recall this fact. In Brady's day film was made by precipitating chemicals on a glass plate and had to be used and developed before it dried out. When Brady went into the middle of a battle to take a picture, he had to be a few steps from his darkroom wagon. There, he made the film, put it in a light-tight holder, rushed to his camera outside, took the picture, and rushed back to develop it. This requires steady nerves if you are within artillery range. Photography ceased being a spectator sport when the chemist George Eastman stabilized a film that would withstand drying and deposited it on flexible cellulose. Even then, it probably would have grown slowly if Eastman had not teamed up with a businessman down the block in Rochester, New York, by the name of Kodak.

elements were found to be radioactive. The elements were *thorium* and *uranium,* the two heaviest elements found in nature. (Not the two densest elements, the two heaviest. Some types of atoms—for example, gold—pack tighter to make a cubic centimeter weigh more.) This work was speeded up when it was found that a radioactive material could cause a charged electroscope nearby to lose its charge. Furthermore, the rate at which the electroscope lost charge was a measure of the strength of the radioactivity.

An extremely interesting point emerged from measuring the radioactivity of thousands of specimens. While all (and only) rocks containing uranium or thorium were radioactive, the original material, pitchblende, an ore of uranium, was the most radioactive one found. Furthermore, other radioactive rocks displayed a level of radioactivity which was directly proportional to the amount of uranium or thorium they contained. Thus, if a chemical analysis showed that one rock had 5 times as much uranium as another, measurement with the electroscope would show it to be 5 times as radioactive. The Curies concluded that radioactivity was a property of particular elements only. Pitchblende was different. It was much more radioactive than its uranium content would have predicted. (1) If radioactivity is a property of elements only; (2) if the only element showing up in a chemical analysis of pitchblende, which was radioactive, was uranium; and (3) if the amount of uranium in pitchblende was too small to account for the level of radioactivity observed, what would you conclude? Take a minute, since the conclusion the Curies were forced to is a bold jump which put them on their way to the Nobel Prize. Ready? The Curies concluded that there must be an undiscovered element in the pitchblende in such small amounts that a standard chemical analysis simply missed it. Furthermore, to have such a strong effect in such small quantity, the new element would have to be much more intensely radioactive that either uranium or thorium. The Curies set out to isolate a new element. After 4 years of hard labor, they isolated not one but two new radioactive elements, *polonium* and then *radium.* Each ton of pitchblende when carefully chemically separated yields a speck of radium smaller than a sand grain. Radium was so incredibly energy-rich it created an immediate sensation. In a dark room the radium gives off a pale blue glow and is visible. It would take 4 tons of pure uranium metal to equal the radioactivity of a single kilogram of radium. Day and night the radium gave off the radiations which could expose film or discharge electroscopes. How can a lump of an element radiate energy day and night in huge amounts? Where does this energy come from? It was like possessing a magic piece of coal that burned day and night without being used up. The image of an inexhaustible energy supply suddenly loomed in the imagination.

Actually, the radium was not really inexhaustible. Careful measurement showed that it was slowing down its output with time. Radium *decays* so that after a period of 1600 years it has lost half of its radioactivity. After another 1600 years it would lose half of the remainder, and so on. This period of time, 1600 years for radium, is referred to as its

The ladies have their day

AIP Niels Bohr Library

Marie Curie

Marie and Pierre Curie (shown above) had the only family that could sit down to breakfast and have a major gathering of past and prospective Nobel laureates. Among them, the family ended up with four Nobel Prizes. Pierre Curie and Marie Sklodovska Curie received the Nobel Prize in physics in 1903. Pierre was killed in a traffic accident 2 years later. However, Marie went on to win the Nobel Prize in chemistry in 1911. Not until 1972 did one person receive two Nobel Prizes in science again: John Bardeen in physics in 1956 and 1972, and Frederick Sanger in chemistry in 1958 and 1980. The Curie's eldest daughter, Irene, went on to win the Nobel Prize in chemistry in 1935. Madame Curie and Irene became the first parent-offspring pair to win separate prizes. This achievement went unmatched by men until Aage Bohr matched his father's 1922 prize in physics with his own in 1975.

Madame Curie was one of the most extraordinary individuals of this century. Higher education generally was denied to women, and the Sorbonne in Paris was almost alone in Europe in admitting female students. Marie worked as a governess to send her sister to France for study. Following this she left her native Poland to complete her studies in Paris. The hardships she overcame, which included abject poverty and male prejudice, offer a study in tenacity. After her marriage to Pierre, they began the attempt to isolate radium. The only space they could find was a shed previously used by the medical school for the storage of cadavers, and which had been abandoned because it was considered too run down for that purpose. With backbreaking labor they crushed tons of ore and proceeded with a careful separation.

Marie had always been an ardent Polish nationalist. At that time, Poland was under the military occupation of Russia. The Russians had been making a determined effort to erase any memory of a separate Polish state. Teaching in Polish was forbidden. Books and newspapers were allowed to be printed only in Russian. All of her life, Marie had fought against this oppression, sometimes too openly. In one bold stroke Marie Sklodovska Curie did more to reassert the legitimate right of Poland to remain on the map and in the minds of people than did the leaders of a generation. The Curies discovered a new element and named it polonium. From every science classroom in the world the name of the oppressed country would stare down at each new generation from its place on the periodic table. The Russians protested at once, and continued to do so until their government fell in the revolution of 1914.

An extreme irony occurred after the discovery of radium. Almost immediately it was found to have an effect on living things. Pierre carried the first sample in a sealed vial in his vest. He developed a skin rash, then sores near the vial. Tests showed that radium left near living cells destroyed them in time. At once, radium found a use in some cases of inoperable cancer. If placed in a tumor, the radium would destroy the cancer cells and could then be removed. The price of radium skyrocketed to over $100,000 per gram (over $1 million for a paper-clip mass if you correct for inflation). All of a sudden, Marie, who had won two Nobel Prizes for her work with radium, was priced out of the market and did not have access to any of the precious stuff to continue her research. Hearing of this, several women's groups in America banded together in a national drive, and in 1921 presented Mme. Curie with her own gram of radium for research. She accepted the gift but gave title to her Radium Institute in Paris so that it would be preserved for research.

One legacy of radium could not be given away. Both Mme. Curie and her daughter Irene spent their adult lives working around intensely radioactive substances. Both developed and died from leukemia in their 60s. During her life she had suffered resentment and vilification from male prejudice. In less than a generation after her death in 1934 she was given a different accord when the newly produced ninety-sixth element was named curium.

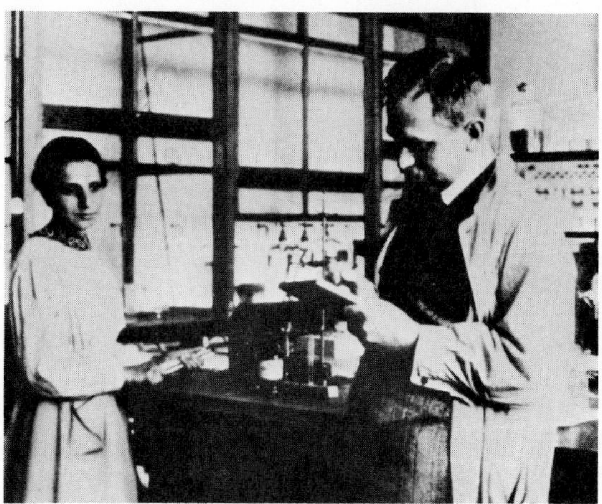

Otto Hahn, "A Scientific Autobiography," New York, Charles Scribner's Sons, 1966

Lise Meitner

While Mme. Curie managed career and family at the same time to the apparent detriment of neither, other female scientists of this century have chosen other paths. Opting for her career, Lise Meitner (above) worked in collaboration with Otto Hahn for a generation. She had left Germany and was in Sweden in 1939 when she pieced together the process of nuclear fission from the data she was working with in Germany.

AIP Niels Bohr Library. Stein Collection

Maria Geopport-Mayer

Unlike either Marie Curie or Lise Meitner, Maria Geopport-Mayer chose to interrupt her career to raise a family. Shown with her first child (above), she remained in contact with theoretical physics through the child-raising years via her husband's work. Beginning her research studies again after her children were grown, she went on to win the 1963 Nobel Prize in Physics for her shell-structure model of the atomic nucleus. Interestingly, she was denied a paid position with her husband at the university until after she had completed the work which was to earn her the prize.

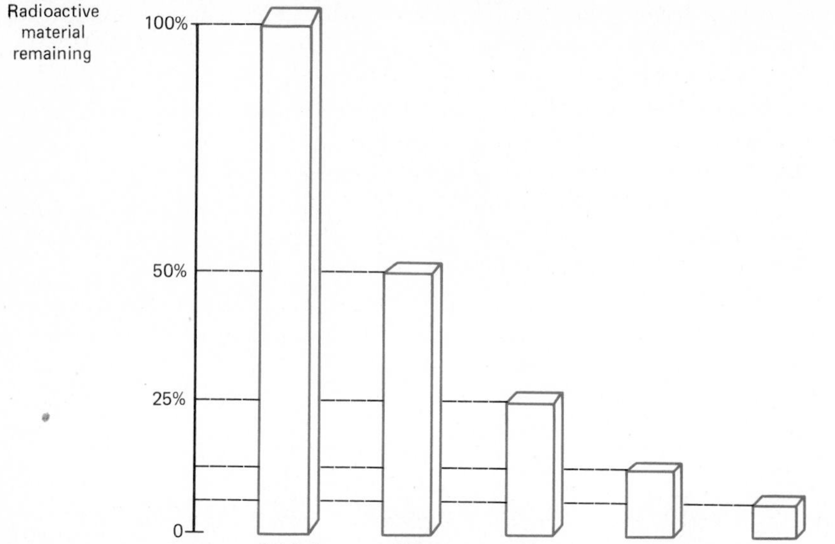

Radioactive material remaining

100%

50%

25%

0

	Original sample	After one half-life	After two half-lives	After three half-lives	After four half-lives
Mass remaining	1	$\frac{1}{2}$	$\frac{1}{4}$	$\frac{1}{8}$	$\frac{1}{16}$
Time	0	T	$2T$	$3T$	$4T$

FIGURE 17-1 The half-life of an element T is the time required for it to lose one-half of its radioactivity.

half-life. All radioactive substances are characterized by a half-life, or a time period during which they lose half the radioactivity they start with (Fig. 17-1). Uranium has a half-life of 4.5 billion years (4.5×10^9 years). Polonium is more active, with a half-life of only 140 days, which presents special problems. If polonium loses half of its activity every 140 days, it should drop to only 16 percent of its original activity in a single year. If we started with 100 tons of pure polonium, in 10 years we would have only 1 g of active polonium. In 100 years, starting with a mass of polonium the size of the earth, we would have a single active atom left. If this is true, how then was it possible for the Curies to isolate an active, measurable sample from rocks that had been untouched in the earth for perhaps millions of years? Clearly, it would not be possible to have a sample left after this time no matter how rich in polonium the rock may have been originally. The conclusion that we are forced to accept is that the polonium is being *made* or *produced* somehow in the rock continually. This view is correct. As radium decays it slowly changes into a radioactive gas called *radon.* Radon decays with a half-life of only 92 h to form polonium. The important point is that radium, radon, and polonium are all distinct and separate elements. *Nature is changing one element into another during the process of radioactive decay.* This process is called natural *transmutation.* Alchemists had searched for centuries to find ways to transmute base (common) metals into gold. Here was nature accomplishing much the same thing.

The Nature of Radioactivity

Radioactive decay revealed an untapped area of fundamental understanding about both matter and energy. A clarification of the nature of radioactive decay was provided by placing a small piece of radium in a well inside of a lead block (Fig. 17-2). This was placed in a vacuum chamber. Radiations given off by the radium would be absorbed by the lead unless they moved down the narrow channel in the lead. In this way a beam of the radiation was obtained for study. The area where this beam emerged into the vacuum chamber was placed between the poles of an electromagnet. If the rays were positively charged, they would be bent by the magnetic field in one direction. If negatively charged, they would be bent in the opposite direction. If the rays were uncharged, they would not be deflected at all by the field. It was found that radium was emitting three distinct types of rays. One bent to the right in the field, one bent to the left, and one was undeflected. Ernest (later Lord) Rutherford (1871–1937) named these the *alpha* (α), *beta* (β), and *gamma* (γ) rays, respectively. Only the undeflected γ ray was truly a massless, chargeless radiation like light. The other two components of the beam were both streams of particles having both mass and charge. The β particle had a charge of -1 and a mass equal to that of the electron. In fact, the β particles were electrons ejected at very high speeds from the radioactive atoms. The α particles were quite different. They had a charge of $+2$ and a mass some 8000 times as great as a β particle. Because the α particles had much less charge for their increased mass, they were bent much less strongly by the magnetic field. However, the α particle, too, was traveling at high speeds of several thousand miles per second. The three rays differed considerably in their ability to penetrate barriers. The α particle penetrated several centimeters in air but was stopped by barriers which the β particles could penetrate. The γ radiations, moving at the speed of light, were the most penetrating form of

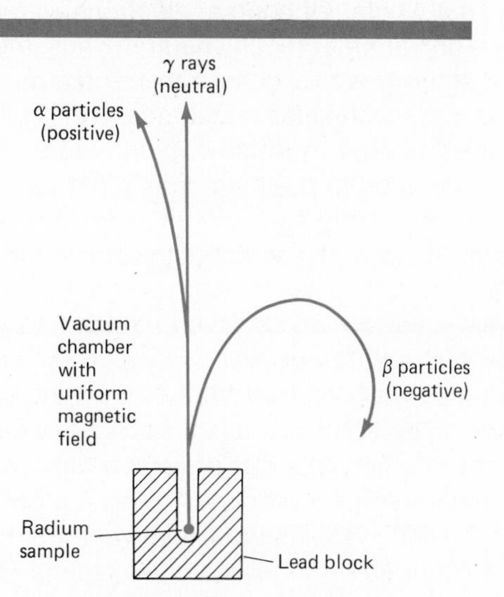

FIGURE 17-2
Allowing a beam of the rays from radium to pass into a vacuum chamber with a magnetic field showed three types of rays to be present—positive alpha rays (α), negative beta rays (β), and neutral gamma rays (γ). Only the latter was truly a ray, like light.

radiation yet observed; they penetrated matter much more easily than x-rays, and shields of dense materials were required to reduce their intensity.

This difference in penetration suggested to Rutherford a clever way to determine the nature of the α particle. An α particle is massive enough to be in the range of atomic masses. A single α particle has about 4 times as much mass as an entire atom of hydrogen. Rutherford sealed a fine piece of radium into a very thin walled glass tube. This tube was then placed in a thick-walled gas discharge tube. When this tube was connected to a high-voltage power supply, the gas trapped in the tube would glow (this is the principle of the neon sign). By examining the spectrum of the light given off, the elements present in the gas could be identified with great precision. Rutherford did this and then put the tube aside to allow the decay of the radium to work for awhile. The α particles should be able to penetrate the thin inner glass wall, but not the thicker outer wall. Thus, the α particles should be trapped between the glass tubes in the gas discharge area where the spectrum is formed. Every second 1 g of radium (about the mass of a thumbtack) gives off roughly 50 billion α particles. The second spectrum of the gas in the tube disclosed the presence of an element which had not been there before. This was a small amount of the inert gas *helium*. α particles became helium atoms when slowed down. They were unlikely to be entire helium atoms, even though they had almost the same mass, since they were much too small. The penetrating ability of the α particle suggested that it was much smaller than an entire atom. [Actually, α particles are about one hundred-thousandth (10^{-5}) of the diameter of a typical atom.]

A key experiment in revealing the nature of the atom involved the use of α particles bombarding a thin metal foil. J. J. Thompson was trying to produce the first model for the structure of the atom. Electrons are tiny in size and of almost negligible mass compared to the atom. It was realized that the electron was a very dense little crumb of matter compared to an atom as a whole. Of course, electrons were negatively charged, yet atoms as a unit are electrically neutral. Thus, the atom also would have to contain some positive charge to be neutral. Thompson imagined the remaining mass of the atom smeared out throughout its volume as a cloud of positive charge in which one or more electrons were imbedded like the plums in a plum pudding. This *plum-pudding model* of the atom was the only reasonable speculation as to the internal structure of atoms at the time. Rutherford was about to stick in his thumb and pull out the Nobel Prize.

The most malleable metal, which could be beaten into the thinnest sheets, was gold. Thus, the *gold-foil experiment* was begun (1911) (Fig. 17-3). A beam of α particles was obtained from radium and shot into the foil. The α particles were detected by tiny flashes of light as each struck a movable fluorescent screen which could be observed with a microscope. With the foil removed, the α particles produced a fine cluster of sparkles on the screen. When the foil was inserted, everyone was quite confident about what should happen—namely, almost nothing. An α particle

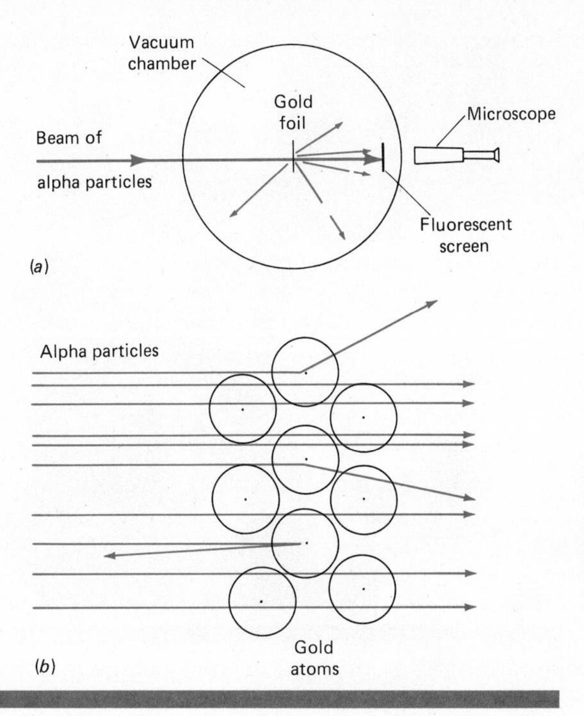

FIGURE 17-3 The gold-foil experiment established the existence of the atomic nucleus. A narrow beam of α particles was allowed to bombard a thin gold foil in a vacuum. The eventual impact of the particles after passing through the foil could be observed on a fluorescent screen with a microscope as flashes of light. While most particles went right through the foil, some were deflected at every angle that could be observed. Rutherford could explain these results by assuming that almost all of the mass and positive charge of the atom was concentrated in a tiny volume of the atom he named the nucleus. Only if an α particle passed very close to a nucleus was its path affected.

moving at high speed should ignore electrons like a cannonball moving through a swarm of mosquitoes. Thompson's cloud of positive charge should be so spread out that it should have even less effect. At best, a very tiny deflection of a few particles which were unlucky enough to encounter many electrons in their path through the gold should occur. This was *not* what happened. True, the vast majority of the α particles passed straight through the foil with no deflection. However, every so often a flash occurred at the edge of the screen visible in the microscope. As the microscope was moved off-axis, these occasional flashes persisted. After moving several degrees away from the central axis, an entirely unexpected phenomenon was observed, which did not stop at a few degrees. Some α particles were found to be scattered at every angle that could be observed. A few particles were even scattered straight back in the direction from which they came! This was strange behavior. If you are firing your cannons into a cloud and all of a sudden a cannonball comes right back at you out of the cloud, you are entitled to several conclusions as you duck. Clearly, the cannonball hit something much more substantial than cloud. Whatever it hit was dense and massive or the cannonball would have torn right through it. The scatter pattern was not the "hard-sphere" pattern you would expect if you shot millions of BB's against a bowling ball; there was too much small-angle scattering. The momentum of the α particles was substantially unchanged. An α particle coming straight back from a collision implied that it had encountered a much more massive object (much as a ball bouncing off a wall). From this experiment, Rutherford pieced together the first accurate glimpse of the internal structure of an atom. A gold atom has a mass (atomic weight)

FIGURE 17-4
Ernest Lord Rutherford (1871–1937) was one of the most skilled experimental physicists in the first part of this century. While very young and working out of a sheepherder's shed in his native New Zealand, he helped establish world electrical standards. His later work in Canada and England established the early ideas of the atomic nucleus and nuclear reactions. The photograph shows Rutherford when he was director of the Cavendish Laboratory. The sign above him requests "Talk Softly Please." While the apparatus was delicate and could be shaken out of adjustment by loud noises, individuals who were there insist the sign was to protect the researchers from being shaken by Rutherford's booming voice. (*Photograph by C. E. Wynn-Williams. Courtesy of AIP Niels Bohr Library.*)

some 197 times that of the hydrogen atom and a positive charge (atomic number) of 79 units.* Rutherford concluded that all of the positive charge of the atom and almost all of its mass had to be packed into a tiny space much smaller than the atom itself (in fact, about the size of the α particle). He called this concentration of mass and positive charge the *atomic nucleus*. α particles did not actually collide with the nucleus of the gold atoms. If the path of the positively charged α particle caused it to approach the positive nucleus too closely, the α particle was repelled from its path. Even when aimed for a head-on collision, the α particle was brought to a complete halt by the repulsion and then repelled back in the direction from which it came. If the charges of nucleus and particle are known, as well as the initial momentum, we can calculate the distance of closest approach, which is also the largest possible size of the nucleus. Rutherford concluded that essentially all of the atom was empty space so far as the moving α particle was concerned. Only the extremely tiny nucleus had any reasonable amount of mass in the atom. The nucleus compared in volume to the atom was on the scale of a dust grain compared to a basketball. The vast volume outside of the nucleus had to be taken up by the only particles left, the electrons. Rutherford imagined these to be looping in orbits about the nucleus in order to "hold up" or "defend" the volume of the atom in the same way that moving fan blades defend an area greater than themselves from intrusion.

* Rutherford's student Moseley determined that the atomic number was the same as the positive charge on the nucleus. Prior to this the atomic number had just listed the order of the elements for the periodic table. Moseley's work established that the atomic number determined which element was which, not atomic weight as had been believed. Moseley is a particularly tragic figure in twentieth-century science. He completed his brilliant work in his 20s, enlisted when World War I broke out, and was killed within a few months of arriving at the front. All death in war is tragic, but Moseley had already qualified himself as a national resource, and one of the few people in any generation capable of reshaping human understanding.

Radioactive Decay and Transmutation

With the discovery of the atomic nucleus in 1911 the nature of radioactive decay began to become clearer. From Fig. 13-3 you will recall that an atom is composed of three basic particles: electrons, protons, and neutrons. The atom is described by two numbers, the first of which is the *atomic number,* or number of positive charges on the nucleus. This is the same as the number of protons, since each proton has a charge of +1 unit (1 unit = 1.6×10^{-19} C). The second number describing an atom is the *atomic mass number,* which is the sum of the protons and neutrons in the nucleus. Each proton or neutron has a mass of about one *atomic mass unit* (1 u = 1.6×10^{-27} kg). Since the electrons in an atom are nearly negligible in mass by comparison to protons and neutrons, the atomic mass number is quite close to the actual mass of the atom in atomic mass units. Different atoms of the same element can and do vary in the number of neutrons they have in the nucleus. Thus, the measured atomic weight of a large collection of atoms will represent the average weight of all these *isotopes* of the particular element in whatever proportion they occur in nature. For example, about 80 percent of naturally occurring magnesium atoms have an atomic mass number of 24, while atoms with mass numbers of 25 and 26 constitute about 10 percent each of the remainder. When the atomic weight of a piece of magnesium is determined, we obtain the average weight of billions of atoms as 24.31 u. This number is the atomic weight reported in the periodic table. When we wish to refer to a particular atom, the atomic weight is meaningless and only the atomic mass number of that particular atom is of interest. Since radioactivity is an internal process of individual atoms and is not influenced by surrounding atoms, only single atomic nuclei are described. We do this by writing the atomic number (Z) in front of and below the symbol of the element, and the atomic mass number (A) above the symbol. Thus, the symbol $^{24}_{12}Mg$ describes the commonest isotope of magnesium. This nucleus contains 12 protons, indicated by Z, and 24 total particles in the nucleus, indicated by A. Of these 24 particles, 12 must be the required protons. The symbols $^{25}_{12}Mg$ and $^{26}_{12}Mg$ describe the other two stable isotopes of magnesium. You may notice that since all magnesium atoms must have 12 protons, no more and no less in order to be magnesium atoms, it is really not necessary to write both Z and the symbol for the element, since they contain the same information. Frequently discussed isotopes are often written without Z. Thus $^{238}_{92}U$ is often written as U 238, or uranium 238; $^{14}_{6}C$ as C 14, or carbon 14 (the isotope used in radioactive dating). The shortened form fits into a printed line with less distraction.

Radium represents the nucleus $^{226}_{88}Ra$, and it decays by emitting an α particle at high speed from its nucleus. However, the α particle is itself a helium nucleus, as we have seen ($^{4}_{2}He$). If the radium nucleus had 88 protons and then ejected 2 of them along with the α particle, the remaining nucleus must have only 86 protons left. Likwise, if the nucleus started with 226 particles and expelled 4, some 222 must remain. We could indicate this as

$$^{226}_{88}\text{Ra} \rightarrow {}^{4}_{2}\alpha + {}^{222}_{86}\text{Rn}$$

Note that the nucleus which has been produced by the α decay, or *daughter* nucleus, is no longer radium, but the new element radon. The radon in turn emits an α particle to become polonium ($^{218}_{84}$Po). While radium and radon decay only by α emission, the polonium can decay by either α or β emission. However, β decay presents a new problem. The β particle is actually an electron ejected at high speed from the nucleus of the atom. Electrons, however, are not found in the nucleus, but outside in the remaining volume of the atom. Where then does the nucleus obtain an electron to discard in β decay? The answer is quite simple. If we could take a neutron and somehow set it on a shelf and watch it we would find out. A neutron in a stable atom normally lasts forever. A free neutron, however, itself decays with a half-life of about 12 min to form three particles, an electron, a proton, and an odd particle called an antineutrino ($\bar{\nu}$):

$$^{1}_{0}n \rightarrow {}^{1}_{1}p + {}^{0}_{-1}e + {}^{0}_{0}\bar{\nu}$$

Note that to have the numbers work, the electron, our β particle, must have a Z of -1, its charge. The important idea is that *β decay represents one of the neutrons of the nucleus changing into a proton.* Thus, in β decay, A does not change, but Z increases by 1. We have one more proton in the nucleus than before the β decay. Applied to polonium we obtain

$$^{218}_{84}\text{Po} \xrightarrow{\nearrow {}^{0}_{-1}\beta} {}^{218}_{85}\text{At}$$

By now it should be apparent that with 218 particles remaining in the nucleus which began as radium, this chain of radioactive decay could continue for some time. When will it end? Actually, any decay chain ends when a stable nucleus is produced. This is any nucleus which is not radioactive, does not decay, and is perfectly capable of sitting out the rest of eternity in its present form. For the chain we have been examining, this point is reached after six further decays in the isotope of lead, $^{206}_{82}$Pb. By examining this process backward we may also determine that this chain begins with the commonest isotope of uranium, $^{238}_{92}$U, some five short steps before we reach radium. The chain begins at U 238 because its half-life (of 4.5 billion years) is so long that it apparently has not had the time to decay away since the formation of the universe. If we assume that all Pb 206 in the universe began as U 238, we could estimate the age of the universe from the ratio of these two isotopes at present. Unfortunately, we cannot determine the amounts of these materials in the deep interior of the earth. We can, and do, use this method to date particular rocks, where all of the Pb 206 produced would be trapped along with the U 238 remaining. A similar method is to compare the amounts in a particular rock of U 235 with its eventual daughter nucleus, Pb 207. Still other "radioactive clocks" for dating samples exist.

Nuclear Stability

The picture of the atomic nucleus presented by Rutherford presents an immediate problem. Protons, each with a positive charge, repel each other. The size of an atomic nucleus ranges from about 1.4×10^{-15} m for hydrogen (a proton) to almost 9×10^{-15} m for the largest nucleus. In uranium, nature packs 92 positively charged protons into this tiny volume. The question is not why nuclei fall apart in radioactive decay, but rather, how can they possibly stay together? We may scale up the repulsive force on a proton in a uranium nucleus to a mass of your size. If we did so, you would have to resist a force of over *a billion billion tons!* The electrical repulsion between protons in a nucleus is irresistibly strong for the forces known in 1900. However, most nuclei are comfortably stable and do not fly apart. Thus, there must be an additional, unknown force which holds the particles of the nucleus together. Actually, there are two *nuclear forces* which remained to be discovered at that time, both stronger than the electric force. While the details of these forces are not necessary for our purposes, one property stands out. Rutherford's scattering of α particles by nuclei in a gold foil could be completely accounted for by the force of electrical repulsion. Even at this close approach of an α particle to the nucleus, no nuclear force was apparent. Thus, *nuclear forces must operate over an extremely short range,* perhaps only the size of the nucleus itself. All elements in the periodic table beyond bismuth (atomic number 83) are radioactive. Perhaps the nucleus becomes so large that particles at the edge occasionally fall outside the range of the nuclear force and are repelled away in radioactive decay. This at least was a reasonable speculation.

It occurred to Rutherford that while an α particle cannot overcome the severe repulsion of a large nucleus, such as gold, it should be able to penetrate against the repulsion of a small nucleus. If an α particle could approach the nucleus closely enough, it might get within the range where the nuclear force would overcome the electrical repulsion. Thus, the bombarding particle might be absorbed by the nucleus, causing it to change. To have this happen, the α particle would have to approach the light nucleus on a direct collision course. Otherwise, it would be deflected to one side or the other and rush past the nucleus. Since such a "direct hit" is very unlikely, it would be necessary to observe the path of an individual particle before and after collision to confirm what interaction had occurred. Rutherford was aided in this pursuit by a device developed by C. T. R. Wilson. The Wilson *cloud chamber* operated on a principle similar to that which produces vapor trails behind some high-altitude jet aircraft today. When air is saturated with a vapor, the vapor will tend to condense and form droplets on any convenient particle nearby. When an α particle passes through the gas in the chamber, it leaves a trail of free electrons and ions. The latter serve nicely as centers around which condensation of the vapor may take place. Thus, the path of the particle is marked by a vapor trail made up of tiny droplets of condensation about the atomic debris caused by the particle. This renders the actual path of the particle visible for study. By placing the chamber between the poles of a magnet, the curvature of the paths allows measurement of the momentum of the α particle, and charge or mass of other particles. A forked Y in the track indicates that the α particle has collided with some nucleus and that fragments were scattered in different directions.

In 1919 Rutherford finally produced the tracks which he had been looking for. He found the record of an α particle making a collision with a nitrogen nucleus from the air. By measurement of the paths of the objects produced he showed them to be a proton and an oxygen atom:

$$^{4}_{2}\alpha + {}^{14}_{7}N \rightarrow {}^{1}_{1}p + {}^{17}_{8}O$$

Rutherford thus became the first to observe the direct evidence of *natural transmutation,* or the changing of one element into another. This was a *natural* transmutation since all that Rutherford had done was to observe and interpret it. He could not produce the α particle on demand; nature did that. The same event had occurred billions of times in the past (and was occurring) as some radioactive nucleus in a rock sent off an α particle which struck a nearby air atom to change it to another element. On the other hand, it was seen at once that transmutation need not be natural if only very high speed particles could be produced at will. Two of Rutherford's students began to work on an *accelerator* for charged particles. This device used the attraction of charged particles for opposite charges to raise particles to speeds (energies) high enough to penetrate the nucleus. If a proton with a charge of 1 unit ($q = 1.6 \times 10^{-19}$ C) is introduced at an opening in a positively charged plate, it will be repelled by the plate. If a negatively charged plate is placed some distance away, the proton will be attracted to it. If the potential difference between the plates is V, the kinetic energy acquired by the proton in accelerating across the gap between the plates is

$$E = \tfrac{1}{2}mv^2 = qV$$

Since the masses of particles on the atomic scale are so tiny, it is convenient to define a new unit of energy on this scale. The energy acquired by an electron accelerating across a potential difference of one volt is called *one electronvolt* (1 eV). Calculations indicated that a light nucleus could be penetrated at will under 1 megaelectronvolt (MeV) (1 MeV = 10^6 eV). This level could be exceeded with a single potential difference of 1 million volts. In 1932 John Cockroft and E. T. S. Walton had produced the first machine to accelerate particles across a high potential difference. Walton became the first human to observe the flashes on a fluorescent screen which indicated an *artificial* or humanmade transmutation was taking place during the machine's trial runs.*

* At the same time an obscure play was running on the London stage in which a typical "mad-scientist" type did something called "splitting the atom" and thereby gained control of the world. When artificial transmutation was announced, front-page newspaper headlines read "British Scientists Split the Atom." The scientists were quite startled by the heavy news coverage they continued to receive until this connection was pointed out. Accelerators have been known as "atom smashers" ever since. The actual reaction was quite modest. A lithium target was bombarded with a beam of protons with energies of only 100,000 eV, producing α particles as products: $^{1}_{1}p + {}^{7}_{3}Li \rightarrow {}^{4}_{2}He + {}^{4}_{2}\alpha$.

FIGURE 17-5
Aerial view of the Fermi National Accelerator Laboratory at Batavia, Illinois. The main accelerator ring is 2 km in diameter. A small town of 10,000 could comfortably fit on the land inside the ring. The ring is large enough to be seen in a spacecraft view of Chicago (Fig. 15-16*a*). Actually, much smaller accelerators begin particles on their journey and then inject them into the main ring where they effectively "ride the crest" of speeding radio waves to be accelerated to energies equivalent to a voltage difference of 5×10^{11} V (500 GeV). (*Photograph courtesy of Fermilab.*)

As the investigation of the nucleus has posed more sophisticated problems, accelerators have become larger. Today, the largest accelerators can comfortably house a small town. Energies of 500 gigaelectronvolts (GeV) (5×10^{11} eV), or millions of times that of the machine of Cockroft and Walton, are being produced. Each time a basic question about the nucleus is resolved, more questions are posed by the data obtained. The machines for continuing work in nuclear research have become so costly that only the largest nations can afford them. After World War II the countries of Western Europe banded together to form a partnership in nuclear research called CERN since they could no longer afford individual, national efforts. Currently CERN has agreed to construct the largest accelerator yet built. The largest operating accelerator is at the National Accelerator Laboratory outside of Chicago (Fig. 17-5).

Mass and Energy—The Theory of Relativity

While many of the details describing nuclear changes have been cleared up, the problem of energy remains puzzling. At the turn of the century prophets were predicting a future of limitless energy if the full secret of nuclear energy could be revealed. The interpretation of the source of nuclear energy came from an entirely unexpected direction along with one of the great revolutions of physical thought. In 1905 the German journal *Annals of Physics* was published with what appeared to be a normal collection of scientific papers. It has since become a collector's item. (The manuscript for one paper sold in 1944 for the highest price ever paid by collectors for a scientific paper—some $4 million.) Among

FIGURE 17-6 Albert Einstein (1879–1955) was certainly the most famous scientist of this century. He erupted on the scientific scene after his publications of 1905 and became a world figure after confirmation of his general theory of relativity published in 1915. While he liked a quiet life of study, sailing, and walking, his fame often involved him in world events. His ardent stand as a pacifist against military conscription led to much ridicule. He fled the antisemitism of Nazi Germany for America. He was offered, and turned down, the presidency of Israel. His profound insights into the nature of the universe made him a legend in his own time and a symbol of advanced thought. (*Photograph courtesy of AIP Niels Bohr Library.*)

the papers were three by one author. This is unusual, but not startling. However, this set of papers constituted one of the most incredible displays of human intellect in history. The author of these papers was an obscure clerk in the Swiss Patent Office by the name of Albert Einstein (Fig. 17-6).

The first paper ended once and for all the argument about the existence of atoms. In 1827 the biologist Robert Brown had observed a constant jiggling of fine particles at high magnification under his microscope. Einstein's mathematical treatment of the phenomenon ended a century of speculation. He demonstrated that this behavior was to be expected for fine particles under the constant bombardment of the still finer molecules of fluid in which they were suspended. In *Brownian motion* we see the visual evidence for the existence of molecules which themselves are too small to be seen using visible light. (Atoms are much smaller than light waves and will not reflect them any more than a dust speck will reflect a sound wave.)

The second paper in the journal contained a detailed experiment to confirm or reject a hypothesis posed earlier. When this *photoelectric effect* was later demonstrated, it paved the way for the last of the great revolutions in physics. Einstein was awarded the Nobel Prize for this contribution to understanding. We will consider it in the next chapter.

The final paper contained the special theory of relativity, and is Einstein's true claim to fame (although, as you might have realized by now, the other two papers would have placed his name in textbooks also). Einstein and relativity will be forever linked in history even though he was an incredibly versatile human being with contributions to many fields of science (as with Newton or Maxwell before him). Our search for an explanation of the source of nuclear energy requires that we examine some of the ideas of relativity. Initially, this will seem like a complete change of topic, but we will end up with the explanation we seek. Relativity is not particularly difficult, but it is very different. The ideas of relativity are harder to believe than to understand.

Relative Motions

You are traveling 10 m/s on a bus and you throw a ball forward at an initial speed of 5 m/s. At the same time you throw a second ball toward the rear of the bus at the same speed. Any observer moving with you on the bus would agree on the speed of each of these balls, and would assign the values of +5 m/s and −5 m/s to them, respectively. An observer standing on the ground outside of the bus would disagree. Since both balls share the forward motion of the bus, this observer assigns speeds of +15 m/s and +5 m/s to each of the balls. Let the speed of the bus be u, and the speed of the ball relative to the bus v. To the person in the bus sharing the motion at speed u, the two balls have speeds of $+v$ and $-v$, respectively. Before they were thrown, the outside observer would see the balls moving past with the bus at speed u. When the two balls are thrown, the additional motion adds or subtracts from this initial speed. The speeds are thus $u + v$ and $u - v$. None of this is very surprising. This idea of the addition of relative speeds was clearly stated by Galileo. The most surprising element of *Galilean relativity,* as this is called, is that nature does not work this way at high speeds.

Consider a somewhat more difficult example. Imagine a river flowing downstream at a uniform speed of c. The river has a width of d. We have two identical motorboats each of which can make a speed of v (Fig. 17-7). One of these boats we will send directly across the river to a pier on the opposite bank and then back. The second boat will move the identical distance d downriver to a pier and then back upriver to the starting point. The question to be answered is simply which of the two boats will complete the round trip in the shorter time? Since the two boats have the same distance to travel and can move at the same speed, your first reaction might be that they would take the same time. This would be correct if not for the current. If the current is not negligible, they would take different times according to the ideas of Galilean relativity. Let us calculate the time for a round trip in each case to show this point.

The boat moving downstream travels at a speed of v. But, of course, the water is moving downstream, too, at a speed of c. Even without the motor, the boat would be carried along by the current. Using the motor, the boat going downstream has a combined speed of $v + c$. Similarly, when moving upstream the speed of the boat relative to the water is v. But, some of this speed is used to fight the current. If the speed of the boat was the same as that of the current, the boat would appear to rest motionless in the water relative to the riverbank. Going upstream the effective speed of the boat is $v - c$. The time to get to the objective is the distance divided by the speed:

$$\Delta t = \frac{\Delta d}{v}$$

For the roundtrip we obtain the total time as

$$t_1 = \underset{\text{(Downstream)}}{\frac{d}{v + c}} + \underset{\text{(Upstream)}}{\frac{d}{(v - c)}} = \frac{2dv}{v^2 - c^2}$$

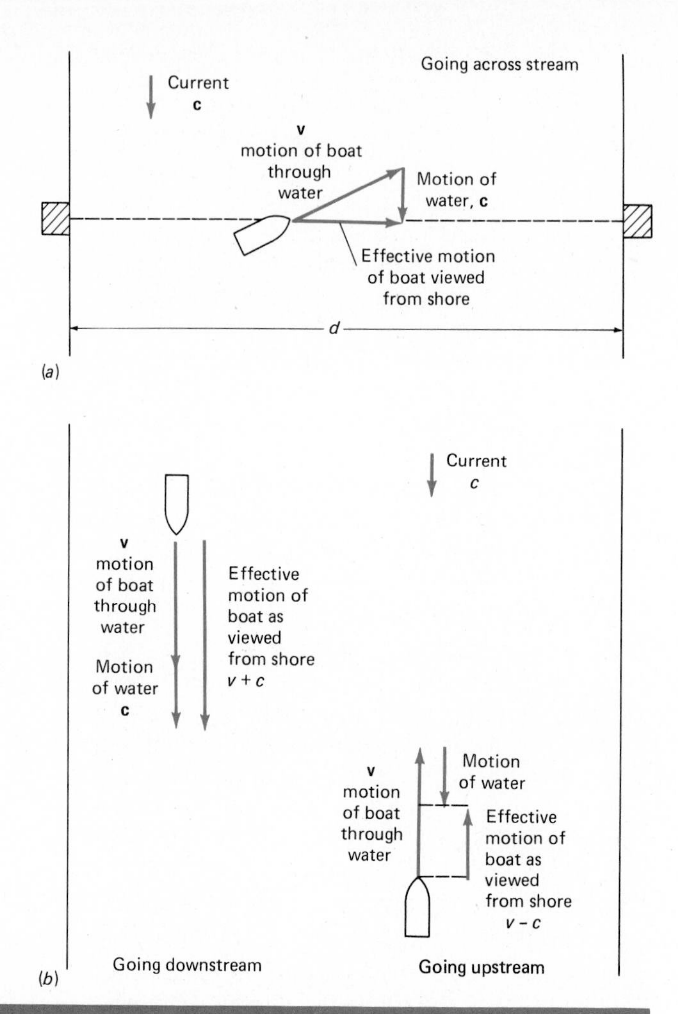

FIGURE 17-7 The effective speed of a boat varies with direction if there is a current. When going across the current (*a*) the boat must be directed upstream so that some of its forward motion may cancel out the downstream drift with the current. Going upstream of downstream (*b*) the current either subtracts from or adds to the boat's motion in the water to yield the effective speed.

(The final form at the right is obtained by placing both fractions over the common denominator and adding them.) We may try this out by inserting some numbers. Imagine that the boat can move with a speed of 5.00 km/h, the current is 3.00 km/h, and the one-way distance is 2.00 km (v = 5.00 km/h; c = 3.00 km/h; d = 2.00 km). Downstream the effective speed is $v + c$ or 8.00 km/h, and the trip takes 15 min, or 0.250 h. Upstream, $v - c$ is only 2.00 km/h and the trip will require 1 h. The total time for the round trip, t_1, is thus 1.25 h. This is what we obtain from using the final form also.

$$t_1 = \frac{2dv}{v^2 - c^2}$$

$$= \frac{2(2.00\ \text{km})(5.00\ \text{km/h})}{(5.00\ \text{h})^2 - (3.00\ \text{h})^2}$$

$$= \frac{20.0\ \text{km}^2/\text{h}}{16.0\ \text{km}^2/\text{h}^2} = 1.25\ \text{h}$$

Crossing the river produces a different result. If the boat was sent directly across the river, the current would carry it considerably downstream from its objective during the crossing. To avoid this the boat must be aimed somewhat upstream. Then, part of the boat's speed cancels the current, and the boat moves directly across the water to its objective. The vector situation which results can be solved to produce a speed of crossing of $\sqrt{v^2 - c^2}$. Using the numbers of the previous example, the speed of crossing is $\sqrt{(5.00 \text{ km/h})^2 - (3.00 \text{ km/h})^2} = 4.00$ km/h. This speed will be the same whether crossing the river on the trip out or the trip back. Thus, the time for the round trip is

$$t_2 = \frac{2d}{\sqrt{v^2 - c^2}}$$

$$= \frac{2(2 \text{ km})}{4.00 \text{ km/h}} = 1.00 \text{ h}$$

Clearly this is a different result than for the boat moving upstream and then downstream. The problem is that while this is the method of computing travel time relative to velocity for boats, it does not work for light.

The earth is moving through space in its orbital motion at a speed of 30 km/s. Classical experience suggests that to produce a wave you must disturb some substance from its state of rest. Whatever the medium for carrying light waves was (it was named the *ether*), at some parts of our orbital motion around the sun we should be plowing through it at great speed. This produces a situation very much like that of a current in a river. The earth should be moving through the ether, or, what really is the same thing, the ether should be flowing past us. The earth's speed is really quite respectable—one-thousandth of a percent of the speed of light. The difference in times noted in the case of the boats should be easily detectable with light waves. This occurred to the two American experimentalists Michelson and Morley in Cleveland. One beam of light was sent out in the direction of the earth's motion, reflected, and returned to the instruments. Part of the same beam was diverted at right angles, sent the same distance, reflected, and returned. If even a small difference in the time required for the round trip existed between the two paths, the beams of light would interfere with each other when mixed after their return. The instrument developed for the measurement, the Michelson *interferometer,* could easily measure a shift in time of travel of the size expected. Yet, it was not there! All of the equipment was mounted on a block of granite floated in a pool of mercury so that it could be rotated. Measurements were tried at every angle, at every time of day, in every season of the year. There simply was *no measurable* difference in the speed of light for the two paths regardless of when or how they were measured.

One explanation of this would be to assume that the ether was "fixed" to the earth somehow and moved with us. Of course, if this were true, it would not be fixed for any other object in the universe, and the earth would be special and unique. Every time in the history of human thought we have assumed a special and unique role for the earth, we have gotten into trouble and have had to backtrack. Yet, what other explanation could there be? Two theorists, Fitzgerald and Lorentz, suggested that the difficulty could be overcome if length shortened in the direction of travel. A contraction of $\sqrt{1 - (v^2/c^2)}$ in the direction of motion would produce identical times of flight for the two beams. This was widely regarded as a wild attempt to "save the pieces," since theory and experiment produced differing results. Einstein, however, arrived at the same conclusion from a widely different direction. (In fact, he was isolated enough to have been unaware of the results of the Michelson-Morley experiment.) Einstein was fond of asking himself fundamental questions, such as: "If you were running alongside a moving light wave, how would it appear to you?" He concluded that Maxwell's electromagnetic theory and Newton's mechanics could work together only after some fundamental modification. As a starting point he assumed that relative motions at constant velocity should not produce different ways of viewing the nature of the universe. This is the *principle of relativity:*

> *The laws of physics are the same for observers in all systems moving at constant velocity relative to each other.*

Along with this principle, Einstein assumed as his second point:

> *The speed of light in vacuum, c, must be a constant independent of the motion of such observers.*

By themselves these two statements do not appear particularly startling, but Einstein continued by exploring their implications. The results were that space (length), mass, and even times were *not absolute* quantities but depended upon the state of motion. We are used to thinking of the length of a meterstick as being fixed and absolute. Or so it will appear to us if it shares our motion. Imagine instead that we measure the length of a meterstick glued to the outside of a spaceship that passes us moving at six-tenths the speed of light, 0.60 c. If the meterstick is glued on perpendicular to the motion of the rocket, it will measure 1.00 m. However, if a second meterstick has its length in the direction of the spaceship's motion, it will not measure 1 m at all. It would come to only 80 cm:

$$\sqrt{1 - \frac{(0.6\,c)^2}{c^2}} = 0.80$$

Furthermore, the whole ship would appear shrunk in its direction of motion to some eight-tenths its length when at rest relative to us. People on the ship, too, would be contracted in their direction of motion. However, if they measured us as they passed we would appear contracted in the direction of motion in the same way. Of course, we have not done this experiment with people yet. Six-tenths the speed of light is 180,000 km/s. The fastest people have moved relative to the earth during the space program has been only about 50 km/s. Subatomic particles are much more convenient than people to accelerate to high speed for the measurement of relativistic effects.

A second prediction that is much easier to check out has to do with time. Just as length is contracted, time is stretched, or *dilated,* with high-speed relative motion. If L_o is the length of an object at rest relative to an observer, its length when in motion at speed v is given by

$$L = L_o\sqrt{1 - (v^2/c^2)}$$

Time is dilated in much the same way. Where T_o is the time interval (between ticks of an accurate clock, for example) when the timepiece is at rest relative to the observer, the interval when moving at speed v is

$$T = \frac{T_o}{\sqrt{1 - (v^2/c^2)}}$$

Should you care to work it out, if $v = 0.6c$, 1 h on the moving spaceship clock would appear to the observer to be 1.25 h (Fig. 17-8). Even though the speed of an earth satellite compared to the speed of light is tiny, comparisons of the timing of very accurate clocks in orbit with others on the ground show precisely the slowdown predicted by relativity. A clock can be any device that marks time intervals. The rate at which human beings live is a clock of sorts. The speed of thinking, the rate of digestion, the pulse rate—all are clocks. All should appear to slow as the human moves faster and faster relative to the observer, just as the human's length should shrink to a tiny fraction of rest length. Radioactive disintegration is also a clock uninfluenced by normal changes in the

FIGURE 17-8 The implications of relativity are different from common experience. Imagine that an individual in a spaceship shoots past a twin on earth traveling at a large fraction of the speed of light. Each holds a meterstick in each hand. The metersticks held perpendicular to the motion ($L_\perp$) appear to be the same length. However, the meterstick held parallel to the traveler's motion ($L_\parallel$) appears shortened—as does the traveler and the entire spaceship in this direction. Furthermore, clocks that were synchronized before the motion began no longer are. The time interval for the spaceship (T') appears to be stretched out or dilated, with time moving more slowly. If we tried to deflect the spaceship by applying a known force, we would also find that its mass had increased greatly. What would the twin on earth appear like to the one streaking by in the spaceship?

environment. If such particles are accelerated to near the speed of light, the rate of decay *does* drop exactly as predicted by Einstein's expression. If we could whirl 1 g of radium in a machine at the speed of light, its decay would stop, since when $v = c$, the time interval for a clock becomes indeterminately large. There are other problems in trying to do this experiment, however. As the relative speed increases, the mass of an object increases also. If m_o is the *rest mass* of an object, its mass measured by an observer when its relative speed is v is

$$m = \frac{m_o}{\sqrt{1 - (v^2/c^2)}}$$

This implies that as the speed of an object is increased to near the speed of light, its mass becomes much larger. This is a well-known effect that must be taken into consideration in designing particle accelerators. A machine bending a beam of fast particles with a magnet will fail completely if the strength of the magnet is designed to handle the rest mass of the particles. As energy is given to the particles to accelerate them to high speed, their mass increases. Stronger forces are needed to overcome their inertia and bend them around corners than if their mass had not changed.

The relationship between mass and energy was one of the most celebrated ideas to come from the theory of relativity. Using the relationships for relativity Einstein recalculated the expression for kinetic energy. He obtained

$$E_k = mc^2 - m_oc^2$$

If the speed of motion is zero, the relativistic mass m is just the rest mass m_o, and the kinetic energy expression becomes zero. For a speed greater than zero but much less than the speed of light, the value of the expression is approximately $\frac{1}{2}m_ov^2$, the classic expression for kinetic energy. If the expression m_oc^2 is taken to be the energy of an object at rest, a simple equation gives the total energy:

$$E = mc^2$$

where m is the relativistic mass. The interpretation of this expression is simple but important. Since c is a universal constant, the relationship implies that the energy contained in a substance is directly proportional to its relativistic mass. *Energy addition implies a mass increase.* If we heat an iron ball, it should weigh more than it did at a lower temperature. In a chemical reaction that liberates energy, the products would have slightly less mass than the reactants. In both of these cases, however, the mass change would be too fine to detect. The speed of light is a very large number (3×10^8 m/s), and even larger when squared. Yet, this is the constant which determines how much mass appears from a given amount of energy.

For example, a barrel contains about 150 kg of oil. Burning of this oil will release about 6.5×10^9 J of energy (1.6×10^9 cal). How much mass disappears as the oil is burned?

$$m = \frac{E}{c^2}$$

$$= \frac{6.5 \times 10^9 \text{ J}}{(3 \times 10^8 \text{ m/s})^2}$$

$$= 7 \times 10^{-8} \text{ kg}$$

This loss of mass is too small to detect directly. However, if we could somehow "burn" the oil so that its entire mass was converted to energy, we could obtain 10 billion times as much from the same barrel of oil. Matter itself represents a supercondensed form of energy. Here then is the answer to the problem of the energy source in radioactive decay. Where does the energy of radium, for example, come from? The particles produced in the radioactive decay have less mass than the original atom. The mass which has been lost has been converted to energy.

Nuclear Reactions and Mass

The relationship between mass and energy requires that careful determinations of mass be made. The first of the devices to perform such precise measurements was the *mass spectrograph* developed in 1920. In this instrument, atoms are ionized by having an electron removed. The ions are accelerated down a tube and deflected by uniform magnetic field. As the mass of an ion is increased, its path is bent less by the field. The amount of bending which does occur provides a precise measure of mass. The first important contribution of this work was to verify the existence of isotopes. An oxygen sample will display ions with masses of 16, 17, or 18 u, but nothing in between these values. Early work confirmed that all atomic masses are quite close to whole number multiples of the atomic mass unit. Finer work revealed some small discrepancies. It became apparent that the mass of an atom was *not quite* the sum of the masses of the particles which made up the atom. The current definition of the atomic mass unit (u) is that it is one-twelfth of the mass of the commonest type of carbon atom, C 12, (or 1.6×10^{-27} kg). On this scale, a proton has a mass of 1.00728 u, a neutron mass is 1.00867 u, and an electron is 0.00055 u. A normal helium atom, ${}^4_2\text{He}$, will have two of each of these particles. If we add the masses of these six particles, we obtain 4.0330 u as the mass of the parts of the helium atom. The actual measured mass of the helium atom is 4.00260 u. Thus the atom is some 0.0304 u less massive than the masses of the parts making it up! This difference is called the *mass defect.* A bowling ball on a housetop possesses gravitational potential energy. If the ball falls to the ground, it has lost that energy. The implication of relativity is that this energy change would show up as a small decrease in the mass of the bowling ball. With weak gravitational forces such mass changes are too small to be detected. This is not so when particles "fall" to a lower potential energy state in a nucleus under the action of nuclear forces. In this case the mass defect corresponding to the decreased energy state is measurable. It is easier to talk about

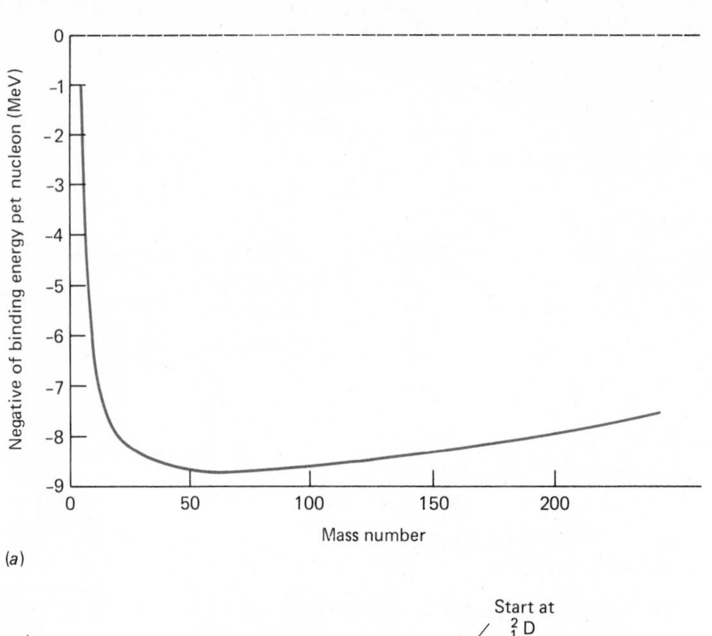

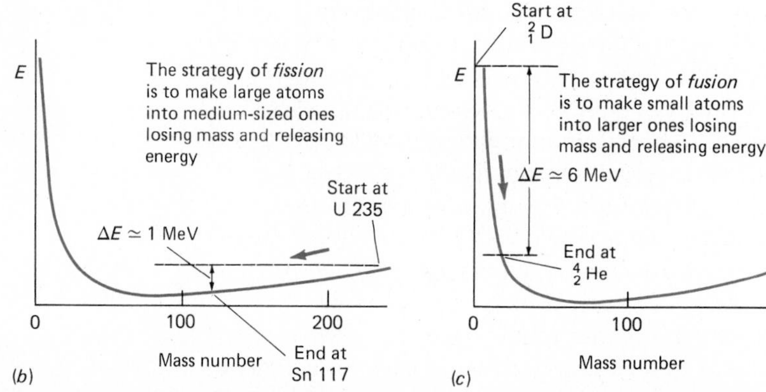

FIGURE 17-9 The mass of a nucleus is less than the mass of the particles (nucleons) making it up. If an iron atom with mass number 50 (Fe 50) could be assembled from bins of separate protons and neutrons, its formation would liberate almost 8.18 MeV per nucleon. This represents an energy of 8×10^{17} J for each kilogram of iron formed, or more than the output of a large electric power plant for a year. Thus an iron atom is almost half a proton's mass short of the sum of the masses of its separate pieces. This "lost" mass shows up as released energy by the relationship $\Delta E = \Delta m\, c^2$ of relativity theory. This also suggests two strategies for the release of energy. Fission splits large atoms to make medium-sized ones and releases energy (*b*). Fusion joins small atoms to make larger ones and releases energy (*c*).

such changes in terms of the negative of the mass defect, which is called the *binding energy* of the nucleus. This binding energy varies from one nucleus to another. To compare various nuclei it is necessary to divide the binding energy of each nucleus by the number of *nucleons* (protons and neutrons) in the nucleus. The result is the curve of Fig. 17-9, which shows atomic mass numbers vs. binding energy per nucleon for the elements of the periodic table. The conclusion this allows is that the intermediate-sized atoms, with A from 40 to 80, would require the greatest amount of energy to separate into individual particles. These nuclei, centered roughly around the iron atom in the periodic table, represent the bottom of the energy pit. Natural transmutation will proceed toward the bottom of this pit only, liberating energy as it occurs.

No strategy is known for the complete conversion of mass to energy in a controlled manner. Such *annihilation* of matter liberates an incredible amount of energy. The annihilation of a single atomic mass unit liberates an energy of 931.44 MeV. The complete conversion of only 2 kg

(4.5 lb) of any kind of matter (say old pop bottles) to energy would liberate enough heat to completely vaporize all of the water in Lake Erie. We have no idea how to tap this gigantic energy source. However, the binding energy curve suggests a strategy to obtain a small amount of this energy stored in matter. If a wholesale method could be found to split apart very heavy atoms into intermediate-sized pieces, a large amount of energy would be liberated. The energy released per nucleon would be proportional to the difference in binding energies in the heavy atom and in the intermediate-sized product. Figure 17-9*b* shows the position of uranium 238 in the chart. If this nucleus could be split into two smaller ones averaging 119 nucleons each, the difference shown by ΔE would be released as energy. This energy difference is on the order of 1 megaelectronvolt per nucleon, or some 240 MeV for all of the nucleons involved in the splitting, or *fission,* of a single uranium atom. This is only about one-thousandth of the energy which is theoretically available from the complete annihilation of the uranium atom, but something is better than nothing. Even in fission the energy is gigantic. It would take the complete fission of about 2 tons of uranium to boil away Lake Erie instead of the 2 kg of matter using annihilation. This amount would still fit conveniently into the back of a truck. All that remained was to figure out how to split big nuclei into intermediate-sized ones. This eventually grew into one of the greatest scientific and industrial efforts of the twentieth century.

A second strategy is apparent from the binding energy curve for the release of energy. If relatively light nuclei can be made into heavier ones, considerable energy will be released. Figure 17-9*c* shows the position of the nucleus of the heavy hydrogen or deuterium nucleus, ${}^2_1H$. If two of these could be welded into a helium nucleus, ${}^4_2He$, almost 6 times as much energy per nucleon would be released as in uranium fission. The process of joining together lighter nuclei to form heavier ones is called *nuclear fusion.* This process is particularly important since it provides for a nearly limitless energy source if it can be done on a large scale. Over 90 percent of the matter of the universe is in the form of hydrogen. The same hydrogen in gasoline hydrocarbons which yields 20 to 30 mi/gal when burned chemically could yield millions of miles to the gallon in fusion. Hydrogen fusion is the main process which stars use to generate their heat and light. The reaction

$$ {}^1_1H + {}^1_1H + {}^1_1H + {}^1_1H \rightarrow {}^4_2He $$

is the essential source of most of the energy we experience, including that stored in fossil fuels. This process almost never proceeds as written, but rather in a series of steps leading to the production of the helium. All that is required is to have the hydrogen nuclei slam together at incredibly high speed to have them fuse into a heavier nucleus. These speeds (temperatures) are much too high for normal matter to survive, so this cannot be done in some special furnace. Research advances in the last decade make it clear that we are winning in the attempt to produce controlled fusion (just as the hydrogen bomb demonstrated that we could trigger an uncontrolled fusion). It seems possible that the commercial production of electric power by fusion will occur in your lifetime, although it is too early to be certain. Aside from the huge fuel supply available, fusion has the advantage that neither products nor reactants are radioactive. Any radioactive wastes associated with fusion will be incidental materials in limited quantities.

The Discovery of Nuclear Fission

E. T. S. Walton, codesigner of the first particle accelerator, recalled a conversation at the Cavendish Laboratory in the late 1930s. All of the electronics specialists had just left for a top-secret project (which was to produce the first radar). A colleague mentioned how sad it was that all of these fine minds had to be diverted into producing weapons for the war that seemed inevitable.He added that it was fortunate that Walton and he were in a field with *no practical implications whatever,* since both were nuclear physicists! In less than a decade both Hiroshima and Nagasaki would be destroyed by nuclear bombs, yet even then the structure of matter appeared a quest of pure science. Today, more electric energy is produced by nuclear power plants than by all the generating stations on earth at the end of World War II. We will offer a brief description of the discoveries that led to this change.

Theory had predicted the existence of the neutron long before its actual discovery by Chadwick in 1932. The problem was that the neutron was an elusive phantom of a particle. Charged particles have to bull their way through matter, throwing charged fragments in every direction and leaving very messy trails. Neutrons are different. They slide through normal matter with the ease of a seasoned burglar, and interact only when they join a nucleus. Neutrons may be produced by the bombardment of beryllium by the α particles from radium. It occurred to several investigators that the neutron is an ideal particle to induce nuclear changes. The neutron may drift quietly into the nucleus while charged particles must fight the fierce electrical repulsion. In particular the bombardment of uranium was of great interest. This is the largest atom known in nature, and there was curiosity as to what would happen when it was made even larger by absorbing an additional neutron. The answer was that the neutron triggered a β decay and the production of a new element. This element in turn underwent β decay to yet another element. Since uranium was named after the planet Uranus, it seemed only natural to call these two higher elements neptunium and plutonium after the next two planets.

$$^{1}_{0}n + ^{238}_{92}\mathrm{U} \longrightarrow ^{239}_{92}\mathrm{U} \xrightarrow{^{0}_{-1}\beta} ^{239}_{93}\mathrm{Np} \xrightarrow{^{0}_{-1}\beta} ^{239}_{94}\mathrm{Pu}$$

While this process was still in doubt, it was a concern of a team of German scientists. Lise Meitner was a physicist working with Otto Hahn, a chemist, on the problem of neutron bombardment of uranium. Hahn, was one of the foremost analytic chemists in the world. He kept turning up minute traces of strange elements in samples of pure uranium bombarded by neutrons. For example, he identified the element barium after bombardment. This was puzzling since barium is not even close to uranium in the periodic table, having an atomic number of only 56 compared to the 92 of uranium. Yet, Hahn knew that none was present in the sample before it was exposed to neutrons. His caution kept him from publishing such an unlikely result until further tests had been conducted. At this time, politics intervened. Lise Meitner was a respected physicist, but she was Jewish. She was informed that it would save both her and the state some embarrassment if she left Germany under Adolf Hitler and worked somewhere else. Meitner left for Sweden and shortly after her arrival the interpretation of the results of uranium bombardment occurred to her. What must be happening to produce the trace of lighter elements was not a reaction of normal uranium atoms at all. Instead it was the result of a neutron striking the rare uranium isotope, U 235. Natural samples of uranium are about 99% U 238 with less than 1% of the lighter isotope, U 235, mixed with it. While U 238 absorbs a neutron to climb the periodic table to plutonium, the reaction of U 235 is quite different. Bombardment of an atom of U 235 with a neutron causes nuclear fission where the U 235 nucleus breaks into medium-sized nuclei. A sample reaction might be

$$^{1}_{0}n + ^{235}_{92}\text{U} \rightarrow ^{236}_{92}\text{U} \rightarrow ^{145}_{56}\text{Ba} + ^{88}_{36}\text{Kr} + 3^{1}_{0}n + \text{energy}$$

The actual nuclei produced vary considerably from fission to fission, but the reaction always liberates considerable energy and several free neutrons. Meitner's ideas were confirmed literally within hours of the accidental announcement of her discovery in America. Then, after a brief flurry of discussion, a self-imposed censorship stopped open discussion of this area of research. The year was 1939, the same year in which the German invasion of Poland began World War II. Perhaps you have noted the key point in uranium fission already. Uranium fission has as one of its products free neutrons, but it was a neutron that triggered the reaction in the beginning. Whenever one of the products of a reaction is the key material that began the reaction in the first place, the substance is capable of a *chain reaction.* By analogy, imagine that people have a special design. If an adult is hit with a bullet, he or she immediately gives off a burst of energy like a firecracker. In the process the adult splits into two children and several randomly aimed, speeding bullets. A lunatic firing a single bullet into a crowded football stadium filled with such adults would trigger utter chaos. As the first adult underwent fission, the speeding bullets liberated would be likely to strike other adults. In a small amount of time the whole stadium would go up in a roar limited only by the number of people present. Worse yet, U 235 spontaneously gives off an occasional neutron to start the event rolling. It was immediately clear that any time and any place enough U 235 was accumulated in a heap, it would detonate in an explosion equivalent to thousands of tons of high explosives. What prevents this in nature? Clearly a natural chain reaction is prevented by the dilution of U 235. Neutrons liberated by a U 235 fission will be absorbed by other atoms long before they encounter another U 235 atom and trigger another fission. Similarly, a small lump of U 235 would not go up in a chain reaction since more neutrons would be lost to the outside through the surface than would be produced inside the lump. It requires a large enough mass of U 235, called the *critical mass,* to ensure that neutrons liberated at the center are not lost but find other U 235 atoms and keep the chain going. While calculations disclosed that this critical mass was not something you could put in your pocket, it was small enough for a single person to carry (in two separate bundles, please) or to be fired in an artillery shell. This was a frightening prospect anytime, but on the eve of the most total war in human history it was horrifying. World War II produced 15,000,000 military deaths and about twice as many civilian deaths to become humanity's most extensive bloodbath. About a quarter million of these were to be nuclear casualties.

Building A Bomb

Disturbing signs appeared immediately. Foreign-born scientists who had emigrated to the United States learned that Germany was rounding up the available uranium supply. German nuclear scientists were being relocated for top-secret work. Yet, as late as 1941 an article was published by an American journalist speculating on the production of an atomic bomb. The situation in Europe looked so definite that Einstein's colleagues induced him to write a letter to President Roosevelt. Einstein was an archpacifist, yet he understood the concern, and his was the one name which could not easily be ignored by the government. Even then, no real commitment was made until 1943, by which time the Germans had a 4-year head start. Then the American program, named the Manhattan Project, took off in full force. The Manhattan Project constituted the first large-scale involvement between scientists and government and changed the nature of this interaction permanently.

The largest problem presented in the production of a bomb was in the separation of U 235 from natural uranium. After all, both U 235 and U 238 are uranium. Whatever one isotope does the other will do since they are atoms of the same element and share every chemical property. The problem is not unlike that of distinguishing two identical twins, alike down to the last mole, except that one weighs 235 lb and the other 238 lb. The obvious solution would be to weigh them individually, but in dealing with atoms this would take an excessive time. Even if a mass spectrograph would separate 1 billion atoms of U 235 per second, it would have to operate continuously for over 1 million years to accumulate enough U 235 for a bomb. The solution for our twins is easy: Have them run a very long and difficult obstacle course. If the course is demanding enough, the small weight advantage of the lighter twin should determine the winner. To translate this into atoms, the world's largest industrial facility was built at Oak Ridge, Tennessee. Uranium was made into a gas, UF_6, uranium hexafluoride. This gas was then pumped at very high pressure into a chamber with one wall made of porous material. On the other side of this wall was another chamber kept at high vacuum. The gas molecules had to diffuse through billions of tiny pores and passageways in the material to proceed from one chamber to the next. The earliest gas molecules to make it through the barrier had a higher percentage of U-235 atoms than the starting mixture. This procedure was repeated with another chamber using only the earliest gas molecules, which further *enriched* the amount of U-235 atoms in the mixture. As this was done over and over again, the U 235 grew in purity until it was suitable for bomb production. (The uranium for nuclear reactor rods needs to be enriched in U 235 also. The enrichment in this case is nowhere near enough to make them suitable for bombs.)

A second strategy for producing a bomb existed. No naturally occurring isotope other than U 235 would enter into a chain reaction, but the humanmade element plutonium would. Since plutonium is an entirely different element, it could be separated easily from the uranium used to produce it. The device to manufacture plutonium was called a *nuclear reactor.* Rods of uranium metal enriched in U 235 are placed near each other, and the chain reaction in U 235 is allowed to proceed at a steady rate. That is, enough fissions take place to keep the number of neutrons shooting around in the reactor at a constant level, neither increasing nor decreasing. Most of the neutrons do not find a uranium-235 nucleus, and many instead strike U-238 nuclei to produce plutonium. However, the neutrons from U-235 fission are moving too fast to be absorbed efficiently by U-238 nucleus. A material called a *moderator* is thus used between the fuel rods to slow the neutrons to an ideal speed. Calculations disclosed two ideal moderators: carbon atoms, as in graphite, and heavy water (or water made with heavy hydrogen, ${}^2_1H$). At this point another prod for speeding up efforts occurred. After the invasion of Norway, the Germans took special interest in the only plant in the world that was producing heavy water. The only use for heavy water in quantity was as the moderator in a nuclear reactor. The only use for a reactor known at this time was to produce plutonium for nuclear bombs. Thus it became clear that the Germans were indeed working on a nuclear bomb (as it turned out, so were the Japanese and the Russians). The test model of a nuclear reactor was first operated under the stadium stands (since torn down) at the University of Chicago. On December 2, 1942, the human race entered the nuclear age in the middle of a city of 3 million people. Later tests were done in appropriately isolated spots. Banks of large reactors were erected in the desert of Washington state near Hanford. The United States (and Great Britain, as we had pooled efforts in 1941) had its first two massive facilities to produce fissionable materials. Actually, these were the two efforts that worked. Many other techniques were tried and abandoned somewhere along the line. Time was precious, so everything reasonable was tried until it ran out of promise. In a war where oil tankers were being torpedoed within sight of the residents from New

York to the Carolinas, and millions of innocent civilians were being slaughtered in Europe, the prospect of being second to develop a nuclear bomb was not pleasant. History had the final ironic twist. By the time the bombs were ready, the scientifically sophisticated Germans were out of the war. Their effort at isotope separation had failed and their reactor program never had gotten far off the ground. President Roosevelt died, and his successor, Harry Truman, was told of the nuclear program only after he became president. Almost immediately he was presented the problem of authorizing its use against Japan. Within 4 months of his becoming president Hiroshima was destroyed by a U-235 weapon and Nagasaki by a plutonium bomb. The nuclear age had begun on a melancholy note, but it had begun, never to leave.

Campus Sociology and Nobel Prizes

Beer halls have a long association with campuses. College students are traditionally an impoverished class barely scratching out the money to survive long years of study. One requirement of social life, therefore, has been that it be cheap. In dry towns or at the high school level this promoted the soda parlor, malt shop, or hamburger joint through time. (Legend has it that the ice cream sundae was developed in Evanston, Illinois, in response to a blue law stating that ice cream sodas could not be sold on Sunday.) At a more mature, but possibly less sober, level is the beer hall. Local custom varies somewhat. Even in the "old days" of a generation or more ago students at the University of Alaska stocked their own supply in the dorm. This was politely overlooked since a 6-mi walk at −50° F could be hazardous. Most campuses, however, had some spot that made up in enthusiasm what it lacked in appearance. A few became accidental legends. Morey's at Yale gained fame by fitting in a song nicely, for example. A typical off-campus, campus institution was the Pretzel Bell at Ann Arbor. Almost totally devoid of the trappings that we lovingly refer to today as "atmosphere," the Pretzel Bell had more atmosphere than today's color-coordinated medieval castles can provide. The interior was done in depression-sterile decor. The main gimmick was a modestly sized ship's bell mounted on a beam. Whenever some fortunate person came of drinking age, the bell was rung loudly to alert the clientele. This individual was then expected to stand on the table and down the first legal mug (for females) or pitcher (for males) of beer with a careless abandon that took W. C. Fields a lifetime of practice. This ritual constituted the only floor show, entertainment, distraction, or sales promotion. Other than this the place was given over to talking (more frequently shouting) with your tablemates. Believe it or not, in this atmosphere serious physics can be done, and possibly with more enjoyment than in the laboratory.

A group of high-energy physicists from the University of Michigan were in the Pretzel Bell recuperating from the slings and arrows of outrageous fortune that nature throws in the path of most research. Physics was reaching an experimental limit. Particle accelerators had grown into massive monsters that speeded up the charged particles in the beam to incredibly high energies. That was the problem. The particle—a proton, for example—would come streaking into the target at near the speed of light and shatter into smithereens the nucleus it encountered. By studying the pieces of the nucleus, physicists managed to learn more and more about

the nucleus itself. To study the pieces you had to be able to detect them and measure their properties. But at the new, higher energies of the machines the particles were recoiling so fast they just didn't stay in the apparatus long enough to have much happen. The particles kicked out of the nucleus were not simple things you could store on a shelf. Sooner or later they would either decay into pieces or come across another particle, interact with it, and be gone from view. This interaction was the most interesting part of the process. The trouble was, they had to do it sooner to be detected—not later. A particle traveling at near the speed of light spends only three-billionths of a second inside a 1-m-long detecting instrument. The instrument used to record the particles was the Wilson cloud chamber, in which the particles left tracks much as a jet leaves a vapor trail in the upper atmosphere. Since the chamber was filled with air, particles weren't slowed down very much in collisions inside the chamber. Furthermore, it was proving very difficult to keep a large volume of air right at the saturation point where cloud formation could occur. Thus, it didn't seem likely that much larger cloud chambers could be built.

Given this background, let's rejoin our physicists at the Pretzel Bell. The conversation was probably typical of such a group—speculations on the ordinary small phenomena of life—the random oscillations of a passerby; the perfect cubic nature of the salt grains on the table. The conversation became serious and got down to beer—more especially, to beer bubbles. Have your ever noted how bubbles form as a pitcher or mug of beer warms? Certain sites on the inner wall develop a bubble. At some critical time the bubble is released from the wall and another forms and rises until a string of rising bubbles in the beer is produced. Once in a while a string of bubbles is produced in the interior of the beer away from the walls, and it was on these that speculation centered. Why does one spot rather than another in all that volume of beer suddenly generate a string of bubbles? One speculation was that a high-energy cosmic ray streaking down through the room interacted with the beer at that particular point releasing its energy. The energy released by this interaction generated enough heat in a tiny volume of the beer to push the area over the edge and encourage the carbon dioxide to come out of solution and form a gas bubble.*

Donald Glaser was one of the group. He had been working on the recording of particle tracks by bubble formation in a liquid. The next day he tried beer. Unfortunately, it did not work, but hydrogen did. The bubble chamber keeps a volume of liquid hydrogen just at the boiling point. When a high-energy particle comes barreling through the hydrogen, some of its kinetic energy is given to the hydrogen atoms it is plowing through. These local atoms thus acquire enough energy to boil, and they are pushed over the brink and form a bubble of hydrogen gas. Thus, the path of the particle is marked by a string of bubbles at the same time its speed is reduced to keep it in the apparatus long enough to observe its decay or interaction. Dr. Glaser was awarded the 1960 Nobel Prize in Physics for the discovery of the bubble chamber and went on to become the chairman of the U.S. Atomic Energy Commission.

* It probably hasn't escaped your notice by this point that this seems to be a very promising area of research. After all, once the material shows any signs of passing its peak and going flat it is no longer of any research value. One wonders at the reaction of the IRS if the vast supplies purchased were suddenly claimed as research expense.

Summary

In 1895 Roentgen discovered a penetrating form of radiation emitted from cathode-ray tubes. He named these radiations *x-rays* and found them to be much like light rays. Becquerel discovered penetrating rays emitted from the rock *pitchblende*. The spontaneous emission of penetrating rays from matter is called *radioactivity* and is a property of the particular elements thorium and uranium. The Curies isolated two additional radioactive elements in polonium and radium. Each radioactive element is characterized by its *half-life,* or the time required for a sample to lose half of its radioactivity. Radioactive decay represents natural *transmutation,* or the changing of one element into another. Analysis of the rays given off by radium with a magnetic field disclosed three types of radiation. *Alpha* (α) *particles* have a positive charge twice the size of the charge of an electron ($e = -1.6 \times 10^{-19}$ C). The α particle has a mass of about 4 times the mass of a proton, or four *atomic mass units (u),* where 1 u = 1.6×10^{-27} kg. *Beta* (β) *particles* are identical to electrons with a charge of -1 (e) and a mass which is almost negligible (1/1836 u) compared with that of a proton. The third radiation is the *gamma* (γ) *ray,* which is a massless, chargeless form of pure energy, like light or x-rays, but much more penetrating than either of these.

Rutherford demonstrated that the α particle became a helium atom ($^4_2$He) when slowed down to normal speeds. He bombarded a thin film of gold atoms with α particles. From the data Rutherford concluded that almost all of the mass and positive charge of the gold atom must be concentrated in a very tiny portion of the atomic volume. He called this concentration the atomic *nucleus.* Nuclei can be represented by two numbers. The *atomic number* (Z) is the number of protons, or positive charges, the nucleus contains. The *atomic mass number* (A) is the total number of particles or nucleons (protons plus neutrons) in the nucleus (each of which has a mass of about 1 u). When a nucleus decays by α *emission,* Z is lowered by 2 and A is lowered by 4 to produce an atom of a different element. In β *decay,* A is not changed, but Z is increased by 1 since a neutron has changed into an additional proton. γ emission affects neither A nor Z. All atoms of the same element must have the same value of Z, but may differ in A; such are referred to as *isotopes* of the same element. The ability of nature to pack many positively charged protons into the small volume of the nucleus allows us to infer the existence of a new type of force. This *nuclear force* must be much stronger than that of electrical repulsion, but must be effective only across very short distances. With sufficient energy (speed) a charged particle may be forced to within range of this nuclear force and absorbed into the nucleus. Rutherford observed the cloud-chamber tracks indicating such a natural *transmutation* in 1919. Later machines called *accelerators* were built to produce bombarding particles for *artificial transmutation.* The *electronvolt* (eV) is the amount of energy acquired by an electron moving through a potential difference of one volt (1 eV = 1.6×10^{-19} J).

Understanding of the energy changes involved in nuclear interactions came from the *theory of relativity.* The special theory of relativity postulated that length, time, and mass were not absolute quantities but depended on the relative velocity separating two systems. A factor, $\sqrt{1 - (v^2/c^2)}$, relates quantities between two moving systems. If the length, mass, and duration of an interval are L_o, m_o, and T_o for an observer at rest with respect to these measurements, the same values for an observer moving at speed v would be

$$L = L_o\sqrt{1 - (v^2/c^2)}$$

$$m = \frac{m_o}{\sqrt{1 - (v^2/c^2)}}$$

$$T = \frac{T_o}{\sqrt{1 - (v^2/c^2)}}$$

The total energy of an object was proportional

to its relativistic mass by the relationship $E = mc^2$. This equivalence of mass and energy implies that a release of energy would be accompanied by a lowering of mass.

Careful measurements of the masses of nuclei indicated that the atomic mass was not equal to the sum of the masses of the particles making up the atom. This *mass defect* was interpreted as being related to the *binding energy* which held the nucleus together. Since intermediate-sized nuclei have the least binding energy, some energy could be released by producing these nuclei. This could be done by splitting larger nuclei (*fission*) or joining lighter nuclei together (*fusion*). On the eve of World War II it was discovered that the relatively rare isotope U 235 was fissionable. This opened the possibility of a *chain reaction*, leading to both bombs and nuclear reactors.

Questions

1 Why did the Curies believe that the rock pitchblende contained an undiscovered element?

2 If a radioactive carbon-14 atom emits a β particle, what does it change into?

3 What would be left if a neutron underwent β decay?

4 If you shot streams of BB's at a bowling ball, would more BB's scatter in the direction back toward you or at small angles from their original forward direction? Why?

5 The text mentions five different methods for detecting the rays given off by radioactive substances. How many of these techniques can you list?

6 The most widely recognized instrument for detecting radiation is the geiger counter, invented by Hans Geiger, a student of Rutherford. As a particle passes through a gas, it ejects thousands of electrons from the shells of the atoms it encounters. These are attracted to a positive wire in the tube by a high positive voltage, giving a momentary current, or "click." Geiger counters effectively count β particles, only work well with α particles if they have very thin "windows" to expose to the sample, and are quite poor at counting γ rays. Can you explain this behavior of the counter?

7 Why are radiation detectors which render the path of the particle visible always placed inside a strong, uniform magnetic field?

8 Nuclear forces are vastly stronger than other forces. How do we know this?

9 The earliest accelerators comfortably fit inside a small room, while the newest ones are large enough to shelter a large town. Why have these machines grown so?

10 Before relativity theory we listed a separate law of conservation of mass and a law of conservation of energy. How must this change after relativity?

11 Some subnuclear particles last for only billionths of a second before decaying to other particles. What is the advantage of studying them at very, very great speeds in an accelerator?

12 A spaceship racing away from the earth at $0.6c$ ($\frac{6}{10}$ the speed of light) flashes a laser message back at the earth. At what speed does the beam arrive at the earth? A second ship approaching the earth at $0.6c$ flashes us a similar message with its laser. What is the measured speed of its light beam at the earth? What would be notably different between these two messages?

Problems

1 The first six steps in the decay of ${}^{238}_{92}U$ are emissions of α, β, β, α, α, α, respectively. Draw this chain of radioactive emissions, indicating the correct atomic mass number and

atomic number of each product. (Symbols of the elements may be found in the periodic table inside the back cover.)

2 An atom of U 238 decays in a series of steps to an atom of Pb 206.

(*a*) How many α emissions will this take?
(*b*) The atomic number of uranium is 92. If it emitted this many α particles and nothing else, what would its new atomic number be?
(*c*) The atomic number of lead is 82. How many β particles must be emitted in addition to the α particles during the total decay of U 238 to Pb 206?

3 The electrons in a TV set are accelerated across a potential difference of 10,000 V. What is the energy of the electrons that emerge to race toward the screen? In electronvolts? In megaelectronvolts? In joules?

4 α particles ejected from radium atoms emerge with a kinetic energy of 4.8 MeV (1 MeV = 10^6 eV = 1.6×10^{-13} J). The α particle has a mass of 4.00278 u, or about 6.6×10^{-24} g. Neglecting any effect due to relativity, at what speed is an α particle ejected from a radium atom?

5 (*a*) We know that 1 lb of U 235 has about 10 mol of U-235 atoms in it. The fission of a single U-235 atom releases about 200 MeV. How many joules of energy may be obtained from the complete fission of 1 lb of U 235?
(*b*) A competition swimming pool is 10 × 35 × 2 m. To vaporize all of the water in this pool would require about 1.5×10^{12} J. How does the energy released by the 1 lb of U 235 compare with this?

6 A particular type of atom has an average life span of 1.0 s before undergoing radioactive decay. What would the average life span of these atoms be if they were traveling in an accelerator at a speed of $0.9c$?

7 A 7-kg bowling ball is brought to a speed of $0.8c$. If we try to deflect its path, what will we find its mass to be?

8 A spaceship leaves earth at a speed of $0.5c$ for α Centauri. An observer on earth believes the trip took 8 years. What will the traveler on the spaceship record as the time for the trip?

CHAPTER EIGHTEEN

The Atomic Realm

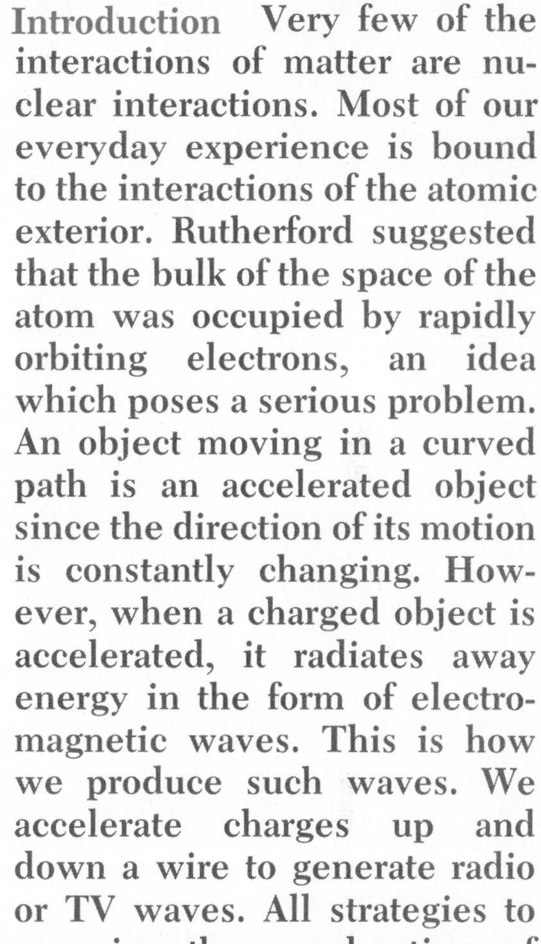

Introduction Very few of the interactions of matter are nuclear interactions. Most of our everyday experience is bound to the interactions of the atomic exterior. Rutherford suggested that the bulk of the space of the atom was occupied by rapidly orbiting electrons, an idea which poses a serious problem. An object moving in a curved path is an accelerated object since the direction of its motion is constantly changing. However, when a charged object is accelerated, it radiates away energy in the form of electromagnetic waves. This is how we produce such waves. We accelerate charges up and down a wire to generate radio or TV waves. All strategies to produce electromagnetic waves require the acceleration of charges, and all observed accelerations of charge result in such waves. Thus we would expect an orbiting electron to be constantly radiating energy in the form of electromagnetic waves. The only apparent source for such energy would be the orbital motion of the electron itself. We would expect that the electron would continually spiral into a lower and lower orbit until it eventually fell into the nucleus. Calculations indicate that atoms with orbiting electrons following the laws of physics known in 1900 should completely collapse in a tiny fraction of a second! If all of the atoms in your house collapsed their electrons into their nuclei, the entire house would fit comfortably within the period at the end of this sentence. Clearly, matter does not collapse and is really quite stable. This must imply that either electrons are not in motion in atoms or our 1900 understanding of the rules was wrong, or, at best, incomplete.

Actually, the last of the great revolutions in physical science had to take place before we could interpret atoms. This was the revolution which produced *quantum mechanics*. While the name sounds imposing, you use the products of its ideas daily, from a transistor radio to a hand calculator. The development of quantum mechanics required the relearning of one of the ideas which should have been learned with relativity. With relativity we found that Newton's ways of summarizing motion did not stretch all the way to the case of extreme speeds. Why should this be surprising? In his life Newton never went half as fast as our current highway speed limit. The marvel is that his laws cover the cases of jet travel and space shots. We may wish for physical laws to be valid in realms entirely beyond the physical experience of the people who first formulated them, but we have no right to expect it. We are about to enter another such realm; that of the ultratiny. It should not be surprising that it will present a very different set of physical behaviors than we might imagine from our experience with huge glops of matter. The behavior of an atom cannot be understood from the behavior of gigantic clusters of atoms, such as dust motes, bacteria, and pieces of lint.

Radiant Energy

Any object above the temperature of absolute zero radiates away some energy. This is obvious for light bulbs or stars but less obvious for people, rocks, and ice cubes. It is, however, a fact. Objects colder than their environs tend to absorb more radiant energy than they themselves emit and thus become warmer. A 1-cm cube of steel slightly above room temperature at 27°C (300 K) will radiate energy at a rate of about $\frac{1}{4}$ W. If we double the absolute temperature of the cube to 327°C we find it radiates at 16 times this rate. In general, the energy radiated by an object is proportional to the fourth power of its absolute temperature, $E = kT^4$. As the temperature increases other changes are apparent also. One of the easiest ways to observe this is to turn the burner of an electric stove on to the "high" position. As the heating coil warms up, no initial change is apparent except a clicking caused by the expansion of the metal. Soon, while no change is visible, your skin informs you heat is being radiated from the coil. As the temperature continues to climb you slowly become aware of a small change in the appearance of the surface of the coil. The color of the metal has switched from a dirty gray-black to a reddish gray. The cause is revealed by switching off the room lights. The coil has achieved the state called *red hot* and is giving off a dull red light. The light will intensify as you watch the continued heating, and it shifts from a dull red through bright red to an orange-red. This is as far as a stove burner will usually allow you to follow the process. The hot wire in a light bulb goes farther (hotter) to orange, then yellow, and eventually yellow-white. *White hot* is about as high in temperature as normal materials can be

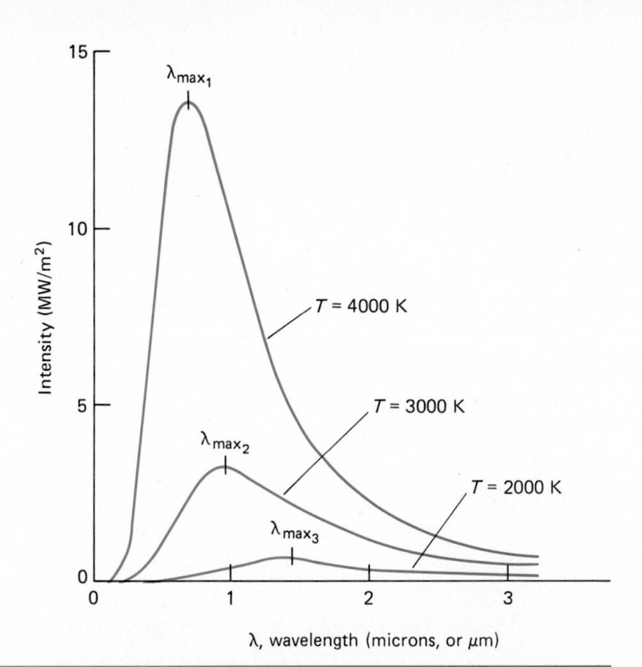

FIGURE 18-1 Intensity of radiation vs. wavelength for objects at three different temperatures. The shapes of these three curves would be quite similar for any three temperatures having the ratio of 2/3/4. Visible light ranges from about 0.35 (violet) to 0.80 (red) μm. Thus, λ_{max} for the object at 4000 K would be a deep red radiation at the limit of visibility. However, each of these sources radiates some light at every visible wavelength and would be incandescent.

heated on the earth. Even the most resistant solids begin to melt after this point is reached. The stars can and do go further yet, to blue hot and beyond. We may conclude from this behavior upon heating that the color of the light produced depends upon temperature. Beyond this we may conclude that a hot object radiates a considerable range of waves. Just because the heating coil is glowing fiercely, it has not stopped producing heat. Furthermore, the fact that white hot can be achieved ensures that a mixture of various wavelengths of light is being given off. Ever since Newton's experiments with light it has been accepted that white is a mixture of various colors of the spectrum. When white hot is achieved, a broad range of wavelengths must be emitted to register at the eye as white. Actually, if we would measure every wavelength being given off and its intensity for a particular temperature, we would obtain a curve such as that shown in Fig. 18-1. If we double the absolute temperature of the object, several things happen to produce the tallest curve shown. First of all, the total energy radiated is given by the area under this curve. As we have seen, the area in the second case must be 16 times as great as before. Secondly, there is a shift in the peak of the curve. The wavelength which is most strongly radiated is designated λ_{max}. If we double the temperature, λ_{max} shortens to half its previous length. As the object becomes hotter it favors the radiation of shorter waves. The relationship of λ_{max} to temperature is particularly simple:

$$\lambda_{max}T = k \qquad \text{where } k = 2.90 \times 10^{-3} \text{ m} \cdot \text{K}$$

For example, the sun radiates essentially all colors of visible light with its peak radiation occurring at about 5.0×10^{-7} m in the yellow range of the spectrum. From this we may calculate the temperature of the sun's surface to be

$$T = \frac{k}{\lambda_{max}}$$

$$= \frac{2.90 \times 10^{-3}\ \text{m} \cdot \text{K}}{5.0 \times 10^{-7}\ \text{m}}$$

$$= 0.58 \times 10^{4}\ \text{K} = 5800\ \text{K}$$

These empirical relationships are reasonably easy to picture and present few surprises. Problems began when physicists tried to predict the shape of the curves in Fig. 18-1 from the laws of Newton and Maxwell. The most careful work predicted the shape for wavelengths longer than λ_{max} quite well. However, these treatments did not predict that the curve would reach a peak and then decline at very short wavelengths. Instead, they predicted that the curve would continue to rise and shoot up off the graph. Since the area under the curve is the energy radiated, theory predicted that an object above absolute zero should radiate an infinite amount of energy. This clearly does not conform to reality.

In 1900 the German theoretical physicist Max Planck (1858–1947) suggested a way out of this difficulty. If one assumption was made about the energy radiated, Planck showed that the exact shape of the curve could be predicted. This assumption was that the atoms radiating energy did not give it off in a continuous stream, but only in "chunks" or "pieces." Planck called such packages of energy *quanta,* and asserted that the amount of energy in a single quantum was not a constant but varied directly with the frequency of the radiation. Thus, if a quantum is radiated, its energy would be

$$E = hf$$

where f is the frequency of the radiation [in hertz (Hz), or cycles per second] and h is a constant, now called Planck's constant ($h = 6.6 \times 10^{-34}\ \text{J} \cdot \text{s}$). This assertion is known as the *quantum hypothesis.* For example, the yellow light given off by a sodium vapor lamp has a wavelength of 5.89×10^{-7} m. Dividing this into the speed of light, we obtain a frequency for this light of 500 billion vibrations per second:

$$f = \frac{c}{\lambda}$$

$$= \frac{2.998 \times 10^{8}\ \text{m/s}}{5.89 \times 10^{-7}\ \text{m}}$$

$$= 5.09 \times 10^{14}\ \text{Hz}$$

The quantum hypothesis would then predict that this light would be radiated *only* in quanta with a fixed energy of

$$E = hf$$

$$= (6.6 \times 10^{-34}\ \text{J} \cdot \text{s})(5.09 \times 10^{14}\ \text{Hz})$$

$$= 3.4 \times 10^{-19}\ \text{J}$$

This yellow light would be given off in chunks, or quanta, and each quantum would have this much energy. If the light were brighter, this would imply that more quanta were being radiated each second, but each would still have this energy.

The quantum hypothesis is really a very revolutionary idea. Coming in chunks or pieces of a certain size is a behavior associated with particles. Having a definite frequency of vibration is a wave characteristic. In the quantum hypothesis we have both a particle property and a wave property bound into the same, simple equation. This was particularly distressing in the case of light since the wave-vs.-particle argument on the nature of light appeared to have been settled decisively a half-century earlier. A decisive experiment was clearly necessary.

The Photoelectric Effect

When light shines on the surface of a metal, it sometimes is capable of ejecting electrons from the metal surface. This is the *photoelectric effect.* If a metal plate is attached to an electroscope, the leaves of the electroscope will disclose the development of an electric charge when light from an intense arc is shined on the metal. The effect works even better when the metal is enclosed in a vacuum tube and a nearby metal grid is maintained at a low positive voltage with respect to the plate (Fig. 18-2). With this arrangement, any ejected electrons are attracted over to the positive grid and do not rejoin the metal plate. In this effect, Einstein perceived the experimental test of the quantum hypothesis. Continuous waves and quanta should behave quite differently in the manner in which they would eject electrons from the metal.

Picture the situation with a continuous light wave washing past an electron in the metal. As the electric field of the wave alternated to the right and then left, for example, the electron would experience an electric force driving it to the left, then to the right. A certain amount of work, W (called the *work function* of the metal), is necessary to break free the electron from the surface. A wave with great amplitude (energy) might be intense enough to do this amount of work on the electron almost immediately and break it free from the metal. A weaker wave might require a longer time to deliver this much energy. The electron could oscillate back and forth with the wave as it absorbed its energy. This oscillation would grow with time until enough energy had been absorbed to break free. Thus, from wave theory, any frequency of light waves should be able to provide electrons with the energy needed, W, to leave the surface. A bright light might do it at once, while a dim light would have to illuminate the surface for some time until enough energy had been delivered. Wave theory thus predicts free electrons for any frequency of light, but with a time delay for a dim source.

The prediction from the quantum hypothesis is quite different. If light energy comes in packages, a single package will either contain enough energy to do the job or not. Even with the most intense light source, the

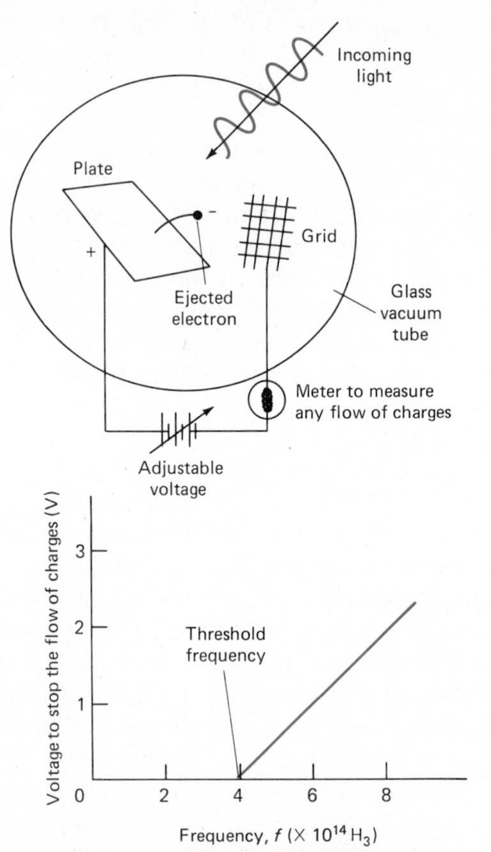

FIGURE 18-2
Light striking a metal plate is capable of ejecting electrons from the metal. This is called the photoelectric effect. If a nearby grid is positively charged, it will attract ejected electrons and a current will flow through the meter. By making the grid negative, as shown, the electrons are repelled back toward the plate. The voltage required to stop all electrons from reaching the grid is a measure of the maximum energy at which the electrons are ejected.

probability of a single electron receiving two quanta at once is so small it may be ignored. A single quantum of light, or *photon* (as Einstein named it), will have to have sufficient energy to allow the electron to break free from the surface. But, since the energy of the photon depends upon the frequency of the light, the color of the light (frequency) is of the greatest importance. Each metal surface would have a particular color of the spectrum, or *threshold frequency,* at which the light photons would have just enough energy to eject electrons. Photons of a color redder than this would not have the required energy. Thus, if the threshold was in the green portion of the spectrum, pure yellow light, even if bright enough to eventually melt the metal plate, would be expected to produce *no* photoelectrons. Meanwhile, a higher frequency than the required green would produce free electrons easily. Furthermore, the instant that a photon with the required energy reaches the plate an electron will be ejected. If the light comes from a blue-green firefly glowing 2 km away from the plate, the instant the photon hits the electron is ejected. The quantum hypothesis predicts no time delay since the photon either has the required energy and ejects electrons at once or does not have the required energy and thus never ejects electrons. Furthermore, if the light is above the threshold frequency, there should be energy left over beyond that required to break free electrons from the surface. This energy should remain with the electron as kinetic energy as it leaves the metal.

The experiment had actually been performed in detail in 1902 but had lacked an interpretation. The results were completely those predicted by Einstein's application of the quantum hypothesis. Light indeed came in packages of a given amount of energy, not as a continuous smear. Quanta were real, not a fiction to help theorists over a problem. Again, however, nature was behaving in a manner that appeared opposed to experience. Picture an analogy. Joshua and his army are in front of the walls of Jericho. At the notches in the walls there are many enemy soldiers. An informer tells Joshua that sound waves can knock the soldiers off the wall and out into the plain before the walls. Joshua arms all the Israelites with large horns, and even borrows amplifiers from all the teenagers. At a signal a deep bass blare is begun that shakes the plain. Some houses in Jericho collapse with the vibration. Chunks of mortar fall from the seams in the walls. But not a single soldier is ejected. Joshua orders a halt to think things over. Was the informer lying? At that moment, a child in the Israelite camp a kilometer back, excited by the festivities, blows a high-pitched flat note on a toy bugle. As the faint wisp of sound reaches the wall, a soldier is shot high into the air and arches out to land with a thud on the plain near Joshua's feet. Would you blame Joshua for thinking that this behavior was opposed to what he had a right to believe on the basis of experience? Since humans have experience with large amounts of energy and big masses, we might forgive Joshua.

Order in the Spectrum

When it comes to the interaction of subatomic masses and radiation, the science of 1900 had only one area of experience, which had not been fully recognized as such—the bright-line spectra of elements. In a low-pressure gas the atoms are usually too far apart to interact with their neighbors. The light of the spectrum emitted by such a gas represents the internal adjustments in energy of single atoms. Only one mathematical regularity had been noted in the patterns of bright lines. The colors emitted by hydrogen atoms seemed to fall into a regular series of frequencies. The formula for this series was discovered by a secondary school teacher named Balmer. It was

$$\frac{1}{\lambda} = R\left(\frac{1}{4} - \frac{1}{m^2}\right)$$

where m could take on the values 3, 4, 5, . . . , and where the set of values of λ produced were the wavelengths observed in the hydrogen spectrum. R was a constant with the value of $1.1 \times 10^7\ \text{m}^{-1}$. Subsequently, other series of spectral lines were discovered for hydrogen in the infrared and ultraviolet portions of the spectrum. All fit the formula

$$\frac{1}{\lambda} = R\left(\frac{1}{n^2} - \frac{1}{m^2}\right) \qquad n, m = \text{integers}, m > n$$

If $n = 1$ the ultraviolet series of lines was obtained; $n = 2$ for the visible Balmer series; $n = 3, 4$ for infrared series of lines. There was no idea why this order should occur in the spacing of the wavelengths emitted by hydrogen. Likewise, there was no explanation for the apparent lack of order in the spectra of other elements. This was about to change.

The Orbits of Electrons

Rutherford had suggested that the bulk of the atomic volume could be filled by orbiting electrons. As pointed out, theory suggested such an accelerated electron should radiate away energy, and the atom should collapse. The person to fill in the remainder of the atom was Niels Bohr (1885–1962). Bohr was both a student of Rutherford and one of his close friends in later life.

Decades or even centuries of human experience are tied up in a scientific law. Clearly, if such a law has passed thousands or millions of scientific tests, it is not to be discarded lightly. Still, it may be that such a law is not an exact statement of how the physical world is behaving, but only an approximation for a certain range of human experience. When a new range of experience opens up, a law may or may not require modification. Bohr worried about this as he was about to discard a tried and true physical law. What are the pitfalls inherent in rejecting an established idea? He summarized it in his *correspondence principle:*

> *A new law governing physical behavior must be such that when translated into the older areas of experience it would produce the same predictions of behavior as the law it seeks to replace, while yielding correct predictions for the new area of experience where the old law breaks down.*

With this principle, Bohr rejected the argument that electrons in orbits should radiate energy with the reply that they obviously do not, so let us look for a new relationship which applies to the world of the atom as well as the realm of common experience. Recent experience had suggested that various commodities came in discrete packages of a given size rather than of any value whatever (such as mass in definite particles, or energy in quanta). Bohr proposed that the *angular momentum* (mvr) of the electron in a hydrogen atom could come only in packages of a given size. (He proposed this theory because it worked; that is, it led to correct predictions. Exactly why it worked he spent his life trying to discover.) The basic package was the Planck constant h divided by 2 times pi ($\pi = 3.14159 \ldots$). The angular momentum could be $h/2\pi$, or twice this, or 3 times this, and so on. In short,

$$mvr = \frac{nh}{2\pi} \qquad \text{where } n = 1, 2, 3, \ldots$$

Most of these values (such as m, the mass of the electron; or h, the

Planck constant) were known. The speed of the electron, v, could be eliminated in calculation. From this, Bohr could work out a value for r, the theoretical radius of the hydrogen atom. Bohr calculated a value of the radius of the hydrogen atom from this formula and obtained $r = 5.29 \times 10^{-11}$ m. For the first time, the probable size of an atom had been predicted from theory. Up to this time scientists had no idea why an atom should be any particular size. Furthermore, the prediction was exactly in agreement with the measured size of the hydrogen atom! More than any other single event, this prediction of a known but otherwise unexplained value set science on the road to a full quantum theory of matter and its interactions. Bohr summarized his findings for electrons in a series of postulates:

1 The electron normally occurs in one of a series of *stationary states* at which no radiation of energy takes place.

2 Energy which is emitted or absorbed causes the electron to make a transition between two of these allowed states. The energy which is then emitted or absorbed is equal to the energy difference between these states.

3 Each allowed state is given by the formula $mvr = nh/2\pi$.

4 The ordinary laws of physics apply except for the transition between stationary states.

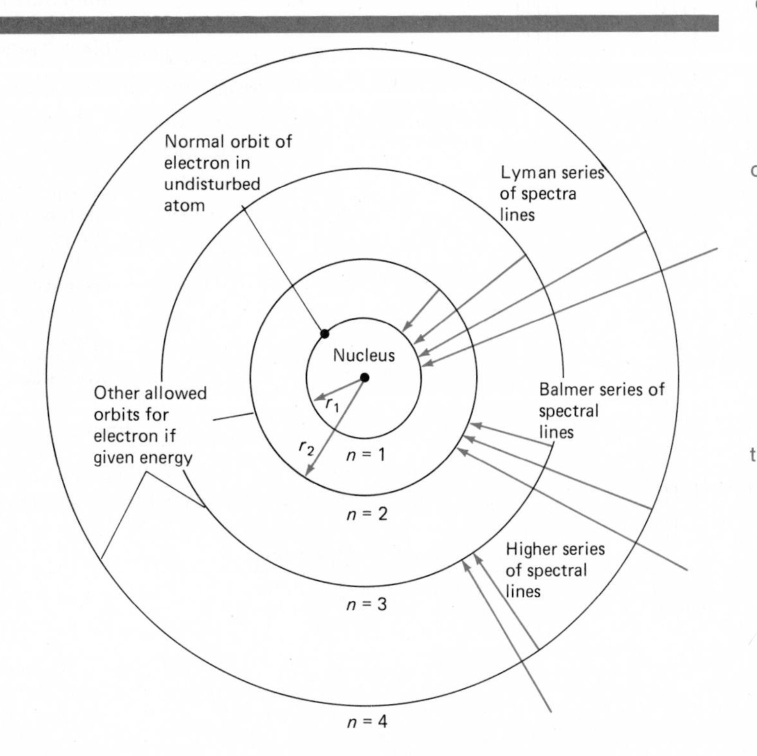

FIGURE 18-3
The Bohr model of the hydrogen atom has the electron normally occupying the orbit closest to the nucleus ($n = 1$). If the electron is "disturbed" by being given energy, it may be raised to a higher orbit. However, only very specific orbits are "allowed," and nothing is allowed between these. When the electron drops from a higher orbit to a lower one, it must release energy as a burst of electromagnetic radiation. For example, the set of all drops from a higher orbit to the orbit where $n = 2$ gives rise to the set of visible spectral lines for hydrogen. The set of electron transitions to the lowest orbit gives rise to the ultraviolet Lyman series of lines. Orbital radius is proportional to the square of n. Thus, the second orbit is 4 times as far from the nucleus as the first orbit.

Bohr immediat
tulates for the hydr
defines a possible
principle quantum
the hydrogen elect
outside of the atom
measurements had
the voltage require
Again, right on! Fi
electron made a tr
For example, the tr
3.03×10^{-19} J. Fro
a frequency of 4.57
this gives a wavel
the visible spectr
6.56279×10^{-7} m.
Bohr's rules for the
dicted every spect
radiation or not.

Electron energy (eV)

The Rutherfords on the left and the Bohrs on the right were close friends in addition to having their names attached on a model of the atom. (Mrs. Oliphant, wife of the photographer, is in the center.) (*Courtesy of AIP Niels Bohr Library.*)

Theoretical physics of the shootout

"The Gunfight at Blegdamsvej Strasse"

Niels Bohr was a legend of sorts in his own time. He served as the head and chief "father figure" of a generation of theoretical physicists. While a great teacher, he was apparently far from being a good lecturer. Most of his formal addresses were ponderous things that had to be endured. At a more personal level, Bohr was superb. His students became part of his "family" at the Institute for Theoretical Physics in Copenhagen. Many a night, the work on the structure of the atom would go on all night with lively discussions in the central meeting room. When he was personally making little progress, Bohr liked to escape from his work with his favorite recreation. He was a devoted fan of American western movies. The reader of today might have to strain a bit to get the correct mental picture. This was before the "adult" westerns, or social commentary movies. These were "shoot-em-ups" with the cavalry charging over the hill to rescue the wagon train from the nasty old Indians. The score at the end usually stood "Good Guys 113; Bad Guys 0." These were the days when the hero wore white and the villain could be identified without half trying by his moustache and black hat (very useful when the sound-system quit). Bohr was an intent viewer and really got involved in the minimal plots.

One night after the movie, Bohr was walking back with some of his students. He broke the silence by asking whether anyone had observed that in the shootouts, the bad guy invariably drew his gun first, but the good guy always won. Well, this statement didn't excite the students very much and they responded with a cautious "so what?" Bohr said he had a theory to propose to explain his observation. His students were puzzled; they didn't see that anything needed explanation. Bohr persisted. He pointed out that the bad guy knows he will have to make the first move. He is therefore involved in making a conscious decision of when to start. On the other hand, the good guy (the one in brilliant target-white, remember) also knows that the first move is up to the bad guy. He therefore is not making a decision. He leaves the decision to the villain and merely reacts to the events. Now, the point of all this, Bohr continued, is that a conscious decision requires processing by the brain and the resulting command to the muscles is sent out along the relatively slow pathways of the nerves executing conscious commands (the central nervous system). Meanwhile, the good guy is not playing the same game. He is set to react to the first visual impulse he receives of the bad guy going for his gun. This can be cycled up to an automatic act not requiring processing in the brain, and use the high-priority nerve paths of the autonomic nervous system—the one controlling reflex acts. (Perhaps this explains why there could be so many mindless heroes in those days, too.). Result—good guy has the villain well ventilated before the villain can even confirm muscular control of the gun.

Bohr's students were astounded! They tried to point out to him that the good guy won because the script called for it! They pointed out that if the good guy didn't win there would be moral outrage and the letters would have swamped Hollywood (they would have, too, in those days). Bohr acknowledged most of the counterarguments, and even the most telling one—that most of the real good guys as well as bad guys in the old West were such rotten shots that if you could have collected the lead thrown around you could have used it to line the casket of the loser. Still Bohr persisted. The basic idea was sound, he asserted, and should work much as he described it. To the casual stroller in Copenhagen it might have been one of the funniest arguments he had ever come across, but the battle lines were drawn now! Neither side was going to give in easily. The students rattled the script and Bohr waved the theory of nerve function. How do you settle such a disagreement? Brahe would have fought a duel! Why not? What does any red-blooded physicist do when confronted with conflicting opinions on a physical situation? Experiment!

It was agreed! The next day, George Gamow was to go out and obtain two holster sets with cap pistols. Bohr would be the good guy, Gamow the villain. They would wear the guns until Gamow, at an appropriate moment of his choosing, would call out Bohr for the showdown. Well, this idea sounds fine on paper, but there were complications. Here was Bohr, winner of a Nobel Prize, probably the most respected man in Denmark after the king, and the director of the Institute for Theoretical Physics at 22 Blegdamsvej Strasse, committed to run around all day in a toy holster and cap pistol. (We haven't mentioned it, but Bohr was a gigantic man. In photos he is the one two heads taller than the pack. He was also very athletic. His doctor made him give up skiing when he was in his 60s. Bohr then took up tennis and wore out opponents. In desperation, his doctor recommended bicycling. Bohr agreed, and bicycled from Copenhagen to Athens, Greece.) The mental picture of this huge man with a toy cap pistol slung over his business suit is incredible. The next day couldn't have been worse for the experiment. A group of visitors were due in for a presentation of Bohr's interpretation of complex spectra.

In spite of all this, the experiment proceeded. With heroic dignity, the visitors managed not to notice or comment on Bohr's strange costume. Bohr, having worn the gun all morning to get used to it, had overlooked it by this time and forgot to enlighten them. Events were proceeding smoothly in the seminar where Bohr was at the blackboard running through several equations. At this point, Gamow chose to step conspicuously into the aisle! Bohr continued talking, switched the chalk to the other hand and turned to face Gamow. Gamow drew! Bohr proceeded to outdraw him, fired the first shot, and turned back to the board without interrupting the sobriety of the occasion. Unfortunately, history has not recorded the reaction of the audience to all of this. If much in the way of theoretical atomic physics was learned after this incident, it must stand as a monument to human concentration and mental discipline!

experimental results. Uhlenbeck and Goudsmit imagined the electron itself to be spinning on its axis as it orbited the nucleus, much as the earth spins on its axis as it orbits the sun. Since the electron is a charged particle, this spin would set up a small magnetic field. This spin was of a constant size but could line up either in agreement with the external field or in disagreement. For this they assigned the *electron-spin* quantum number, m_s, which could have only two values, $+\frac{1}{2}$ or $-\frac{1}{2}$.

Each of these modifications improved the number of spectral lines which could be predicted. However, there were fundamental problems. Each modification or addition was precisely that, a "tack-on" to save the theory and make it agree with experiment. There was no doubt that the idea of quantized behavior of electrons was substantially correct, but there was no hint of why this was the case. Furthermore, while the theory could predict the energy (frequency) at which a spectral line should occur, it could not predict the observed differences in brightness between the lines. A dozen years after Bohr's original breakthrough into the world of the atom a new interpretation emerged.

The de Broglie Hypothesis and Matter Waves

In 1924 a young French scientist, Prince Louis Victor de Broglie (born 1892), completed his doctoral dissertation at the Sorbonne in Paris. The idea he proposed seemed like a wild speculation, yet once it was announced it caused an immediate revision of the way the atom was viewed.

The photoelectric effect demonstrated that light waves have particle-type properties. Could it not be that a particle, such as the electron, would have wave-type properties? Photons were capable of transferring momentum (mv) upon impact just as a particle would. The momentum transferred was inversely proportional to its wavelength ($mv \propto 1/\lambda$). Why should not the moving electron have a wavelength associated with it which was inversely proportional to its momentum? By equating the energy of a moving particle as derived by Einstein ($E = mc^2$) with the energy expression of Planck ($E = hf$), de Broglie derived the simple formula

$$\lambda = \frac{h}{mv}$$

as the wavelength to be associated with a particle moving with momentum mv (h is the Planck constant). Stripped to its bare essentials, de Broglie's suggestion was that *moving particles have wave properties.*

How does the de Broglie hypothesis influence the view of electrons in atoms? Consider the electron in orbit. Let us form a mechanical picture of this situation. Picture something that wiggles up and down while racing around in a circle. Both of these motions are very fast, so the only thing we can perceive is the long-time average disturbance for a particular spot. For simplicity let us "snip" the circular orbit and lay it out along a straight line (Fig. 18-5*a*). If there is no particular relationship between the up-and-down oscillation and the motion along the orbit, each time around finds the oscillation at a different portion of its cycle as it passes a given point on the orbit. For example, the solid wave runs off the diagram at the right, but for a circular orbit, the right end and the left end of the line are the same point. Thus, the dashed line continues the same oscillation on the second pass of the orbit. The second pass does not agree at all with the first, nor will the third pass agree with the first two. If we sampled a point on this orbit we would find every part of the wave occurring at that point at some time, from crest through trough. Since the only thing we can sample is the long-time average disturbance at that point, such a wave would be *self-canceling*. The average disturbance at any point along the path would be zero. Can such a wave be noncanceling? Picture the situation in Fig. 18-5*b*. The wave begins as the line begins and proceeds through crest and then trough. The length of the orbit is just sufficient for one full wave. The second time this oscillation begins it will be back at the left end and will repeat the drawn pattern exactly. Each pass will repeat exactly the pattern drawn. This wave is not self-canceling. Each phase of the oscillation cycle occurs always at the same point along the orbit. The position where the crest is drawn is always a crest, and so on. The long-time average of this wave would not be zero for any point along the orbit (except the two fixed nodes). We may note that the requirement to produce this situation was that the length of the orbit ($2\pi r$, since it is a circle) be exactly one wavelength. The same situation would occur if $2\pi r$ was exactly two wavelengths, or three, or any whole number n. Thus, the condition for a noncanceling wave in orbit becomes

$$2\pi r = n\lambda$$

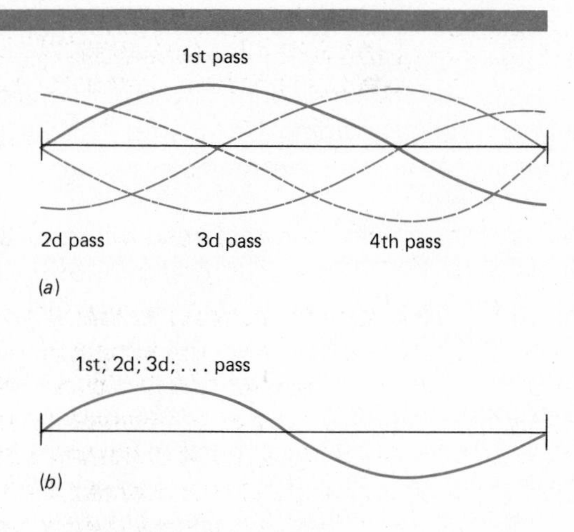

FIGURE 18-5 A wave proceeding around a circular orbit (here cut and laid in a line) produces a different amplitude at each point on each pass. Only if the length of the orbit is exactly one wavelength (or any whole number of wavelengths) will a stable pattern that repeats exactly be formed.

However, according to the de Broglie hypothesis, $\lambda = h/mv$; thus

$$2\pi r = \frac{nh}{mv}$$

By simply rearranging terms in this expression we obtain

$$mvr = \frac{nh}{2\pi}$$

But, this is *exactly* the expression which Bohr had to assume in 1912 to explain the properties of the hydrogen atom. In the dozen years between Bohr's assumption and the de Broglie hypothesis no one had been able to determine why this relationship should be correct or what it was really saying about the design of nature. Only one thing had been certain: that it was essentially a correct description of nature. Suddenly, de Broglie presented an explanation. *Matter in motion has wave properties.* For a huge object, such as a bacterium, the mass is so great that the observed behavior is dominated by those properties we ascribe to particles. For a tiny object, such as an electron, the wave properties can never be negligible. In certain circumstances the particle properties will dominate the observed behavior, and under other conditions the wave properties will dominate. Why are electrons confined to certain given orbits only? These are the only positions where a standing wave of the electron's wavelength may be formed. At these positions the electron is self-reinforcing, while between these positions its wave nature makes it self-canceling.

This interpretation of matter as a dual wave-particle is fascinating, but is it real? If electrons are—What do you call them? "Wavicles"?—they should show the standard wave properties under the correct conditions. Electrons should show refraction and reflection—but they do, since these properties are shared by particles. They especially must display the "pure" wave property of diffraction. Very promptly the conclusion of the laboratories came in—they do! Electrons (and by implication all matter) display wave properties. Similarly, light (and by implication all radiation) displays quantum or particle behavior. Once again nature had revealed one of its subtleties, and it was a brand new ball game. Old laws and ways of viewing things were not wrong; they were right only for a restricted range of masses and motions. Once the atomic realm was entered the wave nature of particles could not be ignored. Our viewpoint would have to be reshaped for a *wave mechanics* or *quantum mechanics,* as it came to be called.

The new science of quantum mechanics was drafted independently mainly by two individuals (Fig. 18-6). Erwin Schrödinger (1887–1961) worked from the classical theory of waves, while Werner Heisenberg (born 1901) used a purely mathematical approach independent of earlier theory. Indeed, it took many years to prove that in addition to supplying the same answers to problems, the resulting methods were identical mathematically. There is no such thing as nonmathematical quantum mechanics; it is very complex and requires the precision of mathematics at every step. This does not imply that its conclusions are of

(a)

(b)

FIGURE 18-6 The individuals who separately produced a mathematical theory of wave mechanics were Erwin Schrödinger (*a*) and Werner Heisenberg (*b*). Their two treatments seemed entirely different but yielded the same answers for the same problems. It took many years to prove that the two theories reduced to forms which were mathematically identical. *(Photographs courtesy of AIP Niels Bohr Library, Bainbridge Collection.)*

interest only to scientists. Every day the typical American uses quantum-mechanical devices that were completely beyond the power of science to understand in the 1920s. Today you can carry more calculation power around in your hand than the largest computers on earth possessed in the 1950s. This miniaturization is part of a quantum-mechanical revolution.

This may all be true, but yet the mind wants to know. If we could sit down in some submicroscopic subway next to an electron, what would it be like?—a super-jiggly bowling ball or a bunch of waves chasing their own tails? Almost at the start, Heisenberg showed that nature may have built in its own way of foiling our ability to find out. This is called the *Heisenberg uncertainty principle.* Basically, this principle states that as we measure one property with greater and greater precision, another becomes fuzzier and fuzzier, leaving us as badly off as we began. Such a set of properties is position and momentum. If Δx is the error in the measurement of position, and Δmv is the error in momentum, Heisenberg showed that $(\Delta x)(\Delta mv) \geq h/2\pi$. Thus, if we really tried to pin down the electron's wavelength (by measuring mv), we would lose all knowledge of where it was. Since this idea was formulated, it has held up quite well in the face of clever experimentation; thus the limitation may well be fundamental in nature.

The Quantum and Chemistry

Just as with the extension of Bohr's theory, the new quantum mechanics describes the electron in an atom in terms of four quantum numbers. These numbers determine the shapes of electron orbitals in atoms. Since the electron is not a tiny billiard ball, it is not possible to draw a discrete

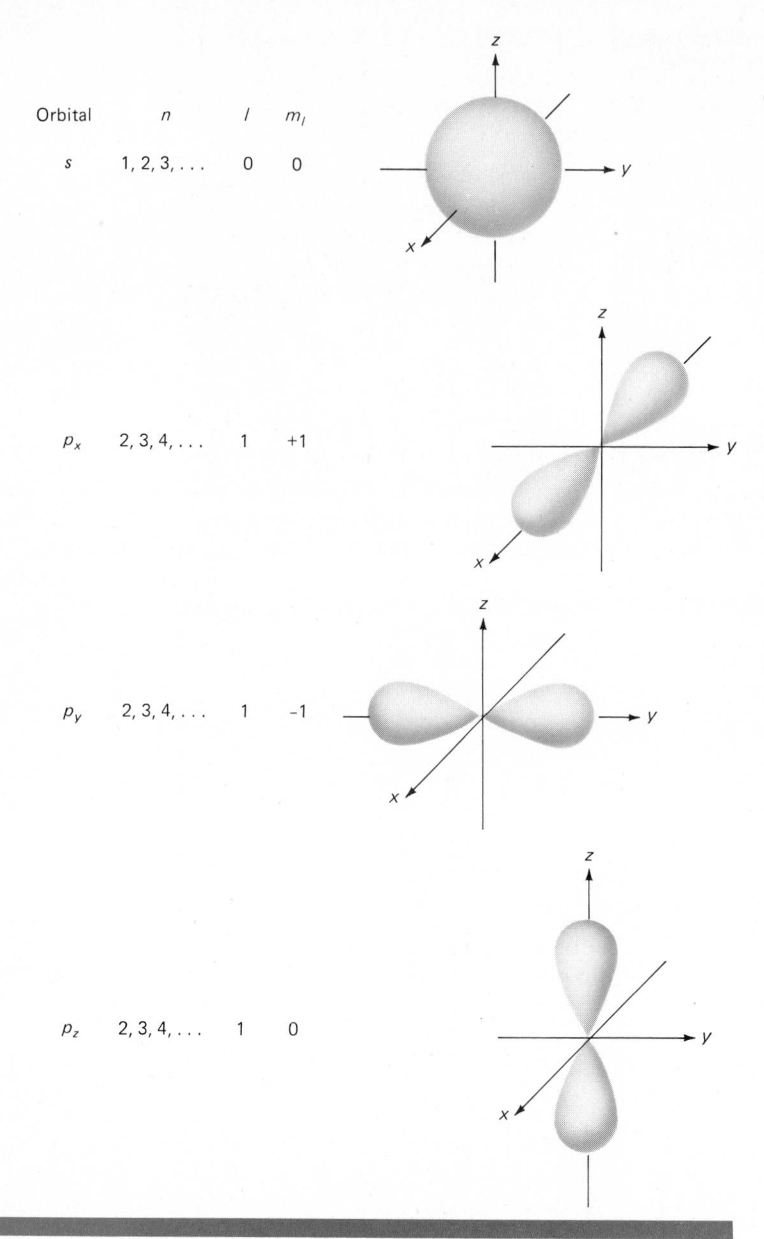

FIGURE 18-7
Calculation of the likelihood (probability) of finding the electron at any position in space around a nucleus yields the electron-cloud representation for electron orbits. The first two electrons in any shell (*s* electrons) are equally likely to be in any direction from the nucleus and have the highest probability of being at a distance which is the orbital radius for that shell. We may represent this as a spherical shell of electron density about the nucleus. After the two *s* electrons, the next six electrons in the same shell—the *p* electrons—do not distribute themselves evenly in space about the nucleus. Rather they distribute about three mutually perpendicular directions in space which we may call the *x*, *y*, and *z* axes, respectively, in a type of figure-eight or "pinched-sausage" shape as shown.

pathway for one in an atom. But, it is possible to calculate the likelihood of the electron's being at any given position at a given time. This leads to the *electron-cloud model of the atom,* which represents the position of the electron averaged over time. For the first orbit ($n = 1$) this is not very surprising, since the cloud is a spherically symmetrical shell centered on the nucleus. But, for $n = 1$, the angular-momentum quantum number l must be zero. Calculations show that whenever $l = 0$, a spherical distribution of the electron about the nucleus will occur. However, when l is not zero, different shapes emerge. When $l = 1$, the distribution is centered on a line through the nucleus, and looks like a figure eight in cross section (Fig. 18-7). Additional electrons in the same shell with $l = 1$ will align their orbitals along lines *perpendicular* to this first one. Thus, there are three possible orbitals for a shell when $l = 1$—one along some line we may call the x axis, one along a y axis, and one along the z axis of space. These three directions correspond to the three possible values of m for the case of $l = 1$, namely, $m = +1, 0, -1$.

At this point it is appropriate to call a time-out to review terminology. The quantum number l determines the geometry of the orbital, and each value of l also has a letter associated with it. These letters came from the various series of spectral lines and are always used in connection with electron orbitals (Table 18-1). Thus, s orbitals are the spherically symmetrical ones that occur when $l = 0$, and p orbitals are the figure eights of $l = 1$, and so on. Sometimes, the p orbitals are separately noted as p_x, p_y, and p_z, respectively, to distinguish their mutually perpendicular orientations in space. The d orbitals and beyond have an even more complex geometry than the p orbitals, but electrons in these orbitals usually are not engaged in interactions with other atoms so we will ignore them.

Wolfgang Pauli (1900–1958) formulated a basic idea concerning the electrons in an atom in terms of quantum numbers. Basically, as an atom gets larger, each new electron joins the atom by going into the lowest energy state available. However, *no two electrons in an atom may have the same set of values for the four quantum numbers* (n, l, m, m_s). This idea cleared up the arrangement of elements in the periodic table, and is called the *Pauli exclusion principle.* With one electron in an atom (hydrogen), the electron will occupy the lowest energy state of $n = 1, l = 0, m = 0, m_s = \frac{1}{2}$ (or 1, 0, 0, $\frac{1}{2}$, for short). Of course, if energy is given to that electron, it may be raised to a higher energy state temporarily. However, it will quickly fall back to this *ground* state in one or more jumps between shells by radiating away bursts of energy. Each possible radiated photon will represent a jump between allowed orbitals and a specific line in the hydrogen spectrum. When a second electron occupies the same atomic volume, as with helium, the set of quantum numbers (1, 0, 0, $\frac{1}{2}$) is already taken. The second electron will have the first three numbers the same, but differ in electron spin (1, 0, 0, $-\frac{1}{2}$). This represents a very minor energy difference from the first electron. Two electrons may share the same orbital provided they have opposite electron spins. However, this does end all possible combinations with $n = 1$. We say that with helium, we have filled the first shell. An *electron shell* thus consists of all orbitals with the same value of n.

With the lithium atom we begin the second electron shell. The first two electrons have the same quantum numbers as helium (1, 0, 0, $\frac{1}{2}$; and

TABLE 18-1
Terminology for Atomic Orbitals

Value of l	0	1	2	3
Letter used	s	p	d	f
Spectral series name	Sharp	Principle	Diffuse	Fundamental

TABLE 18-2
Orbital Arrangement of Electrons for the First 13 Elements*

ATOMIC NUMBER	ELEMENT	QUANTUM NUMBERS (ORBITAL)						
		$n = 1$, $l = 0$ (1s)	$n = 2$, $l = 0$ (2s)	$n = 2$, $l = 1$ (2p)			$n = 3$, $l = 0$ (3s)	$n = 3$, $l = 1$ (3p)
				p_x	p_y	p_z		
1	Hydrogen	↑						
2	Helium	↑↓						(First shell complete)
3	Lithium	↑↓	↑					
4	Beryllium	↑↓	↑↓					
5	Boron	↑↓	↑↓	↑				
6	Carbon	↑↓	↑↓	↑	↑			
7	Nitrogen	↑↓	↑↓	↑	↑	↑		
8	Oxygen	↑↓	↑↓	↑↓	↑	↑		
9	Fluorine	↑↓	↑↓	↑↓	↑↓	↑		
10	Neon	↑↓	↑↓	↑↓	↑↓	↑↓		(Second shell complete)
11	Sodium	↑↓	↑↓	↑↓	↑↓	↑↓	↑	
12	Magnesium	↑↓	↑↓	↑↓	↑↓	↑↓	↑↓	
13	Aluminum	↑↓	↑↓	↑↓	↑↓	↑↓	↑↓	↑

* An upward arrow indicates an electron spin of $+\frac{1}{2}$, while a downward arrow indicates an electron spin of $-\frac{1}{2}$.

1, 0, 0, $-\frac{1}{2}$) to fill the $n = 1$, s orbitals. The third electron will have $n = 2$. While l in this case could be zero for an s orbital, or 1 for a p orbital, the case of $l = 0$ always represents the lowest energy state for any value of n. This means that s orbitals will always fill with electrons before any are added to a p orbital of the same shell. Lithium will have two electrons in its first-shell s orbitals and one electron in the second-shell s orbital. The shorthand for this is $1s^2 2s^1$. Beryllium, the next atom, has the *electron configuration* of $1s^2 2s^2$. Boron, with an atomic number of 5, adds the fifth electron to a p orbital of the second shell ($1s^2 2s^2 2p^1$). The electron configurations for the first 13 atoms are indicated in Table 18-2. If we were to describe that of aluminum, it would be $1s^2 2s^2 2p^6 3s^2 3p^1$. This informs us of what type of orbital each of the 13 electrons of aluminum occupies.

Electrons added to an atom always seek the lowest energy state available, just as a bowling ball dropped into a pit would tend to lodge in the lowest niche available. This implies that a new shell may be started even before the previous one is filled. For example, not all orbitals with $n = 4$ have lower energy than some orbitals with $n = 5$. In general, the order of energy as calculated from quantum mechanics is: 1s, 2s, 2p, 3s, 3p, 4s, 3d, 4p, 5s, 4d, 5p, 6s, 4f,. . . . This would be very difficult to remember, but a simple device allows the order to be determined any time it is needed. Writing the various orbitals in a triangular table of n values vs. l values as in Table 18-3, the order of filling is given by diagonal sweeps from upper right to lower left as shown. The jumps in this order match perfectly with the various disruptions in the periodic table. For example, after the two 4s electrons are added in the potassium and calcium atoms, respectively, the next set of orbitals to be filled is the 3d, which can contain 10 electrons (five possible orbits with a pair of elec-

TABLE 18-3

n Values vs. *l* Values*

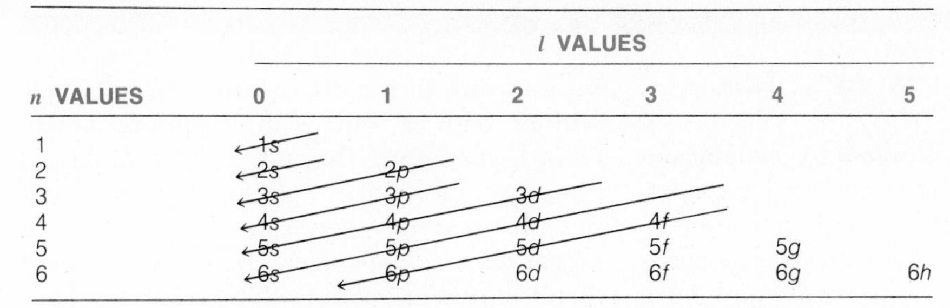

n VALUES	*l* VALUES: 0	1	2	3	4	5
1	1*s*					
2	2*s*	2*p*				
3	3*s*	3*p*	3*d*			
4	4*s*	4*p*	4*d*	4*f*		
5	5*s*	5*p*	5*d*	5*f*	5*g*	
6	6*s*	6*p*	6*d*	6*f*	6*g*	6*h*

* The order of increasing energy of subshells (orbitals) is correctly given by the diagonal sweeps.

trons in each). The periodic table at this point has 10 elements from scandium ($Z = 21$) to zinc ($Z = 30$), all of which are metals and reasonably similar chemically. These are called *transition elements,* and all have two electrons in their outer shell, the 4*s* orbital. The change between elements is that electrons are being added to an inner shell, the 3*d* orbitals, not the outer shell so involved in the chemistry of the atom. Another point of interest about this group of elements may be noted. In general, as an electron is added to an atom and could assume one of several vacant orbitals, it will take a private orbital rather than share an orbital with an existing electron. As we add 10 electrons to the five 3*d* orbitals of this group, by the middle of the group we have many *unpaired* electrons. Paired electrons, with opposite spins, cancel the effects of each other to produce no external magnetic effect. Unpaired electrons, on the other hand, need not cancel and may in fact align their spins to produce a magnetic effect detectable outside of the atom. This external effect may cause nearby atoms to align their spins either with or against those of the first atom. If large groups of atoms align their spins with each other, the result may be a magnetic effect detectable in the space surrounding the matter. This is the origin of permanent magnets, and the transition element of this group most noted for this effect is iron ($Z = 26$).

We could proceed with the development of the periodic table, but by this time you have perceived the moral. Every variation in the ordering of elements of the periodic table is a reflection of changes in orbitals to which electrons are being added for particular elements. The geometry of orbitals has a great deal to do with the geometry of the molecules that result from chemical bonding. For example, consider molecules with two atoms singly bonded to an oxygen atom. Water is a good example. Heating measurements disclose how difficult it is to get such a molecule to rotate. Such measurements indicate that the three nuclei (oxygen and the two other atoms) are never arranged in a straight line, but that the two additional atoms attach to the oxygen at nearly right angles to each other. The orbitals used for bonding by oxygen are two *p* orbitals. Each of these has a very definite direction in space, and the two orbital axes are perpendicular to each other (Fig. 18-8). The geometry of the H_2O molecule is immediately clarified by picturing the orbitals involved.

The university research lab

The majority of the elementary school science programs of the last 15 years have laboratory activities built in as a part of the course. Virtually all high school science curricula feature laboratory programs, and have for generations. It is therefore difficult to realize how new the idea of a school science laboratory really is. Before this time much of the scientific work purused was conducted in kitchens, living rooms, and offices. Davy set up his equipment and discovered the element iodine in a hotel room. Goodyear discovered the process of vulcanization of rubber at his kitchen stove after decades of working out of the family kitchen. Private laboratories were a preserve of the rich and were usually converted kitchens. Newton did most of his work in his apartment. Rutherford worked for a time in a sheepherders' shed in New Zealand. Hall perfected the electrolytic separation of aluminum in the equivalent of the family garage.

When the Cavendish Laboratory became the operating address of the Department of Physics of Cambridge University in 1874, it was a bold innovation. Sixty years after its opening, the Cavendish still required a large loft room under the roof to be reserved for new graduate students. This room was dubbed "the nursery," and it was here that some of the most brilliant minds in physics first learned to handle simple equipment. This experience was almost totally lacking in the students sent to the Cavendish from the leading schools around the world. It seems a bit amusing to picture a future Nobel Prize winner with an outstanding undergraduate record searching through a bin of wood scraps or bits of wire to put together the apparatus for a simple experiment. Today, some elementary schools have better equipment than the nursery could offer, and many high schools have an equipment base that would awe a worker at the Cavendish during its first 50 years. The idea of direct experimentation being closely associated with instruction in the sciences is quite new historically.

Further, the Cavendish *did* attract the future Nobel Prize winners. In the first 75 years of the Nobel Prize, 22 Nobel laureates have been associated with the Cavendish Laboratory. It was here that J. J. Thompson "discovered" the electron, that Rutherford continued his string of discoveries on the atomic nucleus, that Chadwick discovered the neutron, that Cockcroft and Walton built the first particle accelerator and with it produced the first humanmade transmutation of matter; and, more recently, it was here that Watson and Crick decoded the structure of DNA.

It would be hard to deny that the experiment of associating a research laboratory with a university science department has paid off well during the first 100 years of the Cavendish Laboratory in an increase in human understanding of nature. In many ways this association became the model for the science departments at major universities. It would be surprising to find a major school of science today without laboratories, and almost incredible to find a major graduate school without research facilities in science.

Similarly, we would expect the three covalent-bond directions of the nitrogen atom to be nearly mutually perpendicular since p orbitals are involved. When covalent bonding occurs between atoms, the electron clouds of the atoms merge to produce a new cloud geometry. The shape of this new distribution may be calculated from wave mechanics. It dis-

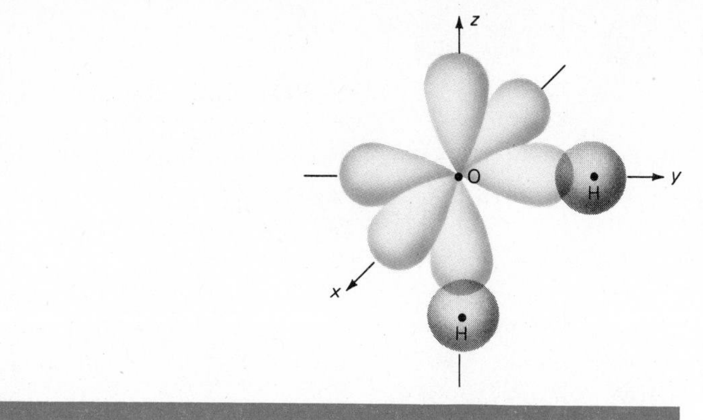

FIGURE 18-8 Bonding in the water molecule may be visualized by the shape of orbitals. The two electrons in the first shell of the oxygen atom ($1s^2$) are not drawn, but would be the shape of and smaller in size than the hydrogen orbitals shown. The next two electrons in the oxygen atoms ($2s^2$), also not shown, would be a spherically symmetric shell centered on the nucleus from which the tips of the p orbitals would protrude. The p_x orbital has a pair of electrons with opposite spins and is thus full and will not form bonds. The p_x and p_z orbitals in the oxygen have only one electron each and will bond with the electrons in the orbitals of the hydrogen atoms. Bonding distorts the orbitals so that the loop nearer the hydrogen grows and the loop away from it shrinks (not shown). The strength of the bond varies with the amount of orbital overlap. This model predicts the two covalent bonds of oxygen will lie at 90° to each other—an experimental fact.

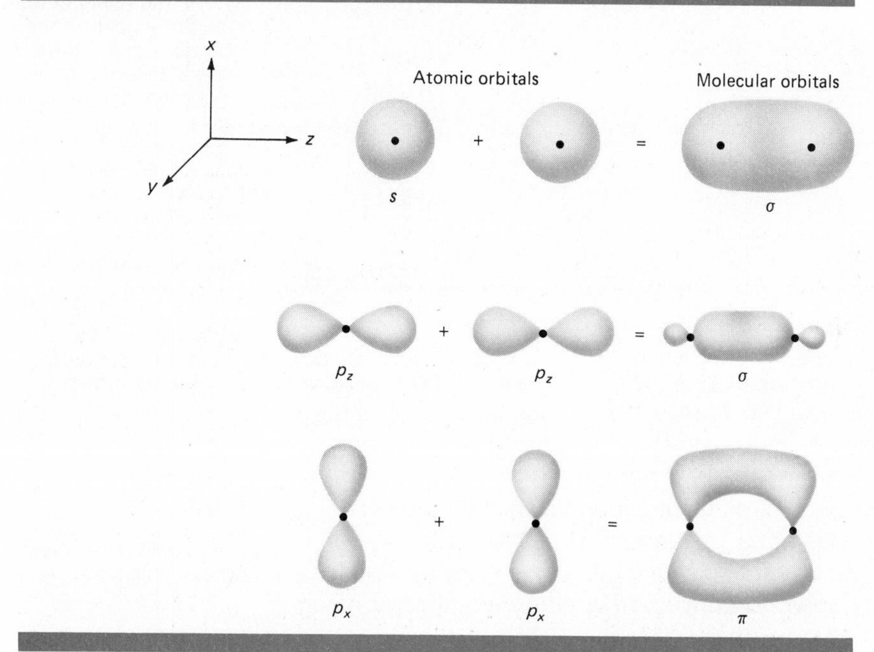

FIGURE 18-9 We may recalculate the probability of electron distribution between atoms after bonding has taken place. Orbitals shared by two atoms of a molecule are called molecular orbitals. Bonding of two electrons in s orbitals gives a shape called a sigma (σ) bond. Bonding of two electrons in p orbitals lying along the same line forms a similar molecular orbital. Bonding of electrons in p orbitals that are parallel is different. The molecular orbitals lie on both sides of the line joining the nuclei, in a pi (π) orbital.

closes that very different results are to be expected in the bonding between p orbitals in two different geometric situations (Fig. 18-9). If the p orbitals of two atoms are along the line of joining of the two atoms, a sigma-type bond (σ) is formed, which is similar to two joined s orbitals. However, if the p orbitals are parallel to each other and perpendicular to the line along which the atoms are joined, the orbitals overlap and join above and below the line of the nuclei in a pi-type bond (π). This difference helps to clear up the problem posed by the benzene molecule, C_6H_6 (Chap. 14). While conventionally drawn as a six-member ring with alternating single and double bonds, this is not the structure (Fig. 18-10). Double bonds are more easily attacked chemically than single bonds, yet all bonds in benzene are equivalent. Once the six carbons are joined in a ring, the p orbitals protrude from each atom above and below the plane of the ring. Should a particular carbon's p orbital be used to bond

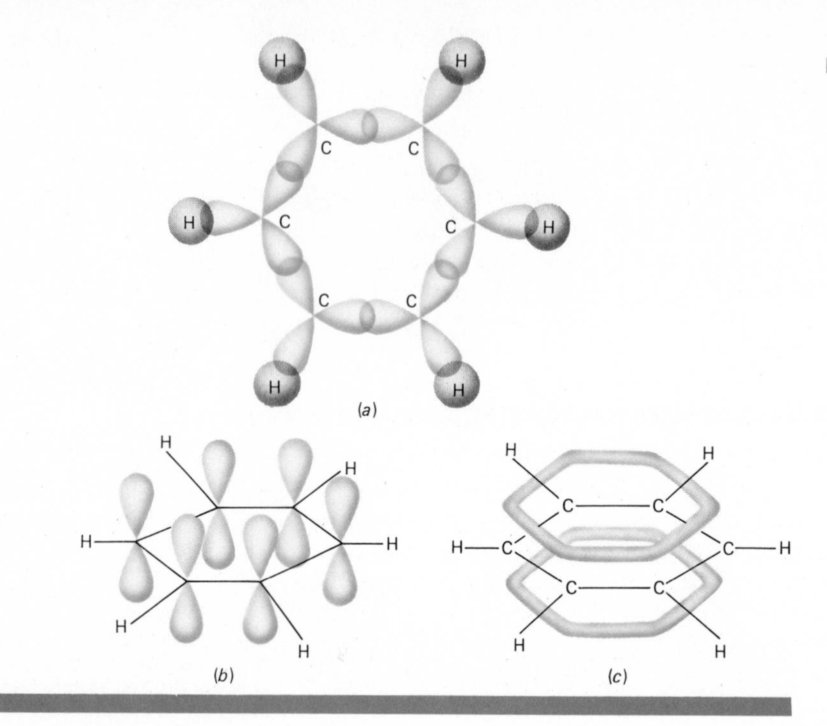

FIGURE 18-10
Molecular orbitals offer a picture of the bonding arrangement in benzene. Drawing (*a*) shows the bonding in the plane of the molecule. Each carbon bonds to two neighboring carbons and a hydrogen atom. Drawing (*b*) indicates that this leaves an electron in a *p*-type orbital available for bonding. Each of these orbitals projects up and down from the plane of the carbon nucleii. These electrons could form π bonds to either of two adjacent electrons in parallel *p* orbitals. Drawing (*c*) indicates the result. The *p* orbitals of all six carbon atoms bond both left and right in π bonds. This produces an independent stream of orbitals fused above and below the plane of the nuclei. The electrons are delocalized, or freed from particular atoms, and occupy these streams or bands of charge. The result is a bonding between carbon atoms that is stronger than a single bond but less than a full double bond, as is the case in benzene.

to the atom to its right or the atom to its left? The calculation indicates that all the *p* orbitals overlap to form a continuous stream of charge above the ring and another below it. Thus, sandwiched between two rings of bonds, the benzene molecule is particularly safe from chemical attack and quite stable. Such streams of π bonds are thought to be important in conveying bond information along the spine of biologically important protein molecules. Disruption of one part of the stream influences all parts of it. This may serve to convey information electrically along nonconducting molecules.

Quantum Levels in Solids

Wave-mechanical ideas may be applied to pure collections of atoms of the same type. Imagine a group of atoms pushed closer and closer until they form a crystal. Separately, each atom has a set of allowed energy states similar to the energy levels of Fig. 18-4 for the hydrogen atom. The Pauli principle asserts that no two electrons may occupy exactly the same energy state. As the atoms are crowded together the same energy levels of individual atoms merge to form a band of allowed energy states. These bands are formed from billions of similar allowed levels now differing by tiny amounts of energy. The highest energy level containing electrons in an atom is called the *valence level*. In a sodium atom, for example, this consists of two very closely spaced energy levels differing only by the direction of the electron's spin. As sodium atoms are crowded together, these levels for a particular atom merge with similar levels for all of the other atoms to form a band of extremely close energy states called the *valence band.* Of course, a sodium atom has only one electron in its valence shell even though there are two closely spaced available energy levels. This means that only half of the allowed levels in the valence band of a collection of sodium atoms are occupied by electrons. A miniscule amount of energy can move an electron from one of these levels to another, and thus promote the motion from one atom to another throughout the mass of sodium atoms. This result is completely general. If a collection of atoms has many vacant levels in its valence band, electrons may move easily from atom to atom, and the material is a metal and a conductor of electricity.

The opposite case is where all the levels of the valence band are filled by electrons. These are nonconductors or *insulators.* The next highest energy level in individual atoms above the valence level also broadens to form a band of energy states as the atoms are merged. In order for an electron to move from atom to atom in a nonconductor, it must receive enough energy to jump the energy gap separating the top of the valence band from this higher band (see Fig. 18-11). Once in this higher band the electron can move from one energy level in it to another with only tiny energy adjustments. Since an electron in this upper band is free to move in the crystal, the band is called the *conduction band.* Typical conductors have resistances of a few millionths of an ohm (Ω) per centimeter, while typical insulators have resistances of millions of millions of ohms per centimeter. Between these extremes lies the small group of semiconductors ($R \simeq 10^6\ \Omega$/cm). In these materials, valence and conduction bands approach each other but do not meet. Very small amounts of impurities added to such a semiconductor may provide available energy levels in the gap below the conduction band (*N*-type) or levels above the valence band (*P*-type). The junction formed when *N*-type and *P*-type semiconductors are connected to form a *diode* is of great importance. A photon of light may be absorbed by an electron in the valence band of the *P*-type material and promote it to the *N*-type conduction band. The electrons emerging may constitute a source of electric current. This device is called a *solar cell* or *photovoltaic source* (Fig.

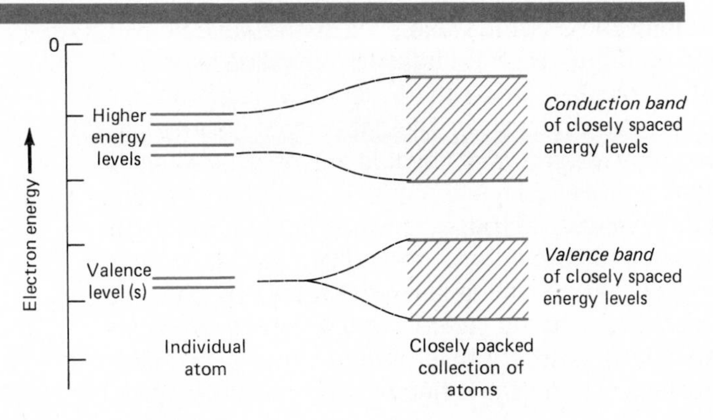

FIGURE 18-11
When atoms are packed together, the energy levels of the single atoms form bands of levels differing by small amounts of energy. If some of the valence levels of the individual atoms were vacant, the valence band would be only partly filled, and electrons may move freely from atom to atom with negligible adjustments in energy. This is the situation in conductors, such as the metals. If all levels in the valence band are filled, the electron must acquire enough energy to "jump" the energy gap to the conduction band in order to be mobile throughout the crystal. This is the situation in nonconductors.

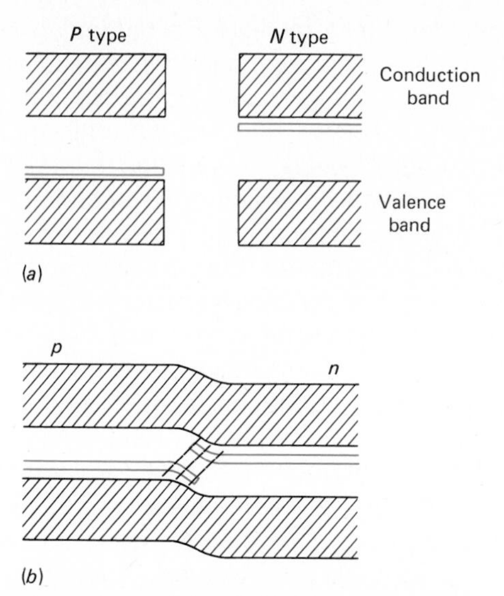

FIGURE 18-12 *P*-type semiconductors have trace impurities added which contribute additional levels above the valence band. *N*-type semiconductors have impurities contributing levels below the conduction band. When joined (*b*), a natural motion of charges occurs which effectively lowers the entire band structure of the *N*-type relative to the *P*-type semiconductor. Electrons in the conduction band can effectively flow from *P* to *N* but can move in the opposite direction only by acquiring energy. An external voltage connected to the materials can raise or lower the bands even more relative to each other. Electrons supplied the *N* material will be encouraged to make the dotted transitions to the valence levels of the *P* material. To do this they must lose energy as a photon of light. This is the principle of the light-emitting diode (LED). In the opposite process, lower electrons in the *P* material may absorb a photon of light to jump to an available level in the *N*-type material. This is the basis of solar cells to generate electricity.

18-12). If the band gap is too large, only exceptionally energetic electrons can make the transition. The opposite process occurs with even more efficiency. If a current is run into the *N*-type material, electrons may jump to the valence level at the *NP* boundary, giving off a photon of light as they do so. This is the source of light in *light-emitting diodes* (LEDs) as are found in calculators and some digital watches. This process may be nearly 100 percent efficient; that is, almost all of the electric energy supplied is converted to light. Efficiency is the serious problem with solar cells. First of all, only certain photons in sunlight are suitable for raising electrons to the conduction band. Secondly, the very high crystal purity required makes the material of the cell expensive—not in material cost but in the energy of processing. Progress over the last decade in improving efficiency of solar cells has exceeded most expectations.

Band theory in materials predicts another well-known effect. When a wave front of light passes electrons in the conduction band, if the frequency is exactly that which would be emitted by the electron dropping to the valence band, it is encouraged to make this drop. The light emission which was stimulated by the passing wave joins on the wave front to make the intensity of the wave greater. This "light amplification by stimulated emission of radiation" or *laser,* for short, has found widespread application from its first production in crystals. The capacity of materials to "lase" or produce high-intensity light waves by this means is not limited to crystals or even to solids. In a laser, electrons are "pumped" to the conduction band by some means. The passing wave then causes them to drop back and release their energy. By bouncing the wave back and forth between two mirrors at the ends of the laser, the wave is allowed to grow to considerable size. Light from a laser is very different from light coming from an ordinary source. Laser light is *coherent;* that is, all of the light emerging at any instant is of the same phase—e.g., all crest. Powerful crests follow powerful troughs. Ordinary light is much like the sound of a crowd at the Super Bowl—every frequency and phase hopelessly mixed together. This coherence of laser light is a powerful tool for several types of interactions and studies. With a laser a cheap plastic disk can be produced for a few cents which will record several hours of television programming. Because the frequencies of light are so much higher than radio waves, a single beam can encode much more information. Laser transmission is already being used in communications.

The premier application of the quantum-band theory of what is called the *solid state* is in the transistor. A *PN* junction may be used to *rectify* a current, that is, to allow electrons to flow only in one direction through the material. Triple junctions, *NPN* or *PNP,* go beyond this. They allow *amplification* of an electric signal. A small electric signal (variable voltage) fed to the central section of the transistor will cause this section to increase or decrease the flow of electricity across the device. It does this by raising and lowering the band levels of the central section relative to the other two. Thus, a small variation fed in, such as the signal from a radio station, can produce a large electrical variation coming out. If the electrical variation out is large enough, it may be used to drive a speaker and broadcast the radio program. Most important perhaps are the logic circuits which allow us to build small calculators and computers. Today, transistors are rarely used singly, but are made by the hundreds in various zones of a single flake of a semiconductor. A single *integrated circuit* (IC) may contain thousands of transistors, diodes, resistors, etc., on a fragment of matter smaller than your fingernail (Fig. 18-13). This has eliminated the need to produce thousands of separate, single-function devices and wire them together to produce a single device, such as a calculator. A single solid chip with no moving parts does just as well at much less cost and with more reliability.

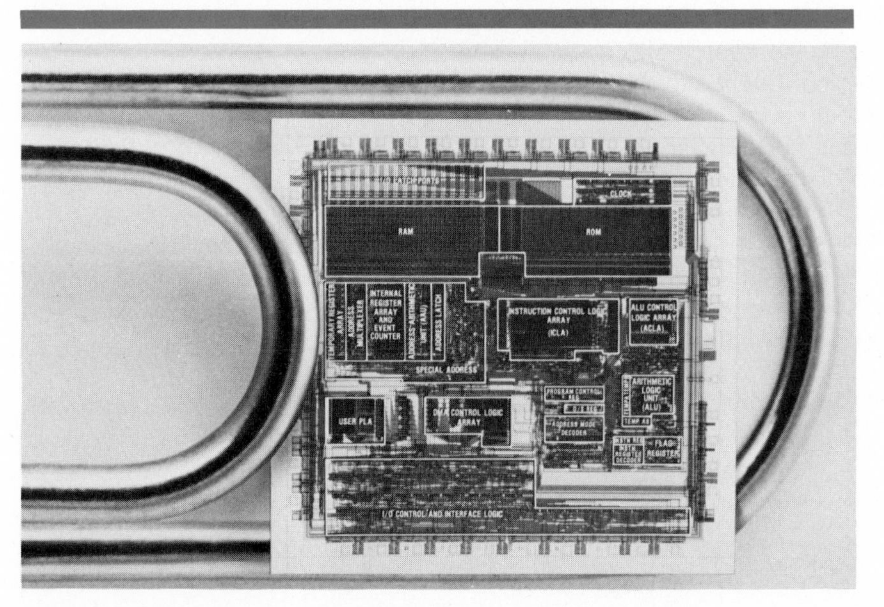

FIGURE 18-13
PN junctions form a diode which allows the control of the flow of current. Used in sets, *NPN* or *PNP*, they make up a transistor, which can also amplify—or make a small current larger. The advantages of these solid-state devices is that they have no moving parts to wear out and they can be made extremely small. The original commercial computers of the mid-1950s took up a room plus an additional room for their air-conditioning system. The photograph shows a computer on a single semiconducting chip. This integrated circuit (IC) has some 20,000 transistors, resistors, diodes, etc., for computer memories, logic and arithmetic units, input, and output. It is shown in a standard-size paper clip. Integrated circuits with twice this number of components are now available, and much more elaborate ones are being designed. (*Photograph courtesy of Bell Laboratories.*)

The Nature of Particles

In 1930, Paul Dirac (born 1902) succeeded in merging both the quantum ideas and relativity with the first relativistic theory of the electron. (Experimentally, these two theories had been wedded earlier. In 1923 Compton showed that the scattering of photons by electrons could be accounted for only by using *both* the quantum theory and relativity.) Dirac's theory implied two possible states. One was the electron. The other was a particle with the mass of the electron and the same size charge, but positive. Should this particle ever encounter an electron (which would be almost unavoidable), the two particles should totally annihilate in a burst of pure energy. This *antiparticle* to the electron was named the *positron* when it was accidentally discovered in 1932. If there are antielectrons, should there be antiprotons and antineutrons? There should be, and there are, but these particles were not detected until the large accelerators were built which could generate particles of enough energy to produce them in collisions. Antiparticles present the image of antiatoms with positive electrons quietly orbiting a nucleus of negative protons and antineutrons. Such antimatter is completely consistent with physical theory. A universe built of antimatter could function just as well as our own. However, whenever matter and antimatter would come together, they would completely annihilate in a burst of pure energy.* Another implication of quantum ideas applied to the nucleus was that the binding force between particles should occur in the form of a particle. Picture two children throwing a ball back and forth while on roller skates. Each time the ball is caught it would impart backward momentum to push the two children farther apart. *Exchange particles* were the opposite. The faster and harder the particles were exchanged the more they should be drawn together. In 1935, Yukawa predicted the existence of such a nuclear exchange particle, called a *meson,* with a mass some 270 times the mass of the electron. This particle should come in positive, negative, and neutral varieties. Of course, these particles exist or we would not be discussing them. However, on the way to their discovery other particles were discovered in nuclear fragments. Later, still more were discovered. The simple atom of protons, neutrons, and electrons dissolved into a sea of particles. When a sufficiently energetic particle is smashed into a nucleus, we may expect any of about 40 subatomic particles to be ejected. All have definite properties, and only our old friends of electron, proton, and neutron will last very long before disappearing in some interaction. Simplicity has gone and we have the equivalent of a periodic table of particles. Missing is the underlying insight that makes sense out of observed order. Why should a proton be 1836 times as massive as an electron? How did nature decide on exactly this number? Why do all charges

* There is no way of telling matter from antimatter at a distance. Every spectral line radiated by hydrogen atoms would also be radiated by antihydrogen atoms. If some star in the sky were made of antimatter, it is difficult to see how we would find out before we got there. There has been some suggestion that meteorites falling to the earth occasionally may have been made of antimatter. Most frequently discussed is the Siberian meteorite of 1908. No fragments of this have ever been found, yet it leveled forests tens of kilometers from its impact point and started fires 100 km distant.

seem to come in the same size chunk, and why this observed size? What, for that matter, is charge?

By the time we enter the field of particle physics we are straining the extensions of the quantum mechanics derived for atomic interactions. New formulations called *field theories* and *group theory* are used, borrowing heavily on ideas derived for atomic behavior. These are quantum theories in the sense that all properties come in discrete chunks rather than in continuous smears. The first semblance of order comes as particles are grouped into categories by their properties. For some time the indirect evidence has been accumulating that heavy particles (called *hadrons,* a class that includes protons, neutrons, and mesons) were themselves made of subparticles called *quarks.* Quarks have not been observed directly in over a decade of searching, but the belief in their reality grows. Quarks carry fractional charges of $+\frac{2}{3}$ or $-\frac{1}{3}$ and occur in groups of three to make up a proton, for example (with charges of $+\frac{2}{3}$, $+\frac{2}{3}$, and $-\frac{1}{3}$, respectively). Each quark is described in terms of properties which may or may not have something to do with a property that may be easily pictured. Thus, the fanciful properties of "flavor" and "color" are given to quarks but have nothing to do with these attributes in large glops of matter. (This tendency came from particle theory, which used properties such as "charm" and "strangeness" to describe particles.) High-speed electrons shot into a single nucleon (a neutron, for example) scatter from separate charge centers within the nucleon entirely in accord with the quark hypothesis. It is not certain whether the quark hypothesis is bringing us to the brink of a very fundamental understanding of the nature of matter or is the start of the unpeeling of another layer of subtlety.

Summary

Objects above absolute zero radiate an amount of energy proportional to the fourth power of their absolute temperature ($E \propto T^4$). This energy is radiated as a mixture of electromagnetic waves, such as light waves, radio waves, etc., depending on the temperature. The most frequently radiated wavelength, λ_{max}, may be determined from the temperature by the relationship: $\lambda_{max}T = k$ (where $k = 2.9 \times 10^{-3}$ m · K). Attempts to predict these facts in terms of classical laws failed. Planck succeeded in explaining the observed radiation pattern if he assumed that energy was radiated only in discrete "packages" which he called *quanta.* In his *quantum hypothesis* he suggested that the amount of energy in a single quantum was proportional to the frequency of the radiation. $E = hf$ (where $h = 6.6 \times 10^{-34}$ J · s). Einstein used this idea of radiant energy coming in quanta of definite sizes to explain the ejection of electrons from a metal surface by light, called the *photoelectric effect.* He named a single quantum of light a *photon.*

Bohr recognized that the rules pertaining to changes inside an atom must be different and not predictable from classical laws. In his *correspondence principle* he stated the rule that should guide a scientist in such a situation. He proceeded by asserting that the angular momentum of an electron (mvr) also came only in quanta of a given size, $mvr = nh/2\pi$ (where n is a whole number). From this he produced a predicted size for hydrogen atoms which was in complete agreement with measurement. He also predicted the energies of the lines in the hydrogen spectrum. Since Bohr's approach was so successful, other scientists began to *quantize* other variables associated with the electron. Each of these improved some predictability without providing insight as to why the approach seemed to work.

In 1924 de Broglie asserted that *moving matter had wave properties.* The wavelength associated with a particle was the Planck constant h divided by the particle's momentum mv, or $\lambda = h/mv$. This predicted that because of its wave nature the electron could reside only in the *stationary states* which were assumed by Bohr. This wave interpretation of matter led to *wave mechanics,* or *quantum mechanics.* The condition of an electron in an atom could be described in terms of four *quantum numbers* (n, l, m, m_s), and each combination of these predicted a different geometry for the average electron position. Pauli formulated the *exclusion principle,* which asserted that no two electrons in the atom could have the same set of values for these numbers. This led to the first theoretical explanation of the regularities among elements found in the periodic table.

The successful techniques for describing the situation in single atoms was extended to crystalline aggregates of atoms in *solid-state physics.* This predicted the broadening of electron energy levels of single atoms into bands. Interactions between the *valence band* and the *conducting band* in collective matter allowed the prediction of several phenomena. Practical devices resulting from this include solar cells, transistors, lasers, and light-emitting diodes (LEDs) as found in calculators and watches.

Many new nuclear particles have been predicted by theory and then discovered. Most importantly the *quark* theory asserts that nucleons are composed of sets of three particles called quarks. Quarks have fractional charges and have never been observed separately. The evidence favors the quark theory and seems to provide the best pathway for further understanding of the nucleus.

CHAPTER NINETEEN

The Modern View of the Universe

Introduction The date July 20, 1969, will always remain one of the landmarks of this century. On that date astronauts Armstrong and Aldrin announced the coming of age of the civilization by setting foot on the moon (Fig. 19-1). Whether future generations treat the event quite like October 12, 1492, or not, remains to be seen. The last generation has produced more information on the solar system than all of prior human history. Better maps are available of Mars or Mercury today than existed for Brazil in 1950, or for the United States in 1900. This has been the most incredibly exciting century in human history in almost every way. No single field has had more of its share of discovery and further expanded horizons than that of astronomy.

The Moon The moon is really too big compared to the earth to be considered a satellite. Most treatments today consider us a two-planet system. The moon has about one-fourth the diameter of the earth and one-eightieth of its mass. No other moon of the solar system has even a tenth of the diameter of its planet (even though four of those moons are larger than "our" moon). Since ancient times it has been known that the moon is 30 earth diameters away from us. Figure 19-2 shows the earth, the moon, and their separation to the correct scale. On this scale, Venus would be the same size as the earth but some 20 m off the page at *closest approach.* The sun would be a ball almost 1 m in diameter and nearly the length of a football field away, while the next nearest star would be on the other side of the United States.

FIGURE 19-1 Humankind realized a long dream on July 20, 1969, when the first human footprints were placed on the debris of the moon's surface. The technology that had to be developed to accomplish this has had considerable fallout in everyday life. Nevertheless, a decade of space exploration made the incredible appear to be routine and lost the public imagination. Succeeding generations will have to make the human toehold in space permanent and secure. (*NASA.*)

Observation of the moon from the earth allows us to divide its surface into separate parts. For example, the dark, relatively flat areas called *maria* (seas) by Galileo stand out well from cratered uplands. Mountain ranges are rare on the moon, although deep trenches and faultlike formations may be seen. The first surprises from lunar exploration came before any landing took place. The moon always keeps one face turned toward the earth, although various "wobbles" allow us to see some 59 percent of the surface. The back surface of the moon is different from the face that we see. The dark plains of the maria make up almost half of the surface visible from the earth. The back surface of the moon is notably deficient in maria, being almost entirely heavily cratered uplands. The far side of the moon also has many more large craters than the near side. Deflection in the paths of satellites orbiting the moon disclosed dense areas of rock underlying the maria on the near side. This imbalance of mass causes the denser side to point toward the earth.

The data obtained after landing presents a complex picture of the

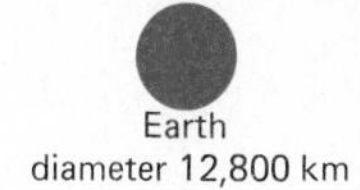

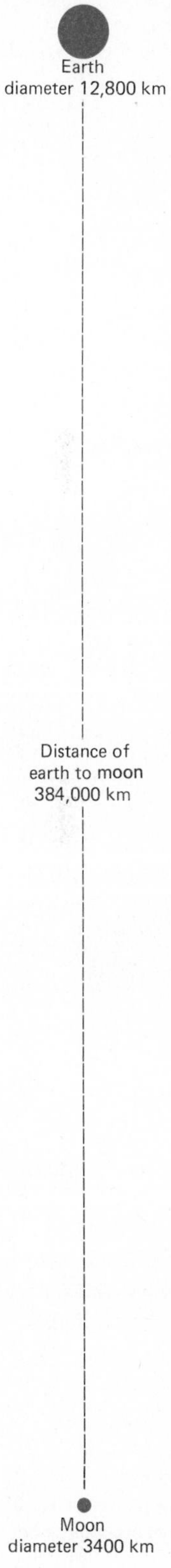

FIGURE 19-2
The earth, the moon, and their separation drawn to scale.

moon. Little change has occurred during the last 3 billion years, but, prior to that, the moon went through several cycles. As expected, the moon has no atmosphere and thus weathering and erosion effects. Moon rocks show differences in conditions and time of origin. The ages of samples range from 4.4 to 3.1 billion years old. The older age is close to the best estimate of the age of the solar system. Thus, the moon underwent much of its changes early and has since remained very stable. After a molten origin the material cooled and was extensively cratered. Later, release of heat in the interior remelted much of the rock and gave the moon a period of lava activity after many features had formed. Seismographs placed on the moon disclose a structure resembling the interior of the earth. An outer crust exists twice as thick under the extensive uplands as under the maria of the near side. Under this is a mantle extending downward from about 70 km. Data suggest a core which may be metallic. Moonquakes occur centered many hundreds of kilometers below the surface and are mainly caused by the tidal forces of the earth.

A View of the Planets

Mercury is difficult to observe from the earth because it is always seen close to the sun. The first satellite pictures in 1974 were suggestive of the heavily cratered far side of the moon (Fig. 19-3). The detail seemed less sharply etched than on the moon, and the dark, contrasting maria were absent. Mercury has a magnetic field about one-hundredth as strong as that of the earth. This probably indicates an iron core and helps to account for the high density of the planet. Crossing the planet are a series of interconnecting low ridges called *scarps.* These may be "shrinkage marks," like the ridges on a prune, caused by a contraction of the core. Measured surface temperatures reach a high of nearly 800 K, almost sufficient to melt the metal lead and bring the element mercury to its boiling point. The bare, dark rock of Mercury reflects light more poorly than any other planet. Only 6 percent of the light falling on the surface is reflected, compared with 76 percent for Venus.

Venus is the closest of the planets to the earth in size, mass, and distance. (See Table 19-1.) Because of its highly reflective cloud cover, science-fiction stories often imagined it a world of oceans and swamps (Fig. 19-4). Venus is much too hot for liquid water to form, much as liquid oxygen would not form on earth. Before a series of space probes visited Venus, beginning in 1961, there was disagreement about the density and composition of its atmosphere and its surface temperature. The reality, however, outdid most of the wilder ideas which had been suggested. The atmosphere of Venus is almost 95% carbon dioxide, with 5% nitrogen, 1% oxygen, and a smaller amount of water vapor. However, the surface pressure of the atmosphere is 100 times that of the earth, or

FIGURE 19-3
This composite photo of Mercury may be easily mistaken for the moon. Mercury has roughly 3 times the volume of the moon but presents the same heavily cratered surface. The largest craters are about 200 km across and would comfortably hold Connecticut. (*NASA.*)

TABLE 19-1
The Planets

PLANET	PLANETARY RADIUS (km)	PLANETARY RADIUS (EARTH = 1)	MASS (EARTH = 1)	ORBITAL RADIUS (km)	ORBITAL RADIUS (AU)	ORBITAL PERIOD	ROTATION PERIOD
Mercury	2,440	0.38	0.06	57.9	0.39	88^d	59^d
Venus	6,100	0.95	0.81	108	0.72	225^d	244^d
Earth	6,380	1	1	149.6	1	1 yr	1^d
Mars	3,400	0.53	0.11	228	1.52	1.88 yr	24^h37^m
Jupiter	71,400	11.2	317	778	5.20	11.9 yr	9^h50^m
Saturn	60,400	9.4	95	1,430	9.54	29.5 yr	10^h26^m
Uranus	25,400	4.0	15	2,870	19.2	84.1 yr	12^h
Neptune	24,300	3.8	17	4,500	30.1	165 yr	16^h
Pluto	2,500	0.4	0.1 (?)	5,900	39.5	249 yr	6^d9^h

FIGURE 19-4
In visible light, Venus presents very little detail. This ultraviolet image shows a cloud swirl centered on the planet's south pole. Radar images give us the best coverage of the surface. Venus has been one of the largest surprises of space exploration, being very different from the best earth-based guesses. (*NASA.*)

roughly 1000 tons/m^2. The Union of Soviet Socialist Republics has sent eight probes into the atmosphere of Venus, beginning in 1966. The earliest of these were destroyed by the totally unexpected conditions, but Venera 9 in 1975 lasted for almost an hour on the surface and sent back the first picture of a planetary surface. Days later, Venera 10 sent back a second picture when it landed several thousand miles away. Since the great density of the atmosphere had been established by earlier probes, a well-worn, rounded surface was expected. Instead, the first picture shows a field of sharply edged boulders and fragments. The United States has had three probes photograph and measure Venus as they flew by, the largest effort being the set of atmospheric probes in 1978 to 1979. Venus seems to present a geologically active surface. Radar maps have disclosed large objects resembling volcanoes, as well as the ever-present craters.

Mars always has seemed the most likely candidate for life besides the earth. Telescopes reveal a tantalizing change of color with season on

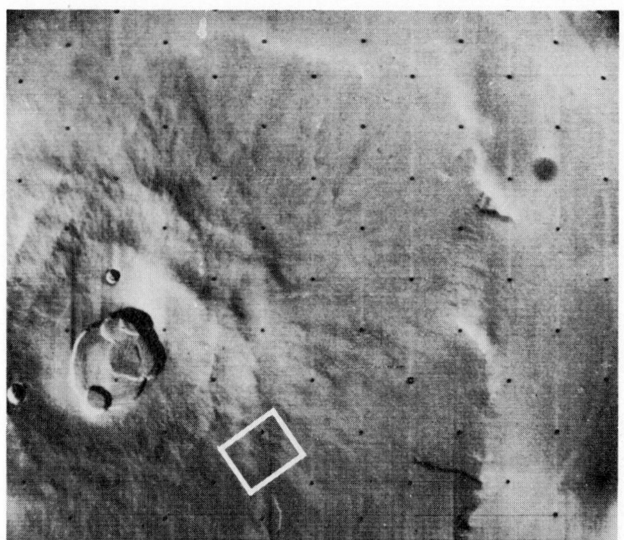

FIGURE 19-5 Twice the height of Mount Everest, the multiple crater of Olympus Mons looks spaceward from Mars. Most of the volcano is seen in the upper picture, including the cliff at which it ends on the right. The lower picture shows the detail of its lava streamers and cracks from the area in the rectangle at the top. (*NASA.*)

Mars. In the spring the clearly defined polar cap on Mars begins to shrink and the deep reddish areas nearer the equator turn to a gray-green. This was very suggestive of vegetation on Mars. The first space probes revealed a surface much more interesting than the fanciful ones produced by earthbound observers. Channels seem to argue for the past presence of running water on the surface. Prominent from very far in space are the large volcanoes of Mars. Olympus Mons (Mount Olympus), the largest of these, is well over twice as tall as any mountain on earth (Fig. 19-5). It ends in a fascinating cliff falling to the plains far below. An astronaut on the plain would have the spectacular sight of a lava wall rising as much as 5 times the height of the walls of the Grand Canyon. This wall circles the volcano and encloses a base with an area larger

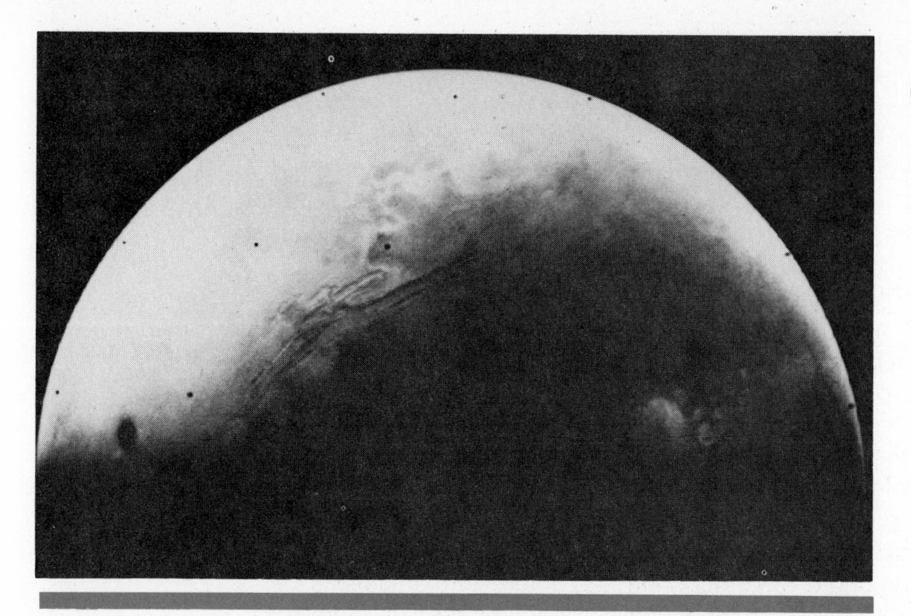

FIGURE 19-6 The Grand Canyon of Mars shows clearly in this approach photograph of the Viking I spacecraft. Some 5000 km long and up to 5 km deep, this canyon represents a scale not observed in earth canyons. Several European countries would fit on the floor of the canyon. The dark round spot is one of the major volcanoes of Mars (Mount Ascraeus), all of which are easily visible from deep in space. (*NASA.*)

than the size of Pennsylvania. The entire island of Oahu (Honolulu, Pearl Harbor, and all) would fit comfortably into the crater at the top of Olympus Mons with enough room for a sizable lake surrounding it. If some space-age entrepreneur could manage snow from the pink Martian sky, with a pressure only 1 percent that of the earth, the result would be a bonanza. What skier could resist a constant downhill slope as long as the distance from New York to Washington, D.C.? For hikers, there is a canyon broader and deeper than the Grand Canyon and as long as the United States is wide! (See Fig. 19-6.) To become a tourist paradise, Mars would have to change considerably. Currently, it is too cold (20°C on the equator on a warm day, down to −120°C), and the atmosphere is too thin and of the wrong gases. Mars has half the earth's radius but only one-tenth the mass. This gives Mars an acceleration due to gravity at the surface of only 3.7 m/s^2. This is too weak a gravitational pull to retain light gases in its atmosphere. A gas with heavy molecules, such as CO_2 with a molecular weight of 44, is favored and makes up most of the Martian atmosphere. Water vapor, with a molecular weight of 18, has been retained in only small quantities, and oxygen (molecular weight = 32) is present only in trace amounts. On Mercury, the "baking off" of any atmosphere was rapid and inevitable. A molecule eventually acquires enough speed from random collisions to be carried entirely away from the planet. On Mars the process was much slower since the absolute temperature is only half that of Mercury, but it nevertheless has lost most of its gases.

The Viking spacecraft both photographed Mars from orbit and sent landers to the surface in 1976. Each returned to earth a marvelously detailed picture of the Martian surface. Experiments in each lander tested for the presence of life by two methods. The first method was to incubate a soil sample, add a little water, and check for the gases evolved by life forms as we know them. In both cases such gases were produced. Before anyone could celebrate, the results of the second method came in. This was a chemical analysis to test for complex organic molecules always associated with life on earth. The results of both of these tests were completely negative—no complex carbon molecules to the limit of measurement. While all hope has not been abandoned for some life forms on Mars, it faded a great deal when inorganic reactions of rock were demonstrated that could produce the results of the first tests. Simple forms of earth life could probably adapt to the pressure and temperature on Mars in time. The scarcity of water would be a serious problem, but the worst problem might be exposure to the sun's ultraviolet radiation without the protective cover of an atmosphere to filter it out. Some future astronaut can wear clothes to protect from severe sunburn and possible skin cancer. What would you suggest for a plant?

Past Mars and before Jupiter lies a vast separation of space. This is inhabited by thousands of bodies called *asteroids.* The largest of these is about 800 km across. Most are in nearly circular orbits, but a few of the asteroids do not belong to the group. The most interesting of these is Icarus, the only object, besides comets, that approaches the sun more closely than Mercury. At one time it was fashionable to speculate that the asteroids might represent the fragments of a broken-up planet. This is interesting, but unlikely. The combined mass of the asteroids would make up only a good-sized moon. Furthermore, satellites passing through the asteroid belt on the way to Jupiter did not record an increase in the dust encountered in this region. More likely, the asteroids represent fragments that never managed to coalesce into a planet.

Jupiter is the giant of the planets in the sun's family (Fig. 19-7). In mass and in volume it is larger than all of the rest of the planets put together. It has about a tenth of the sun's radius and a thousandth of its volume. With Jupiter we may begin the division of planets into two classes. Mars is only 1.5 times as far from the sun as is the earth, but Jupiter is over 5 times as far. This is a notable jump or skip in the spacing of the planets filled only by the asteroids. The planets from Mercury to Mars are called the *inner planets,* and from Jupiter outward they are called the *outer planets.* Equally important is that with Jupiter the basic nature of the planets changes. Before Jupiter planets are relatively small and dense, and afterward they tend to be big and fluffy. We call the first group the *terrestrial planets* (after *terra,* Latin for earth). The second group is called the *Jovian planets,* after Jupiter. These are sometimes also called the *gas giants* since their low densities indicate they they must be composed of mainly atmosphere rather than rock as we know it. Jupiter has a density only one-fourth that of the earth and only a third higher than water. Saturn has such a low density it would float in water. Jupiter has 14 moons, two of which are about the same size as the planet Mercury (one larger and one slightly smaller). These two, Ganymede and Callisto, along with the smaller Io and Europa, were the four satellites of Jupiter first seen by Galileo and are called the *Galilean satellites.* Spectacular photographs of these moons were sent back to earth by the two

FIGURE 19-7
A composite photo of Jupiter showing the Great Red Spot and many secondary swirling disturbances in its atmosphere. Features as small as 150 km across may be seen. The Great Red Spot has been visible since telescopic observations of Jupiter began. Why an atmospheric feature so big and distinct should be a stable feature in an atmosphere for centuries is puzzling. (Compare this view with the best earth-based photograph of Jupiter in Fig. 6-9.) (*NASA*.)

Voyager space probes of 1979. The most startling discovery was that of active volcanoes on Io, photographed during eruptions (Fig. 19-8). Jupiter is a huge blob of light gases like hydrogen and helium and may have no solid surface at all beneath the clouds, but may condense to a liquid hydrogen zone. The visible surface of the planet is composed of cloud tops and indicates the fastest rotation of any planet in spite of its size. Portions toward the equator rotate more rapidly than near the poles, and this may result in the division of the atmosphere into obvious bands. The most fascinating marking on the planet is the so-called Great Red Spot, which has been there since telescopes large enough have observed it. This too is made of cloud tops, but they stand a little above the general surface of clouds. While many suggestions have been offered to explain this spot, none is widely accepted, and most are not tenable. The spot is roughly twice the size of the earth.

Jupiter has a surprisingly high magnetic field which almost burned out the instruments aboard the Pioneer 10 spacecraft in 1973. This is the

FIGURE 19-8 An active volcano in eruption on Io photographed by Voyager I. Materials are being ejected to an altitude of 100 km with velocities up to 1 km/s (2000 mi/h). Several volcanoes were photographed in simultaneous eruptions on Io, each more violent than major eruptions on earth. (*NASA.*)

largest magnetic field of the planets that have been measured, and much stronger than that of the earth (which is second highest).

From Jupiter outward we suffer the huge disadvantage of not having sent our instruments there. A Voyager spacecraft received a gravitational "whip effect" from Jupiter for the trip to Saturn in 1979. The use of this type of boosting of energy for trips to distant planets saves a great deal of effort. This effect was used to fling the Pioneer satellites completely out of the solar system into interstellar space.

Saturn, of course, is a beautiful planet because of its rings. A set of five similar rings was discovered around Uranus in 1977 as the planet slid across our view of a star. Variations in the star's brightness before and after its *occultation* (blocking out of light) by the planet indicated the presence of the rings. A faint ring was photographed around Jupiter by a Voyager spacecraft in 1979. The closer an object approaches a planet, the faster it should move in orbit. For a large object, such as a moon, the closest edge should be orbiting more rapidly than the far edge. If a moon approached the planet too closely, the resulting gravitational shearing stress would become too great for rock structures to resist and the moon would be pulled apart. The distance within which a large object will be torn apart by this tidal effect of gravity is called the *Roche limit.* The rings of Saturn, as well as the more recently discovered ones of Uranus and

Jupiter, are all within this limit. Thus, the rings are believed to be a fine rock debris of moons wandering too close to their gravitational master. The rings certainly are not solid. The fainter ones allow light to pass through them, and the speed of rotation of each ring is greater the nearer it is to the planet.

Uranus has another distinction beyond its rings. It is the only planet whose axis of rotation lies in the plane of its orbit. In general, the various rotations and revolutions in the solar system are all counterclockwise as viewed from far above the North Pole of the earth out in space (the slow rotation of Venus being an exception). Uranus, however, nearly points its North Pole at the sun in one part of its orbit. Half a Uranian year later (42 earth years later) it would point its South Pole nearly directly at the sun. Uranus is a gas giant 4 times our diameter, 19 times as far from the sun, and probably −200°C (−300°F). Thus, this inclination of its axis to its orbit should make little difference. Imagine what would happen to the seasons on earth if we had an inclination like that of Uranus. Uranus is unique among the four gas giants in that it does not have a really large moon—one approaching the size of our moon. Saturn has the largest moon in the solar system. Titan (radius 2500 km) is larger than Mercury and definitely possesses an atmosphere containing methane (CH_4). This is slightly larger than Jupiter's moon, Ganymede, which does not appear to have much of an atmosphere but may have a polar cap of some sort. Neptune's largest satellite is Triton, which is some 15 percent larger in radius than our moon. That leaves Uranus alone among the giants with four moons all under a third the radius of ours.

Finally we come to Pluto. Just as every junior high school class seems to develop a class clown, the solar system has its Pluto. Small, in a region where planets are big; dense, where planets are least dense; dim, in a region where planets are highly reflective—Pluto is a misfit. Pluto is earth-sized or smaller and as dense as Mars but is out beyond the gas giants. The size ratio of planets from Jupiter to Pluto is that of a large medicine ball, a smaller medicine ball, two basketballs, to a tennis ball. The four gas giants have nearly circular orbits, while Pluto's orbit is the most flattened elliptical orbit in the solar system. Pluto's disregard for traffic control is so bad that it crossed inside of Neptune's orbit in 1979 and will remain closer to the sun than Neptune until the end of the century. Then it will recross Neptune's orbit and swing farther out beyond Neptune's orbit than Saturn is from the sun. Not only is this ill-mannered for a planet, but while the rest of the orbits of the planets are very near to the same plane in space, Pluto's orbit is tilted in space. From the earth outward, all orbits are within $2\frac{1}{2}°$ of being coplanar, except Pluto's, which is 17° off the plane of the ecliptic. For some time it was reasonable to regard Pluto as a moon somehow escaped from Neptune. This is tempting since Pluto is more moonlike than it is similiar to the giant planets. Recently, however, Pluto has been shown to have a moon of its own. This makes the escaped-moon hypothesis unlikely. Pluto is very difficult to observe well enough to measure with much certainty any characteristics other than its orbit.

The Distances of the Stars

CHAPTER NINETEEN

Parallax is the apparent shift in nearby star positions due to the motion of the earth in its orbit. Measurement of this effect had been sought since the time of ancient Greece. Bessel announced the first measured parallax in 1836 for the star Sirius. Sirius is the brightest star in the sky after the sun, and might be expected to be one of the closest stars. Struve measured the parallax of Alpha Centauri at about the same time, but published it much later (he did his work in the Southern Hemisphere, but journals were published in the Northern Hemisphere, and the only connection was by sailing ships). Each degree on a protractor is divided into 60 minutes of arc (60′), and each minute into 60 seconds of arc (60″). A second of arc is a very small angle. Imagine that you set up a post in the middle of a large desert. You then walk some 29 km (18 mi) directly away from the post. Looking back in the clear desert air you can see the post in the distance against the silhouette of the distant mountains. Sighting the post with your right eye you can spot its position against the mountains. Without moving your head, if you close your right eye and open your left eye, the image of the post will shift *one second of arc* against the more distant background mountains; we define this shift as *one parallax-second,* using as a baseline the distance between our eyes. Of course, astronomers use a much larger baseline between two observations. If we obtain a view of a star today, in 6 months the earth will be entirely on the other side of the sun and halfway around in its orbit. The earth is 1.5×10^8 km (9.3×10^7 mi) from the sun, which is called the *astronomical unit* (AU). Sighting from a base of one astronomical unit, any star which showed a shift of one second of arc would be at a distance of one parallax-second from the earth, or one *parsec* (pc). Astronomers always measure large distances in parsecs since the distance may be obtained directly from the parallax: $d = 1/p''$, where p'' is the parallax in seconds of arc and d is the distance in parsecs. It would take a beam of light traveling at 300,000 km/s (186,000 mi/s) some 3.26 years to move a distance of 1 pc. Thus, 1 pc is 3.26 light-years.

There is no star with a parallax this large; Struve found the closest star with a parallax of 0″.76, while Bessel could manage only 0″.38 for Sirius. This implies that Alpha Centauri and Sirius are 1.3 and 2.6 pc away (4.3 and 8.6 light-years), respectively. Yet these are two of the *closest* of all stars to the solar system. (Alpha Centauri is a triple star. Three other stars are closer than Sirius, all of which are too dim to be seen by the naked eye. Sirius itself is a double star, that is, two stars orbiting about each other.*) Astronomers can measure parallaxes which are only $\frac{1}{500}$″ today, provided a star is bright enough to be seen with a telescope over this distance. While most of the few thousand stars which

* Students who wish to obtain a feeling for the immediate stellar vicinity of the solar system are encouraged to obtain a copy of the game "Starforce: Alpha Centauri" (Simulations Publications, Inc., 44 E. 23d St., New York, NY 10010). You may enjoy the game, but more importantly the gameboard is an accurate replica of the positions of star systems within a volume of about 6 pc of the solar system, some 70 star systems. If you do play the game, where each hexagon has a diameter of 1 light-year, remember that it would take us 26,000 years to cover one hexagon at the greatest speed humans have attained in rocket travel!

may be seen with the naked eye are within this range, the vast majority of stars which may be seen with the telescope have *no* measurable parallax whatever—*they are simply too far away.* Thus, if the distance between your eyes represents the space between earth and sun, no star would be within 37 km (23 mi), and the vast majority of stars would be well beyond 1700 km (1000 mi). Space is big indeed, with only a miniscule fragment of it occupied by anything. Yet, we do have hope of interpreting it, since there are thousands of stars within measurable range.

The basic information sent to the earth by the stars is their radiation, notably light. If we wish to interpret the universe, we need to know what that light implies. The intensity of light varies with the inverse of the square of the distance ($I \propto 1/d^2$). Thus, if a given star is 10 times as far away, it will only be one one-hundredth as bright. Ever since ancient Greece (Ptolemy's book, in fact) astronomers have rated stars by a system of *magnitudes* (Table 19-2). When magnitude was defined precisely, it was agreed that a difference of 5 magnitudes would mean a brightness ratio of 100/1. Each visual magnitude represents about 2.5 times the brightness. The scale works backward, with the brightest stars being assigned the lowest magnitude. Alpha Centauri has an *apparent magnitude* (m) of about zero, which means that it appears 2.5 times as bright as Pollux ($m = 1$), one of the twins in the group called Gemini. Of course, apparent magnitude has little to do with real brightness. If a star is very close it will appear bright, like the sun. The sun actually gives off almost the same amount of light as Alpha Centauri, which is 1.3 pc from the earth. The sun is only 0.00004 pc from the earth and has an apparent magnitude of -26.8, 50 billion times the apparent brightness of Alpha Centauri ($2.5^{26.8}$), even though both stars radiate the same heat and light. To rate stars in comparison to each other we must not depend on how far they happen to be from us. The *absolute magnitude* (M) of a star is a measure of the radiations you would receive at a standard distance of 10 pc. On this scale, the sun would have an M of 4.9. That is, if we were

TABLE 19-2
Magnitudes of Stars

APPARENT MAGNITUDE,* m	RELATIVE BRIGHTNESS (m OF 0 = 100)	EXAMPLES
−1	250	Sirius (−1.6)
0	100	Alpha Centauri, Vega, Rigel
+1	40	Betelgeuse, Altair, Pollux
+2	16	Polaris, Dubhe (brightest star in the Big Dipper)
+3	6.3	Delta Ursae Majoris (dimmest star in the Big Dipper)
+4	2.5	Epsilon Eridani (nineth closest star system)
+5	1.0	Epsilon Indi (fourteenth closest star system)
+6	0.40	
+7	0.16	

* For determining the apparent magnitudes from $m = 3$ to $m = 7$ it is handiest to view the Pleiades (the Seven Sisters) in the constellation of Taurus. This open galactic cluster of stars might well be renamed "the eyechart of the sky." The group contains one star with magnitude 3, five with $m = 4$, one with $m = 5$, three with $m = 6$, and fourteen with $m = 7$. Thus, it is possible to define the limiting magnitude for seeing the stars simply by counting the number visible in the Pleiades in the winter sky.

10 pc distant from the sun, it would appear to be a star with a magnitude of 4.9. Since the dimmest stars that can be seen with the naked eye have a magnitude of about 7, this makes the sun clearly visible. At 10 pc (32.6 light-years) the sun would not be as bright as any of the stars of the Big Dipper are at the earth, but you could see it if it were pointed out. On the other hand, the second brightest star in the sky, Canopus (visible only from the Southern Hemisphere), is a distance of 30 pc yet has an apparent magnitude of -0.72. Canopus thus must be 1600 times as bright as the sun, and there are stars much brighter still. Part of the difference in brightness is due to temperature. From the last chapter you may recall that the color of an incandescent object is a measure of its temperature ($\lambda_{max}T = k$). Canopus is 1000 K hotter than the sun in terms of its color. Even so, it should radiate only about twice as much energy per square meter. This suggests that Canopus has 800 times the surface area to radiate heat and light as does the sun, or that Canopus has over 9 times the diameter of the sun. This is information that we cannot obtain directly.

Separate techniques for the indirect measurement of stellar diameters exist. These do show some stars to be much larger than the sun. Indeed, for the very largest stars, if they replaced the sun, the orbit of both the earth and Mars would be entirely inside of the star. Such stars were originally detected because they were both very bright (high M) and, very cool (reddish). Such stars are known as *red giants*. The only way their great brightness is achieved at low temperature is with a very great surface area to radiate heat and light to space. The opposite condition also exists, of stars that are very hot but not very bright. This occurs only if the star is quite small in surface area. Each square meter of surface radiates a great deal of heat and light, but there are relatively few square meters. Such stars are called *white dwarfs,* and may be of about the same size as the earth. The nearest example of this is the white dwarf Sirius B. Sirius A is 23 times as bright as the sun; while Sirius B, which orbits Sirius A, is only some eight-thousandths the brightness of the sun despite its being hotter. The presence of this white dwarf was inferred in 1844 by Bessel, even though it was not seen until 1862. Such a *binary star system* is not unusual. About half of all known stars belong to such a multiple star system, which leaves over half of the visible stars as singles.

Motions of the Stars

Stars have independent motions through space. The easiest motion to interpret is that which occurs *across our line of sight* to the star. This is called *proper motion* and shows up as the position of the star is compared year after year. The largest proper motion of any star viewed from the solar system is that of Barnard's star (Fig. 19-9). This is also our closest star after Alpha Centauri at a distance of 1.8 pc. The annual shift in position of Barnard's Star of 10″.3 is 20 times the size of its parallax even though it has the second highest parallax of the stars. Every 360 years, Barnard's Star appears to move a full degree in the sky. This is a

FIGURE 19-9 The proper motion of Barnard's star clearly shows up in these photographs taken 6 months apart. While none of the other stars appears to have changed position relative to one another, Barnard's star has shown considerable motion. Even though it is the second closest star system to the sun, this star is too faint to be visible to the naked eye, as are most of the stars in the portion of the sky in the photograph. (*Lick Observatory photograph.*)

huge amount and would certainly have been noticed in a star of respectable magnitude; however, Barnard's Star is a red dwarf with a visual magnitude of 9.5 and would have to be 10 times as bright as it is to be visible on earth to the naked-eye observer.

Of course, a star may have a very large motion relative to the solar system and not appear to change its position among the stars at all. This would be the case if the motion was *along our line of sight* and the star was moving either toward us or away from us. Such motion is called a *radial motion,* and a motion of this sort must be determined from the light received from the star. If a source of waves is moving toward an observer, more crests arrive in a given time than are emitted by the source as the distance between waves is reduced by the motion of the source. If the motion is away from the observer, fewer waves arrive per unit time

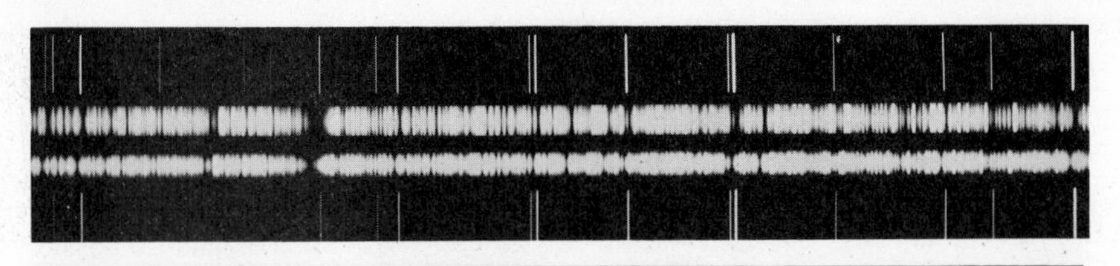

FIGURE 19-10 The motion of the earth in its orbit about the sun shows up in these spectra of the star Arcturus. The bright-line spectra at top and bottom are comparison spectra used to measure any doppler shift. The upper-star spectrum reveals dark lines shifted to the right (blue) from the comparison spectrum. These yield an approach of 18 km/s for Arcturus. The second spectrum taken 6 months later shows a red shift of −32 km/s. The difference is due entirely to the orbital motion of the earth toward or away from Arcturus. Incidentally, the portion of the spectra shown represents the yellow-orange portion of the spectrum and only one-fortieth of the range of visible colors. (*Hale Observatories.*)

than are emitted. This is the *doppler shift* which was discussed in Chap. 12. Thus, radial motion shows up as an apparent shift in the wavelength of waves (Fig. 19-10). Details in the star's spectrum, such as dark absorption lines, will be shifted toward the red end of the spectrum if the star is receding, or toward the blue if the star is approaching. The speed of the radial motion can be calculated easily once the size of the shift is measured. A doppler shift may be encountered which is both to the red and to the blue end of the spectrum at once. The spectral line, instead of being a fine line, *broadens* to a greater width than normal. Clearly, parts of the star are both approaching and receding from us. This is exactly the effect we expect for a *rapid rotation.* Light from the part of the star moving toward us during this rotation is blue-shifted, while the opposite limb is carried away from us and its light is red-shifted. Thus the width of stellar spectral lines may be a measure of their rate of rotation. Of course, if the star is rotating with its axis pointed toward the solar system, no broadening of its spectral lines occurs.

The measurement of large numbers of radial motions and proper motions discloses a bias in each. In one portion of the Northern Hemisphere of the sky the stars *average* a blue shift, indicating a rate approach of 20 km/s (12 mi/s). Exactly halfway around the sky, stars have an average red shift, indicating a speed of recession of 20 km/s. In these numbers we are not seeing some generalized drift of the stars but the motion of the entire solar system through the local star field. The sun with its planets is sliding smoothly through space in roughly the direction of the bright star Vega. Vega is the brightest star in the northern half of the sky. If you live in the northern United States and look up some summer night to see a very bright blue-white star almost directly overhead, you have identified the direction in which the solar system is moving relative to the stars.

Spectral Type and the Classification of Stars

Spectra provide information beyond the motions of the stars. Most stellar spectra represent a continuous spectrum with dark absorption lines. While some of these are moderately sunlike, most differ widely from the solar spectrum. Some spectra are dominated completely by strong lines due to hydrogen or helium. In the sun's spectrum, these are relatively weak, but many metallic lines (for example, lines of calcium) are prominent. Still other spectra display wide bands due to molecules such as titanium dioxide. Astronomers began to sort these spectra into classes almost at once. The current system uses seven main classes and three branch classes for most stars. The class into which a star fits is given by a letter known as its *spectral type*. The order of spectral types, from hottest and bluest is O, B, A, F, G, K, M, R, N, S (Table 19-3). (This would be difficult to remember without the code sentence: "Oh Be A Fine Girl; Kiss Me Right Now, Smack!" Somehow, with this sentence generations of students have mastered the order of spectral types immediately.) Each class may be further subdivided using the numerals zero (bluest, hottest) to 9. Thus, the sun is a G2 star, as is the brightest member of the Alpha Centauri Triple (the other two are a K5 and the type M red dwarf called Proxima Centauri because it swings out closest to us in its orbit). Of the 50 stars *nearest to us* after the sun, we have: one A (Sirius, the brightest star in the sky); one F (Procyon, eighth brightest star); two G's (Alpha Centauri A and Tau Ceti); six K's (all visible to the naked eye, none spectacular); thirty-five M's (two visible to the naked eye); and five unclassified white dwarfs. This survey of near interstellar space is complete out to about 5 pc (16 light-years). We have no really nearby stars of type O or B, and this is probably fortunate. There are no type O stars for a very considerable distance, yet they stand out because they are incredibly hot and bright. Of the 25 brightest stars as seen from the earth, 11 are B stars. The B star Rigel is so far that it is at our limit to detect and measure its parallax. Yet for every star within this measurable volume of space, there are thousands of stars visible beyond it. Spectral types aided in piercing this limiting volume of direct measurement. Figure 19-11 shows a graph of spectral type vs. the absolute magnitude for the stars. This is called a *Hertzsprung-Russell diagram* (or H-R diagram) after its originators. Such a diagram essentially plots surface temperature (spectral type) vs. brightness (absolute magnitude). Since absolute magnitude is used, it is necessary to be able to measure the distance to each star to determine its true brightness. Thus, only nearby stars to a distance of perhaps 200 pc can be graphed. The majority of stars fall into a broad band called the *main sequence*. At the upper right is a diffuse cluster of stars which are reddish (type M) and have cool surfaces, but have very high absolute magnitudes. These are the red giants. Below the main sequence is a cluster of stars which are quite hot but not very bright. This is the group of the white dwarfs.

TABLE 19-3
Spectral Types

SPECTRAL TYPE	MASS (SUN = 1)	RADIUS (SUN = 1)	TEMPERATURE, K	ABSOLUTE MAGNITUDE, M	LUMINOSITY (SUN = 1)
O	40	18	50,000	−6.0	21,000
B	7	4	16,000	−1.1	230
A	2.2	1.8	8,000	+2.1	12
F	1.4	1.2	6,700	+3.4	3.6
G	0.9	0.93	5,500	+5.2	0.70
K	0.7	0.74	4,300	+8.0	0.052
M	0.2	0.32	3,000	+12.3	0.0010

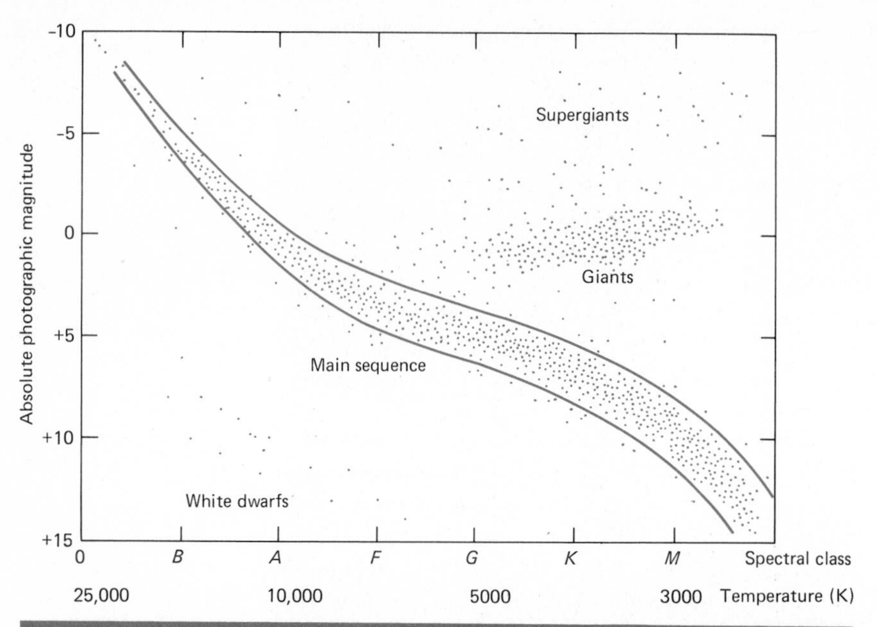

FIGURE 19-11 The H-R diagram is a graph of absolute magnitude of stars vs. their spectral type—effectively, true brightness vs. surface temperature. If thousands of stars are plotted, the dots fall into a band called the main sequence, along with two zones of scattered dots—red giants and white dwarfs.

Interpretation of this pattern took some time. At first it was guessed that stars might begin hot and cool down as they aged, or vice versa. Other schemes were proposed also, but how does one decide which is likely to be correct? The key idea was to examine the H-R diagrams of groups of stars which were of the same age. How do you tell when stars are of the same age? This task can be hard enough with people, but with stars it seemed impossible. The answer was to use *clusters* of stars which were gravitationally associated yet isolated from other stars (Fig. 19-12). Such clusters, with hundreds to tens of thousands of stars, are fairly common. Since they are proceeding through time and space as an associated group, it is logical to guess that they had a common origin. The H-R diagram of the stars in a cluster should lack older stars if the cluster is relatively young, or vice versa. However, we would not expect the H-R diagram of a cluster to be the same as the normal population of mixed stars of all types. We find that young clusters have O, B, and A stars on the main sequence, but that at some point stars slip off to the right of their main sequence position. Since early spectral types, such as O or B, are very bright, very hot, and very large stars, they have a relatively short life span before they exhaust their fuel supply for fusion. Thus, a cluster with these stars could not be very old. We believe that the stars to the right of the main sequence are in the process of attaining the form where they will spend the majority of their lifetimes. If this is correct, *early* spectral types (O, B, A) get onto the main sequence faster than do *later* spectral types (G, K, M). This difference would be almost completely due to the difference in *masses* of these stars. Stars apparently condense out of large volumes of dust and gas in space. Gravitational attraction pulls this gas and dust together into the forming star. The

FIGURE 19-12
The globular cluster of stars M3 in the constellation of Canes Venatici is barely visible to the naked eye. Such a cluster is literally a "ball of stars" more tightly packed as they near the center. Even at the center, however, individual stars are light-years apart. Most such clusters are off the main portions of space with large numbers of stars. Globular clusters are quite common. Messier's catalog of the 1770s names this cluster as object number three, hence the designation M3. Of 109 objects listed in Messier's catalog some 29 are globular clusters. (*Lick Observatory photograph.*)

greater the mass and density of the original gas cloud, the faster the particles will be pulled together and the larger the star that will form. At some time the huge pressures inside a condensing sphere of matter will become high enough to start nuclear reactions in the interior of the mass, and the star is "turned on." Calculations show that the pressures at the center of the planet Jupiter are not too far from those required to begin nuclear fusion. Jupiter is nearly as big a planet as we would expect to find in the universe. If the mass of Jupiter could be increased to perhaps a tenth that of the sun, it could become a star, and the solar system would be a binary star system. The great mass of an O or B star is accumulated rapidly, before the radiated light of the star itself can "blow" the rest of the gas and dust of the cloud away into space. A type O star may exhaust its fuel and evolve off the main sequence before a type G star, such as our sun, has achieved enough central pressure from contraction to ignite. Once the fusion reaction begins in a star like the sun, it would rapidly become incandescent, but it would be redder than it would be in its final form. The star would change slowly as its surface became hotter until it had joined the main sequence. Once on the main sequence, a star like our sun would be stable for billions of years.

The examination of an older cluster shows all of the late spectral types in place on the main sequence. However, the early spectral types are gone. As these stars disappear from a cluster, the scattering of stars in the red-giant area of the diagram increases. From this we could conclude that stars might enlarge and become redder after their span on the main sequence is over. Actually, however, the later history of stars depends very much upon the mass of the star at each stage in its development. Atomic physics applied to the conditions of stellar interiors predicts several possible fates for a star depending largely on its mass. After the interior of a star reaches the pressures necessary for nuclear fusion to take place, the star is ready to slip onto its place on the main sequence. Prior to this time, the star has not been a cold, nonluminous object. The heat associated with gravitational contraction raises the temperature of the star to the million degrees or so required for fusion. Thus, the star becomes hot long before the stable mechanism of fusion begins, which will operate during most of its life span. The basic energy process in a star is the conversion of hydrogen to helium, which may be done directly or with several intermediate steps depending upon the temperature. Once fusion begins, there is an intense radiation pressure from the zone of fusion pushing the matter outward. This is balanced by the gravitational attraction pulling materials toward the center. At some point these forces balance, and a stable rate of energy production is maintained in the interior of the star.

Our sun should have a life span in this state on the order of 10 billion years. Materials in the stellar interior, however, seem to mix very little. Stars begin as mostly hydrogen, the commonest material in the universe. As fusion takes place inside the star, a *core* begins to develop, which is almost pure helium produced in fusion. As more and more hydrogen is converted to helium, the size of this core grows. The zone where fusion occurs, which began at the center of the star, moves outward with the surface of the core. Once this shell of hydrogen fusion starts to reach the outer portions of the star, the outer layers will be expanded (which drops their temperature) while more energy is being radiated. The star will move toward becoming a red giant. With a star of the sun's mass another process will occur—the core temperature will reach the point where nuclear reactions of the helium will begin to form heavier elements (generally carbon). Thus, two zones of nuclear reaction can occur at once. Eventually, both will run out of fuel to continue, and a long collapse will occur. Pulsations in this process may blow away the outer portion of the star in a series of explosions, but the collapse is inevitable. Gravitational force eventually crushes the materials together into a contracting white-hot mixture of particles. The pressures are beyond those of normal atoms to withstand, so a collapsed state of matter occurs, and we have arrived at the white dwarf. As the contraction of this matter ceases, it will radiate the last of its energy and become dark. In this collapsed state, atoms are crushed and matter packs very tightly. The white dwarf circling Sirius, called Sirius B, is smaller in size than the earth but has a mass equivalent to that of the sun, a million times the earth in volume. This would imply that every cubic centimeter of Sirius B has a mass of many tons.

The process of stellar evolution does not have to end as it will for the sun. If the mass of the collapsing star is greater than 1.4 times the mass of the sun, another process may take place. Nuclear reactions may occur in the center, producing elements as heavy as iron. However, such a process is unstable and a collapse is inevitable. In the event we call a *supernova,* an iron core may reach a temperature at which it will engage in nuclear reactions. Chapter 17 pointed out that iron nuclei are as stable as nuclei ever become, and energy may be obtained by producing them from either lighter or heavier nuclei. Thus, when an iron core reacts, it *absorbs* energy rather than liberating it. The core collapses and the upper layers fall in upon it. A shock wave completely blows away the outer layers of the star while matter streaming into the core is crushed. Pressures may reach the point where electrons are crushed into protons to form neutrons. A sphere of crushed neutrons would pack much more tightly than a normal mixture of particles, and a ball small enough to fit into a small crater on the moon could be produced. Such a *neutron star* would have very strange properties. A cubic centimeter of one neutron star would have a mass greater than all of the automobiles on earth. A neutron star could do something that an average star could not do—change very rapidly. The fastest that the sun could display any large change would be limited by how fast a signal could go from one part of it to another. In 1967 stars were discovered which were pulsing, or varying their output at a very fast rate. These *pulsars* were measured with radio telescopes and were sending constant timed pulses of energy with periods as short as $\frac{1}{4}$ s. This observation was consistent with the idea of a rotating neutron star, but impossible for a large object like a normal star. The evidence was not conclusive, and so the search turned to the constellation of Taurus. The last brilliant supernova explosion in our immediate part of the universe occurred in 1054 in Taurus. The fragments of that explosion have been identified with the object called the Crab Nebula. If a pulsar is a neutron star formed from the core of a supernova, where better to look for one than at our own local supernova? ("Local" is a very relative term. The Crab Nebula is estimated to be 1300 pc or 4200 light-years distant.) Even at this distance it was brighter than Venus in the sky for 2 years before it faded. The expanding luminous filaments of matter of the nebula are about 1 pc across. Of course, they are really much larger by now, since what we see is how the Crab Nebula would have appeared 4200 years ago. The central star in the Crab Nebula is a pulsar, but it has a period of only $\frac{3}{100}$ s. This established the star as much smaller than the earth, for example. If the earth rotated at this rate, the equator would have to move several times the speed of light. Inertial forces would shred the earth if we rotated at any sizable fraction of this speed. The pulsation rate of the central pulsar of the Crab Nebula established it as a very small object. Even more incredible, it was pulsing not only in the radio spectrum but in visible light as well! It is not coincidence that the newest pulsar is also the fastest in rotation. Careful measurements indicate that all pulsars are slowing down in time. In the case of the Crab Nebula, the energy lost in slowing is equal to the radiation given off by the entire nebula. Astronomers had long wondered where filaments of materials separated from a supernova for 900 years got the energy to glow.

One final state of matter is predicted by theory. If the mass of the collapsing core of a supernova is large enough, say 3 times the mass of the sun, the collapse should not stop at the neutron star stage. The gravitational pressures should be so intense that the neutrons could be crushed into each other and the collapse could continue. Continue to what? The theory does not imply that it collapses *to* anything, merely that the collapse continues and the mass compacts into a smaller and smaller volume with greater and greater density. Where does this stop? Nothing seems to imply that it must stop. This does not mean that you could watch the matter shrinking past the size of the earth, Florida, Manhattan, a basketball, and so on, because of the gravitational field set up by the mass which is contracting. At a certain distance from the center of collapse, the gravitational field would be so intense that nothing could get out, *not even light.* For this reason, such an object is called a *black hole* in space. Such an object would "appear" to be a gravitational field centered on no visible object. Close up (but not too close, or you could never get out) there would be a sphere of darkness of a very definite radius determined by the mass inside. By itself, a black hole might never be detected, but as part of a binary star system detection should be relatively easy. Recall that only stars of large mass can form black holes. Thus, they should display a large gravitational effect, and may even dominate an orbiting star. If you walked onto a field and saw an elephant being swung around in loops on the end of a rope, but the rope ended in nothing, your feeling for how physical laws are operating might seem violated. If astronomers find a star whirling around, orbiting about nothing as it proceeds through space, they will have our first black hole.

The Expansion of Techniques—Radio Astronomy

Different materials are transparent to different types of electromagnetic waves. The atmosphere of the earth allows visible light through. Longer waves, in the infrared region, are absorbed. This absorption is done mainly by carbon dioxide molecules and gives rise to the greenhouse effect which keeps the earth's surface at a livable temperature. Shorter waves than light are absorbed mainly by ozone molecules high in the atmosphere. This prevents dangerous amounts of ultraviolet radiation from harming us with perpetual sunburn or more serious complications. While both of these limits are useful, they allow astronomers only a small "window" of possible radiations with which to study the universe. Clouds of dust in space absorb the rays in this range and pull down the shade on the window for observing objects behind them. It appeared as though whole regions of space would be blocked from direct observation forever.

The first half of this century produced a large expansion of systems to service the consumer. Gas pipes had served individual homes, but then came electric lines, water, telephone lines, and so on. Larger and larger areas were served, and problems in distribution and transmission had to be solved. A young engineer at Bell Telephone Laboratories was assigned the task of identifying the sources of the static that hissed and crackled on long-distance telephone lines. In 1931 Karl Jansky built a highly directional, steerable radio antenna to track down the source of the noise. Looking like the wings ripped off a house-sized biplane, the antenna rotated on four wheels to point to any part of the horizon. Jansky rapidly determined that much of the high-pitched static was from thunderstorms, but there was a part to the noise that was different. This noise seemed to grow most intense once each day, so at first Jansky thought it might be associated with the sun. More work made it clear that it was not. Finally, in rechecking his records Jansky realized that the peaks did not come every 24 h, but every 23 h 56 min 4 s. This realization might send shivers up the spine of an astronomy buff. This time, you may recall, is the *sidereal day,* or the time that it takes for the heavens to appear to make one complete revolution about the earth (see Chap. 4). Jansky had been listening in on the waves from the center of the Milky Way galaxy itself, right through spiral arms and dust clouds. Astronomers had a second window through which to "view" the universe. Actually, the radio window had been tried decades before on the sun with no success. This was because stars normally emit very little radio disturbance except during flares. Even so, Janksy's discovery caused very little stir among astronomers. His paper gathered dust in the same libraries where another paper by Fleming described an obscure mold called *penicillium* that produced a bacteria-destroying chemical. Both papers eventually produced very large changes.

Radio waves are so much larger than light waves that a much larger detector is required to produce a sharp image. Even a small detector of light such as the eye can produce a sharply resolved, nonblurry image. It takes a radio telescope several blocks in diameter to equal the resolution of the eye with light. Single telescopes this size are too big to be steered on earth. The largest fully steerable radio telescope today has a parabolic dish as wide across as a football field (with end zones). For a nonsteerable telescope, an entire valley in Puerto Rico was lined as a reflector with the detector for the reflected waves supported on cables high above the valley. The largest compound radio telescope will be the very large array (VLA) radio telscope, which uses a desert floor in New Mexico (Fig. 19-13). Twenty-seven separate, steerable parabolic antennas will be spaced in a Y some 27 km (17 mi) across. Signals from these antennas will be merged by a computer to equal the resolving power of a single telescope of this size. Each of the separate "dishes" is

FIGURE 19-13 The very large array (VLA) radio-telescope complex under construction in New Mexico. Signals from all 27 radio telescopes will be merged by the computer to generate a radio-wave image nearly as sharp as optical images. The problem with radio waves is not "brightness," but resolving the precise direction from which they come. Built with National Science Foundation funding, the VLA will be able to resolve direction as well as a single antenna 27 km across. (*National Radio Astronomy Observatory.*)

the size of a 10-story building and a major structure by itself. The operation of the VLA, scheduled for 1981, should bring to radio astronomy a crispness of image (resolution) approaching that of optical telescopes.

Appropriately, we may end this section back where it began, in New Jersey at the Bell Telephone Laboratories. In 1965, using a special horn-type antenna, Penzias and Wilson were still working on the problem presented to Jansky. They were trying to locate every possible source of static. They eliminated sources down to a persistent, small background that was found in all directions in space. This background of static corresponded to the radiation emitted by a black body at a temperature of only 3 K. To even detect this low a background all of the sensitive detecting parts of the equipment had to be kept as cold as possible. The *thermal noise* in a quality amplifier at room temperature would sound like the clash of Attila the Hun with a Roman legion at this level. This thermal noise comes from the radiations associated with the jiggling of the atoms in the equipment themselves. To prevent this, the equipment was cooled by liquid helium at 4 K to provide as sensitive a detector of radiation as possible. The 3-K background was real. Theorists showed it to be the residual radiation of the creation of the universe. To many astronomers this was the confirmation of a particular theory for the origin of the universe, and it earned the 1978 Nobel Prize in Physics for Penzias and Wilson. Ironically, it also made possible an experiment which came to a spectacular nothing 90 years earlier. Small shifts in the 3-K background indicate the motion of the sun in its rotation about the galaxy. On top of this another shift suggests the motion of the entire Milky Way galaxy in the direction of the constellation Virgo. In 1887 the Michelson-Morley experiment failed to detect an absolute motion of the earth through space based on the time of travel of light (Chap. 17). That result swung physical thought in a direction that culminated in the theory of relativity.

The Organization of Stars—The Galaxy

The first suggestion of order in the nighttime sky is the band of faint light called the Milky Way that circumnavigates the sky. When Galileo turned his telescope to the Milky Way, he saw more stars than he had imagined could exist. Hershel quantified this by counting the number of stars of each magnitude visible in various directions. Around 1800 he announced that the numbers of faint stars decreased with distance in the directions perpendicular to the Milky Way. He concluded that we were near the center of a disc of stars roughly 4 times as wide as thick. Not everything beyond the planets was a star, however. During the first trip around the world, Magellan's crew had puzzled over two large patches of dim light in the southern sky (Fig. 19-14). These were nowhere near the Milky Way and could not be isolated patches of it. They were called the Magellanic Clouds. Later, the astronomer Messier, an ardent comet-watcher, became annoyed by fuzzy patches in the sky which he kept mistaking for comets. He decided to make a list of all such objects in the sky for himself and fellow comet-seekers to avoid. Ironically, the Messier objects in his catalog of the 1770s became the most studied objects in the sky. (For example, the first Messier object, M1, is the Crab Nebula; M13 is the great cluster in Hercules; and M31 is the great galaxy in Andromeda. Messier objects have come to include a list of "must" viewing for small telescope users.)

An important breakthrough came with the work of Henrietta Leavitt at Harvard in 1912. She was studying variable stars of the type known as *Cepheid variables*. These stars go through a regular variation in brightness in a pattern which took a few days to a month or more. No Cepheids occur within range for direct measurement of their distance, hence their absolute magnitude. However, the lesser Magellanic Cloud had been shown to be a gigantic swarm of stars (an irregular galaxy) and it presented numerous Cepheids. Leavitt studied 25 of these and discovered that the brightness of a Cepheid depended on the period of light variation. The longer the star took in its variation, the brighter it was at maximum. The dimmest Cepheids, with a period of about a day, have an absolute magnitude near zero. Thus, the brightest ones, with the longest periods, are very bright stars which can be seen across large distances (the lesser Magellanic Cloud itself is a great distance away, over 40 kpc

FIGURE 19-14 The greater Magellanic Cloud in the southern hemisphere of the sky is really entirely outside of our own grouping (galaxy) of stars. It is composed of several billion stars some 50 kpc (160,000 light-years) distant. Along with the lesser Magellanic Cloud in the same portion of the sky it is an irregularly shaped satellite galaxy of our own which is several dozen times as large. (*Lick Observatory photograph.*)

or 160,000 light-years). Cepheid variables provided a way to measure very large distances.

During this same decade, a new model for our star system was introduced. Previous models had implied that the star system, or galaxy, was centered on the solar system. This had been suggested by the method of star counts in various directions. These counts fell off faster with distance than they should if the stars were uniformly distributed in the plane of the Milky Way. Mostly, this effect was due to the absorption of light by the gas and dust of interstellar space, but this was not known. In 1917 Harlow Shapley proposed a new model based on the distribution of globular clusters of stars. These are not randomly distributed in the sky but center on an area in the direction of the constellation Sagittarius. Shapley suggested that the sun is about two-thirds of the way out from the center of a huge system of stars some 300,000 light-years across. While this idea was by no means accepted at the time, it has shown to be essentially correct.

The Milky Way galaxy is a wheel-shaped disk composed of some 200 billion stars with associated gas and dust (Fig. 19-15). The center of

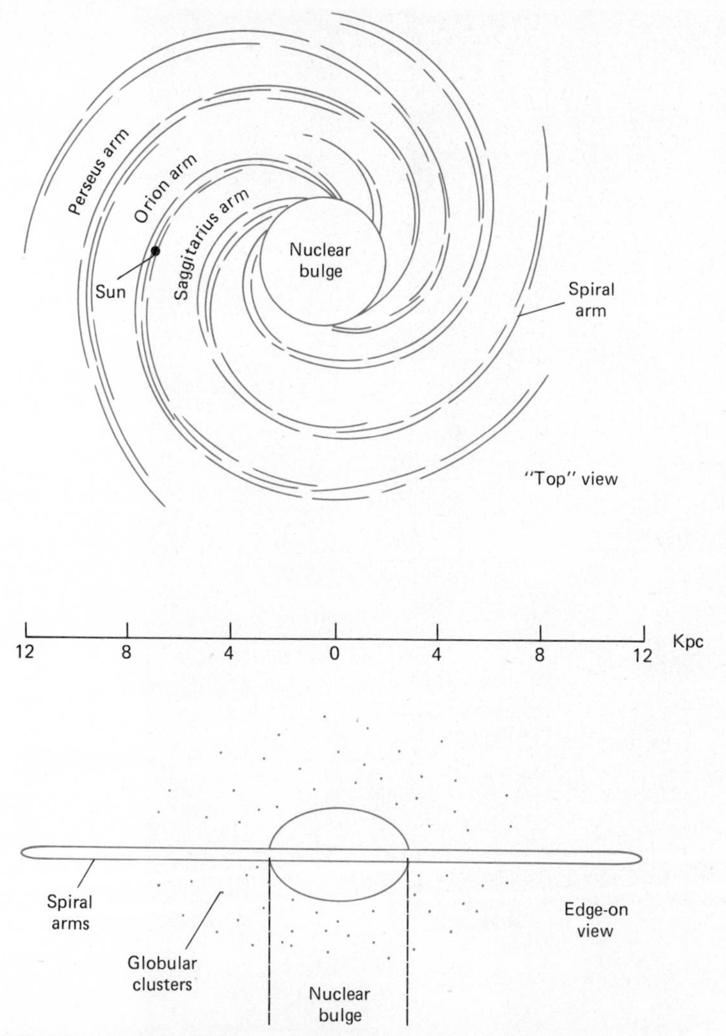

FIGURE 19-15 The Milky Way galaxy, to which the sun belongs, represents a flattened disk of over 200 billion stars. The sun is located on the third spiral arm out from the nuclear bulge of the galaxy. The vast majority of all the stars visible in the night sky are located within the dot marking the sun's position.

the galaxy swells to form a flattened spherical *nuclear bulge* some 4000 pc across. The sun is about 8000 pc from the center in the region of the *spiral arms.* This disk is over 500 pc thick and of varying stellar density. We are on the inside edge of a spiral arm running in the direction of the constellation Cygnus, sometimes called the Orion arm. About 2 kpc closer to the nucleus is the Sagittarius arm of the galaxy and a similar distance outward is the Perseus arm. We cannot see much of this structure directly because of the absorption of light by interstellar dust. However, radio telescopes can operate with waves that are not absorbed and that show a detailed spiral structure. To see this structure in a different sense, it is necessary only to look at other spiral galaxies visible to us outside of our own. If at the dawn of human recorded history we had begun to travel to the center of our galaxy at the speed of light, we still would have over three-fourths the way to go to arrive. This structure is so

FIGURE 19-16
Our "sister galaxy" in the local group of galaxies is M31, the great galaxy in Andromeda. Slightly larger than our own Milky Way galaxy, M31 likewise shows smaller satellite galaxies clustering near it. A small smudge to the naked eye (if you know where to look), the small telescope reveals a blurred area. Only a large telescope begins to resolve the light of M31 into the hundreds of billions of separate stars that make it up. The separate stars visible are members of our own galaxy and incredibly closer to us than to M31. We must look through this veil of neighbors, across a vast starless void, to see our companion galaxy. (*Lick Observatory photograph.*)

vast that it taxes the human imagination to picture a bit of it. Even so, the galaxy does not fill a very large part of the void that is intergalactic space. The spiral galaxy which is easiest to observe in the night sky is the great galaxy in Andromeda (Fig. 19-16). This may be seen with the naked eye on a clear night as a blur in the constellation of Andromeda. The blur becomes a smudge in binoculars and a structured smudge in a small telescope. A large telescope reveals the spiral structure to a photographic plate. This galaxy is about 20 galaxy-lengths away from us, and, like the Milky Way galaxy, it has two dwarf galaxies as satellites. The satellite galaxies of the Milky Way galaxy are the two Magellanic Clouds, which are irregular in size and shape. The satellites of the Andromeda galaxy are dwarf elliptical galaxies. This packing of galaxies in space is tighter than the average because all of these objects are part of the *local group* of galaxies. This is a linked group of perhaps two dozen galaxies with the Milky Way and Andromeda spirals being about as far apart as any. At least one other large spiral galaxy belongs to the group, M33 in Triangulum. Others may lurk near the plane of the Milky Way where the dust of our own spiral would make them difficult to observe. The light output of the Andromeda galaxy (M31) expressed as an absolute magnitude is −21. This is about one magnitude, or 2.5 times, greater than the calculated brightness of the Milky Way galaxy. The third spiral, M33, has an absolute magnitude of −18.9. In mass, M33 is only one-fifth our mass. The large Magellanic Cloud is only about one-tenth the mass of M33, although it nearly equals it in brightness.

The Expanding Universe

Beyond the local group we see more and more galaxies as far as the telescope can collect the faint light into images. One important fact emerges about this light. The farther a galaxy is the more its light is shifted toward the red end of the spectrum. The interpretation of this *red shift* is that it indicates that the objects emitting the light are moving away from us. But, red-shifted galaxies occur in all directions. If we look twice as far, the red shift is twice as great and the galaxies are moving away at twice the speed. Just because all these galaxies are moving away from us does not imply we are the center of expansion. The view of galactic recession would be the same from any galaxy. The best analogy is to picture a polka-dotted balloon. As the balloon is blown up each dot would see all of the others getting farther away. Like the balloon, *the universe is expanding.*

If all galaxies are moving away from each other, does this not imply that at some time in the past they must have been packed tightly together? Yes, this is implied. The rate of expansion is difficult to measure with precision. This value is known as *Hubble's constant H,* and its value is estimated at about 30 to 60 km/s of recession for every 1 million pc of distance. If we ran all of the galaxies backward at this rate, they would converge into the primordial fireball of creation some 10 to 20 billion years ago. At this time all of the mass of the universe would be packed into a superdense blob of interacting photons. At some instant at that time the gigantic expansion began which we can see continuing today. Astronomers refer to this instant as the *big bang.* Radiation would dominate such a fireball until thousands of years of expansion dropped the temperature to the point where ordinary matter could form. This matter would not form huge gravitational clouds for another billion years of expansion. Finally, these large clouds would break into clusters, these into galaxies, these into individual stars in a cycle of smaller and smaller condensations. The exact time of the big bang depends upon the value of the Hubble constant, and, of course, the entire hypothesis depends upon the red shift indicating a recession of the galaxies. While many alternative schemes have been proposed, none has survived tests based on the data.

The big bang interpretation of the universe has several interesting implications. Our largest optical telescopes can gather enough light to resolve the images of galaxies over 2 billion light-years away. The light the telescope detects began traveling toward us over 2 billion years ago and is just now arriving. If we had a larger telescope capable of resolving images some 20 billion light-years distant, could we watch the big bang take place? A simpler question would be, as we look farther and farther and more and more into the past appearance of the universe, do we notice the galaxies becoming more tightly packed in space? The answers depend very much on the type of universe we live in. Einstein's general theory of relativity predicts that the universe is curved, finite, and unbounded. (An example involving area may help. The area of a sphere is curved, finite, and unbounded. Einstein predicts the same sort of relationship for the volume of space.) This appears to be correct. As the universe expands, the gravitational attraction of the masses of the galaxies for each other tend to slow the expansion down. Will this attraction be enough to stop the expansion and pull all of the matter back together? Current measures suggest that it will not, and the expansion will probably continue forever.

Currently, the farthest known objects do not seem to be galaxies but strange quasistellar objects, or *quasars.* Their images are small and starlike except for tantalizing smudges of detail. Yet, some of these objects have red shifts indicating incredible speeds of recession of over 85 percent the speed of light. Quasars cannot be galaxies since some vary with time periods much too short for a galaxy to react; thus they cannot be much larger than giant stars. If the red shift of quasars is entirely due to the expansion of the universe, they can be almost twice as far away as the galaxies at our limit of measurement. This is the fundamental question about the quasar. How can an object not much larger than the stars produce enough radiation for us to detect at a distance where whole galaxies are too dim to be seen? Quasars might be some strange superenergetic form that could exist 4 or 5 billion years after the big bang but are now extinct. (Just as nature could, and did, produce natural nuclear reactors on earth a few billion years ago but could not do so today.) One suggestion is that quasars may represent black holes in galactic nuclei "sucking in" gas and stars at the center of a galaxy. Another suggests zones of matter-antimatter annihilation. The nature of quasars is far from being understood.

Astronomy represents the oldest of the sciences. For centuries it required of each of its practitioners thousands of tedious hours of measurement. Within the last two generations the accumulated data, the new instruments, and the theoretical development in our understanding of matter all have allowed an explosion of astronomical knowledge to take place. There is no reason to expect this process to change, or even slow down, in the foreseeable future.

Summary

The recent past has produced gigantic increases in our understanding of astronomy. Astronaut landings on the moon and satellite studies of the planets have expanded our information. Volcanoes on bodies such as Mars and Venus disclose geologic activity in the recent past, while eruptions were observed on Io. Most surfaces observed show extensive cratering, indicating a great age of the planetary surfaces without serious modification. More has been learned in the past decade about the bodies of the solar system than was known previously.

Early studies of the stars had measured the distance to several thousand stars by using *parallax,* or the shift in the stars' apparent position caused by the motion of the earth. Astronomers measure the distances to stars in *parsecs* (pc) or parallax-seconds, where 1 pc is the distance light travels in 3.26 years. Brightness is measured in *magnitudes,* where lower numbers indicate brighter stars, and each full magnitude is about 2.5 times as bright as 1 magnitude higher. The *apparent magnitude* (*m*) of a star measures how bright it appears from the earth, while the *absolute magnitude* (*M*) rates the star's brightness if placed at a standard distance of 10 pc. Stars vary in color, or surface temperature, and they may be grouped into *spectral types* which largely measure this variable. A plot of spectral type vs. absolute magnitude is known as an H-R (Hertzsprung-Russell) diagram. Most stars fall on a band in this diagram called the *main sequence.* Hot, dim stars (called *white dwarfs*) appear below this band; and cool, bright stars (called *red giants*) appear above it in the diagram. While most stars spend the majority of their life cycle on the main sequence, these other cases represent phases in stellar evolution.

Stars are grouped into gigantic associations called *galaxies.* The Milky Way galaxy to which we belong is some 25 kpc (300,000 light-years) across and contains some 200 billion stars. Distribution of globular clusters in space implied that the center of our galaxy is in the direction of the constellation Sagittarius. This cannot be seen from earth because of many clouds of dust and gas in space which obscure it, but we can detect its radio-wave emissions. Since we can measure the distance of only a very few of the nearest stars directly, determination of the scale of the galaxy required the development of indirect methods to determine distances. The first of these was the use of *Cepheid variables,* or stars whose brightness changes regularly. These are very bright stars whose absolute magnitude may be measured directly by the period of variation. The majority of galaxies are so far that no individual stars may be resolved in their images. However, as the distances of galaxies increased, it was found that their spectrum shifted toward the red. The only consistent interpretation of this is that galaxies are moving apart, and the universe is expanding. If this is true, at one time in the past, the matter in the galaxies must have been very tightly crowded together. This suggests that the matter in the universe had a particular origin some 10 to 20 billion years ago, "blew apart," and has been expanding ever since. This view of the universe is known as the *big bang theory.*

Questions

1 Why is a heavily cratered surface an indicator of the great age of the moon's surface?

2 Why should the crust of the moon be twice as thick as that of the earth?

3 Can you give a reason why any cloud pattern in an atmosphere should remain about the same for decades of time? Are there any such semistable features in the earth's atmosphere?

4 Jupiter's moon Io is the only solid body of the solar system yet found with a geologically active surface that is actively reshaping

itself, like the earth's. If Io has been around for some 4 billion years since the formation of the solar system, where would it get the energy necessary for volcanic eruptions? Where does the energy for such activity come from on earth?

5 The earth is tilted on its axis by $23\frac{1}{2}°$, giving us our seasons. Describe what the year would be like if the earth's tilt was nearly 90°, like that of Uranus.

6 The bright stars Arcturus and Pollux both have their absolute magnitudes and apparent magnitudes about the same. What can we infer about these two stars?

7 The star Rigel in Orion has an absolute magnitude of −6.2 and an apparent magnitude of 0.3. What can you tell from these numbers?

8 Observation of the spectrum of a star reveals that some of the spectral lines remain in position. Other lines, however, shift first toward the red, then reverse and shift toward the blue end of the spectrum, and repeat this pattern at regular intervals. How would you interpret this pattern?

9 Whatever rotation is in a dust and gas cloud before it contracts to become a star remains with it after contraction—much as a skater pulls in his or her arms and legs to spin faster. Thus, from conservation of angular momentum (Chap. 6) we expect most stars to rotate rapidly. Early spectral types (O, B, A) do just this, but from spectral type F5 through later types most stars have very slow rotations. How could you account for this—where could the angular momentum have gone? (*Hint:* Where is most of the angular momentum of the solar system?)

10 RR-Lyrae-type variable stars are of spectral type A and brighten and darken in a period of 12 h by about 1 full magnitude. At brightest, they all have an absolute magnitude of about zero. Several thousand have been identified, mainly in star clusters. What use could be made of such information (and has been)?

11 It is very difficult to obtain a completely reliable value for Hubble's constant. Why is this the case?

12 When just the stars of globular cluster M3 (Fig. 19-12) are plotted in an H-R diagram the later spectral types (F, G, M,) are present on the main sequence. There are no early spectral types (O, B, A) present. Instead, the plotted stars curl off the main sequence to a scattering of red giants. What does this pattern suggest about M3 and the evolution of stars?

Problems

1 The Roche limit for a satellite is 2.44 times the planetary radius. How closely could the moon approach the earth before breaking up into a ring?

2 The star Tau Ceti is frequently mentioned as a sunlike star which is a likely candidate for planets and possibly life forms. Its parallax is 0″.276 (seconds of arc). How far is it from us in parsecs; in light-years? What is its likely spectral type from its description?

3 Two stars with the same absolute magnitude are seen in the night sky. One appears 5 magnitudes brighter than the second. How many times as far away is the second star as the first?

4 The resolving power of a telescope is determined by the width of its collecting surface compared to the wavelength it is detecting. The human eye has a lens opening of about 0.5 cm compared to the Hale telescope on Mount Palomar with an "opening" of 500 cm. Thus, the telescope has 1000 times the resolving power. This is sufficient to see two candles 1 m apart some 20,000 km away as

two separate candles. If the wavelength of light used by the telescope is 5×10^{-7} m, how large an antenna would a radio telescope have to be to achieve the same resolution if it used the 21-cm waves of neutral hydrogen?

5 Barnard's Star is of spectral class M5 with an apparent magnitude of 9.5 and an absolute magnitude of 13.2. Its parallax is 0".543 with a radial velocity of −108 km/s and a proper motion of 10".3. Write a short description of Barnard's Star, translating this numerical information into words.

6 The red star Betelgeuse in Orion has an absolute magnitude of −5.6, over 10 magnitudes brighter than the sun.

(*a*) About how many times as much heat and light does Betelgeuse radiate compared to the sun?
(*b*) The surface of Betelgeuse is about half the absolute temperature of the sun's surface. How much energy is radiated by each square meter of its surface compared to the square meter of the sun? (*Recall:* $I \propto T^4$)
(*c*) If the total radiation from Betelgeuse is this many times as much as from the sun [part (*a*)], yet each unit of area radiates only such a small fraction of energy as the sun [part (*b*)], then how many times as much surface area as the sun must Betelgeuse have?
(*d*) The radius of Betelgeuse is how many times as large as the radius of the sun? (The area of a sphere is: $A = 4\pi r^2$.)

7 Write down a column of 10 numbers. Begin with zero. The next number is 3, and each successive number is just twice the previous one (6, 12, . . .). Add 4 to each number to make a new column. Divide each of these numbers by 10 to get a final column of numbers. Compare this final set of numbers to the orbital radius column (in AU) of Table 19-1. This procedure is referred to as Bode's "law" for the spacing of the planets and has been known since before the American Revolution. No one knows why it works as well as it does.

(*a*) According to Bode's law, how many planets out from the sun should Jupiter be?
(*b*) According to the numbers from Bode's law, at what distance from the sun is there a planet "missing"?
(*c*) Which planet doesn't seem to fit Bode's law at all? Does Table 19-1 give any other numbers which do not seem to fit for this planet?

8 A typical nitrogen molecule of the air at room temperature (300 K) is traveling at a speed of 450 m/s (100 mi/h).

(*a*) How fast would such a molecule be traveling on the daytime surface of Mercury? (*Recall: T* is a measure of average kinetic energy.)
(*b*) The velocity an object must have to be able to coast away from the earth and not be pulled back by gravitational attraction is 11.2 km/s (6.94 mi/s). For Mercury, this escape velocity is 3.9 km/s. Would it appear that the nitrogen molecule above would be able to leave the surface?
(*c*) Actually, while the average speed of a nitrogen molecule at room temperature is about 450 m/s, some are moving at much greater speeds. Astronomers calculate that the *average* molecule must have a speed under 0.2 the escape velocity for a planet to retain its atmosphere for a billion years. Based on this criterion, could Mercury retain an atmosphere of nitrogen molecules?
(*d*) Based on the data above for the nitrogen molecule (molecular mass = 28), what would be the average speed of a hydrogen molecule (molecular mass = 2) in the earth's atmosphere at the same temperature? Does this speed make it likely for the hydrogen molecule to escape from the earth?

CHAPTER TWENTY

Science and Modern Life

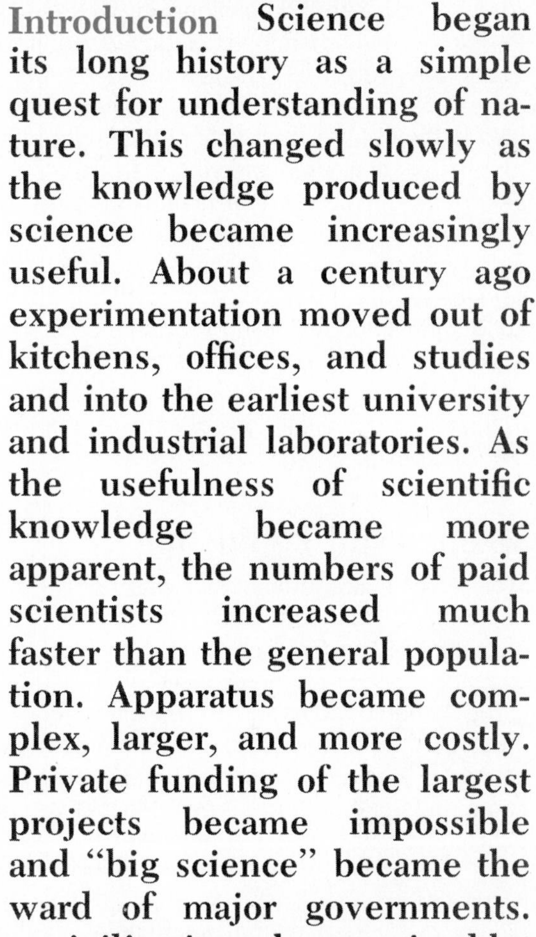

Introduction Science began its long history as a simple quest for understanding of nature. This changed slowly as the knowledge produced by science became increasingly useful. About a century ago experimentation moved out of kitchens, offices, and studies and into the earliest university and industrial laboratories. As the usefulness of scientific knowledge became more apparent, the numbers of paid scientists increased much faster than the general population. Apparatus became complex, larger, and more costly. Private funding of the largest projects became impossible and "big science" became the ward of major governments. Over the last few centuries, Western civilization, characterized by its developed science, has "squeezed out" the other civilizations on earth. This process began as simple military expansion boosted by a superior technology. Later, when the Western world lost its appetite for colonialism, the expansion still continued, spurred by the visible benefits brought by Western science and technology. A few centuries have transformed science from the private pursuit of the intellectually curious to one of the underpinnings of society. The speed and power of this movement has no parallel in human history. Over 80 percent of all the scientists that have ever lived are alive today! Some 30,000 scientific journals report new findings at the rate of 100,000 pages *per day!* Despairing of keeping up with this mass of information, most scientists retreat into narrow areas of specialization. The public, only vaguely aware of this flow, tries to cope with the bewildering pace of change in all aspects of life. People born in a horse-and-buggy world live out their old age watching space travel on their TV screens. Individuals from generations who have always had electronic communications, jet travel, and miracle drugs look at their elders as reactionaries in their attempt to adapt institutions to the present while preserving elements of protection learned from past experience. It seems that human physical existence in advanced societies gets easier while psychological existence becomes more difficult to cope with. The word "science" comes from the Latin word *scienca*—knowledge. Past societies confronted by serious problems sought the solution in wisdom. Every poll today indicates that the public expects solutions not from wisdom but from knowledge—science. But knowledge applied unwisely can create as many problems as it solves. In this chapter we will look at some of the problems confronting us and the relationship of science to them.

Science and Population

When your average instructor was born, it was onto an earth with only half of its present population. Recent retirees were born on a planet with only one person for each four now alive. World population is doubling every 35 years or so. In the past there were three great checks on the size of the human population. These were disease, famine, and war. The next time you look around at your classmates it is sobering to realize that *all of you* have already exceeded the average human life span in Europe for some parts of the Middle Ages. Most of you do not have the experience of brothers and sisters, or playmates, dying of illnesses. None of you has lived through an epidemic killing people more rapidly than they could be buried. If you are fortunate enough to have very old members of your family alive, chat with them about their recollections of the influenza epidemic during World War I (which killed more Americans than the war did). In a more recent time most people of your parent's generation will recall friends and acquaintances crippled for life by polio (infantile paralysis). Your world, the world of today, is vastly different from any of the past in that it has little experience with widespread, serious childhood illness, death, and physical deformity. When so-called Legionnaires' disease appeared on the American scene in the late 1970s, it was headline news for weeks and months while the death toll barely reached 100. Such diseases were simply overlooked in the massive mortality rates of yesteryear.

The attack on sanitation- and nutrition-related diseases began in earnest a little more than a century ago. However, the second quarter of this century produced the first real human ability to combat disease. We have not fully recovered from the implications of this revolution in human health care. By midcentury, childhood mortality rates and epidemics were down all over the world. Social systems accustomed to producing many children so that some would survive their parents into adulthood did not change so rapidly. The population of the earth at the time of

Christ was an estimated 300 million. By the time of Newton (1660) it had doubled to 600 million. In the 200 years to Maxwell (1860) it had redoubled to exceed a billion (Fig. 20-1A). In the past 120 years it has doubled *and* redoubled so that we approach the 4.5 billion mark. More people are alive today than have ever died. This fundamental change impacts almost every area of our life. Since its founding the United States has shown a similar population trend (Fig. 20-1B).

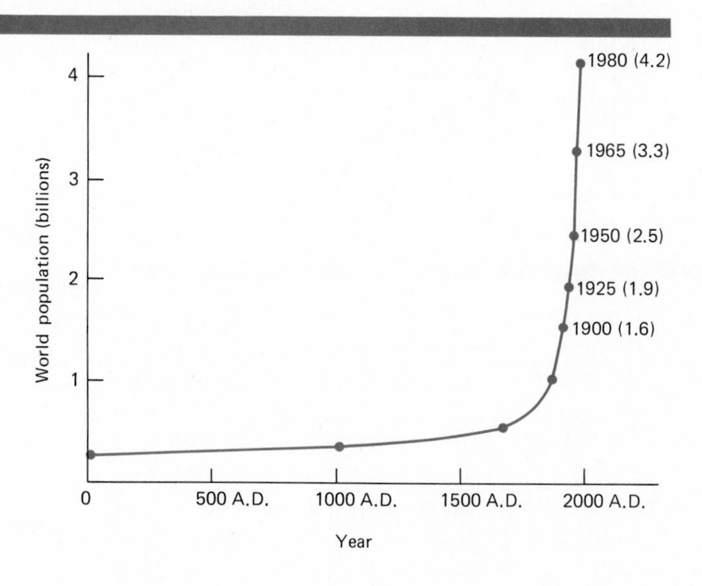

FIGURE 20-1A The human population of the world represents a runaway growth curve which will eventually reach some limit imposed by available resources, natural forces, or human action. Projections of the curve to the present lead to many "standing-room-only" statements for conditions in the next century.

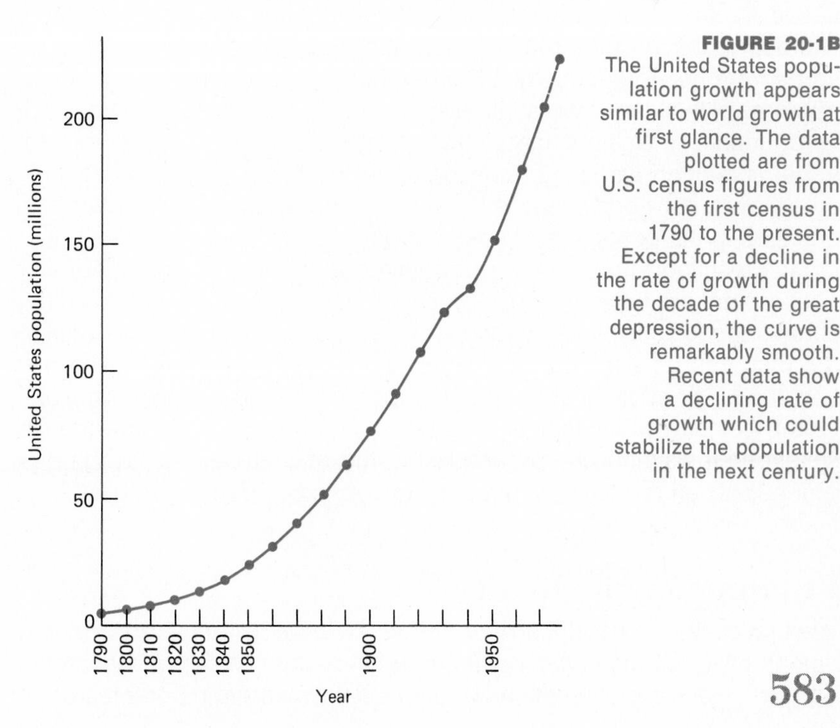

FIGURE 20-1B The United States population growth appears similar to world growth at first glance. The data plotted are from U.S. census figures from the first census in 1790 to the present. Except for a decline in the rate of growth during the decade of the great depression, the curve is remarkably smooth. Recent data show a declining rate of growth which could stabilize the population in the next century.

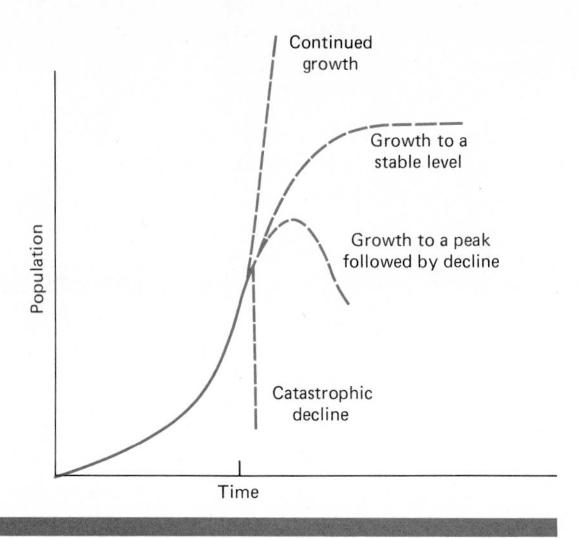

FIGURE 20-1C
The future trend of population is difficult to predict. The option of continued exponential growth through the next century is limited by the simple dimensions of the earth. Expansion away from the earth's surface could ease the limits, but probably not rapidly enough to allow the expansion to continue unchecked. Growth tending toward a stable level is quite likely, but what that level is or its implications for the human condition cannot be predicted. Growth to a peak followed by a decline is suggested by pessimists who foresee this as consequence of expanded natural resources. Some optimists see this as a conscious choice of an enlightened future race to improve living conditions. Least desirable is the catastrophic decline as the result of nuclear war or our inability to cope with a universal natural disaster.

As disease and epidemic problems came slowly under control, and the infant mortality rate dropped, pressures on the food supply grew. Famine asserted itself as the second control on population. With thousands dying of starvation daily, science and technology were called upon to work the miracle in the underdeveloped world that they had accomplished in the developed nations. Large efforts were made to increase agricultural productivity: new plants, drought-resistant crops, disease-resistant strains, high-yield hybrids—all were introduced. In some cases the productivity per acre was increased sixfold, in many cases fourfold. Yet, rather than produce stability and improve living standards, these gains did not change living patterns, and births absorbed the new productivity. Whereas before thousands were starving, now tens of thousands were in the same condition. The early moral of the introduction of Western medical and agricultural knowledge into the underdeveloped world was that gains in the living standards of these areas were not necessarily promoted. In the absence of fundamental changes in the political and social structures of these societies any increase resulted in a yet larger population at the same level of squalor. Thus, science is not a universal cure for the ills of humankind. Given reasonable good sense and a chance for ordered development, the future of the race can be bountiful in ways which can only be imagined today. Given despair at the host of difficult problems facing us and a refusal to modify customs which lead to long-term instability (such as American insistence on wasting energy and using the environment as a dumping ground), science can only prolong an inevitable decline (Fig. 20-1C).

Environmental Quality

The first book on air pollution was written almost 400 years ago, and conditions have not improved in this area since then. Weather forecasters rattle off levels of pollutants as glibly as if they were a natural feature of the air like its moisture content and temperature. *Fumifugiam,* written in 1661, explored the effects of thousands of coal fires in London on its inhabitants. These effects were worst when the city was suffering one of its famous fogs. It was not known then, but this situation is characterized by an atmospheric *temperature inversion* (Fig. 20-2). Normally, the air gets colder as you ascend. On occasion, a warmer upper layer of air can "seal in" a pocket of colder air on the surface in an inversion. Temperature inversions prevent the vertical mixing of the air, and pollutants generated at the surface are trapped in this lower layer of air. As the generation of pollutants continues, their concentration in this thin pocket of air grows to dangerous levels. The most famous incident of this type was the London "killer fog" of 1952. During the period of the inversion from December 5 to 9, almost 4000 more deaths than normal (*excess deaths*) were reported in London. Repeated incidents in 1956 and 1962 each produced nearly 1000 excess deaths. Similar situations have occurred in New York City, Washington, Philadelphia, Chicago, and many other major cities around the world. While temperature inversions cause an immediate crisis, they act only to dramatize a constant problem. Pollutants which are always present and largely unnoticed aggravate respiratory ills and produce a steady trickle of excess deaths at other times.

Major air pollutants include carbon monoxide, hydrocarbons, nitrogen oxides, sulfur oxides, and small particles, or *particulates.* Generally, urban air has 25 times as much carbon monoxide, 10 times as much CO_2 and dust particles, and 5 times as much sulfur dioxide as rural air. The oxides of nitrogen are seldom found at dangerous levels themselves; however, they interact with sunlight to produce the secondary pollutants known as *smog.* Up to two-thirds of the carbon monoxide and half the hydrocarbons and oxides of nitrogen were introduced to the atmosphere by automobile and truck engines prior to 1967. Exhaust

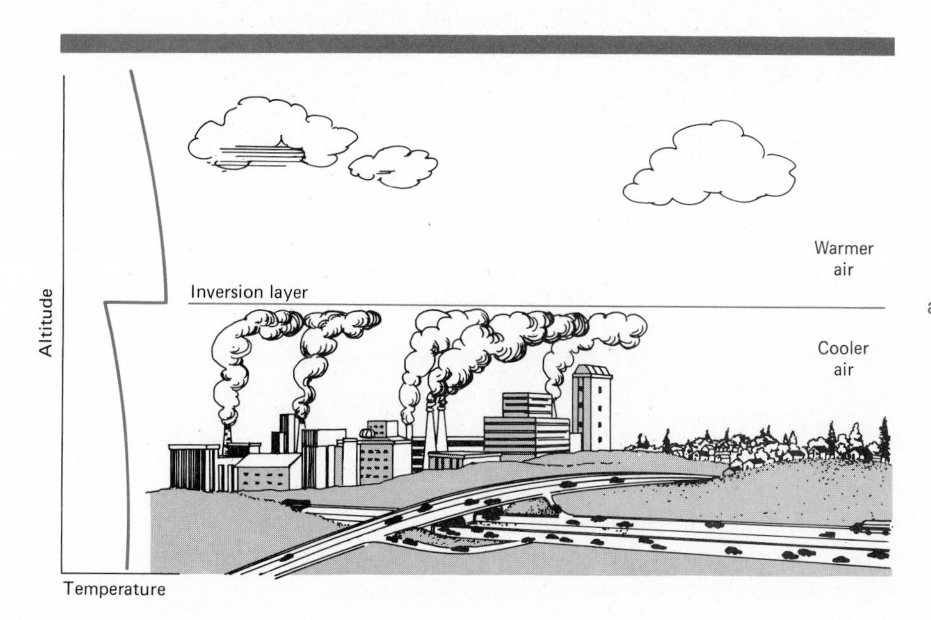

FIGURE 20-2
Temperature normally decreases with altitude. An inversion occurs when warmer air at some height seals in a pocket of cooler air underneath it. This prevents the vertical mixing of the air, and pollutants generated at the surface begin to concentrate in the trapped volume of air.

TABLE 20-1 Air Quality Criteria

POLLUTANT	LEVEL AT WHICH ADVERSE HEALTH EFFECTS OBSERVED	1971 AIR QUALITY STANDARD	
		AVERAGE VALUE	ONCE PER YEAR
Carbon monoxide	15 mg/m³	—	10 mg/m³
Hydrocarbons	200 μg/m³	—	160 μg/m³
Nitrogen oxides	120 μg/m³	100 μg/m³	—
Particulates	80 μg/m³	75 μg/m³	260 μg/m³
Sulfur dioxide	115 μg/m³	80 μg/m³	365 μg/m³

Source: Condensed from Emil Chanlett, *Environmental Protection,* McGraw-Hill, New York, 1973, pp. 270–271, table 5-24.

standards adopted then required the lowering of the levels of all three of these pollutants by over 90 percent in a series of steps over the next decade. The Clean Air Amendments of 1970 lowered allowed hydrocarbon emissions from 17 g/mi down to 0.41 g/mi (Table 20-1). While everyone agrees on the desirability of reductions in pollution levels, time and again extensions of deadlines and exceptions were sought to this law. Ultimately, its provisions went into effect and still stronger standards were set for the 1980s. Compliance with these standards mandated serious changes for American automobiles. Engines run better with excess fuel for the oxygen available, but this favors the formation of both carbon monoxide and hydrocarbon emissions. High-compression engines deliver more power for their size, but this favors the formation of oxides of nitrogen. Ultimately, most cars used chambers with a catalyst to remove hydrocarbons and CO after it was produced, and dropped compression to generate less nitrogen oxides. This in turn resulted in a 27 percent drop in the miles per gallon achieved by vehicles. Furthermore, since the catalyst would be rendered inactive (*poisoned*) by lead, unleaded gas had to be introduced. Tetraethyl lead in gasoline acts to slow down the burning in an engine and thus reduces engine "knock." Without lead a higher-octane gasoline must be used which requires more refining. By 1970 over 200,000 tons of carbon monoxide and 45,000 tons of hydrocarbons were being dumped into the air by vehicle exhausts *each day.* Furthermore, without emission controls both of these numbers would have doubled by 1980 (Fig. 20-3).

The control of sulfur dioxide is a different problem. Most is released at power plants using high-sulfur coal and oil, and amounts to over 30 million tons per year. Some of this may be removed by using devices called *stack gas scrubbers.* These are quite costly to install and maintain, do not remove all of the pollutants, and generate their own solid wastes which must be disposed of. Sulfur dioxide is not only a human health hazard but is very damaging to structures and finishes. Since the introduction of automobile emission standards, sulfur dioxide has become the most bothersome air pollutant. Part of the reluctance to greatly expand the use of coal for energy stems from the lack of a solution for the elimination of sulfur dioxide. Estimates of the excess deaths due to coal pollutants have increased steadily as more data become available. Coal

(*a*)

(*b*)

FIGURE 20-3 Considerable progress has been made in cleaning the air from extreme levels of pollutants. (*a*) The top photograph is a view of downtown Pittsburgh from the exit to a tunnel immediately across the river. The photograph was taken at midday in 1940, and particulates emitted by local industries and homes have completely obscured the large buildings. (*b*) By 1972, a number of local programs aimed at cleaning the air had produced the scene below taken from the same position as a generation and a half earlier. By this time the main threat to health had switched from particulates from coal fires to less visible automobile exhaust emissions. (*Courtesy of Allegheny County Department of Works.*)

now accounts for less than a fifth of United States energy use. In the absence of highly effective means for reducing coal-related pollutants, expansion of its use to satisfy a majority of U.S. energy needs would be environmentally unacceptable.

The Need for Energy

In the 1960s experts were arguing about problems in energy delivery. By 1970 some of these problems began to affect consumers, and by 1980 everyone knew that serious energy problems lay ahead of the nation and the world. Details of the crisis in energy may be debated, but several basic facts underlie the problem. From the time of the Industrial Revolution onward, developed societies have been powered mainly by *fossil fuels*—first coal and more recently oil and natural gas. Nature takes extremely long periods of time to produce fossil fuels, so long that we must consider them *nonrenewable*. When they are used up they are gone—essentially forever. Viewed from the perspective of centuries, fossil fuels are nature's gift for the start-up of industry and technology until more sophisticated, renewable energy sources may be developed. Fossil fuels will dissappear from the world energy picture—the only question is when. While population has been leaping upward during this century, energy use has been growing much faster. For example, the average American uses twice as much energy today as at the start of World War II, and there are twice as many of us. Today, the United States alone consumes as much energy each year as the entire world did (Americans included) in 1953. Figure 20-4 shows the growth of American use of electric energy during this century. These trends rush us toward the expenditure of world fossil fuels at breakneck speed. In the 1970s as much fuel was used as in all of previous human history. Simple

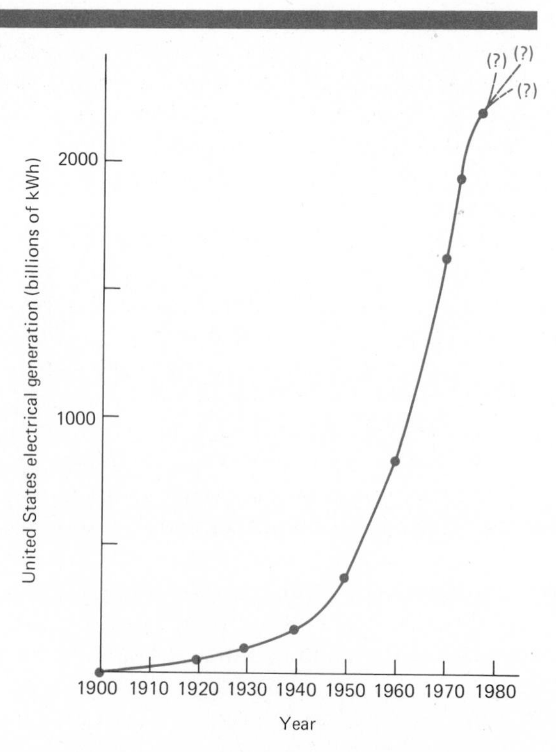

FIGURE 20-4
The growth in the use of electric energy during this century has outstripped even the population growth. Generation of electricity in the 1980s will reach 100 times the rate of 1900, while population will have expanded only 4 times. This compounded growth is true of many energy systems, but none is as extreme as electrical generation. We have expanded from 50 kWh per person per year in 1900 to 10,000 kWh per person per year currently.

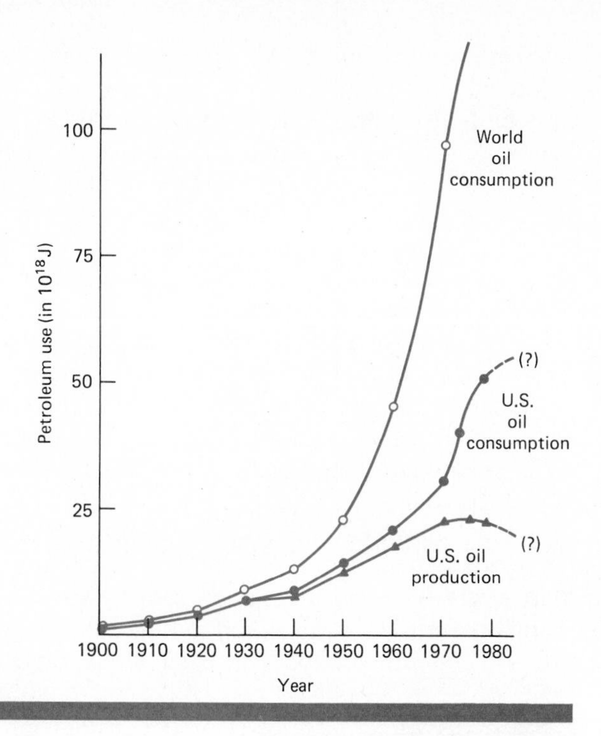

FIGURE 20-5 The record of world and U.S. petroleum production and consumption shows that the United States dominated both for the first half of the century. Following that, world demand outgrew ours. The decade beginning in 1970 saw an increase in American use just when American production reached its peak. The growing gap was filled by imports. The Arab oil embargo of 1973 produced a change in the rate of American growth of demand, but toward the end of the decade we were importing more than half of our needs and paying over 10 times as much per barrel as at the start of the decade.

arithmetic based on this rate indicates that between 1980 and the year 2000 we would have to find, mine, drill, pump, develop, and use 7 times as much energy as was used by the entire world up to the year 1980 to maintain this rate of growth. In the case of oil, it is debatable whether 7 gal remain in the ground for every gallon already used. Even if it does, using it up at this accelerated rate deprives future generations of products more valuable than fuel—such as lubricants, chemicals for plastics, synthetic fibers, and so on. At some point, our use of these materials simply must change.

While the energy crisis affects the entire world, it particularly impacts the United States. As a nation we have been endowed with more than our fair share of every important fuel—oil, natural gas, oil shale, uranium, and especially coal. However, our use of energy has outrun our own resources in the case of oil, and we are dipping heavily into the world supply. American production of oil peaked in the 1970s and will probably decline in the future (Fig. 20-5). At the same time, American use expands annually. We use 6 times as much energy per person as the world average, and we tend to favor the cleanest, most convenient forms of fuels. At the start of World War II about half of American energy came from coal, and essentially all fuels came from within the country. Since that time coal use has expanded very little but now makes up only one-sixth of our energy supply. Oil is relied on for 3 times the energy supplied by coal, and natural gas for twice as much. Of some 3 billion watts of power used, oil supplies half. About half of this, or one-fourth of our total energy use, was supplied by imported oil from foreign sources in 1980. This in turn creates grave problems for national security and the economy. Additionally, other nations resent the insatiable American demand for oil since it tends to drive up the world price of oil. Underdeveloped countries, for example, simply cannot afford the oil needed for the development of their industry and agriculture. Developed countries resent the American demand in the face of their own efforts to limit dependence on imported oil. Such countries, lacking their own oil supplies, have achieved much lower use rates than the United States in an effort to stabilize supply and price. While Americans complained about $1-per-gallon gasoline in 1980, roughly 3 times the price of 1970, they neglected to note that this was about half the price in oil-dependent, developed countries around the world. Americans are the great energy gluttons of the world. National policies for many years have undervalued the price of energy and thus encouraged its use. Such policies made it cheaper, for example, to burn additional fuel each year rather than to properly insulate a house. Now that we can no longer keep up with our own demands for energy, serious changes are inevitable for the future.

A definite relationship exists between the standard of living of a population and its per capita energy use. The graph in Fig. 20-6 illustrates the amount of goods and services produced (gross national prod-

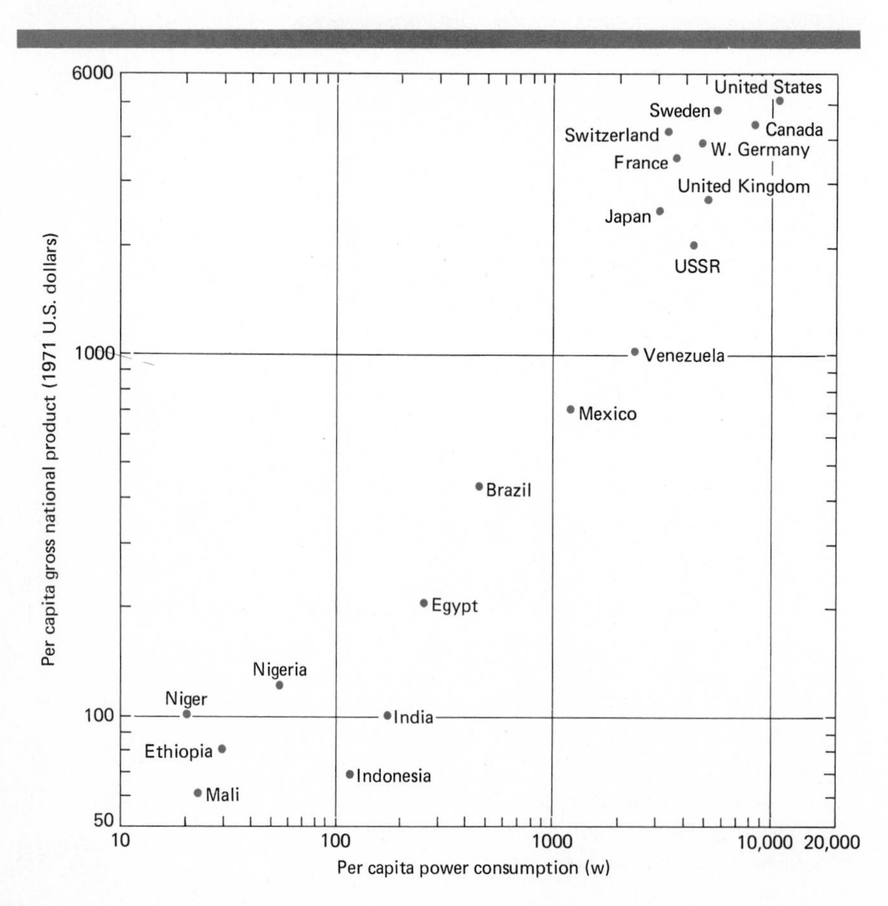

FIGURE 20-6
A plot of per capita gross national product (total goods and services produced in a country divided by its population) vs. the per capita power consumption. Note the relationship between energy use and productivity.

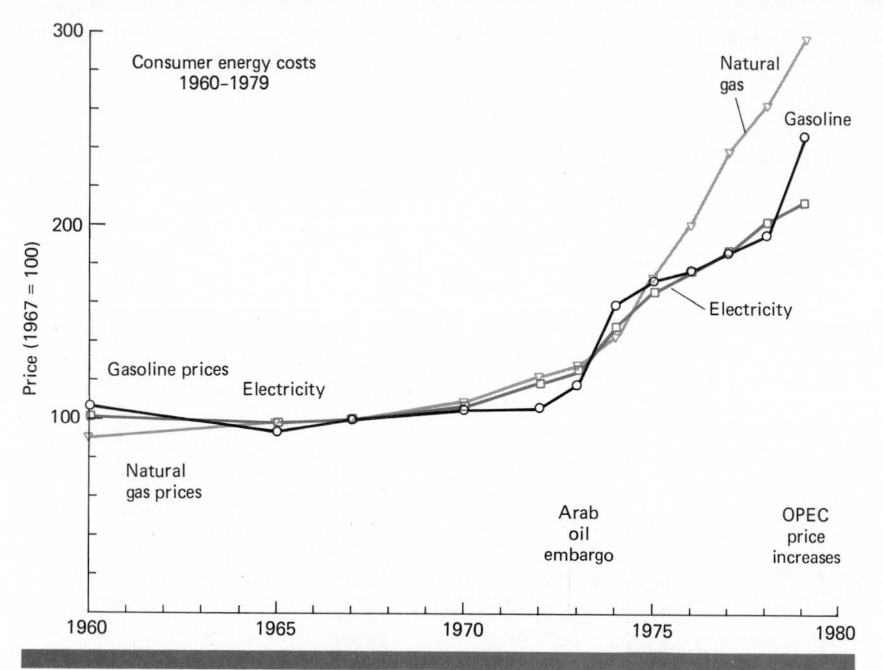

FIGURE 20-7
Energy prices were much more stable than the consumer price index (CPI) until 1973. Indeed, the cost of electricity declined each year until 1969. Since 1969 energy cost increases have outpaced the rise in the general cost of living (CPI) and, indeed, fed its growth. Inflationary rise in the CPI was definitely not fed by energy prices before 1973. During the decade 1967–1977, all prices rose 184 percent while energy costs increased 211 percent.

uct or GNP) per person vs. average energy use for 1971. Clearly power consumption is related to productivity. For example, from the graph we may read that the average American consumes about 5 times as much power as the average Venezuelan (10,200 W vs. 1120 W) and produces about 5 times as much ($5000 vs. $1000). Considerable discussion questions whether this must be the case. Could productivity be maintained while energy use is sharply cut? A number of practices are clearly wasteful of energy and do not contribute to the standard of living—such as failure to insulate homes. Industry has found that it can cut energy use by up to 30 percent in some cases without reducing its output. As energy continues to get more expensive we will either have to maintain productivity with less energy or face a decline in the standard of living. If energy costs make up a larger and larger portion of the price of things we buy, we will have no choice as a nation but to buy less (Fig. 20-7). Since this is not a very happy prospect, it is important to reduce the amount of energy going into products and services.

Patterns of Energy Use

Energy consumption in the United States can be divided into four basic sectors of the economy: residental use, industrial use, transportation, and commerce. Figure 20-8 indicates the total amount of energy devoted to each of these sectors and the source of the energy. Over half of all the oil used is consumed in the transportation sector, since other fuels make almost no contribution to transportation. Of this oil, over 60 percent is

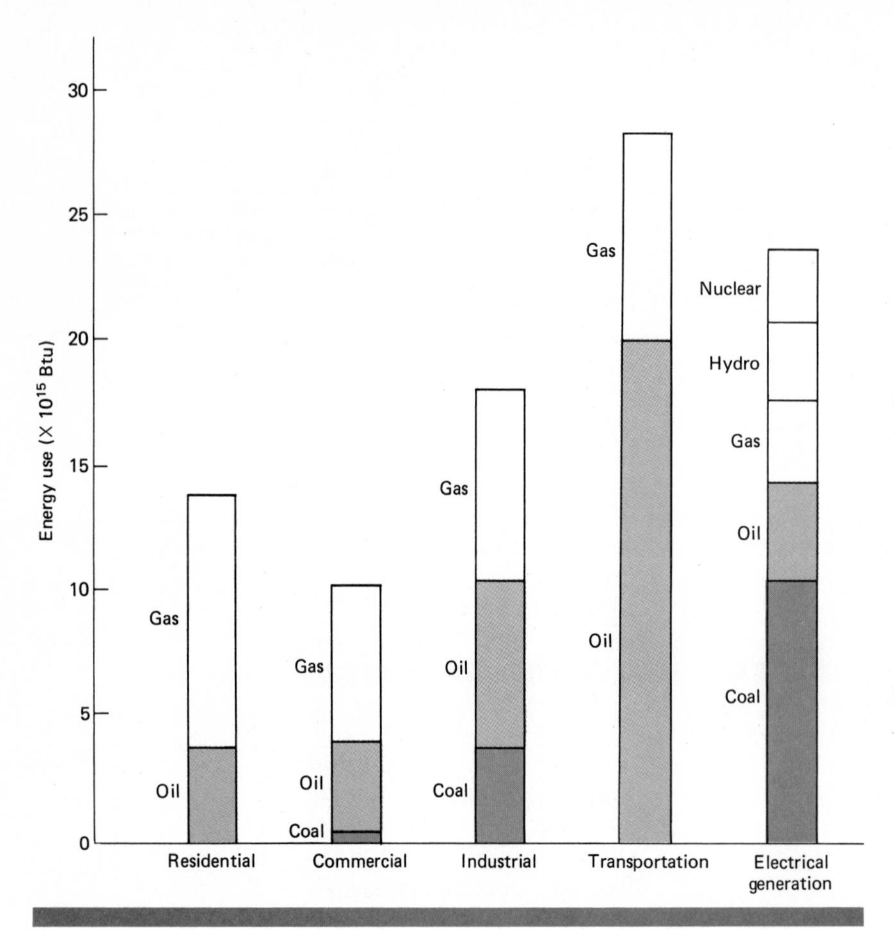

FIGURE 20-8
Energy use in various sectors of the United States' economy in 1978. Total energy use was 93.8 mQ (93.8×10^{15} Btu) in 1978, up from 72.8 mQ in 1973. However, every sector shows strong shifts in energy use. The first three sectors shown actually declined in energy use during 1973–1978. Two large shifts in energy use were notable: natural gas provided 4 percent of the energy used in transportation in 1973, but well over 25 percent of the energy used in 1978; nuclear energy and hydroelectricity together accounted for only 10 percent of the energy used in electrical generation in 1973, but by 1978 these sources accounted for 26 percent of the energy used for this purpose. (*From Statistical Abstract of the United States, 1979.*)

used to move people (55 percent by automobile) and about 30 percent is used to move goods (23 percent by trucks). Thus, any solution to our problems with oil supply must focus on the largest single user of oil, the American automobile. Simply put, solutions boil down to using less oil, producing more oil, and switching to alternative fuels. A reasonable approach must incorporate all three, although in the very long view only the last is a real solution provided the fuels selected are renewable and not fossil fuels.

Using less oil in transportation can mean cutting the total passenger miles traveled, shifting to more fuel-efficient modes of transportation, or both. However, the amount of oil used in transportation almost exactly equals the total level of imports. To eliminate the use of imported oil just by cuts in transportation would require that all transportation of people and goods cease. Thus, gains in fuel efficiency alone cannot solve our oil problems. However, without substantial gains in fuel efficiency the problem would go from serious to intolerable in a few years. In 1978 we had close to 150 million registered vehicles consuming an average of 860 gal of fuel per year apiece. Total registrations were growing at the rate of about 50 million per decade in the same time period. Just to stay

TABLE 20-2

Energy Efficiency of Various Forms of Transportation

TRANSPORTATION	PASSENGER MILES PER GALLON*
Bicycle	1500
Walking	500
Intercity bus (near capacity load)	225
Volkswagen sedan (at capacity load)	120
Motorcycle (one person)	60
Jet plane (near capacity load)	40
Light private airplane	20
Automobile, full-sized (transporting one person around town)	10
Ocean liner (at capacity load)	10

* Per gallon of gasoline or energy equivalent in other fuels.
Source: After calculations in R. H. Romer, *Energy: An Introduction to Physics,* 1976, W. H. Freeman, San Francisco.

even with fuel use, by 1988 the average fuel used per vehicle must drop to 650 gal/year. To gain on the problem cuts in use would have to be even more extensive. By law, fuel economy for new cars, which averaged 14 mi/gal in 1975, must reach an *average* of 27.5 mi/gal in 1985. The redesigning and retooling efforts for this will cost $80 billion, or more than *all* corporate profits in 1975. However, federal mileage requirements for new cars in the 1980s are designed to keep the problem from getting worse but will not reduce total oil use unless fewer new automobiles are produced than expected. One of the main problems in trying to change oil use in the transportation sector is that the American life-style and much of our economy is tied to the private automobile. The use of smaller, more fuel-efficient cars and elimination of single-driver commuting could cut use by 40 and 30 percent, respectively, if fully implemented. Mass transportation systems could replace private cars for a similar 30 percent savings in fuel if fully used by the public. Most implemented changes have produced fuel savings at a much slower rate than additional vehicle registrations have expanded fuel use. Improvements in the technology of moving people and goods cannot reduce total fuel use without fundamental changes in the established patterns of American driving, either by choice, by the increased price of fuels, or by law. Most programs today nibble at the edges of the problem rather than attacking it directly (Table 20-2).

Ultimately, of course, we will need to replace oil as a fuel entirely. This will come slowly as the increased cost of oil stimulates the use of alternative fuels. The replacement for drilled and pumped oil may be "manufactured oil." This can be oil tied to rock structures and not in a liquid form, as in the tar sands of Canada or the shale oil of the United States. Since these materials must be processed, usually by heating, to release their petroleum content, they could never compete in price with liquid oil already in a usable form. Indeed, the actual cost of producing oil in Saudi Arabia (exploration, drilling, pumping, storing, and loading on ships) is about 40 cents per barrel. By contrast it will take at least $15 per barrel to get oil from shale ready for shipment, and possibly more. Thus, short-term oil discoveries could always destroy a shale-oil industry just by releasing a large amount of oil onto the market and dropping the price below shale processing costs. Largely for this reason, tar sands and shale oils have not been developed extensively until recently. Although there is a great amount of oil in shales around the world, the vast majority would not be economically recoverable at any foreseeable time in the future. We can probably obtain half as much oil from shales as that obtained from wells in the United States, or nearly 100 billion barrels. This would have been a 40-year supply at the 1950 rate of use or a 20-year supply at the 1970 use rate. By 1990, when shale oil should be in wide use, the entire amount of high-quality shales will constitute a 10-year supply. Beyond this point the oil remaining in shales probably will be much harder to recover. The energy for recovery might easily exceed the fuel value of the oil. Furthermore, shale oil will have a serious environmental impact. Processing shales will require large amounts of water in the already water-limited western United States. Once the oil is removed from the shale, the disposal of the waste may become a serious problem since the rock expands some 20 percent on processing. While we must work on shale-oil development, full-scale use of this source is at best a temporary solution of the oil supply problem, and it poses several additional problems which must be solved.

Both natural gas and oil may be made from coal. In processing coal into other fuels some of the energy value of the coal is lost, but this might be compensated for by the desirability of having liquid and gaseous fuels. However, as with shale oil, the environmental impact could be serious. Particularly noted is the likelihood of increased release of several carcinogens in using coal oil. The economics of the process is uncertain, and a coal liquefication or gasification industry would be very vulnerable to a temporary drop in the world price of these fuels. The only operating pilot plant in the United States in 1975 converted 30 tons of coal per day to 100 barrels of oil at a cost of some $30 per barrel. Several larger pilot plants are under construction. Since the United States has over one-fifth of the world's known coal reserves, the development of this source of fuels is logical. Rates of coal use in the United States have not expanded significantly from the rates attained during World War I. The main difficulty in expanding coal use is in the high levels of pollution traditionally associated with coal. Major improvements are necessary in this area.

Two long-range candidates for our main transportation fuel are alcohol and hydrogen. Alcohol is attaining prominence currently as a partial substitute for gasoline in "gasohol," a 9/1 gasoline/alcohol mixture. While automobile engines can run perfectly well on pure alcohol, they require minor adjustments to do so. The gasahol mixture works well in unmodified engines. Alcohol may be produced from most plant products by fermentation. Small industries producing alcohol from grains and from wood chips already exist and could be expanded. The difficulty is that after fermentation the alcohol must be distilled to a high degree of purity. Today, this requires as much energy as we obtain from the alcohol and thus provides no gain in energy. For example, if oil were used to distill the alcohol, it would take the equivalent of a gallon of gasoline

FIGURE 20-9
Almost any mention of the use of hydrogen as a fuel rekindles discussion of the Hindenburg disaster of May 7, 1937. Compared with modern air disasters, this was a relatively mild occurrence. Many more passengers and crew escaped successfully than died, and several of the 31 immediate deaths were due to panic jumps from this height before the Zeppelin began to settle. Modern tests indicate that hydrogen is a much safer fuel in most respects than liquid hydrocarbons, yet the dramatic impact of the Hindenburg disaster remains after four decades. (*United Press International.*)

to make a gallon of alcohol. A second problem with alcohol is that the most desirable feedstocks in its production are foods. For example, it is estimated that every 1 percent of the corn crop devoted to alcohol production will raise the price of corn 2.5 percent. This would raise costs in the beef industry and reduce the amount of food available for export. The great advantage of alcohol is that it is renewable, since a fuel crop can be grown annually.

Hydrogen gas can be produced from water and used as a fuel (Fig. 20-9). When burned, the only combustion product is water vapor. Thus, practically no pollution problems are associated with the use of hydrogen as a fuel. The separation of hydrogen from water may be accomplished electrically or by using high temperatures. The energy for this process could come from solar power, nuclear reactors, coal, or some similar source. Hydrogen has the advantage that it could use the existing network of natural gas mains and lines into individual homes. For use in transportation the hydrogen could be stored in pressurized gas cylinders. However, considerable weight is saved and safety is improved if the gas is absorbed by metal chips under pressure. Hydrogen is at least as safe to use as the fuels it replaces, and probably is safer. Several problems exist with hydrogen, one of the largest being public acceptance. Kilogram for kilogram, hydrogen stores more energy than any fossil fuel. Since it is not found uncombined in nature, free hydrogen

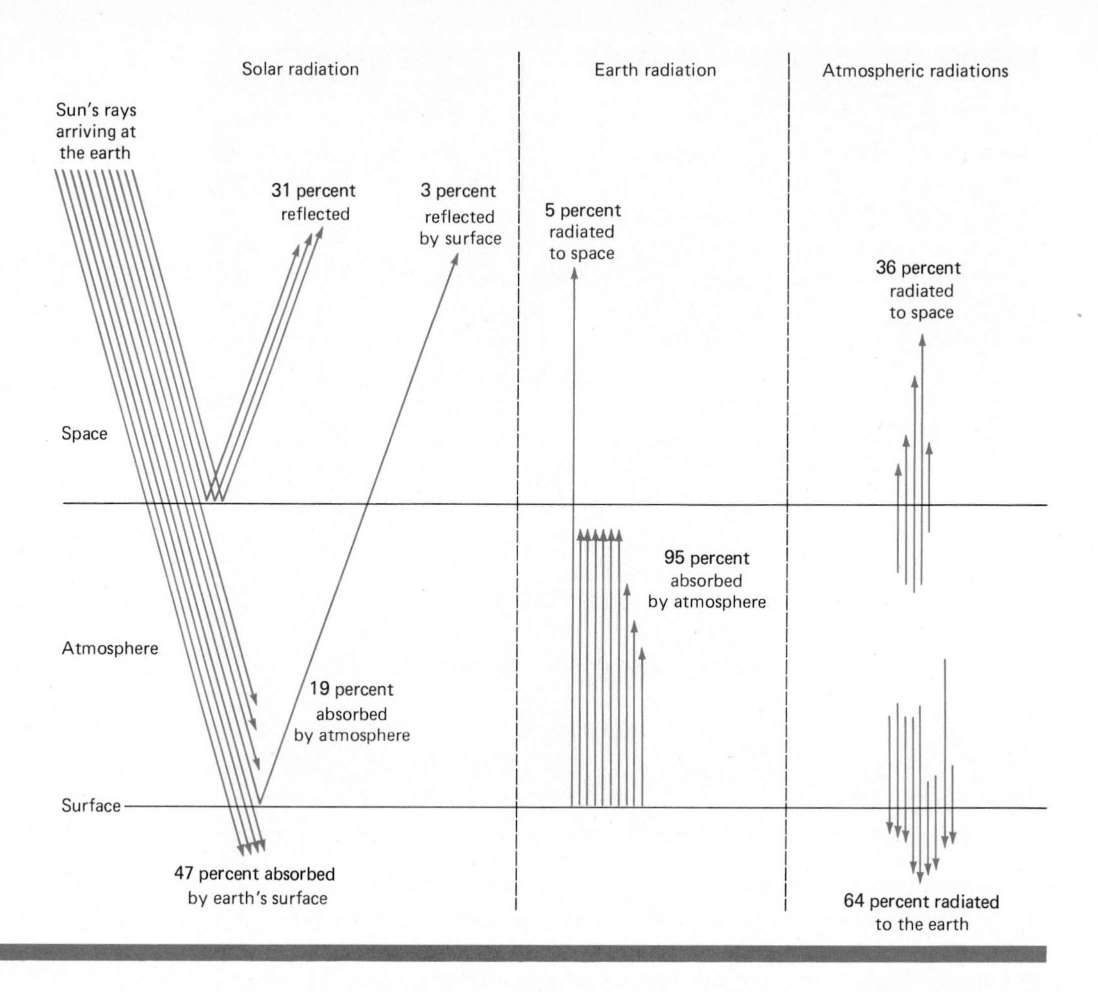

FIGURE 20-10
The fate of radiations from three sources illustrates the greenhouse effect. Short-wavelength solar radiation mostly penetrates to be absorbed by the surface and atmosphere. Longer-wavelength earth radiations are mostly absorbed by the atmosphere, whose radiations in turn are mostly back to the earth. In absolute terms, the earth receives twice as much heat from the atmosphere as from the sun. In effect, atmosphere and surface are "playing catch" with the bulk of the energy while the sun makes up their losses.

would have to be produced. Thus it would be only the storage mechanism for energy from whatever source was used to produce it.

Alcohol and hydrogen share what may become the single most important feature in a fuel for the future. Neither of these fuels results in the long-term increase of carbon dioxide in the atmosphere. Carbon dioxide in the atmosphere tends to absorb radiated heat from the earth in the *greenhouse effect* (Fig. 20-10). Short-wavelength radiation from the sun is allowed through the atmosphere to warm up the earth. However, CO_2 prevents longer-wavelength earth radiation from escaping to cool the earth down. The result of increased CO_2 in the atmosphere could be disastrous, since much of the surface of the continents is quite close to sea level. During most of geologic history less land surface was above water than at present. Most of our low-lying, fertile plains near sea level are dry land simply because the level of the sea is kept low by 7 million cubic miles of water being tied up in the polar ice caps of the earth. Promoting any serious change in the earth's climate is flirting with catastrophe. The current growth rate in the use of fossil fuels of 4.3 percent per year does just this. The earth has two great repositories for excess carbon. The first is in sedimentary rocks such as limestone, $CaCO_3$, or dolomite, $CaMg(CO_3)_2$; and the second is in the beds of fossil fuels. At the current growth rate in the use of fossil fuels, we will double the concentration of CO_2 in the atmosphere by the year 2035. This should be sufficient to raise the average temperature of the earth some 2 to 3°C (4 to 5°F). A massive use of synthetic fuels (shale oil, etc.) would speed up the process since CO_2 is released both in producing the fuels and in using the fuels. A crash program in synthetic fuel production could double CO_2 concentration within 30 years. Unfortunately, any increase in the earth's temperature would not be even across the earth. Projections indicate that the effect at the poles is likely to be 3 or 4 times as great as that at the equator. It is not impossible that in your lifetime the greatest cooperative effort in the history of humankind might have to be taken to stop fossil fuel use or else watch entire countries disappear beneath the sea. Certainly the next decade or so will define the likelihood of this crisis occurring.

Solar Energy

Probably the most important renewable source of energy in our future is solar energy. The earth receives some 6×10^{24} J/year of energy from the sun. This is roughly 20,000 times as much energy as is required for all human uses at present. Every square meter aimed directly at the sun on earth can collect up to 1000 W of power (1 kW/m^2). This is about 860 kcal/h, or 3.6×10^6 J/h for each square meter in direct sunlight. Enthusiasm with numbers such as these is already causing the same problems found in the early days of nuclear power. At one time a politician claimed for nuclear power that "electricity will be so cheap we won't bother to meter it." Lately, one Pennsylvania politician made the similar statement, "Let's turn on the sun for energy, its free!" Actually, the energy may be free, but collecting and using it certainly will not be. The main problem with sunlight as a power source is that it is not intensive but *extensive,* that is, it is spread out thinly over a great area. After all, standing in full, direct sunlight on a cold winter's day you are warmed by the sun, but you can still freeze to death. Your body does not have sufficient area to collect as much solar energy as you need in such a situation. A collector is required to gather enough energy for a particular use.

Humankind has used some forms of solar energy since the beginnings of history. Wood fires release the trapped energy of the sun gathered by the solar collectors of the tree's leaves. Sailboats and windmills use the sun's energy that has become motion of the air. Hydroelectric dams rely on the sun to raise water vapor from the oceans and deposit it on the uplands as rain and snow. While these are all solar energy systems, a much greater variety of ways to tap the power of the sun are available today. Direct sunlight may be trapped by a glass-enclosed absorber and used to heat water. Such a solar heating system can provide hot water for washing and bathing or for the heating of homes. While 1 kW/m^2 is the maximum rate of delivery of energy from the sun, the

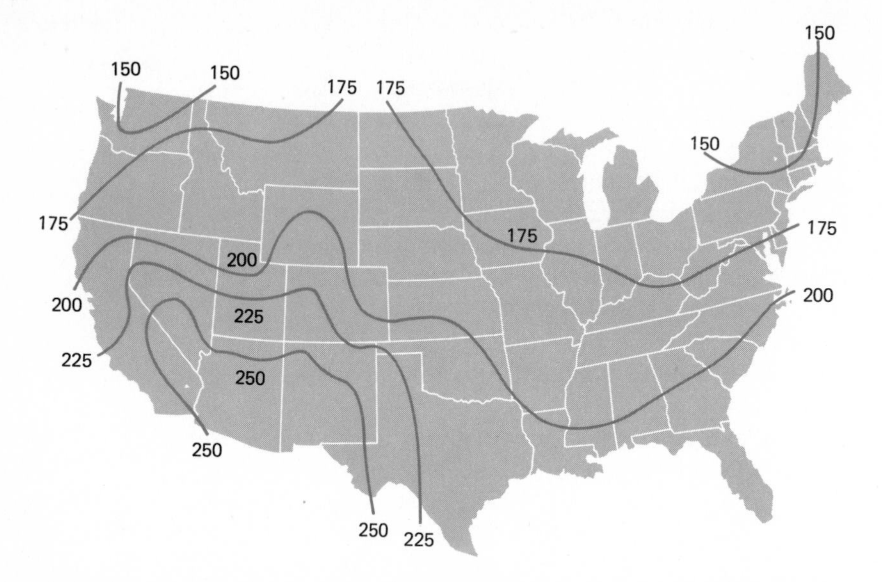

FIGURE 20-11 The ideal radiation received by 1 m² of collector pointed at the sun is 1 kW. The actual value varies. The map indicates the average annual wattage of radiation on a horizontal square meter of surface for various locations. Local variations can change the values for various zones by large amounts, mainly downward.

average delivery rate is much lower (Fig. 20-11). No sunlight is available at night, or during bad weather. Furthermore, it is costly and complicated to steer collectors so that they always present a flat face to the sun. A fixed horizontal surface in the desert southwest may average 250 W/m² for the year, while along the northern tier of states the average is closer to 150 W/m². Local climatic differences influence these numbers by as much as 20 percent. Seasonal differences cause these numbers to change greatly during the year. Thus, northern states receive only a fifth as much energy in the winter, when they need it most, as in the summer—typically, 50 to 60 W/m² in December to 250 to 270 W/m² in June. This is, of course, precisely the root cause of why it is cold in the winter. Still, the amount of energy delivered to a large surface by the sun is very respectable. A collector pointed at the sun can achieve as much as 290 kcal/m² · h if operating at 30 percent efficiency and ideal weather conditions (Fig. 20-12). This is enough to warm 1 gal of water (4 kg) about 62°C, or from 50 to 180°F, in 1 h. Thus, on an ideal day, 1 m² (11 ft²) of collector can produce about the amount of hot water used by the average middle-class American in that day. Of course, it could also produce no hot water on a densely cloudy day. It is usually not widely appreciated, but hot water heating in homes is the second largest user of energy in the residential sector, consuming some 2.5 percent of all energy used in the United States. We use more energy to heat water than in refrigeration and cooking combined.

The largest portion of residential use is in the heating of our homes, requiring one-tenth of all energy used (Table 20-3). A large portion of this could be conserved simply by better insulation and the elimination of any leaks to the outside. Solar energy can be used in the same manner as in water heating to provide heat to our homes. To do so, a much larger

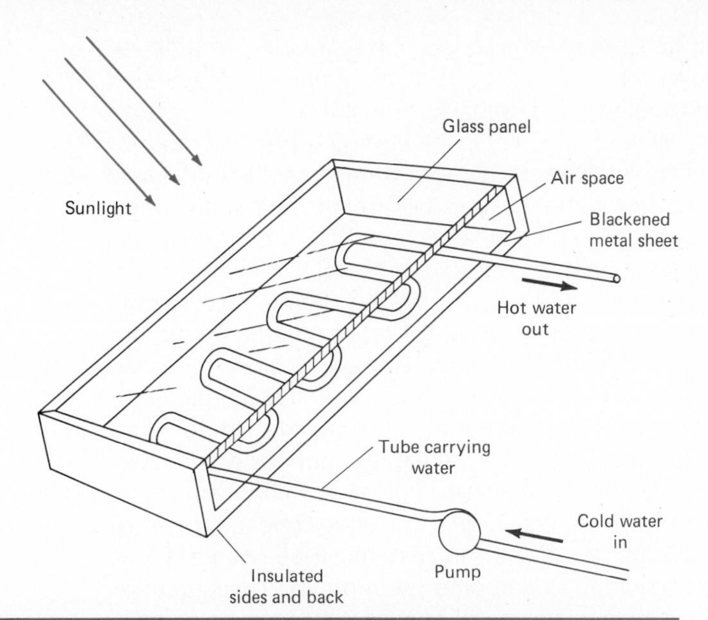

FIGURE 20-12
The solar collector uses the greenhouse effect to retain the energy of sunlight. The sun's rays pass through the glass and are absorbed by the blackened metal surface. The longer-wavelength rays from the metal are reflected by glass, trapping the energy. Tubes welded to the metal circulate water which absorbs the heat gained by conduction. This can provide a warm water source for a household, or, in larger installations, enough hot water to keep the house itself warm.

TABLE 20-3
Subdivision of Residential Energy Consumption

USE	PERCENT CONSUMPTION
Space heating	55.4*
Hot water	13.1
Lighting	7.8
Refrigeration	5.5
Cooking	4.6
Air conditioning	4.0
Television	3.0
Freezers	2.1
Clothes drying	1.9
Misc. appliances	1.6
Diswashers	0.4
Washing machines	0.3
Total	100.0

* Includes waste heat in electric energy generation where appropriate.
Source: Statistical Abstracts of the United States.

array of solar collectors is used, providing hot water to a storage system usually in the home's basement. As heat is needed, hot water can be pumped from this storage reservoir or air circulated around it to provide warmth. Storage is essential since the system must deliver heat during the night. It is usually most economical to rely on a solar heating system for a major portion of the heat needed but not for all. The additional collectors and heat storage needed to provide for the coldest, cloudiest stretch of winter can make the cost of such a system prohibitive. Currently solar heating systems can compete in costs with conventional heating systems in only a limited portion of the country. Large-scale manufacturing will undoubtedly cut the costs of these systems in time.

However, it remains to be seen whether costs can be made competitive with fuel-based systems in very cloudy or cold areas. Conventional systems using fuel are relatively inexpensive to install but pose annual fuel bills for their users. With solar systems the energy is free, but the initial cost of the system is much higher. Compared over the 20- to 30-year life spans of each of these systems, costs would be fairly similar except for the interest expense of a high initial investment. Rising fuel costs improve the position of solar heating as do governmental tax incentives or low-interest loans for installing such systems. One problem frequently neglected in discussing solar energy is that the large areas required by collectors use large amounts of materials. Current designs employ glass, aluminum, and copper in large amounts. One study criticizing solar energy asserts that the conversion of all new construction to solar energy by the year 2000 would require us to expand copper production to 28 times the current level. Even if this were possible, without stringent safeguards smelting activities can be highly polluting, releasing to the air poisons in the ores such as arsenic. In general, it seems as if the problems of solar heating can be overcome by improved designs and mass production of components, and it will play an ever-increasing role in the future.

Electricity also may be produced directly from sunlight by photovoltaic cells as discussed in the previous chapter. In the mid-1970s such solar electricity was well over 100 times as expensive to generate as that produced by conventional means. Since then, progress in improving the efficiency of solar cells and lowering their costs have exceeded projections. It seems quite likely that the mid-1980s will produce solar electrical systems which can begin to compete with systems using ever more costly fuels. For major high-technology systems, big installations usually imply significantly greater efficiencies. However, solar cells work quite well at any scale of use. This already favors their use in remote areas if costs can be brought into line with other systems. A problem always present with solar devices is that of energy storage for periods when the sun is not shining. Small solar electrical systems could solve this by the use of batteries. For large systems this would be too expensive and several other schemes have been suggested. One of these would be to separate water into hydrogen and oxygen by electrolysis. The hydrogen could then be burned as a fuel to regain the energy when needed, or the hydrogen and oxygen could be recombined at a controlled rate in a device called a *fuel cell,* which produces the electricity back with high efficiency.

Such a great variety of ways exist to tap solar energy that even a listing of them is very extensive. All suffer at the moment from cost problems due in no small part to our past practice of seriously undervaluing the price of our fuels. Like any developing technology we can expect growing pains in some areas of solar use. However, there is no question but that solar energy will grow to produce a sizable portion if not a majority of our energy needs in the far future.

A problem exists with wide-scale use of small solar power systems.

Will such small users retain the right to stay connected to the power grid of electric utilities? If they do, they must be supplied with their electric power during long, cloudy periods or when winds are not blowing. Thus, the utilities must maintain enough generating capacity to supply regular users plus the occasional demand of solar users when their source of energy lags. This requires the industry to maintain excess generating capacity which may only rarely be needed. Furthermore, this capacity would have to be maintained to supply the individuals who used the system the least and thus paid the smallest amount for its construction and maintenance. Also, generating stations that can start up promptly to provide "peak-power" demands, yet are infrequently used, are the most expensive stations to run. "Base-load" stations which run continually are the least expensive source of electric energy. Extensive use of small solar systems with regular power backup would make these problems quite serious.

Fact vs. Fiction— A Personal Statement about Nuclear Power

For some time the nation has been arguing the merits and risks of nuclear power generation. It is fairly safe to assert that much of this discussion has proceeded away from any factual base and into the realm of the purely emotional response. In no small part the confusion of the well-meaning citizen stems from the handling of nuclear information in the press. Controversy makes a good news story and confrontations produce good TV coverage. Time and again sensational charges make headlines nationally, while the reports of commissions and panels investigating these charges to find them unfounded are lucky to get a note on page 79 of the same papers. The decision whether to rely on nuclear plants for the generation of electric power is, and properly should be, a politico-economic decision and not a scientific one. However, more nonsense and misinformation has been advanced as scientific fact in this discussion than any other in modern memory. To say anything good about nuclear power has resulted increasingly in being branded "pronuclear," even when what is said is merely factual. Regardless of any emotional bias each of us may have, we have an obligation to do our homework on the issue and argue from the facts. The confused mothers crying at licensing hearings over the safety of their children deserve this of us all.

RADIATION, CANCER, AND MUTATION

Nuclear reactions release energy, mainly in the form of heat. The heat produced can replace a wood or coal fire under a boiler producing steam. This process describes what happens in a nuclear power plant. A nuclear reactor, however, represents a large concentration of radioactive materials, and numerous products of the fission process are even more radioactive. To understand the hazard this represents we must discuss the effects of radiation on living matter.

Radiation harms living material in the process of its being absorbed. If a cosmic ray, for example, passes through your body and does not interact with the matter in you releasing energy, it has no effect on you. This process is occurring continually. When a particle is absorbed, its kinetic energy is released at the point of absorption, potentially breaking bonds between atoms and severing molecules. For key molecules in the body this can be serious. In particular, when the DNA molecule in a gene is broken, it may not reassemble in exactly the original pattern. Since the genes contain nature's blueprint for the key structures of life, any change could produce an incorrect blueprint. This *radiation damage* can produce certain forms of *cancer* and can lead to *mutation,* or a change in the form or chemical functioning of an offspring. For example, increased exposure to sunlight, in particular the ultraviolet portion of sunlight, is related to an increased incidence of skin cancer. The actual effect of an absorbed ray or particle is a matter of chance. A single absorbed particle in one individual could perhaps produce a serious change, while trillions upon trillions of absorbed particles in another individual may produce no observable adverse effect. To discuss such a situation we must talk about the likelihood or probability of adverse effects for a given amount of radiation.

The basic unit in which absorbed dose of radiation is measured is called the *gray,* which is one joule of absorbed energy per kilogram of matter (1 gray = 1 J/kg). More commonly used is the *rad,* which is one-hundredth of a gray. In discussing human damage these units are not adequate since different radiations with the same energy may have different effects. α particles, for example, are 10 times as likely to cause damage to human cells as are γ rays of the same energy. To account for this difference, the dose in rads is multiplied by the *relative biological efficiency* of the type of radiation present to obtain the standard measure of human dose, the *rem.* Thus, *one rem of exposure* to γ rays would imply 10^{-2} J of energy absorbed per kilogram of matter exposed, or 10^{-3} J/kg if

TABLE 20-4

Population Exposure to Radiation

SOURCE	mrem/YEAR
Natural	
Cosmic rays (at sea level, the exposure increases greatly with altitude)	44
Rocks and building materials	40
Radioactive materials in the body	18
Subtotal	102
Artificial	
Medical and dental x-rays	73
Fallout from nuclear weapons testing	4
Miscellaneous products and activities	2
Occupational exposures	0.8
Nuclear power plant releases	0.003
Subtotal	80.
Total	182

α particles were causing the exposure. Normal doses received are much smaller than this and are usually measured in *millirems* (1 mrem = 10^{-3} rem).

Human beings are constantly exposed to radiation since it is a natural part of the physical world. Radiation comes from several sources (Table 20-4). *Cosmic rays* from space cause an average exposure of about 44 mrem/year at sea level. Persons living at high altitudes easily can double this, while living in a jet at 30,000 ft could increase this exposure 100 times. Equally variable is exposure from *rocks* and *building materials.* Average exposure amounts to some 40 mrem/year. This amount is lower for frame houses and higher for brick or stone. An interesting illustration of levels of radiation occurred at a briefing to acquaint senators with radiation. A scientist took a background count and reported, "It's 400 counts per minute, which comes to 250 mrem/year, which is about $2\frac{1}{2}$ times the usual background." Senator Glenn of Ohio asked incredulously, "We are getting more right here than they got downwind from Three Mile Island?" The response was "You sure are!"* The Senate Office Building in Washington, D.C., is made of granite. Such monumental stone buildings frequently have high levels of *background radiation* due to radioactive materials in the stone. In this case, the annual dose is about one-seventh the *maximum permissible* dose allowed the general population.

The final source of natural background radiation is from materials inside the human body itself. You have seen that a certain fraction of the carbon in all living beings is radioactive carbon 14 (C 14). Actually, quite a few different radioactive isotopes are found mixed in nature with atoms the body takes up. The largest exposure here comes from potassium 40, which contributes most of the 18 mrem/year received from sources in the body.

The sources listed thus far represent an inescapable part of life on earth and always have. They add to just over 100 mrem/year. To this we must now add the radiation exposure caused by human activity. The largest contribution, by far, comes from the medical and dental use of *x-rays.* This adds an average exposure of some 73 mrem/year to the population, but, of course, may be much higher when an extensive series of x-rays is required or for persons receiving radiation treatments. The dual role of radiation as potentially damaging on one hand yet powerfully helpful in diagnosis and treatment of disorders on the other leads to a problem in *trade-offs.* Generally, the possible medical benefits of such use very much outweigh the potential risks.† Fallout from past nuclear weapons testing contributes less than 4 mrem/year to the general population, and this exposure has been declining since the treaty halting

* Quoted from *The New York Times,* Sunday, July 1, 1979, sec. 1, p. 28.

† A much discussed exception recently was routine mammography, or x-ray breast scans of presumably healthy young women done periodically to detect the presence of small tumors. Since the benefit did not overwhelmingly outweigh the risk involved, the practice is no longer recommended routinely for women under 40. For younger women, some uncertainty or suspicion of tumor is required for x-ray mammography to be used.

above-ground testing. Short-term effects of such tests can be a health hazard. For example, the isotope strontium 90 in fallout is taken up by plants, concentrated by cows in milk, and may be reconcentrated by children drinking the milk into their developing bone structure. (Strontium is similar to calcium in its chemistry.) This *biological concentration* of the radioactive material occurs with other isotopes also but most have quite short half-lives, and withholding affected food products from the market for a short time reduces the activity to a small fraction of the natural background radiation.

An additional 2 mrem/year comes from the products around us and our way of life—for example, radium-dial wristwatches (no longer marketed but still usable), radiation from TV sets, additional cosmic radiation from airplane flights, and so on. The added exposure of technicians, doctors, nurses, dentists, nuclear workers, and the like average to another 0.8 mrem/year for the population. Finally, we currently receive an added 0.003 mrem/year from releases by all of the nuclear installations in service. If there is a tripling of nuclear power plants and use of reprocessing plants for nuclear fuel rods, this is expected to increase to as much as 0.4 mrem/year by the year 2000.

All of this adds to some 102 mrem/year from natural sources and an additional 80 mrem/year from artificial sources, for a total average exposure of 182 mrem/year. This much is the result of measurement and is simply factual. We know a great deal about the health implications of large doses of radiation. A dose of 500 rem (500,000 mrem) delivered in a short time period over the entire body is the human *half-fatal dose*—that is, half the people exposed to this dose die, the other half survive. Death usually occurs several days after exposure. Exposures at about half this level usually produce no symptoms for about 2 weeks. The individual then suffers a series of temporary afflictions such as diarrhea, vomiting, loss of hair, bleeding from the gums, etc. This level of exposure is sometimes reached in medical and dental treatment but is rarely administered over the entire body. Radiation workers are limited to an occupational exposure of 5 rem/year, except for pregnant females who are limited to 0.5 rem during the gestation of the fetus (because the rapid cell divisions of the fetus make it particularly susceptible to radiation risks).

Less easy to deal with are long-term effects of radiation such as the inducing of cancer and mutation. At large doses the association of delayed cancers and radiation exposure is well documented. The earliest of these to develop is leukemia (which was the cause of Mme. Curie's death). Generally, the probability of cancer appearing within 20 years of exposure is about 100 cases per million people per 1.0 rem of exposure. The difficulty is that this rate can be confirmed only for doses very large compared to the natural background exposure. The normal incidence of cancer in the American population is 160,000 cases per million people over their life spans. The vast majority of these are definitely *not* associated with radiation. Any effect of small doses of radiation disappears into this normal background of naturally occurring cancer cases and is near impossible to detect. Thus, the exact risk of low-level radiation is simply not known. For example, if risk shrinks directly with exposure (one-tenth the exposure means one-tenth as many cancers), we might expect one additional cancer case due to the Three Mile Island nuclear accident. If risk shrinks less rapidly than exposure, as many nuclear critics maintain, we may expect perhaps 10 additional cancer cases over the life span of the exposed population. If the body largely "shrugs off" very low exposures, as some nuclear advocates maintain, no additional cancers are to be expected. Since the population exposed in this instance numbered some 2 million people and may be expected to develop some 320,000 normal cases of cancer over their life spans, it is difficult to see how we can determine which of these three choices is correct. To be safe, the operating rule of cautious scientists has been to presume the worst case consistent with what is known. Such *worst-case analysis* is one of the reasons there is a great argument over nuclear safety. Worst-case analyses simply are not done for other types of human activity. One source of data about the effects of low-level radiations comes from populations that live in areas where the soil is quite highly radioactive. A large patch of India, for example, has a natural background of over 2000 mrem/year from the soil. Millions of humans have lived on this land over millennia. Studies of this population do not indicate unusual cancer rates or mutation effects.

The problem of human mutations is distinct from the inducing of cancer and is much more closely woven into our subconscious. We have all been brought up on a steady diet of radiation-induced monsters in the movies and on TV. Every time Hollywood needs a new monster they seem to reach for a radioactive shaker. This they sprinkle on anything handy (ants, crickets, grasshoppers . . . probably people-eating earthworms) like fairydust from Tinker Bell to produce an instant monster. This is the heritage of the speculations following the nuclear bombing of Hiroshima and Nagasaki. Then, the level of mutations induced by radiation was not known. Then, as now, scientists accepted the proposition that radiation can and does induce mutations in every living species. What was not known were the numbers—how many mutations for a given exposure? Oddly enough the very bombings which stimulated a generation of radioactive mutants in movies provided the data to obtain the numbers. Survivors of Hiroshima and Nagasaki, some exposed to radiation up to the human fatal dose, have been studied for over 30 years. The children and grandchildren of these individuals have been carefully examined for mutations. Conclusion—*the mutation rate in these offspring of nuclear bomb survivors is indistinguishable from the mutation rate in a normal human population.*

NUCLEAR SAFETY

It makes abundant sense to make every reasonable effort to limit human exposure to radiation. The key questions center on the word "reasonable." There has been no attempt to evacuate the Kerala area of India simply because living there gives people many times the dose annually

FIGURE 20-13 A photograph looking into the core of a small research reactor. The core of enriched uranium fuel rods is suspended deep in a large pool of water. The long rods reaching to the core are control rods, which absorb neutrons to stop the reaction, and devices to insert and remove fuel rods. The core is luminous and glows with a blue light emitted when high-speed particles enter the water from the fuel assembly. This view is not possible in a power reactor, which generates enough energy to convert the surrounding water into high-temperature steam. The research reactor is cooled by a natural convection in the pool of water. Power reactors separate the areas above and below the core and force water through the core with high-pressure pumps. (*Courtesy of the Breazeale TRIGA Reactor Group, Penn State College of Engineering.*)

of any citizen exposed at Three Mile Island. The benefits to these people of being able to use the land simply outweigh the risks of the increased exposure. Thousands of American business executives face a similar *risk-vs.-benefit decision* in their widespread use of high-altitude jet travel (although the risk of death in an airplane accident, while small, is far larger than the risk posed by increased radiation). Thus, the main effort in a nuclear power plant is to contain the radiation in a single place and away from people. *It is a scientific impossibility for a nuclear reactor to explode in a nuclear chain reaction.* It is essential to emphasize this point since polls indicate that this is not understood by the public. Under no circumstances whatever can a nuclear reactor detonate like a nuclear bomb since the uranium fuel is not sufficiently rich in fissionable isotopes. The containment system thus does not have to cope with this extreme case. Actually, the containment is threefold (Fig. 20-13). Nuclear fuel is clad in zirconium metal alloys to prevent it from contacting cooling water. The cooling water itself does not leave the containment area but delivers its heat to a separate supply of water in a heat exchanger. The reactor itself is located in a large steel pressure vessel which is typically 5 in or more thick. This resembles a multistory seamless pot with a large bolt-on lid. Toward the top of this structure large pipes conduct water flows into and out of the reactor. Through the lid are extended sensors, control rods, and the like. All of this structure along with emergency systems and the heat exchangers are located in the reinforced concrete containment building. This is a large, pressurized structure designed to withstand extreme forces from within or

FIGURE 20-14 The Diablo Canyon (California) reactor containment during construction. Several layers of steel reinforcing rods are being fitted prior to pouring the concrete walls of the containment. This building completely surrounds the reactor vessel and all systems containing the water passing through the reactor. The Nuclear Regulatory Commission requires the building to survive the impact of a 747 jumbo jet without a breach. (*Courtesy of the Atomic Industrial Forum.*)

from without and entered only by large vault doors. American standards require the containment building to withstand the direct impact of a fully fueled 747 jumbo jet and the resulting fire without developing a crack. Some European standards are even higher than this since terrorism is viewed more seriously in Europe (Fig. 20-14).

There are two ways for radiation to escape this multiple containment. The first is by brute force in penetrating these layers of defense, and the second is by sneaking through in some preventable way as happened at Three Mile Island. While the second of these is much more probable, it is the first case which is always discussed in worst-case analysis. Reactors generate a great amount of heat and this is their weapon to breach the system of containment. If the large pipes conducting water to the reactor, or steam from the reactor, were to break suddenly, the steam would be vented into the containment building. With the loss of this coolant the core of the reactor, where heat is generated, would begin to rise in temperature. If the level of water drops for any reason, it is the job of the emergency core cooling system (ECCS) to keep the reactor flooded with water. Presuming these standby systems fail, the core may come uncovered and begin to melt. Such a meltdown implies serious damage to the reactor and threatens the containment structure if it continues. As meltdown continues tons of material would fall to the floor of the steel containment vessel. In an extreme case the heat being generated is adequate to melt through the steel and keep going to melt into the several feet of concrete floor. This is the beginning of the now famous "China Syndrome" discussed by engineers for some time. Serious danger can occur if the core fragments melt through the building to the level of groundwater. The clouds of steam generated would carry vents of radioactive fission particles up outside of the containment and into the atmosphere. Such a nightmare has not occurred, of course, but it is referred to as the *maximum credible nuclear accident* (MCNA)—that is, if everything possible goes wrong, this is the worst case imaginable.

The questions now become, how much death and destruction could such an event produce and how likely is it to occur? There is some agreement on the first question. Again, presuming everything is the worst possible—the reactor is near a population center, the winds carry the radioactive particles and steam over this population, a thermal inversion of the atmosphere prevents the spread and dilution of the radiation, etc.—the estimate is chilling. Perhaps 400 immediate deaths and an additional 10,000 cancer cases over the life span of the population affected. (If this terrifying spectacle unnerves you, remember it is worst-case analysis. For comparison, nuclear advocate Ralph Lapp suggests a worst-case analysis for an airplane accident. A Boeing 747 jet loaded with 400,000 lb of jet fuel for a nonstop flight to Tokyo takes off from Los Angeles International Airport on New Year's Day only to crash into the Rose Bowl packed with 100,000 football fans.)

The second question raised is how likely (probable) is the MCNA? The most extensive study of this question was the Rasmussen report, widely criticized by antinuclear groups. This report concluded that such an accident was thousands to millions of times less likely than other causes of mass loss of life such as dam failures, airplane crashes, and the like. Indeed, the only event the report found with an equal probability of occurring was for a similar group of people to be wiped out by a large meteorite striking the earth. The Rasmussen report has been under attack ever since it was issued. Serious critics claimed the probability of MCNA was 10 to 100 times more likely than claimed. Overlooked in this conclusion is that reactors would still be the safest large installations on earth at this level. In early 1978 the Nuclear Regulatory Agency abandoned the executive summary portion of the report as the basis for making nuclear decisions, although not the technical report itself. Undoubtedly the debate of how likely is very unlikely will continue for the foreseeable future, so let us examine what data we have.

The most serious accident to date in a commercial nuclear reactor was at the Three Mile Island (TMI) nuclear power station near Harrisburg, Pennsylvania. This caused some one-half billion dollars of damage to the installation and the possible future loss of life as explained earlier. This was an extremely serious accident compounded of a series of mistakes, malfunctions, bureaucratic foul-ups, administrative short-sightedness, operator errors, and more, leaving sufficient blame for almost everyone involved. While far from the worst-case MCNA nightmare, it was bad enough. Furthermore, it demonstrated to everyone's satisfaction that in spite of layer after layer of defenses, very serious nuclear power accidents could indeed happen. Prior to TMI nuclear advocates had been making stronger and stronger claims of nuclear safety in reaction to similarly extreme claims from nuclear critics.

The U.S. Navy has almost 2000 reactor-years of operating experience to the present, and the civilian power industry has over half of this. (In 1979 there were 71 operating nuclear power plants in the United States.) As a very rough guess we might assume that we would have a TMI-size accident, or worse, every 3000 reactor-operating-years. Thus, with 300 reactors in operation an accident of TMI size might be expected every 10 years. Of these, some fraction might be the worst-case meltdown with widespread death and destruction. How can we judge the safety of nuclear reactors closer than this? In terms of *absolute safety,* we cannot, since operating experience is still too limited for a solid conclusion. Even relatively neutral expert opinion varies. Nuclear advocates insist the MCNA is an event with the probability of occurring once in a million reactor-operating-years. With 1000 reactors in operation they would expect one such catastrophe per millennium. Extreme nuclear critics claim one MCNA per 200 reactor-years, by which reckoning we should have had 15 such events already. What data are available suggest reactors to be less safe than advocates maintain but much safer than the extreme critics suggest.

Judging the *relative safety* of nuclear power is much more appropriate and informative than trying to assess its *absolute safety.* An early estimate of the National Academy of Science on the dangers of coal burning put the level at 25 deaths, 60,000 excess respiratory infections and some $12 million of property damage annually for every 1000 MW of generating capacity. Worse yet, an official congressional estimate of the pollution hazard in the United States due to coal burning was released in June 1979. This put excess deaths attributable to the pollutants from coal combustion at 48,120 in the year 1975 alone and destined to rise to 55,800 by 1985. This is the equivalent of *five* maximally credible nuclear accidents *every year* (or a minimum of 5000 Three Mile Island accidents per year!). This is occurring quietly and without much comment right now. Oil and gas, if we could get enough, tend to be less dangerous than coal but still more of a hazard than nuclear power.

Do not mistake the moral of all of this! At present we *must* continue to burn coal and oil and gas, and we must continue to operate our reactors. *The option of not having the power available is much more hazardous to human health and welfare than the present dangers.* At the same time we must seek to evaluate and reduce risk in every form of energy use. *Absolute safety is simply not possible in any facet of human life.* Risks cannot be reduced to zero for any form of energy delivery system but we must seek to minimize them. The argument over nuclear power can (and has) filled many books. It is part of your responsibility as an educated citizen to read on both sides of this issue. Neutral voices have been becoming rarer as time goes on. Not the least important consideration in the argument is that of costs. Escalating energy costs work a disproportionate hardship on the poor of the nation and the world. Every detailed analysis this author has seen makes operating nuclear plants the

FIGURE 20-15 Two main strategies are being pursued to attain nuclear fusion. As in a hydrogen bomb, once hydrogen atoms are at a high enough temperature (moving fast enough), they will fuse to form heavier nuclei and liberate energy. The problem is that no device can contain such a temperature without its walls vaporizing. The first strategy holds a superhot mix of hydrogen ions confined in a strong magnetic field without touching any walls. A second strategy is shown here in the Shiva laser fusion project. A building full of large power lasers delivers simultaneous beams from several dozen directions to the target chamber at the center. These beams compress and heat a tiny hydrogen pellet at the center of the chamber, causing its temperature to reach millions of degrees in an instant. The resulting fusion produces a burst of energy which is delivered as heat to the walls of the chamber. When firing, the Shiva laser beams contain as much energy as all power plants are producing on earth. In spite of this, a laser fusion system would have to be many times as big to deliver useful amounts of energy. (*University of California, Lawrence Livermore Laboratory, and the U.S. Department of Energy.*)

least expensive source known—even before the large OPEC price increases of 1978 to 1979. For example, the one-third of electric energy supplied by nuclear plants to consumers in Vermont saved an average of $106 per household in 1978 alone. In the next 6 months, oil prices had already increased some 60 percent above the data of this comparison.

Ultimately, we may hope for the replacement of fisson reactors with nuclear fusion reactors (Fig. 20-15). These have very little associated radioactivity and have a virtually inexhaustible supply of fuel in the hydrogen of the oceans. The major problem is that an operating fusion reactor has yet to produce useful power. The final years of the 1970s produced considerable progress and the scientific community remains quite optimistic that success in the laboratory is likely in the 1980s. Even then, however, it will take 20 to 40 years by most estimates to produce a commerical plant delivering useful energy. Thus, this is a possible solution for the next century's energy needs, but not our own.

A Note to End On

Much of the science-related news of the last decade related to health has been depressing. For example, substance after substance is reported as a potential cancer-causing agent (carcinogen). Chemical dumps have caused miscarriages, malformed babies, and even deaths. Up to 20 percent of cancers may be caused by chemical exposure on the job. Insect sprays have caused sterility, miscarriage, and deaths. The wastes of society degrade the environment. Mercury, asbestos, arsenic, DDT, and other products of the technology threaten us. All of this has been widely reported in the press and is true. What is not mentioned is that the explosion of knowledge and techniques in the sciences has led to the ability to measure and assess risks much more finely than ever before. Practices accepted for decades and centuries as safe sometimes reveal high risks. The practice has not changed, but the ability to measure has. Adjustments to this increased ability to measure are painful and can cause the nonastute to falsely conclude that the world is falling apart. Yet, consider the facts on human health alone. Human beings are living longer than ever before in history. Furthermore, fewer spend old age as senile, dependent individuals but retain vigor to great age. In the decade of the 1970s the average American life span increased almost $\frac{1}{2}$ year each year—the greatest rate of increase in history. (If it would ever reach 1 year per year we would live forever.) Clearly, looking at such facts, we cannot be doing everything wrong. Any age in the history of the human race would be overjoyed to change places with us. Recall that the next time you tend to be panicked by a headline of imminent threat or danger. We should not overlook or minimize health hazards—they are real—but neither should we enlarge them beyond their actual size.

The future belongs to all of us. Science can help us in limping along using old patterns to solve the problems of the day. Science can also contribute mightily to any future civilization which our collective will wants to attain. The decision on the pattern of development is not up to science, nor should it be. We are the public! Collectively, we as citizens must rationally decide on the broad evolution of our society. Trying to duck this responsibility and trust others to let us muddle through somehow earns us whatever we get, for better or for worse.

Questions

1 From the data of Fig. 20-1A, what was the rate of growth of the human population ($R = \Delta p/\Delta t$, where p stands for population and t for time) for the period:

(*a*) Between the time of Christ (0 A.D.) and 1860 A.D?
(*b*) Between 1860 and 1900?
(*c*) Between 1900 and 1950?
(*d*) Between 1950 and 1980?
(*e*) Most people take the term "the population explosion" to mean simply that the population is growing. The eye-opening meaning of the phrase "population explosion" is best revealed by the set of numbers for parts *a* through *d* of this question. Try to express the implication of this set of numbers in words.

2 Construct a graph of the rate of growth of the human population vs. time for the period 1680 to present. Each set of data points in Fig. 20-1A will yield a rate of growth (as in the previous problem). Plot this rate of growth vs. the date which is the midpoint in time of the interval used.

3 The statement is made in the caption of Fig. 20-1B that the population of the United States appears to be heading for stability in the next century. In view of the continued growth displayed in the curve, how can this statement be made? (*Hint:* Consider the previous two questions.)

4 How many 100-W light bulbs would a person have to keep lit to use 10,000 kWh in a year (as asserted in the caption of Fig. 20-4)?

5 Compare the curves of Figs. 20-1A and 20-1B and Fig. 20-4 by estimating the number of years it requires near the right side of the graph for the y variable to double in value. What are the doubling times for world population, U.S. population, and U.S. electric energy use, respectively?

Problems

1 *Trade-Offs* Airlines used to have a policy requiring stewardesses to be unmarried, and they lost their in-flight job when wed. Women's groups saw in this practice a sexual discrimination designed to please predominantly male business passengers with a corps of attractive and eligible females. While women's rights groups successfully lobbied against this practice, it had its origin in radiation safety standards. Occupational radiation standards for a developing fetus limit exposure to 0.5 rem (500 mrem) during its entire 9 months of gestation. (Nonoccupational exposure standards are set much lower.) The background radiation due to cosmic rays at 30,000 ft in an airplane averages 4000 mrem/year.

(*a*) How many hours would have to be spent at this altitude to acquire a 500-mrem dose of radiation?
(*b*) If a stewardess works 6 h/day at this altitude with no breaks, how many work days would be required to absorb a 500-mrem dose?
(*c*) Presuming that a female becomes aware that she is pregnant $1\frac{1}{2}$ months after conception, will the fetus have exceeded the limiting exposure of radiation in this time?
(*d*) At 60,000 ft the average background rises to 13,000 mrem/year; would flying at this level change the answer to part *c*?
(*e*) During periods after a flare on the sun the background rate of radiation may rise to 1000 rem/year and remain there for several hours. How many hours in flight at this rate could expose the fetus beyond the allowed level?
(*f*) If working during a period of maximum sunspot activity with frequent solar flares, is it likely that a stewardess could expose her fetus to a dose of radiation beyond the maximum permissible dose of 500 mrem *before she knew she was pregnant?*
(*g*) Since the human fetus is much more succeptible to radiation damage than the adult, and only women carry a fetus, is a rule like that of the airlines requiring only stewardesses to be unmarried (and presumably unpregnant) sexual discrimination or not?
(*h*) Are you as likely to arrive at a rapid decision as to whether this past airline policy was fair or not once the science behind the policy is explained to you? Would a typical newspaper or TV article on a stewardess protest rally be likely to explain the background of airline policy?
(*i*) The Equal Employment Opportunity

Commission (EEOC) has the legal power to demand equal employment of married stewardesses. The Nuclear Regulatory Commission (NRC) has the power to forbid pregnant stewardesses from occupational flying once they have reached the maximum permissible dose for the fetus. Since both the EEOC and the NRC can bring legal actions against the airlines, where does this leave them if these two agencies cannot cooperate?

2 The maximum permissible dose (MPD) is set as low as practical to avoid possibly dangerous doses yet allow reasonable medical and occupational exposure to radiation. The MPD is well below any demonstrated medical effect. To put the American MPD of 0.5 rem for the fetus in perspective consider the following.

(*a*) What is the average dose absorbed by a fetus conceived, carried, and born in the Kerala region of India? How does this compare to the MPD?

(*b*) The average background from radioactive materials in a particular tourist beach in Brazil is 17.5 rem/year. How many days could a pregnant woman spend at this beach before her fetus absorbed the MPD? (By the way, the highest naturally occurring backgrounds measured at various areas of the earth's surface run as high as 90 rem/year, or almost 1000 times as high as the average background radiation, and nearly 20 times the adult occupational MPD.)

3 Airline passengers receive an annual dose of approximately 330 million rems from high-altitude flying annually. About how many cancers would you expect this to induce among the flying population?

4 While the sun can deliver 1000 W/m^2 to a surface perpendicular to its rays, nighttime, bad weather, etc., reduce the average value at a fixed collector to about 250 W/m^2.

(*a*) At this rate, what area of collectors would be required to receive the 2.5×10^{12} W of power currently used from all sources in the United States?

(*b*) 1 km^2 is 10^6 m^2, and there are 2.6 km^2 in 1 mi^2. How many square miles is the answer to part *a*?

(*c*) If the solar collectors are only 33 percent efficient, and thus convert only one-third of the sunlight to a usable form of energy, how does this increase the total collector area needed? What is this new area?

(*d*) The area of the United States (excluding Alaska and Hawaii) is roughly 3 million square miles (3,022,387 mi^2). What percentage of the surface of the country would have to be lined with solar collectors at 33 percent efficiency to provide all of our energy needs?

(*e*) What would this percentage drop to if the efficiency of the collectors could be increased to 100 percent?

APPENDIX ONE

Mathematical Review

Fractions and Decimals

Sometimes it is convenient to keep a number as a fraction in a calculation. Whenever it is necessary to convert a fraction to a decimal, the implied division of the fraction is carried out. For example, $\frac{2}{5}$ literally means "2 divided by 5." Actually doing the division yields the decimal 0.40. A fraction such as $\frac{17}{38}$ is almost always reduced to a decimal, in this case 0.45, at once. The most common problem students have with dividing out fractions is in reporting the result to too many significant figures, such as 0.44736842, simply because the calculator yields these digits. Study the rules on significant figures later in this appendix to avoid this error.

Compound fractions consist of one fraction divided by another and are encountered frequently, particularly in the handling of units. Consider the problem of $\frac{1}{5}$ divided by $\frac{3}{8}$. To reduce this to a simple fraction it is necessary to get the denominator, $\frac{3}{8}$, equal to 1. We may do this by multiplying the denominator by $\frac{8}{3}$. However, if we do this with the denominator we must also multiply the numerator, $\frac{1}{5}$, by $\frac{8}{3}$ to retain the value of the expression:

$$\frac{\frac{1}{5}}{\frac{3}{8}} \times \frac{\frac{8}{3}}{\frac{8}{3}} = \frac{\frac{1}{5} \times \frac{8}{3}}{\frac{3}{8} \times \frac{8}{3}} = \frac{\frac{8}{15}}{\frac{24}{24}} = \frac{\frac{8}{15}}{1} = \frac{8}{15}$$

This procedure is certainly justified since the quantity we multiplied by, $\frac{8}{3}/\frac{8}{3}$, is a number over itself and must be equal to 1. We may always multiply by 1 without changing the value of a quantity. This entire procedure is frequently remembered by the slogan "invert the divisor and multiply." Thus,

$$\frac{\frac{1}{5}}{\frac{3}{8}} = \frac{1}{5} \times \frac{8}{3} = \frac{8}{15}$$

which has the same effect as the longer procedure.

Percentage means "percentum" or "per hundred" and literally reports the number of parts per hundred parts. Thus, 15 percent (15%) means 15 parts per hundred, or 15/100 or 0.15. To change a decimal into percentage it is necessary only to move the decimal point two digits to the right. If the question is asked, "$\frac{3}{8}$ is what percentage?," the fraction is first reduced to a decimal, 0.375, and then the decimal point is shifted to produce 37.5 percent. Percentage is frequently used in science to indicate the presumed amount of error in an experimental measurement. The *percentage error* (% error) of a measurement is the difference between the measurement and the accepted value of the quantity being measured expressed as a percentage:

$$\%\ \text{error} = \frac{(\text{experimental value} - \text{accepted value})}{\text{accepted value}} \times 100$$

For example, a measurement of the height of a flagpole using the length of its shadow and angle to the sun might yield 11.4 m. If the actual height is 12.0 m, the percentage error would be

$$\%\ \text{error} = \frac{11.4\ \text{m} - 12.0\ \text{m}}{12.0\ \text{m}} \times 100 = -5.0\%$$

(The negative sign indicates only that the measured value was less than the accepted value in this case.)

Combining fractions seems to cause confusion for many students. The slogan usually taught to ease this confusion is that "you can't add oranges and apples." If two fractions are to be added or subtracted they must have a *common denominator,* or represent parts of a whole divided in the same way. Thus, $\frac{1}{3} + \frac{1}{2}$ may not be added directly. The first number represents one part of an object divided into three parts, while the second is one part in two. We may find a common denominator by multiplying the one denominator by the other, in this case $3 \times 2 = 6$. One part in three is two parts out of six, while one part in two is three parts in six. Now we may add:

$$\frac{1}{3} + \frac{1}{2} = \left(\frac{1}{3} \times \frac{2}{2}\right) + \left(\frac{1}{2} \times \frac{3}{3}\right) = \frac{2}{6} + \frac{3}{6} = \frac{5}{6}$$

Frequently, the fractions encountered in science are so complex that it is simpler to convert them to decimals and then add. Consider the formula for electrical resistances in parallel:

$$\frac{1}{R_P} = \frac{1}{R_1} + \frac{1}{R_2}$$

If $R_1 = 15\ \Omega$ and $R_2 = 42\ \Omega$, what is R_P?

$$\frac{1}{R_P} = \frac{1}{15} + \frac{1}{42} = 0.067 + 0.024 = 0.091$$

If

$$\frac{1}{R_P} = 0.091$$

then

$$\frac{R_P}{1} = \frac{1}{0.091} = 11$$

This could also have been worked as a problem in combining fractions:

$$\frac{1}{R_P} = \frac{1}{15}\left(\frac{42}{42}\right) + \frac{1}{42}\left(\frac{15}{15}\right) = \frac{42}{630} + \frac{15}{630} = \frac{57}{630}$$

If

$$\frac{1}{R_P} = \frac{57}{630}$$

then

$$\frac{R_P}{1} = \frac{630}{57} = 11$$

Equations

An equation is a statement that two expressions are numerically equal. The expression to the right of the equals sign will yield the same number on evaluation as the expression to the left of the equals sign. Any mathematical procedure that changes the value of only one side of the equation destroys this equality. This is exactly like masses on the pans of a balance. If the masses are initially in balance, any mass added or subtracted from one pan will only destroy this balance.

ALLOWED OPERATIONS

Thus, operations with equations are allowed only if they affect both sides of the equation in the same way. In particular we may:

1 *Add (or subtract) the same quantity to both sides of an equation.* For example,

$$
\begin{array}{r}
°F = \frac{9}{5}°C + 32 \\
-\,32 \qquad\quad -\,32 \\
\hline
°F - 32 = \frac{9}{5}°C
\end{array}
$$

By subtracting the same number (32) from both sides, we are assured that the result retains the original equality.

2 *Multiply (or divide) both sides of the equation by the same quantity.* For example,

$$°F - 32 = \frac{9}{5}°C$$

$$\frac{5}{9}(°F - 32) = \frac{5}{9}(\frac{9}{5}°C)$$

$$\frac{5}{9}(°F - 32) = °C$$

Notice that in this example we must multiply the entire left side by $\frac{5}{9}$, not just the first term. Expanded, this would be

$$\frac{5}{9}°F - \frac{5}{9}(32) = °C$$

(While this form is correct, the expression is usually seen in the form we derived first. This is because it is easier to subtract 32 from the Fahrenheit reading before multiplying by $\frac{5}{9}$ to obtain degrees Celsius, rather than multiplying degrees Fahrenheit by $\frac{5}{9}$ and then subtracting 17.8.)

3 *Square (or take the square root) of both sides of an equation, or multiply each side of the equation by itself as many times as desired as long as the same is done to the other side.* Thus, in the equation

$$v = \sqrt{4.9h}$$

we may square the equation or multiply each side by itself, yielding

$$v \times v = \sqrt{4.9h} \times \sqrt{4.9h}$$

$$v^2 = 4.9h$$

4 *Divide (or multiply) one equation by another* (since if equals are divided by equals, the quotients are equal). If $pV = Nkt$ for an enclosed gas, where N and k are unknown constants, we may still use the expression to evaluate changes in pressure, volume, and absolute temperature.

In the first case: $p_1V_1 = NkT_1$

In a later case: $p_2V_2 = NkT_2$

We may divide the right side of the first equation by the right side of the second, and similarly divide the left sides to yield

$$\frac{p_1V_1}{p_2V_2} = \frac{NkT_1}{NkT_2} \quad \text{or} \quad \frac{p_1V_1}{p_2V_2} = \frac{T_1}{T_2}$$

(after the constants have divided out to 1). Such a use of equations is quite common when we wish to compare one situation involving a particular mathematical expression with another use of the same expression. Such *comparative* use of equations may lead to terms that have not changed being eliminated from consideration. However, any equation may be divided in this manner by any other, whether similar or not.

SOLVING AN EQUATION FOR A PARTICULAR VARIABLE

One of the commonest operations with equations is to rearrange the equation to allow it to be used to find the value of a particular variable. The expression $A = LW$ may be said to be *solved for* the area A. This implies that A is all by itself on one side of the equation and is to the first power. Assume we had the same expression and wanted to solve for W, the width. In this case, W is not by itself on one side of the equation but is multiplied by L. To eliminate this, we need only divide both sides by L:

$$\frac{1}{L}A = LW\left(\frac{1}{L}\right) \quad \text{or} \quad \frac{A}{L} = W$$

On the right side we obtain L/L, which is equal to 1. Thus, W is alone on the right side and to the first power. We have solved for W.

Example

Solve $E = \frac{1}{2}mv^2$ for v.

Step 1 Multiply both sides by 2 to remove $\frac{1}{2}$.

$$E = \tfrac{1}{2}mv^2$$

$$2E = mv^2$$

Step 2 Divide both sides by m to remove the m.

$$2E\left(\frac{1}{m}\right) = mv^2\left(\frac{1}{m}\right) = \left(\frac{m}{m}\right)v^2 = v^2$$

Now we have the expression

$$\frac{2E}{m} = v^2$$

Step 3 Take the square root of both sides to reduce v to the first power, yielding

$$v = \sqrt{\frac{2E}{m}}$$

The original equation has been solved for v.

Practice Problems

Solve the following for the variable indicated:

$H = CMT$ for M $\qquad$ $E = mc^2$ for c

$F = G\dfrac{m_1m_2}{R^2}$ for G $\qquad$ $F = Q\dfrac{q_1q_2}{R^2}$ for R

$d = v_0t + \frac{1}{2}at^2$ for v_0, for a

Significant Figures

The reasons for the concern in science on how a number is reported is discussed at the beginning of Chap. 2. If a measurement of 2.3 is multiplied by a measurement of 3.7, the product is reported as 8.5 and not 8.51 even though the calculator yields this additional figure. To understand how to decide how many figures to report in an answer we must first know when a figure is significant.

1 Any digit is a significant figure if it is the direct result of a measurement.

2 Any nonzero digit reported is significant.

3 The digit zero is significant whenever it occurs between nonzero digits. Thus, in 203 the zero is a significant figure.

4 The digit zero is not significant when written to the right of a number to indicate place. Thus, in 57,000, the digits 5 and 7 are significant but the three zeros are not. The zeros merely indicate the answer to the question "57 what?": They answer "thousands." With the number 230, the measured part is 23, while the zero asserts these were 23 tens of something. [Should a measurement indicate exactly 230 (not 231 or 229, but 230) of something, we place a decimal point after the zero to indicate that it is significant—230.]

5 Zeros to the left of the first nonzero digit are never significant figures. Thus, 0.0031 has two significant figures, 3 and 1.

6 Terminal zeros to the right of the decimal point are always significant, as in 1.2300.

Practice Problems

How many significant figures do the following figures have?

(*a*) 207 (*b*) 2700 (*c*) 10.300
(*d*) 14,070 (*e*) 1.003 (*f*) 0.000310

MULTIPLICATION AND DIVISION

The rules for significant figures in multiplication and division are particularly simple. No product (or quotient) may be known better than the data going into the calculation. Therefore, in multiplying or dividing, the answer may never have more significant figures than the number with the least significant figures entering the calculation. For example, $2.31 \times 4.0076201 = ?$ In this case the first number, 2.31, is known to three significant figures, while the second is known to eight significant figures. The weaker number controls, and the product is reported to *three* significant figures, as 9.26. This is true even for a complex chain of operations:

$$\frac{3.0016 \times 27.934 \times 3.14159}{2.1 \times 17.873} = 7.0$$

Here the accuracy to which the answer may be reported is controlled by the weakest number, 2.1, which is known to only two significant figures. Thus, the answer may have only this number of significant figures, and no more.

The only time when a number may appear which does not bear on the significant figures of an answer is when it is *not* the result of measurement, but of counting. For example, when doubling something you multiply by 2, and may write it that way. This, however, is considered as taking exactly twice—a counting process—and the 2 involved may be ignored in determining the number of significant figures. In the formula, $E_K = \frac{1}{2}mv^2$, the $\frac{1}{2}$ likewise has no bearing on the number of significant figures.

Practice Problems

How many significant figures will the answer to the following problems have? (Do not work the calculations.)

(*a*) 3.72×4.1

(*b*) $\frac{87.95}{42}$

(*c*) $\frac{1.0031 \times \pi(2.7)^2}{14.61}$

(*d*) $\frac{1}{2}(31.7)(24.9)^2$

ADDITION AND SUBTRACTION

The operations of addition and subtraction present a different problem in determining the number of significant figures. In general, the answer may be quoted only to the decimal place where the weakest number runs out, beginning from the left. Thus, if 270,000 is to be added to 927, the weakest number is the first—27 ten thousands.

$$\begin{array}{r} 27|0{,}000 \\ +\quad |\ 927 \\ \hline 27|0{,}000 \end{array}$$

Since the second number doesn't influence the ten thousands' column at all, the answer is 270,000.

For example, you have \$2.37 and your friends have about \$19 between them. What do you all have together? The second number gives no information about the dimes or pennies available from your friends; therefore, the result may be quoted to the nearest dollar only:

\$19 + \$2.37 = \$21

Consider that \$19 to the nearest dollar is anything from \$18.50 to \$19.49. Thus, if we knew to the penny the money available from your friends, the answer would range between a low of \$20.87 and a high of \$21.86. The best summary of an answer somewhere within this range is the answer given, \$21.

Practice Problems

To what place may the answers to the following operations be quoted?

(*a*) 10 + 234 + 6
(*b*) 241.7 + 0.03 + 0.20
(*c*) (31 + 294) − 14.9
(*d*) 3.0 m + 70 cm + 215 mm
(*e*) 0.00036 + 8.00041
(*f*) 27.6321 − 0.029

A Combined Example

The combination of these rules is sometimes confusing. Consider the following: Jane performs an experiment and gets a result of 82.4. The accepted value of the constant Jane measured is 81.92. What is the percentage error in her measurement? From the formula of the preceding section, we substitute

$$\% \text{ error} = \frac{82.4 - 81.9}{81.92} \times 100 = \frac{0.5}{81.92} \times 100 = ?$$

Numerically this works out to 0.61035. . . . Where do we stop in the reporting of the answer? The "100" in the formula is a counting number and does not figure into determining the significant figures, which is instead determined by the difference in the numerator. The number 0.5 has only one significant figure and thus the answer may have only one significant figure. The error is 0.6 percent, not 0.61 percent, and so on.

Scientific Notation

To handle numbers easily (especially very big numbers and very small ones) we usually express them in scientific notation. This rewrites any number as a number between 1 and less than 10, times 10 to an appropriate power. For example, the number 2300 could be rewritten as 2.3 × 1000. To write the correct power of 10 recall what an exponent on a number means: 3^4 means 3 multiplied by itself 4 times, or 3 × 3 × 3 × 3. Thus $10^3 = 10 \times 10 \times 10 = 1000$. Thus, our original number could be written as 2.3×10^3.

For positive powers of 10 the exponent indicates how many zeros the 1 is followed by in writing out the number. Thus, 10^6 = 1,000,000; 10^7 = 10,000,000; and so on.

For negative powers of 10 the exponent indicates in which place after the decimal point the first nonzero digit occurs. Thus $10^{-3} = 0.001$. This is illustrated in the following series:

$$10^4 = 10 \times 10 \times 10 \times 10 = 10{,}000$$

$$10^3 = 10 \times 10 \times 10 = 1{,}000$$

$$10^2 = 10 \times 10 = 100$$

$$10^1 = 10 = 10$$

$$10^0 = \quad 1 = \quad 1$$

$$10^{-1} = \quad \tfrac{1}{10} = \quad 0.1$$

$$10^{-2} = \quad \tfrac{1}{10} \times \tfrac{1}{10} = \quad 0.01$$

$$10^{-3} = \quad \tfrac{1}{10} \times \tfrac{1}{10} \times \tfrac{1}{10} = \quad 0.001 \quad \text{etc.}$$

Examples

Regular number		*Scientific notation*
27,900	=	2.79×10^4
0.00482	=	4.82×10^{-3}

Calculations in scientific notation are quite easy once the basic principles are understood. In general, numbers in science will have three parts: the numerical part, 10 to a power, the unit of measurement. Scientific notation allows us to group each part and treat each separately in most calculations. Consider the problem

$$(2.3 \times 10^6 \text{ m})(4.1 \times 10^{-2} \text{ m})(2.1 \times 10^{-3} \text{ m}) = ?$$

We may regroup each type of information separately to produce

$$(2.3 \times 4.1 \times 2.1) \quad (10^6 \times 10^{-2} \times 10^{-3}) \quad (\text{m} \times \text{m} \times \text{m}) = ?$$

$$(20.) \qquad (10^1) \qquad (\text{m}^3) \qquad = 20. \times 10^1 \text{ m}^3$$

$$= 2.0 \times 10^2 \text{ m}^3$$

The part of this which may be novel to you is the operations with powers of 10.

Consider

$$10^2 \times 10^3$$

This means

$$(10 \times 10) \times (10 \times 10 \times 10) = 10 \times 10 \times 10 \times 10 \times 10 = 10^5$$

In general:

$$10^A \times 10^B = 10^{A+B}$$

To multiply two powers of 10, you need only to add the exponents. Or, to divide powers of 10 (as in $10^3/10^2$), you only subtract the exponent of the divisor from the exponent of the dividend.

$$\frac{10^3}{10^2} = 10^{3-2} = 10^1$$

In general:

$$\frac{10^A}{10^B} = 10^{A-B}$$

The greatest caution must be exercised when adding or subtracting numbers in scientific notation. Recalling the slogan "you can't add apples and oranges," we see that numbers must be expressed in the same power of 10 for addition (or subtraction).

$$(2.3 \times 10^5) + (5.9 \times 10^4) = ?$$

$$(23 \times 10^4) + (5.9 \times 10^4) = 29 \times 10^4$$

$$= 2.9 \times 10^5$$

The change of the first number was legal since 2.3×10^5 means $(2.3 \times 10 \times 10 \times 10 \times 10 \times 10)$. We can certainly multiply the 2.3 by one of these 10s to produce 23, leaving a string of only four 10s to establish the power of the answer. Thus

$$2.3 \times 10^5 = 23 \times 10^4 = 230 \times 10^3 = 2300 \times 10^2$$

$$= 23{,}000 \times 10 = 230{,}000 = 2{,}300{,}000 \times 10^{-1} \qquad \text{etc.}$$

The first expression may be changed to any appropriate power of 10 for addition.

Consider a problem: The smallest atom, hydrogen, is about 1×10^{-10} m across. A light ray travels at a speed of 3.0×10^8 m/s. How long does it take a ray of light to pass a hydrogen atom? (If the problem looks useless to you, reword it: "How long does a hydrogen atom have to 'grab' a light ray as it passes?")

$$v = \frac{\Delta d}{\Delta t} = \frac{1 \times 10^{-10}\text{ m}}{3.0 \times 10^8\text{ m/s}} = \tfrac{1}{3} \times (10^{-10-8}) \times \left(\frac{\text{ms}}{\text{m}}\right) = 0.3 \times 10^{-18}\text{ s}$$

Now, ask yourself the question, "If I had worked this problem out longhand, what is the likelihood that I would have made an error?

$$v = \frac{0.000000000\,1\text{ m}}{300000000\text{ m/s}} = 0.0000000000000000003\text{ s}$$

Suppose that you slipped a single zero someplace. The answer would become 10 times too large or one-tenth the size it should be. Ten times too large is a 1000 percent error! When working with numbers of this type, most individuals will commit occasional errors in the placement of the decimal point. Using scientific notation tends to eliminate this problem, and is thus valuable for preserving accuracy as well as reducing writer's cramp.

APPENDIX TWO

Answers and Solutions to Odd-Numbered Problems

CHAPTER 1

1 The amount of earth and water combined equals the amount of fire.

5 (*b*), (*c*), (*d*).

7 $D = \frac{M}{V} = \frac{200\ \text{g}}{10.4\ \text{cm}^3} = 19.2\ \text{g/cm}^3$; silver: $\frac{200\ \text{g}}{19.0\ \text{cm}^3} = 10.5\ \text{g/cm}^3$.

CHAPTER 2

1 88 km/h

3 From Prob. 2-1; 1.0 mi = 1.609 km = 1609 m.
Therefore, *1 mi is greater than 1500 m by 109 m.* 3 m 54 s = 234 s is the time to run the mile. To determine the time for 1500 m at this speed, we set up the proportion

$$\frac{1500\ \text{m}}{1609\ \text{m}} = \frac{t}{234\ \text{s}} \qquad \text{or} \qquad t = 234\ \text{s}\,\frac{1500\ \text{m}}{1609\ \text{m}} = 218\ \text{s} = 3\ \text{m}\ 28\ \text{s}$$

5 $-227°F$.

7 104 cm^2 without considering significant figures, or 1.0×10^2 cm^2 when considering them. The precise area agrees with the latter but not the former number.

CHAPTER 3

3 7.069 m/s; 7.014 m/s. The record for the 1500-m race represents the higher average speed.

---

5 $$a = \frac{\Delta v}{\Delta t} = \frac{80 \text{ km/h}}{8.6 \text{ s}} = \frac{80{,}000 \text{ m/h}}{8.6 \text{ s}}\left(\frac{1 \text{ h}}{60 \text{ m}}\right)\left(\frac{1 \text{ m}}{60 \text{ s}}\right) = 2.6 \text{ m/s}$$

The instantaneous acceleration is not likely to be this at any given moment since the car's drive train varies in its ability to deliver twisting force with speed. This is why a car has a first gear, second gear, etc., and you feel slight jerks as it shifts from one to another.

---

7 -9.8 m/s; -9.8 m/s; where the negative signs indicate downward speeds.

9 -4.9 m; -44 m.

---

11 $d = 3$ km; $v_o = 7.7$ m/s. To find a:

$$v_f^2 - v_o^2 = 2ad \qquad \text{or} \qquad a = \frac{v_f^2 - v_o^2}{2d}$$

$$a = \frac{0^2 - (7.7 \text{ m/s})^2}{2(3000 \text{ m})} = -0.0098 \text{ m/s}^2$$

To find the time to stop:

$$d = \tfrac{1}{2}at^2 \qquad \text{or} \qquad t^2 = \frac{2d}{a} \qquad t = \sqrt{\frac{2(3000 \text{ m})}{-0.0098 \text{ m/s}^2}} = 6.1 \times 10^5 \text{ s}$$

This time is almost 7 days to come to a stop! Fortunately, the water helps by exerting a drag on the ship, which makes the acceleration much larger at the start than the number above indicates.

---

13 Segment I; 2.5 m/s.

15 Segment V, -6 m/s; 10.1 m/s is the record for the 100-m dash. Ralph's speed is a respectable fraction of the record—he is in good physical condition.

17 20 m/s at $t = 15$ s.

19 Segment B is likely at 15 m/s (34 mi/h); otherwise, Janice's maximum speed of 20 m/s (45 mi/h).

21 3 m/s^2.

23 (*a*) 2.0 h; (*b*) 3.0 h; (*c*) 1.7 h; (*d*) yes.

CHAPTER 4

1 (90° − your latitude) + $23\frac{1}{2}$°; (90° − your latitude) − $23\frac{1}{2}$°.

3 About 6 A.M., 6 P.M.; about 6 P.M., 6 A.M.; about noon, midnight.

7 (*a*) $29\frac{1}{2}$ days (*b*) 365.2 days (*c*) yes (*d*) $29\frac{1}{2}$ days (*e*) no.

CHAPTER 5

1 180 s

---

5 Yes. $\frac{1}{T_{syn}} = \frac{1}{T_{in}} - \frac{1}{T_{out}}$ or $\frac{1}{T_{out}} = \frac{1}{T_{in}} - \frac{1}{T_{syn}}$

$$\frac{1}{T_{out}} = \frac{1}{60\text{ s}} - \frac{1}{90\text{ s}} = \frac{3}{180\text{ s}} - \frac{2}{180\text{ s}} = \frac{1}{180\text{ s}}$$

If $\frac{1}{T_{out}} = \frac{1}{180\text{ s}}$, then $\frac{T_{out}}{1} = \frac{180\text{ s}}{1} = 180\text{ s}$

This is also the answer given in Prob. 1.

---

13 About 1°/day.

17 $T = 84$ years.

19 $T_{syn} = 587$ days.

CHAPTER 6

1 (*a*) 9.8 N; (*b*) 18.6 N.

3 980 N; 0.20 m/s²; 1.2 m/s².

---

5 Wagon plus boy = 25 kg; 10 kg of bricks thrown off at 2.0 m/s. Momentum is conserved, or, momentum of wagon plus boy equals momentum of bricks in opposite direction.

$$(mv)_{wagon} = (mv)_{bricks}$$

$$(25\text{ kg})v_W = (10\text{ kg})(2.0\text{ m/s})$$

$$v_W = \frac{20\text{ kg}\cdot\text{m/s}}{25\text{ kg}} = 0.80\text{ m/s}$$

CHAPTER 7

1 78 N; 78 J.

3 4.0 m/s.

---

5 (*a*) $E_p = mgh = (1000\text{ kg})(9.8\text{ m/s}^2)(3.0\text{ m}) = 29{,}000$ J.
(*b*) 29,000 J, since it may convert all of its potential energy.

(*c*) A watt is a joule per second, so it falls in 1 s. The power is $\frac{29{,}000\text{ J}}{1\text{ s}}$ or 29,000 W or 29 kW.

CHAPTER 8

1 $-40°$; $-459.69°F$.

3 1.0°C.

7 630 kcal.

---

9 The 11 lb of fat have a mass of 5.0 kg and, according to the table, must represent 44,500 kcal of stored food energy (5.0 kg × 8900 kcal/kg). The body requires 2000 kcal/day, but we will provide only 1000 kcal/day in food. The additional 1000 kcal/day must come from the stored fat.

$$\frac{44{,}500 \text{ kcal}}{1000 \text{ kcal/day}} = 44.5 \text{ days to expend all 11 lb}$$

---

13 (*a*) 3200 kcal/person; (*b*) 5.3 m^2; (*c*) 53,000 m^2; (*d*) 13; (*e*) $2,650,000; (*f*) $87,500; (*g*) $56/day; (*h*) $11,200/year; (*i*) $133,000/year based on 20-year useful life and excluding interest charges.

CHAPTER 9

1 7.5 m/s.

3 25/s.

5 12.4 m; 0.081 m.

---

7 For open-end resonance in a tube, $L = \frac{1}{2}\lambda = \frac{1}{2}\frac{V}{f}$ (1 Hz = 1/s).

$$L_1 = \tfrac{1}{2}\left(\frac{340 \text{ m/s}}{250\text{/s}}\right) = 0.68 \text{ m} \qquad \text{[Handel]}$$

$$L_2 = \tfrac{1}{2}\left(\frac{340 \text{ m/s}}{274\text{/s}}\right) = 0.62 \text{ m} \qquad \text{[Sousa]}$$

Thus, Handel's flute would be 6 cm longer than Sousa's.

CHAPTER 10

3 5000 N/C; 50,000 V.

5 1800 W; 2.3 hp.

CHAPTER 12

1 ZnO; $AlCl_3$; NaF; $CaCl_2$ (every other formula).

3 $2Li + 2H_2O \rightarrow 2LiOH + H_2$.

5 H_2O—18; Na_2O—62; CaF_2—78 (every other formula).

7 11%.

9 About 2/1; yes, roughly.

11 $H_2 + Cl_2 \rightarrow 2HCl$; 2.0; 4.0.

13 Al_2O_3, MW = 102.

$$2\ \text{Al} \longrightarrow (2 \times 27) = 54$$

$$3\ \text{O} \longrightarrow (3 \times 16) = \underline{48} \quad \text{molecular weight}$$
$$102$$

(*a*) $\frac{54}{102}$ is Al or 53%. 0.53 × 1000 kg = 530 kg, the mass of aluminum in a ton of bauxite ore.
(*b*) If 530 kg are Al, 470 kg must be oxygen. The molecular mass of oxygen gas (O_2) is 32:

$$\frac{470\ \text{kg}}{32\ \text{g/mol}} = \frac{470{,}000\ \text{g}}{32\ \text{g/mol}} = 15{,}000\ \text{mol}$$

Each mole occupies 22.4 liters at STP, so

$$(15{,}000\ \text{mol})\ (22.4\ \text{liters/mol}) = 3.4 \times 10^5\ \text{liters.}$$

15 $\frac{1}{238}$ mol, or 2.5×10^{21} atoms.

CHAPTER 17

3 10.000 eV; 10 MeV; 1.6×10^{-12} J.

5 (*a*) 10 mol × (6×10^{23} atoms/mol) = 6×10^{24} atoms.

$$(6 \times 10^{24}\ \text{atoms})(200\ \text{MeV/atom}) = 12 \times 10^{26}\ \text{MeV}$$

but, 1.0 MeV = 1.6×10^{-13} J. Thus, the energy released is

$$(12 \times 10^{26}\ \text{MeV})(1.6 \times 10^{-13}\ \text{J/MeV}) = 1.9 \times 10^{14}\ \text{J}$$

(*b*) $\frac{1.9 \times 10^{14}\ \text{J available}}{1.5 \times 10^{12}\ \text{J/pool}}$ = 130 pools could be vaporized per pound of uranium.

7 12 kg.

CHAPTER 19

1 15,500 km.

3 10 times as far.

7 (*a*) 5 mag ⇒ 100 times as bright
10 mag ⇒ 100 × 100 times, or 10,000 times as bright.
(*b*) $I \propto T^4$; if $T_B = \frac{1}{2}T_{sun}$ the intensity of its surface radiation is $(\frac{1}{2})^4 = \frac{1}{16}$ of the intensity of the sun's radiation per square meter of surface.
(*c*) (Radiation/m²) × (number of m²) = total radiation, or

$$R_{tot} = IA$$

Writing this equation for the sun and Betelgeuse, and dividing one equation by the other yields:

$$\frac{R_{\text{sun}}}{R_B} = \frac{l_{\text{sun}}}{l_B}\frac{A_{\text{sun}}}{A_B}$$

Rearranging:

$$A_B = A_{\text{sun}}\left(\frac{l_{\text{sun}}}{l_B}\right)\left(\frac{R_B}{R_{\text{sun}}}\right)$$

$$A_B = A_{\text{sun}}\left(\frac{32}{1}\right)\left(\frac{10{,}000}{1}\right) = 32{,}000\ A_{\text{sun}}$$

(*d*) $\dfrac{A_B}{A_{\text{sun}}} = \dfrac{4\pi R_B^2}{4\pi R_{\text{sun}}^2}$ or $\dfrac{A_B}{A_{\text{sun}}} = \dfrac{R_B^2}{R_{\text{sun}}^2}$

$$R_B^2 = \left(\frac{A_B}{A_{\text{sun}}}\right) R_{\text{sun}}^2 = 32{,}000\ R_{\text{sun}}^2$$

$$R_B = 180\ R_{\text{sun}}$$

Thus, Betelgeuse has 180 times the radius of our sun, which is over half again as large as the earth's entire orbit around the sun.

CHAPTER 20

1 (*a*) $\frac{1}{8}$ year = 1100 h.
(*b*) 180 days.
(*c*) No.
(*d*) $\dfrac{500}{1300} = \dfrac{1}{26}$ year = 337 h = 14 days. Yes.
(*e*) This is 114 mrem/h, requiring 4.4 h for the MPD.
(*f*) Extremely likely.

3 (330×10^6 rems) × (100 cancers/10^6 rems) = 33,000 cancers.

APPENDIX THREE

Glossary

aberration, stellar The apparent shift of stars in the sky caused by the differing direction of the earth's motion in its orbit around the sun.

absolute magnitude The actual brightness of stars based on how each would appear in magnitudes if placed at a standard distance of 10 parsecs (pc).

acceleration The rate of change of velocity, or the velocity change divided by the length of time it takes to make the change ($a = \Delta v/\Delta t$).

accelerator A device for bringing charged particles to high speed for penetrating atomic nuclei and studying the structure of matter.

acid Taken from the latin word for "sour," this is any member of a group of substances which tend to liberate hydrogen ions (H^+) in solutions. If this tendency is powerful, the acid may be a dangerous corrosive such as sulfuric acid or muriatic (hydrochloric) acid. Weak acids include citric or boric acids, or aspirin.

acid rain The rainfall with unusually low (acid) pH caused by oxides of sulfur and nitrogen forming sulfuric and nitric acids in clouds—usually attributed to the emissions of coal burning.

activation energy The energy which must be supplied to begin a reaction, frequently needed to break temporary bondings of molecules, such as in the striking of a match before it bursts into flame.

activity series A listing of elements in terms of increasing chemical activity.

alpha (α) particle A helium nucleus consisting of two protons and two neutrons ejected at high speed from an atomic nucleus in the process of radioactive decay.

amino group Chemical group of the formula NH_2 where the nitrogen atom retains the ability to form one additional covalent bond to another atom.

ampere Unit of current, or electric flow, equal to one unit of charge passing a given point per second: 1.0 ampere (A) = 1.0 coulomb/second (C/s).

amplification Increasing the size (intensity) of a wave, by any of a number of means.

amplitude With a single wave, the size of the displacement of a particular portion of the wave from the rest position of the material carrying the wave. In general, the maximum disturbance of the wave—the loudness of the sound, the brightness of the light, and so on.

angle of incidence The angle at which a wave or ray comes into a surface, measured between the line of travel of the ray and the perpendicular to the surface at the point where the ray strikes it.

angle of refraction The angle between a perpendicular or normal to the surface at the point of refraction and the refracted ray.

angular momentum The product of mass times velocity times radius for spinning or rotating objects (*mvr*). Always a constant in the absence of unbalanced twisting forces. Thus, as a skater straightens from a spinning, crouched position to reduce the average radius of the rotating mass, the speed of rotation increases such that *mvr* is constant.

annihilation The union of a particle and its antiparticle causing them to disappear as matter in a pure burst of energy.

annular eclipse An eclipse of the sun when the sun appears slightly larger than the moon in the sky and a bright rim of sunlight is seen around the moon even at total eclipse.

anode A positively charged plate or electrode.

anticline An upward folded ripple in the crust of the earth.

antinodes Position of maximum vibration in a standing wave pattern, where the crossing waves alternately produce a large crest, then a large trough, and so on.

antiparticle "Mirror-image" particles predicted by theory and found in nature or produced in laboratories, such as antielectrons or antiprotons. Whenever matter is produced from pure energy, it is as a particle and its corresponding antiparticle.

apparent magnitude The brightness of stars rated on how they appear from the earth.

aquifer A porous rock containing water in the spaces between its particles and allowing the movement of this water within its structure.

Archimedes' principle An object immersed in a fluid loses an amount of weight equal to the weight of the fluid displaced.

artificial transmutation The changing of one element to another by the action of a human being.

astronomical unit The average distance of the earth from the sun, used for measuring distances within the solar system. (Abbreviated AU.)

atom The smallest part of an element; the smallest constituent of normal matter as it is ordinarily encountered.

atomic mass number The number of particles, protons plus neutrons, in the nucleus.

atomic mass unit A mass originally taken as nearly that of the proton or neutron and equal to 1.6×10^{-27} kg. Now defined more precisely as one-twelfth of the mass of a carbon-12 atom. (Abbreviated u.)

atomic number The number of positive charges, or protons, in the nucleus of an atom. Responsible for determining what element an atom is.

Avogadro hypothesis Asserts that equal volumes of all gases under the same conditions contain the same number of molecules.

Avogadro number The number of atoms in a gram atom of a substance or molecules in a mole. $N_o = 6.02 \times 10^{23}$ molecules per mole.

azimuth The angle to any point on the horizon measured from the due north point on the horizon sweeping eastward (clockwise).

background radiation The radiation from the environment to which all individuals are exposed.

base (alkali) Any material which tends to liberate hydroxide radicals (OH^-) in solution.

beta (β) particle An electron ejected at high speed from an atomic nucleus in the process of radioactive decay.

binding energy The energy required to hold the various particles of the nucleus together.

biological concentration Increasing the concentration of some material present in the environment in trace amounts by the action of a living system.

black hole A super-collapsed state of matter where gravitational forces have crushed atomic particles into one another with such force that the matter occupies essentially no space. This concentration of mass creates a gravitational pull so strong that not even a light beam can escape from it—hence the name.

Boltzmann's constant (k) A fundamental constant relating large-scale (macroscopic) behavior to atomic events in several expressions, especially $pV = NkT$, where $k = 1.38 \times 10^{-23}$ J/K.

bond Any mechanism that holds one atom to another.

bright-line spectrum The light and radiations given off by specific elements excited as a low-pressure vapor. Such spectra are characterized by only certain specific frequencies of light and no others.

caloric A hypothetical, self-repulsive fluid that was heat itself according to the incorrect caloric theory.

calorie The basic unit of heat equal to the amount of heat required to raise the temperature of one gram of water one degree Celsius. (Abbreviated cal; not to be confused with the dietary Calorie, which is the scientific kilocalorie and 1000 times as large.)

carbohydrate Large group of foodstuffs providing the majority of energy requirements for life forms. Chemically, molecules having carbon, hydrogen, and oxygen with the latter two in the proportion of water molecules (H_2O) and approximately one of these for each carbon atom, as in the sugar glucose, $C_6H_{12}O_6$.

carboxyl group Also known as organic acid group. Chemical group of the formula —COOH, where the carbon atom shares a double covalent bond with one oxygen atom, a single covalent bond with the hydroxyl (—OH) group, and retains the ability to attach to some other atom in a single covalent bond. The weak attachment of the H in this group tends to make compounds with it mildly acidic, as in vinegar or in citric acid.

catalyst Any substance which acts to speed up the rate at which a chemical reaction takes place but which is not itself used up in the reaction.

catastrophism The discredited belief that the appearance of the earth's crust was due to violent catastrophes rather than to earth processes working as they are now.

cathode A negatively charged plate or electrode.

cathode ray A stream of particles (electrons) found to be coming from the cathode, or negative plate, inside an evacuated tube with plates on opposite sides connected to a high voltage.

celestial coordinate system A system for describing the position of objects in the sky centered on the celestial equator and using two coordinates, declination and right ascension.

celestial equator The collection of all points in the sky directly overhead at some point on the equator. The projection of the earth's equator onto the sky. (Abbreviated CE.)

centrifugal force Any force directed away from the center of a circle—usually used incorrectly in everyday discourse since such forces almost never exist except in highly contrived situations.

centripetal force Any force directed toward the center of a circle tending to accelerate objects inward.

Cepheid variable A particular type of quite bright variable star in which the period of variation of brightness is directly related to the brightness at maximum, thus a powerful tool for measuring distances indirectly.

chain reaction Any reaction in which one of the products is the commodity required to start the reaction in the first place, which allows the reaction to spread and grow. For example, heat is a product of the burning of paper but is also what is needed to begin the burning, allowing the next part of the paper to burn.

circular acceleration The acceleration required to produce uniform circular motion and equal to either $(4\pi^2 R)/T^2$ or v^2/R, where R is the radius of the circle and T is the period of the motion.

circumference The distance around a sphere or circle given by $c \times 2\pi R$, where $\pi \times 3.14159 \ldots$ and R is the radius of the circle.

closed end An end to a medium carrying a wave, not capable of vibrating in the manner required by the wave, as in a fastened-down violin string.

cloud chamber A device for studying the tracks of particles by producing vapor trails (or clouds) along the atomic debris left in the paths of the particles.

compound A pure substance with very specific properties made up of two or more elements chemically united in a particular way, as in water (H_2O).

compression A situation of forces acting to push the molecules of a material closer together.

compressions That portion of a sound wave where the air pressure rises above normal air pressure.

concentric Having the same center, as with circles.

conduction A process for the transfer of heat and/or electric charge by the passing of heat and/or charge from atom to atom through the body of the material.

conductor A material substance which offers very little resistance to the passage of heat and/or electricity through it.

configuration The arrangement of electrons in the various electron shells of an atom.

conservation law A statement about the behavior of nature, asserting that the total of some commodity or property does not change during interactions.

containment vessel The large reinforced concrete structure around a nuclear reactor designed to contain the steam and radiation which would be released in the event of a catastrophic failure of the system.

continental drift The theory that continental blocks of the earth have separate motions relative to each other and at one time were joined.

continuous spectrum A range of wavelengths representing all colors of visible light.

convection A circulation of a fluid usually set up by temperature differences. A process for the transfer of heat requiring the flow of matter from place to place as in ocean currents, air circulation from a heat vent, and so on.

converge To come together, as with parallel rays of light after passing through a convex lens.

convex A surface bulging outward at the center. In a lens, one thicker through the center than at the edges.

core Innermost zone of the earth beginning at a depth of 2900 km and extending to the center of the earth.

coriolis force A ficticious (apparent but not real) force seeming to deflect objects on the earth moving toward the equator to the west, and those moving away toward the east. Caused by the earth's rotation and the inertia of masses.

coulomb The unit of electric charge in the SI system of metric units. In terms of the more fundamental charges found on the particles in atoms, 1.0 coulomb (C) = 6.25×10^{18} times the charge of an electron or a proton.

Coulomb's law The basic law of electrical attraction and repulsion of charges; the force (in newtons) between charges q_1 and q_2 (in coulombs) separated by a distance d (in meters) is $F = Q\frac{q_1q_2}{d^2}$, where Q is a constant of value $8.99 \times 10^9\ N \cdot m^2/C^2$.

covalent bonding Bonds formed between atoms by the sharing of electrons in pairs between the atoms.

crest The portion of a wave with the greatest positive displacement from the rest condition of the medium carrying the wave. With sound waves the zone of greatest air pressure (compression); with water waves the highest portions of the water surface.

critical angle The largest possible angle to the perpendicular to a surface at which a light ray can escape from a material with a higher index of refraction to a material with a lower index of refraction. At this angle, the ray escapes and just skims the surface. Beyond this angle the ray reflects at the boundary and remains in the original material.

crust The outermost portion of the earth containing rocks as we know them and ending at a well-defined boundary to a lower layer, the mantle, at depths of about 8 km under the oceans or 30 km under the continents.

crystal A regular geometric structure formed by repeating units of identical chemical composition at the molecular level, as with quartz crystals or a chunk of rock candy (sugar crystals).

current electricity Electrical situations characterized by a sustained and systematic flow of electric charges.

dark-line spectrum A continuous spectrum representing all frequencies (colors) except for very specific colors so greatly reduced in intensity as to appear to be missing.

declination The separation in the sky, reported as an angle, between an object and the celestial equator.

definite proportions Having constant composition of each element by mass; for example, all pure water is 89 percent oxygen and 11 percent hydrogen by mass.

dehydration The removal of water, especially the chemical removal of water, from molecules, usually resulting in the remaining parts of these molecules joining in a bond.

diffraction The process of the bending of a wave as it passes the edge of a barrier or through a fine opening to spread out into the "shadow" area not accessible to the wave if it traveled in a straight line.

dispersion The separation of light by the slightly different bending of each color (wavelength) during refraction.

dissolve To break apart into separate molecules or even parts of molecules (ions) and mix intimately with the molecules of a dissolving fluid called the solvent; for example, sugar or salt dissolving in water.

diverge To spread apart, as with parallel rays of light after passing through a concave lens.

doppler shift An apparent shift in the wavelength (and frequency) of a source caused by the motion of the source toward or away from the observer, resulting in a shorter or longer apparent wavelength, respectively.

eccentricity A measure of the deviation of an orbit from a circle, which has zero eccentricity. Technically, the distance between the foci of the ellipse divided by the greatest distance across the ellipse.

ecliptic The apparent path of the sun through the sky. Also, the projection of the plane of the earth's orbit onto the heavens.

elastic Capable of returning all of the energy used to disturb a given material when it is allowed to return to its original shape or position and not use the energy to permanently deform itself, as with a spring. In collisions, any collision that conserves the kinetic energy of the particles involved.

electric circuit A complete pathway for the flow of electric charges which closes back on itself to return charges to their point of origin. Circuits usually contain elements, such as a battery or generator, to provide electric potential energy (voltage) to set up a flow of charges, as well as elements which resist the passage of charges by requiring work to be done in passing.

electric field A region of space in which an electric force exists on a charged particle. The electric field vector: **E** is defined as the force acting on a unit positive charge placed at any point in space. This is frequently summarized by drawing lines of electric force, where the strength of **E** increases as the lines draw closer and the direction of **E** is always parallel to the lines in any region of space.

electric potential difference Voltage.

electrochemical cell A device converting chemical potential energy into an electric potential difference. A cell consists of two dissimilar conductors partially imbedded in an electric conducting fluid. The electric potential, or voltage, develops between the exposed ends of the conductors. Two or more such cells when connected are called a battery.

electrolysis The separation of a chemical compound by the application of electricity.

electromagnetic spectrum The range of detectable electromagnetic waves, including long waves such as radio waves or electric power waves, through microwaves and radar waves, into infrared rays, visible light, and shorter waves such as x-rays and γ rays.

electromagnetic wave A transverse wave produced by the acceleration of an electric charge consisting of variation of electric- and magnetic-field information and which moves through a vacuum at 300,000 km/s; for example, light or radio waves.

electron A subatomic particle found in all atoms with a negative charge of 1 unit (-1.6×10^{-19} C) and a mass almost negligible (1/1836) compared to the proton or neutron. The particle involved in all chemical reactions between atoms, and whose actual motion produces all normal electrical effects.

electronvolt A unit of energy useful in describing atomic actions, equal to the energy acquired by a free electron when accelerated across a potential difference of one volt. (Abbreviated eV.)

element The simplest form of matter, which cannot be reduced into more basic parts by any chemical means, and from which all normal matter is composed.

elementary charge The smallest charge yet measured in nature, the charge on the electron ($-$) or the proton ($+$) of some 1.60×10^{-19} C.

endothermic Any reaction which absorbs heat during the reaction.

enriched Uranium in which the percentage of the isotope U 235 has been increased above the level occurring in nature.

entropy A measure of the amount of randomness or disorder of a system fundamental to the second law of thermodynamics, which asserts that the entropy of the universe is increasing.

epicycle In the Ptolemaic System, a smaller circle in which a planet moves and which has a center that moves around a larger circle in the basic orbit of the planet.

equant A scheme in the Ptolemaic System for adjusting the observed variable speeds of planets in moving along their orbits. The motions were at constant speed as seen from a point not at the center of the orbit, and the earth was displaced from the center an equal amount in the opposite direction.

equilibrium (chemical) A state of balance in a chemical reaction where products break down at the same rate as reactants produce them so that the concentration of each substance remains constant with time .

equilibrium constant A number describing the state at which equilibrium occurs between the reactants and the products of the reaction. Numerically, the concentrations of the products multiplied together divided by the concentrations of the reactants multiplied together.

equivalent resistance The single electrical resistance which could replace some combination of resistances in a circuit and have no effect on currents outside of the area replaced.

era The largest division into which geologic time is subdivided.

erosion The process of transporting materials of the crust from one place to another, as in the washing away of topsoil, or a rockslide.

ether A name given to the hypothetical medium carrying light waves.

excess deaths Deaths beyond the level normally expected in a population and attributed to some special effect.

exothermic Any reaction liberating heat or other energy.

extrapolation Projecting the value of a variable beyond the region for which data are available, as beyond the points on a graph, or in forecasting.

family (chemical) A group of elements with similar combining tendencies and chemical properties. One column in the periodic table.

fault A crack in the rock structure of the earth

along which a motion of blocks on opposite sides of the crack may occur.

fission A nuclear reaction in which a larger nucleus splits or divides into two smaller nuclei of somewhere near half the size of the original nucleus, each liberating energy in the process, as in an atomic bomb.

fluorescence A property of some forms of matter to absorb energy and reradiate it as visible light.

focal length The distance from a lens or mirror at which parallel rays of light converge to a point.

focus To bring light rays into a sharp image with crisply outlined edges, as in focusing a projector or a camera.

fold A bending in the rock layers of the crust.

force A push or a pull acting on a material object. Measured by the rate at which the motion of the mass is changed ($F = ma$) in newtons (N): $1.0\ \text{N} = 1.0\ \text{kg} \cdot \text{m/s}^2$.

fossil Any record of a life form in the rocks of the earth's crust, such as imprints or casts of organisms, shells, bone remains, or even footprints.

frequency The number of full actions or cycles of a repetitive event completed in one time unit, as in one-twenty-fourth of a rotation per hour for the earth with respect to the sun. The reciprocal of the period ($f \times 1/T$).

fulcrum The position at which a material object is supported and about which it is free to rotate, as in the pivot point of a lever.

fundamental mode The lowest-frequency standing wave which may be set up on a vibrating system—for musical instruments, the basic note being sounded.

fusion A nuclear reaction in which smaller nuclei unite to form a single larger nucleus liberating energy in the process, as in a hydrogen bomb.

galaxy The largest known grouping of stars, gas, and dust in the universe, containing up to several hundred billion stars in a single galaxy.

gamma (γ) ray A quantum of electromagnetic energy emitted by an atomic nucleus, like light, but much shorter in wavelength.

geomagnetic reversal A geologically abrupt shifting of the earth's entire magnetic field from one orientation to approximately the opposite—so that a north-seeking pole would point southward. Such reversals have occurred periodically in the past, and one may be due.

geosyncline An extremely large downward folded portion of the earth capable of filling with enough sediment to distort the earth by accumulated weight.

glacier A mass of ice and compacted snow of sufficient thickness to become plastic at its base and flow.

gram atom (mole) The mass of a molecule calculated from the atomic masses listed in the periodic table translated into grams (g): e.g., 18 g of water is 1 mole (mol) and therefore contains 6.02×10^{23} molecules of H_2O.

greenhouse effect A shielding effect of the earth's atmosphere, allowing short-wave solar radiation to reach the surface but absorbing long-wave earth radiation, thus retaining heat at the earth. Similar to the heating of a closed car sitting in a parking lot in direct sunlight, causing the interior temperature to rise.

ground state The lowest energy position of an atom, in which its electrons are normally found.

group Any covalently bonded assemblage of atoms that tends to be found with similar structure attached to a wide variety of molecules and which confers particular chemical properties on that molecule, as with —COOH, the acid group, or —OH-producing alcohols.

half-fatal dose The amount of radiation at which one-half of the exposed population dies and the other half survives, about 500 rem for human beings.

half-life The length of time required for the radioactivity of a sample to decline to one-half of its initial value.

harmonics The series of frequencies that can set up standing waves on the same medium at the same time—for example, the set of all waves making up a particular note on a violin string.

heat of fusion The amount of heat required for a given mass of a substance to change from a solid to a liquid at the same temperature. Usually given in calories per gram or kilocalories per kilogram, as with ice to water, where $H_f = 80$ cal/g.

heat of reaction The heat liberated or absorbed in a chemical reaction. Usually stated in kilocalories per mole of product; positive if heat is absorbed or negative if heat is given off by the reaction.

heat of vaporization The amount of heat required for a given mass of a substance to change from a liquid to a vapor at the same temperature. Usually given in calories per gram or kilocalories per kilogram, as with the value for water to steam, where $H_v = 540$ cal/g.

horizontal component That portion of a vector acting in a horizontal plane and having no vertical, or up-and-down, part to it.

horsepower A unit of power attributed to James Watt and supposedly equal to the sustained work output of a horse: 1.0 horsepower = 746 watts (W) = 550 ft · lb/s.

H-R diagram A graph of the masses of stars vs. their brightness (or spectral type vs. surface temperature).

Hubble constant A number relating the distance of galaxies to their red shift, hence, speed of recession from us.

hydrocarbon Any compound formed of carbon and hydrogen, as in methane, CH_4, or benzene, C_6H_6.

hydrogen bond A weak electric bond formed between partially charged zones of molecules that are electrically neutral as a whole. Of great importance in explaining the ability of water to dissolve a wide variety of materials.

hydrolysis The chemical addition of water to a structure, causing its breakdown into smaller molecules, as in the hydrolysis of peptide bonds of proteins in the stomach in the digestion of meats.

hydrophilic "Water-loving"; a molecule or a part of a molecule that tends to dissolve in water or be attracted by its polar bonds.

hydrophobic "Water-fearing"; a molecule or a part of a molecule which is not attracted by the polar bonds of water and thus tends to separate from it, as with a bead of oil.

hydroxyl group Chemical group of the formula —OH, where the oxygen atom retains the ability to attach to some other atom in a single covalent bond.

igneous extrusion Any of the processes of earth-building in which magma is forced to the surface of the earth, as in volcanoes or lava flows.

igneous intrusion Any process of earth-building in which the crust is formed by the action of molten rock below the surface but does not appear at the surface.

igneous rock A rock resulting from the cooling of a molten mass of rock, as in granite or basalt.

image distance The distance from a lens or mirror at which the image of an object is formed.

incandescence The emission of light by a substance heated to a high temperature sufficient to glow.

inert Not taking part in any chemical combinations.

inertia The tendency of matter to persist in whatever motion it has and to resist a change in this motion. A measure of this inertial tendency is called mass.

infrared A range of wavelengths similar to visible light but of longer wavelength and not visible to the human eye.

insolation The total heat and light received from the sun—a maximum of about 1 kW/m² for a surface on earth pointing directly at the sun.

insulator A material substance which offers very high resistance to the passage of heat and/or electricity through it.

intensity A measure of how densely packed the energy of a system is, as with the brightness of a light or the loudness of a sound.

interference A property of waves allowing two or more of them to add together at particular times and places to produce anything from a very large effect (constructive interference) to no disturbance at all (destructive interference); as with two sound waves combining to produce silence.

ion Any charged particle formed from the giving or taking of electrons from a neutral atom.

ionic bonding The chemical joining of atoms by the physical transfer of electrons from one atom to another and the subsequent electrical attraction of the resulting ions.

isomers Chemical compounds with identical chemical formulas but different chemical structures and thus different chemical and physical properties.

isostasy The principle that each portion of the earth's crust is essentially floating on the mantle underneath and rises or sinks to a level determined by its density.

isotopes Atoms of the same element (same atomic number) but with differing numbers of neutrons, hence different atomic mass numbers.

joule The basic metric unit (abbreviated J) of work or energy. The work done when a force of one newton is exerted through a distance of one meter. Also the energy required to lift a mass of 1.0 kg about 10.2 cm at the earth's surface. $1\ J = 1\ N \cdot m = 1\ kg \cdot m^2/s^2$.

kinetic energy Energy stored in the motion of a

mass and equal to the work required to bring the mass to its current speed, or $E_k = \frac{1}{2}mv^2$.

latitude A measure of distance north or south of the equator on earth reported as an angle as seen at the center of the earth between the position and the closest point on the equator.

leaching The removal of soluble materials from rock and soil by the action of water; particularly the carrying of these chemicals away from the surface to greater depths by percolating rainwater.

lens A shaped piece of a material transparent to a wave used to bend and shape wave fronts, especially those used to focus light as in eyeglasses or projector lenses.

linear source (of waves) A source which generates a wave disturbance along a line that then proceeds away from the source as a linear wave front.

local group The cluster of galaxies to which our own belongs.

longitudinal wave A type of wave passed from place to place by a motion of the particles of the material carrying the wave back and forth in the direction the wave is traveling, as with sound waves.

machine Any device to transform the size and/or direction of forces.

magma Molten rock, usually called lava if it appears at the earth's surface, but the term includes all lavas.

magnetic field A region of space in which a magnetic force exists on poles or moving charges. As normally drawn, the strength of the force is indicated by the spacing of the lines of force, and the direction of the force is parallel to the lines.

magnitude A scale of the brightness of stars developed in ancient times and still used. The lower the magnitude, the brighter the star; a difference of 5 magnitudes implies a brightness difference of 100 to 1.

main sequence The band in a graph of the spectral type vs. surface temperature of stars within which most stars fall.

mantle Zone of the earth under the crust beginning at the moho, some 5 to 40 km deep, and extending to the core.

mass The amount of matter in a substance, measured by the object's inertia, or tendency to resist a change in its motion.

mass spectrograph A device to separate isotopes based on their masses by giving them a charge and shooting them through a magnetic field where the path of more massive particles are bent or deflected less.

mechanical advantage A number describing the factor by which a machine multiplies the force applied to it. (Abbreviated MA.)

metal Chemically, any element which tends to lose electrons in a reaction. Usually physically characterized by being shiny, a good conductor of electricity, and deformable with the application of forces without breaking (nonbrittle).

metamorphic rock A rock which has been substantially changed in form and/or pressure, as in the production of marble from limestone.

mineral Any pure material in the earth's crust, distinguished by a definite chemical composition and uniform properties, as in a fleck (crystal) of mica imbedded in a granite sample.

mirror Any good reflector of light with a smooth surface that does not randomize the direction of reflected rays.

mixture Two or more materials (elements or compounds) thrown together in any proportions without any necessary chemical union, resulting in a material with variable properties which may be separated physically into separate components; for example, a bag of garbage or a lump of granite.

moho (Mohorovičić discontinuity) The abrupt transition zone between the crust and the mantle revealed worldwide by a change in seismic waves and occurring at an average depth of 5 km under oceans or 33 km under the continents.

molarity (of solutions) A number for reporting the concentration of solutions in terms of the number of moles of the dissolved material per liter of solution. For example, if 1 mol of salt (58 g) is dissolved in 1 liter of solution, it would be a 1.0 molar saline solution.

mole (gram atom, gram molecular mass) The mass of the molecule calculated from the atomic masses of the periodic table translated into grams; for example, 32 g of O_2 is 1 mol, and contains therefore 6.02×10^{23} molecules.

momentum The product of mass times velocity (*mv*) having direction as well as size. Always a constant in the absence of unbalanced forces.

moraine A ridge of unsorted (of all sizes) rock debris left at the side of (lateral moraine) or end of (terminal moraine) a glacier.

multiple proportions When elements form more

than one compound, the percentages of a given element in each of the compounds form simple whole number ratios by mass.

neutron Subatomic particle normally found in the nucleus with mass nearly one atomic mass unit and no electric charge.

neutron star A star made of incredibly dense collapsed matter where the electrons of the atoms have been crushed into the nucleus to join with protons, producing "wall-to-wall" neutrons.

node A position of zero disturbance as two or more waves cross each other.

nonmetal A chemical substance which tends not to lose electrons in a chemical reaction, especially atoms which tend to gain electrons.

nuclear force Either of the two forces which are dominant in nuclear interactions; each is much stronger than either electric forces or gravitational forces, the other two types known in nature.

nuclear reactor A device for the controlled release of nuclear energy by sustaining a chain reaction at a given level.

nucleon Any particle (proton or neutron) normally residing in the nucleus.

nucleus A concentration of mass and positive charge at the center of an atom containing all of its protons and neutrons.

object distance The distance of the source of light rays, the object, from a lens or mirror which forms an image of that object.

observer system A system of coordinates in astronomy to report position in the sky as a set of angles. *Azimuth* is the clockwise angle around from due north measured along the horizon. *Altitude* is the angle of the object above the nearest point on the horizon.

ohm Unit of electrical resistance indicating the potential energy which must be converted per unit of charge for each unit of current which flows: 1.0 ohm (Ω) = 1.0 volt/ampere (V/A).

Ohm's law The basic relationship between voltage, current, and resistance in electric circuits, $V = IR$. Applicable to an entire circuit, or any small part of a circuit separately.

opaque The property of not allowing a wave, especially light, to pass through—as with a brick wall to light, or a metal sheet to a radio wave.

open end An end to a medium carrying a wave capable of vibrating in the manner required by the wave, as the end of a fiberglass fishing pole.

optics, law of For mirrors or lenses, the relationship between focal length f, image distance D_I, and object distance D_O is

$$\frac{1}{f} = \frac{1}{D_O} + \frac{1}{D_I}$$

ore Any rock from which a component may be recovered economically (at a profit).

organic Anything pertaining to carbon and its compounds, such as DDT, kerosene, axle grease, or any life form. (Used entirely differently in general usage, so be careful.)

oscillation or vibration The periodic back-and-forth displacement of a system from its normal, stable condition.

overtone The set of higher-frequency waves also set up when a particular note, the fundamental, is sounded on a string or pipe.

oxidation Uniting with oxygen, as in the burning of gasoline. More generally, any reaction tending to increase the positive charge on a particle, such as: $Na^0 \rightarrow Na^+ + e^-$.

parallax The apparent shift of a nearer object against a more distant background as the observer's point of view is changed.

parsec One parallax second (abbreviated pc), a standard measure of astronomical distance. The distance at which a star would have a parallax of one second of arc if viewed from two points separated by one astronomical unit. About 3.26 light-years.

peptide bond Bond found between amino acids producing the structure in all proteins. Formed by removing a hydrogen atom from the amino group of the first amino acid and a hydroxyl group from the other to produce a separate molecule of water and to join the two amino acids.

period The time required to complete one full cycle or action of a repetitive event (T) as in 24 h, the period of the earth's rotation with respect to the sun. The reciprocal of frequency ($T = 1/f$).

period (chemical) One set of elements of increasing atomic number occurring before an element is encountered with extreme similarity to one listed. One row in the periodic table.

period (of geologic time) The largest subdivision of an era of geologic time.

periodic law In any listing of elements by increasing atomic number, a specific set of chemical behaviors tends to repeat in another element at some distance down the list, one period later.

periodic waves Any wave disturbance that repeats exactly in space and/or time—a repeating wave pattern.

perpendicular At right angles, or 90°, to.

pH A number describing whether a solution is acid, neutral, or basic (alkaline). A pH of 7.0 is neutral, with lower numbers increasingly acidic and higher numbers increasingly alkaline.

phase The position represented within a periodic or repeated cycle, as with the moon in the full-moon phase, or a wave at crest.

phlogiston A hypothetical "fire element" used in the past to explain the behavior of burning. In fires, phlogiston was supposedly liberated. Replaced by the oxygen theory of burning.

phosphorescence The absorption of energy by a material and the later radiation of this energy as visible light.

photoelectric effect The ejection of electrons from a metal by the action of light falling on the metal.

photon A quantum or "package" of light, having a fixed energy proportional to its frequency.

photovoltaic source (solar cell) A device that converts some of the energy of incoming light to a voltage difference between two electric connections.

plate (crustal) A large slab of the earth's crust which moves on the surface as a unit. There are six major crustal plates with separate motions and many minor ones.

plate tectonics The interpretation of the deformations of the earth in terms of the independent motions of the plates of the earth's surface.

point source (of waves) A source which disturbs the medium carrying the wave at a single point, generating waves which proceed away from the source in every direction in circular or spherical wave fronts.

polar molecule A molecule which is electrically neutral overall but which has zones with a partial excess or deficiency of electrons causing charged zones on the molecule.

polarization A property of transverse waves allowing the medium to vibrate in a particular plane only, say up and down for a wave traveling sideways.

potential energy Energy stored in the situation or condition of matter, as in a raised weight or compressed spring, or bottle of nitroglycerin. For raised weights, $E_p = mgh$.

power The rate at which work is done (or the rate at which energy is transformed). Measured in watts, or joules per second.

pressure The force acting divided by the area over which it is applied.

propagation The process of passing a disturbance from place to place through a material, as in the propagation of waves.

proper motion The part of the motion of a star perpendicular to the line of sight of the observer which shows up as a motion against the more distant background of stars.

protein A joining of amino acid molecules, frequently dozens or hundreds, by peptide bonds to form long molecular chains.

proton Subatomic particle found in the nucleus of the atom, having a mass of nearly one atomic mass unit and with positive charge.

pulsar A star varying in brightness very, very rapidly due to its rapid rotation, producing flashes similar to those of a rotating light on a police car, but much more frequent.

Pythagorean theorem For any triangle with a right angle (90°), the length of one side making up the right angle squared plus the length of the other side making up the right angle squared equals the length of the side across from the right angle squared: $A^2 + B^2 = C^2$.

quantum The smallest "chunk" or "package" in which radiant energy is transferred, with energy proportional to the frequency of the radiation.

quantum hypothesis The assertion that radiant energy does not come in continuous streams but is transferred in "packages" (quanta) of energy, $E = hf$, where h is called the Planck constant and f is the frequency.

quantum number A number describing in which of several possible states a particular particle is; for example, which shell or subshell an electron is in within an atom.

quark One of the hypothetical subparticles thought to make up subatomic particles like electrons or protons.

quasar A "quasistellar" source of light appearing quite starlike on a photographic plate but incredibly far away and thus radiating much more energy than entire nearby galaxies.

rad A unit of radiation exposure equal to one-hundredth of a joule of energy released per kilogram of substance.

radial motion (of stars) The part of the motion of a star directly toward or away from the observer. Indicated by a shift in the spectrum of the source toward the blue for approach, or toward the red for recession.

radiation (1) Anything that is radiated; especially massless, chargeless rays such as light, x-rays, etc., or high-speed particles ejected by a nucleus. (2) A process for the transfer of energy from place to place usually via electromagnetic waves, particularly infrared rays for heating effects.

radical Two or more atoms attached in a particular way which enter into chemical reactions as a group and require other materials to attach to in order to form a stable compound; e.g., nitrate radical (NO_3^-), ammonium ion (NH_4^+), etc.

radioactivity The spontaneous emission of energy and high-speed particles from an atomic nucleus.

rarefaction That portion of a sound wave during which the pressure of the particles falls below normal air pressure.

reactant Any chemical substance entering into a chemical reaction.

real image An image formed by reassembling the light rays from each point on an object into another point in space. If a screen is placed at this image position, the reassembled rays will produce a visual image on the screen, as with projectors.

red giant A star which is at low temperature (red) but is very bright and thus must have a very large size.

red shift The distortion of a spectrum by shifting the light emitted toward the red end of the spectrum caused by the rapid motion of the source away from the observer.

reduction (1) Separating oxygen from an oxide to produce the metal. (2) Any reaction tendency to lower the positive charge on a particle, as in $Fe^{2+} + 2e^- \rightarrow Fe^0$.

reflection The "bouncing" of a wave or particle from the surface to a new material.

reflector A telescope using a mirror as its largest (perhaps only) optical element to form the image.

refraction The bending of the direction of a wave's motion as it encounters the boundary to a medium in which it travels at a different speed.

refractor A telescope using a lens (or set of lenses) as the largest optical element to form its image.

relative biologic efficiency A rating of a type of radiation in terms of its efficiency in producing biologic damage relative to electromagnetic radiations such as x-rays or γ rays.

relativistic Calculated according to the theory of relativity, as in relativistic mass, which increases with speed.

rem A unit of delivered radiation dose equal to the rad times the relative biologic efficiency of the type of radiation—x-rays, β particles, neutrons, etc. One rem of x-rays releases $\frac{1}{100}$ J per kilogram of body exposed.

resistance Any device requiring that potential energy be converted (work be done) in order for electric charges to pass through it. Measured in ohms or volts per ampere.

resonance A condition of constructive interference with waves which occurs when the energy input is timed precisely to the natural vibrational frequency of the substance, allowing the vibration to grow in time to a large size.

resonance (chemical) The rapid shifting of a chemical bond on one atom back and forth between two other atoms.

rest mass The mass of an object when at rest relative to the observer.

retrograde motion Any apparent westward motion of an object in the sky with respect to the stars.

reversibility The general tendency of a light ray to follow the same path backward through an optical system if its direction is reversed.

reversible reaction A chemical reaction capable of proceeding not only in the direction written but in the backward direction as well (products react to yield the original reactants). In principle, all reactions are reversible.

right ascension The east-west position in the sky of an object measured along the celestial equator proceeding eastward from the vernal equinox point, reported in hours of rotation required to bring it to the position of the vernal equinox.

scalar A quantity possessing size only (but not direction), such as work, energy, and so on.

sediment Any material that is deposited as the result of settling out from the fluid which transported it, as air-deposited sand dunes or mud layers in a delta.

sedimentary rock Any rock formed from materials which were originally deposited in layers by the action of a fluid; as in the formation of shale from clay.

seismograph A device to record shock waves transmitted through the rock structure of the earth.

semiconductor A material offering more electrical resistance than a typical conductor but much less than a typical insulator.

shell (or orbit) The distinct zone of an atom in which one or more electrons are allowed to be present.

sial Any of a group of igneous rocks rich in relatively low density and light-colored minerals, notably those of silicon and aluminum.

sidereal Anything pertaining to or with reference to the stars. Thus, the sidereal day of $23^h56^m4^s$ is the time the stars appear to rotate once around the earth.

silica The basic mineral combination of silicon and oxygen, formed of a silicon atom bonded to four oxygen atoms arranged evenly in space, each of which is also attached to a similar silicon atom in another such unit.

sima Any of a group of igneous rocks rich in dense and dark minerals, notably those of silicon and magnesium.

smog Originally a mixture of smoke and fog. A thick mist formed by the action of sunlight on pollutants in the air, particularly hydrocarbons and oxides of nitrogen.

soil A complex mass of weathered rock, plant debris, and other fragments occurring at the earth's surface.

solubility A number describing the ability of a substance to dissolve in water.

specific heat A number comparing the heating and cooling of a substance to the behavior of liquid water, which requires 1.0 cal/g to increase its temperature 1°C.

spectral type A classification system for stars based on the appearance of their spectra, and essentially a measure of stellar mass.

spectrum A range of frequencies, especially that represented in the light given off by some particular source, produced by passing light through a prism or using a diffraction grading of very closely spaced slits.

standing wave A pattern that results when a wave and its reflection cross over each other to produce fixed positions where the two cancel at all times, called nodes, and between which the waves constructively interfere to produce a large disturbance.

state A specifiable condition of a substance or a system, as in the *states of matter*—usually listed as solid, liquid, or gas.

static electricity Electrical situation in which charges have no sustained and systematic motion, and while they may momentarily rearrange themselves, are essentially at rest.

stationary states The set of allowed, stable conditions (orbits) of an electron in an atom.

STP Standard conditions for chemical description consisting of 1 atmosphere (atm) of pressure and a temperature of 0°C.

subduction zone A region of the earth's crust where two plates meet and one is pulled under the surface into the mantle and to its destruction. Frequently masked by ocean trenches.

supernova The explosion of a star in a gigantic burst of energy, blowing the outer layers of the star away into space and compressing its core into a neutron star.

syncline A down-folded ripple in the crust of the earth.

synodic period The time required between successive "meetings" or lineups of astronomical objects; for example, 780 days between events for the sun, earth, and Mars to be in a line with the earth in the middle.

synthesis The "putting together" or building of a structure from diverse parts.

tectonics The study of the structural deformation of the earth.

temperature inversion A relatively unusual situation in which the temperature of the air does not decrease with altitude but may increase to some boundary with colder upper air. This condition tends to trap emissions in a pocket of air near the ground, resulting in levels potentially dangerous to human health.

tension A situation with forces acting to pull the molecules of a material apart.

terminal velocity The maximum velocity that can be attained by a particular mass driven by a given force through a resisting material, as a falling object through air.

thermodynamics The science of thermal energy and its transfer.

total internal reflection The reflection of light from the inside surface of a transparent material with a greater index of refraction than the material surrounding it. This occurs when the ray strikes the surface at an angle greater than the critical angle, and is used in prism binoculars, camera viewfinders, etc., to preserve all of the light for the image.

transform fault A fault or crack in the earth's crust along which a sideways, horizontal motion of the crust on either side of the fault occurs.

transistor A device using the quantum properties of atoms in a regularly arranged solid to control the direction and size of electric currents.

translucent Allowing light to pass through but with sufficient distortion so that a clear image of what is on the other side of a given material is not provided, as with frosted glass.

transmit To pass along from place to place, as in the transmission of a wave.

transmutation The changing of one element into another.

transparent Allowing a wave, especially light, to pass through with little or no randomization of its direction, thus allowing a clear image to be formed, as with a windowpane.

transverse wave A type of wave passed along a material by a motion of the material at right angles to the direction the wave is traveling, as with a hump shaken into a taut clothesline.

trough The position in a repeating wave with the greatest negative displacement from the rest condition of the medium carrying the wave. With sound waves, the zones with the lowest air pressure (rarefactions); with water waves, the lowest zones of the water surface.

umbra The deepest part of a shadow where all direct illumination is blocked.

uniform circular motion Motion at constant speed in a perfect circle of constant radius.

uniformitarianism The belief that geologic processes are essentially constant in time, and that the same processes which can be seen working today are the ones which have shaped the earth's crust through time.

valence A number describing the combining tendency of an element (or radical) in reacting with other substances.

valence band The semicontinuous set of electron states in a crystal separated by very fine energy differences, which results as the electron orbits in the valence shell of individual atoms "merge" as the atoms are packed together in the crystal.

valence shell The outermost electron shell of an atom, the only portion of the atom normally contacting the outside world and thus largely responsible for its chemical behavior.

vector Any quantity having both a size and a direction, such as force, velocity, etc.

velocity A measure of the rate of motion, the size of which is called speed, and having a particular direction ($v = \Delta d/\Delta t$).

vernal equinox point The intersection point in the sky of the celestial equator and the path of the sun (ecliptic) as the sun ascends from the southern sky to the northern half. The sun occupies this point on or about March 21, marking the end of winter and the beginning of spring. The starting point for the east-west coordinates of the sky in astronomy.

vertical component That portion of a vector acting in a pure up-and-down direction.

vibration A motion of a substance when disturbed back and forth from its normal position of rest.

virtual image An image wherein light rays appear to be coming from a particular position in space but in actuality may never go near that location, as with the image in a bathroom mirror. Such an image may never be captured on a screen.

visible light The range of wavelengths of electromagnetic radiation detected by the human eye, ranging from about 3.5 to 7.8×10^{-8} m.

vitamin A particular chemical structure needed by the body but not used in the body as a source of energy. Usually not produced in the body and required in trace amounts in foods.

volatile Any material which is easily vaporized.

volt Metric unit (abbreviated V) of electric potential difference between two points equal to the work, in joules, required to move one coulomb of charge from one point to the other: 1.0 V = 1.0 J/C.

voltage A measure of electric potential energy difference per unit of charge between two points of space, measured in volts, or joules per coulomb. For example, if a battery is rated at 12 V, this implies that each coulomb of charge delivered by the battery can do 12 J of work in flowing around a circuit including the battery.

watt The basic metric unit (abbreviated W) of power equivalent to one joule of work per second.

white dwarf A star which is quite hot (white) but not very bright and thus must be very small, perhaps earth-sized.

work A measure of accomplishment given by the product of the force exerted times the distance the object is moved in the direction of the force. Measured in joules, or newton-meters.

worst-case analysis Calculating the consequences of an accident based on the worst possible sequence of events to increase its seriousness.

zenith The point in the sky directly overhead for any observer.

zodiac A band of 12 constellations in the sky through which the ecliptic passes and thus in which the sun, the moon, and the planets are always found.

Index

Abberation:
  of lenses, 310
  of starlight, 136–138
    illustrated, 137
Absolute magnitude (M), 559, 563
Absolute zero, 204
Abyssal plains, 450
Acceleration:
  definition of, 59
  due to gravity (g), 59–60
  of the earth toward the sun, 161
  empirical measurement of, 42–45
  of the moon, 161
  motion when constant, 70–72
  as slope of v-vs.-t graph, 67
Accelerator, atomic, 496–497
Acid, 374–377
Acid group (carboxyl group), 389–390
Acidity:
  and plants, 376
  of rain (natural), 422
Activation energy, 355
Activity, chemical, 342–346
Activity series, 343–344

Adams, John Couch, 164–166
Age:
  of earth, 434–435
  of rocks, 429–430, 435
Air, 415
  and cosmic rays, 444
  urban vs. rural, 585
Air pollution, 584–587
Alchemy, 328–329
Alcohol, 389
  as fuel, 594–595
Alexandria, 15, 16, 98
Alkali metals, 362–363
Alkalis (bases), 347, 374–377
Alloy, 328
Almagest (Ptolemy), 110
Alpha Centauri, 558–560, 563
Alpha helix, 410
Alpha particle:
  discovery of, 489
  and discovery of nucleus, 490–491
  emission of, 493–494
  nature of, 490
  properties of, 489

Altitude of object in sky, 81, 83
Amino acids, 401
  essential, 402
    table of, 403
Amino group, 390
Ammonia, 428–429
Ampere (unit of current), 276
Amplification, 543
Amplitude, 238
Analemma, 85
Analysis, chemical, 332
Andromeda galaxy (M31), 574–575
Angular momentum, 157–159
  of electron, 524
Angular speed, 87
Annihilation of matter, 506
Annular eclipse, 100–101
  illustrated, 101
Antimatter, 544
Antiparticle, 544
Apogee, 184
Apparent magnitude (*m*), 559
Aquifer, 476
Archimedes, 11–13
Archimedes' principle, 12–13
Arctic Circle, 88–89, 96
Arcturus, 562
Aristarchus, 9–11
Aristotelian System, 13, 17–23
Aristotle, 8, 13, 17–23, 264
  vs. Galileo, 55–56
Arrhenius, Svante, 368*n*.
Aspirin, 390–391
Asteroids, 554
  motion of, illustrated, 103
Astrology, origins of, 5–6
Astronomical observatory, 125
  illustrated, 123, 124
Astronomical unit (AU), 558
Astronomy, revolution in (*see* Copernican revolution)
Atmosphere:
  air pressure, 294
  transparency, 568
ATP (adenosine triphosphate) molecules, 275
  illustrated, 398
"Atom smashers," 496*n*.
Atomic mass number (A), 360, 493
Atomic mass unit (u), 505
Atomic number (*Z*), 360
  and nuclear charge, 492*n*., 493
Atomic symbol, 335
Atomic theory:
  ancient, 329
  modern, 334
Atomic weights, determination of, 335–336
Atoms:
  combining tendencies of, 355
  structure of, 359–361
Avogadro, Amadeo, 336
Avogadro hypothesis, 336
Avogadro number, 339
Azimuth, 83

Babylon, science of, 5, 81, 101
Bacteria, action of, 395, 397
Balmer series of spectral lines, 523–524, 526
Barnard's Star, 560–561
Basalt, 419, 420
Bases (alkalis), 347, 374–377
Bases, genetic, 409–411
Batholith, 472–473
Battery, electrical, 275–277
Becher, Joachim, 329
Becquerel, Henri, 484
Beer, 512–513
Benzene:
  bond lengths, 387–388
  molecular orbitals, 539–540
  resonance of, 388
  structure of, 387
Bessel, Friedrich, 138, 558, 560
Beta decay of uranium, 508
Beta particle:
  discovery of, 489
  emission of, 494
  properties of, 489
Big bang, 575–576
Binary star system, 560
Binding energy, 506–507
Black, Joseph, 202, 329, 330
Black Hills of South Dakota, 472, 473
Black holes, 568
Blood, human, 404
Bohr, Niels, 524–530, 532–533
Bond, chemical, 335
    covalent, 369–372
    hydrogen, 373–374
    ionic, 366–370
    pi orbital, 539
    relative strengths of, 373–374
    sigma orbital, 539
Bonding:
  atomic separation in, 374
  of carbon atoms, 387–388
  chemical, simple model, 355–356
  molecular orbitals for, 538–540
  of silicon atoms, 417–418
Bones, 176–177
Bonneville, Lake, 442–443
Boundary, wave behavior at, 234–236
Boyle, Robert, 203
Bradley, James, 136–138
Brahe, Tycho, 122–128
  biography of, 125–128
  illustrations of, 123, 125
Broglie, Prince Louis Victor de, 530
Brownian motion, 498
Bubble chamber, 513
Buckland, William, 434
Buoyancy, 11

Calendar, 98–99
Caloric, 200
Caloric theory, 199–202

Calorie (dietary), 197*n*.
table of, for fast foods, 198
Calorie (scientific):
definition of, 197
and work in humans, example of, 207
Cambrian Period, 435, 436
Canadian Shield, 436
Cancer, radiation and, 602, 604, 605
Canopus, 560
Carbohydrates, 392–399
Carbon:
atomic structure of, 369–370
combining tendency of, 369–370, 381–383
simple compounds, table of, 382
Carbon dioxide:
and carbon monoxide, 334, 340
discovery of, 329
and greenhouse effect, 596
Carbon-14 dating, 443–444
Carboniferous Period, 435, 437
Carboxyl (acid) group, 389–390
Carnot, Sadi, 215
Cartilage:
and friction of joints, table of, 186
and joints, 176
Catalyst, 377–378
automobile exhaust, 586
Catastrophism, 433
Cathode-ray tube, 357–359
and x-rays, 483
Cavendish, Henry, 167, 329, 330, 333, 349, 356
Cavendish Laboratory, 537–538
Celestial coordinate system, 87
Celestial equator (CE), 85–90, 93
determining position of, 90
from latitude, 89
on star map, 95
Cenozoic Era, 435–436
Centrifugal force, 145–147
Centripetal force, 145–147
Cepheid variables, 571–572
Chadwick, Sir James, 443, 508
Chain reaction, 509
Change of state, 196
Change of variables method in graphing, 37–45
steps in, 38–39
Charge, electrical (*see* Electrical charge)
Chemical bonding, simple model, 355–356
(*See also* Bond)
Chemical composition of the earth, 415–417
Chemical equations, 336–338
Chemical reaction:
decomposition, 346
displacement, 344–345
and energy, 364–366
rate of, 377–378
Chicago, 438, 476–477, 497, 511
Chile saltpeter, 422, 428
Chlorophyll, illustrated, 405
Cholesterol, 400
Chromosomes, 409
Cinder cones, 468
Circular motion, 159–160
acceleration in, 159
speed in, 159
Circular wave front, 229
Classical science, end of, 481–483
Clocks, 84–85
radioactive, 494
Clothing, heat and, 190–194
Cloud chamber, 495, 513
Clusters of stars, 559*n*., 564–566
distribution in space, 572–573
Coal:
formation of, 437
problems in use of, 586
use in the United States, 589
Cockcroft, John, 496–497
Collisions, 156–157
Colorado Plateau, 464
Combining volumes of gases, 336–337
Combustion in automobile, 385–386
Compound, chemical, 332
Compression of wave, 231, 245
Condensation, chemical (*see* Dehydration)
Conduction:
electrical charging by, 267
of heat, 190–194
Conduction band, 541, 542
Conjunction, planetary, 118
Conservation of energy, 180–184, 216–217
Conservation law, 156
Constellations, 91, 94
on star map, 95
Continental drift, 458, 463
best fit of continents, illustrated, 466
rate of, 462
Continents:
with change in sea level, 449–450
through geological time, 437
as plateaus, 449–450
shape of, 437
Convection, 190–192
Copernican revolution, 78, 110
significance of, 122
Copernican System, 110, 117–122, 143
assumptions of, 119
problems of, 122
and retrograde motion, 120–121
significance of, 122
Copernicus, Nicolaus, 118
illustration of, 117
Core of the earth, 452
Coriolis force, 148–149
Correspondence principle, 524
Cosmic rays, 443, 513, 602, 603
Coulomb, Charles, 269
Coulomb (unit of charge), 270
Covalent bonding, 369–372
Crab Nebula, 567–568

Crater Lake, 469
Crick, Francis, 410
Critical angle, 306
Crookes, William, 357
Crust of the earth, 416
  thickness of, 452, 454–455
Crystal, 418
Curie, Irene, 486, 487
Curie, Marie, 484–488
  biography of, 486–487
Curie, Pierre, 484–486, 488
Current, electric, 276
    and magnetic field, 282
    from magnetic field, 285–288
Cuvier, Georges, 432–434

Dalton, John, 334
Dating, radiocarbon, 443–444
Dating of rocks, 429–430
  geological eras, table of, 435
  modern dates, 435
  relative, 430–434
Davy, Sir Humphrey, 262–263, 345*n*., 356
Day:
  length of, 84–85
  length of daylight period, 97–98
  sidereal, 93
Declination, illustrated, 86
Decomposition reaction, 346
Deferent (in Ptolemaic System), 114
Definite proportions, law of, 333–334
Dehydration:
  in carbohydrates, 394–395
  in fats, 399
  in proteins, 401
Democritus, 9, 328–329
Density, 11–13, 17
  of the earth, 416–417
  of rock types, 419
Diabetes, 411
Diffraction:
  of light, 316
  patterns of diffracted rays, illustrated, 407
  and protein structure, 406–407
Diffusion, 209
Dikes, 472
Dinosaurs:
  age of, 436
  discovery of, 434
Dirac, Paul, 544
Disease, 395–397
  ancient view of, 20–21
  and population, 582–584
Dispersion, 307
Displacement reaction, 344–345
Distance:
  object and image, 300
  of stars, 558–559
  with uniform acceleration, 70–73
  from *v*–vs.–*t* curve, 68, 69
DNA (deoxyribonucleic acid):
  components of, 409–410
DNA (deoxyribonucleic acid) *(Cont.)*:
  discovery of, 409
  structure of, 410–411
  variation in, 410
Doppler shift:
  of stars, 561–562
  of waves, 252

Ear, human, hearing and, 245
Earth:
  chemical composition of, 415–417
  crust of, 416
  density of, 416–418
  internal structure of, 452–453
  orbital data, 550
  radiation balance of, 596
  shape of, 9
  size of, 11, 16–17
Earthquake waves, 220*n*., 241*n*., 450–453
Eccentric (in Ptolemaic System), illustrated, 116
Eclipses, 100–102
  first prediction of, 8
  illustrated, 101
Ecliptic, definition of, 93
Efficiency:
  of engines, 215
  of machines, 173
Egyptian science, 2–5, 98
Einstein, Albert, 27, 498, 521–523, 576
Electric charge, 265, 358
  force on, in magnetic field, 285–286
  unit of, 269–270
Electric circuit, 275
Electric field, 271–272
Electric motor, 283–284
Electricity:
  dynamic or current, 273–281
  magnetic effects of, 281–288
  static, 264–273
Electrochemical cells, 276–277
Electrodes, 345–346, 356, 358
Electrolysis, 357
Electromagnetic induction, 285
Electromagnetic radiation, x-rays, 483–484
Electromagnetic spectrum, 323
Electromagnetic waves, 320–322
Electromagnetism, 263
Electron:
  allowed orbits in hydrogen, 524–527
  and alpha particles, 491
  angular momentum of, 524
  arrangement in atoms, 360–363, 490
  change of orbit, 525–527
  discovery of, 358–359
  emission from surface by light, 521–523
  energy of orbits, illustrated, 526
  ground state, 535
  mobility in metals, 541
  motion in atoms, 517–518
  orbitals (suborbits), 534–540
  stationary states of, 525

Electron *(Cont.)*:
wave properties of, 530–532
wavelength of, 530–531
Electron-cloud model of atom, 533–534
Electron spin, 527–528
Electronvolt (eV), 496
Electroscope, 266–267
Elements:
ancient, 17–19
dates of discovery of, 332, 333
electrical separation of, 345–346
periodic law of, 348
Elevation of earth's surface, illustrated, 450
Ellipse, 129–130
illustrated, 130
Empirical laws, Kepler's 132–133
Empirical process, 25–26
Energy:
of annihilation, 506
binding, of nucleii, 506–507
in chemical reactions, 364–366
cost of, 591
crisis in, 588–592
definition of, 175
of earthquake, 453*n*.
efficiency in transportation, 593
of fission, 506–507
of fuels, 221
of fusion, 506–507
and GNP, 590, 591
growth in use of, 588–590
of hydrocarbons, 386
and mass, 504–505
of nuclear explosion, 453*n*.
of nuclear reactions, 505–507
in photosynthesis, 392
production of, in body cells, 395, 398–399
radiated by matter, 518–520
relativistic, 504–505
residential use of, table of, 599
solar (*see* Solar energy)
of spreading wave, 228–230
in stars, 565–568
use of, in the United States, 591–592
Enrichment of uranium, 511
Entropy, 218–219
Enzymes, 397, 398
Epicycles, 112–116
"do-it-yourself guide to," 112–113
Equal-area law, 131–132
Equant (in Ptolemaic System), illustrated, 116
Equation of straight line, 33–36
Equation of time, 84
Equator, sun's motions at, 79, 82
Equatorial bulge, 162–163
Equilibrium, chemical, 375–377
Equinox, 94–96
precession of, 102–103
Eratosthenes, 11, 16–17
Erie, Lake, 442, 443
Erosion, 420–422
by glaciers, 438–440
Erosion *(Cont.)*:
of Niagara Falls, 443
and soil production, 426–427
and weathering, 424–429
Eruptions, 468–472
Essential amino acids, 402
illustrated, 403
Evaporation of perspiration, 192–193, 201
Excess deaths, 585, 586, 609
Exchange particles, 544
Exclusion principle, 535
Expanding universe, 575–576
Expansion with heating, 187–189
interpretation of, 200
Explosives, 428
Extrapolation, 204
Eyes, defenses of, 397

Falling objects, analysis of, 42–45
Family, chemical, 348–349
Famine, population and, 584
Faraday, Michael:
biography of, 262–263
and electrolysis, 356
and electromagnetism, 285
and magnetic fields, 260–261
Faraday (unit of charge), 357
Fats, 393, 399–400
Faults, 473–475
Lewis overthrust, 475
San Andreas, 451, 470, 473
Teton, 474
thrust, 473–475
transform, 459–460
Fertilizers, chemical, 428–429
Fiber optics, 306
Fictitious forces, 146, 148
Field:
electric, 271–272
magnetic (*see* Magnetic field)
Fission, nuclear, 506–512
Fizeau, H., 310, 322
Fluorescence with x-rays, 483
Focal length:
of curved mirror, 299
of lens, 308
Folding, 473, 476–478
Food:
adequacy of, 402
component of, 393
Forces(s), 145–163
and acceleration, 151
ancient ideas of, 19
centrifugal, 145–147
centripetal, 145–147
on charge due to magnetic field, 285–286
electrical, 285–286
fictitious, 146, 148
gravitational, 160–163
human, 176–177
between magnets, 258–260
and mass, 149

Force(s) *(Cont.)*:
  nuclear, 495
  and orbits, 146–147
  in rocks, 461, 478
  sets of, 154–155
  unit of (Newton), 151
Formula, chemical, 336–340
Fossils, 432–437
  dinosaurs, 434, 436
  evidence for continental drift, 463–466
  by geologic era, 435, 437
  index, 433–434
Foucault, Jean, 138, 320, 322
Fraunhofer lines, 312, 313
Frequency:
  threshold, 522
  of waves, 242
Frequency range of sound, 245–246
Fresnel, Auguste, 318
Friction, 54, 175*n*., 179–180
  coefficient of, table, 186
Fusion reaction in stars, 565–567
Fusion reactors, 506, 507, 610

Galaxies, 571–576
Galileo Galilei:
  and air pressure, 294
  and change in science, 318
  and the Copernican System, 133–135
  Dialogue on the Great World Systems, 135
  experiments with ramps, 25, 53–56
  on falling objects, 25, 42
  moons of Jupiter, 133, 554
  on motion, 53–56
  and Newton, 144
  and pendulum, 84
  phases of Venus, 134
  principle of inertia, 145
  and relative motions, 499
  on role of mathematics, 27
  The Starry Messenger, 133–135
  and sunspots, 134
  and thermometer, 187
  trial and condemnation of, 135
Galvani, Luigi, 274
Gamma ray:
  discovery of, 489
  properties of, 489–490
Ganges basin, 453, 454
Gases, behavior of, 202–204
Gay-Lussac, Joseph, 336
Generator, electric, 287–288
Genetic engineering, 411–412
Genetic material, 408–412
  DNA structure of, 409–411
  "lost patterns," 408–409
Geologic time, table of, 435
Geomagnetic reversal, 264, 460
Geometry, 26
Geopport-Mayer, Maria, 488
Geosyncline, 457–458
Gilbert, William, 258
Glacial lakes, 441, 445
Glacier National Park, 475
Glaciers, 437
  action of, 438–440
  dating of, 445–446
  effect on sea level, 438
  effect of weight, 443–445, 456
  evidence for, 438
  and the Great Lakes, 441, 443, 445
  maximum extent, 445
  motion of, 439
  mountain, illustrated, 440, 441
  thickness of, 437
Glasser, Donald, 513
Glucose:
  chemical breakdown in body, 395, 398, 399
  structure of, illustrated, 393
  use in body, 394
Gnomon, 81–86
Gold-foil experiment, 490–492
Goodyear, Charles, 52
Graben, 3
  illustrated, 4, 473
Grand Canyon, 429, 430, 464–465
Grand Teton Mountains, 473, 474
Granite, 419
  weathering of, 421–424
Graphing:
  axes, 32
  change of variables, 37–45
  intercept, 34
  origin of, 32
  scales, 32
  slope, 33
  slope-intercept form, 33–35
  variables, 31–32
Gravitation:
  effect of mountains on, 454–455
  law of, 160–163
  and stars, 564–568
Great Lakes, 440–445
Greek science:
  inadequacies of, 23–24
  origins of, 6–7
  periods of, 13
  problems of, 13–15
  time chart, 14
Greenhouse effect, 596
Grounding, electrical, 273

Haber, Fritz, 428–429
Haber process, 428–429
Hahn, Otto, 488, 508, 509
Half-life, 485–486
  of C 14, 444
  of radium, 485
  of uranium, 486
Halley, Sir Edmund, 434–435
Halley's Comet, illustrated, 165
Hanford, Washington, 511
Harmonic law of orbits, 132–133, 143
Harmonics, 247

Hawaii, 467–469
Hearing, range of, 245
Heat, 187
  of fusion, 197
  mechanical equivalent of, 205–206
  and motion, 205–212
  of reaction, 365
  vs. temperature, 194–196
  transfer of, 190–194
  unit of (Kcal), 196
  of vaporization, 197
Heat engines, 211–215
Heavy water, 361
Heisenberg, Werner, 532–533
Heliocentric (*see* Sun-centered universe)
Helium, discovery of, 313
Heme group, 404
  illustrated, 405
Hemoglobin, 404–408
Henry, Joseph, 285
Herschel, Sir William, 163–164, 310, 571
Hertz, Heinrich, 322
Hertzsprung-Russell (HR) diagram, 563–564
Himalayan mountain chain, 453–455
Hipparchus, 9, 11, 102, 110, 113*n*.
Hiroshima, 512
Horsepower, 213*n*.
Horst, 473
Hubble's constant H, 575
Human body:
  conversion of energy by, 207
  motion of, 176–177
  temperature of, 191–193
Human life span, 611
Hutton, James, 433
Huygens, Christian, 313–314, 320
Hydrocarbons:
  alkane series, 383–384
  aromatic, 387–388
  combustion of, 385–386
  energy content of, 386
  fuel value of, 386
  as pollutants, 585–587
  saturated, 383–384
  separation of, 385
Hydrogen:
  discovery of, 329
  energy in fusion, 506–507
  as fuel, 595–596
  isotopes of, 361–362
Hydrogen bonding, 373–374
Hydrogenization of fats, 400
Hydrolysis:
  in carbohydrates, 394
  in fats, 399
  in proteins, 401
Hydrophilic compounds, 389
Hydrophobic compounds, 389

Ice sheets (*see* Glaciers)
Ideal case, 54, 56
Ideal-gas law, 205
Igneous activity, 466–473
Igneous rocks, 419–421
Illinois Basin, 476–477
Image:
  in curved mirror, 299–303
  from lenses, 308–310
  in plane mirror, 297–298
  real, 298
  virtual, 298
Incidence, angle of, 237–238, 296, 303
Inclined plane, 171, 172
Index of refraction, 304
  and speed of light, 315
Index fossils, 433–434
Induction, electrical charging by, 268
Inert gases, 349
Inertia:
  of air masses and weather, 148–149
  definition of, 145
  and orbits, 146–147
  principle of, 54, 145–146
Infrared radiation, 310, 568
Inorganic chemistry, 381
Insolation, 97
Insulator:
  electric, 541
  of heat, 194
Insulin, 411–412
Integrated circuit (IC), 543
Intensity:
  of solar radiation, 597
  of thermal radiation, 518–519
Interference:
  of light, 315–318
  of waves, 239–241
Interferometer, 501
Interferon, 412
Internal combustion engine, 214
Ionia, 8
Ionic bonding, 366–370
Ions, 366–370
Isolated system, 182–183
Isomers, 384
Isostasy, 455–456
Isotopes:
  and atomic weight, 493
  definition of, 360
  of hydrogen, 361–362

Jansky, Karl, 569
Joints, human, 176
    friction of, 186
Joule, James Prescott, 203, 205–208
Joule (unit of work or energy), 172
Jupiter:
  apparent motion of, 106
  atmosphere, 555
  data table, 550
  magnetic field, 555–556
  moons, 554–557
  ring, 557

Kekulé, Friedrich, 388
Kelvin scale (of temperature), 204
Kepler, Johannes, 125–133
  biography of, 126–128
  laws of astronomy, 128–133, 143
Kilauea, 467, 469
Kilocalorie, 196–197
Kilogram atom (kilomole), 338
Kinematics, definition of, 53
Kinetic energy, 178–180
  and potential energy, 180–184
Kinetic theory of matter, 211
Krakatoa, 468, 469
Krebs cycle, 400

Laboratories in science, 537–538
Laccolith, 472
Laser, 542, 543
Latitude:
  from height of North Star, 89
  from rising of sun, 79
Lava flows, 467
Lava volcanoes, 468–469
Lavoisier, Antoine Laurent:
  biography of, 330–331
  chemical definitions, 332, 339
  and elements, 334
  and oxidation, 333
Leaching, 421–422
Leavitt, Henrietta, 571
Le Chatelier's principle, 375
LED (light-emitting diode), 542
Length, contraction of, 502, 503
Lenses, 308–310
Lever, 176–177
Leverrier, Urbain, 164–166
Lewis overthrust, 475
Leyden jar, 273
Libby, Willard, 444
Light:
  behavior of materials with, 295–296
  divergence from source, 297
  nature of, 293–294
  production of, 321–322
Light-years, 558
Limestone, 422, 423
Linear wave front, 229
Lines of force, 260
Long Island, New York, 477
Longitudinal waves, 231
Lunar eclipse, 100
  illustrated, 101
Lunar phases, 101–102
  illustrated, 102
Lyell, Charles, 433–434
Lysozyme, 397

Machine, definition of, 170
Magellanic Clouds, 571, 572, 574
Magma, 420
Magnetic field, 260–264
  from electric current, 282–285
  to produce a current, 285–288
Magnetic flux, 286
Magnetic materials, types of, 259
Magnetism, 258–260
  of ocean floor, 460
  source of, 283
Magnifying glass, 308–309
Magnitude of stars, 559
Main sequence, 563
Manhattan Project, 510
Mantle of the earth, 452
Mars:
  apparent motion of, 104–106
  atmosphere, 553
  canyon, 553
  color change, 551–552
  Ptolemaic interpretation of, 114–116
    illustrated, 115
  retrograde loop, 105
  synodic period, 105
  tests for life, 553–554
  variation in brightness, 106
  volcanoes, 552–553
Mass:
  and black holes, 568
  definition of, 54
  relativistic, 504
  of stars, 563, 564, 567
  subatomic, 505
  and supernova, 567–568
  vs. weight, 152–153
Mass defect, 505–506
Mathematics, role of, in science, 23, 26, 51–52
    Galileo on, 27
    da Vinci on, 52
Mauna Kea, 468
Mauna Loa, 469
Maximum credible nuclear accident (MCNA), 608, 609
Maxwell, James Clerk, 263, 320–322
Mazama, Mount, 469
Mechanical advantage (MA), definition of, 172
Mechanical energy, 182
Mechanistic determinism, 482
Medical practice, changes, 395–397
Meitner, Lise, 488, 508, 509
Mendel, Gregor, 409
Mendeleev, Dmitri, 347–349
Mercury:
  apparent motion of, 103
  data, 550
  problems with orbit, 166
  surface of, 549
    illustrated, 550
Meson, 544
Mesozoic Era, 435, 436, 458
Messier, Charles, 571
Metabolic rate of humans, 190
Metallurgy, 328
Metals:
  atomic structure, 367–368
  and conduction of heat, 193
  oxidation of, 342–344

Metals (Cont.):
  production from ores, 328
  properties of, 342
Metamorphic rocks, 420–421, 423
Metric system, 29
  prefixes, table of, 29
Michelson, Albert, 481, 501
Michelson-Morley experiment, 501–502
Michigan, Lake, 438, 445
Midoceanic ridges, 459
Milky Way galaxy:
  brightness of, 575
  and local group, 574
  motion in space, 571
  size and shape of, 572–574
Minerals, 417–419
Mirror:
  curved, 298–303
  plane, 297–298
  telescope, 303
Model (scientific), 356
  of the atom, 359–361
Mohorovičić discontinuity (moho), 452
Mole (gram molecular mass), 338
Molecular weight of hemoglobin, 404
Molecule, 334
Momentum:
  angular, 157–158
  conservation of, 155–158
  definition of, 155
Moon:
  age of, 549
  distance of, 9
  eclipses, 100–101
    illustrated, 101
  internal structure, 549
  landing on, 547, 548
  length of month, 98
  lunar calendar, 98–99
  motions of, 98–102
  nodes of orbit, 100–101
  phases of, 101–102
    illustrated, 102
  size and mass of, 547, 549
  solar month, 99
  surface of, 548
  synodic month, 99
Moons of Jupiter, discovery of, 133, 143
Moraine, glacial, 439
    illustrated, 440
Moseley, Henry Gwyn-Jeffreys, 492
Motion:
  ancient ideas of, 19–23
  astronomical, 110–117
  of the earth, confirmations of, 135–138
  energy of (*see* Kinetic energy)
  and force, 147–153
    first law of, 145–147
    second law of, 147–153
    third law of, 153–155
  "perfect," 22, 23, 111
  rotational, illustrated, 158
  of solar system in space, 562
  of stars: proper, 560–561
    radial, 561–562

Motion (Cont.):
  translational, illustrated, 158
  uniform circular, 159–160
  of water in wave, 231–232
Multiple proportions, law of, 335
Mysticism, 9, 127

Nagasaki, 512
Natural philosophy, 8
Neptune:
  data table, 550
  discovery of, 164–166
  and Pluto, 557
Nerve impulses, speed of, 529
Neutralization, chemical, 376–377
Neutron:
  and chain reaction, 508–509
  discovery of, 508
  and radiocarbon, 443–444
Neutron bombardment, 508
Neutron star, 567
Newton, Isaac, 25, 51, 143
  biography of, 144–145
  and dispersion, 307
  and geological dating, 434
  laws of motion by, 145–156
  and nature of light, 313–314
  and telescope, 164, 303
Newton (unit of force), 151
Niagara Falls, 443
Nitrogen fixation, 428–429
Nodes:
  of moon's orbit, 100
  regression of, 101
Nonmetals, chemical, 342
Nuclear fission, 506–512
Nuclear force, inferred, 495
Nuclear fusion, 506, 507
  in stars, 565–567
Nuclear fusion reactors, 610
Nuclear power, 601–610
Nuclear reactions in stars, 565–568
Nuclear reactors:
  costs of, 609–610
  and explosion, 606
  meltdown, 607
  relative safety of, 609
  safety systems in, 606–607
Nucleus, atomic, 360
    discovery of, 490–492

Oak Ridge, Tennessee, 511
Oersted, Hans, 282
Ohm, Georg Simon, 278
Ohm (unit of resistance), 278
Ohm's law, 278
Oil:
  from coal, 594
  effect on atmosphere, 597
  U. S. production of, 589
  use of, 589, 590
    in transportation, 592, 593
Olympus Mons (on Mars), 552–553

Operational definition, 57
Optics, law of, 301, 309
Orbitals (suborbits), 533–540
  and bond directions, 538–539
  illustrated, 534
  and magnetic properties, 536
  molecular, 539–540
  and periodic table, 535–536
Orbits:
  energy of, 183–184
  shape of, 129–131
Organic chemistry, 381–413
  carbohydrates, 392–399
  DNA (*see* DNA)
  hydrocarbons (*see* Hydrocarbons)
  lipids or fats, 393, 399–400
  proteins, 393, 400–408
Organic groups, 388–391
  table of, 390
Overtones, 247–248
Oxygen, 329
Oxygen theory of combustion, 331

Pain, relief of, 390–391
Paleozoic Era, 435, 436
Pangaea, 458
Paracelsus, 329
Parallax, 124, 136
  definition of, 10–11
  of stars, 558
    of Alpha Centauri, 558
    limit of, 558–559
    of Sirius, 558
Parallel cells, 277
Parsec (pc), 558
Particle physics, 544–545
Particles of rock, table of, 424
Particles of a system, 202
Particulates, 585–587
Pauli, Wolfgang, 535
Pendulum, energy of, 182–183
Penzias, A., 570
"Per," definition of, 57
Perigee, 184
Periodic law of elements, 348
Periodic waves, 241–244
Perturbation, 162
pH, table, 376, 377
Phlogiston, 329
Phosphorescence, 483
Photoelectric effect, 498, 521–523
Photography, 484*n*.
Photon, 522, 530
Photosynthesis, 392
Photovoltaic source, 541
Pi bond, 539
Pipes, waves in, 249–251
Pitchblende, 484–485
Planck, Max, 520
Planck's constant (*h*), 520
Planetary motion, equal-area law, 131–132
  illustrated, 131
Planets:
  apparent motions of, 102–106
  classification of, 554
  data table, 550
  modern view of, 549–557
Plate tectonics, 458–466
  continental drift, 458
  fossil evidence, 463, 465
  Gorda plate, 470–471
  major plates, illustrated, 462
  midoceanic ridges, 459
  plate boundaries, 461
  rate of motion, 462
  sea-floor spreading, 460–461
  subduction zones, 462
Pleistocene Epoch, 435, 437–440, 445
Plum-pudding model of atom, 490
Pluto:
  crossing Neptune's orbit, 557
  data table, 550
  discovery of, 165–166
  moon of, 557
  orbit, 557
Plutonium, 508, 511
Polar molecules, 373, 389
Polaris (North Star), 86–88
  illustrated, 92
Polarization, 318–319
Poles, magnetic, 259
Pollutants, air, 584–587
    standards for, table of, 586
Pollution deaths, 585, 609
Polonium:
  decay of, 494
  discovery of, 484
  half-life of, 485
  naming of, 486
Population, human, 582–584
Position, energy of (*see* Potential energy)
Positron, 443
Potential energy, 174–178
  and kinetic energy, 180–184
Power, definition of, 174
Precambrian Era, 435–436
Precession of the equinoxes, 102–103
Precision, astronomical, 122–124
Pressure:
  definition of, 187
  measurement of, 37
  and volume of gas, 37–38, 203–205
Priestley, J., 329–330
Principia (Newton), 118, 145
Problems in science of 1900s, 481–483
Propagation of waves, 230–232
Proper motion of stars, 560–561
Proportion:
  direct, 19, 35
  inverse, 37–38
Protein content of fast foods, 198–199
Proteins, 393, 400–408
  hemoglobin, 404–408
Proton, acceleration of, 496
Ptolemaic System, 110–117, 143
  assumptions of, 111–112

Ptolemaic System *(Cont.)*:
- difficulties with, 117
- epicycles, 112–116

Ptolemy, Claudius, 15, 110–117, 559
Pulley, 171
Pulsars, 567–568
Pythagoras, 9
Pythagorean theorem, 9

Quantization of variables, 520–521, 524–525, 527
Quantum (quanta), 520
- of angular momentum, 524
- of light (photon), 522
- of radiant energy, 520–521

Quantum hypothesis, 520
- experimental verification, 521–523

Quantum levels in solids, 540–543
Quantum mechanics, 518, 532
Quantum numbers, 527–528
- for atoms, 534–537

Quarks, 545
Quasar, 576

Radial motion of stars, 561–562
Radiation, atomic:
- of heat, 190–192
- of space, 570–571
- thermal, 518–520
- wavelength of, 519–520
- x-rays, 483–484

Radiation, nuclear:
- background, 603
- biological concentration, 604
- biological effects of, 601–605
- effects of large doses, 604
- measures of dose, 602
- and mutations, 605
- natural, 602, 603
- from radium, 489, 490

Radiation detection:
- biological effect of, 486–487
- cloud chamber, 495
- electroscope, 485
- fluorescent screen, 483
- photographic film, 484

Radicals, chemical, 340–341
Radio telescope, 569–570
Radioactive clocks, uranium, 494
Radioactive dating, 435, 443–444
Radioactive decay of C 14, 444
Radioactivity, discovery of, 484
Radiocarbon dating, 443–444
Radium:
- and cancer, 488
- discovery of, 485
- effect on living cells, 486–487
- half-life, 485
- radiation from, 489–490

Ramsay, William, 349
Randomization, 208
- of energy, 218

Rarefaction, 231, 245
Rate of distance changes, 57
Real image, 298
Rectification, 543
Red giants, 560, 566
Red shift of galaxies, 575
Reduction, chemical, 344
Reflection:
- of light, 296–303
  - by curved mirrors, 298–303
  - by plane mirror, 297–298
  - by surfaces, 295–296
- of waves, 233

Refraction:
- angle of, 238, 296, 303
- index of, 238
- of light, 303–309
- Snell's law of, 304
- of waves, 236–238

Refrigerator, 190
Relative motion, 499–502
Relativity, 498–505
- assumptions about, 502
- and energy, 504–505
- length contraction, 502–503
- and mass, 504
- principle of, 502
- time dilation, 503–504

Rem (unit of radiation dose), 602
Resistors, electrical, 277–281
- in parallel, illustrated, 279
- in series, 279–281
  - illustrated, 279

Resolution of radio telescope, 569–570
Resonance, 249
- in tubes, 250–251

Retrograde motion:
- definition of, 103
- illustrated, 105
- of Mars, 105–106

Reversible reactions, 342
Revolutionibus Orbium Coelestium, *De* (Copernicus), 118
Rhyolite, 467
Right ascension, 87, 94
- illustrated, 95

Risk-vs.-benefit decision, 606
Rock cycle, 420–421
Roentgen, Wilhelm, 483
Römer, Olaus, 84, 136–137, 143
- illustration of, 137

Royal Institution, 262–263
Rubber, discovery of vulcanization, 52
Rumford, Count (Benjamin Thompson), 202, 205
Rutherford, Ernest, 489–490, 528
- and accelerator, 496
- discovery of nucleus, 492
- gold-foil experiment, 490–491
- and natural transmutation, 495, 496
- nature of alpha particle, 490
- nature of radiation, 489

St. Helens, Mount, 470–471
Salt, composition of, 341
San Andreas fault, 451, 470, 473
Saturn:
  apparent motion of, 106
  data table, 550
  density of, 554
  moons, 557
  rings, 556–557
Scattering:
  of alpha particles, 490–492
  of light, 296
Scheele, 329–330
Schrödinger, Erwin, 532–533
Science:
  definition of, 1–2
  expansion of, 581–582
Scientific method, 25–26
Scientific revolution, 51, 78
Sea-floor spreading, 460–461
Sea level:
  changes in, 437
    effect of, 449–450
Seasons, 78–81, 94–98
Sedimentary rock, 420–423
Sedimentary strata, 429
  distorted, 431–432
  illustrated, 430
Sediments, 420–422, 425
  table of, 424
  thickness of, 456
Seismograph, 451
Semiconductors, 541, 542
Separation of hydrocarbons, 385
Series-connected cells, 277
Shale oil, 593
Shapley, Harlow, 572
Shells, electron, 360–361
Shield volcanoes, 468–469
SI (System Internationale) units, 29
Sidereal day, 93
  and radio telescope, 569
Sidereal period of planets, table, 141
Sierra Nevada, 473, 475
Sigma bond, 539
Significant figures, 29–31
Silica, 417
Silicon, bonding of, 417–418
Sills, 472
Sine of an angle, 304
Sinusoidal waves, 243
Sirius, 86, 558–560, 563, 566–567
Slope-intercept form of equation of straight line, 33–35
Smog, 585
Snake River plain, 467
Snell, Willebrod, 304
Snell's law of refraction, 304
Soil:
  acid balance of, 376–377
  erosion of, 424–428
  fertility of, 426–427
  fertilizers for, 428–429
  types of, 426–427
Solar cell, 541
Solar eclipse, 100
Solar energy, 597–601
  annual average, 597–598
  collectors, 599
  to electricity, 600
  intensity of, 597
  problems of, 600–601
  storage of, 599–600
Solar month, 99
Solar system, motion in space, 562, 571
Solid-state physics, 543
Sonic boom, 253
Sound, 245–250
Sound intensity, human ear and, 228*n*.
Specific heat, 197
Spectrum, 307, 310–313
  and atomic structure, 523, 525–527
  of hydrogen, 523
    illustrated, 525, 526
Speed:
  average, 58, 68–69
  and *d*–vs.–*t* graphs, 65–67
  definition of, 57
  instantaneous, 58
  of light, 322
  vs. velocity, 60–61
  of waves, 227
    and wavelength, 242
Standard conditions(sc), chemical, 339
Standards, air quality (1971), 586
Standing waves, 244
  electron orbits as, 531
Star map, 95
Starry Messenger, *The* (Galileo), 133–135
Stars:
  apparent motion of, 91–94
  clusters of (*see* Clusters of stars)
  collapsed, 567–568
  core of, 566
  energy production, 565–568
  evolution of, 564–568
  masses of, 563, 564, 567
  multiple, 558, 560
  neutron, 567
  proper motion, 560–561
  pulsations, 566–568
  radial motion, 561–562
  spectral types of, 562–565
  supernova, 567, 568
  temperatures, 563, 566
State of matter, 195–196
  and heating, 200–202
Steam engine, 211–215
Steam turbine, 215
Steno, Nicolaus, 429–430
Stone Age, 327–328
Strength of bones, 177
Streuve, O., 138, 558
Strings, waves on, 247–248
Subatomic particle, discovery of, 358
Subduction zones, 462
Sugars, simple, 392–395
    double, 394–395
    and sweetness, 394

Sulfur dioxide as pollutant, 585–586, 588
Sun:
  apparent motions of, 78–81
  distance of, 9
  insolation, 97
  motion seen from earth, 78–80
  rate of motion, 87
  and seasons, 94–98
  temperature from radiation, 519–520
Sun-centered universe:
  ancient theory of, 9–10
  (*See also* Copernican System)
Sunspots, discovery of, 134
Superior, Lake, 442, 443
Supernova, 567–568
Superposition, principle of, 240
Superstition, 6, 7
Systems:
  Aristotelian, 13, 17–23
  Copernican (*see* Copernican System)
  Ptolemaic (*see* Ptolemaic System)

Tartaglia, 25
Tectonics, 466
  (*See also* Plate tectonics)
Temperature:
  and depth in earth, 452
  vs. heat, 194–196
  of human body, 191–193
  meaning of, 209
  of sun's surface, 190
Temperature change with heating, 194–198
Temperature conversions (°F and °C), 34–35
Temperature inversion, 585
Temperature scales:
  absolute or kelvin, 203–204
  Celsius scale, 188–189
  Fahrenheit scale, 188
  relative vs. absolute, 188
Tendons, 176
Terminal velocity, 180
Tertiary Period, 435, 437
Thales of Miletus, 8, 101, 264
Theoretical process, definition of, 26, 51–52
Thermal energy, 210, 216
Thermal radiation, 518–520
Thermodynamics, 216–219
Thermometer, 187–189
Thermostat, 188
  illustrated, 189
Thompson, J. J., 358–359, 483, 490
Thorium, 485
Three Mile Island (TMI) nuclear reactor, 606–608
Tidal waves (tsunamis), 241*n*.
Time dilation, 503–504
Tombaugh, Clyde, 165
Torque:
  in electric motor, 284–285
  illustrated, 158
Torsional wave, 232
Total internal reflection, 306
Trade-offs, 603
Transform faults, 459–460
Transistor, 543
Transverse waves, 231
Transmutation, 328, 486, 496
Transverse waves, 231
  nature of light, 318–319
Traveling waves, 243
Trilobite, 436
Tropic of Cancer, 88–89, 96
Tsunamis (tidal waves), 241*n*.
Turbine, steam, 215

Ultraviolet radiation, 310–568
Uncertainty principle, 533
Unconformity, 464
Uniformitarianism, 434
Uniformly accelerated motion, 59–60
  analysis of, 70–72
Units, English system of, illustrated, 28
Uranium:
  for atomic bombs, 510–512
  bombardment of, 508–510, 512
  chain reaction of, 510
  decay of, 494
  energy in fission, 506–507
  half-life of, 486
  radioactivity in, 485
Uranium isotopes, 509
  separation of, 510–511
Uranus:
  data table, 550
  discovery of, 163
  moons, 557
  rings, 556
  rotation of, 557
Ussher, Bishop, 434

Valence, 339–341
Valence band, 541, 542
Valence shell of electrons, 361
Vectors, 60–63
Vega, 562
Velocity, 57
  and *d*–vs.–*t* graphs, 65–67
  vs, speed, 60–61
  terminal, 180
Venus:
  apparent motion of, 103–104
  atmosphere, 549, 551
  data, 550
  greatest elongation, 104
  phases of, 134
  Ptolemaic interpretation, 113–114
    illustrated, 115
  surface of, 551
  synodic period, 104
Vinci da, and human structure, 176
Virtual image, 298
Vitamin C, 409*n*.
Vitamins, 393

Volcanoes, 467–472
Volt, 272–273
Volta, Alessandro, 275–276

Walton, E. T. S., 496–497
Water:
  action on rock, 421–422
  behavior with heating, 194–197
  chemical formulation, 364–365
  dissociation of, 374–375
  dissolving ability, 372–373
  formula of, 334–336
  heat of fusion, 197
  heat of vaporization, 197
  ionization of, 374–375
  molecular orbitals in, 538–539
  specific heat, 197
  structure of molecule, 371–373
  underground, 476–477
Water-suface waves, 225–226, 228–229
  breaking behavior, 239
  and depth of water, 236
  motion of water, 231–232
Watson, James, 410, 411*n.*
Watt, James, 213
Watt (unit of power), 174
Wave nature of matter, 532–533
Wave properties of electron, 530–532
Wavelength, 242
  of light, 317–318, 323
  maximum for radiation, 519–520
  of radiant energy, 519–520
Waves, 225–253
  definition of, 227
  earthquake, 220*n.*, 241*n.*, 450–453
  longitudinal, 231
  moving sources, 251–253
  periodic, 241–244
  im pipes, 249–250
  sinusoidal, 243
  sound, 245–250
  standing, 244, 531
  on strings, 247–248
  tidal, 241*n.*
  torsional, 232
  transmission of, 234
Waves *(Cont.)*:
  transverse, 231, 318–319
  traveling, 243
  types of, 230–232
  water-surface (*see* Water-surface waves)
Weather patterns, 148–149
Weathering, 420–422
  and erosion, 424–429
  of granite, 421–424
Wegner, Alfred, 458
Weight:
  at equator vs. poles, 162–163
  vs. mass, 152–153
White dwarfs, 560, 566–567
Wilson, C. T. R., 495
Wilson, R. W., 570
Winds, 148–149
Women in science, 487–488
Work:
  and change of state, 210
  conservation of, 173
  definition of, 170–174
  done by an expanding gas, 211
  to lift a weight, 174–175
    (*See also* Potential energy)
  and power, 174
  to speed up a mass, 178–180
    (*See also* Kinetic energy)
  to transfer electric charge, 272–273
  unit of (Joule), 172
Work function, 521

X-rays, 483–484

Year, length of, 98–99
Yellowstone National Park, 420
Yellowstone Plateau, 467
Young, Thomas, 315–318
Young's experiment, 315–318

Zenith, 81, 87
Zion National Park, 431
Zodiac, 94